The
AutoCAD® 2004
Tutor for
Engineering Graphics

The AutoCAD® 2004 Tutor for Engineering Graphics

ALAN J. KALAMEJA

autodesk®
press

THOMSON
DELMAR LEARNING

Australia • Canada • Mexico • Singapore • Spain • United Kingdom • United States

THOMSON

DELMAR LEARNING

autodesk®
press

The AutoCAD® 2004 Tutor for Engineering Graphics
Alan J. Kalameja

Autodesk Press Staff

**Vice President, Technology
and Trades SBU:**
Alar Elken

Editorial Director:
Sandy Clark

Senior Acquisitions Editor:
James DeVoe

Senior Development Editor:
John Fisher

Editorial Assistant:
Mary Ellen Martino

Executive Marketing Manager:
Cynthia Eichelman

Channel Manager:
Fair Huntoon

Marketing Coordinator:
Sarena Douglass

Production Director:
Mary Ellen Black

Production Manager:
Andrew Crouth

Production Editor:
Tom Stover

Art and Design Specialist:
Mary Beth Vought

**Library of Congress
Cataloging-in-Publication Data**
Kalameja, Alan J.
  The AutoCAD 2004 tutor for engineering graphics / Alan J. Kalameja.
      p. cm.
  ISBN 0-4018-5082-0

604.20285
K14a

CONTENTS

INTRODUCTION

Engineering graphics is the process of defining an object graphically before it is constructed and used by consumers. Previously, this process for producing a drawing involved the use of drawing aids such as pencils, ink pens, triangles, T-squares, etc. to place an idea on paper before making changes and producing blue-line prints for distribution. The basic principles and concepts of producing engineering drawings have not changed even when using the computer as a tool.

This text uses the basics of engineering graphics to produce 2D drawings and 3D computer models using AutoCAD and a series of tutorial exercises that follow each chapter. Following the tutorials in most chapters, problems are provided to enhance your skills in producing engineering drawings. A brief description of each chapter follows:

CHAPTER 1 – GETTING STARTED WITH AUTOCAD

This first chapter introduces you to the following fundamental AutoCAD concepts: Screen elements; Use of function keys; Opening up an existing drawing file; Basic drawing techniques using the LINE, CIRCLE, and PLINE commands; Understanding absolute, relative, and polar coordinates; Using the Direct Distance mode for drawing lines; Using Object snaps and the AutoSnap feature, polar and object tracking techniques; Using the Erasing command; Right-click mouse customizations; Saving a drawing. Drawing tutorials follow at the end of this chapter.

CHAPTER 2 – DRAWING ORGANIZATION

This chapter introduces the concept of drawing in real world units through the setting of drawing units and limits. The importance of organizing a drawing through layers is also discussed through the use of the Layer Properties Manager dialog box. Color, linetype, and lineweight are assigned to layers and applied to drawing objects. Layer filtering, layer states, and how to create template files are also discussed in this chapter.

CHAPTER 3 – AUTOCAD DISPLAY AND SELECTION OPERATIONS

This chapter discusses the ability to magnifying a drawing using numerous options of the ZOOM command. The PAN command is also discussed as a means of staying in a zoomed view and moving the display to a new location. Productive uses of realtime zooms and pans along with the effects a wheel mouse has on zoom and pan are included. All object selection set modes are discussed such as Window, Crossing, Fence, and All to name a few. Finally this chapter discusses the ability to save the image of your display and retrieve the saved image later through the View Control dialog box.

CHAPTER 4 – MODIFY COMMANDS

The full range of Modify commands are discussed in this chapter and include ARRAY, BREAK, CHAMFER, COPY, EXPLODE, EXTEND, FILLET, LENGTHEN, MIRROR, MOVE, OFFSET, PEDIT, ROTATE, SCALE, STRETCH, and TRIM. Various tutorial exercises are provided at the end of this chapter to emphasize these very important commands.

CHAPTER 5 – PERFORMING GEOMETRIC CONSTRUCTIONS

This chapter discusses how AutoCAD commands are used for constructing geometric shapes. The following drawing related commands are included in this chapter: ARC, DONUT, ELLIPSE, MLINE, POINT, POLYGON, RAY, RECTANG, SPLINE, and XLINE. A tutorial exercise is provided at the end of this chapter.

CHAPTER 6 – ADDING TEXT TO YOUR DRAWING

Use this chapter for placing text in your drawing. Various techniques for accomplishing this task include the use of the DTEXT and MTEXT commands. The creation of text styles and the ability to edit text once it is placed in a drawing is also included. A tutorial exercise is included at the end of this chapter.

CHAPTER 7 – OBJECT GRIPS AND CHANGING THE PROPERTIES OF OBJECTS

The topic of grips and how they are used to enhance the modification of a drawing is presented. The ability to modify objects through the Properties window is discussed in great detail. A tutorial exercise is included at the end of this chapter to reinforce the importance of changing the properties of objects.

CHAPTER 8 – SHAPE DESCRIPTION/MULTIVIEW PROJECTION

Describing shapes and producing multiview drawings using AutoCAD are the focus of this chapter. The basics of shape description are discussed along with proper use of linetypes, fillets, rounds, chamfers, and runouts. A tutorial exercise is available at the end of this chapter and outlines the steps used for creating a multiview drawing using AutoCAD.

CHAPTER 9 – BEGINNING DRAWING LAYOUTS

This chapter deals with the creation of layouts before a drawing is plotted out. A layout takes the form of a sheet of paper and is referred to as Paper Space. A wizard to assist in the creation of layouts is also demonstrated. Once a layout of an object is created, scaling through the Viewports toolbar is discussed. The creation of numerous layouts for the same drawing is also introduced including a means of freezing layers only in certain layouts. Various exercises are provided throughout this chapter to reinforce the importance of layouts.

CHAPTER 10 – PLOTTING YOUR DRAWING

Beginning and advanced of outputting your drawings to a print or plot device are discussed in this chapter through a series of tutorial exercises. One tutorial demonstrates the use of the Add-A-Plotter Wizard to configure a new plotter. Plotting from a layout is discussed through a tutorial. This includes the assignment of a sheet size. Tutorial exercises are also provided to create a color-dependent plot style. Plot styles allow you to control the appearance of your plot. Other tutorial exercises available in this chapter include ePlot capabilities and Plotting to the Web.

CHAPTER 11 – DIMENSIONING BASICS

Basic dimensioning rules are introduced in the beginning of this chapter before all AutoCAD dimensioning commands are discussed. Placing linear, diameter, and radius dimensions are then demonstrated. The powerful QDIM command allows you to place baseline, continuous and other dimension groups in a single operation. A tutorial exercise is provided at the end of this chapter.

CHAPTER 12 – THE DIMENSION STYLE MANAGER

A thorough discussion of the use of the Dimension Styles Manager dialog box is included in this chapter. The ability to create, modify, and override dimension styles are discussed. A detailed tutorial exercise is provided at the end of this chapter.

CHAPTER 13 – ANALYZING 2D DRAWINGS

This chapter provides information on analyzing a drawing for accuracy purposes. The AREA, ID, LIST, and DIST commands are discussed in detail in addition to how they are used to determine the accuracy of various objects in a drawing. A tutorial exercise follows that allows the user to test their drawing accuracy.

CHAPTER 14 – SECTION VIEWS

The basics of section views are described in this chapter including full, half, assembly, aligned, offset, broken, revolved, removed, and isometric sections. Hatching techniques through the use of the Boundary Hatch dialog box are also discussed. The ability to apply a gradient hatch pattern is also discussed. Tutorial exercises follow at the end of the chapter that deal with the topic of section views.

CHAPTER 15 – AUXILIARY VIEWS

Producing auxiliary views using AutoCAD is discussed in this chapter. Items discussed include rotating the snap at an angle to project lines of sight perpendicular to a surface to be used in the preparation of the auxiliary view. One tutorial exercise is provided in this chapter.

CHAPTER 16 – BLOCK CREATION, AUTOCAD DESIGNCENTER, AND MDE

This chapter covers the topic of creating blocks in AutoCAD. Creating local and global blocks such as doors, windows, and pipe symbols will be demonstrated. The Insert dialog box is discussed as a means of inserting blocks into drawings. The chapter continues by explaining the many uses of the AutoCAD DesignCenter. This feature allows the user to display a browser containing blocks, layers, and other named objects that can be dragged into the current drawing file. The use of tool pallets is also discussed as a means of dragging and dropping blocks and hatch patterns into your drawing. This chapter also discusses the ability to open numerous drawings through the Multiple Document Environment and transfer objects and properties between drawings. A tutorial exercise can be found at the end of this chapter.

CHAPTER 17 – ATTRIBUTES

This chapter introduces the purpose for creating attributes in a drawing. A series of four commands step the user to a better understanding of attributes. The first command is ATTDEF and is used to define attributes. The ATTDISP command is used to control the display of attributes in a drawing. Once attributes are created and assigned to a block, they can be edited through the ATTEDIT command. Finally, attribute information can be extracted using the ATTEXT command. Extracted attributes can then be imported into such applications as Microsoft Excel and Access. Various tutorial exercises are provided throughout this chapter to become better acquainted with this powerful feature of AutoCAD.

CHAPTER 18 – EXTERNAL REFERENCES AND IMAGE FILES

The chapter begins by discussing the use of External References in drawings. An external reference is a drawing file that can be attached to another drawing file. Once the referenced drawing file is edited or changed, these changes are automatically seen once the drawing containing the external reference is opened again. Performing in-place editing of external references will also be demonstrated. Importing image files is also discussed and demonstrated in this chapter. A tutorial exercise follows at the end of this chapter to practice with external references.

CHAPTER 19 – MULTIPLE VIEWPORT DRAWING LAYOUTS

This very important chapter is designed to utilize advanced techniques used in laying out a drawing before it is plotted. The ability to layout a drawing consisting of various images at different scales will also be discussed. The ability to create user-defined rectangular viewports through the MVIEW command and is discussed. The creation of non-rectangular viewports will also be demonstrated. A tutorial exercise follows to practice this advanced layout technique.

CHAPTER 20 – SOLID MODELING FUNDAMENTALS

This chapter begins with a comparison between isometric, extruded, wireframe, surfaced, and solid model drawings. The chapter continues with a detailed discussion on the use of the User Coordinate System and how it is positioned to construct objects in 3D. The display of 3D images through the 3DORBIT, VPOINT and 3DCORBIT commands is discussed. Creating various solid primitives such as boxes, cones and cylinders is discussed in addition to the ability to construct complex solid objects through the use of the Boolean operations of union, subtraction, and intersection. The chapter continues on by discussing extruding and rotating operations for creating solid models in addition to filleting and chamfering solid models. Analyzing solid models is also covered in this chapter. Because of the importance of this design paradigm taking you from 2D to 3D, two tutorial exercises follow at the end of this chapter.

CHAPTER 21 – EDITING SOLID MODELS

The ALIGN and ROTATE3D commands are discussed as a means of introducing the editing capabilities of AutoCAD on solid models. Modifications can also be made to a solid model through the use of the SOLIDEDIT command, which is discussed in this chapter. The ability to extrude existing faces, imprint objects, and create thin wall shells will be demonstrated throughout this chapter. A tutorial exercise can be found at the end of this chapter.

CHAPTER 22 – CREATING ORTHOGRAPHIC VIEWS FROM A SOLID MODEL

Once the solid model is created, the SOLVIEW command is used to layout 2D views of the model and the SOLDRAW command is used to draw the 2D views. Layers are automatically created to assist in the annotation of the drawing through the use of dimensions. A tutorial exercise is available at the end of this chapter along with instructions on how to apply the techniques learned in this chapter to other solid models.

CHAPTER 23 – PRODUCING PHOTOREALISTIC RENDERINGS

This chapter introduces you to the uses and techniques of producing renderings from 3D models in AutoCAD. A brief overview of the rendering process is covered along with detailed information about placing lights in your model, producing scenes from selected lighting arrangements, loading materials through the materials library supplied in AutoCAD, attaching materials to your 3D models, how to create your own custom materials, applying a background to your rendered image, and experimenting with fog and how it affects the 3D model.

CHAPTER A – ISOMETRIC DRAWINGS *(Available on the CD in the back of the book in PDF format. Requires Acrobat Reader)*

This chapter discusses constructing isometric drawings with particular emphasis on using the SNAP command along with the Style option in AutoCAD used to create an isometric grid. Methods of toggling between right, top and left isometric modes will be explained. In addition to isometric basics, creating circles and angles in isometric are also discussed. Tutorial exercises on producing isometric drawings follow along with additional problems at the end of this chapter.

CHAPTER B –   EXPRESS TOOLS *(Available on the CD in the back of the book in PDF format. Requires Acrobat Reader)*

This appendix covers numerous Express Tools that are supplied with AutoCAD. Express Tools consist of a library of productivity tools designed to assist in the following categories: Layers, Blocks, Text, Layout Tools, Dimension, Selection Tools, Modify, Draw, and File Tools.

ONLINE COMPANION

This new edition contains a special Internet companion piece. The Online Companion is your link to AutoCAD on the Internet. Monthly updates include a command of the month, FAQs, and tutorials. You can find the Online Companion at:

> http://www.autodeskpress.com/resources/olcs/index.html

ACKNOWLEDGEMENTS

I wish to thank the staff at Autodesk Press for their assistance with this document, especially John Fisher, Sandy Clark, Jim DeVoe, Mary Beth Vought, and Tom Stover. Thanks also go out to John Shanley of Phoenix Creative Graphics for his part in the desktop publishing aspects of this document. I would also like to thank Rob Bolus, Gary Crafts, and Kevin Lang of Trident Technical College for performing the technical edit on the entire manuscript.

The publisher and author would like to thank and acknowledge the many professionals who reviewed the manuscript to help us publish this AutoCAD text. A special acknowledgement is due to the following instructors who reviewed the chapters in detail:

Wen. M. Andrews, J. Sargeant Reynolds Community College, Richmond, VA

Kirk Barnes, Ivy Tech State College, Bloomington, IN

Paul Ellefson, Hennepin Technical College, Brooklyn Park, MN

Rajit Gadh, University of Wisconsin, Madison, WI

Gary J. Masciadrelli, Springfield Technical Community College, Springfield, MA

ABOUT THE AUTHOR

Alan J. Kalameja is the Department Head of Computer-Integrated Manufacturing at Trident Technical College located in Charleston, South Carolina. He has been at the College for over 22 years and has been using AutoCAD since 1984. He has authored the AutoCAD Tutor for Engineering Graphics in Release 14, AutoCAD 2000 and AutoCAD 2004. All of his books are published by Delmar/Autodesk Press.

CONVENTIONS

All tutorials in this publication use the following conventions in the instructions:

Whenever you are told to enter text, the text appears in boldface type. This may take the form of entering an AutoCAD command or entering such information as absolute, relative or polar coordinates. You must follow these and all text inputs by pressing the ENTER key to execute the input. For example, to draw a line using the LINE command from point 3,1 to 8,2, the sequence would look like the following:

 Command: **L** *(For* LINE*)*
Specify first point: **3,1**
Specify next point or [Undo]: **8,2**
Specify next point or [Undo]: *(Press* ENTER *to exit this command)*

Instructions for selecting objects are in italic type. When instructed to select an object, move the pickbox on the object to be selected and press the pick button on the mouse or digitizing puck.

If you enter the wrong command for a particular step, you may cancel the command by pressing the ESC key. This key is located in the upper left hand corner of any standard keyboard.

Instructions in this tutorial are designed to enter all commands, options, coordinates, etc., from the keyboard. You may use the same commands by selecting them from the screen menu, digitizing tablet, pulldown menu area, or from one of the any floating toolbars.

NOTES TO THE STUDENT AND INSTRUCTOR CONCERNING THE USE OF TUTORIAL EXERCISES

Various tutorial exercises have been designed throughout this book and can be found at the end of each chapter. The main purpose of each tutorial is to follow a series of steps towards the completion of a particular problem or object. Performing the tutorial will also prepare you to undertake the numerous drawing problems also found at the end of each chapter.

As you work on the tutorials, you should follow the steps very closely, taking care not to make a mistake. However, most individuals rush through the tutorials to get the correct solution in the quickest amount of time only to forget the steps used to complete the tutorial. A typical comment made by many is "I completed the tutorial...but I don't understand what I did to get the correct solution."

It is highly recommended to both student and instructor that all tutorial exercises be performed two or even three times. Completing the tutorial the first time will give you the confidence that it can be done; however, you may not understand all of the steps involved. Completing the tutorial a second time will allow you to focus on where certain operations are performed and why things behave the way they do. This still may not be enough. More complicated tutorial exercises may need to be performed a third time. This will allow you to anticipate each step and have a better idea what operation to perform in each step. Only then will you be comfortable and confident to attempt the many drawing problems that follow the tutorial exercises.

The CD-ROM in the back of the book contains AutoCAD drawing files for the Try It! exercises. To use drawing files, copy files to your hard drive, then remove their read-only attribute. Files cannot be used without AutoCAD. Files are located in the /Drawing Files/ directory. Also included are complete PDF files of the AutoCAD 2004 Tutor for Engineering Graphics Project Manual, drawing files for the tutorial exercises, and Chapters A and B of this textbook.

SUPPLEMENTS

e.resource™–This is an educational resource that creates a truly electronic classroom. It is a CD-ROM containing tools and instructional resources that enrich your classroom and make your preparation time shorter. The elements of *e.resource* link directly to the text and tie together to provide a unified instructional system. Spend your time teaching, not preparing to teach.

ISBN 0-4018-5084-7

Features contained in *e.resource* include:

- **Syllabus:** Lesson plans created by chapter. You have the option of using these lesson plans with your own course information.

- **Chapter Hints:** Objectives and teaching hints that provide the basis for a lecture outline that helps you to present concepts and material.

- **PowerPoint® Presentation:** These slides provide the basis for a lecture outline that helps you to present concepts and material. Key points and concepts can be graphically highlighted for student retention. There are over 300 slides, covering every chapter in the text.

- **Exam View Computerized Test Bank:** Over 600 questions of varying levels of difficulty are provided in true/false and multiple-choice formats. Exams can be generated to assess student comprehension or questions can be made available to the student for self-evaluation.

- **Drawing Solutions:** These drawing files are the solutions for end of chapter problem exerises.

- **Video and Animation Resources:** These AVI files graphically depict the execution of key concepts and commands in drafting, design, and AutoCAD and let you bring multi-media presentations into the classroom.

Spend your time teaching, not preparing to teach!

Getting Started with AutoCAD

Chapter 1 begins with an explanation of the components that make up a typical AutoCAD display screen. You will learn various methods of selecting commands; some from the pull-down menu, others from toolbars, and still others by entering the command from the keyboard. You will learn how to begin a new drawing. Once in a drawing, you will construct line segments with the LINE command in addition to drawing lines accurately with the direct distance mode, absolute coordinates, relative coordinates, and polar coordinates. Object snap will be discussed as a way to construct all types of objects with greater precision. Additional drawing aids such as object snap tracking and polar tracking will be discussed. Other basic drawing commands such as constructing circles and polylines will be shown. You will also be introduced to the ERASE command for removing drawing objects. Other topics included in this chapter are saving your drawing, getting questions answered through the HELP command, using the Active Assistant, creating a clean drawing screen, status bar menu controls, right click customization, and exiting an AutoCAD drawing session.

THE AUTOCAD DRAWING SCREEN

When you first launch AutoCAD, you will see the drawing screen illustrated in Figure 1–1. At the top of this screen is the menu bar, with the pull-down menus containing various commands under such categories as File, Edit, View, and Insert (to name a few). Directly beneath the menu bar is the Standard toolbar containing buttons for such commands as NEW and OPEN in addition to two flyout buttons supporting the Zoom and Object Snap features. The Object Properties toolbar (below the Standard toolbar) is designed for manipulating layers, color, linetypes, and line weights. Two additional toolbars appear in Figure 1–1. The Draw toolbar appears on the left side and contains buttons for such commands as LINE, CIRCLE, ARC, and BHATCH. The Modify toolbar appears on the right side and contains buttons for FILLET, CHAMFER, ERASE, and ARRAY, to name just a few. The User Coordinate System icon is displayed in the lower corner of the drawing editor to alert you to the coordinate system currently in use.

At the lower part of the display screen is the command prompt area, where AutoCAD prompts you for input depending on which command is currently in progress. At the very bottom of the screen is the status bar. Use this area to toggle on or off the following modes: Coordinate Display, Snap, Grid, Ortho, Polar Snap, Object Snap, Object Tracking, Line Weight, and Model/Paper. Click the button once to turn the mode on or off. Scroll bars are present just below and to the right of the drawing editor screen. You can use these to pan to a different screen location, especially when the screen has been magnified using one of the ZOOM command options. You can use the graphics cursor to pick points on the screen or select objects to edit.

Tool palettes consist of tabbed areas that allow you to organize blocks and hatches in an efficient manner. Once these blocks and hatches are organized, you can easily insert them into your drawing through drag-and-drop techniques. Tool palette use will be covered in greater detail in chapters 14 (Section Views) and 16 (Block Creation, AutoCAD DesignCenter, and MDE).

When you begin a new drawing, you perform the drawing inside Model mode. To plot the drawing out, you switch to one of the Layout modes presented by tabs at the bottom of the display screen.

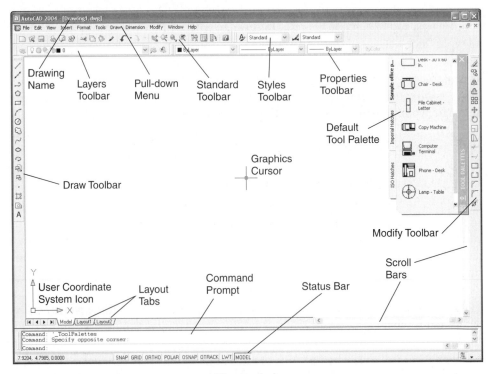

Figure I–I

USING KEYBOARD FUNCTION KEYS

Figure 1–2

Once you open a drawing file, you have additional aids to control such settings as Grid and Snap. Illustrated in Figure 1–2 is a group of function keys, which are found above the alphanumeric keys of a typical keyboard. These function keys are labeled F1 through F12; only keys F1 through F11 will be discussed here. Software companies commonly program certain functions into these keys to assist you in their application, and AutoCAD is no exception. Most keys act as switches, which turn functions on or off (see Table 1–1).

Table 1–1

Function Key	Definitions
F1	AutoCAD Help Topics
F2	Toggle Text/Graphics Screen
F3	Object Snap settings On/Off
F4	Toggle Tablet Mode On/Off
F5	Toggle Isoplane Modes
F6	Toggle Coordinates On/Off
F7	Toggle Grid Mode On/Off
F8	Toggle Ortho Mode On/Off
F9	Toggle Snap Mode On/Off
F10	Toggle Polar Mode On/Off
F11	Toggle Object Snap Tracking On/Off

When you press F1, the AutoCAD Help Topics dialog box is displayed.

Pressing F2 takes you to the text screen consisting of a series of previous prompt sequences. This may be helpful for viewing the previous command sequence in text form.

Pressing F3 toggles the current Object Snap settings on or off. This will be discussed later in this chapter.

Use F4 to toggle Tablet mode on or off. This mode is only activated when the digitizing tablet has been calibrated for the purpose of tracing a drawing into the computer.

Pressing F5 scrolls you through the three supported Isoplane modes used to construct isometric drawings (Right, Left, and Top).

F6 toggles the coordinate display, located in the lower left corner of the status bar, on or off. When the coordinate display is off, the coordinates are updated when you pick an area of the screen with a mouse or digitizer puck. When the coordinate display is on, the coordinates dynamically change with the current position of the cursor.

F7 turns the display of the grid on or off. The actual grid spacing is set by the GRID command and not by this function key.

F8 toggles Ortho mode on or off. Use this key to force objects such as lines to be drawn horizontally or vertically.

Use F9 to toggle Snap mode on or off. The SNAP command sets the current snap value.

Use F10 to toggle the Polar Tracking on or off. Use F11 to toggle Object Snap Tracking on or off. Both these function keys will be discussed in greater detail in this chapter.

The status bar, illustrated in Figure 1–3, holds most commands controlled by the function keys. For example, click the OSNAP button once to turn the Object Snap on; click the OSNAP button again to turn the Object Snap off. You can control Object Snap (F3), the coordinate display (F6), Grid (F7), Ortho (F8), Snap (F9), Polar Tracking (F10), and Object Snap Tracking (F11) in the same way. In addition, the LWT button turns Lineweight mode on or off. Clicking the MODEL button enters the Page Setup mode. Both the LWT and the MODEL buttons will be discussed in greater detail in the chapters that follow.

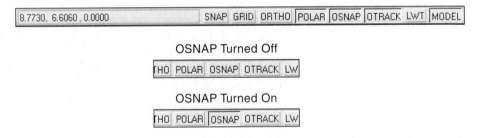

Figure 1–3

Right-clicking one of the buttons displays the shortcut menu in Figure 1–4. Choose Settings to access various dialog boxes that control certain features associated with the button.

Figure 1–4

METHODS OF CHOOSING COMMANDS

AutoCAD provides numerous ways to enter or choose commands for constructing or editing drawings: pull-down menus, toolbars, dialog boxes, right-click shortcut menus, icon menus, and command aliases for direct entry from the keyboard.

PULL-DOWN MENUS

The Draw pull-down menu, shown in Figure 1–5, is very popular with many users. It allows you to choose most AutoCAD 2004 commands.

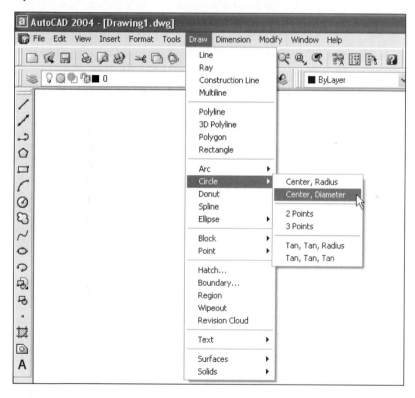

Figure 1–5

TOOLBARS

Figure 1–6 illustrates a toolbar, an additional method for initiating commands. The toolbar illustrated gives you access to Zoom commands. You can find other toolbars by choosing Toolbars from the View pull-down menu. By default, the toolbar is displayed flat. When the cursor rolls over a tool, a 3D border is displayed, along with a tooltip that explains the purpose of the command (see Figure 1–7).

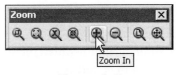

Figure 1–6   **Figure 1–7**

Another method of accessing commands in a toolbar is through a flyout, which is identified by a small triangle located in the lower right corner of the button. Pressing and holding your mouse down on the toolbar button displays other buttons (see Figure 1–8). Note that when you select a button from a flyout, that button replaces the one originally at the top.

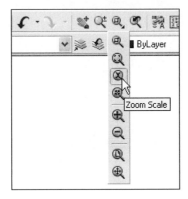

Figure 1–8

ACTIVATING TOOLBARS

Many toolbars are available to assist you in using other types of commands. Choosing Toolbars from the View pull-down menu, as shown in Figure 1–9, displays the Customize dialog box, illustrated in Figure 1–10, with the Toolbars tab displayed.

By default, six toolbars are already active or displayed in all drawings: the Draw, Layers, Modify, Properties, Standard, and Styles toolbars. To make another toolbar active, click in the empty box next to the name of the toolbar. This will display the toolbar and allow you to preview it before making it a part of the display screen. If this toolbar is correct, click the Close button. This will close the Customize dialog box but leave the selected toolbar. Illustrated in Figure 1–11 is the Dimension toolbar, which is activated when a check is placed in the box next to Dimension on the Toolbars tab of the Customize dialog box.

A quicker way of activating toolbars is by moving your cursor over any command button and then pressing the right mouse button. A shortcut menu appears that displays all toolbars, as shown in Figure 1–12 . Placing a check beside the toolbar name displays this toolbar.

Figure 1–9

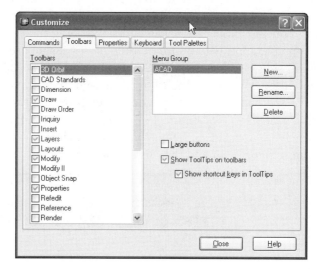

Figure 1–10

Figure 1–11

Figure 1–12

DOCKING TOOLBARS

It is considered good practice to line the side edges of the display screen with toolbars. The method of moving toolbars to the sides of your screen is called docking. Press down on the toolbar title strip and slowly drag the toolbar to the screen edge until the toolbar appears to jump. Letting go of your mouse button will dock the toolbar to this side of the screen. Practice this by docking various toolbars to your screen. To prevent docking, press the CTRL key as you drag the toolbar. This will allow you to move the toolbar into the upper or lower portions of your display screen without the toolbar docking (see Figure 1–13). Also, if a toolbar appears to disappear, it might actually be alongside or below toolbars that already exist. You might have to turn off toolbars in order to find the missing one.

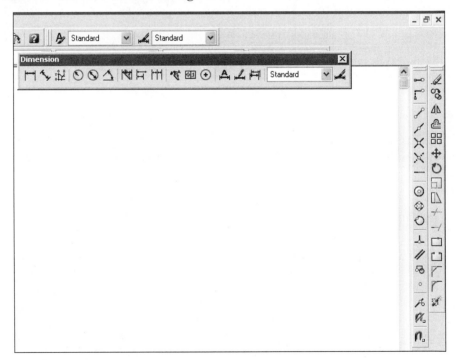

Figure 1–13

DIALOG BOXES

Settings and other controls can be changed through dialog boxes. Illustrated in Figure 1–14 is the Open dialog box, which will be discussed later in this chapter. You can increase the size of certain dialog boxes by moving your cursor over the border of the dialog box, and this is true of the Open dialog box. Two arrows will appear that allow you to hold down the pick button of your mouse (usually the left button) and stretch the dialog box in that direction. If you move your cursor to the corner of the dialog box, you can stretch the box in two directions. These methods can also be used to make the dialog box smaller, although there is a default size for the dialog boxes, which limit smaller sizes. The dialog box cannot be stretched if no arrows appear when you move your cursor over the border of the dialog box. Practice this by activating the Open dialog box with the OPEN command and stretching this dialog box larger or smaller.

Figure 1–14

RIGHT-CLICK SHORTCUT MENUS

Many shortcut or cursor menus have been developed to assist with the rapid access to commands. Clicking the right mouse button activates these commands. The Default shortcut menu is illustrated in Figure 1–15. It is displayed whenever you right-click in the drawing area and no command or selection set is in progress.

Figure 1–16 shows an example of the Edit shortcut menu. This shortcut consists of numerous editing and selection commands. This menu activates whenever you right-click in the display screen with an object or group of objects selected but no command is in progress.

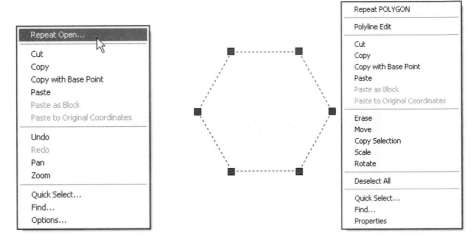

Figure 1–15   **Figure 1–16**

Right-clicking in the Command prompt area of the display screen activates the shortcut menu in Figure 1–17. This menu provides quick access to the Options dialog box. Also, a record of the six most recent commands is kept, which allows you to select from this group of previously used commands.

Illustrated in Figure 1–18 is an example of a Command-Mode shortcut menu. When you enter a command and right-click, this menu will display options of the command. This menu supports a number of commands. In the figure, the 3P, 2P, and Ttr (tan tan radius) listings are all options of the CIRCLE command.

Figure 1–17

Figure 1–18

ICON MENUS

Icon menus display graphical representations of commands or command options. The icon menu in Figure 1–19 shows numerous hatch patterns. The graphic images of the patterns allow you preview the pattern before you place it in a drawing. Icon menus allow you to select items quickly because of their graphical nature.

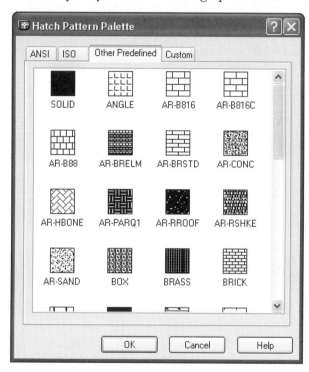

Figure 1–19

COMMAND ALIASES

You can enter commands directly from the keyboard. This practice is popular for users who are already familiar with the commands. However, you must know the exact command name, including its exact spelling. To assist with the entry of AutoCAD commands from the keyboard, numerous commands are available in shortened form, and these shorter forms are called *aliases*. The list in Table 1–2 shows most of the command aliases that are located in the support file ACAD.PGP. Once you are comfortable with the keyboard, command aliases provide a fast and efficient method of activating AutoCAD commands.

Table 1–2   Command Aliases

Alias	Command	Alias	Command
3DO	3DORBIT	DOR	DIMORDINATE
A	ARC	DR	DRAWORDER
AA	AREA	DRA	DIMRADIUS
AL	ALIGN	DST	DIMSTYLE
-AR	ARRAY	DT	DTEXT
AR	Array Dialog	DV	DVIEW
ATT	ATTDEF	E	ERASE
ATE	ATTEDIT	ED	DDEDIT
ATTE	-ATTEDIT	EL	ELLIPSE
B	BLOCK	EX	EXTEND
BH	BHATCH	EXT	EXTRUDE
BO	BOUNDARY	F	FILLET
BR	BREAK	G	GROUP
C	CIRCLE	H	BHATCH
CH	PROPERTIES	HE	HATCHEDIT
CHA	CHAMFER	HI	HIDE
CP	COPY	I	INSERT
D	DIMSTYLE	IM	IMAGE
DAL	DIMALIGNED	IN	INTERSECT
DAN	DIMANGULAR	L	LINE
DBA	DIMBASELINE	LA	LAYER
DCE	DIMCENTER	LE	QLEADER
DCO	DIMCONTINUE	LEN	LENGTHEN
DDI	DIMDIAMETER	LI	LIST
DED	DIMEDIT	LS	LIST
DI	DIST	LT	LINETYPE
DIV	DIVIDE	LTS	LTSCALE
DLI	DIMLINEAR	M	MOVE
DO	DONUT	MA	MATCHPROP

Alias	Command	Alias	Command
ME	MEASURE	RR	RENDER
MI	MIRROR	S	STRETCH
ML	MLINE	SC	SCALE
MO	PROPERTIES	SCR	SCRIPT
MT	MTEXT	SHA	SHADE
MV	MVIEW	SN	SNAP
O	OFFSET	SP	SPELL
OP	OPTIONS	SPL	SPLINE
OS	OSNAP	ST	STYLE
P	PAN	SU	SUBTRACT
PE	PEDIT	T	MTEXT
PL	PLINE	TOL,	*TOLERANCE
PO	POINT	TOR	TORUS
POL	POLYGON	TR	TRIM
PRE	PREVIEW	UN	UNITS
PRINT	PLOT	UNI	UNION
PS	PSPACE	V	VIEW
PU	PURGE	VP	DDVPOINT
R	REDRAW	W	WBLOCK
RE	REGEN	WE	WEDGE
REC	RECTANGLE	X	EXPLODE
REG	REGION	XL	XLINE
REV	REVOLVE	XR	XREF
RO	ROTATE	Z	ZOOM

STARTING A NEW DRAWING

To begin a new drawing file, select the NEW command using one of the following methods:

From a toolbar button (Insert the file b_std_new.PCX here)

From the pull-down menu (File > New)

From the keyboard (New)

Command: **NEW**

Entering the NEW command displays the dialog box illustrated in Figure 1–20.

The Template selection displays a list of various files that have an extension of .DWT. These represent template files and are designed to conform to various standard drawing sheet sizes. Associated with each template file is a corresponding title block that is displayed in the Preview area. You can scroll through the various template files and get a glimpse of the title block matched to the template file.

You can also elect to start a drawing from scratch by clicking on the arrow next to Open, located in the lower-right corner of the dialog box. The three options include opening a drawing with the default settings, opening with imperial units, and opening with metric units. These types of drawings consist of a number of default settings, such as a pre-defined sheet size or preset units of measure.

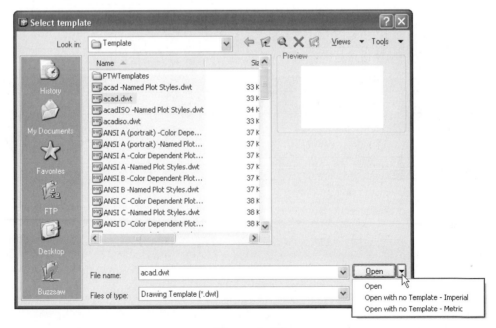

Figure 1–20

ACTIVATING THE TRADITIONAL STARTUP DIALOG BOX

Choosing Options from the Tools pull-down menu displays the Options dialog box shown in Figure 1–21. Click the System tab and locate the Startup heading located under General Options. Selecting "Show traditional startup dialog" will display the dialog box in Figure 1–22 whenever you issue the new command. This dialog box contains four buttons designed to open a drawing, start a drawing from scratch, start a drawing using a template, and use a wizard to make changes to drawing settings before creating. This wizard will be discussed in greater detail in chapter 2, "Drawing Setup and Organization."

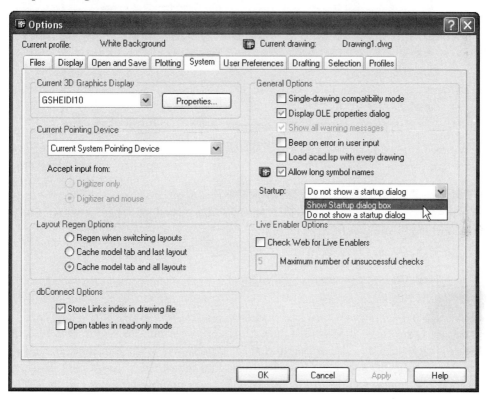

Figure 1–21

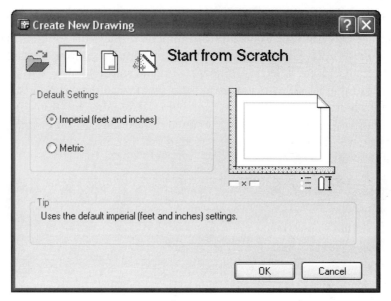

Figure 1–22

OPENING AN EXISTING DRAWING

The OPEN command is used to edit a drawing that has already been created. Select this command from one of the following:

From a toolbar button (Insert the file b_std_open.PCX here)

From the pull-down menu (File > Open)

From the keyboard (Open)

When you select this command, a dialog box appears similar to Figure 1–23. Listed in the edit box area are all files that match the type shown at the bottom of the dialog box. Because the file type is .DWG, all drawing files supported by AutoCAD are listed. To choose a different folder, use standard Windows file management techniques by clicking in the Look in edit box. This will display all folders associated with the drive. Clicking the folder will display any drawing files contained in it.

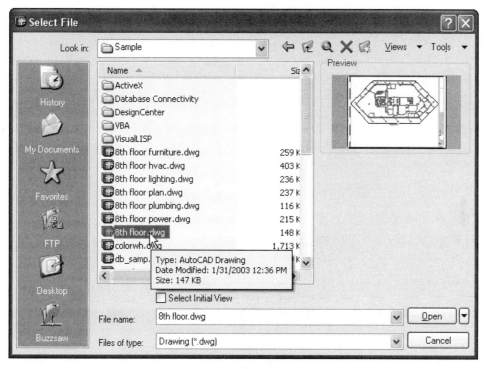

Figure 1–23

 Note: If a drawing file does not appear in the preview window, it was last opened in a version prior to AutoCAD Release 13. This drawing must be opened in AutoCAD 2004 and then saved before it appears in the Preview area.

POINTING TO YOUR FAVORITE FOLDERS

One of the more time-consuming tasks involved in CAD is finding where you saved certain files. Even when you know the location of the files, you must perform a series of selections by browsing through various folders to get to the desired file. This is the purpose of identifying your most commonly used folders under the heading of Favorites. In Figure 1–24, suppose you wanted to create a favorite out of the folder called Sample. First click the Sample folder. Then click the down arrow in the Tools area of the dialog box. Choosing the Add to Favorites option allows you to "Add Selected Item to Favorites."

Once this is accomplished, click the Favorites button located along the left side of the Select File dialog box. Notice the folder Sample has been added as one of the favorites in Figure 1–25. You can now select this folder and all drawing files listed in this folder will be displayed.

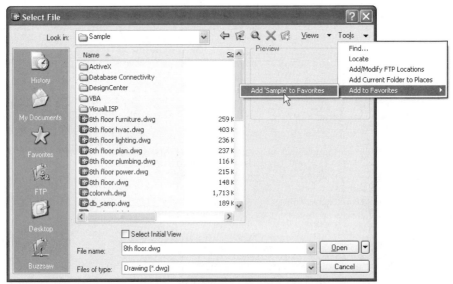

Figure 1–24

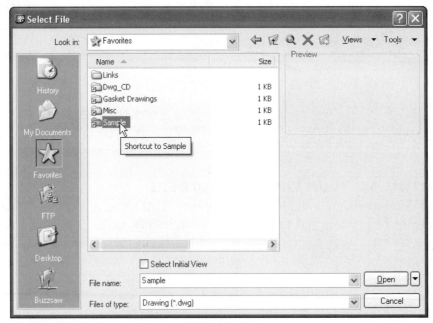

Figure 1–25

BASIC DRAWING TECHNIQUES

The following sections discuss some basic techniques used in creating drawings. These include drawing lines, circles, and polylines, using object snap modes and tracking, and erasing objects.

THE LINE COMMAND

Use the LINE command to construct a line from one endpoint to the other. Choose this command from one of the following:

> From a toolbar button (Insert the file b_draw_line.PCX here)
>
> From the pull-down menu (Draw > Line)
>
> From the keyboard (L or Line)

As the first point of the line is marked, the rubber-band cursor is displayed along with the normal crosshairs to help you see where the next line segment will be drawn. The LINE command stays active until you either use the Close option or issue a null response by pressing ENTER at the prompt "To point."

Try It! – Create a new drawing from scratch. Study Figure 1–26 and follow the command sequence for using the LINE command.

 Command: **L** *(For LINE)*
Specify first point: *(Pick a point at "A")*
Specify next point or [Undo]: *(Pick a point at "B")*
Specify next point or [Undo]: (Pick a point at "C")
Specify next point or [Close/Undo]: *(Pick a point at "D")*
Specify next point or [Close/Undo]: *(Pick a point at "E")*
Specify next point or [Close/Undo]: *(Pick a point at "F")*
Specify next point or [Close/Undo]: **C** *(To close the shape and exit the command)*

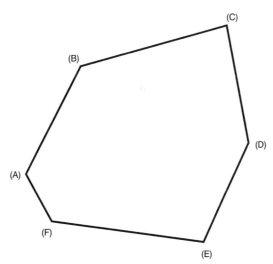

Figure 1–26

From time to time, you might make mistakes in the LINE command by drawing an incorrect segment. As illustrated in Figure 1–27, segment DE is drawn incorrectly. Instead of exiting the LINE command and erasing the line, you can use a built-in Undo within the LINE command. This removes the previously drawn line while still remaining in the LINE command. Refer to Figure 1–27 and the following prompts to use the Undo option of the LINE command.

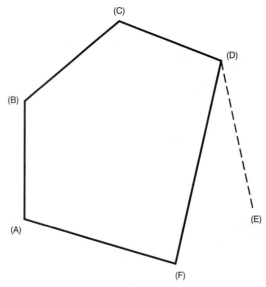

Figure 1–27

⬜ Command: **L** *(For LINE)*
Specify first point: *(Pick a point at "A")*
Specify next point or [Undo]: *(Pick a point at "B")*
Specify next point or [Undo]: *(Pick a point at "C")*
Specify next point or [Close/Undo]: *(Pick a point at "D")*
Specify next point or [Close/Undo]: *(Pick a point at "E")*
Specify next point or [Close/Undo]: **U** *(To undo or remove the segment from "D" to "E"*
    and still remain in the LINE command)
Specify next point or [Close/Undo]: *(Pick a point at "F")*
Specify next point or [Close/Undo]: **End** *(For Endpoint mode)*
of *(Select the endpoint of the line segment at "A")*
Specify next point or [Close/Undo]: *(Press ENTER to exit this command)*

Another option of the LINE command is the Continue option. The dashed line segment in Figure 1–28 was the last segment drawn before the LINE command was exited. To pick up at the last point of a previously drawn line segment, type the LINE command and press ENTER twice. This will activate the Continue option of the LINE command.

Command: **L** *(For LINE)*

Specify first point: *(Press ENTER to activate Continue Mode)*
Specify next point or [Undo]: *(Pick a point at "B")*
Specify next point or [Undo]: *(Pick a point at "C")*
Specify next point or [Close/Undo]: End *(For Endpoint mode)*
of *(Select the endpoint of the vertical line segment at "A")*
Specify next point or [Close/Undo]: *(Press ENTER to exit this command)*

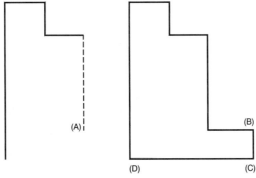

Figure 1–28

THE DIRECT DISTANCE MODE FOR DRAWING LINES

Another method is available for constructing lines, and it is called drawing by Direct Distance mode. In this method, the direction a line will be drawn in is guided by the location of the cursor. You enter a value, and the line is drawn at the specified distance at the angle specified by the cursor. This mode works especially well for drawing horizontal and vertical lines. Figure 1–29 illustrates an example of how the Direct Distance mode is used.

Try It! – Create a new drawing from scratch. Turn Ortho mode On. Then use the following command sequence to construct the line segments using the Direct Distance mode of entry.

 Command: **L** *(For LINE)*

Specify first point: **2.00,2.00**
Specify next point or [Undo]: *(Move the cursor to the right and enter a value of 7.00 units)*
Specify next point or [Undo]: *(Move the cursor up and enter a value of 5.00 units)*
Specify next point or [Close/Undo]: *(Move the cursor to the left and enter a value of 4.00 units)*
Specify next point or [Close/Undo]: *(Move the cursor down and enter a value of 3.00 units)*
Specify next point or [Close/Undo]: *(Move the cursor to the left and enter a value of 2.00 units)*
Specify next point or [Close/Undo]: **C** *(To close the shape and exit the command)*

22

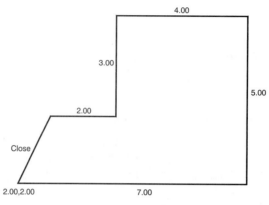

4.00

3.00

2.00

5.00

Close

2.00,2.00   7.00

Figure 1–29

Figure 1–30 shows another example of an object drawn with Direct Distance mode. Each angle was constructed from the location of the cursor. In this example, Ortho mode is turned off.

 Try It! – Create a new drawing from scratch. Be sure Ortho mode is turned Off.

Then use the following command sequence to construct the line segments using the direct distance mode of entry.

 Command: **L** *(For LINE)*
Specify first point: *(Pick a point at "A")*
Specify next point or [Undo]: *(Move the cursor and enter 3.00)*
Specify next point or [Undo]: *(Move the cursor and enter 2.00)*
Specify next point or [Close/Undo]: *(Move the cursor and enter 1.00)*
Specify next point or [Close/Undo]: *(Move the cursor and enter 4.00)*
Specify next point or [Close/Undo]: *(Move the cursor and enter 2.00)*
Specify next point or [Close/Undo]: *(Move the cursor and enter 1.00)*
Specify next point or [Close/Undo]: *(Move the cursor and enter 1.00)*
Specify next point or [Close/Undo]: **C** *(To close the shape and exit the command)*

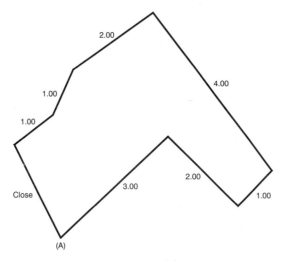

Figure 1–30

USING OBJECT SNAP FOR GREATER PRECISION

The most important mode used for locking onto key locations of objects is Object Snap. Figure 1–31 gives an example of the construction of a vertical line connecting the endpoint of the fillet with the endpoint of the line at "A." The LINE command is entered and the Endpoint mode activated. When the cursor moves over a valid endpoint, an Object Snap symbol appears along with a tooltip telling you which OSNAP mode is currently being used. Another example of the use of Object Snap is in dimensioning applications where exact endpoints and intersections are needed.

 Try It! – Open the drawing file 01_Endpoint. Follow the illustration in Figure 1–31 and the command sequence below to draw a line segment from the endpoint of the arc to the endpoint of the line.

Command: **L** *(For LINE)*
Specify first point: **End** *(For Endpoint mode)*
of *(Pick the endpoint of the fillet illustrated in Figure 1–31)*
Specify next point or [Undo]: **End** *(For Endpoint mode)*
of *(Pick the endpoint of the line at "A")*
Specify next point or [Undo]: *(Press ENTER to exit this command)*

Perform the same operation to the other side of this object using the Endpoint mode of OSNAP.

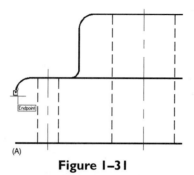

(A)

Figure 1–31

You can choose Object Snap modes in a number of different ways. Illustrated in Figure 1–32 is the standard Object Snap toolbar. Figure 1–33 shows the Object Snap modes that can be activated when you hold down SHIFT and press the right mouse button while within a command such as LINE or MOVE. This shortcut menu will appear wherever the cursor is currently positioned. Another method of activating Object Snap modes is from the keyboard. When you enter Object Snap modes from the keyboard, only the first three letters are required. The following pages give examples of the application of all Object Snap modes.

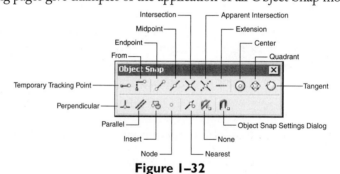

Figure 1–32

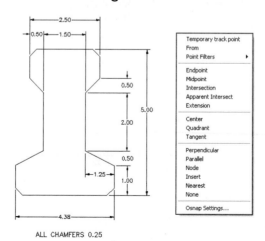

ALL CHAMFERS 0.25

Figure 1–33

OBJECT SNAP MODES

OBJECT SNAP CENTER (CEN)

Use the Center mode to snap to the center of a circle or arc. To accomplish this, activate the mode by clicking the Object Snap Center button and moving the cursor along the edge of the circle or arc similar to Figure 1–34. Notice the AutoSnap symbol appearing at the center of the circle or arc.

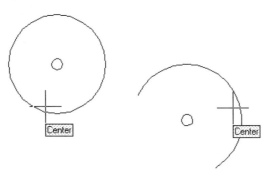

Figure 1–34

OBJECT SNAP ENDPOINT (END)

The Endpoint mode is one of the more popular Object Snap modes; it is helpful in snapping to the endpoints of lines or arcs. One application of Object Snap Endpoint is during the dimensioning process, where exact distances are needed to produce the desired dimension. Activate this mode by clicking the Object Snap Endpoint button, and then move the cursor along the edge of the object to snap to the endpoint. In the case of the line or arc shown in Figure 1–35, the cursor does not actually have to be positioned at the endpoint; favoring one end automatically snaps to the closest endpoint.

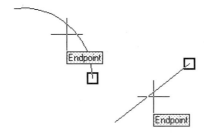

Figure 1–35

OBJECT SNAP EXTENSION (EXT)

When you select a line or an arc, the Object Snap Extension mode creates a temporary path that extends from the object. Picking the end of the line at "A" in

Figure 1–36 acquires the angle of the line. A tooltip displays the current extension distance and angle. Picking the end of the arc at "B" acquires the radius of the arc and displays the current length in the tooltip.

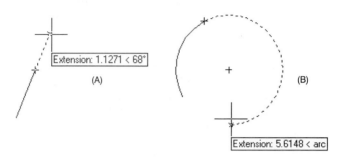

Figure 1–36

 OBJECT SNAP FROM (FRO)

Use the Object Snap From mode along with a secondary Object Snap mode to establish a reference point and construct an offset from that point. Open the drawing file 01_Osnap From. In Figure 1–37, the circle needs to be drawn 1.50 units in the X and Y directions from point "A." The CIRCLE command is activated and the Object Snap From mode is used in combination with the Object Snap Intersection mode. The From option requires a base point. Identify the base point at the intersection of corner "A." The next prompt asks for an offset value; enter the relative coordinate value of @1.50,1.50 (This identifies a point 1.50 units in the positive X-direction and 1.50 units in the positive Y-direction). This completes the use of the From option and identifies the center of the circle at "B." Study the following command sequence to accomplish this operation:

Try It! – Open the drawing file 01_Osnap From. Use the illustration and prompt sequence below for constructing a circle inside the shape with the aid of the Object Snap From mode.

Command: **C** *(For CIRCLE)*
Specify center point for circle or [3P/2P/Ttr (tan tan radius)]: **From**
Base point: **Int** *(For Intersection Mode)*
of *(Select the intersection at "A" in Figure 1–37)*
<Offset>: **@1.50,1.50**
Specify radius of circle or [Diameter]: **D** *(For Diameter)*
Specify diameter of circle: **1.25**

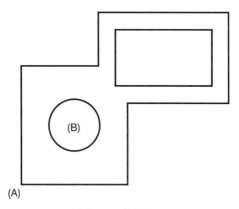

Figure 1–37

OBJECT SNAP INSERT (INS)

The Object Snap Insert mode snaps to the insertion point of an object. In the case of the text object in Figure 1–38, activating the Object Snap Insert mode and positioning the cursor anywhere on the text snaps to its insertion point, in this case at the lower left corner of the text at "A." The other object illustrated in Figure 1–38 is called a block. It appears to be constructed of numerous line objects; however, all objects that make up the block are considered to be a single object. Blocks can be inserted in a drawing. Typical types of blocks are symbols such as doors, windows, bolts, and so on—anything that is used several times in a drawing. In order for a block to be brought into a drawing, it needs an insertion point, or a point of reference. The Object Snap Insert mode, when you position the cursor on a block, will snap to the insertion point of a block.

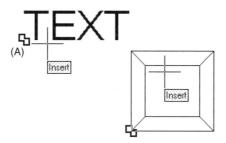

Figure 1–38

OBJECT SNAP INTERSECTION (INT)

Another popular Object Snap mode is Intersection. Use this mode to snap to the intersection of two objects. Position the cursor anywhere near the intersection of two objects and the intersection symbol appears. See Figure 1–39.

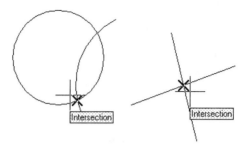

Figure 1–39

OBJECT SNAP EXTENDED INTERSECTION (INT)

Another type of intersection snap is the Object Snap Extended Intersection mode, which is used to snap to an intersection not considered obvious from the previous example. The same Object Snap Intersection button is for performing an extended intersection operation. Figure 1–40 shows two lines that do not intersect. Activate the Object Snap Extended Intersection mode and pick both lines. Notice the intersection symbol present where the two lines, if extended, would intersect.

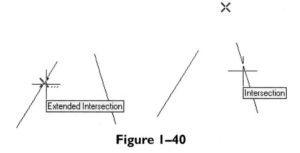

Figure 1–40

OBJECT SNAP MIDPOINT (MID)

The Object Snap Midpoint mode snaps to the midpoint of objects. Line and arc examples are shown in Figure 1–41. When activating the Object Snap Midpoint mode, touch the object anywhere with some portion of the cursor; the midpoint symbol will appear at the exact midpoint of the object.

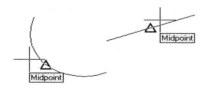

Figure 1–41

OBJECT SNAP NEAREST (NEA)

The Object Snap Nearest mode snaps to the nearest point it finds on an object. Use this mode when you need to grab onto an object for the purposes of further editing. The nearest point is calculated based on the closest distance from the intersection of the crosshairs perpendicular to the object or the shortest distance from the crosshairs to the object. In Figure 1–42, the appearance of the Object Snap Nearest symbol helps to show where the point identified by this mode is actually located.

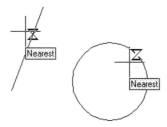

Figure 1–42

OBJECT SNAP NODE (NOD)

The Object Snap Node mode snaps to a node or point. Touching the point in Figure 1–43 snaps to its center. The Object Snap Nearest mode shown in Figure 1–42 can also be used to snap to a point.

Figure 1–43

OBJECT SNAP PARALLEL (PAR)

Use the Object Snap Parallel mode to construct a line parallel to another line. In Figure 1–44, the LINE command is started and a beginning point of the line is picked. The Object Snap Parallel mode is activated and an existing line is highlighted at "A"; the Object Snap Parallel symbol appears. Finally, move your cursor to the position that makes the new line parallel to the one just selected. The line snaps parallel along with the presence of the tracking path and tooltip giving the current distance and angle.

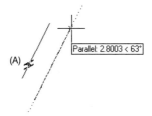

Figure 1–44

OBJECT SNAP PERPENDICULAR (PER)

The Object Snap Perpendicular mode is helpful for snapping to an object normal (or perpendicular) from a previously identified point. Figure 1–45 shows a line segment drawn perpendicular from the point at "A" to the inclined line "B." A 90° angle is formed with the perpendicular line segment and the inclined line "B." With this mode, you can also construct lines perpendicular to circles.

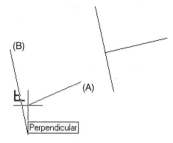

Figure 1–45

OBJECT SNAP QUADRANT (QUA)

Circle quadrants are defined as points located at the 0°, 90°, 180°, and 270° positions of a circle, as in Figure 1–46. Using the Object Snap Quadrant mode will snap to one of these four positions as the edge of a circle or arc is selected. In the example of the circle in Figure 1–46, the edge of the circle is selected by the cursor location. The closest quadrant to the cursor is selected.

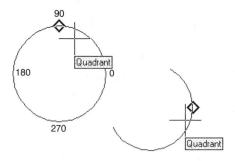

Figure 1–46

OBJECT SNAP TANGENT (TAN)

The Object Snap Tangent mode is helpful in constructing lines tangent to other objects such as the circles in Figure 1–47. In this case, the Object Snap Tangent mode is being used in conjunction with the LINE command. The point at "A" is anchored to the top of the circle using the Object Snap Quadrant mode. When dragged to the next location, the line will not appear to be tangent. However, with Object Snap Tangent mode activated and the location at "B" picked, the line will be tangent to the large circle near "B."

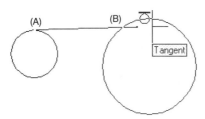

Figure 1–47

 Try It! – Open the drawing file 01_Tangent. Follow this command sequence for constructing a line segment tangent to two circles:

 Command: **L** *(For LINE)*
Specify first point: **Qua** *(For Quadrant mode)*
from *(Select the circle near "A")*
Specify next point or [Undo]: **Tan** *(For Tangent mode)*
to *(Select the circle near "B")*

OBJECT SNAP DEFERRED TANGENT (TAN)

When a line is constructed tangent from one point and tangent to another, the Deferred Tangent mode is automatically activated (see Figure 1–48). This means that instead of the rubber-band cursor being activated where a line is present, the deferred tangent must have two tangent points identified before the line is constructed and displayed.

Try It! – Open the drawing file 01_Deferred Tangent. Follow this command sequence for constructing a line segment tangent to two circles:

Command: **L** *(For LINE)*
Specify first point: **Tan** *(For Tangent mode)*
from *(Select the circle near "A")*
Specify next point or [Undo]: **Tan** *(For Tangent mode)*
to *(Select the circle near "B")*

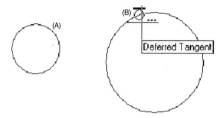

Figure 1–48

 Try It! – Open the drawing file 01_Osnap. Various objects consisting of lines, circles, arcs, points, and blocks need to be connected up with line segments at their key locations. Follow the prompt sequence and Figure 1–49 below for performing this operation.

Command: **L** *(For LINE)*
Specify first point: **End**
of *(Pick the endpoint at "A")*
Specify next point or [Undo]: **Nod**
of *(Pick the node at "B")*
Specify next point or [Undo]: **Tan**
to *(Pick the circle at "C")*
Specify next point or [Close/Undo]: **Int**
of *(Pick the intersection at "D")*
Specify next point or [Close/Undo]: **Int**
of *(Pick the line at "E")*
and *(Pick the horizontal line at "F")*
Specify next point or [Close/Undo]: **Qua**
of *(Pick the circle at "G")*
Specify next point or [Close/Undo]: **Cen**
of *(Pick the arc at "H")*
Specify next point or [Close/Undo]: **Mid**
of *(Pick the line at "J")*
Specify next point or [Close/Undo]: **Per**
to *(Pick the line at "K")*
Specify next point or [Close/Undo]: **Ins**
of *(Pick on the I-Beam symbol near "L")*
Specify next point or [Close/Undo]: **Nea**
to *(Pick the circle at "M")*
Specify next point or [Close/Undo]: *(Press ENTER to exit this command)*

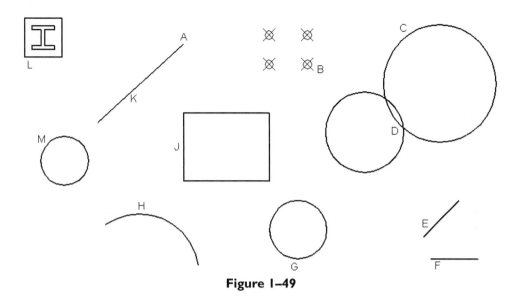

Figure 1–49

CHOOSING RUNNING OBJECT SNAP

So far, all Object Snap modes have been selected from the shortcut menu or entered at the keyboard. The problem with these methods is that if you're using a certain mode over a series of commands, you have to select the Object Snap mode every time. It is possible to make the Object Snap mode or modes continuously present through Running Osnap. Select Drafting Settings from the Tools pull-down menu in Figure 1–50 to activate the Drafting Settings dialog box and the Object Snap tab. You could also right click the OSNAP button located in the Status bar located at the bottom of your screen, as shown in Figure 1–51. Selecting Settings from the shortcut menu will activate the Drafting Settings dialog box illustrated in Figure 1–52. By default, the Endpoint, Center, Intersection, and Extension modes are automatically selected. Whenever your cursor lands over an object supported by one of these four modes, a yellow symbol appears to alert you to the mode. It is important to know that when you make changes inside this dialog box, the changes are applied to all drawing files. Even if you quit a drawing session without saving, the changes made to the Object Snap modes are automatic. You can select other Object Snap modes by checking their appropriate boxes in the dialog box; removing the check disables the mode.

Figure 1–50

Figure 1–51

Figure 1–52

These Object Snap modes remain in effect during drawing until you click the OSNAP button illustrated in the Status Bar in Figure 1–53; this turns off the current running Object Snap modes. To reactivate the running Object Snap modes, click the OSNAP button again and the previous set of Object Snap modes will be back in effect. You can also activate and reactivate running Object Snap by pressing the F3 function key.

SNAP	GRID	ORTHO	POLAR	OSNAP	OTRACK	LWT	MODEL

Figure 1–53

OBJECT SNAP TRACKING

In Figure 1–54, a hole represented in the top view as a series of hidden lines needs to be displayed as a circle in the front view. The hole will be placed in the center of the rectangle identified by "A" and "B." Tracking mode provides a quick and easy way to locate the center of the rectangle without drawing any construction geometry.

When you start tracking and pick a point on the screen, AutoCAD forces the next point to conform to an orthogonal path that could extend vertically or horizontally from the first point. The first point may retain an X or Y value and the second point may retain separate X or Y values. Use the following command sequence to get an idea of the operation of Tracking mode. For best results, be sure Ortho mode is turned off.

 Try It! – Open the drawing file 01_Tracking. Follow the prompt sequence and illustrations below for completing this operation.

Command: **C** *(For CIRCLE)*
Specify center point for circle or [3P/2P/Ttr (tan tan radius)]: **TK** *(For Tracking)*
First tracking point: **Mid**
of *(Pick the midpoint at "A" in Figure 1–54)*
Next point (Press ENTER to end tracking): **Mid**
of *(Pick the midpoint at "B" in Figure 1–54)*
Next point (Press ENTER to end tracking): *(Press ENTER)*
Specify radius of circle or [Diameter]: **D** *(For Diameter)*
Specify diameter of circle: **0.50**

The result is illustrated in Figure 1–55, with the hole drawn in the center of the rectangular shape. You could also set Running OSNAP to Midpoint, turn OSNAP on and turn OTRACK on to achieve the same results as in the command sequence above. This will be discussed later in this chapter.

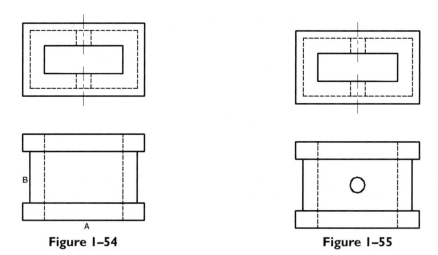

Figure 1–54   Figure 1–55

POLAR TRACKING

Previously in this chapter, the Direct Distance mode was highlighted as an efficient means of constructing orthogonal lines without the need for using one of the coordinate modes of entry. However it did not lend itself to lines at angles other than 0°, 90°, 180°, and 270°. Polar tracking has been designed to remedy this. This mode allows the cursor to follow a tracking path that is controlled by a preset angular increment. The POLAR button located at the bottom of the display in the Status area turns this mode On or Off. Choosing Drafting Settings from the Tools pull-down menu in Figure 1–50 displays the Drafting Settings dialog box shown in Figure 1–56 and the Polar Tracking tab.

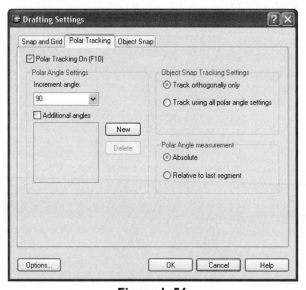

Figure 1–56

Before continuing, you need to be familiar with a few general terms:

Tracking Path – This is a temporary dotted line that can be considered a type of construction line. Your cursor will glide or track along this path (see Figure 1–57).

Tooltip – This displays the current cursor distance and angle away from the tracking point (see Figure 1–57).

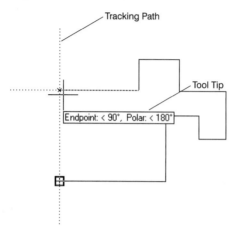

Figure 1–57

 Tip: Both POLAR and ORTHO modes cannot be turned on at the same time. Once you turn POLAR on, ORTHO automatically turns off, and vice versa.

 Try It! – To see how the Polar Tracking mode functions, construct an object that consists of line segments at 10° angular increments. Create a new drawing starting from scratch. Then, set the angle increment through the Polar Tracking tab of the Drafting Settings dialog box, shown in Figure 1–58.

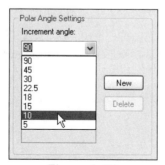

Figure 1–58

Start the LINE command, anchor a starting point at "A," move your cursor to the upper right until the tooltip reads 20°. Enter a value of 2 units for the length of the line segment (see Figure 1–59).

Move your cursor up and to the left until the tooltip reads 110° as in Figure 1–60 and enter a value of 2 units. (This will form a 90° angle with the first line.)

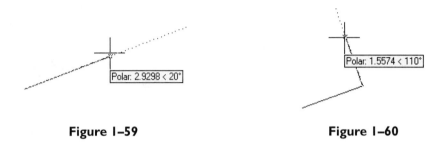

Figure 1–59   **Figure 1–60**

Move your cursor until the tooltip reads 20° as in Figure 1–61, and enter a value of 1 unit.

Move your cursor until the tooltip reads 110° as in Figure 1–62, and enter a value of 1 unit.

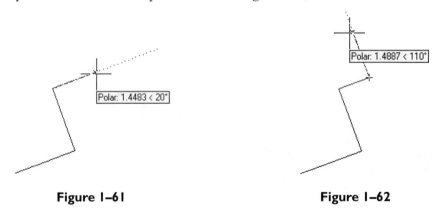

Figure 1–61   **Figure 1–62**

Move your cursor until the tooltip reads 200° as in Figure 1–63, and enter a value of 3 units.

Move your cursor to the endpoint as in Figure 1–64, or use the Close option of the LINE command to close the shape and exit the command.

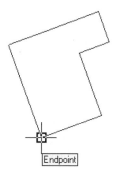

Figure 1–63   **Figure 1–64**

SETTING A POLAR SNAP VALUE

An additional feature of using polar snap is illustrated in Figure 1–65. Clicking the Snap and Grid tab of the Drafting Settings dialog box displays the dialog box in the figure. Clicking the Polar Snap option found along the lower right corner of the dialog box allows you to enter a polar distance. When SNAP and POLAR are both turned on and you move your cursor to draw a line, not only will the angle be set but your cursor will also jump to the next increment set by the polar snap value.

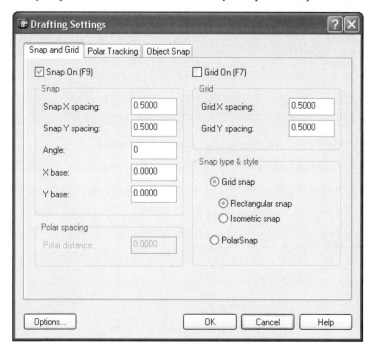

Figure 1–65

Try It! – Open the drawing file 01_Polar. Set the polar angle to 30 degrees and a polar snap distance to 0.50 unit increments. Be sure POLAR and SNAP are both turned On in your status bar and that all other modes are turned off. Begin constructing the object in Figure 1–66 using the command prompt sequence below as a guide.

Command: **L** *(For LINE)*
Specify first point: **7.00,4.00**
Specify next point or [Undo]: *(Move your cursor down until the tooltip reads Polar: 2.5000<270 and pick a point)*
Specify next point or [Undo]: *(Move your cursor right until the tooltip reads Polar: 1.5000<0 and pick a point)*
Specify next point or [Close/Undo]: *(Polar: 2.0000<30)*
Specify next point or [Close/Undo]: *(Polar: 2.0000<60)*
Specify next point or [Close/Undo]: *(Polar: 2.5000<90)*
Specify next point or [Close/Undo]: *(Polar: 3.0000<150)*
Specify next point or [Close/Undo]: *(Polar: 1.5000<180)*
Specify next point or [Close/Undo]: *(Polar: 2.5000<240)*
Specify next point or [Close/Undo]: *(Polar: 2.5000<120)*
Specify next point or [Close/Undo]: *(Polar: 1.5000<180)*
Specify next point or [Close/Undo]: *(Polar: 3.0000<210)*
Specify next point or [Close/Undo]: *(Polar: 2.5000<270)*
Specify next point or [Close/Undo]: *(Polar: 2.0000<300)*
Specify next point or [Close/Undo]: *(Polar: 2.0000<330)*
Specify next point or [Close/Undo]: *(Polar: 1.5000<0)*
Specify next point or [Close/Undo]: *(Polar: 2.5000<90)*
Specify next point or [Close/Undo]: **C** *(To close the shape)*

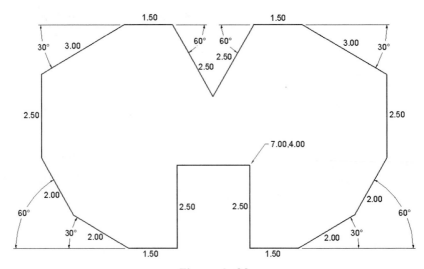

Figure 1–66

MORE ON OBJECT SNAP TRACKING MODE

Object Snap Tracking works in conjunction with Object Snap. Before you can track from an Object Snap point, you must first set an Object Snap mode or modes from the Object Snap tab of the Drafting Settings dialog box. Object Snap Tracking can be toggled on or off with the OTRACK button located in the Status bar in Figure 1–67.

| SNAP | GRID | ORTHO | POLAR | OSNAP | OTRACK | LWT | MODEL |

Figure 1–67

The advantage of using Object Snap Tracking is in the ability to choose or acquire points to be used for construction purposes. Care must be taken when acquiring points that you in fact do not pick these points. They are used only for construction purposes. For example, two line segments need to be added to the object in Figure 1–68 to form a rectangle. Here is how to perform this operation using Polar Tracking.

 Try It! – Open the drawing file 01_Otrack Lines. Notice in the status bar that POLAR, OSNAP, and OTRACK are all turned on. Be sure that Running Osnap is set to Endpoint mode. Enter the line command and pick a starting point for the line at "A." Then, move your cursor directly to the left until the tooltip reads 180°. The starting point for the next line segment is considered acquired (see Figure 1–69).

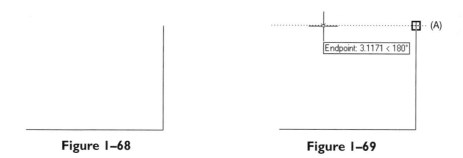

Endpoint: 3.1171 < 180°   (A)

Figure 1–68   **Figure 1–69**

Rather than enter the length of this line segment, move your cursor over the top of the corner at "B" to acquire this point (be careful not to pick the point here). Then move your cursor up until the tooltip reads 90° (see Figure 1–70).

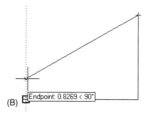

(B)   Endpoint: 0.8269 < 90°

Figure 1–70

Move your cursor up until the tooltip now reads angles of 90° and 180°. Also notice the two tracking paths intersecting at the point of the two acquired points. Picking this point at "C" will construct the horizontal line segment (see Figure 1–71). Finally, slide your cursor to the endpoint at "D" to complete the rectangle (see Figure 1–72).

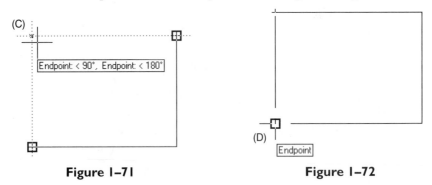

Figure 1–71   Figure 1–72

 Tip: Pausing the cursor over an existing acquired point a second time will remove the tracking point from the object.

 Try It! – Open the drawing file -1_Otrack Pipes. Set Running Osnap mode to Midpoint. Set the polar angle to 90 degrees. Be sure POLAR, OSNAP, and OTRACK are all turned on. Enter the line command and connect all fittings with lines illustrated in Figure 1–73.

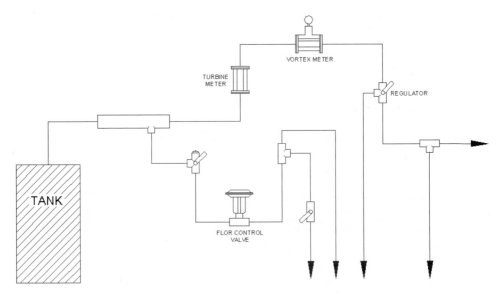

Figure 1–73

USING TEMPORARY TRACKING POINTS

A very powerful construction tool includes the ability to use an extension path along with the Temporary Tracking Point tool to construct objects under difficult situations, as illustrated in Figure 1–74.

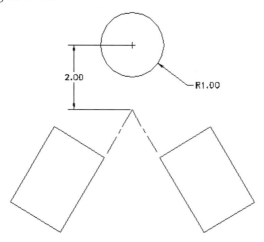

Figure 1–74

An extension path is similar to a tracking path except that it is present when the Object Snap Extension mode is activated. To construct the circle in relation to the two inclined rectangles, follow the next series of steps.

 Try It! – Open the drawing file 01_Temporary Point. Set Running Osnap to Endpoint, Intersection, Center, and Extension. Check to see that OSNAP and OTRACK are turned on and all other modes are turned off. Activate the circle command; this will prompt you to specify the center point for the circle. Move your cursor over the corner of the right rectangle at "A" to acquire this point. Move your cursor up and to the left, making sure the tooltip lists the Extension mode (see Figure 1–75).

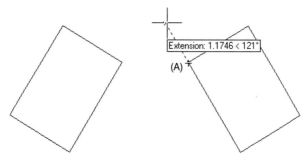

Figure 1–75

With the point acquired at "A," move your cursor over the corner of the left rectangle at "B" and acquire this point. Move your cursor up and to the right, making sure the tooltip lists the Extension mode (see Figure 1–76).

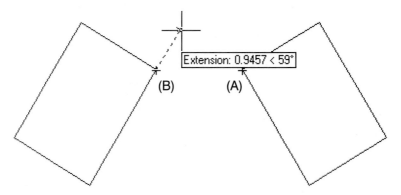

Figure 1–76

Keep moving your cursor up until both acquired points intersect as in Figure 1–77. The center of the circle is located 2 units above this intersection. Click the Object Snap Temporary Tracking button and pick this intersection.

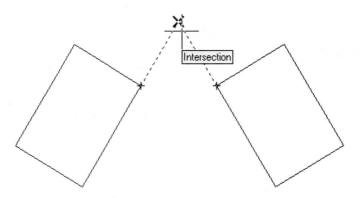

Figure 1–77

Next, move your cursor directly above the temporary tracking point as in Figure 1–78. The tooltip should read 90°. Entering a value of 2 units identifies the center of the circle.

The completed construction operation is illustrated in Figure 1–79.

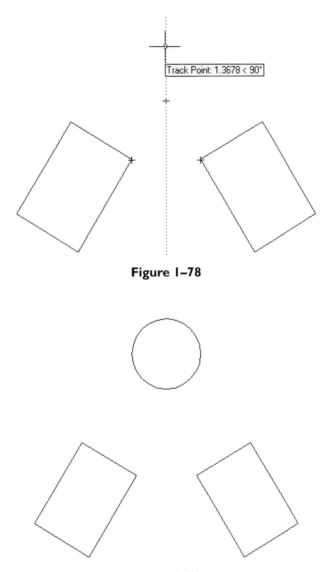

Figure 1–78

Figure 1–79

ALTERNATE METHODS USED FOR PRECISION DRAWING: CARTESIAN COORDINATES

Before drawing precision geometry such as lines and circles, you must have an understanding of coordinate systems. The Cartesian or rectangular coordinate system is used to place geometry at exact distances through a series of coordinates. A coordinate is made up of an ordered pair of numbers identified as X and Y. The coordinates are then plotted on a type of graph or chart. The graph, shown in Figure 1–80, is made up of two perpendicular number lines called coordinate axes. The horizontal axis is called the X-axis. The vertical axis is called the Y-axis.

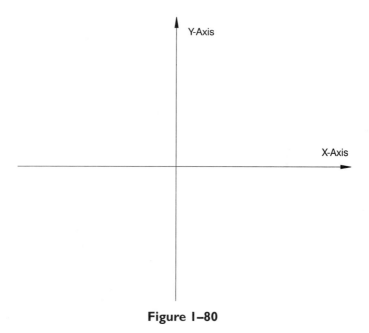

Figure 1–80

As shown in Figure 1–81, the intersection of the two coordinate axes forms a point called the origin. Coordinates used to describe the origin are 0,0. From the origin, all positive directions move up and to the right. All negative directions move down and to the left.

The coordinate axes are divided into four quadrants that are labeled I, II, III, and IV, as shown in Figure 1–82. In Quadrant I, all X and Y values are positive. Quadrant II has a negative X value and positive Y value. Quadrant III has negative values for X and Y. Quadrant IV has positive X values and negative Y values.

For each ordered pair of (X,Y) coordinates, X means to move from the origin to the right if positive and to the left if negative. Y means to move from the origin up if positive and down if negative. Figure 1–83 shows a series of coordinates plotted on the number lines. One coordinate is identified in each quadrant to show the positive and negative values. As an example, coordinate 3,2 in the Quadrant I means to move 3 units to the right of the origin and up 2 units. The coordinate -5,3 in Quadrant II means to move 5 units to the left of the origin and up 3 units. Coordinate -2,-2 in Quadrant III means to move 2 units to the left of the origin and down 2. Lastly, coordinate 2,-4 in Quadrant IV means to move 2 units to the right of the origin and down -4.

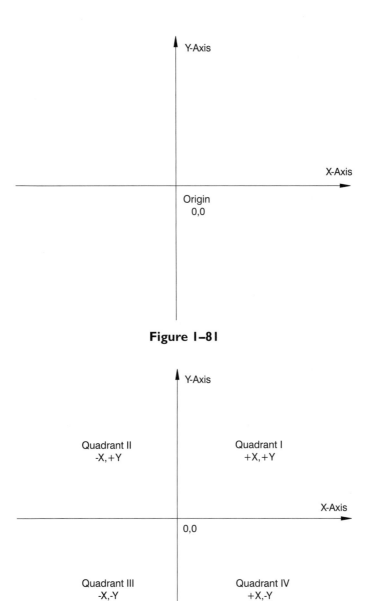

Figure 1–81

Figure 1–82

When you begin a drawing in AutoCAD, the screen display reflects Quadrant I of the Cartesian coordinate system in Figure 1–84. The origin 0,0 is located in the lower left corner of the drawing screen.

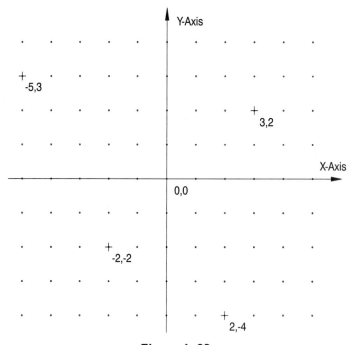

Figure 1–83

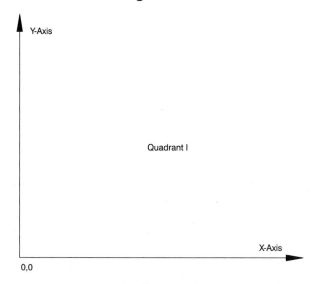

Figure 1–84

ABSOLUTE COORDINATE MODE FOR DRAWING LINES

When drawing geometry such as lines, you must use a method of entering precise distances, especially when accuracy is important. This is the main purpose of using coordinates. The simplest and most elementary form of coordinate values is absolute coordinates. Absolute coordinates conform to the following format:

$$X,Y$$

One problem with using absolute coordinates is that all coordinate values refer back to the origin 0,0. This origin on the AutoCAD screen is usually located in the lower left corner when a new drawing is created. The origin will remain in this corner unless it is altered with the LIMITS command.

 Try It! – Create a new drawing file starting from scratch. Use the following line command prompts and illustration in Figure 1–85 to construct the shape.

Command: **L** *(For LINE)*
Specify first point: **2,2** *(at "A")*
Specify next point or [Undo]: **2,7** *(at "B")*
Specify next point or [Undo]: **5,7** *(at "C")*
Specify next point or [Close/Undo]: **7,4** *(at "D")*
Specify next point or [Close/Undo]: **10,4** *(at "E")*
Specify next point or [Close/Undo]: **10,2** *(at "F")*
Specify next point or [Close/Undo]: **C** *(To close the shape and exit the command)*

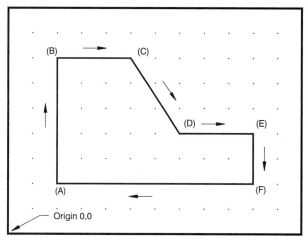

Figure 1–85

As you can see, all points on the object make reference to the origin at 0,0. Even though absolute coordinates are useful in starting lines, there are more efficient ways to construct lines and draw objects.

RELATIVE COORDINATE MODE FOR DRAWING LINES

With absolute coordinates, the origin at 0,0 must be kept track of at all times in order for the correct coordinate to be entered. With complicated objects, this is sometimes difficult to accomplish and as a result, the wrong coordinate may be entered. It is possible to reset the last coordinate to become a new origin or 0,0 point. The new point would be relative to the previous point, and for this reason, this point is called a relative coordinate. The format is as follows:

$$@X,Y$$

In this format, we use the same X and Y values with one exception: the At symbol or @ resets the previous point to 0,0 and makes entering coordinates less confusing.

 Try It! – Create a new drawing file starting from scratch. Use the following line command prompts and illustration in Figure 1–86 to construct the shape.

Command: **L** *(For LINE)*
Specify first point: **2,2** *(at "A")*
Specify next point or [Undo]: **@0,4** *(to "B")*
Specify next point or [Undo]: **@4,2** *(to "C")*
Specify next point or [Close/Undo]: **@3,0** *(to "D")*
Specify next point or [Close/Undo]: **@3,-4** *(to "E")*
Specify next point or [Close/Undo]: **@-3,-2** *(to "F")*
Specify next point or [Close/Undo]: **@-7,0** *(back to "A")*
Specify next point or [Close/Undo]: *(Press ENTER to exit this command)*

 Tip: In each command, the @ symbol resets the previous point to 0,0.

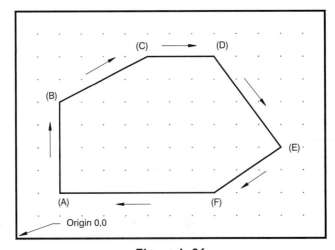

Figure 1–86

POLAR COORDINATE MODE FOR DRAWING LINES

Another popular method of entering coordinates is the polar coordinate mode. The format is as follows:

@Distance<Direction

As the preceding format implies, the polar coordinate mode requires a known distance and a direction. The @ symbol resets the previous point to 0,0. The direction is preceded by the < symbol, which reads the next number as a polar or angular direction. Figure 1–87 describes the directions supported by the polar coordinate mode.

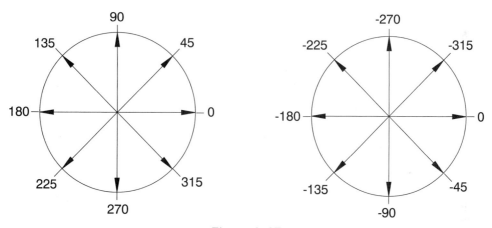

Figure 1–87

Try It! – Create a new drawing file starting from scratch. Use the following line command prompts and illustration in Figure 1–88 to construct the shape.

Command: **L** *(For LINE)*
Specify first point: **3,2** *(at "A")*
Specify next point or [Undo]: **@8<0** *(to "B")*
Specify next point or [Undo]: **@5<90** *(to "C")*
Specify next point or [Close/Undo]: **@5<180** *(to "D")*
Specify next point or [Close/Undo]: **@4<270** *(to "E")*
Specify next point or [Close/Undo]: **@2<180** *(to "F")*
Specify next point or [Close/Undo]: **@2<90** *(to "G")*
Specify next point or [Close/Undo]: **@1<180** *(to "H")*
Specify next point or [Close/Undo]: **@3<270** *(to close back to "A")*
Specify next point or [Close/Undo]: *(Press ENTER to exit this command)*

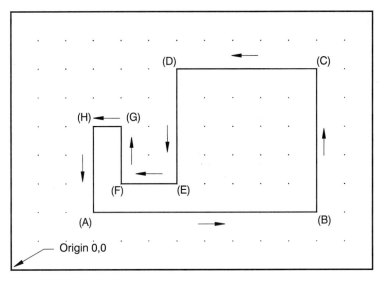

Origin 0,0

Figure 1–88

COMBINING COORDINATE MODES FOR DRAWING LINES

So far, the preceding pages concentrated on using each example of coordinate modes (absolute, relative, and polar) separately to create geometry. At this point, we do not want to give the impression that once you start with a particular coordinate mode, you must stay with the mode. Rather, you create drawings using one, two, or three coordinate modes in combination with each other. In Figure 1–89, the drawing starts with an absolute coordinate, changes to a polar coordinate, and changes again to a relative coordinate. It is your responsibility as a CAD operator to choose the most efficient coordinate mode to fit the drawing.

 Try It! – Create a new drawing file starting from scratch. Use the following line command prompts and illustration in Figure 1–89 to construct the shape.

Command: **L** *(For LINE)*
Specify first point: **2,2** (at "A")
Specify next point or [Undo]: **@3<90** *(to "B")*
Specify next point or [Undo]: **@2,2** *(to "C")*
Specify next point or [Close/Undo]: **@6<0** *(to "D")*
Specify next point or [Close/Undo]: **@5<270** *(to "E")*
Specify next point or [Close/Undo]: **@3<180** *(to "F")*
Specify next point or [Close/Undo]: **@3<90** *(to "G")*
Specify next point or [Close/Undo]: **@2<180** *(to "H")*
Specify next point or [Close/Undo]: **@-3,-3** *(back to "A")*
Specify next point or [Close/Undo]: *(Press ENTER to exit this command)*

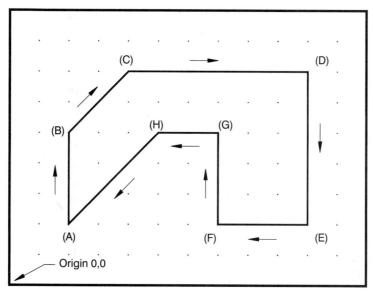

Figure 1–89

THE CIRCLE COMMAND

The CIRCLE command constructs various types of circles. This command can be selected from any of the following:

> From a toolbar button (Insert the file b_draw_circle.PCX here)
>
> From the pull-down menu (Draw > Circle)
>
> From the keyboard (C or Circle)

Choosing Circle from the Draw pull-down menu displays the cascading menu shown in Figure 1–90. All supported methods of constructing circles are displayed in the list. Circles may be constructed by radius or diameter. This command also supports circles defined by three points and construction of a circle tangent to other objects in the drawing. These last two modes will be discussed in chapter 5, "Performing Geometric Constructions."

54

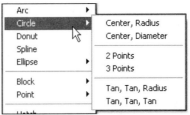

Figure 1–90

CIRCLE BY RADIUS MODE

Use the CIRCLE command and the Radius mode to construct a circle by a radius value that you specify. After selecting a center point for the circle, you are prompted to enter a radius for the desired circle.

 Try It! – Create a new drawing file starting from scratch. Use the following circle command prompts and illustration in Figure 1–91 to construct a circle by radius.

 Command: **C** *(For CIRCLE)*

Specify center point for circle or [3P/2P/Ttr *(tan tan radius)*]: *(Mark the center at "A")*
Specify radius of circle or [Diameter]: **1.50**

CIRCLE BY DIAMETER MODE

Use the CIRCLE command and the Diameter mode to construct a circle by a diameter value that you specify. After selecting a center point for the circle, you are prompted to enter a diameter for the desired circle.

 Try It! – Create a new drawing file starting from scratch. Use the following circle command prompts and illustration in Figure 1–92 to construct a circle by diameter.

 Command: **C** *(For CIRCLE)*

Specify center point for circle or [3P/2P/Ttr (tan tan radius)]: *(Mark the center at "A")*
Specify radius of circle or [Diameter]: **D** *(For Diameter)*
Specify diameter of circle: **3.00**

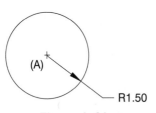

Figure 1–91

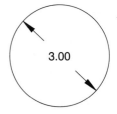

Figure 1–92

USING THE PLINE COMMAND

Polylines are similar to individual line segments except that a polyline can consist of numerous segments and still be considered a single object. Width can also be assigned to a polyline, unlike regular line segments; this makes polylines perfect for drawing borders and title blocks. Polylines can be constructed by selecting any of the following:

> From a toolbar button (Insert the file b_draw_pline.PCX here)
>
> From the pull-down menu (Draw > Polyline)
>
> From the keyboard (PL or Pline)

Study Figures 1–93 and 1–94 and both command sequences that follow to use the PLINE command.

 Try It! – Create a new drawing file starting from scratch. Follow the command prompt sequence and illustration below to construct the polyline.

 Command: **PL** *(For PLINE)*
Specify start point: *(Pick a point at "A" in Figure 1–93)*
Current line-width is 0.0000
Specify next point or [Arc/Close/Halfwidth/Length/Undo/Width]: **W** *(For Width)*
Specify starting width <0.0000>: **0.10**
Specify ending width <0.1000>: *(Press ENTER to accept the default)*
Specify next point or [Arc/Close/Halfwidth/Length/Undo/Width]: *(Pick a point at "B")*
Specify next point or [Arc/Close/Halfwidth/Length/Undo/Width]: *(Pick a point at "C")*
Specify next point or [Arc/Close/Halfwidth/Length/Undo/Width]: *(Pick a point at "D")*
Specify next point or [Arc/Close/Halfwidth/Length/Undo/Width]: *(Pick a point at "E")*
Specify next point or [Arc/Close/Halfwidth/Length/Undo/Width]: *(Press ENTER to exit this command)*

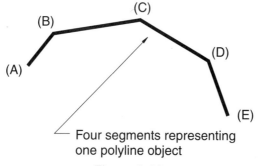

Four segments representing one polyline object

Figure 1–93

 Try It! – Create a new drawing file starting from scratch. Follow the command prompt sequence and illustration below to construct the polyline object. The Direct Distance mode of entry could also be used to construct this object.

Command: **PL** *(For PLINE)*
Specify start point: *(Pick a point at "A" in Figure 1–94)*
Current line-width is 0.0000
Specify next point or [Arc/Close/Halfwidth/Length/Undo/Width]: **@1.00<0** *(To "B")*
Specify next point or [Arc/Close/Halfwidth/Length/Undo/Width]: **@2.00<90** *(To "C")*
Specify next point or [Arc/Close/Halfwidth/Length/Undo/Width]: **@0.50<0** *(To "D")*
Specify next point or [Arc/Close/Halfwidth/Length/Undo/Width]: **@0.75<90** *(To "E")*
Specify next point or [Arc/Close/Halfwidth/Length/Undo/Width]: **@0.75<180** *(To "F")*
Specify next point or [Arc/Close/Halfwidth/Length/Undo/Width]: **@2.00<270** *(To "G")*
Specify next point or [Arc/Close/Halfwidth/Length/Undo/Width]: **@0.50<180** *(To "H")*
Specify next point or [Arc/Close/Halfwidth/Length/Undo/Width]: **@2.00<90** *(To "I")*
Specify next point or [Arc/Close/Halfwidth/Length/Undo/Width]: **@0.75<180** *(To "J")*
Specify next point or [Arc/Close/Halfwidth/Length/Undo/Width]: **@0.75<270** *(To "K")*
Specify next point or [Arc/Close/Halfwidth/Length/Undo/Width]: **@0.50<0** *(To "L")*
Specify next point or [Arc/Close/Halfwidth/Length/Undo/Width]: **C** *(To close the shape and exit the PLINE command)*

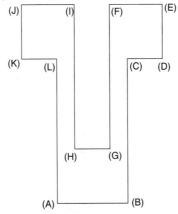

Figure 1–94

ERASING OBJECTS

Throughout the design process, as objects such as lines and circles are placed in a drawing, changes in the design will require the removal of objects. The ERASE command will delete objects from the database. The ERASE command is selected from any of the following:

From a toolbar button (Insert the file b_modify_erase.PCX here)

From the pull-down menu (Modify > Erase)

From the keyboard (E or Erase)

In Figure 1–95, line segments "A" and "B" need to be removed in order for a new line to be constructed closing the shape. Two ways of erasing these lines will be introduced here.

When first entering the ERASE command, you are prompted to "Select objects:" to erase. Notice that your cursor changes in appearance from crosshairs to a pick box, shown in Figure 1–96. Move the pick box over the object to be selected and pick this item. Notice that it will be highlighted as a dashed object to signify it is now selected. At this point, the "Select objects:" prompt appears again. If you need to pick more objects to erase, select them now. If you are finished selecting objects, press ENTER. This will perform the erase operation.

 Command: **E** *(For ERASE)*

Select objects: *(Pick line "A" in Figure 1–96)*
Select objects: *(Press* ENTER *to perform the erase operation)*

Another method of erasing is illustrated in Figure 1–97. Instead of using the ERASE command, highlight the line by selecting it with your crosshairs from the "Command:" prompt. Square boxes also appear at the endpoints and midpoints of the line. With this line segment highlighted, press DELETE on your keyboard, and this will remove the line from the drawing.

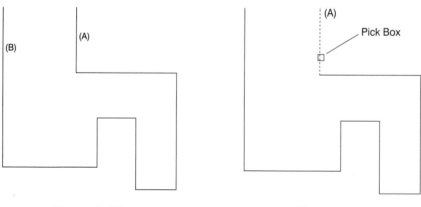

Figure 1–95   **Figure 1–96**

With both line segments erased, a new line is constructed from the endpoint at "A" to the endpoint at "B" as in Figure 1–98.

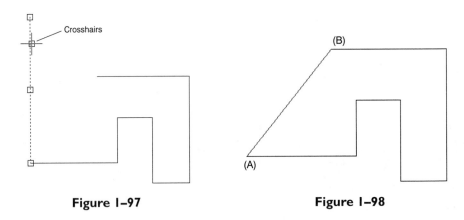

Figure 1–97   Figure 1–98

💾 SAVING A DRAWING FILE

You can save drawings using the QSAVE, SAVE, and SAVEAS commands. The QSAVE command can be selected from the following:

> From a toolbar button (Insert the file b_std_qsave.PCX here)
>
> From the pull-down menu (File > Save)
>
> From the keyboard (New)

The SAVEAS command can be selected from any of the following:

> From the pull-down menu (File > Save As)
>
> From the keyboard (Saveas)

Two of these three commands are found on the File menu, as shown in Figure 1–99.

💾 SAVE

This command is short for QSAVE, which stands for Quick Save. If a drawing file has never been saved and this command is selected, the dialog box shown in Figure 1–100 is displayed. Once a drawing file has been initially saved, selecting this command performs an automatic save and no longer displays the Save Drawing As dialog box.

SAVE AS

Using the Save As command always displays the dialog box shown in Figure 1–100. Simply click the Save button or press ENTER to save the drawing under the current name, displayed in the edit box. This command is more popular for saving the current drawing under an entirely different name. Simply enter the new name in place of the highlighted name in the edit box. Once a drawing is given a new name through this command, it also becomes the new current drawing file.

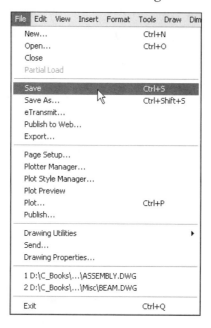

Figure 1–99

The ability to exchange drawings with past releases of AutoCAD is still important to many industry users. When you click in the Save As type edit box in Figure 1–100 a drop-down list appears. Use this list to save a drawing file in AutoCAD 2000/LT 2000 formats. You can also save a drawing file as a Drawing Standard (.dws), a Drawing Template (.dwt), and as a Drawing Interchange Format (.dxf) (see Figure 1–101). The Drawing Interchange Format is especially useful with opening up an AutoCAD drawing in a competitive CAD system.

To ensure that drawing files are saved in a timely manner, you can use the automatic save mechanism. This setting is found by first right clicking on your screen to display the shortcut menu in Figure 1–102. Clicking on Options… will display the Options dialog box, shown in Figure 1–103. The Automatic Save setting is found under the Open and Save tab. By default, AutoCAD will automatically save your drawing every 10 minutes. Adjust this value depending on how often you wish AutoCAD to perform an automatic save on your drawing.

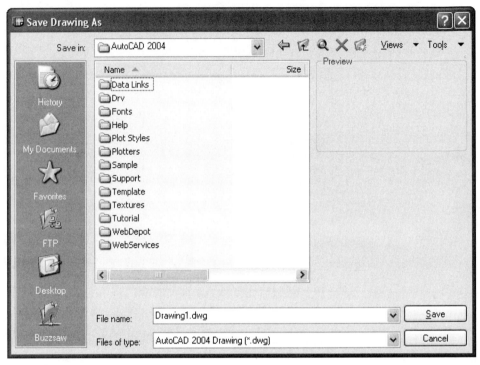

Figure 1–100

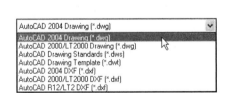

Figure 1–101

Figure 1–102

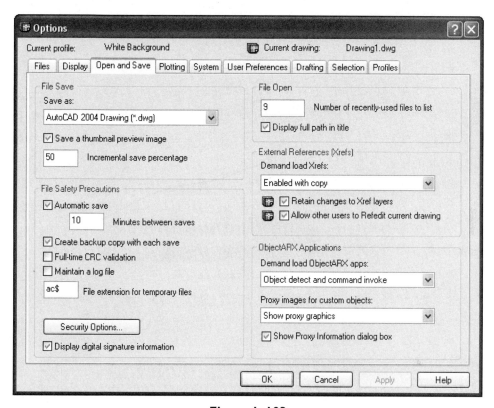

Figure 1–103

THE HELP COMMAND

The HELP command is very useful for seeking help on a number of AutoCAD topics. This command can be selected from the following:

From a toolbar button (Insert the file b_std_help.PCX here)

From the pull-down menu (Help > Help)

From the keyboard (Help)

Selecting Help from the Help menu, shown in Figure 1–104, activates the AutoCAD Help System dialog box, shown in Figure 1–105. Five tabs—Contents, Index, Search, Favorites, and Ask Me—can be used to navigate around Help. When you first start Help, the Contents tab is displayed, giving various topics to select.

Figure 1–104

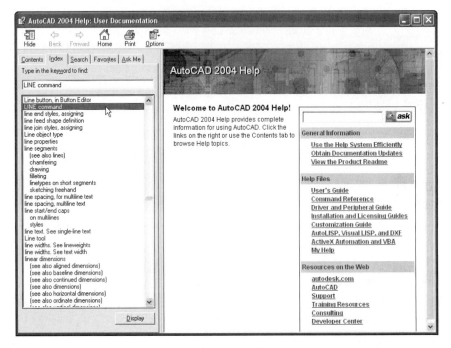

Figure 1–105

The Index tab in Figure 1–105 allows you to enter any AutoCAD command at the edit box and receive help on this command. On the left side of the dialog box, enter the LINE command in the Index edit box. Then click the Display button. When the Topics Found dialog box appears, as shown in Figure 1–106, decide which help topic you want to go to. Pick this help topic and then click the Display button. The results are displayed in Figure 1–107. The right side of the dialog box contains information specific to the LINE command, complete with graphical examples. Using the AutoCAD Help System can prove very beneficial in answering basic questions about any command.

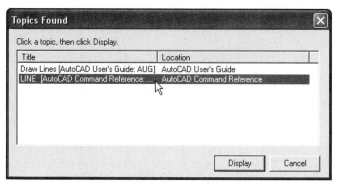

Figure 1–106

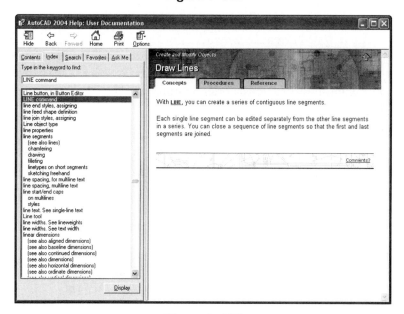

Figure 1–107

THE ACTIVE ASSISTANCE WINDOW

Active Assistance is a new AutoCAD Help feature that monitors each AutoCAD command you enter. Information about that command is displayed in the Active Assistance window. By default, Active Assistance is set for providing help on all commands. For instance, as you enter the LINE command, Active Assistance displays information about this command, as shown in Figure 1–108.

Figure 1–108

Suppose, however, that you would like to control when Active Assistance displays information about your current command. Moving your cursor inside the Active Assistance window and pressing the right mouse button displays the shortcut menu in Figure 1–109. Selecting the Settings option displays the Active Assistance Settings dialog box in Figure 1–110. Here is a brief description of each option:

Figure 1–109

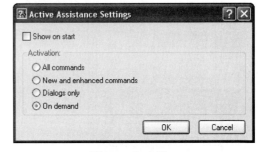

Figure 1–110

SHOW ON START

Active Assistance starts any time you start AutoCAD. If you clear this option, the next time you restart an AutoCAD session and want to display Active Assistance, type ASSIST at the command prompt or click on Active Assistance under Help on the Standard toolbar.

The following additional options allow you to control when the Active Assistance window is displayed.

ALL COMMANDS

By default, Active Assistance is automatically displayed whenever any command is activated.

NEW AND ENHANCED COMMANDS

If this option is selected, Active Assistance will open only when a new or enhanced command is activated.

DIALOGS ONLY

Selecting this option will display the Active Assistance window whenever a dialog box is activated.

ON DEMAND

This option allows you to control when the Active Assistance window is displayed. Selecting this option will turn off the Active Assistance window.

When the Active Assistance is used for the first time, an icon appears in the system tray as shown in Figure 1-111. This provides a convenient spot for reactivating the Active Assistance on a command.

Figure 1–111

 Tip: For working with this book, please set Active Assistance to On Demand. This will allow you to control when to request information about a command.

PRODUCTIVITY AND PERFORMANCE TOOLS

CLEAN SCREEN

To give you maximum area for drawing, a Clean Screen feature is available. This tool hides everything in the display window except for the drawing area, menu bar, and command line. Choose this command from the View pull-down menu as shown in Figure 1-112.

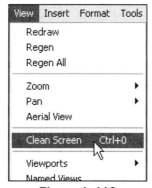

Figure 1–112

After using clean screen, your display should appear similar to Figure 1-113. Notice how all toolbars have been removed to increase the amount of drawing area.

Figure 1–113

Note: Pressing CTRL+0 (zero) will toggle you between the current user interface and the clean screen.

STATUS BAR MENU CONTROLS

You can control the display of buttons on the status bar by turning off those that you do not use very often. To perform this task, click the arrow located at the far right end of the status bar and click Status Bar Menu. This menu will display as shown in Figure 1-114. The presence of a check next to the button descriptor indicates that this item is presently displayed in the status bar.

Removing the check mark adjacent to Snap and Grid will remove these tools from the status bar as shown in Figure 1-115.

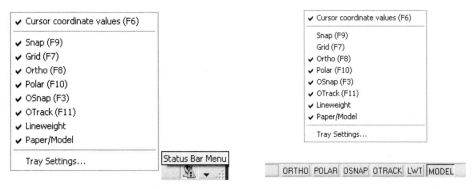

Figure 1–114   **Figure 1–115**

TRAY SETTINGS

For the purpose of providing you with various notifications such as product updates, a series of tray settings is also available through the Status Bar Menu as shown in Figure 1-116.

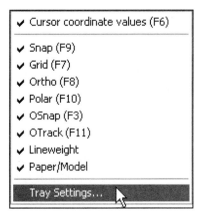

Figure 1–116

Clicking on Tray Settings in Figure 1-116 displays the Tray Settings dialog box as shown in Figure 1-117. Various buttons are available to control the display of icons and notifications located at the far right end of the status bar. The following items will be explained in greater detail:

Display Icons from Services

When this option is selected, a tray of service icons will display at the far right end of the status bar. When this option is cleared, the tray will not display.

Display Notifications from Services

When this option is selected, notifications from various services will display. A typical service may be a notification about future product updates. When this option is cleared, you will not receive any notifications from services.

Through the Tray Settings dialog box, you can set the time that a notification is displayed. There is also a setting that requires you to click the Close button in order to dismiss a notification that displayed on the screen.

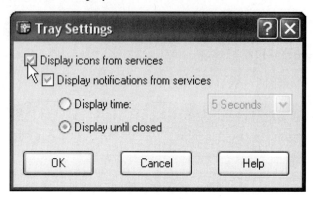

Figure 1-117

RIGHT CLICK CUSTOMIZATION

Depending on how fast or slow you click on the right mouse button, you can get different results. A special control is available that when checked allows you to activate the ENTER key when quickly clicking on the right mouse button. Pressing this button more slowly activates a short cut menu. In order to activate this feature (called time-sensitive right click behavior), follow the next series of steps.

Click Options from the Tools pull-down menu to activate the Options dialog box in Figure 1-118.

While in the Options dialog box, select the User Preferences tab followed by the Right-Click Customization button as shown in Figure 1-118.

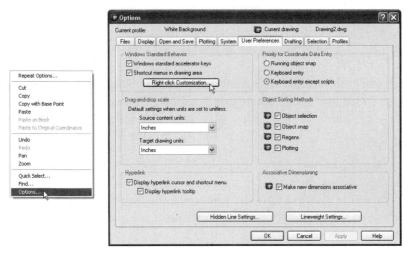

Figure 1-118

Place a check in the box adjacent to Turn on time-sensitive right-click as shown in Figure 1-119; this will activate or turn on this feature. You can also change the duration of the click in milliseconds; the default is 250. You will need to experiment with right clicking before making changes to this setting.

When you are finished, click the Apply & Close button. Then click the OK button in the main Options dialog box to return to your display screen.

Now experiment with the right click capabilities of your mouse. When you quickly click on the right mouse button, AutoCAD should signal this action as the ENTER key. If you press a little slower on this button, a short cut menu appears instead. This is a great way to perform two different tasks by right clicking on the mouse button.

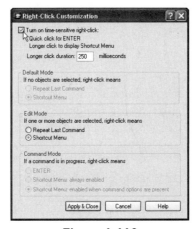

Figure 1-119

EXITING AN AUTOCAD DRAWING SESSION

It is considered good practice to properly exit any drawing session. One way of exiting is by choosing the Exit option from the File menu (see Figure 1–120). Another command that exits AutoCAD is QUIT. Use this command with extreme caution because QUIT exits the drawing file without saving any changes. This command is useful when you are just browsing drawing files where no changes were made. If you are a new user of AutoCAD, do not use this command until you have gained sufficient experience. You can also use the CLOSE command to end the current AutoCAD drawing session.

Whichever command you use to exit AutoCAD, a built-in safeguard gives you a second chance to save the drawing before exiting, especially if changes were made and a Save was not performed. You may be confronted with three options, illustrated in the AutoCAD alert dialog box shown in Figure 1–121. By default, the Yes button is highlighted.

Figure 1–120

Figure 1–121

If you made changes to the drawing but did not save them, and now you want to save them, click the Yes button before exiting the drawing. Changes to the drawing will be saved and the software exits back to the operating system.

If you made changes but do not want to save them, click the No button. Changes to the drawing will not be saved and the software exits back to the operating system.

If you made changes to the drawing and chose the Exit option mistakenly, choosing the Cancel button cancels the Exit option and returns you to the current drawing.

TUTORIAL EXERCISE: 01_PATTERN.DWG

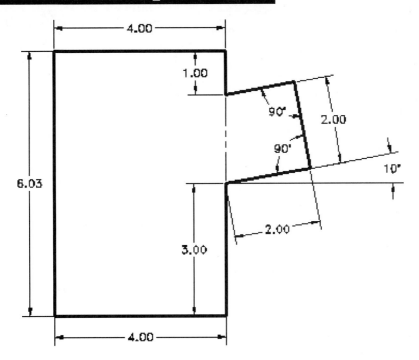

Figure 1–122

Purpose

This tutorial is designed to allow you to construct a one-view drawing of the Pattern using Polar Tracking techniques. See Figure 1–122.

System Settings

Use the current default settings for the limits of this drawing, (0,0) for the lower left corner and (12,9) for the upper right corner.

Layers

Create the following layer with the format:

Name	Color	Linetype
Object	Green	Continuous

Suggested Commands

Open the drawing file called 01_Pattern. The LINE command will be used entirely for this tutorial in addition to the Polar Tracking mode. Running Object Snap should already be set to the following modes; Endpoint, Center, Intersection, and Extension.

Whenever possible, substitute the appropriate command alias in place of the full AutoCAD command in each tutorial step. For example, use "CP" for the COPY command, "L" for the LINE command, and so on. The complete listing of all command aliases is located in Table 1–2.

STEP 1

Open the drawing file 01_Pattern. Activate the Drafting Settings dialog box, click on the Polar Tracking tab, and change the polar tracking angle setting to 10° (see Figure 1–123). Verify that POLAR, OSNAP, and OTRACK are all turned on.

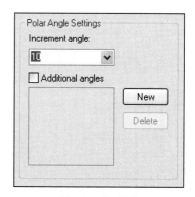

Figure 1–123

STEP 2

Activate the LINE command, select a starting point, move your cursor to the right, and enter a value of 4 units (see Figure 1–124). Notice that your line is green because the current layer, Object, has been assigned the green color.

 Command: **L** *(For LINE)*

Specify first point: **3,1**

Specify next point or [Undo]: *(Move your cursor to the right and enter 4)*

Figure 1–124

STEP 3

While still in the LINE command, move your cursor directly up, and enter a value of 3 units (see Figure 1–125).

Specify next point or [Undo]: *(Move your cursor up and enter 3)*

STEP 4

While still in the LINE command, move your cursor up and to the right until the tooltip reads 10°, and enter a value of 2 units (see Figure 1–126).

Specify next point or [Close/Undo]: *(Move your cursor up and to the right at a 10° angle and enter 2)*

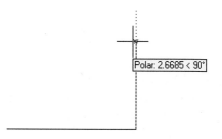

Figure 1–125

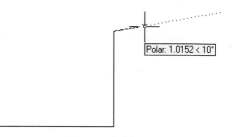

Figure 1–126

STEP 5

While still in the LINE command, move your cursor up and to the left until the tooltip reads 100°, and enter a value of 2 units (see Figure 1–127).

Specify next point or [Close/Undo]:
  (Move your cursor up and to the left at a 100° angle and enter 2)

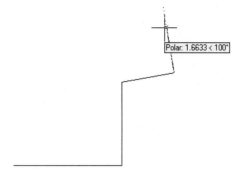

Figure 1–127

STEP 6

While still in the LINE command, first acquire the point at "A." Then move your cursor below and to the left until the Polar value in the tooltip reads 190°, and pick the point at "B" (see Figure 1–128).

Specify next point or [Close/Undo]:
  (Acquire the point at "A" and pick the new point at "B")

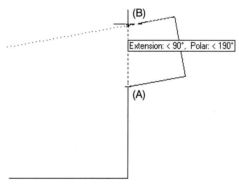

Figure 1–128

STEP 7

While still in the LINE command, move your cursor directly up, and enter a value of 1 unit (see Figure 1–129).

Specify next point or [Close/Undo]:
  (Move your cursor up and enter 1)

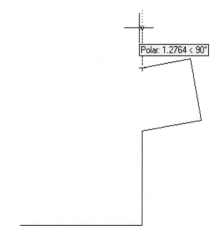

Figure 1–129

STEP 8

While still in the LINE command, first acquire the point at "C." Then move your cursor to the left until the tooltip reads Polar: < 180°, and pick the point at "D" (see Figure 1–130).

Specify next point or [Close/Undo]:
    (Acquire the point at "C" and pick the new point at "D")

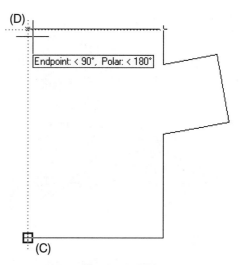

Figure 1–130

STEP 9

While still in the LINE command, complete the object by closing the shape (see Figure 1–131).

Specify next point or [Close/Undo]: C
    (To close the shape and exit the LINE command)

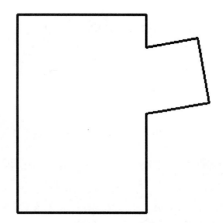

Figure 1–131

TUTORIAL EXERCISE: 01_POLYGON.DWG

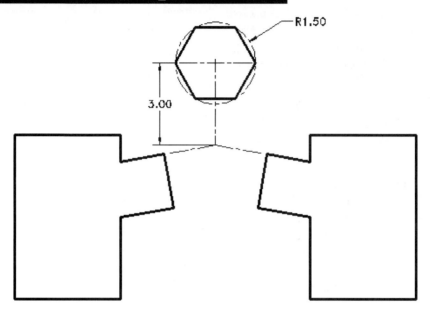

Figure 1–132

Purpose

This tutorial is designed to allow you to construct a polygon using existing geometry as a guide. Polar tracking and Object Snap tracking will be used to accomplish this task (see Figure 1–132). This drawing is available on the CD under the file name 01_Polygon.Dwg.

System Settings

No special system settings need to be made to this drawing file.

Layers

The drawing file 01_Polygon.Dwg has the following layers already created:

Name	Color	Linetype
Center	Yellow	Center
Dimension	Yellow	Continuous
Object	Green	Continuous
Defpoints	White	Continuous

Suggested Commands

The POLYGON command is used to start this tutorial exercise. To identify the center of the polygon, Polar Tracking will be used along with the Osnap Extension mode and the Temporary Tracking Point tool.

Whenever possible, substitute the appropriate command alias in place of the full AutoCAD command in each tutorial step. For example, use "CP" for the COPY command, "L" for the LINE command, and so on. The complete listing of all command aliases is located in Table 1–2.

STEP I

Open the drawing file 01_Polygon.Dwg and activate the POLYGON command. Polar tracking and Object Snap tracking will be used to identify the center of the polygon (see Figure 1–133).

Command: **POL** (For POLYGON)
Enter number of sides <4>: **6**

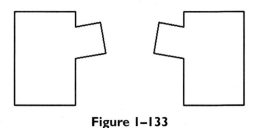

Figure 1–133

STEP 2

At the prompt "Specify center of polygon or [Edge]:", begin locating the center of the polygon by acquiring the endpoint at "A" in Figure 1–134. This creates a tracking path at the same angle as the line selected. Move your cursor above and to the right as illustrated in this figure.

Specify center of polygon or [Edge]:
    (Acquire the point at "A")

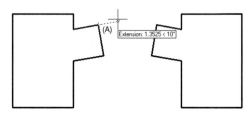

Figure 1–134

STEP 3

While still at the prompt "Specify center of polygon or [Edge]:", acquire the endpoint at "B" in Figure 1–135. This creates another tracking path at the same angle as the line selected. Move your cursor above and to the left as illustrated in this figure.

(Acquire the point at "B" in Figure 1–135)

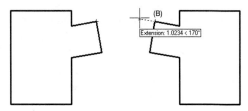

Figure 1–135

STEP 4

Now move both temporary tracking points to meet at their intersection in Figure 1–136 but do not pick a point at this location. The polygon needs to be constructed 3 units above this point. Click the Temporary Tracking Point Osnap mode and then pick a point at the intersection of "C." Notice the following prompt sequence that appears:

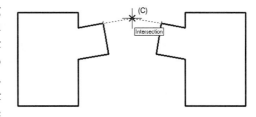

Figure 1–136

Specify temporary OTRACK point: *(Pick a point at the intersection at "C")*

STEP 5

Move your cursor directly above this tracking point in Figure 1–137 and enter a value of 3 units. This will locate the center of the polygon 3 units directly above the last tracking point. Complete the construction by following the remaining POLYGON command prompts:

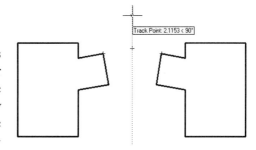

Figure 1–137

(Move your cursor up and enter 3)
Enter an option [Inscribed in circle/ Circumscribed about circle] <I>: *(Press* ENTER *to accept this default value)*
Specify radius of circle: **1.50**

The completed construction task is illustrated in Figure 1–138.

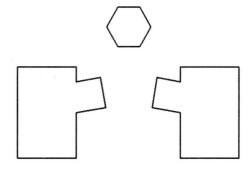

Figure 1–138

Open the Exercise Manual PDF file for Chapter 1 on the accompanying CD for more tutorials and exercises.

If you have the accompanying Exercise Manual, refer to Chapter 1 for more tutorials and exercises.

CHAPTER 2

Drawing Setup and Organization

Chapter 2 will cover a number of drawing set up commands. You will learn how to set up different units of measure with the UNITS command. You can also increase the default sheet size on your display screen with the LIMITS command. The Quick and Advanced Setup wizards will be demonstrated as a means of setting up your drawing given a number of steps. Controlling the grid and snap will be briefly discussed through the Snap and Grid tab located in the Drafting Settings dialog box. The major topic of this chapter is the discussion of layers. All options of the Layer Properties Manager dialog box will be demonstrated along with the ability to assign color, linetype, and lineweight to layers. The Layer Control box and Properties toolbar will provide easy access to all layers, colors, linetypes, and lineweights used in your drawing. Controlling the scale of linetypes through the LTSCALE command will also be discussed. Advanced layer tools such as Filtering Layers, Layer States, and Translating Layers will be introduced in this chapter. More information can be found on the CD supplied with this text or in the accompanying Projects Manual. This chapter concludes with a section on creating template files.

THE DRAWING UNITS DIALOG BOX

The Drawing Units dialog box is available for interactively setting the units of a drawing. Choosing Units... from the Format pull-down menu as shown in Figure 2–1 activates the dialog box illustrated in Figure 2–2.

By default, decimal units are set along with four-decimal-place precision. The following systems of units are available: Architectural, Decimal, Engineering, Fractional and Scientific (see Figure 2–3). Architectural units are displayed in feet and fractional inches. Engineering units are displayed in feet and decimal inches. Fractional units are displayed in fractional inches. Scientific units are displayed in exponential format.

Figure 2–1

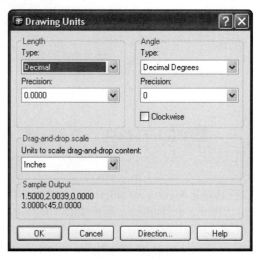

Figure 2–2

Figure 2–3

Methods of measuring angles supported in the Drawing Units dialog box include Decimal Degrees, Degrees/Minutes/Seconds, Grads, Radians, and Surveyor's Units (see Figure 2–4). Accuracy of decimal degree for angles may be set between zero and eight places.

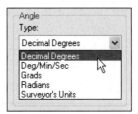

Figure 2–4

Selecting Direction… in the Drawing Units dialog box in Figure 2–2 displays the Direction Control dialog box, shown in Figure 2–5. This dialog box is used to control the direction of angle zero in addition to changing whether angles are measured in the counterclockwise or clockwise direction. By default, angles are measured in the counterclockwise direction.

Figure 2–5

ENTERING ARCHITECTURAL VALUES FOR DRAWING LINES

The method of entering architectural values in feet and inches is a little different from decimal places. To designate feet, you must enter the apostrophe symbol (') from the keyboard after the number. For example, "ten feet" would be entered as (10') in Figure 2–6. When feet and inches are necessary, you cannot use the space bar to separate the inch value. For example, nineteen feet seven inches would be entered as (19'7) in Figure 2–6. If you do use the space bar after the (19') value, this is interpreted as the ENTER key and your value is accepted as (19'). If you have to enter feet, inches and fractions of an inch, use the hyphen (-) to separate the inch value from the fractional value. For example, to draw a line twenty feet eleven and one-quarter inches, you would enter the following value in at the keyboard: (21'11-1/4) See Figure 2–6. Placing the inches symbol (") is not required since all numbers entered without the foot mark are interpreted as inches.

10'

19'7

21'11-1/4

Figure 2–6

 Try It! – Open the drawing file 02_Architectural. Verify that you are in architectural units by activating the Drawing Units dialog box. Using Figure 2–7 as a guide, follow the prompt sequence below to enter numeric values in feet and inches. The command sequence illustrates the use of polar coordinates. You could also use the Direct Distance mode of entry to construct the shape using architectural values.

Command: **L** *(For LINE)*
Specify first point: **4',2'**
Specify next point or [Undo]: **@10'<0**
Specify next point or [Undo]: **@7'6<90**
Specify next point or [Close/Undo]: **@13'7-1/2<0**
Specify next point or [Close/Undo]: **@4'9-1/4<90**
Specify next point or [Close/Undo]: **@7'<180**
Specify next point or [Close/Undo]: **@8'<90**
Specify next point or [Close/Undo]: **@16'7-1/2<180**
Specify next point or [Close/Undo]: **C** *(To close the shape and exit the LINE command)*

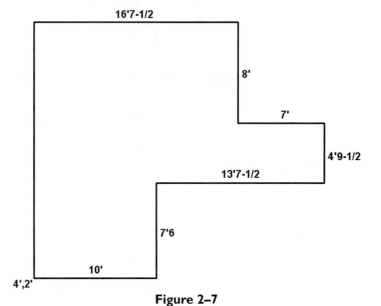

Figure 2–7

THE LIMITS COMMAND

By default, the size of the drawing screen in a new drawing file measures 12 units in the X direction and 9 units in the Y direction. This size may be ideal for small objects, but larger drawings require more drawing screen area. Use the LIMITS command for increasing the size of the drawing area. You can select this command by picking Drawing Limits from the Format pull-down menu (see Figure 2–8); you can also enter this command directly at the command prompt. Illustrated in Figure 2–9 is a section view drawing. This drawing fits in a screen size of 24 units in the X direction and 18 units in the Y direction. Follow the next command sequence to change the limits of a drawing.

Figure 2–8

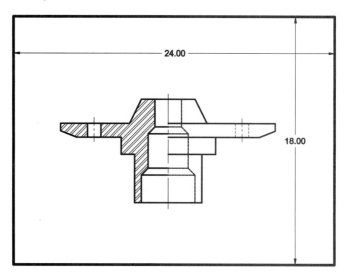

Figure 2–9

Command: **LIMITS**
ON/OFF/<Lower left corner>: <0.0000,0.0000>: *(Press* ENTER *to accept this value)*
Upper right corner <12.0000,9.0000>: **24,18**

Before continuing, perform a ZOOM-All to change the size of the display screen to reflect the changes in the limits of the drawing. You can find ZOOM-All on the View menu. It is also found on the Zoom toolbar, which you can activate by choosing Toolbars... from the View pull-down menu and placing a check in the box next to Zoom.

 Command: **Z** *(For Zoom)*
All/Center/Dynamic/Extents/Left/Previous/Vmax/Window/<Scale(X/XP)>: **A** *(For All)*

CALCULATING THE LIMITS OF THE DRAWING FOR MODEL SPACE

Before you construct any drawing, it is considered good practice to first calculate the limits or sheet size of the drawing. Two items are needed to accomplish this task: the scale of the drawing and the paper size the drawing will be plotted out on. Most design professionals who use CAD systems in industry already have some idea as to the scale of the drawing and whether it will fit on a certain size sheet of paper.

For the purposes of this example, the following steps will illustrate the creation of drawing limits based on a D-size sheet and at a scale of 1/4" = 1'0". This scale is commonly used to construct residential house plans. Because the following steps use this scale and sheet, the method works for calculating limits at any scale and on any sheet of paper.

STEP 1

Determine that a "D" size drawing sheet measures 36" x 24".

STEP 2

Determine the multiplication factor based on the drawing scale. In Computer-Aided Design, most drawings are created full size. The limits set a full-size drawing sheet used to match the full size drawing. The multiplication factor for the scale 1/4" = 1'0" is 48 (1' divided by 1/4). Use this multiplier to increase the size of the paper, enabling an operator to produce a full-size drawing on the computer. This is vastly different from manual drawing practices, where an object had to be reduced to 1/48th of its normal size in order to fit on the D-size paper at a scale of 1/4" = 1'0".

STEP 3

Using the multiplication factor of 48, increase the size of the D-size sheet of paper:

Paper Size from the Plot dialog box	36	24
Multiplication Factor	48	48
	1,728	1,152

Convert the previous values to feet:

1,728 / 12 = 144'   1,152 / 12 = 96'

The limits of the drawing become 144' wide and 96' high.

STEP 4

Use the Drawing Units dialog box to set the current units to Architectural (Decimal mode does not accept feet as valid units). Next, use the LIMITS command and enter the lower left corner as 0,0 and the upper right corner as 144',96'. Once the limits have been set, use the ZOOM command and the All option to display the new sheet of paper. To view the active drawing area, use the RECTANG command and construct a rectangular border around the drawing using 0,0 as the lower left corner and 144',96' as the upper right corner. Construct the drawing inside the rectangular area.

Decimal Scales

SCALE	SCALE FACTOR	ANSI "A" 11"x 8.5"	ANSI "B" 17"x11"	ANSI "C" 22"x17"	ANSI "D" 34"x22"	ANSI "E" 44"x34"
.125=1	8	88,68	136,88	176,136	272,176	352,272
.25=1	4	44,34	68,44	88,68	136,88	176,136
.50=1	2	22,17	34,22	44,34	68,44	88,68
1.00=1	1	11,8.5	17,11	22,17	34,22	44,34
2.00=1	.50	5.5,4.25	8.5,5.5	11,8.5	17,11	22,17

Architectural Scales

SCALE	SCALE FACTOR	ANSI "A" 11"x 8.5"	ANSI "B" 17"x11"	ANSI "C" 22"x17"	ANSI "D" 34"x22"	ANSI "E" 44"x34"
1/8"=1'-0"	96	88',68'	136',88'	76',136'	272',176'	352',272'
1/4"=1'-0"	48	44',34'	68',44'	88',68'	136',88'	176',136'
1/2"=1'-0"	24	22',17'	34',22'	44',34'	68',44'	88',68'
3/4"=1'-0"	16	14.7',11.3'	22.7',14.7'	29.3',22.7'	45.3',29.3'	58.7',45.3'
1"=1'-0"	12	11',8.5'	17',11'	22',17'	34',22'	44',34'
2"=1'-0"	6	5.5',4.25'	8.5',5.5'	11',8.5'	17',11'	22',17'

Metric Scales

SCALE	SCALE FACTOR	ANSI "A" 11"x 8.5"	ANSI "B" 17"x11"	ANSI "C" 22"x17"	ANSI "D" 34"x22"	ANSI "E" 44"x34"
1=1	25.4 mm	279,216	432,279	559,432	864,559	1118,864
1=10	10 cm	110,85	170,110	220,170	340,220	440,340
1=20	20 cm	220,170	340,220	440,340	680,440	880,680
1=50	50 cm	550,425	850,550	1100,850	1700,1100	2200,1700
1=100	100 cm	1100,850	1700,1100	2200,1700	3400,2200	4400,3400

THE QUICK SETUP WIZARD

Starting a new drawing and clicking on the Use a Wizard button, shown in Figure 2–10, and then choosing the Quick Setup mode displays the Quick Setup dialog box, as shown in Figure 2–11, consisting of two categories. The Units category (see Figure 2–11) is used to graphically control the units of the drawing. Five units of measure are available through this tab, namely Decimal, Engineering, Architectural, Fractional, and Scientific. Clicking on the radio button displays a sample of the units' appearance in the drawing. When you have completed setting the units of the drawing, click on the Next> button to display the next dialog box in the Quick Setup sequence.

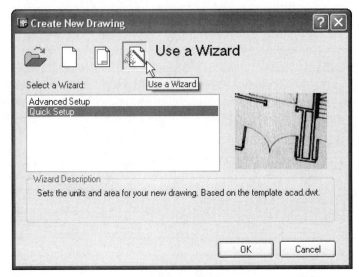

Figure 2–10

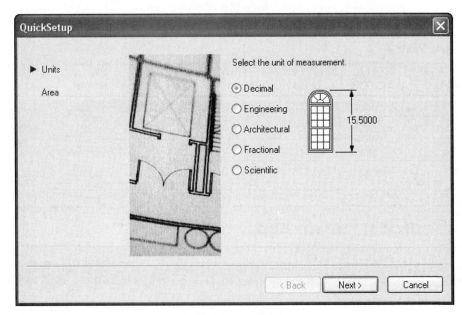

Figure 2–11

The Area category deals with the drawing area and is given as a Width and a Length value (see Figure 2–12). Setting this is comparable to using the LIMITS command. As you enter the width and length of the drawing area, the sample area image updates to reflect the changes in the values. At times, operators mistakenly substitute the width value for the length and vice versa. The image will allow you to preview the limits or drawing area before returning to the drawing screen. If the drawing area is incorrect, you can make changes, and the sample area image will update to show the latest changes. At any point, you may elect to return to the Units category by clicking on the <Back button. When you complete the setting of the drawing units and area, click on the Finish button to return to the drawing editor, where the units and area will be updated to reflect the changes you made in the Quick Setup Wizard.

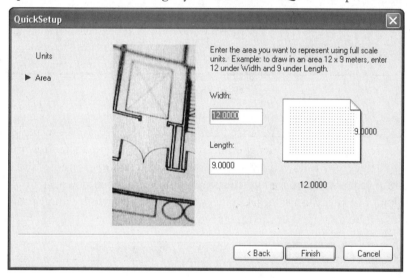

Figure 2–12

THE ADVANCED SETUP WIZARD

When you activate the Quick Setup Wizard, you are guided through two categories to make changes in the units and area of the drawing. Clicking on Advanced Setup in the Create New Drawing dialog box displays the Advanced Setup dialog box illustrated in Figure 2–13. The Advanced Wizard contains five categories for making changes to the initial drawing setup. The first category deals with the units of the drawing. This tab is almost identical to the dialog box found in the Quick Setup Wizard; it allows you to choose among Decimal, Engineering, Architectural, Fractional, and Scientific units. When you change the units, they will preview in the sample units area. An additional control for units in the Advanced Setup Wizard allows you to change the precision of the main units. In Figure 2–13, the precision for the decimal units is four decimal places. Changing the precision will update the Sample Units area, as shown in Figure 2–13.

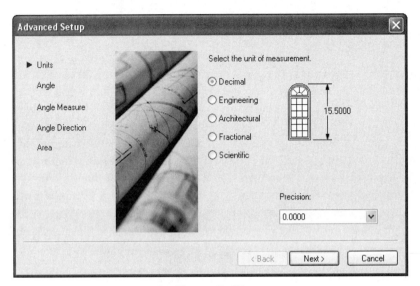

Figure 2–13

The second category of the Advanced Setup Wizard deals with how angles will be measured (see Figure 2–14). Five methods of angle measurements include the default of Decimal Degrees followed by Degrees/Minutes/Seconds, Grads, Radians, and Surveyor. A precision box for the measurement of angles allows you to change to different angular precision values. Changes to the measurement of angles and angular precision will be displayed in the image icon supplied.

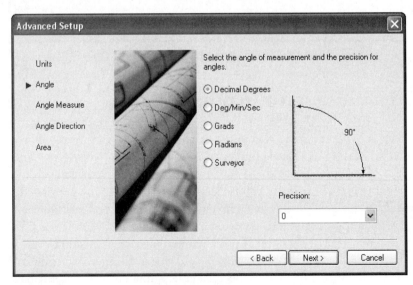

Figure 2–14

The third category controls the direction for angle measurement. By default, angles are measured starting with East for an angle of 0, North for an angle of 90°, West for 180°, and South for 270°. You have the option of changing the direction of zero; to make these changes, update the Angle Zero Direction area illustrated in Figure 2–15.

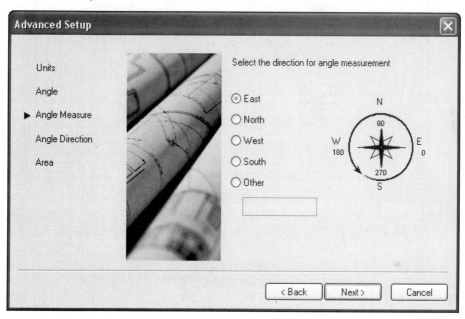

Figure 2–15

The fourth category of the Advanced Setup dialog box deals with Angle Direction. All angles are by default measured in the counterclockwise direction. Use the dialog box illustrated in Figure 2–16 to change from counterclockwise measurement of angles to clockwise angular measurements. Throughout this book, all examples dealing with angular measurements will keep the default setting of counterclockwise for angular direction.

The fifth category of the Advanced Setup dialog box is used to change the drawing area (see Figure 2–17). Enter values for the Width and Length in the edit boxes provided, and watch the Sample Area image update to show the new drawing area.

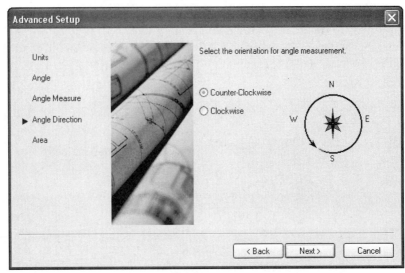

Figure 2–16

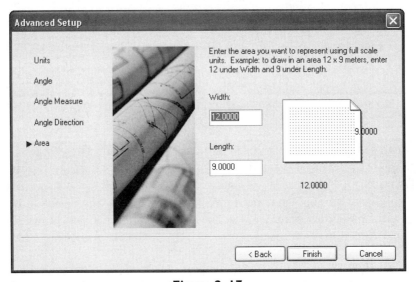

Figure 2–17

THE GRID COMMAND

Use grid to get a relative idea as to the size of objects. Grid is also used to define the size of the display screen originally set by the LIMITS command. The dots that make up the grid will never plot out on paper even if they are visible on the display screen. You can turn grid dots on or off by using the GRID command or by pressing F7, or by single-clicking on GRID, located in the status bar at the bottom of the display screen (see Figure 1–3). By default, the grid is displayed in 0.50-unit intervals similar to Figure 2–18. Illustrated in Figure 2–19 is a grid that has been set to a value of 0.25, or half its original size.

 Try It! – Create Open the drawing file 02_Grid. Use the following command sequence and illustrations below for using the GRID command.

Command: **GRID**
Specify grid spacing(X) or [ON/OFF/Snap/Aspect] <0.5000>: **On**

Command: **GRID**
Specify grid spacing(X) or [ON/OFF/Snap/Aspect] <0.5000>: **0.25**

While a grid is a useful aid for construction purposes, it may reduce the overall performance of the computer system. If the grid is set to a small value and it is visible on the display screen, it takes time for the grid to display. If a very small value is used for grid, a prompt will warn you that the grid value is too small to display on the screen.

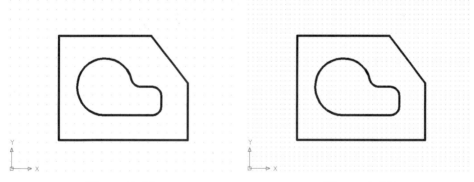

Figure 2–18   Figure 2–19

THE SNAP COMMAND

It is possible to have the cursor lock onto or snap to a grid dot as illustrated in Figure 2–20; this is the purpose of the SNAP command. By default, the current snap spacing is 0.50 units. Even though a value is set, the snap must be turned on for the cursor to be positioned on a grid dot. You can accomplish this by using the SNAP command as shown in the following sequence, by pressing F9, or by single-clicking on SNAP in the status bar at the bottom of the display screen (see Figure 1–3).

Some drawing applications require that the snap be rotated at a specific angular value (see Figure 2–21). Changing the snap in this fashion also affects the cursor. Use the following command sequence for rotating the snap.

 Try It! – Open the drawing file 02_Snap. Use the following command sequence and the illustration in Figure 2–21 below for using the SNAP command.

Command: **SN** *(For SNAP)*
Specify snap spacing or [ON/OFF/Aspect/Rotate/Style/Type] <0.5000>: **On**

Command: **SN** *(For SNAP)*
Specify snap spacing or [ON/OFF/Aspect/Rotate/Style/Type] <0.5000>: **R** *(For Rotate)*
Specify base point <0.0000,0.0000>: *(Press ENTER to accept this value)*
Specify rotation angle <0>: **30**

One application of rotating snap is for auxiliary views where an inclined surface needs to be projected in a perpendicular direction. This will be explained and demonstrated in chapter 15.

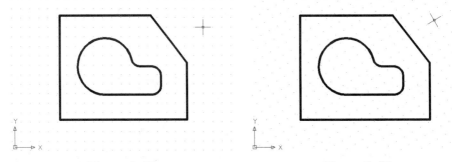

Figure 2–20   **Figure 2–21**

 Tip: To affect both the grid and the snap, set the grid to a value of zero (0). When setting a new snap value, this value is also used for the spacing of the grid.

THE SNAP AND GRID TAB OF THE DRAFTING SETTINGS DIALOG BOX

Choosing Drafting Settings… from the Tools pull-down menu, as shown in Figure 2–22, displays the Drafting Settings dialog box, shown in Figure 2–23. Use this dialog box for making changes to the grid and snap settings. The Snap type & style area controls whether isometric grid is present or not.

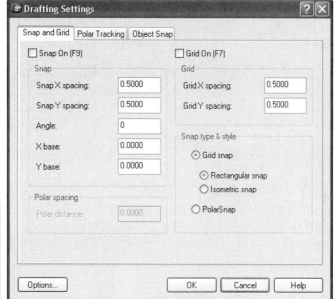

Figure 2–22   **Figure 2–23**

THE ALPHABET OF LINES

Before you construct engineering drawings, the quality of the lines that make up the drawing must first be discussed. Some lines of a drawing should be made thick; others need to be made thin. This is to emphasize certain parts of the drawing and it is controlled through a line quality system. Illustrated in Figure 2–24 is a two-view drawing of an object complete with various lines that will be explained further.

The most important line of a drawing is the Object line, which outlines the basic shape of the object. Because of their importance, Object lines are made thick and continuous so they stand out among the other lines in the drawing. It does not mean that the other lines are considered unimportant; rather, the Object line takes precedence over all other lines.

The Cutting Plane line is another thick line; it is used to determine the placement in the drawing where an imaginary saw will cut into the drawing to expose interior details. It stands out by being drawn as a series of long dashes separated by spaces. Arrowheads determine how the adjacent view will be looked at. This line will be discussed in greater detail in chapter 14, "Section Views."

The Hidden line is a medium weight line used to identify edges that become invisible in a view. It consists of a series of dashes separated by spaces. Whether an edge is visible or invisible, it still must be shown with a line.

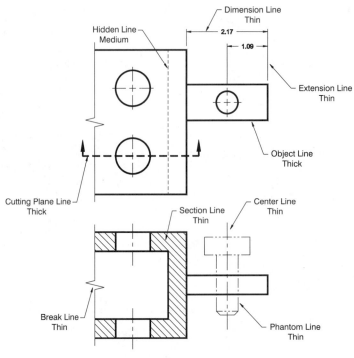

Figure 2–24

The Dimension line is a thin line used to show the numerical distance between two points. The dimension text is placed within the dimension line, and arrowheads are placed at opposite ends of the dimension line.

The Extension line is another thin continuous line, used as a part of the overall dimension. Extension lines show the edge being dimensioned. In Figure 2–24, the vertical extension lines are used to indicate the horizontal dimension distance.

When you use the Cutting Plane line to create an area to cut or slice, the surfaces in the adjacent view are section lined using the Section line, a thin continuous line.

Another important line used to identify the centers of circles is the Center line. It is a thin line consisting of a series of long and short dashes. It is a good practice to dimension to center lines; for this reason center lines, extension lines, and dimensions are made the same line thickness.

The Phantom line consists of a thin line made with a series of two dashes and one long dash. It is used to simulate the placement or movement of a part or component without actually detailing the component.

The Break line is a thin line with a "zigzag" symbol used to establish where an object is broken to simulate a continuation of the object.

ORGANIZING A DRAWING THROUGH LAYERS

As a means of organizing objects, a series of layers should be devised for every drawing. You can think of layers as a group of transparent sheets that combine to form the completed drawing. Figure 2–25 shows a drawing consisting of object lines, dimension lines, and border. Organizing these three drawing components by layers is illustrated in Figure 2–26. Only the drawing border occupies a layer that could be called "Border." The object lines occupy a layer that could be called "Object," and the dimension lines could be drawn on a layer called "Dim." At times, it may be necessary to turn off the dimension lines for a clearer view of the object. Creating all dimensions on a specific layer will allow you to turn off the dimensions while viewing all other objects on layers still turned on.

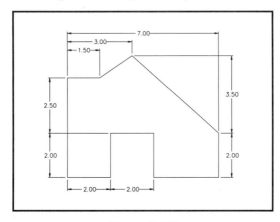

Figure 2–25

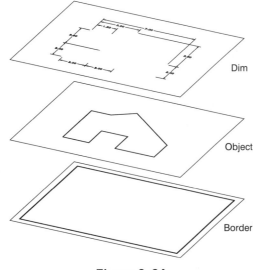

Figure 2–26

THE LAYER PROPERTIES MANAGER DIALOG BOX

An efficient way of creating and managing layers is through the Layer Properties Manager dialog box. Choose Layer…from the Format menu, as in Figure 2–27, to display the Layer Properties Manager dialog box in Figure 2–28.

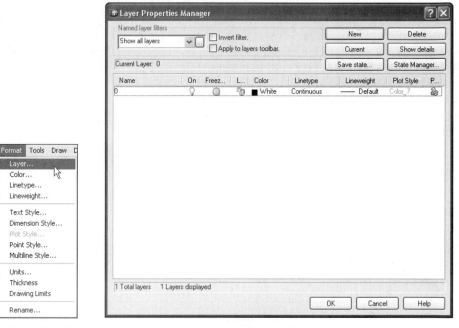

Figure 2–27   **Figure 2–28**

Another way of activating this dialog box is by picking the Layers button located in the Object Properties toolbar in Figure 2–29.

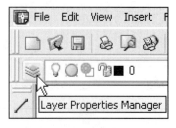

Figure 2–29

Once the dialog box displays, buttons located along the right side of this dialog box are designed for creating new layers, making an existing layer current to draw on, deleting a layer that is unused, or showing a series of details regarding a selected layer (Layer 0 cannot be deleted).

In addition to these buttons, a number of layer states exist that enable you to perform the following operations: turning layers on or off, freezing or thawing layers, locking or unlocking layers, assigning a color, linetype, lineweight, and plot style to a layer or group of layers and whether to plot a layer or group of layers (see Figure 2–30). A brief explanation of each layer state is described below:

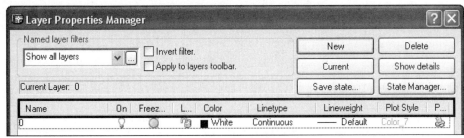

Figure 2–30

On/Off – Makes all objects created on a certain layer visible or invisible on the display screen. The On state is symbolized by a yellow light bulb. The Off state has a light bulb icon shaded black.

Freeze – This state is similar to the Off mode; objects frozen will appear invisible on the display screen. Freeze, however, is considered a major productivity tool used to speed up the performance of a drawing. This is accomplished by not calculating any frozen layers during drawing regenerations. A snowflake icon symbolizes this layer state.

Thaw – This state is similar to the On mode; objects on frozen layers will reappear on the display screen when they are thawed. The sun icon symbolizes this layer state.

Lock – This state allows objects on a certain layer to be visible on the display screen while protecting them from accidentally being modified through an editing command. A closed padlock icon symbolizes this layer state.

Unlock – This state unlocks a previously locked layer and is symbolized by an open padlock icon.

Color – This state displays a color that is assigned to a layer and is symbolized by a square color swatch along with the name of the color. By default, the color White is assigned to a layer.

Linetype – This state displays the name of a linetype that is assigned to a layer. By default, the Continuous linetype is assigned to a layer.

Lineweight – This state sets a lineweight to a layer. An image of this lineweight value will be visible in this layer state column.

Plot Style – A plot style allows you to override the color, linetype, and lineweight settings made in the Layer Properties Manager dialog box. Notice how this area is grayed out. When working with a plot style that is color dependent, you cannot change the plot style. Plot styles will be discussed in greater detail later in this book.

Plot – This layer state controls which layers will be plotted. The presence of the printer icon symbolizes a layer that will be plotted. A printer icon with a red circle and diagonal slash signifies a layer that will not be plotted.

CREATING NEW LAYERS

Clicking on the New button of the Layer Properties Manager dialog box creates a new layer called Layer1, which displays in the layer list box as shown in Figure 2–31. Since this layer name is completely highlighted, you may elect to change its name to something more meaningful, such as a layer called Object or Hidden to hold all object or hidden lines in a drawing.

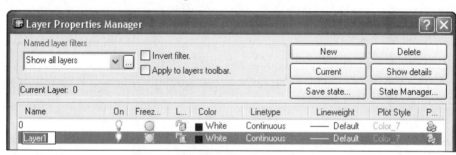

Figure 2–31

Illustrated in Figure 2–32 is the result of changing the name Layer1 to Object. You could also have entered "OBJECT" or "object" and these names would appear in the dialog box.

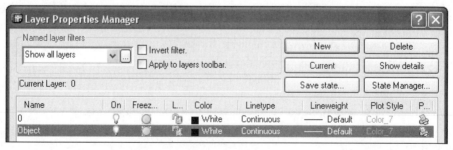

Figure 2–32

You can also be really descriptive with layer names. In Figure 2–33, a layer has been created called "Section" (This layer is designed to hold crosshatch lines)." You are allowed to add spaces and other characters in the naming of a layer. Because of space limitations, the entire layer name will not display until you move your cursor over the top of the layer name. As a result, the layer name will appear truncated as in the layer "Object(Thi...object lines)" in Figure 2–33.

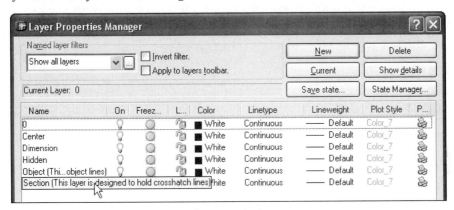

Figure 2–33

If more than one layer needs to be created, it is not necessary to continually click on the New button. Because this operation will create new layers, a more efficient method would be to perform either of the following operations: After creating a new "Layer1", change its name to Dim followed by a comma (,). This will automatically create a new Layer1 (see Figure 2–34). Change its name followed by a comma and a new layer is created, and so on. You could also have pressed ENTER twice, which would create another new layer.

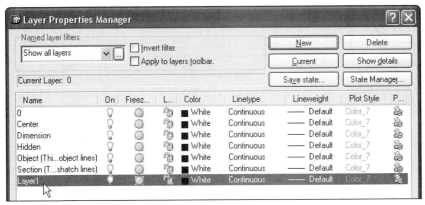

Figure 2–34

ASSIGNING COLOR TO LAYERS

Once you select a layer from the list box of the Layer Properties Manager dialog box and the color swatch is selected in the same row as the layer name, the Select Color dialog box shown in Figure 2–35 is displayed. Three tabs allow you to select three different color groupings. The three groupings (Index Color, True Color, and Color Books) are described as follows.

INDEX COLOR TAB

This tab, shown in Figure 2–35, allows you to make color settings based on 255 AutoCAD Color Index (ACI) colors. Standard, Gray Shades, and Full Color Palette areas are available for you to choose colors from.

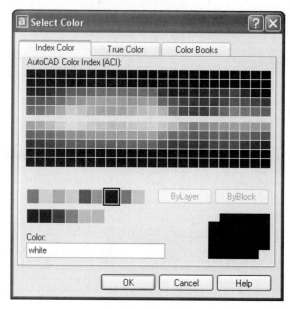

Figure 2–35

TRUE COLOR TAB

Use this tab to make color settings using true colors, also known as 24-bit color. Two color models are available for you to choose from, namely Hue, Saturation, and Luminance (HSL), or Red, Green, and Blue (RGB). Through this tab, you can choose from over sixteen million colors (see Figure 2–36).

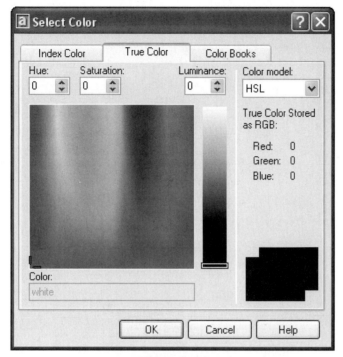

Figure 2–36

COLOR BOOKS TAB

Use the Color Books tab to select colors that use third-party color books (such as PANTONE®) or user-defined color books. You can think of a color book as similar to those available in hardware stores when selecting household interior paints. When you select a color book, the name of the selected color book will be identified in this tab (see Figure 2–37).

 Note: The Index tab will be used throughout this text. However you are encouraged to experiment with the True Color and Color Books tabs.

Figure 2–37

ASSIGNING LINETYPES TO LAYERS

Selecting the name Continuous next to the highlighted layer activates the Select Linetype dialog box, shown in Figure 2–38. Use this dialog box to dynamically select preloaded linetypes to be assigned to various layers. By default, the Continuous linetype is loaded for all new drawings. To load other linetypes, click on the Load button of the Select Linetype dialog box; this displays the Load or Reload Linetypes dialog box, shown in Figure 2–39.

Once the Load or Reload Linetypes dialog box is displayed, as in Figure 2–39, use the scroll bars to view all linetypes contained in the file ACAD.LIN. Notice that, in addition to standard linetypes such as HIDDEN and PHANTOM, a few linetypes are provided that have text automatically embedded in the linetype. As the linetype is drawn, the text is placed depending on how it was originally designed. Notice also three variations of Hidden linetypes; namely HIDDEN, HIDDEN2, and HID-DENX2. The HIDDEN2 represents a linetype where the distances of the dashes and spaces in between each dash are half of the original HIDDEN linetype. HIDDENX2 represents a linetype where the distances of the dashes and spaces in between each dash of the original HIDDEN linetype are doubled. Click on the desired linetypes to load. When finished, click on the OK button.

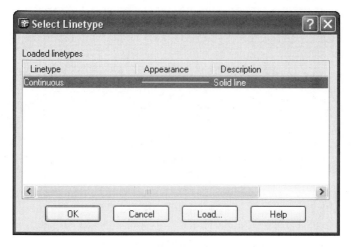

Figure 2–38

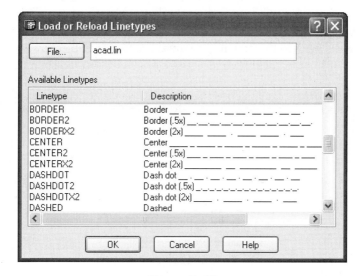

Figure 2–39

The loaded linetypes now appear in the Select Linetype dialog box, as in Figure 2–40. It must be pointed out that the linetypes in this list are only loaded into the drawing and are not assigned to a particular layer. Clicking on the linetype in this dialog box will assign this linetype to the layer currently highlighted in the Layer Properties Manager dialog box.

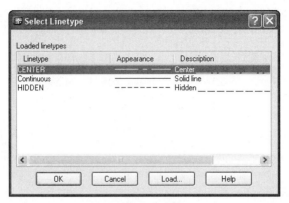

Figure 2–40

ASSIGNING LINEWEIGHT TO LAYERS

Selecting the name Default under the Lineweight heading of the Layer Properties Manager dialog box activates the Lineweight dialog box, shown in Figure 2–41. Use this to attach a lineweight to a layer. Lineweights are very important to a drawing file—they give contrast to the drawing. As stated earlier, the object lines should stand out over all other lines in the drawing. A thick lineweight would then be assigned to the object line layer.

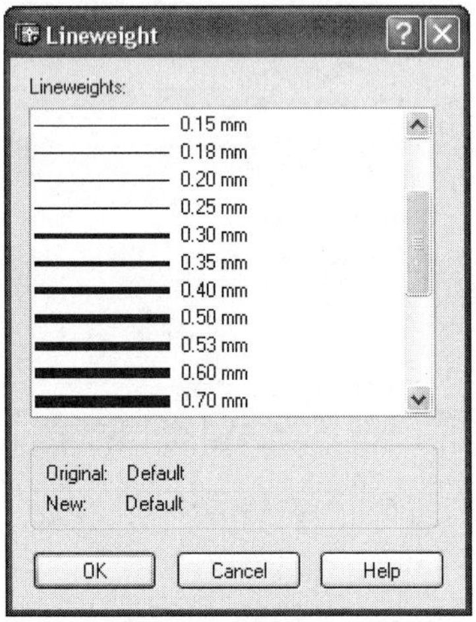

Figure 2–41

In Figure 2–42, a lineweight of 0.50 mm has been assigned to all object lines and a lineweight of 0.30 has been assigned to hidden lines. However, all lines in this figure appear to consist of the same lineweight. This is due to the lineweight feature being turned off. Notice in this figure the LWT button in the Status bar. Use this button to toggle on or off the display of assigned lineweights.

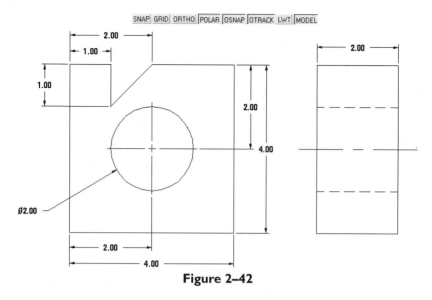

Figure 2–42

Clicking on the LWT button in Figure 2–43 to turn the lineweight function on. The results are displayed in the illustration.

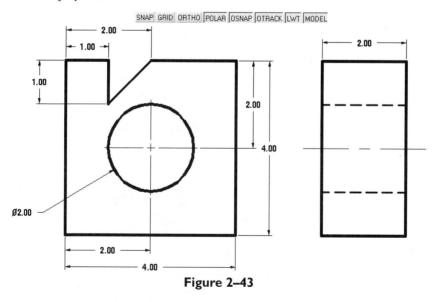

Figure 2–43

THE LINEWEIGHT CONTROL DIALOG BOX

You have probably already noticed that when turning on the lineweights through the LWT button, all lineweights appear very thick. In a complicated drawing, the lineweights could be so thick that it would be difficult to interpret the purpose of the drawing. To give your lineweights a more pleasing appearance, a dialog box is available to control the display of your lineweights. Clicking on Lineweight in the Format pull-down menu in Figure 2–44 will display the Lineweight Settings dialog box in Figure 2–45. Notice the position of the slider bar in the Adjust Display Scale area of the dialog box. Sliding the bar to the left near the Min setting will reduce the width of all lineweights. Try experimenting with this on any lineweight assigned to a layer. If you think your lineweights appear too thin, slide the bar to the right near the Max setting and observe the results. Continue to adjust your lineweights until they have a pleasing appearance in your drawing.

Figure 2–44

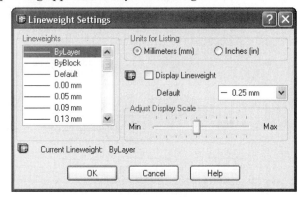

Figure 2–45

SHOWING LAYER DETAILS

Once layers have been created along with color, linetype, and lineweight assignments, the display of the Layer Properties Manager dialog box will be similar to Figure 2–46. Initially, when layers are created, they are placed in the dialog box in the exact order in which they were created. If you close the dialog box by clicking on the OK button and then revisit it at a later time, all layer names are reordered and displayed in alphabetical order (numbers preceding letters).

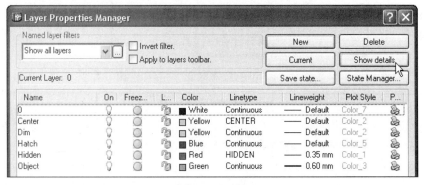

Figure 2–46

Clicking on the Show details button (see Figure 2–46) expands the bottom of the dialog box to include more detailed information about the selected layer (see Figure 2–47). Name, Color, Lineweight, and Linetype assignments are isolated in individual edit boxes. Also, the properties of the layer (On/Off, Lock/Unlock, Plot/No Plot, and Freeze/Thaw in all viewports) are displayed in a column to the right of the layer name. Clicking on the Hide details button compresses the dialog box again so that it appears like Figure 2–46.

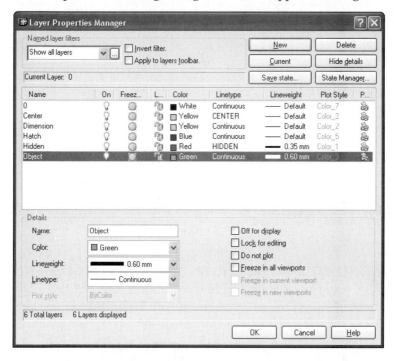

Figure 2–47

THE LINETYPE MANAGER DIALOG BOX

Choosing Linetype... from the partial Format pull-down menu, in Figure 2–48, displays the Linetype Manager dialog box in Figure 2–49. This dialog box is designed mainly to pre-load linetypes.

Figure 2–48

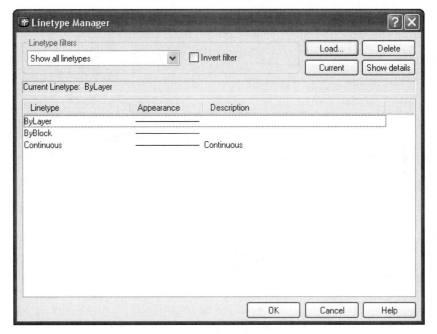

Figure 2–49

Clicking on the Load… button activates the Load or Reload Linetypes dialog box in Figure 2–50. You can select individual linetypes in this dialog box or load numerous linetypes at once by pressing CTRL and clicking on each linetype. Clicking the OK button will load the selected linetypes into the dialog box in Figure 2–51.

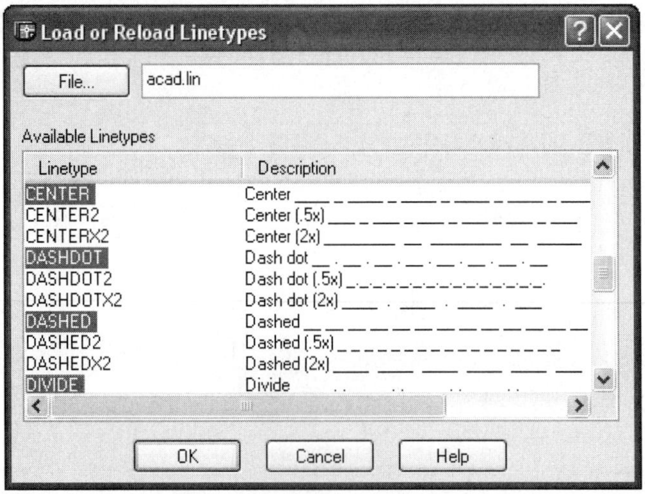

Figure 2–50

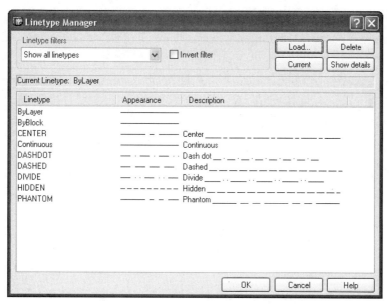

Figure 2–51

THE LAYERS TOOLBAR

The Layers toolbar provides an area to better control the layer properties or states. This toolbar is illustrated in Figure 2–52. The Layer Control area will now be discussed in greater detail.

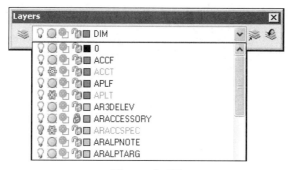

Figure 2–52

CONTROL OF LAYER PROPERTIES

Clicking in the long edit box next to the Layer button cascades all layers defined in the drawing in addition to their properties, identified by symbols (see Figure 2–53). The presence of the light bulb signifies that the layer is turned on. Clicking on the light bulb symbol turns the layer off. The sun symbol signifies that the layer is thawed. Clicking on the sun turns it into a snowflake symbol, signifying that the layer is now frozen. The padlock symbol controls whether a layer is locked or unlocked. By default, all layers are unlocked. Clicking on the padlock changes the symbol to display the image of a locked symbol, signifying that the layer is locked.

Study Figure 2–54 for a better idea of how the symbols affect the state of certain layers.

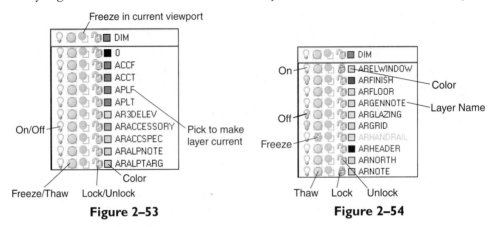

Figure 2–53

Figure 2–54

THE PROPERTIES TOOLBAR

In addition to the Layers toolbar, a Properties toolbar is also available. This toolbar provides three areas to better access options for the control of Colors, Linetypes, and Lineweights. All three areas are illustrated in Figure 2–55.

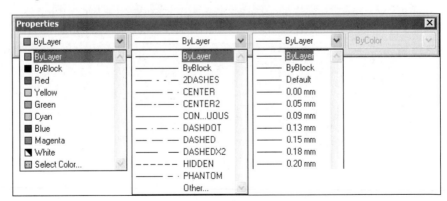

Figure 2–55

 Note: While the Properties toolbar allows you to change color, linetype, and lineweight on the fly, it is considered poor practice to do so. Check to see that each category in the Properties toolbar reads "ByLayer," which means that layers control this category.

MAKING A LAYER CURRENT

Various methods can be employed to make a layer current to draw on. Select a layer in the Layer Properties Manager dialog box and then click on the Current button to make the layer current.

Picking a layer name from the Layer Control box in Figure 2–54 will also make the layer current.

The Make Object's Layer Current button, shown in Figure 2–56, allows you to make a layer the new current layer by just clicking on an object in the drawing. The layer is now made current based on the layer of the selected object.

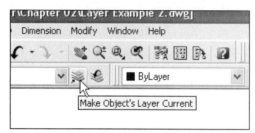

Figure 2–56

USING THE LAYER PREVIOUS COMMAND

The Layer Previous button as shown in Figure 2-57 is used to undo changes that you have made to layer settings such as color or lineweight. You could even turn a number of layers off and use the Layer Previous command to have the layers turned back on in a single step. This command can also be entered in from the keyboard in the following prompt sequence:

Command: **LAYERP**

There are a few exceptions to the use of the Layer Previous command:

- If you rename a layer and change its properties, such as color or lineweight, issuing the Layer Previous command will restore the original properties but will not change back to the original layer name.

- Purged or deleted layers will not be restored by using Layer Previous.

- If you add a new layer, issuing the Layer Previous command will not remove the new layer.

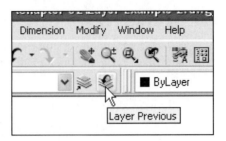

Figure 2–57

RIGHT-CLICK SUPPORT FOR LAYERS

While inside the Layer Properties Manager, right-clicking inside the layer information area displays the shortcut menu in Figure 2–58. Use this menu to make the selected layer current, to make a new layer based on the selected layer, select all layers in the dialog box or clear all layers. You can even select all layers except for the current layer. This shortcut menu provides you with easier access to commonly used layer manipulation tools.

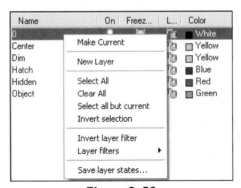

Figure 2–58

CONTROLLING THE LINETYPE SCALE

Once linetypes are associated with layers and placed in a drawing, you can control their scale with the LTSCALE command. In Figure 2–59, the default linetype scale value of 1.00 is in effect. This scale value acts as a multiplier for all linetype distances. In other words, if the hidden linetype is designed to have dashes 0.125 units long, a linetype scale value of 1.00 displays the dash of the hidden line at a value of 0.125 units. The LTSCALE command displays the following command sequence:

Command: **LTS** *(For LTSCALE)*
Enter new linetype scale factor <1.0000>: *(Press ENTER to accept the default or enter another value)*

Try It! – Open the drawing file 02_LTScale. Follow the directions, command prompts, and illustrations below for using the LTSCALE command.

In Figure 2–60, a linetype scale value of 0.50 units has been applied to all linetypes. As a result of the 0.50 multiplier, instead of all hidden line dashes measuring 0.125 units, they now measure 0.0625 units.

Command: **LTS** *(For LTSCALE)*
Enter new linetype scale factor <1.0000>: **0.50**

In Figure 2–61, a linetype scale value of 2.00 units has been applied to all linetypes. As a result of the 2.00 multiplier, instead of all hidden line dashes measuring 0.125 units, they now measure 0.25 units. Notice how a large multiplier displays the center lines as what appears to be a continuous linetype.

Command: **LTS** *(For LTSCALE)*
Enter new linetype scale factor <0.5000>: **2.00**

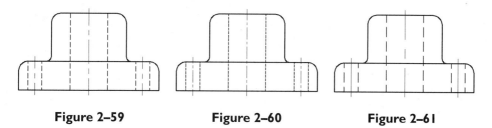

Figure 2–59   **Figure 2–60**   **Figure 2–61**

 Try It! – Open the drawing file 02_Floor Plan illustrated in Figure 2–62. This floor plan is designed to plot out at a scale factor of 1/8"=1'0". This creates a multiplier of 96 units (found by dividing 1' by 1/8). A layer called "Dividers" was created and assigned the Hidden linetype to show all red hidden lines as potential rooms in the plan. However, the hidden lines do not display. For all linetypes to show as hidden, this multiplier should be applied to the drawing through the LTSCALE command.

Since the drawing was constructed in real-world units or full size, the linetypes are converted to these units beginning with the multiplier of 96 in Figure 2–63 through the LTSCALE command.

Command: **LTS** *(For LTSCALE)*
Enter new linetype scale factor <1.0000>: **96**

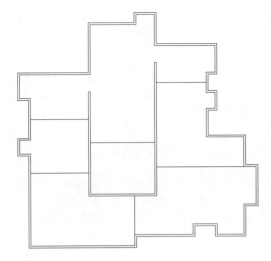

Figure 2–62

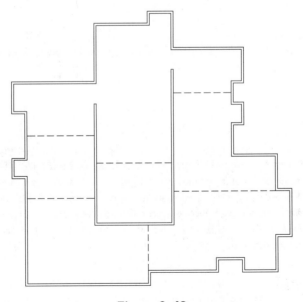

Figure 2–63

Expanding the Linetype Manager dialog box in Figure 2–64 displays a number of linetype details. The Global scale factor value has the same effect as the LTSCALE command. Also, setting a different value in the Current object scale box can scale the linetype of an individual object.

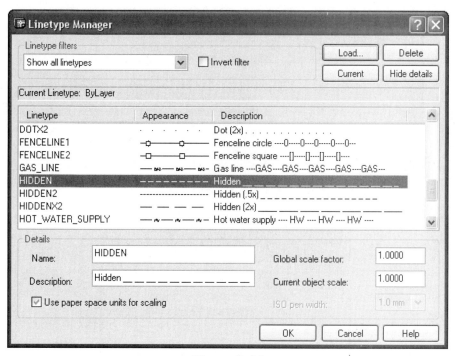

Figure 2–64

ADVANCED LAYER TOOLS

FILTERING LAYERS

Some drawings have so many layers that it is difficult to perform a search of the layers by using the scroll bars. When confronted with numerous layers in a drawing file, filter out all layers except those you want to view. For example, if you would like to list all layers that are assigned the color Red and the Hidden linetype, this technique would work. To activate layer filters, click in the outlined area of the Layer Properties Manager dialog box as illustrated in Figure 2–65. The Named Layer Filters dialog box will appear, similar to Figure 2–66. Through this dialog box, layers can be listed based on whether they are On or Off, Frozen or Thawed, Locked or Unlocked, by Color, or Linetype, or any combination of these parameters. Once a filter set has been created, you can even name this set for later use in the drawing design process.

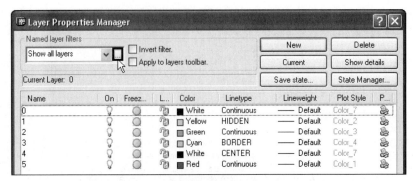

Figure 2–65

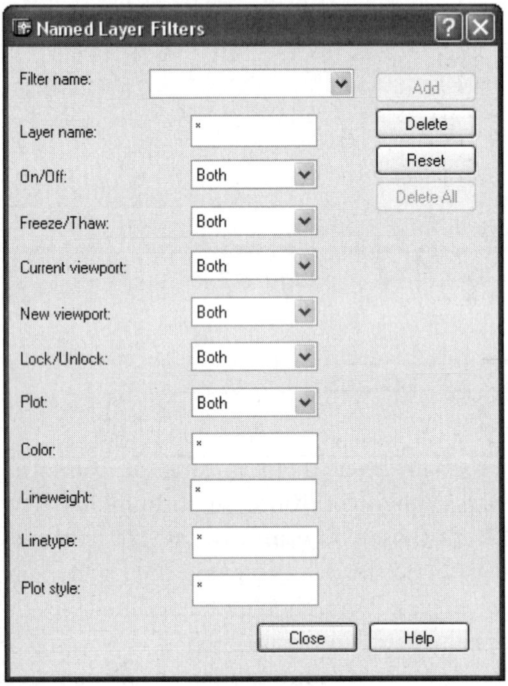

Figure 2–66

LAYER STATES

To manage the settings made in the Layer Properties Manager dialog box, layer states, as shown in Figure 2-67 are saved and then recalled. This option of the Layer Properties Manager dialog box allows you to save and restore layer configurations into "Layer States". You first set the individual layer states normally and then save the states to a unique name as shown in Figure 2-68. These saved layer states then are restored as shown in Figure 2-69.

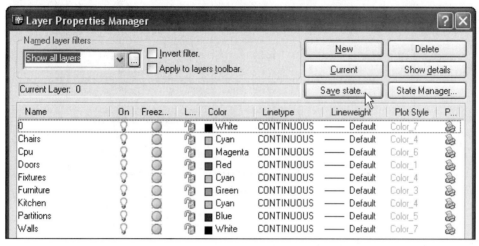

Figure 2–67

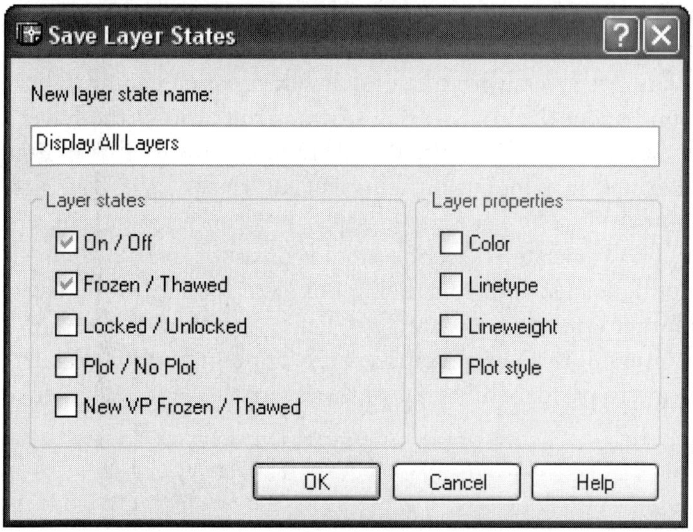

Figure 2–68

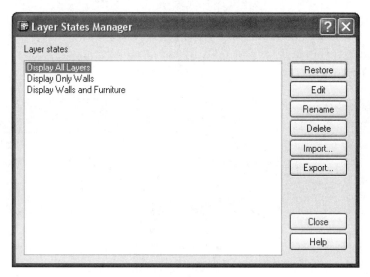

Figure 2–69

TRANSLATING LAYERS

In an effort to improve on the standardization of drawings, a Layer Translator can be used. Here is how it works. Suppose you receive a drawing from a consultant consisting of various layers whose names are given as numbers. You also have your own drawing that has the layers called out by name and function. You can translate or convert the numbered layers into layers with names that carry more meaning. Clicking on CAD Standards found under the Tools... pull-down menu exposes the Layer Translator in Figure 2–70. You could also activate the CAD Standards toolbar in Figure 2–70 and click on the Layer Translator button. This will launch the Layer Translator dialog box shown in Figure 2–71. The Translate From list box contains all the layers in your current drawing. The Translate To list box should consist of layers contained in a drawing that must first be loaded into this dialog box. You highlight one layer to Translate From and another layer to Translate To. Clicking on the Map button displays both old and new layers in the Layer Translation Mappings list box. Finally, click on the Translate button to permanently change the layers.

Figure 2–70

Figure 2–71

 Note: A series of tutorial exercises designed around the use of layer filters, layer states, and translating layers is available in chapter 2 of the accompanying projects manual.

CREATING TEMPLATE FILES

By default, when you start a new drawing from scratch, AutoCAD uses a template. Two templates are available depending on whether you are drawing in inches or millimeters; they are Acad.Dwt (inches) and Acadiso.Dwt (millimeters). Various settings are made in a template file. For instance, in the Acad.Dwt template file, the drawing units are set to decimal units with 4 decimal place precision. The drawing limits are set to 12,9 units for the upper right corner. Only one layer exists; namely layer 0. The grid in this template is set to a spacing of 0.50 units.

If you have to create several drawings that have the same limits and units settings and use the same layers, rather than start with the Acad.Dwt file and make changes to the settings at the start of each new drawing file, a better technique would be to create your own template. Templates can be made up of entire drawings or just have a company border already placed in the drawing. Other templates may appear empty of objects such as lines or circles. However the limits of the drawing may already be set along with grid and snap settings. The following settings are usually assigned to a drawing template:

- Drawing units and precision
- Drawing limits
- Grid and Snap Settings

- Layers, Linetypes, and Lineweights
- Linetype scale factor depending on the final plot scale of the drawing.

 Try It! – Create a new drawing file starting from scratch. You will be creating a template designed to draw an architectural floor plan. Make changes to the following settings:

- In the Drawing Units dialog box, change the units type from Decimal to Architectural.

- Set the limits of the drawing to 17',11' for the upper right corner and perform a ZOOM All. Change the grid and snap to 3" for the X and Y values.

- Create the following layer names:

  - Floor Plan

  - Hidden

  - Center

  - Dimension

  - Doors

  - Windows

- Make your own color assignments to these layers.

- Load the Hidden and Center linetypes and assign these to the appropriate layers.

- Assign a lineweight of 0.70mm to the Floor Plan layer.

- Finally, use the LTSCALE command and set the linetype scale to 12.

Now it is time to save the settings in a template format. Follow the usual steps for saving your drawing. However, when Save Drawing As dialog box appears, first click the arrow on the other side of "Files of type:" in Figure 2–72 to expose the file types. Click on the field that states "AutoCAD Drawing Template File (*.dwt)". AutoCAD will automatically take you to the folder in Figure 2–73 that holds all template information. It is at this time that you enter the name of the template as AEC_B 1=12. The name signifies an architectural drawing for the B size drawing sheet at a scale of 1"=1'-0".

File name:	Drawing2.dwg
Files of type:	AutoCAD 2004 Drawing (*.dwg)

AutoCAD 2004 Drawing (*.dwg)
AutoCAD 2000/LT2000 Drawing (*.dwg)
AutoCAD Drawing Standards (*.dws)
AutoCAD Drawing Template (*.dwt)
AutoCAD 2004 DXF (*.dxf)
AutoCAD 2000/LT2000 DXF (*.dxf)
AutoCAD R12/LT2 DXF (*.dxf)

Figure 2–72

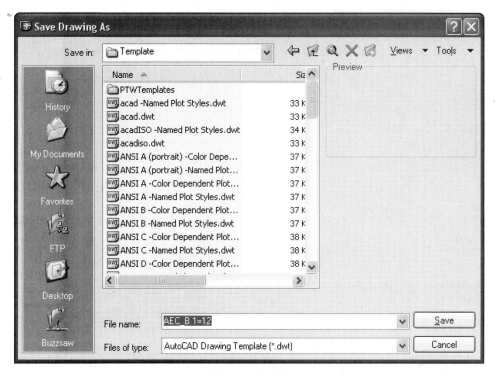

Figure 2–73

Before the template file is saved, you have the opportunity to document the purpose of the template in the Template Description dialog box in Figure 2–74. It is always considered good practice to create this documentation especially with others will be using your template.

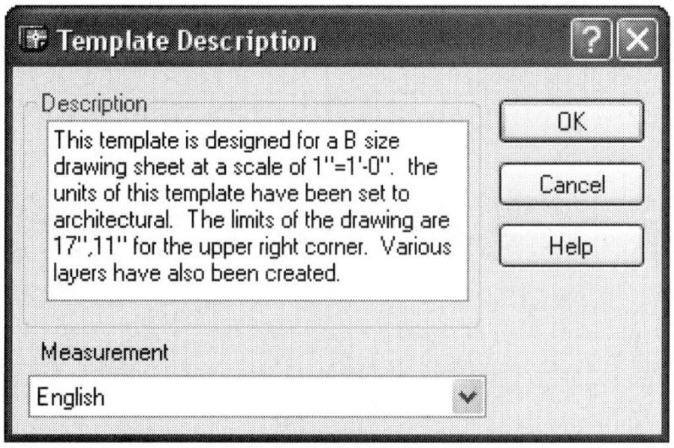

Figure 2–74

TUTORIAL EXERCISE:

Creating Layers Using the Layer Properties Manager Dialog Box

Layer Name	Color	Linetype	Lineweight
Object	White	Continuous	0.60
Hidden	Red	Hidden	0.30
Center	Yellow	Center	Default
Dim	Yellow	Continuous	Default
Section	Blue	Continuous	Default

Purpose

Use the following steps to create layers according to the above specifications.

STEP 1

Activate the Layer Properties Manager dialog box by clicking on the Layer button located on the Object Properties toolbar as shown in Figure 2–75.

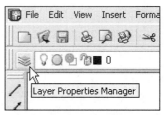

Figure 2–75

STEP 2

Once the Layer Properties Manager dialog box displays in Figure 2–76, notice on your screen that only one layer is currently listed, namely Layer 0. This happens to be the default layer, the layer that is automatically available to all new drawings. Since it is considered poor practice to construct any objects on Layer 0, new layers will be created not only for object lines but for hidden lines, center lines, dimension objects, and section lines as well. To create these layers, click on the New button and notice that a layer is automatically added to the list of layers. This layer is called Layer1 (see Figure 2–77). While this layer is highlighted, enter the first layer name, "Object."

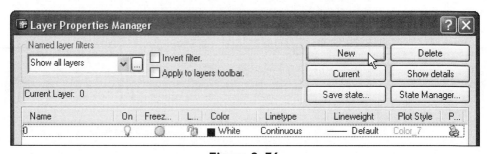

Figure 2–76

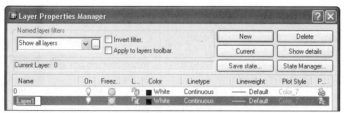

Figure 2–77

Entering a comma after the name of the layer allows more layers to be added to the listing of layers. Once you've entered the comma after the layer "Object" and the new layer appears, enter the new name of the layer as "Hidden." Repeat this procedure of using the comma to create multiple layers for "Center," "Dim," and "Section." Press ENTER after typing in "Section." The complete list of all layers created should be similar to the illustration shown in Figure 2–78.

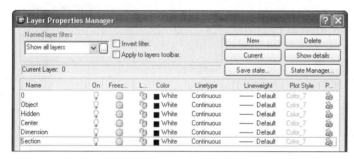

Figure 2–78

STEP 3

As all new layers are displayed, the names may be different, but they all have the same color and linetype assignments (see Figure 2–79). At this point, the dialog box comes in handy in assigning color and linetypes to layers in a quick and easy manner. First, highlight the desired layer to add color or linetype by picking the layer. A horizontal bar displays, signifying that this is the selected layer. Click on the color swatch identified by the box in Figure 2–79 and follow the next step to assign the color "Red" to the "Hidden" layer name.

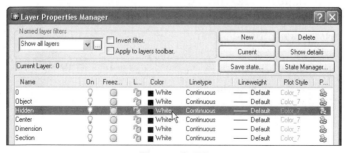

Figure 2–79

STEP 4

Clicking on the color swatch in the previous step displays the Select Color dialog box shown in Figure 2–80. Select the desired color from one of the following areas: Standard Colors, Gray Shades, or Full Color Palette. The standard colors represent colors 1 (Red) through 9 (Gray). On display terminals with a high-resolution graphics card, the full color palette displays different shades of the standard colors. This gives you a greater variety of colors to choose from. For the purposes of this tutorial, the color "Red" will be assigned to the "Hidden" layer. Select the box displaying the color red; a box outlines the color and echoes the color in the bottom portion of the dialog box. Click on the OK button to complete the color assignment; if you select Cancel, the color assignment will be removed, requiring this operation to be used again. Continue with this step by assigning the color Yellow to the Center and Dim layers; assign the color Blue to the Section layer.

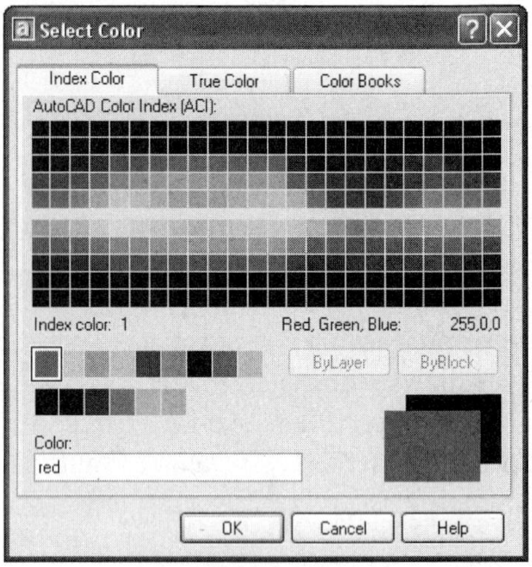

Figure 2–80

STEP 5

Once the color has been assigned to a layer, the next step is to assign a linetype, if any, to the layer. The "Hidden" layer requires a linetype called "HIDDEN." Click on the "Continuous" linetype that is highlighted in Figure 2–79 to display the Select Linetype dialog box illustrated in Figure 2–81. By default, Continuous is the only linetype loaded. Clicking on the Load button displays the next dialog box, illustrated in Figure 2–82.

Figure 2–81

Choose the desired linetype to load from the Load or Reload the Linetype dialog box shown in Figure 2–82. Scroll through the linetypes until the "Hidden" linetype is found. Click on the OK button to return to the Select Linetype dialog box.

Once you're back in the Select Linetype dialog box, shown in Figure 2–83, notice the "Hidden" linetype listed along with "Continuous." Because this linetype has just been loaded, it still has not been assigned to the "Hidden" layer. Click on the Hidden linetype listed in the Select Linetype dialog box, and click on the OK button. Once the Layer Properties Manager dialog box reappears, notice that the "Hidden" linetype has been assigned to the "Hidden" layer. Repeat this procedure to assign the "Center" linetype to the "Center" layer.

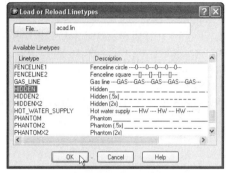

Figure 2–82

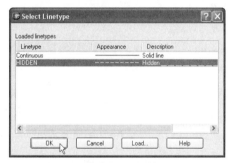

Figure 2–83

STEP 6

Another requirement of the "Hidden" layer is that it uses a lineweight of 0.30 mm to have the hidden lines stand out when compared with the object in other layers. Clicking on the highlighted default Lineweight in Figure 2–79 displays the Lineweight dialog box in Figure 2–84. Click on the 0.30 mm lineweight followed by the OK button to assign this lineweight to the "Hidden" line. Use the same procedure to assign a lineweight of 0.60 mm to the Object layer.

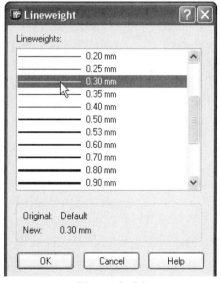

Figure 2–84

STEP 7

Once you've completed all color and linetype assignments, the Layer Properties Manager dialog box should appear similar to Figure 2–85. Click on the OK button to save all layer assignments and return to the drawing editor.

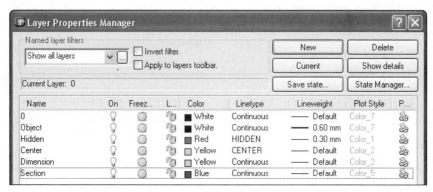

Figure 2–85

STEP 8

When layers are first created, they are listed in the order they were entered. When the layers are saved and the Layer Properties Manager dialog box is displayed again, all layers are reordered alphabetically (see Figure 2–86.)

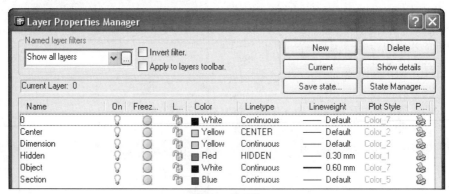

Figure 2–86

STEP 9

Clicking in the Name box at the top of the list of layers reorders the layers. Now, the layers are listed in reverse alphabetical order (see Figure 2–87). This same effect occurs when you click in the On, Freeze, Color, Linetype, Lineweight, and Plot header boxes. Figure 2–88 displays the results of clicking on the Color header where all colors are reordered starting with Red (1), Yellow (2), Blue (5), and White (7).

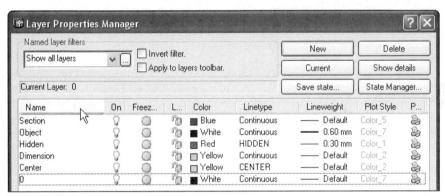

Figure 2–87

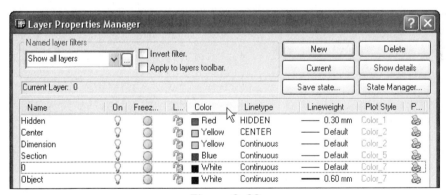

Figure 2–88

 Open the Exercise Manual PDF file for Chapter 2 on the accompanying CD for more tutorials and exercises.

 If you have the accompanying Exercise Manual, refer to Chapter 2 for more tutorials and exercises.

CHAPTER 3

AutoCAD Display and Basic Selection Operations

This chapter introduces you to the various ways of viewing your drawing. A number of options of the ZOOM command will be explained first, followed by a review of panning and understanding how a wheel mouse is used with AutoCAD. User defined portions of a drawing screen called views will also be introduced in this chapter. You will be introduced to various Object Selection methods such as window, crossing, crossing polygon, window polygon, fence, and previous, just to name a few.

THE ZOOM COMMAND

The ability to magnify details in a drawing or reduce the drawing to see it in its entirety is a function of the ZOOM command. It does not take much for a drawing to become very busy, complicated, or dense when displayed in the drawing editor. Therefore, use the ZOOM command to work on details or view different parts of the drawing. One way to select the options of this command is from the View pull-down menu as shown in Figure 3–1. Choosing Zoom displays the various options of the ZOOM command. These options include zooming in real time, zooming the previous display, using a window to define a boxed area to zoom to, dynamic zooming, zooming to a user-defined scale factor, zooming based on a center point and a scale factor, or performing routine operations such as zooming in or out, zooming all, or zooming the extents of the drawing. All of these modes will be discussed in the pages that follow.

Figure 3–1

A second way to choose options of the ZOOM command is through the Standard toolbar area at the top of the display screen. Figure 3–2 shows a series of buttons that perform various options of the ZOOM command. Four main buttons initially appear in the Standard toolbar, namely Realtime PAN, Realtime ZOOM, ZOOM-Window, and ZOOM-Previous. Pressing down on the ZOOM-Window option displays a series of additional zoom option buttons that cascade down. See Figure 3–2 for the meaning of each button. Click on the button to perform the desired zoom operation.

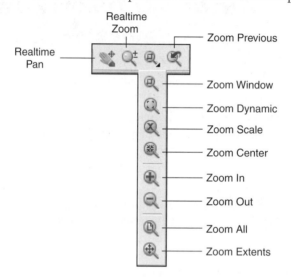

Figure 3–2

Figure 3–3 shows a third method of performing ZOOM command options using the dedicated Zoom floating toolbar. This toolbar contains the same buttons found in the Standard toolbar; it differs in the fact that the toolbar in Figure 3–3 can be moved to different positions around the display screen. You can also activate the ZOOM command from the keyboard by entering either ZOOM or the letter Z, which is its command alias.

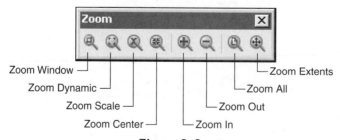

Figure 3–3

Figure 3–4 is a complex drawing of a part that consists of all required orthographic views. To work on details of this and other drawings, use the ZOOM command to magnify or reduce the display screen. The following are the options of the ZOOM command:

Command: **Z** *(For ZOOM)*
Specify corner of window, enter a scale factor (nX or nXP), or
[All/Center/Dynamic/Extents/Previous/Scale/Window] <real time>: *(enter one of the listed options)*

Executing the ZOOM command and picking a blank part of the screen places you in automatic ZOOM-Window mode. Selecting another point zooms in to the specified area. Refer to the following command sequence to use this mode of the ZOOM command on the object illustrated in Figure 3–4.

Command: **Z** *(For ZOOM)*
Specify corner of window, enter a scale factor (nX or nXP), or
[All/Center/Dynamic/Extents/Previous/Scale/Window] <real time>: *(Mark a point at "A")*
Other corner: *(Mark a point at "B")*

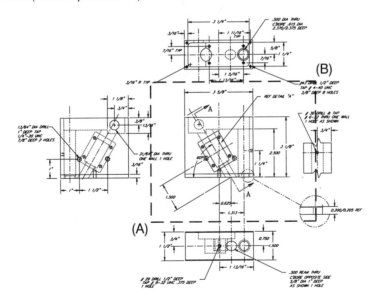

Figure 3–4

The ZOOM-Window option is automatically invoked once you select a blank part of the screen and then pick a second point. The resulting magnified portion of the screen appears in Figure 3–5.

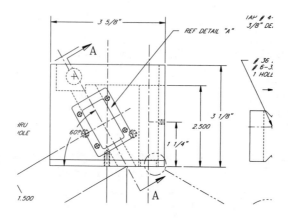

Figure 3–5

 ZOOMING IN REAL TIME

A powerful option of the ZOOM command is performing screen magnifications or reductions in real time. This is the default option of the command. Issuing the Real-time option of the ZOOM command displays a magnifying glass icon with a positive sign and a negative sign above the magnifier icon. Identify a blank part of the drawing editor, press down the Pick button of the mouse (the left mouse button), and move in an upward direction to zoom in to the drawing in real time. Identify a blank part of the drawing editor, press down the Pick button of the mouse, and move in a downward direction to zoom out of the drawing in real time. Use the following command sequence and see also Figure 3–6.

 Command: **Z** *(For ZOOM)*
Specify corner of window, enter a scale factor (nX or nXP), or
[All/Center/Dynamic/Extents/Previous/Scale/Window] <real time>: *(Press* ENTER *to accept Realtime as the default)*

Identify the lower portion of the drawing editor, press and hold down the Pick button of the mouse, and move the Realtime cursor up; notice the image zooming in.

Once you are in the Realtime mode of the ZOOM command, press the right mouse button to activate the shortcut menu shown in Figure 3–7. Use this menu to switch between Realtime ZOOM and Realtime PAN, which gives you the ability to pan across the screen in real time. The ZOOM Window, Original (Previous), and Extents options are also available in the cursor menu in Figure 3–7.

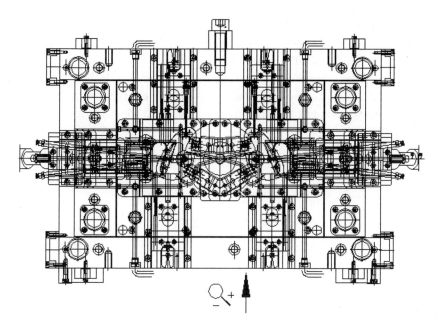

Figure 3–6

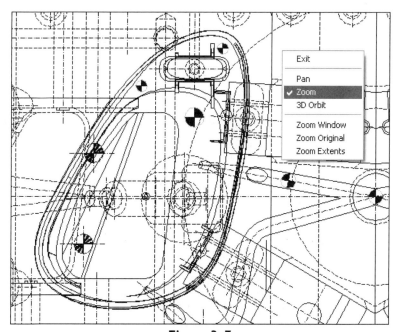

Figure 3–7

USING AERIAL VIEW ZOOMING

Another dynamic way of performing zooms is through the Aerial View option, which is selected from the View pull-down menu in Figure 3–8. Choosing Aerial View activates a dialog box in the lower-right corner of the screen displaying a smaller image of the total drawing (see Figure 3–9). You create a window inside the Aerial View dialog box, which performs the zoom operation in the background of the main AutoCAD screen.

Figure 3–8

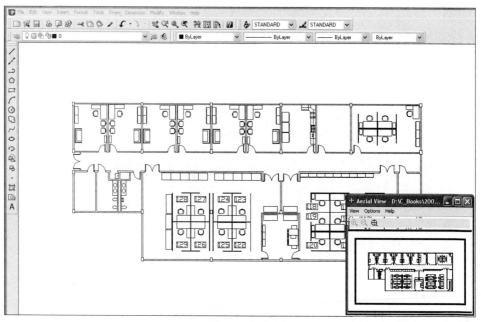

Figure 3–9

The first time you click inside the Aerial View image, a box similar to Figure 3–10 appears. This box represents the current zoom window. Notice that, as you move this box around, your drawing pans to different locations of the screen. The trick to performing zooms with this box is to make it smaller— by single-clicking on the left mouse button, your pick button. When you perform this operation, the box changes to an image of a new box as in Figure 3–11. An arrow is present along the right edge of the rectangular box. Moving the mouse to the left or right allows you to increase or decrease the size of the rectangle. Making the box smaller zooms in to the drawing. Single-clicking again with the pick button of the mouse returns the rectangle to the image illustrated in Figure 3–12. Here the rectangle is smaller; as you move the rectangle around the screen, you can notice the magnified image panning around in the background. To complete the operation, right-click the mouse button to anchor the small rectangle to the image in the Aerial View dialog box.

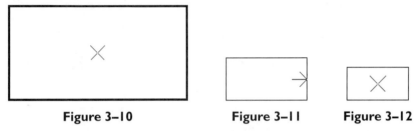

Figure 3–10   **Figure 3–11**   **Figure 3–12**

In Figure 3–13, a small rectangular box has been created in the Aerial View dialog box. The current position of this rectangle in the drawing is displayed in the background of the main AutoCAD screen. Right-clicking the mouse button makes the rectangular box stationary in the Aerial View dialog box, as in Figure 3–14. This frees you up to activate the main AutoCAD screen and work on your drawing in its zoomed-in state.

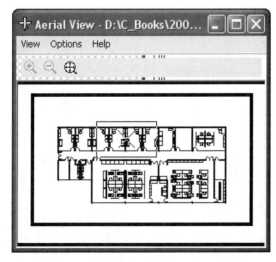

Figure 3–13

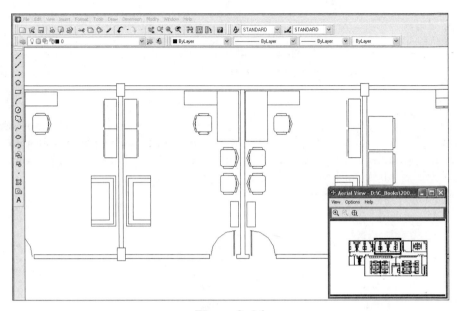

Figure 3–14

USING ZOOM-ALL

Another option of the ZOOM command is All. Use this option to zoom to the current limits of the drawing as set by the LIMITS command. In fact, right after the limits of a drawing have been changed, issuing a ZOOM-All updates the drawing file to reflect the latest screen size. To use the ZOOM-All option, refer to the following command sequence.

Command: **Z** *(For ZOOM)*

Specify corner of window, enter a scale factor (nX or nXP), or
[All/Center/Dynamic/Extents/Previous/Scale/Window] <real time>: **A** *(For All)*

In Figure 3–15, the top illustration shows a zoomed-in portion of a part. Use the ZOOM-All option to zoom to the drawing's current limits in Figure 3–16.

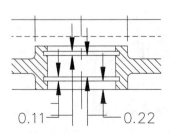

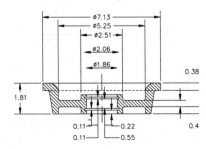

Figure 3–15   **Figure 3–16**

USING ZOOM-CENTER

The ZOOM-Center option allows you to specify a new display based on a selected center point (see Figure 3–17). A window height controls whether the image on the display screen is magnified or reduced. If a smaller value is specified for the magnification or height, the magnification of the image is increased (you zoom in to the object). If a larger value is specified for the magnification or height, the image gets smaller, or a ZOOM-Out is performed (see Figure 3–17).

 Try It! – Open the drawing file 03_Zoom Center. Follow the illustrations and command sequence below to perform a zoom based on a center point.

 Command: **Z** *(For ZOOM)*
Specify corner of window, enter a scale factor (nX or nXP), or
[All/Center/Dynamic/Extents/Previous/Scale/Window] <real time>: **C** *(For Center)*
Center point: *(Mark a point at the center of circle "A" shown in Figure 3–17)*
Magnification or Height <7.776>: **2**

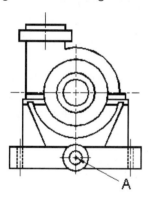
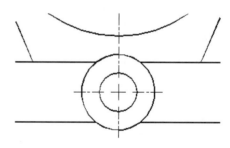

Figure 3–17

 ## USING ZOOM-EXTENTS

The left image of the pump in Figure 3–18 reflects a ZOOM-All operation. Use this option to display the entire drawing area based on the drawing limits even if the objects that make up the image appear small. Instead of performing a zoom based on the drawing limits, ZOOM-Extents uses the extents of the image on the display screen to perform the zoom. The right image in Figure 3–18 shows the largest possible image displayed as a result of using the ZOOM command and the Extents option.

 Try It! – Open the drawing file 03_Zoom Extents. Follow the illustrations and command sequence below to perform a zoom based on the drawing limits (All) and the objects in the drawing (Extents).

Command: **Z** *(For ZOOM)*

Specify corner of window, enter a scale factor (nX or nXP), or
[All/Center/Dynamic/Extents/Previous/Scale/Window] : **A** *(For All)*

 Command: **Z** *(For ZOOM)*

Specify corner of window, enter a scale factor (nX or nXP), or
[All/Center/Dynamic/Extents/Previous/Scale/Window] : **E** *(For Extents)*

Zoom All
Based on Drawing Limits

Zoom Extents
Based on Drawing Objects

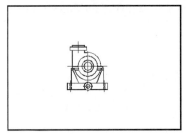 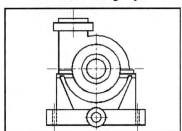

Figure 3–18

 ## USING ZOOM-WINDOW

The ZOOM-Window option allows you to specify the area to be magnified by marking two points representing a rectangle, as in the left image of Figure 3–19. The center of the rectangle becomes the center of the new image display; the image inside the rectangle is either enlarged (see the right image of Figure 3–19) or reduced.

 Try It! – Open the drawing file 03_Zoom Window. Follow the illustrations and command sequence below to perform a zoom based on a window.

 Command: **Z** *(For ZOOM)*

Specify corner of window, enter a scale factor (nX or nXP), or
[All/Center/Dynamic/Extents/Previous/Scale/Window] <real time>: **W** *(For Window)*
First corner: *(Mark a point at "A")*
Other corner: *(Mark a point at "B")*

By default, the window option of zoom is automatic; in other words, without entering the Window option, the first point you pick identifies the first corner of the window box; the prompt "Other corner:" completes ZOOM-Window as indicated in the following prompts:

Command: **Z** *(For ZOOM)*
Specify corner of window, enter a scale factor (nX or nXP), or
[All/Center/Dynamic/Extents/Previous/Scale/Window] <real time>: *(Mark a point at "A")*
Other corner: *(Mark a point at "B")*

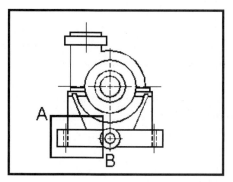
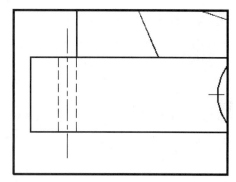

Figure 3–19

USING ZOOM-PREVIOUS

After magnifying a small area of the display screen, use the Previous option of the ZOOM command to return to the previous display. The system automatically saves up to ten views when zooming. This means you can begin with an overall display, perform two zooms, and use the ZOOM-Previous command twice to return to the original display. Zoom-Previous is also less likely to create a drawing regeneration (see Figure 3–20).

 Command: **Z** *(For ZOOM)*
Specify corner of window, enter a scale factor (nX or nXP), or
[All/Center/Dynamic/Extents/Previous/Scale/Window] <real time>: **P** *(For Previous)*

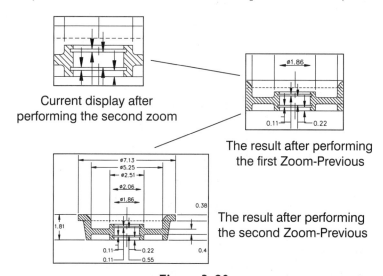

Current display after
performing the second zoom

The result after performing
the first Zoom-Previous

The result after performing
the second Zoom-Previous

Figure 3–20

USING ZOOM-SCALE

 Try It! – Open the drawing file 03_Zoom Scale. Follow the command sequences and illustration in Figure 3–21 for performing a zoom based on a scale factor.

If a scale factor of 0.50 is used, the zoom is performed into the drawing at a factor of 0.50, based on the original limits of the drawing. Notice the image gets smaller.

 Command: **Z** *(For ZOOM)*

Specify corner of window, enter a scale factor (nX or nXP), or
[All/Center/Dynamic/Extents/Previous/Scale/Window] <real time>: **0.50**

If a scale factor of 0.50X is used, the zoom is performed into the drawing again at a factor of 0.50; however, the zoom is based on the current display screen. The image gets even smaller.

 Command: **Z** *(For ZOOM)*

Specify corner of window, enter a scale factor (nX or nXP), or
[All/Center/Dynamic/Extents/Previous/Scale/Window] <real time>: **0.50X**

Enter a scale factor of 0.90. The zoom is again based on the original limits of the drawing. As a result, the image displays larger.

 Command: **Z** *(For ZOOM)*
Specify corner of window, enter a scale factor (nX or nXP), or
[All/Center/Dynamic/Extents/Previous/Scale/Window] <real time>: **0.90**

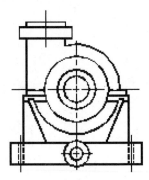

Figure 3–21

USING ZOOM-IN

Clicking on this button automatically performs a zoom-in operation at a scale factor of 0.5X; the "X" uses the current screen to perform the zoom-in operation.

 USING ZOOM-OUT

Clicking on this button automatically performs a zoom-out operation at a scale factor of 2X; the "X" uses the current screen to perform the zoom-in operation.

 THE PAN COMMAND

As you perform numerous ZOOM-Window and ZOOM-Previous operations, it becomes apparent that it would be nice to zoom in to a detail of a drawing and simply slide the drawing to a new area without changing the magnification; this is the purpose of the PAN command. In Figure 3–22, the Top view is magnified with ZOOM-Window; the result is shown in Figure 3–23. Now, the Bottom view needs to be magnified to view certain dimensions. Rather than use ZOOM-Previous and then ZOOM-Window again to magnify the Bottom view, use the PAN command.

Command: **P** *(For PAN)*

Press ESC or ENTER to exit, or right-click to display shortcut menu.

Issuing the PAN command displays the Hand symbol. In the illustration in Figure 3–23, pressing the Pick button down at "A" and moving the hand symbol to the right at "B" pans the screen and displays a new area of the drawing in the current zoom magnification.

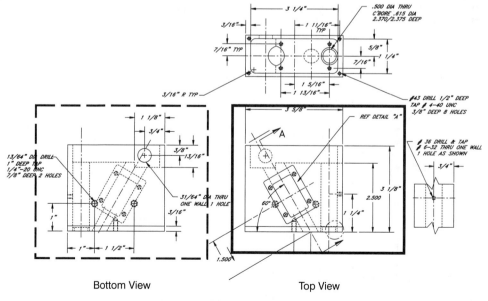

Figure 3–22

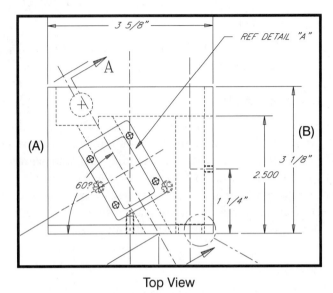

Top View

Figure 3–23

In Figure 3–24, the Bottom view is now visible after the drawing is panned from the Top view to the Bottom view, with the same display screen magnification. PAN can also be used transparently; that is, while in a current command, you can select the PAN command, which temporarily interrupts the current command, performs the pan, and restores the current command.

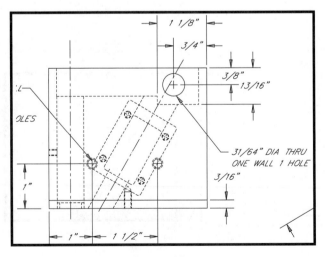

Bottom View

Figure 3–24

You can choose Pan from the View pull-down menu, shown in Figure 3–25, or you can enter P at the keyboard; P is the command alias for the PAN command. Also, don't forget about the scroll bars at the bottom and right side of the AutoCAD display screen. They also allow you to pan across the screen (see Figure 1-1).

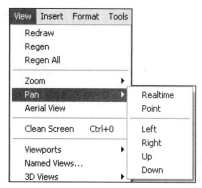

Figure 3–25

SUPPORT FOR WHEEL MOUSE

AutoCAD provides extra support for zooming and panning operations through the Wheel Mouse. This device consists of the standard Microsoft two-button mouse with the addition of a wheel, as illustrated in Figure 3–26. Rolling the wheel forward zooms in to or magnifies the drawing. Rolling the wheel backward zooms out or reduces the drawing.

Pressing and holding the wheel down, as in Figure 3–27, places you in realtime pan mode. The familiar hand icon on the display screen identifies this mode.

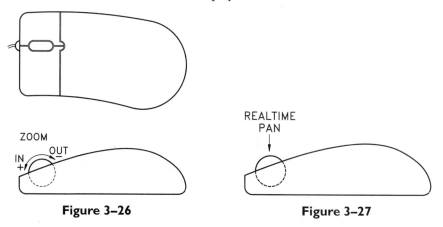

Figure 3–26   **Figure 3–27**

Pressing CTRL and then depressing the wheel places you in Joystick Pan mode (see Figure 3–28). This mode is identified by a pan icon similar to that shown in Figure 3–29. This icon denotes all directions in which panning may occur. Moving the mouse in any direction with the wheel depressed displays the icon in Figure 3–30, which shows the direction of the pan.

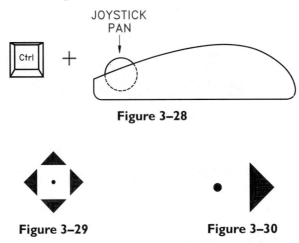

Figure 3–28

Figure 3–29   **Figure 3–30**

The wheel can also function like a mouse button. Double-clicking on the wheel, as in Figure 3–31, performs a ZOOM-Extents and is an extremely productive method of performing this operation. As you can see, the Wheel Mouse provides for realtime zooming and panning with the additional capability of performing a ZOOM-Extents, all at your fingertips.

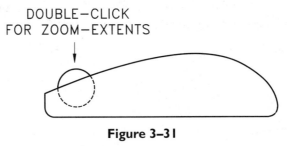

Figure 3–31

CREATING NAMED VIEWS WITH THE VIEW DIALOG BOX

An alternate method of performing numerous zooms is to create a series of views of key parts of a drawing. Then, instead of using the ZOOM command, restore the named view to perform detail work. This named view is saved in the database of the drawing for use in future editing sessions. Selecting Named Views from the View pull-down menu as shown in Figure 3–32 activates the View dialog box, shown in Figure 3–33. You can activate this same dialog box through the keyboard by entering the following at the command prompt:

 Command: **V** *(For VIEW)*

Figure 3–32

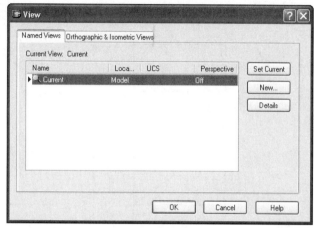

Figure 3–33

 Try It! – Open the drawing file 03_Views. Follow the next series of steps and illustrations used to create a view called "FRONT".

Clicking the New button in the View dialog box activates the New View dialog box shown in Figure 3–34. Use this dialog box to guide you in creating a new view. By definition, a view is created from the current display screen. This is the purpose of the Current Display radio button. Many views are created with the Define Window radio button, which will create a view based on the contents of a window that you define. Choosing the Define View Window button returns you to the display screen and prompts you for the first corner and other corner required to create a new view by window.

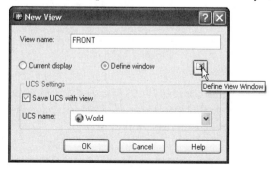

Figure 3–34

146

In Figure 3–35, a rectangular window is defined around the Front view using points "A" and "B" as the corners.

When the window is created, the Define New View dialog box redisplays. Clicking OK saves the view name in the View dialog box in Figure 3–36.

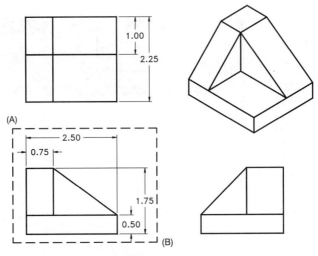

Figure 3–35

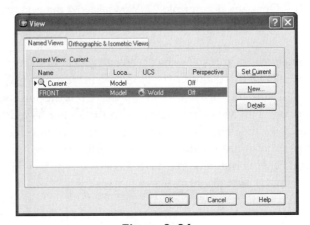

Figure 3–36

 Try It! – Open the drawing file 03_Views Complete. A series of views have already been created inside of this drawing. Activate the View dialog box and experiment restoring a number of these views.

The image in Figure 3–37 illustrates numerous views created from the drawing in Figure 3–35.

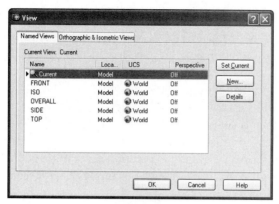

Figure 3–37

Clicking on a defined view name, as in Figure 3–38, followed by right-clicking the mouse displays the shortcut menu used to set the view current. You can also rename, delete, or obtain details of the view through this shortcut menu. Clicking on Set Current displays the Front view, as in Figure 3–39.

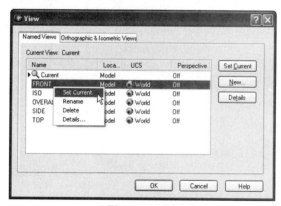

Figure 3–38

View Name = FRONT

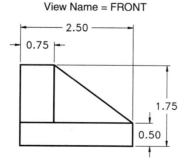

Figure 3–39

CREATING OBJECT SELECTION SETS

Selection sets are used to group a number of objects together for the purpose of editing. Applications of selection sets are covered in the following pages in addition to being illustrated in the next chapter. Once a selection set has been created, the group of objects may all be moved, copied, or mirrored. These operations supported by selection sets will be covered in Chapter 4, "Modify Commands." An object manipulation command supports the creation of selection sets if it prompts you to "Select objects." Any command displaying this prompt supports the use of selection sets. Selection set options (how a selection set is made) appear in Figure 3–40. Figures 3–41 through 3–47 present a few examples of how selection sets are used for manipulating groups of objects.

Add
All
CPolygon
Crossing
Fence
Last
Previous
Remove
Window
WPolygon
Undo

Figure 3–40

SELECTING OBJECTS BY INDIVIDUAL PICKS

When AutoCAD prompts you with "Select objects", a pickbox appears as the cursor is moved across the display screen. Any object enclosed by this box when picked will be considered selected. To show the difference between a selected and unselected object, the selected object highlights on the display screen.

 Try It! – Open the drawing file 03_Select. Enter the ERASE command and at the Select Objects prompt, pick the arc segment labeled "A" in Figure 3–41. To signify that the object is selected the arc highlights.

Command: **E** *(For ERASE)*
Select objects: *(Pick the object at "A")*
Select objects: *(Press ENTER to execute the ERASE command)*

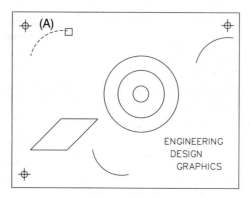

Figure 3–41

SELECTING OBJECTS BY WINDOW

The individual pick method previously outlined works fine for small numbers of objects. However, when numerous objects need to be edited, selecting each individual object could prove time-consuming. Instead, you can select all objects that you want to become part of a selection set by using the Window selection mode. This mode requires you to create a rectangular box by picking two diagonal points. In Figure 3–42, a selection window has been created with point "A" as the first corner and "B" as the other corner. When you use this selection mode, only those objects completely enclosed by the window box are selected. The window box selected four line segments, two arcs, and two points (too small to display highlighted). Even though the window touches the three circles, they are not completely enclosed by the window and therefore are not selected.

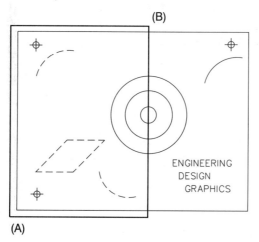

Figure 3–42

SELECTING OBJECTS BY CROSSING WINDOW

In the previous example of producing a selection set by a window, the window selected only those objects completely enclosed by it. Figure 3–43 is an example of selecting objects by a crossing window. The Crossing window option requires two points to define a rectangle as does the window selection option. In Figure 3–43, a dashed rectangle is used to select objects using "C" and "D" as corners for the rectangle; however, this time the crossing window was used. The highlighted objects illustrate the results. All objects that are touched by or enclosed by the crossing rectangle are selected. Because the crossing rectangle passes through the three circles without enclosing them, they are still selected by this object selection mode.

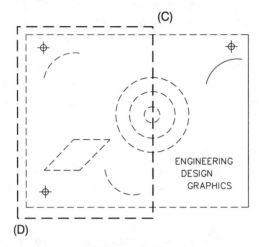

Figure 3–43

SELECTING OBJECTS BY A FENCE

Use this mode to create a selection set by drawing a line or group of line segments called a fence. Any object touched by the fence is selected. The fence does not have to end exactly where it was started. In Figure 3–44, all objects touched by the fence are selected, as represented by the dashed lines.

 Try It! – Open the drawing file 03_Fence. Follow the command sequence below and the illustration to select a group of objects using a fence.

 Command: **E** *(For ERASE)*
Select objects: **F** *(For Fence)*
First fence point: *(Pick a first fence point)*
Specify endpoint of line or [Undo]: *(Pick a second fence point)*
Specify endpoint of line or [Undo]: *(Press* ENTER *to exit fence mode)*
Select objects: *(Press* ENTER *to execute the ERASE command)*

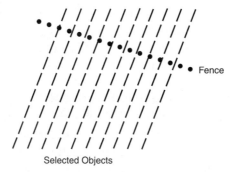

Selected Objects

Figure 3–44

REMOVING OBJECTS FROM A SELECTION SET

All of the previous examples of creating selection sets have shown you how to create new selection sets. What if you select the wrong object or objects? Instead of canceling out of the command and trying to select the correct objects, you can use the Remove option to remove objects from an existing selection set. In Figure 3–45, a selection set has been created and made up of all of the highlighted objects. However, the large circle was mistakenly selected as part of the selection set. The Remove option allows you to remove highlighted objects from a selection set. To activate Remove, press SHIFT and pick the object you want removed; this only works if the "Select objects" prompt is present. When a highlighted object is removed from the selection set, as shown in circle "A" in Figure 3–46, it regains its original display intensity.

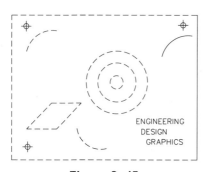

Figure 3–45

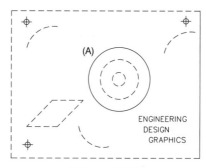

Figure 3–46

SELECTING THE PREVIOUS SELECTION SET

When you create a selection set of objects, this grouping is remembered until another selection set is made. The new selection set replaces the original set of objects. Let's say you moved a group of objects to a new location on the display screen. Now you want to rotate these same objects at a certain angle. Rather than select the same set of objects to rotate, you would pick the Previous option or type "P" at the "Select objects" prompt. This selects the previous selection set. The buffer holding the selection set is cleared whenever you use the U command to undo the previous command.

SELECTING OBJECTS BY A CROSSING POLYGON

When you use the Window or Crossing Window mode to create selection sets, two points specify a rectangular box for selecting objects. At times, it is difficult to select objects by the rectangular window or crossing box because in more cases than not, extra objects are selected and have to be removed from the selection set.

 Try It! – Open the drawing file 03_Select CP. Figure 3–47 shows a mechanical part with a "C"-shaped slot. Rather than use Window or Crossing Window modes, you can pick the Crossing Polygon mode (CPolygon) or type "CP" at the "Select objects" prompt. You simply pick points representing a polygon. Any object that touches or is inside the polygon is added to a selection set of objects. In Figure 3–47, the crossing polygon is constructed using points "1" through "5" A similar but different selection set mode is the Window Polygon (WPolygon). Objects are selected using this mode when they lie completely inside the Window Polygon, which is similar to the regular Window mode.

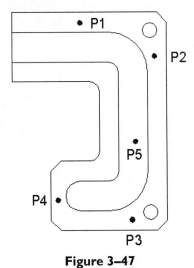

Figure 3–47

APPLICATIONS OF SELECTING OBJECTS WITH THE ERASE COMMAND

Figures 3–48 and 3–49 illustrate the use of the Window and Crossing options of deleting the circle and center marker with the ERASE command.

ERASE—WINDOW

When erasing objects by Window, be sure the objects to be erased are completely enclosed by the window, as in Figure 3–48.

 Command: **E** *(For ERASE)*

Select objects: *(Pick a point at "A" and move the cursor to the right; this automatically invokes the window option)*
Other corner: *(Mark a point at "B" and notice the objects that highlight)*
Select objects: *(Press* ENTER *to execute the ERASE command)*

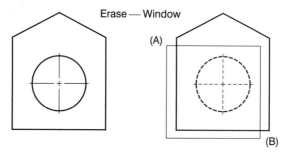

Figure 3–48

ERASE—CROSSING

Erasing objects by a crossing box is similar to the Window Box mode; however, any object that touches the crossing box or is completely enclosed by the crossing box is selected. See Figure 3–49.

 Command: **E** *(For ERASE)*

Select objects: *(Pick a point at "A" and move the cursor to the left; this automatically invokes the crossing option)*
Other corner: *(Mark a point at "B" and notice the objects that highlight)*
Select objects: *(Press* ENTER *to execute the* ERASE *command)*

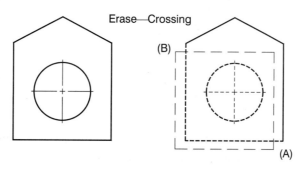

Figure 3–49

Figure 3–50 displays the objects that were selected by the Window option. However, the group of objects on the far right and left have mistakenly been selected and should not be erased. These objects need to be removed from the current selection set of objects with the Remove option of "Select objects."

Command: **E** *(For ERASE)*
Select objects: *(Pick a point at "A" in Figure 3–50 and move the cursor to the right; this automatically invokes the window option)*
Other corner: *(Mark a point at "B" in Figure 3–50 and notice the objects that highlight)*

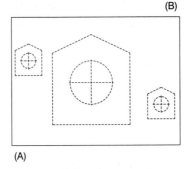

Figure 3–50

Before performing the ERASE command on the objects in Figure 3–50, issue the Remove option to deselect the group of objects shown in Figure 3–51. You can accomplish this easily by holding down SHIFT and picking the objects you do not want to erase.

Select objects: *(While pressing the SHIFT key, begin picking all objects in Figures at "A" and "B" in Figure 3–51. Notice the objects are no longer highlighted)*
Select objects: *(Press ENTER to execute the ERASE command)*

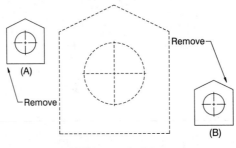

Figure 3–51

By removing objects through Shift-Pick, you remain in the command instead of having to cancel the command for picking the wrong objects and starting over. See Figure 3–52.

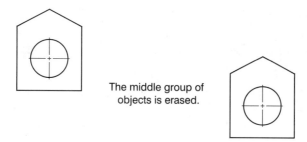

The middle group of
objects is erased.

Figure 3–52

CYCLING THROUGH OBJECTS

At times, the process of selecting objects can become quite tedious. Often, objects lie directly on top of each other. As you select the object to delete, the other object selects instead. To remedy this, press and hold CTRL when prompted to "Select objects." This activates Object Selection Cycling and enables you to scroll through all objects in the vicinity of the pickbox. A message appears in the prompt area when you begin scrolling alerting you that cycling is on; you can now pick objects until the desired object is highlighted. Pressing ENTER not only accepts the highlighted object but toggles cycling off. In Figure 3–53 and with selection cycling on, the first pick selects the line segment; the second pick selects the circle. Keep picking until the desired object highlights.

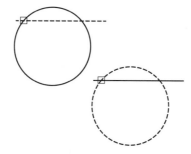

Figure 3–53

PRE-SELECTING OBJECTS

You can bypass the "Select Objects:" prompt by pre-selecting objects at the Command: prompt. Notice that when you pick an object, not only does it highlight but a series of blue square boxes also appear. These objects are called grips and will be discussed in greater detail in chapter 7. To cancel or de-select the object, press the ESC key and notice even the grips disappear. You could also pick a blank part of your screen at the Command: prompt. Moving your cursor to the right and picking a point has the same effect as using the Window option of Select Objects. Only objects completely inside of the window will be selected. If you pick a blank part of your screen at the Command: prompt and move your cursor to the left, this has the same effect as using the Crossing option of Select Objects where any items touched by or completely enclosed by the box are selected.

 Tip: Pressing CTRL + A at the Command: prompt selects all objects in the entire drawing and displays the blue grip boxes. This will even select objects that are on a layer that has been turned off, but will not select those on a layer that is frozen or locked.

THE QSELECT COMMAND

Yet another way of creating a selection set is by matching the object type and property with objects currently in use in a drawing. This is the purpose of the QSELECT command (Quick Select). This command can be chosen from the Tools pull-down menu, as in Figure 3–54. Choosing Quick Select displays the dialog box in Figure 3–55.

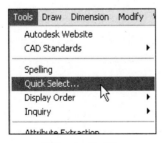

Figure 3–54   **Figure 3–55**

This command works only if objects are defined in a drawing; the Quick Select dialog box does not display in a drawing file without any objects drawn. Clicking in the Object type edit box displays all object types currently used in the drawing. This enables you to create a selection set by the object type. For instance, to select all line segments in the drawing file, click on Line in the Object type edit box in Figure 3–56. Clicking the OK button at the bottom of the dialog box returns you to the drawing and applies the object properties to objects in the drawing.

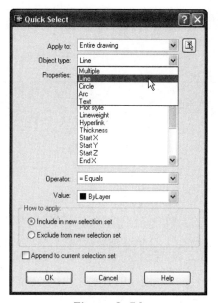

Figure 3–56

Notice, in Figure 3–57, that all line segments are highlighted. (The square boxes positioned around the drawing are called grips and will be discussed later in Chapter 7.) Other controls of Quick Select include the ability to select the object type from the entire drawing or from just a segment of the drawing. You can narrow the selection criteria by adding various properties to the selection mode such as Color, Layer, and Linetype, to name a few. You can also create a reverse selection set. Clicking on the radio button to exclude from the new selection set, shown in Figure 3–55, would create a selection set of all objects not counting those identified in the Object type edit box. The Quick Select dialog box lives up to its name—it enables you to create a quick selection set.

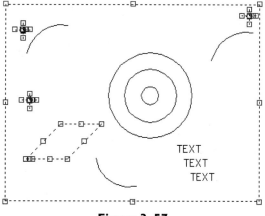

Figure 3–57

TUTORIAL EXERCISE: 03_SELECT OBJECTS.DWG

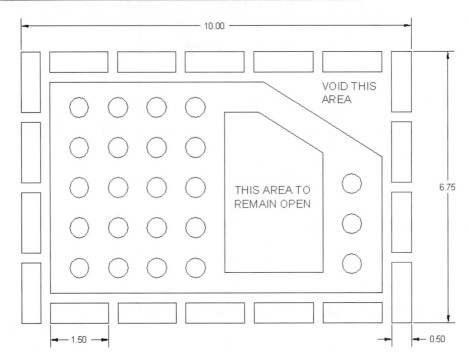

Figure 3–58

Purpose

The purpose of this tutorial is to experiment with the object's selection modes on the drawing shown in Figure 3–58.

System Settings

Keep the current limits settings of (0,0) for the lower-left corner and (16,10) for the upper-right corner. Keep all remaining system settings.

Layers

Layers have already been created for this tutorial exercise.

Suggested Commands

This tutorial utilizes the ERASE command as a means of learning the basic object selection modes. The following selection modes will be highlighted for this tutorial: Window, Crossing, Window Polygon, Crossing Polygon, Fence, and All. The effects of locking layers and selecting objects will also be demonstrated.

Whenever possible, substitute the appropriate command alias in place of the full AutoCAD command in each tutorial step. For example, use "CP" for the COPY command, "L" for the LINE command, and so on. The complete listing of all command aliases is located in Table 1–2.

STEP I

Open the drawing *03_Select Objects.dwg* and activate ERASE at the Command prompt. At the Select objects prompt, pick a point on the blank part of your screen at "A". At the next Select objects prompt, pick a point on your screen at "B". Notice that a solid rectangular box is formed (see Figure 3–59). The presence of this box signifies the Window option of selecting objects.

 Command: **E** (*For ERASE*)
Select objects: (*Pick a point at "A"*)
Specify opposite corner: (*Pick a point at "B"*)

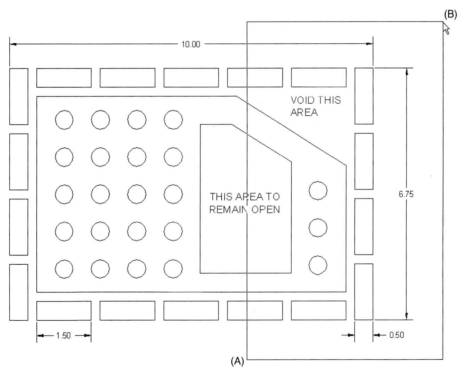

Figure 3–59

The result of selecting objects by a window is illustrated in Figure 3–60. Only those objects completely surrounded by the window are highlighted. Pressing ENTER will perform the erase operation. Before continuing on to the next step, issue the U (UNDO) command, which will negate the previous ERASE operation.

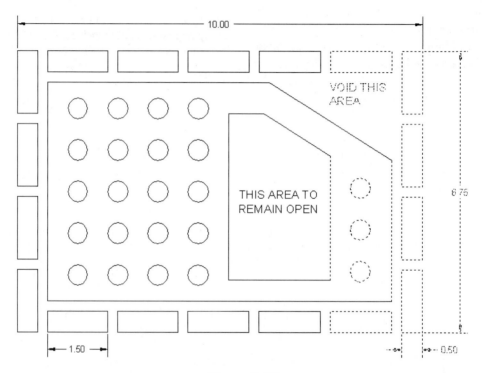

Figure 3–60

STEP 2

With the entire object displayed on the screen, activate the ERASE command again. At the Select objects prompt, pick a point on the blank part of your screen at "A". At the Specify opposite corner prompt, pick a point on your screen at "B". Notice that the rectangular box is now dashed in appearance (see Figure 3–61). The presence of this box signifies the Crossing option of selecting objects.

Command: **E** (*For ERASE*)
Select objects: (*Pick a point at "A"*)
Specify opposite corner: (*Pick a point at "B"*)

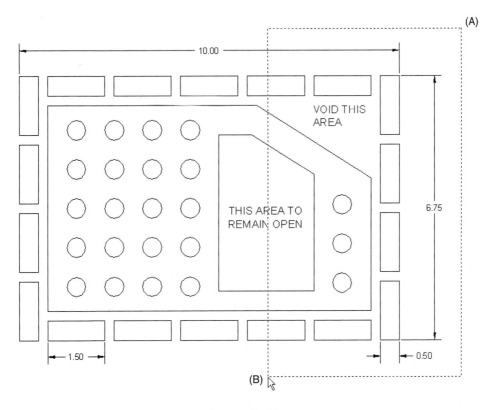

Figure 3–61

The result of selecting objects by a crossing window is illustrated in Figure 3–62. Notice that objects touched by the crossing box, as well as those objects completely surrounded by the crossing box are highlighted. Pressing ENTER will perform the ERASE operation. Before continuing on to the next step, issue the U (UNDO) command, which will negate the previous ERASE operation.

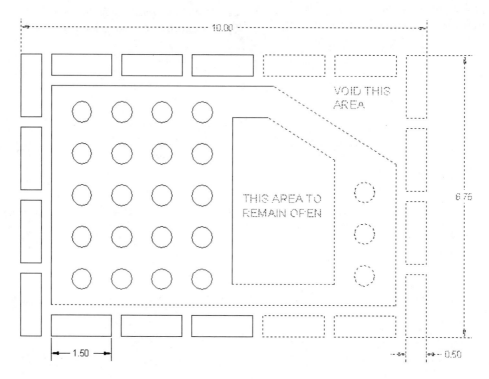

Figure 3–62

STEP 3

Activate the ERASE command again. At the Select objects prompt, pick a point on the blank part of your screen at "A". At the Specify opposite corner prompt, pick a point on your screen at "B" (see Figure 3–63). The presence of this box signifies the Window option of selecting objects.

Command: **E** *(For ERASE)*
Select objects: *(Pick a point at "A")*
Specify opposite corner: *(Pick a point at "B")*

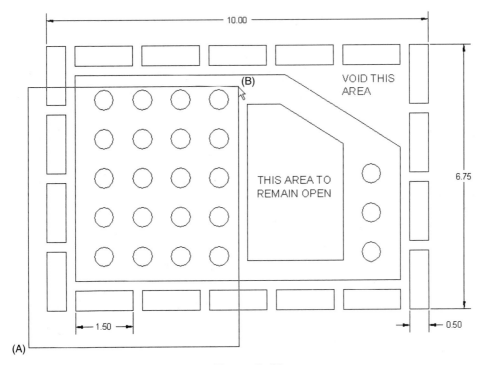

Figure 3–63

The result of selecting objects by a window is illustrated in Figure 3–64. Notice that a number of circles and rectangles have been selected along with a single dimension. Unfortunately, these rectangles and dimension were selected by mistake. Rather than cancel the ERASE command and select the objects again, press and hold down the SHIFT key and click on the highlighted rectangles and dimension. Notice how this operation deselects the objects (see Figure 3–65). Pressing ENTER will perform the ERASE operation. Before continuing on to the next step, issue the U (UNDO) command, which will negate the previous ERASE operation.

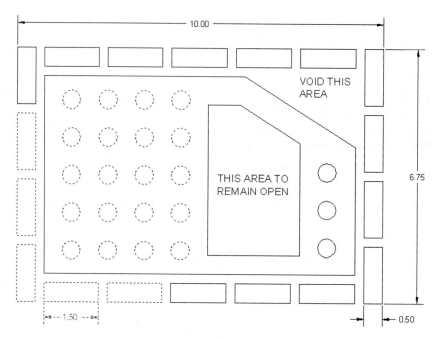

Figure 3–64

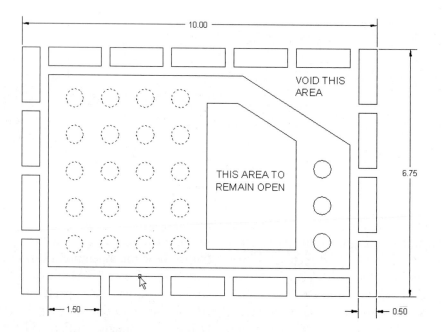

Figure 3–65

STEP 4

Yet another way to select objects is by using a Window Polygon. This method allows you to construct a closed irregular shape for selecting objects. Again, activate the ERASE command. At the Select objects prompt, enter WP for Window Polygon. At the First polygon point prompt, pick a point on your screen at "A". Continue picking other points until the desired polygon is formed, as shown in Figure 3–66.

 Note: The entire polygon must form a single closed shape; edges of the polygon cannot cross each other.

 Command: **E** *(For ERASE)*

Select objects: **WP** *(For Window Polygon)*
First polygon point: *(Pick a point at "A")*
Specify endpoint of line or [Undo]: *(Pick at "B")*
Specify endpoint of line or [Undo]: *(Pick at "C")*
Specify endpoint of line or [Undo]: *(Pick at "D")*
Specify endpoint of line or [Undo]: *(Pick at "E")*
Specify endpoint of line or [Undo]: *(Pick at "F")*
Specify endpoint of line or [Undo]: *(Pick at "G")*
Specify endpoint of line or [Undo]: *(Press ENTER)*
11 found

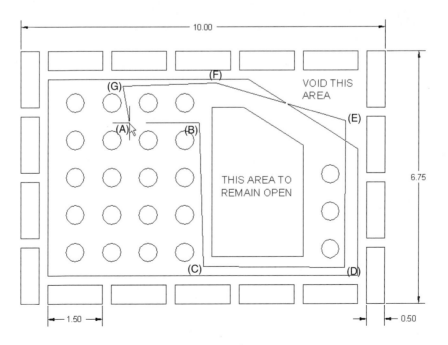

Figure 3–66

The result of selecting objects by a window polygon is illustrated in Figure 3–67. As with the window selection mode, objects must lie entirely inside of the window polygon in order for them to be selected. Pressing ENTER will perform the ERASE operation. Before continuing on to the next step, issue the U (UNDO) command, which will negate the previous ERASE operation.

 Note: A Crossing Polygon (CP) mode is also available to select objects that touch the crossing polygon or are completely surrounded by the polygon.

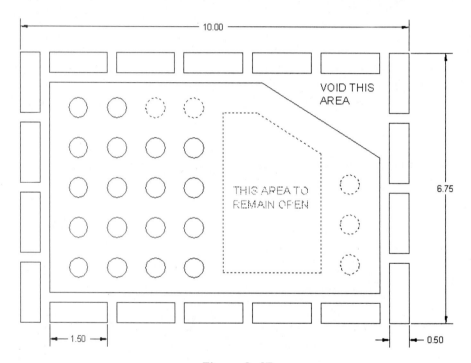

Figure 3–67

STEP 5

Objects can also be selected by a Fence. This is represented by a crossing line segment. Any object that touches the crossing line is selected. You can construct numerous crossing line segments. You cannot use the Fence mode to surround objects as with the Window or Crossing modes. Activate the ERASE command again. At the Select objects prompt, Enter F for Fence. At the First fence point prompt, pick a point on your screen at "A". Continue constructing crossing line segments until you are satisfied with the objects being selected (see Figure 3–68).

 Command: **E** *(For ERASE)*

Select objects: **F** *(For Fence)*

First fence point: *(Pick at "A")*

Specify endpoint of line or [Undo]: *(Pick at "B")*

Specify endpoint of line or [Undo]: *(Pick at "C")*

Specify endpoint of line or [Undo]: *(Pick at "D")*

Specify endpoint of line or [Undo]: *(Pick at "E")*

Specify endpoint of line or [Undo]: *(Press ENTER to create the selection set)*

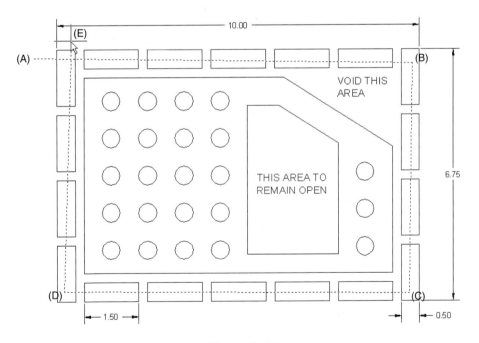

Figure 3–68

The result of selecting objects by a Fence is illustrated in Figure 3–69. Notice that objects touched by the fence are highlighted. Pressing ENTER will perform the ERASE operation. Before continuing on to the next step, issue the U (UNDO) command, which will negate the previous ERASE operation.

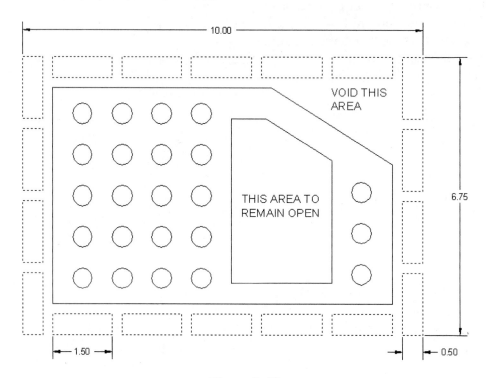

Figure 3–69

STEP 6

When situations require you to select all objects in the entire database of a drawing, the All option is very effecient. Activate the ERASE command again. At the Select objects prompt, Enter All. Notice that in Figure 3–70 all objects are selected by this option.

 Command: **E** *(For ERASE)*

Select objects: **All**

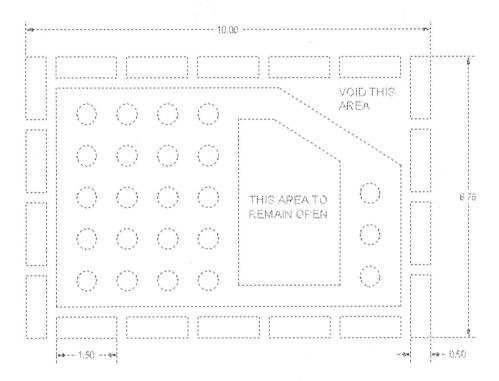

Figure 3–70

Pressing ENTER will perform the ERASE operation. Before continuing on to the next step, issue the U (UNDO) command, which will negate the previous ERASE operation.

 Note: The All option will also select objects even on layers that are turned off. The All option will not select objects on layers that are frozen.

STEP 7

Before performing another ERASE operation on a number of objects, activate the Layer Properties Manager dialog box, select the "Circles" layer, and click the lock icon in Figure 3–71. This operation locks the "Circles" layer.

Name	On	Freez...	L...	Color	Linetype	Lineweight	Plot Style	P...
0				■ White	CONTINUOUS	—— Default	Color_7	
Circles				■ 11	CONTINUOUS	—— Default	Color_11	
Defpoints				■ White	CONTINUOUS	—— Default	Color_7	
Dimension				□ 81	CONTINUOUS	—— Default	Color_81	
Lines				□ 140	CONTINUOUS	—— Default	Color_140	
OBJECT				□ Yellow	CONTINUOUS	—— Default	Color_2	
Rectangles				□ 50	CONTINUOUS	—— Default	Color_50	
Text				■ 240	CONTINUOUS	—— Default	Color_240	

Figure 3–71

To see what effect this has on selecting objects, activate the ERASE command again. At the Select objects prompt, pick a point on the blank part of your screen at "A". At the Specify opposite corner prompt, pick a point on your screen at "B". You have once again selected a number of objects by a crossing box (see Figure 3–72).

Command: **E** *(For ERASE)*
Select objects: *(Pick a point at "A")*
Specify opposite corner: *(Pick a point at "B")*
20 were on a locked layer.

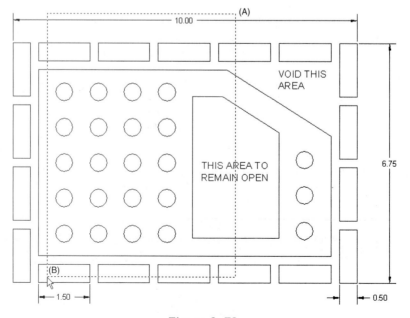

Figure 3–72

The result of selecting objects by a crossing window is illustrated in Figure 3–73. Notice that even though a group of circles was completely surrounded by the crossing window, they do not highlight because they belong to a locked layer. Pressing ENTER will perform the ERASE operation. Before continuing on to the next step, issue the U (UNDO) command, which will negate the previous ERASE operation.

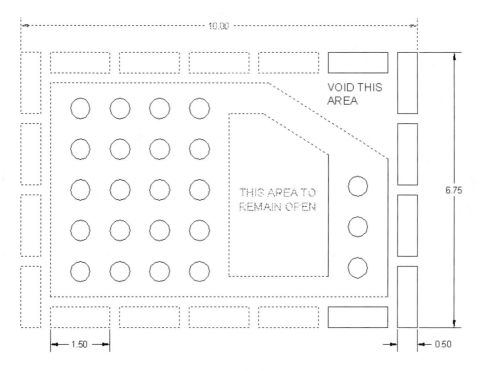

Figure 3–73

STEP 8

This final step illustrates the use of the ZOOM-All and ZOOM-Extents commands and how they differ. First activate the ZOOM command and use the All option (a button is available for performing this operation).

Command: **Z** *(For ZOOM)*
Specify corner of window, enter a scale factor (nX or nXP), or
[All/Center/Dynamic/Extents/Previous/
Scale/Window] <real time>: **A** *(For All)*

The result is illustrated in Figure 3–74. Performing a ZOOM-All zooms the display based on the current drawing limits (the limits of this drawing have been established as 0,0 for the lower-left corner and 16,10 for the upper-right corner).

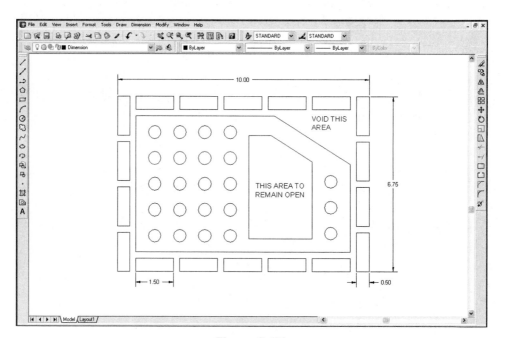

Figure 3–74

Now activate the ZOOM command and use the Extents option (a button is also available for performing this operation).

 Command: **Z** *(For ZOOM)*

Specify corner of window, enter a scale factor (nX or nXP), or

[All/Center/Dynamic/Extents/Previous/Scale/ Window] <real time>: **E** *(For Extents)*

The result is illustrated in Figure 3–75. This zoom is performed to the largest possible display based on the extremities of the objects currently in the drawing.

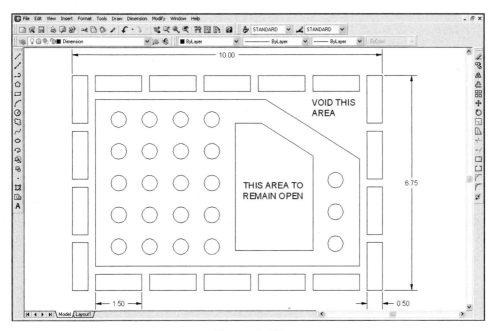

Figure 3–75

 Open the Exercise Manual PDF file for Chapter 3 on the accompanying CD for more tutorials and exercises.

 If you have the accompanying Exercise Manual, refer to Chapter 3 for more tutorials and exercises.

CHAPTER 4

Modify Commands

The heart of any CAD system is its ability to modify and manipulate existing geometry, and AutoCAD is no exception. Many modify commands relieve the designer of drudgery and mundane tasks, and this allows more productive time for conceptualizing the design. This chapter will concentrate on all major editing commands found in AutoCAD, such as Array, Break, Chamfer, Copy, Extend, Fillet, Lengthen, Mirror, Move, Offset, Rotate, Scale, and Stretch. A number of small exercises accompanies each command in order to reinforce the importance of its use.

METHODS OF SELECTING MODIFY COMMANDS

As with all commands, you can find the main body of modify commands on the Modify pull-down menu, shown in Figure 4–1. Another convenient way of selecting modify commands is through the Modify toolbar shown in Figure 4–2.

Figure 4–1

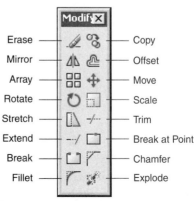

Figure 4–2

You can also enter all modify commands directly from the keyboard either using their entire name or through command aliasing as in the following examples.

> Enter E for the ERASE command
> Enter M for the MOVE command

For a complete listing of other commands and aliases, refer to table 1–2 in chapter 1.

 ## CREATING RECTANGULAR ARRAYS WITH THE ARRAY DIALOG BOX

METHOD #1

The Array dialog box allows you to arrange multiple copies of an object or group of objects in a rectangular or polar (circular) pattern. When creating a rectangular array, you are prompted to enter the number of rows and columns for the array. A row is a group of objects that are copied vertically in the positive or negative directions. A column is a group of objects that are copied horizontally, also in positive or negative directions.

 Try It! – Open the drawing file 04_Array Rectangular1. Suppose the object in Figure 4–3 needs to be copied in a rectangular pattern consisting of three rows and three columns. You also want the rows to be spaced 0.50 units away from each other and the columns to be spaced 1.25 units apart. The result is illustrated in Figure 4–4.

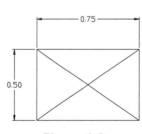

Figure 4–3

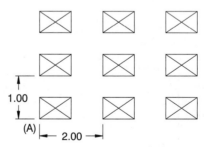

Figure 4–4

Clicking Array in the Modify pull-down menu in Figure 4–5 displays the Array dialog box. When it appears, make the following changes. Be sure the Rectangular Array option is selected at the top left corner of the dialog box. Enter 4 for the number of rows and 3 for the number of columns. For Row Offset, enter a value of 1.00 units. For Column Offset, enter a value of 2.00 units. Click the Select Objects button in the upper-right corner of the dialog box. This will return you to the drawing. Pick the rectangle and the two diagonal line segments. Pressing ENTER when finished will return you to the Array dialog box. The Array dialog box should now appear similar to Figure 4–6. Observe the pattern in the preview image and notice it has been updated to reflect three rows and three columns. Notice also that the Preview< button in the lower-right corner if the dialog box is active. Click this button to preview what the rectangular pattern will look like in your drawing. Clicking the Accept button in the Array alert box in Figure 4–7 will complete the array operation. If you notice an error in the array results, click the Modify button. This will return you to the Array dialog box and allow you to make changes to any value. Clicking the Cancel button will abort the array operation and return you to the command prompt.

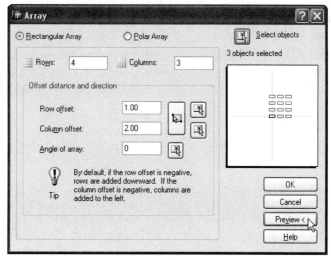

Figure 4–5

Figure 4–6

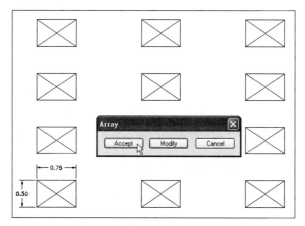

Figure 4–7

Tip: When rectangular-shaped objects are arrayed, a point in the lower-left corner of the shape acts as a reference point where you may calculate the spacing between rows and columns. Not only must the spacing distance be used, but also the overall size of the object plays a role in determining the spacing distances. With the total height of the original object at 0.50 and a required spacing between rows of 0.50, both object height and spacing result in a distance of 1.00 between rows. In the same manner, with the original length of the object at 0.75 and a spacing of 1.25 units between columns, the total spacing results in a distance of 2.00.

PERFORMING RECTANGULAR ARRAYS

METHOD #2

In Figure 4–7, the rectangular array illustrates a pattern that runs to the right and above of the original figure. At times these directions change to the left and below the original object. The only change occurs in the distances between rows and columns, where negative values dictate the direction of the rectangular array. See Figure 4–8.

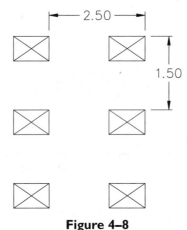

Figure 4–8

Try It! – Open the drawing file 04_Array Rectangular2. Follow the illustration in Figure 4–8 and the command prompt sequence below for performing this operation.

Activate the Array dialog box in Figure 4–9. While in Rectangular array mode, set the number of rows to 3 and the number of columns to 2. Because the directions of the array will be to the left and below the original object, both row and column distances will be negative values. When finished, preview the array; if the results are similar to the illustration in Figure 4–8, click the OK button to create the rectangular array pattern.

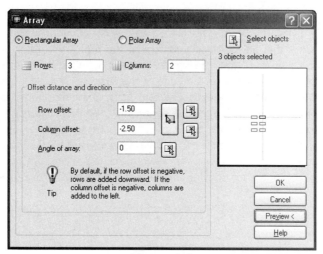

Figure 4–9

Try It! – Open the drawing 04_Array I-Beam and create a rectangular pattern consisting of four rows and five columns. Create a space of 20' between both rows and columns. Dimensions may be added at a later time. The results are displayed in Figure 4–10.

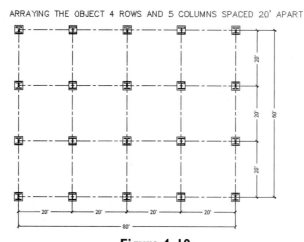

ARRAYING THE OBJECT 4 ROWS AND 5 COLUMNS SPACED 20' APART

Figure 4–10

Note: The I-Beam and footing were converted to a block object with the insertion point located at the center of the footing. For this reason, the I-Beams can be arrayed from their centers.

CREATING POLAR ARRAYS WITH THE ARRAY DIALOG BOX

Polar arrays allow you to create multiple copies of objects in a circular or polar pattern.

Try It! – Open the drawing file 04_Array Polar1 in Figure 4–11 and activate the Array dialog box. Be sure to select the Polar Array option at the top of the dialog box. Click the "Select objects" button at the top of the dialog box and pick the box in Figure 4–12 to array. For the "Center point" of the array, click the "Select objects" button and pick the edge of the large circle at "C" (this works only if the Center OSNAP mode is checked). Enter 8 for the "Total number of items" and be sure the "Angle to fill" is set to 360 (degrees). Also, verify that the box is checked next to "Rotate items as copied." The Array dialog box should similar to the illustration in Figure 4–13. Observe the results in the image icon and click the Preview< button to see the results. Click the Accept button to complete the array operation. Your drawing should appear similar to Figure 4–14.

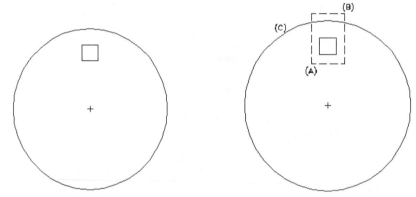

Figure 4–11   **Figure 4–12**

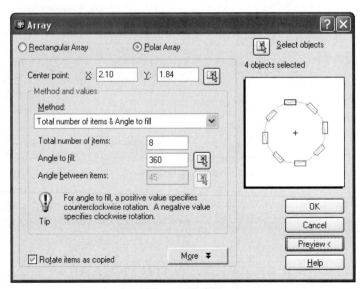

Figure 4–13

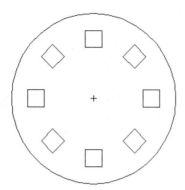

Figure 4–14

 Try It! – Open the drawing file 04_Array Polar2. To array rectangular or square objects in a polar pattern without rotating the objects, first convert the square or rectangle to a block with an insertion point located in the center of the square (this process will be covered in detail in chapter 16). Now all squares lie an equal distance from their common center. Enter the following settings in the Array dialog box in Figure 4–15 to accomplish this operation.

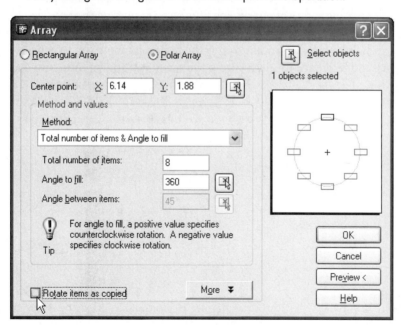

Figure 4–15

Your results should appear similar to the illustration in Figure 4–16.

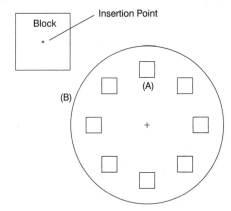

Figure 4–16

BOLT HOLE PATTERNS—METHOD #1

From the previous example, you have seen how easily the Array dialog box can be used for making multiple copies of objects equally spaced around an entire circle. What if the objects are copied only partially around a circle, such as the bolt holes in Figure 4–17? The Array dialog box can be used to copy the bolt holes in 40° increments. Follow the next command sequence for performing the operation shown in Figure 4–17.

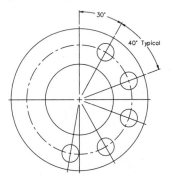

Figure 4–17

 Try It! – Open the drawing file 04_Bolthole2. Using the Array dialog box, select the circle and centerline as the objects to array in Figure 4–19 and pick the center of the circle as the center of the array. Copy the selected objects at 40° increments using the following information entered in at the dialog box. Be sure the "Angle to fill & Angle between items" method is selected in the dialog box in Figure 4–18. This will activate the proper fields and allow you to enter the correct information.

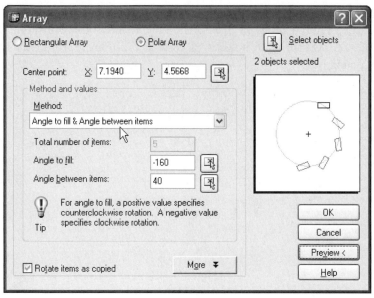

Figure 4–18

The result is illustrated in Figure 4–20. This method of identifying bolt holes shows how you can control the number by specifying the total angle to fill and the angle between items. A negative value is entered in the "Angle to fill" field to force the array to be performed in the clockwise direction.

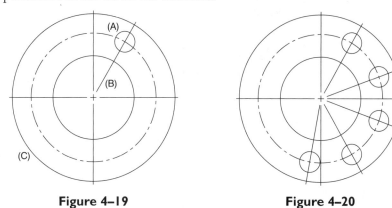

Figure 4–19   **Figure 4–20**

BOLT HOLE PATTERNS—METHOD #2

The object in Figure 4–21 is similar to the object shown in "Bolt Hole Patterns—Method #1." A new series of holes is to be placed 20° away from each other using the Array dialog box.

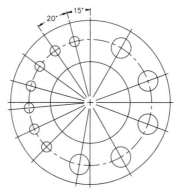

Figure 4–21

 Try It! – Open the drawing file 04_Bolthole3. Using the Array dialog box, select the circle and centerline in Figure 4–23 as the objects to array and pick the center of the circle as the center of the array. Copy the selected objects at 20° increments using the following information entered in the dialog box. Again, be sure the "Angle to fill & Angle between items" method is selected in the dialog box in Figure 4–22. This will activate the proper fields and allow you to enter the correct information.

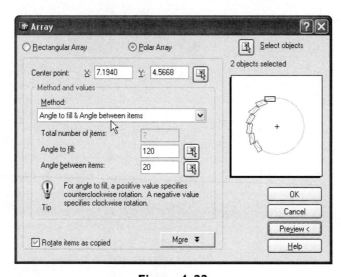

Figure 4–22

Because the direction of rotation for the array is in the counterclockwise direction, all angles are specified in positive values as shown in Figure 4–22. The results are shown in Figure 4–24.

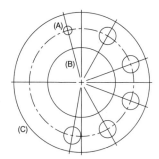

Figure 4–23

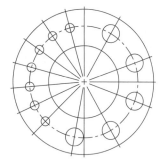

Figure 4–24

THE BREAK COMMAND

The BREAK command is used to partially delete a segment of an object. Choose this command from one of the following:

> the Modify toolbar
>
> the pull-down menu (Modify > Break)
>
> the keyboard (BR or BREAK)

The following command sequence and Figure 4–25 show how the BREAK command is used.

 Try It! – Open the drawing file 04_Break1. Turn off Running OSNAP prior to conducting this exercise. Use the following prompts and illustrations to break the line segment.

 Command: **BR** *(For BREAK)*
Select objects: *(Select the line at "A" in Figure 4–25)*
Specify second break point or [First point]: *(Select the line at "B" in Figure 4–25)*

 Try It! – Open the drawing file 04_Break2. Utilize the First option of the BREAK command along with OSNAP options to select key objects to break. This option resets the command and allows you to select an object to break followed by two different points that identify the break. The following command sequence and Figure 4–26 demonstrate using the First option of the BREAK command:

⊡ Command: **BR** *(For BREAK)*
Select object: *(Select the line shown in Figure 4–26)*
Specify second break point or [First point]: **F** *(For First)*
Specify first break point: **Int**
of *(Pick the intersection of the two lines at "A")*
Specify second break point: **End**
of *(Pick the endpoint of the line at "B")*

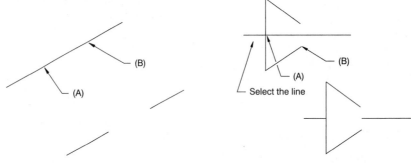

Figure 4–25   **Figure 4–26**

 Try It! – Open the drawing file 04_Break3. Breaking circles is always accomplished in the counterclockwise direction. Study the following command sequence and Figure 4–27 for breaking circles.

⊡ Command: **BR** *(For BREAK)*
Select objects: *(Select the circle in Figure 4–27)*
Specify second break point or [First point]: **F** *(For First)*
Specify first break point: *(Select "A" in either circle in Figure 4–27)*
Specify second break point: *(Select "B" in either circle in Figure 4–27)*

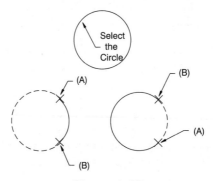

Figure 4–27

BREAK AT POINT

You can also break an object at a selected point. The break is so small that you cannot find it no matter how much you zoom into the break point. In Figure 4–28 on the left, the line is highlighted to prove that it consists of a continuous object. Clicking on the Break at Point tool in the Modify toolbar activates the following command sequence:

Command: **BR** *(For BREAK)*
Select object: *(Select the line anywhere at Figure 4–28 on the left)*
Specify second break point or [First point]: **F** *(For First point)*
Specify first break point: **Mid**
of *(Pick the midpoint of the line in Figure 4–28)*
Specify second break point: **@** *(For previous point)*

The results are illustrated in Figure 4–28 on the right. Here the line is again selected. Notice that only half of the line selects because the line was broken at its midpoint.

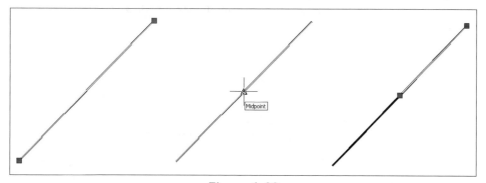

Figure 4–28

THE CHAMFER COMMAND

Chamfers represent a way to finish a sharp corner of an object. The CHAMFER command produces an inclined surface at an edge of two intersecting line segments. Distances determine how far from the corner the chamfer is made. Figure 4–29 is one example of an object that has been chamfered along its top edge.

The CHAMFER command is designed to draw an angle across a sharp corner given two chamfer distances. Choose this command from one of the following:

 the Modify toolbar

 the pull-down menu (Modify > Chamfer)

 the keyboard (CHA or CHAMFER)

The most popular chamfer involves a 45° angle, which is illustrated in Figure 4–30. Even though this command does not allow you to specify an angle, you can control the angle by the distances entered.

 Try It! – Open the drawing file 04_Chamfer Distances. In the example in Figure 4–30, if you specify the same numeric value for both chamfer distances, a 45° chamfer will automatically be formed. As long as both distances are the same, a 45° chamfer will always be drawn. Study the illustration in Figure 4–30 and the following prompts:

 Command: **CHA** *(For CHAMFER)*
(TRIM mode) Current chamfer Dist1 = 0.5000, Dist2 = 0.5000
Select first line or [Polyline/Distance/Angle/Trim/Method/mUltiple]: **D** *(For Distance)*
Specify first chamfer distance <0.5000>: **0.15**
Specify second chamfer distance <0.1500>: *(Press ENTER to accept the default)*
Select first line or [Polyline/Distance/Angle/Trim/Method/mUltiple]: *(Select the line at "A")*
Select second line: *(Select the line at "B")*

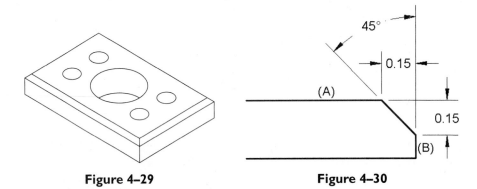

Figure 4–29   **Figure 4–30**

 Try It! – Open the drawing file 04_Chamfer Angle. Figure 4–31 illustrates the use of the CHAMFER command by setting one distance and identifying an angle. Chamfer angles other than 45° are commonly called a beveled edge.

 Command: **CHA** *(For CHAMFER)*
(TRIM mode) Current chamfer Dist1 = 0.5000, Dist2 = 0.5000
Select first line or [Polyline/Distance/Angle/Trim/Method/mUltiple]: **A** *(For Angle)*
Specify chamfer length on the first line <0.1500>: **0.15**
Specify chamfer angle from the first line <60>: **60**
Select first line or [Polyline/Distance/Angle/Trim/Method/mUltiple]: *(Select the line at "A")*
Select second line: *(Select the line at "B")*

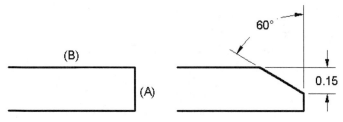

Figure 4–31

 Try It! – Open the drawing file 04_Chamfer Corner. With non-intersecting corners, you could use the CHAMFER command to connect both lines. The CHAMFER command distances are both set to a value of 0 to accomplish this task. See Figure 4–32.

 Command: **CHA** *(For CHAMFER)*
(TRIM mode) Current chamfer Dist1 = 0.5000, Dist2 = 0.2500
Select first line or [Polyline/Distance/Angle/Trim/Method/mUltiple]: **D** *(For Distance)*
Enter first chamfer distance <0.5000>: **0**
Enter second chamfer distance <0.0000>: *(Press ENTER to accept the default)*
Select first line or [Polyline/Distance/Angle/Trim/Method/mUltiple]: *(Select the line at "A")*
Select second line: *(Select the line at "B")*

 Try It! – Open the drawing file 04_Chamfer Pline. Because a polyline consists of numerous segments representing a single object, using the CHAMFER command with the Polyline option produces corners throughout the entire polyline. See Figure 4–33.

 Command: **CHA** *(For CHAMFER)*
(TRIM mode) Current chamfer Dist1 = 0.00, Dist2 = 0.00
Select first line or [Polyline/Distance/Angle/Trim/Method/mUltiple]: **D** *(For Distance)*
Enter first chamfer distance <0.00>: **0.50**
Enter second chamfer distance <0.50>: *(Press ENTER to accept the default)*
Select first line or [Polyline/Distance/Angle/Trim/Method/mUltiple]: **P** *(For Polyline)*
Select 2D Polyline: *(Select the Polyline in Figure 4–33)*

 Note: A Multiple option of the chamfer command allows you to chamfer edges that share the same chamfer distances without exiting and reentering the command.

 Try It! – Open the drawing file 04_Chamfer No-Trim. The CHAMFER command supports a Trim/No trim option, enabling a chamfer to be placed with lines trimmed or not trimmed as in Figure 4–34. Follow the prompt sequence carefully to set the No trim option.

190

Command: **CHA** *(For CHAMFER)*
(TRIM mode) Current chamfer Dist1 = 0.2000, Dist2 = 0.2000
Select first line or [Polyline/Distance/Angle/Trim/Method/mUltiple]: **T** *(For Trim)*
Trim/No trim <Trim>: **N** *(For No trim)*
Polyline/Distance/Angle/Trim/Method/<Select first line>: **D** *(For Distance)*
Enter first chamfer distance <0.0000>: **1.00**
Enter second chamfer distance <1.0000>: *(Press ENTER to accept the default)*
Select first line or [Polyline/Distance/Angle/Trim/Method/mUltiple]: **P** *(For Polyline)*
Select 2D Polyline: *(Select the polyline at "A")*

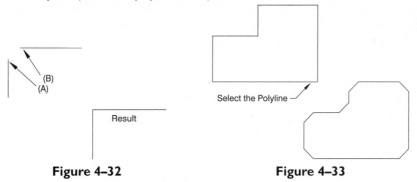

Figure 4–32 **Figure 4–33**

All edges are chamfered without being trimmed.

 Note: Reset the Trim setting in the CHAMFER command back to "Trim" from "No trim." This can be done from the keyboard by entering the command TRIMMODE and setting it to 1 (On).

Command: **TRIMMODE**
Enter new value for TRIMMODE <0>: **1**

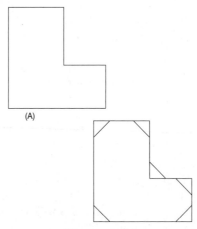

Figure 4–34

Try It! – Open the drawing file 04_Chamfer Beam. Using the illustration provided in Figure 4–35, follow these directions: Apply equal chamfer distances of 0.25 units to corners "AB," "BC," "DE," and "EF." Set new equal chamfer distances to 0.50 units and apply these distances to corners "GH" and "JK." Set a new first chamfer distance to 1.00; set a second chamfer distance to 0.50 units. Apply the first chamfer distance to line "L" and the second chamfer distance to line "H." Complete this object by applying the first chamfer distance to line "M" and the second chamfer distance to line "K."

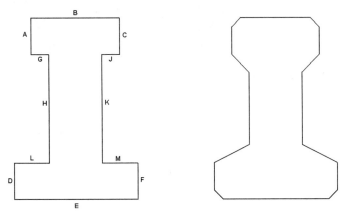

Figure 4–35

THE COPY COMMAND

The COPY command is used to duplicate an object or group of objects. Choose this command from one of the following:

> the Modify toolbar
>
> the pull-down menu (Modify > Copy)
>
> the keyboard (CP or COPY)

In Figure 4–36, the Window mode is used to select all objects to copy. Point "C" is used as the base point of displacement, or where you want to copy the objects to be copied from. Point "D" is used as the second point of displacement, or the destination for copied objects.

 Command: **CP** *(For COPY)*
Select objects: *(Pick a point at "A")*
Other corner: *(Pick a point at "B")*
Select objects: *(Press* ENTER *to continue)*
Specify base point or displacement, or [Multiple]: *(Select the endpoint of the corner at "C")*
Specify second point of displacement or <use first point as displacement>: *(Select a point near "D")*

You can also duplicate numerous objects while staying inside the COPY command. Figure 4–37 shows a group of objects copied with the Multiple option of the COPY command. See the following command sequence for using the Multiple option of the COPY command:

 Command: **CP** *(For COPY)*
Select objects: *(Select all objects that make up the object at "A")*
Select objects: *(Press ENTER to continue)*
Specify base point or displacement, or [Multiple]: **M** *(For Multiple)*
Specify base point: *(Select the endpoint of the corner at "A")*
Specify second point of displacement or <use first point as displacement>: *(Select a point near "B")*
Specify second point of displacement or <use first point as displacement>: *(Select a point near "C")*
Specify second point of displacement or <use first point as displacement>: *(Select a point near "D")*
Specify second point of displacement or <use first point as displacement>: *(Select a point near "E")*
Specify second point of displacement or <use first point as displacement>: *(Select a point near "F")*
Specify second point of displacement or <use first point as displacement>: *(Press ENTER to exit this command)*

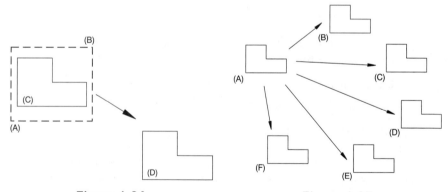

Figure 4–36   Figure 4–37

Try It! – Open the drawing file 04_Copy Multiple. Follow the command sequence in the previous example to copy the three holes a multiple of times. Use the intersection of "A" as the base point for the copy. Then copy the three holes to the intersections located at "B," "C," "D," "E," "F," "G," "H," and "J" (see Figure 4–38).

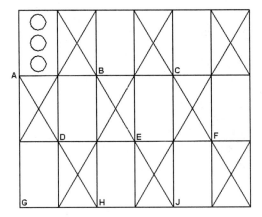

Figure 4–38

THE EXPLODE COMMAND

Using the EXPLODE command on a polyline, dimension, or block separates the single object into its individual parts. Choose this command from one of the following:

> the Modify toolbar
>
> the pull-down menu (Modify > Explode)
>
> the keyboard (X or EXPLODE)

Figure 4–39 is a polyline that is considered one object. Using the EXPLODE command and selecting the polyline breaks the polyline into four individual objects.

Command: **X** *(For EXPLODE)*

Select objects: *(Select the polyline in Figure 4–39)*
Select objects: *(Press* ENTER *to perform the explode operation)*

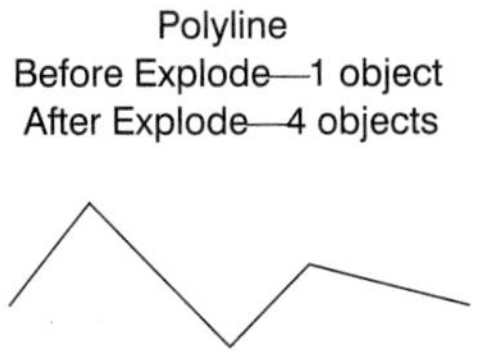

Polyline
Before Explode—1 object
After Explode—4 objects

Figure 4–39

Dimensions consist of extension lines, dimension lines, dimension text, and arrowheads, all grouped into a single object. (Dimensions will be covered in chapter 11, "Dimensioning Basics.") Using the EXPLODE command on a dimension breaks the dimension down into individual extension lines, dimension lines, arrowheads, and dimension text. This is not advisable, because the dimension value will not be updated if the dimension is stretched along with the object being dimensioned. Also the ability to manipulate dimensions with a feature called grips is lost. Grips will be discussed in chapter 7.

Command: **X** *(For EXPLODE)*
Select objects: *(Select the dimension object shown in Figure 4–40)*
Select objects: *(Press ENTER to perform the explode operation)*

Using the EXPLODE command on a block converts the block to individual objects, lines, and circles. If the block had yet another block nested in it, you would need to perform an additional explode operation to break this object into individual objects.

Command: **X** *(For EXPLODE)*
Select objects: *(Select the block shown in Figure 4–41)*
Select objects: *(Press ENTER to perform the explode operation)*

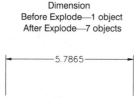

Dimension
Before Explode—1 object
After Explode—7 objects

5.7865

Figure 4–40

Block
Before Explode—1 object
After Explode—15 objects

Figure 4–41

Try It! – Open the drawing file 04_Explode. You can also use the EXPLODE command to explode non-uniformly scaled block objects as shown in Figure 4–42. Use the EXPLODE command by clicking individually on all objects in the drawing. All blocks in this figure will be broken into individual objects without the block being redefined.

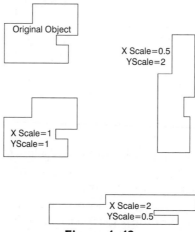

Original Object

X Scale=0.5
YScale=2

X Scale=1
YScale=1

X Scale=2
YScale=0.5

Figure 4–42

THE EXTEND COMMAND

The EXTEND command is used to extend objects to a specified boundary edge. Choose this command from one of the following:

> the Modify toolbar
>
> the pull-down menu (Modify > Extend)
>
> the keyboard (EX or EXTEND)

In Figure 4–43, select the large circle "A" as the boundary edge. After pressing ENTER to continue with the command, select the arc at "B," the line at "C," and the arc at "D" to extend these objects to the circle. If you select the wrong end of an object, use the Undo feature, which is an option of the command, to undo the change and repeat the procedure at the correct end of the object.

 Try It! – Open the drawing file 04_Extend1. Follow the illustration and command sequence below for accomplishing this task.

 Command: **EX** *(For EXTEND)*
Current settings: Projection=UCS Edge=None
Select boundary edges ...
Select objects: *(Select the large circle at "A")*
Select objects: *(Press ENTER to continue)*
Select object to extend or shift-select to trim or [Project/Edge/Undo]: *(Select the arc at "B")*
Select object to extend or shift-select to trim or [Project/Edge/Undo]: *(Select the line at "C")*
Select object to extend or shift-select to trim or [Project/Edge/Undo]: *(Select the arc at "D")*
Select object to extend or shift-select to trim or [Project/Edge/Undo]: *(Press ENTER to exit this command)*

 Tip: An alternate method of selecting boundary edges is to press ENTER in response to the "Select objects" prompt. This automatically creates boundary edges out of all objects in the drawing. When you use this method however, the boundary edges will not highlight.

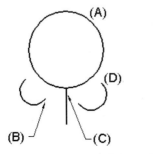

Figure 4–43

Try It! – Open the drawing file 04_Extend2. To extend multiple objects such as the sixteen line segments shown in Figure 4–44, select the line at "A" as the boundary edge and use the Fence mode to create a crossing line from "B" to "C." This will extend all sixteen line segments to intersect with the boundary.

 Command: **EX** *(For EXTEND)*
Current settings: Projection=UCS Edge=None
Select boundary edges ...
Select objects: *(Select the line at "A" in Figure 4–44)*
Select objects: *(Press* ENTER *to continue)*
Select object to extend or shift-select to trim or [Project/Edge/Undo]: **F** *(For Fence)*
First fence point: *(Pick a point at "B")*
Specify endpoint of line or [Undo]: *(Pick a point at "C")*
Specify endpoint of line or [Undo]: *(Press* ENTER *to end the Fence and execute the extend operation)*
Select object to extend or shift-select to trim or [Project/Edge/Undo]: *(Press* ENTER *to exit this command)*

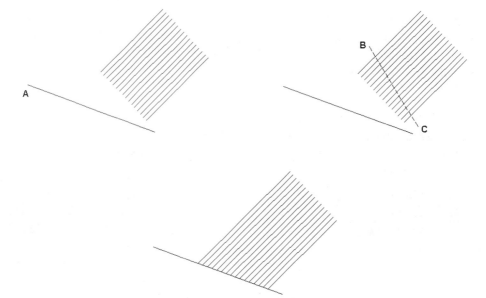

Figure 4–44

 Try It! – Open the drawing file 04_Extend3. Certain conditions require the boundary edge to be extended where an imaginary edge is projected, enabling objects not in direct sight of the boundary edge to still be extended. Study Figure 4–45 and the following prompts on this special case involving the EXTEND command.

Command: **EX** *(For EXTEND)*
Current settings: Projection=UCS Edge=None
Select boundary edges ...
Select objects: *(Pick line "A")*
Select objects: *(Press ENTER to continue)*
Select object to extend or shift-select to trim or [Project/Edge/Undo]: **E** *(For Edge)*
Enter an implied edge extension mode [Extend/No extend] <No extend>: **E** *(For Extend the edge)*
Select object to extend or shift-select to trim or [Project/Edge/Undo]: *(Pick line "B")*

Continue picking the remaining lines to be extended to the extended boundary edge. You could also use the Fence mode to select all line segments to extend. After all lines are extended, press ENTER to exit the command.

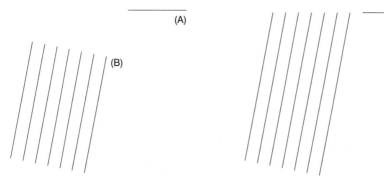

Figure 4–45

Note: Reset the Edge setting in the EXTEND command back to "No Extend" from "Extend." This can be done from the keyboard by entering the command EDGEMODE and setting it to 0 (Zero).

Command: **EDGEMODE**
Enter new value for EDGEMODE <1>: **0**

Tip: While inside the EXTEND command, you can easily toggle to the TRIM command by holding down the SHIFT key at the following command prompt:

Select object to extend or shift-select to trim or [Project/Edge/Undo]: *(Pressing SHIFT while picking objects activates the TRIM command.)*

Try It! – Open the drawing file 04_Extend Pipe. Enter the EXTEND command and press ENTER when the "Select objects:" prompt appears. This will select all objects as boundary edges. Select the ends of all yellow lines representing pipes as the objects to extend. They will extend to intersect with the adjacent pipe fitting. Your finished drawing should appear similar to Figure 4–46.

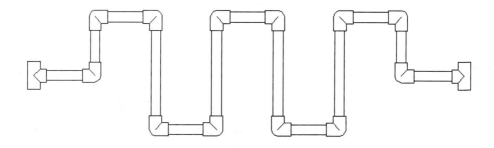

Figure 4–46

THE FILLET COMMAND

Many objects require highly finished and polished surfaces consisting of extremely sharp corners. Fillets and rounds represent the opposite case, where corners are rounded off, either for ornamental purposes or as required by design. Generally a fillet consists of a rounded edge formed in the corner of an object, as illustrated in Figure 4–47 at "A." A round is formed at an outside corner, similar to "B." Fillets and rounds are primarily used where objects are cast or made from poured metal. The metal will form more easily around a pattern that has rounded corners instead of sharp corners, which usually break away. Some drawings have so many fillets and rounds that a note is used to convey the size of them all, similar to "All Fillets and Rounds 0.125 Radius."

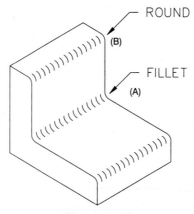

Figure 4–47

AutoCAD provides the FILLET command, which allows you to enter a radius followed by the selection of two lines. The result will be a fillet of the specified radius to the two lines selected. The two lines are also automatically trimmed, leaving the radius drawn from the endpoint of one line to the endpoint of the other line. Choose this command from one of the following:

the Modify toolbar

the pull-down menu (Modify > Fillet)

the keyboard (F or FILLET)

Illustrated in Figure 4–48 are examples of the use of the FILLET command. Because sometimes the command is used over and over again, a Multiple option is available that automatically repeats the command.

 Try It! – Open the drawing file 04_Fillet. Follow the illustration in Figure 4–48 and command sequence below to use the MULTIPLE and FILLET commands.

 Command: **F** *(For FILLET)*
Current settings: Mode = TRIM, Radius = 0.0000
Select first object or [Polyline/Radius/Trim/mUltiple]: **R** (For Radius)
Specify fillet radius <0.0000>: **0.25**
Select first object or [Polyline/Radius/Trim/mUltiple]: **U** (For Multiple)
Select first object or [Polyline/Radius/Trim/mUltiple]: (Select at "A")
Select second object: (Select at "B")
Select first object or [Polyline/Radius/Trim/mUltiple]: (Select at "B")
Select second object: (Select at "C")
Select first object or [Polyline/Radius/Trim/mUltiple]: (Select at "C")
Select second object: (Select at "D")
Select first object or [Polyline/Radius/Trim/mUltiple]: (Press ENTER to exit this command)

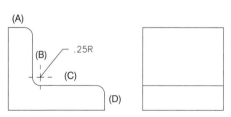

Figure 4–48

 Try It! – Open the drawing file 04_Fillet Corner1. A very productive feature of the FILLET command is its use as a cornering tool. To accomplish this, set the fillet radius to a value of 0. This produces a corner out of two non-intersecting objects. Follow the illustration in Figure 4–49 and the command sequence below for performing this task.

Command: **F** *(For FILLET)*

Current settings: Mode = TRIM, Radius = 0.5000
Select first object or [Polyline/Radius/Trim/mUltiple]: **R** *(For Radius)*
Enter fillet radius <0.5000>: **0**
Select first object or [Polyline/Radius/Trim/mUltiple]: *(Select line "A")*
Select second object: *(Select line "B")*

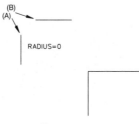

Figure 4–49

 Try It! – Open the drawing file 04_Fillet Corner2. Activate the FILLET command and verify that the radius is set to 0. Use the Multiple option to make the command repeat. Then click on the corners until your object appears similar to the illustration in Figure 4–50.

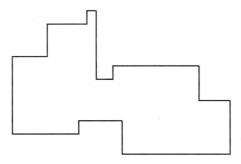

Figure 4–50

 Try It! – Open the drawing file 04_Fillet Pline. Using the FILLET command on a polyline object produces rounded edges at all corners of the polyline in a single operation. Follow the illustration in Figure 4–51 and the command sequence below for performing this task.

 Command: **F** *(For FILLET)*

Current settings: Mode = TRIM, Radius = 0.0000
Select first object or [Polyline/Radius/Trim/mUltiple]: **R** *(For Radius)*
Enter fillet radius <0.0000>: **0.25**
Select first object or [Polyline/Radius/Trim/mUltiple]: **P** *(For Polyline)*
Select 2D polyline: *(Select the polyline in Figure 4–51)*

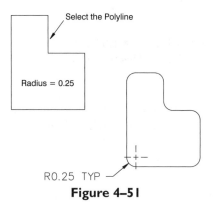

Figure 4–51

The FILLET command can also be used to control whether or not to trim the excess corners after a fillet is placed. Figure 4–52 shows a typical fillet operation where the polyline at "A" is selected. However, instead of the corners of the polyline automatically being trimmed, a new Trim/No trim option allows the lines to remain. The following prompts illustrate this operation.

 Try It! – Open the drawing file 04_Fillet No-Trim. Follow the command sequence and illustration in Figure 4–52 for using the No-Trim option of the FILLET command.

 Command: **F** *(For FILLET)*
Current settings: Mode = TRIM, Radius = 0.0000
Select first object or [Polyline/Radius/Trim/mUltiple]: **T** *(For Trim)*
Enter Trim mode option [Trim/No trim] <Trim>: **N** *(For No trim)*
Select first object or [Polyline/Radius/Trim/mUltiple]: **R** *(For Radius)*
Specify fillet radius <0.0000>: 0.50
Select first object or [Polyline/Radius/Trim/mUltiple]: **P** *(For Polyline)*
Select 2D polyline: *(Select polyline "A")*
6 lines were filleted

 Note: Reset the Trim setting in the FILLET command back to "Trim" from "No trim." This can be done from the keyboard by entering the command TRIMMODE and setting it to 1 (On).

Command: **TRIMMODE**
Enter new value for TRIMMODE <0>: **1**

 Try It! – Open the drawing file 04_Fillet Parallel. Filleting two parallel lines in Figure 4–53 automatically constructs a semicircular arc object connecting both lines at their endpoints. When performing this operation, it does not matter what the radius value is set to. Use the illustration in Figure 4–53 and the command sequence below for performing this task.

Command: **F** *(For FILLET)*
Current settings: Mode = TRIM, Radius = 1.0000
Select first object or [Polyline/Radius/Trim/mUltiple]: *(Select line "A")*
Select second object: *(Select line "B")*

Continue filleting the remaining parallel lines to complete all slots. Quicken the process by using the Multiple option when prompted to "Select first object".

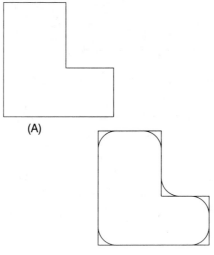

(A)

Figure 4–52

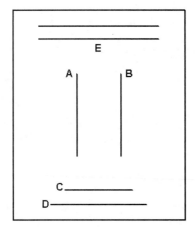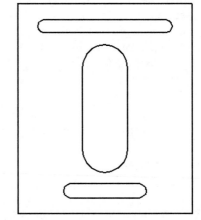

Figure 4–53

THE LENGTHEN COMMAND

The LENGTHEN command is used to change the length of a selected object without disturbing other object qualities such as angles of lines or radii of arcs. Choose this command from one of the following:

the pull-down menu (Modify > Lengthen)

the keyboard (LEN or LENGTHEN)

 Try It! – Open the drawing file 04_Lengthen1. Use the illustration in Figure 4–54 and the command sequence below for performing this task.

Command: **LEN** *(For LENGTHEN)*
Select an object or [DElta/Percent/Total/DYnamic]: *(Select line "A")*
Current length: 12.3649
Select an object or [DElta/Percent/Total/DYnamic]: **T** *(For Total)*
Specify total length or [Angle] (1.0000)>: **20**
Select an object to change or [Undo]: *(Select the line at "A")*
Select an object to change or [Undo]: *(Press ENTER to exit)*

 Tip: After supplying the new total length of any object, be sure to select the desired end to lengthen when you are prompted to select an object to change.

 Try It! – Open the drawing file 04_Lengthen2. When you use the LENGTHEN command on an arc segment, both the length and included angle information are displayed before you make any changes, as shown in Figure 4–55.

Command: **LEN** *(For LENGTHEN)*
Select an object or [DElta/Percent/Total/DYnamic]: *(Select arc "B")*
Current length: 8.0109, included angle: 87
Select an object or [DElta/Percent/Total/DYnamic]: **T** *(For Total)*
Specify total length or [Angle] (1.0000)>: 21
Select an object to change or [Undo]: *(Pick the arc at "B")*
Select an object to change or [Undo]: *(Press ENTER to exit)*

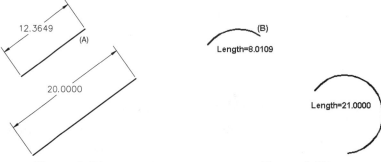

Figure 4–54   **Figure 4–55**

THE MIRROR COMMAND—METHOD #1

The MIRROR command is used to create a mirrored copy of an object or group of objects. When performing a mirror operation, you have the option of deleting the original object, which would be the same as flipping the object, or keeping the original object along with the mirror image, which would be the same as flipping and copying. Choose this command from one of the following:

    the Modify toolbar

    the pull-down menu (Modify > Mirror)

    the keyboard (MI or MIRROR)

Try It! – Open the drawing file 04_Mirror Copy. Refer to the following prompts along with Figure 4–56 for using the MIRROR command:

 Command: **MI** *(For MIRROR)*
Select objects: *(Select a point near "X")*
Specify opposite corner: *(Select a point near "Y")*
Select objects: *(Press ENTER to continue)*
Specify first point of mirror line: *(Select the endpoint of the centerline at "A")*
Specify second point of mirror line: *(Select the endpoint of the centerline at "B")*
Delete source objects? [Yes/No] <N>: *(Press ENTER for default)*

Because the original object needed to be retained by the mirror operation, the image result is shown in Figure 4–57. The MIRROR command works well when symmetry is required.

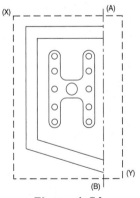
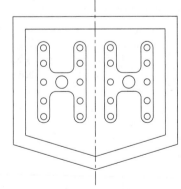

Figure 4–56   **Figure 4–57**

THE MIRROR COMMAND—METHOD #2

The illustration in Figure 4–58 is a different application of the MIRROR command. It is required to have all items that make up the bathroom plan flip but not copy to the other side. This is a typical process involving "what if" scenarios.

Try It! – Open the drawing file 04_Mirror Flip. Use the following command prompts to perform this type of mirror operation. The results are displayed in Figure 4–59.

 Command: **MI** *(For MIRROR)*

Select objects: **All** *(This will select all objects shown in Figure 4–58)*
Select objects: *(Press ENTER to continue)*
Specify first point of mirror line: **Mid**
of *(Select the midpoint of the line at "A")*
Specify second point of mirror line: **Per**
to *(Select line "B," which is perpendicular to point "A")*
Delete source objects? [Yes/No] <N>: **Y** *(For Yes)*

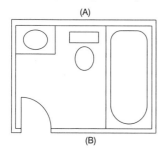

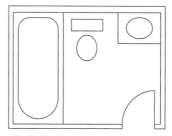

Figure 4–58   **Figure 4–59**

THE MIRROR COMMAND—METHOD #3

Situations sometimes involve mirroring text as in Figure 4–60. A system variable called MIRRTEXT controls this occurrence—by default it is set to a value of 1 or On.

Try It! – Open the drawing file 04_Mirror Text. Use the following prompts to see the results of the MIRROR command on text.

 Command: **MI** *(For MIRROR)*

Select objects: *(Select a point near "X")*
Specify opposite corner: *(Select a point near "Y")*
Select objects: *(Press ENTER to continue)*
Specify first point of mirror line: *(Select the endpoint of the centerline at "A")*
Specify second point of mirror line: *(Select the endpoint of the centerline at "B")*
Delete source objects? [Yes/No] <N>: *(Press ENTER for default)*

Notice that with MIRRTEXT turned on, text is mirrored, which makes it unreadable (see Figure 4–61). Close this drawing file and do not save changes to it.

Try It! – Re-open the drawing file 04_Mirror Text. To mirror text and have it read properly, set the MIRRTEXT system variable to a value of 0 or Off as in the following command sequence. See Figure 4–62.

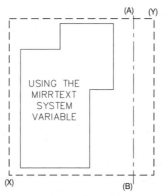

Figure 4–60

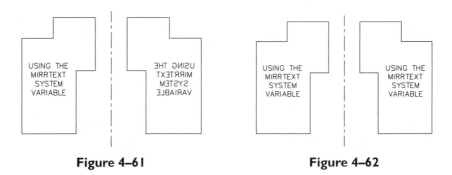

Figure 4–61   **Figure 4–62**

Command: **MIRRTEXT**
New value for MIRRTEXT <1>: **0**

Using the MIRROR command with the MIRRTEXT system variable set to a value of 0 results in text being legible, similar to Figure 4–62. Follow this command sequence to accomplish this.

 Command: **MI** *(For MIRROR)*
Select objects: *(Create a selection set similar to Figure 4–60)*
Select objects: *(Press ENTER to continue)*
Specify first point of mirror line: *(Select the endpoint of the centerline at "A")*
Specify second point of mirror line: *(Select the endpoint of the centerline at "B")*
Delete source objects? [Yes/No] <N>: *(Press ENTER for default)*

 Try It! – Open the drawing file 04_Mirror Duplex. Set the MIRRTEXT system variable to 0, which will turn it off. Use the MIRROR command and create a mirror image of the Duplex floor plan using line "AB" as the points for the mirror line. Do not delete the source objects. Your finished results should be similar to Figure 4–63.

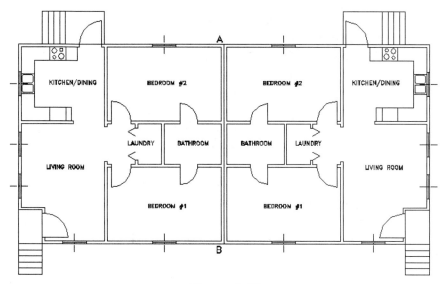

Figure 4–63

 ## THE MOVE COMMAND

The MOVE command repositions an object or group of objects at a new location. Choose this command from one of the following:

   the Modify toolbar

   the pull-down menu (Modify > Move)

   the keyboard (M or MOVE)

Once the objects to move are selected, a base point of displacement (where the object is to move from) is found. Next, a second point of displacement (where the object is to move to) is needed (See Figure 4–64)

 Command: **M** *(For MOVE)*

Select objects: *(Select all dashed objects in Figure 4–64)*
Select objects: *(Press* ENTER *to continue)*
Specify base point or displacement: *(Select the endpoint of the line at "A")*
Specify second point of displacement or <use first point as displacement>: *(Mark a point at "B")*

 Try It! – Open the drawing file 04_Move. The slot shown in Figure 4–65 is incorrectly positioned; it needs to be placed 1.00 unit away from the left edge of the object. You can use the MOVE command in combination with a polar coordinate or Direct Distance mode to perform this operation. Use the illustration in Figure 4–65 and the following command sequence for performing this operation.

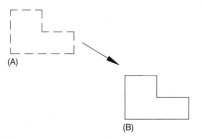

Figure 4–64

Command: **M** *(For MOVE)*

Select objects: *(Select the slot and all centerlines in Figure 4–65)*
Select objects: *(Press ENTER to continue)*
Specify base point or displacement: *(Select the edge of arc "A")*
Specify second point of displacement or <use first point as displacement>: **@0.50<0**

As the slot is moved to a new position with the MOVE command, a new horizontal dimension must be placed to reflect the correct distance from the edge of the object to the centerline of the arc. See Figure 4–66. Another command that affects a group of objects along with the dimension will be explained later.

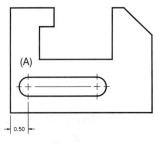

Figure 4–65

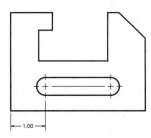

Figure 4–66

THE OFFSET COMMAND

The OFFSET command is commonly used for creating one object parallel to another. Choose this command from one of the following:

    the Modify toolbar

    the pull-down menu (Modify > Offset)

    the keyboard (O or OFFSET)

One method of offsetting is to identify a point to offset through, called a through point. Once an object is selected to offset, a through point is identified. The selected object offsets to the point shown in Figure 4–67.

Try It! – Open the drawing file 04_Offset1. Refer to Figure 4–67 and the following command sequence to use this method of the OFFSET command.

 Command: **O** *(For OFFSET)*

Specify offset distance or [Through] <1.00>: **T** *(For Through)*
Select object to offset or <exit>: *(Select the line at "A")*
Specify through point: **Nod**
of *(Select the point at "B")*
Select object to offset or <exit>: *(Press ENTER to exit this command)*

Another method of offsetting is by a specified offset distance. In Figure 4–68, an offset distance of 0.50 is set. The line segment "A" is identified as the object to offset. To complete the command, you must identify a side to offset to give the offset a direction in which to operate.

 Try It! – Open the drawing file 04_Offset2. Refer to Figure 4–68 and the following command sequence for using this method of offsetting objects.

 Command: **O** *(For OFFSET)*

Specify offset distance or [Through] <1.00>: **0.50**
Select object to offset or <exit>: *(Select the line at "A")*
Specify point on side to offset: *(Pick near "B")*
Select object to offset or <exit>: *(Press ENTER to exit this command)*

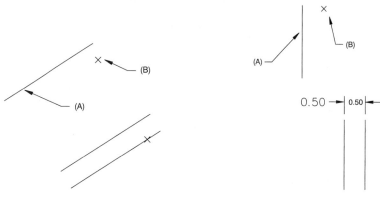

Figure 4–67   Figure 4–68

Another method of offsetting is shown in Figure 4–69, where the objects need to be duplicated at a set distance from existing geometry. The COPY command could be used for this operation; a better command would be OFFSET. This allows you to specify a distance and a side for the offset to occur. The result is an object parallel to the original object at a specified distance. All objects in Figure 4–69 need to be offset 0.50 toward the inside of the original object.

210

 Try It! – Open the drawing file 04_Offset3. See the command sequence and Figures 4–69 and 4–70 that follow to perform this operation.

 Command: **O** *(For OFFSET)*
Specify offset distance or [Through] <1.00>: **0.50**
Select object to offset or <exit>: *(Select the horizontal line at "A")*
Specify point on side to offset: *(Pick a point anywhere on the inside near "B")*

Repeat the preceding procedure for lines "C" through "J."

Notice that when all lines were offset, the original lengths of all line segments were maintained. Because all offsetting occurs inside, the segments overlap at their intersection points (see Figure 4–70). In one case, at "A" and "B," the lines did not meet at all. The CHAMFER command may be used to edit all lines to form a sharp corner. You can accomplish this by assigning chamfer distances of 0.00 units.

 Command: **CHA** *(For CHAMFER)*
(TRIM mode) Current chamfer Dist1 = 0.5000, Dist2 = 0.5000
Select first line or [Polyline/Distance/Angle/Trim/Method/mUltiple]: **D** *(For Distance)*
Specify first chamfer distance <0.5000>: **0**
Specify second chamfer distance <0.0000>: *(Press ENTER to accept the default)*
Select first line or [Polyline/Distance/Angle/Trim/Method/mUltiple]: *(Select the line at "A" in Figure 4–70)*
Select second line: *(Select the line at "B")*

Repeat the above procedure for lines "B" through "I" in Figure 4–70.

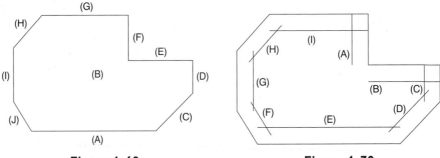

Figure 4–69   **Figure 4–70**

Using the OFFSET command along with the CHAMFER command produces the result shown in Figure 4–71. The chamfer distances must be set to a value of 0 for this special effect. The FILLET command produces the same result when set to a radius value of 0.

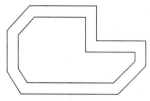

Figure 4–71

📐 THE PEDIT COMMAND

Editing of polylines can lead to interesting results. A few of these options will be explained in the following pages. Choose this command from one of the following:

> the Modify II toolbar
>
> the pull-down menu (Modify > Object > Polyline)
>
> the keyboard (PE or PEDIT)

Figure 4–72 is a polyline of width 0.00. The pedit command was used to change the width of the polyline to 0.10 units.

 Try It! – Open the drawing file 04_Pedit Width. Refer to the following command sequence to use the PEDIT command with the Width option.

📐 Command: **PE** *(For PEDIT)*
Select polyline or [Multiple]:*(Select the polyline at "A")*
Enter an option [Open/Join/Width/Edit vertex/Fit/Spline/Decurve/Ltype gen/Undo]: **W** *(For Width)*
Specify new width for all segments: **0.10**
Enter an option [Open/Join/Width/Edit vertex/Fit/Spline/Decurve/Ltype gen/Undo]: *(Press ENTER to exit this command)*

It is possible to convert regular objects to polylines. In Figure 4–73, the arc segment and individual line segments may be converted to a polyline. The circle cannot be converted unless part of the circle is broken, resulting in an arc segment.

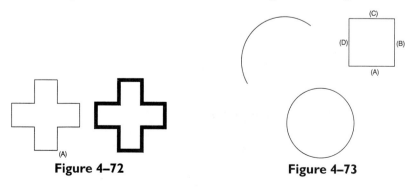

Figure 4–72   **Figure 4–73**

Try It! – Open the drawing file 04_Pedit Join1. Refer to Figure 4–73 and the following prompts for converting the line segments to a polyline.

 Command: **PE** *(For PEDIT)*
Select polyline or [Multiple]:*(Select the line at "A")*
Object selected is not a polyline
Do you want to turn it into one? <Y> *(Press ENTER to accept the default)*
Enter an option [Close/Join/Width/Edit vertex/Fit/Spline/Decurve/Ltype gen/Undo]: **J**
   (For Join)
Select objects: *(Select lines "B" through "D")*
Select objects: *(Press ENTER to join the lines)*
3 segments added to polyline
Enter an option [Open/Join/Width/Edit vertex/Fit/Spline/Decurve/Ltype gen/Undo]:
   (Press ENTER to exit this command)

In the previous example, regular objects were selected individually before being converted to a polyline. For more complex objects, use the Window option to select numerous objects and perform the PEDIT Join operation, which is faster.

 Note: The PEDITACCEPT variable can be used to assist with the editing of polylines. By default, this variable is turned off (0). When turned on (1), and you select a line to be converted into a polyline, the series of prompts alerting you that the object is not a polyline and if you want to turn it into one is suppressed. The line is automatically converted into a polyline. This variable must be entered in from the command prompt as in the following example:

Command: **PEDITACCEPT**
Enter new value for PEDITACCEPT <0>: **1** *(To turn on)*

 Try It! – Open the drawing file 04_Pedit Join2. Refer to the following command sequence and Figure 4–74 to use this command.

 Command: **PE** *(For PEDIT)*
Select polyline or [Multiple]:*(Select the line at "A")*
Object selected is not a polyline
Do you want to turn it into one? <Y> *(Press ENTER for default)*
Enter an option [Close/Join/Width/Edit vertex/Fit/Spline/Decurve/Ltype gen/Undo]: **J**
   (For Join)
Select objects: *(Pick a point at "B")*
Specify opposite corner: *(Pick a point at "C")*
Select objects: *(Press ENTER to join the lines)*
56 segments added to polyline
Enter an option [Open/Join/Width/Edit vertex/Fit/Spline/Decurve/Ltype gen/Undo]:
   (Press ENTER to exit this command)

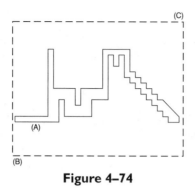

Figure 4–74

The polyline in Figure 4–75 will be used as an example of the use of the PEDIT command along with various curve-fitting utilities. In Figures 4–76 and 4–77, the Spline option and Fit Curve option are shown. The Spline option produces a smooth fitting curve based on control points in the form of the vertices of the polyline. The Fit Curve option passes entirely through the control points, producing a less desirable curve. Study Figures 4–77 and 4–78, which illustrate both curve options of the PEDIT command.

Figure 4–75

SPLINE CURVE GENERATION

 Try It! – Open the drawing file 04_Pedit Spline Curve. Refer to the following command sequence and Figure 4–76 to use this command.

 Command: **PE** *(For PEDIT)*

Select polyline or [Multiple]:*(Select the polyline at "A")*

Enter an option [Close/Join/Width/Edit vertex/Fit/Spline/Decurve/Ltype gen/Undo]: **S** *(For Spline)*

Enter an option [Close/Join/Width/Edit vertex/Fit/Spline/Decurve/Ltype gen/Undo]: *(Press ENTER to exit this command)*

The original polyline frame is usually not displayed when a spline is created and it is shown only for illustrative purposes.

Figure 4–76

FIT CURVE GENERATION

 Try It! – Open the drawing file 04_Pedit Fit Curve. Refer to the following command sequence and Figure 4–77 to use this command.

 Command: **PE** *(For PEDIT)*

Select polyline or [Multiple]:*(Select the polyline at "B")*

Enter an option [Close/Join/Width/Edit vertex/Fit/Spline/Decurve/Ltype gen/Undo]: **F**
  (For Fit)

Enter an option [Close/Join/Width/Edit vertex/Fit/Spline/Decurve/Ltype gen/Undo]:
  (Press ENTER to exit this command)

The Linetype Generation option of the PEDIT command controls the pattern of the linetype from polyline vertex to vertex. In the polyline at "C" in Figure 4–78, the hidden linetype is generated from the first vertex to the second vertex. An entirely different pattern is formed from the second vertex to the third vertex, and so on. The polyline at "D" has the linetype generated throughout the entire polyline. In this way, the hidden linetype is smoothed throughout the polyline.

 Try It! – Open the drawing file 04_Pedit Ltype Gen. Refer to the following command sequence and Figure 4–78 to use this command.

 Command: **PE** *(For PEDIT)*

Select polyline or [Multiple]:*(Select the polyline at "D" in Figure 4–78)*

Enter an option [Close/Join/Width/Edit vertex/Fit/Spline/Decurve/Ltype gen/Undo]: **Lt**
  (For Ltype gen)

Enter polyline linetype generation option [ON/OFF] <Off>: **On**

Enter an option [Close/Join/Width/Edit vertex/Fit/Spline/Decurve/Ltype gen/Undo]:
  (Press ENTER to exit this command)

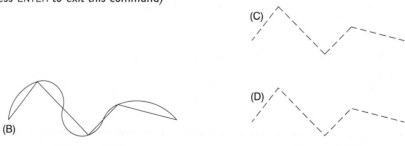

Figure 4–77   **Figure 4–78**

The object shown in Figure 4–79 is identical to that of Figure 4–69. Also in Figure 4–69, each individual line had to be offset to copy the lines parallel at a specified distance. Then the CHAMFER command was used to clean up the corners. There is an easier way to perform this operation.

Try It! – Open the drawing file 04_Pedit Offset. First convert all individual line segments to one polyline using the PEDIT command.

 Command: **PE** *(For PEDIT)*
Select polyline or [Multiple]:*(Select the line at "A")*
Object selected is not a polyline.
Do you want it to turn into one? <Y> *(Press ENTER for default)*
Enter an option [Close/Join/Width/Edit vertex/Fit/Spline/Decurve/Ltype gen/Undo]: **J**
   (For Join)
Select objects: *(Pick a point at "X")*
Specify opposite corner: *(Mark a point at "Y")*
Select objects: *(Press ENTER to continue)*
8 lines added to polyline
Enter an option [Close/Join/Width/Edit vertex/Fit/Spline/Decurve/Ltype gen/Undo]:
   (Press ENTER to exit this command)

The OFFSET command is used to copy the shape 0.50 units to the inside. Because the object was converted to a polyline, all objects are offset at the same time. This procedure bypasses the need to use the CHAMFER or FILLET command to corner all intersections. See Figure 4–80.

 Command: **O** *(For OFFSET)*
Specify offset distance or [Through] <1.0000>: **0.50**
Select object to offset or <exit>: *(Select the polyline at "A")*
Specify point on side to offset: *(Select a point anywhere near "B")*
Select object to offset or <exit>: *(Press ENTER to exit this command)*

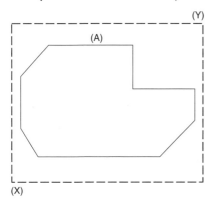

Figure 4–79

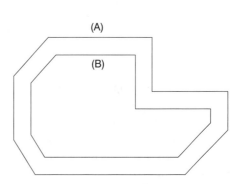

Figure 4–80

MULTIPLE POLYLINE EDITING

Multiple editing of polylines allows for multiple objects to be converted to polylines. This is accomplished with the PEDIT command and the Multiple option.

 Try It! – Open the drawing file 04_Pedit Multiple1. Use the prompt sequence below and Figure 4–81 to illustrate how the MPEDIT command is used.

 Command: **PE** *(For PEDIT)*

Select polyline or [Multiple]: **M** *(For Multiple)*

Select objects: *(Select the arc and line segments in Figure 4–81)*

Select objects: *(Press Enter to continue)*

Convert Lines and Arcs to polylines [Yes/No]? <Y> **Y** *(For Yes)*

Enter an option [Close/Open/Join/Width/Fit/Spline/Decurve/Ltype gen/Undo]: **W** *(For Width)*

Specify new width for all segments: **0.05**

Enter an option [Close/Open/Join/Width/Fit/Spline/Decurve/Ltype gen/Undo]: *(Press Enter to exit this command)*

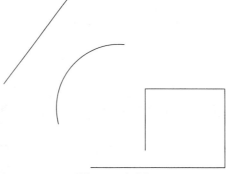

Figure 4–81

As a general rule when joining polylines, you cannot have gaps present or overlapping occurring when performing this operation. This is another feature of using the Multiple option of the PEDIT command. This option works best when joining two objects that have a gap or overlap. After selecting the two objects to join, you will be asked to enter a fuzz factor. This is the distance used by this command to bridge a gap or trim overlapping lines. You could measure the distance between two objects to determine this value. Study Figure 4–82 to better understand the concept of a fuzz factor. Also in Figure 4–82, the thick lines represent the polylines to join. The lines to join can also consist of polylines or regular line segments.

Try It! – Open the drawing file 04_Pedit Multiple2. Study Figure 4–82 and the following command sequences for performing this operation.

 Command: **PE** *(For PEDIT)*
Select polyline or [Multiple]: **M** *(For Multiple)*
Select objects: *(Pick lines "A" and "B" in Figure 4–82)*
Select objects: *(Press ENTER to continue)*
Convert Lines and Arcs to polylines [Yes/No]? <Y> **Y** *(For Yes)*
Enter an option [Close/Open/Join/Width/Fit/Spline/Decurve/Ltype gen/Undo]: **J** *(For Join)*
Join Type = Extend
Enter fuzz distance or [Jointype] <0.0000>: **0.44**
1 segments added to polyline
Enter an option [Close/Open/Join/Width/Fit/Spline/Decurve/Ltype gen/Undo]: *(Press ENTER to exit this command)*

 Command: **PE** *(For PEDIT)*
Select polyline or [Multiple]: **M** *(For Multiple)*
Select objects: *(Pick lines "C" and "D" in Figure 4–82)*
Select objects: *(Press ENTER to continue)*
Convert Lines and Arcs to polylines [Yes/No]? <Y> **Y** *(For Yes)*
Enter an option [Close/Open/Join/Width/Fit/Spline/Decurve/Ltype gen/Undo]: **J** *(For Join)*
Join Type = Extend
Enter fuzz distance or [Jointype] <0.4400>: **0.92**
1 segments added to polyline
Enter an option [Close/Open/Join/Width/Fit/Spline/Decurve/Ltype gen/Undo]: *(Press ENTER to exit this command)*

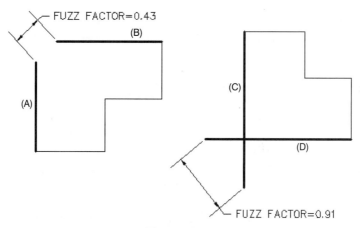

Figure 4–82

⟳ THE ROTATE COMMAND

The ROTATE command changes the orientation of an object or group of objects by identifying a base point and a rotation angle that completes the new orientation. Choose this command from one of the following:

the Modify toolbar

the pull-down menu (Modify > Rotate)

the keyboard (RO or ROTATE)

Figure 4–83 shows an object complete with crosshatch pattern, which needs to be rotated to a 30° angle using point "A" as the base point.

 Try It! – Open the drawing file 04_Rotate. Use the following prompts and Figure 4–83 to perform the rotation.

⟳ Command: **RO** *(For ROTATE)*
Current positive angle in UCS: ANGDIR=counterclockwise ANGBASE=0
Select objects: **All**
Select objects: *(Press ENTER to continue)*
Specify base point: *(Select the endpoint of the line at "A")*
Specify rotation angle or [Reference]: **30**

ROTATE—REFERENCE

At times it is necessary to rotate an object to a desired angular position. However, this must be accomplished even if the current angle of the object is unknown. To maintain the accuracy of the rotation operation, use the Reference option of the ROTATE command. Figure 4–84 shows an object that needs to be rotated to the 30° angle position. Unfortunately, we do not know the angle the object currently lies in. Entering the Reference angle option and identifying two points creates a known angle of reference. Entering a new angle of 30° rotates the object to the 30° position from the reference angle.

 Try It! – Open the drawing file 04_Rotate Reference. Use the following prompts and Figure 4–84 to accomplish this.

⟳ Command: **RO** *(For ROTATE)*
Current positive angle in UCS: ANGDIR=counterclockwise ANGBASE=0
Select objects: *(Select the object in Figure 4–84)*
Select objects: *(Press ENTER to continue)*
Specify base point: *(Pick either the edge of the circle or two arc segments to locate the center)*
Specify rotation angle or [Reference]: **R** *(For Reference)*
Specify the reference angle <0>: *(Pick either the edge of the circle or two arc segments to locate the center)*

Specify second point: **Mid**
of *(Select the line at "A" to establish the reference angle)*
Specify the new angle: **30**

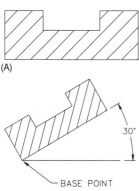

(A)

30°

BASE POINT

Figure 4–83

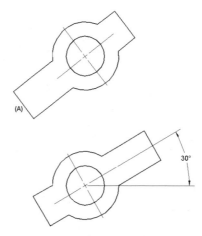

(A)

30°

Figure 4–84

 THE SCALE COMMAND

Use the SCALE command to change the overall size of an object. The size may be larger or smaller in relation to the original object or group of objects. The SCALE command requires a base point and scale factor to complete the command. Choose this command from one of the following:

the Modify toolbar

the pull-down menu (Modify > Scale)

the keyboard (SC or SCALE)

The object in Figure 4–85 will be scaled to the different sizes shown in Figures 4–86 and 4–87.

 Try It! – Open the drawing file 04_Scale1. With a base point at "A" and a scale factor of 0.50, the results of using the SCALE command on a group of objects are shown in Figures 4–86.

Command: **SC** *(For SCALE)*
Select objects: **All**
Select objects: *(Press ENTER to continue)*
Specify base point: *(Select the endpoint of the line at "A")*
Specify scale factor or [Reference]: **0.50**

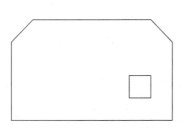

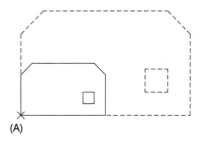

Figure 4–85

Figure 4–86

 Try It! – Open the drawing file 04_Scale2. The example in Figure 4–87 shows the effects of identifying a new base point in the center of the object.

Command: **SC** *(For SCALE)*
Select objects: **All**
Select objects: *(Press ENTER to continue)*
Specify base point: *(Pick a point near "A")*
Specify scale factor or [Reference]: **0.40**

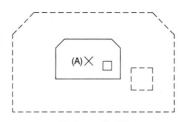

Figure 4–87

SCALE—REFERENCE

Suppose you are given a drawing that has been scaled down in size. However, no one knows what scale factor was used. You do know what one of the distances should be. In this special case, you can use the Reference option of the SCALE command to identify endpoints of a line segment that act as a reference length. Entering a new length value could increase or decrease the entire object proportionally.

 Try It! – Open the drawing file 04_Scale Reference. Study Figure 4–88 and the following prompts for performing this operation.

 Command: **SC** *(For SCALE)*
Select objects: *(Pick a point at "A")*
Specify opposite corner: *(Pick a point at "B")*
Select objects: *(Press ENTER to continue)*
Specify base point: *(Select the edge of the circle to identify its center)*

Specify scale factor or [Reference]: **R** *(For Reference)*
Specify reference length <1>: *(Select the endpoint of the line at "C")*
Specify second point: *(Select the endpoint of the line at "D")*
Specify new length: **2.00**

Because the length of line "CD" was not known, the endpoints were picked after the Reference option was entered. This provided the length of the line to AutoCAD. The final step to perform was to make the line 2.00 units, which increased the size of the object while also keeping its proportions.

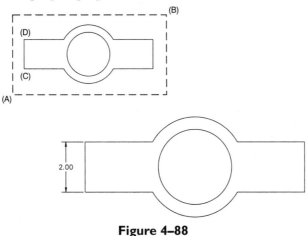

Figure 4–88

THE STRETCH COMMAND

Use the STRETCH command to move a portion of a drawing while still preserving the connections to parts of the drawing remaining in place. Choose this command from one of the following:

    the Modify toolbar

    the pull-down menu (Modify > Stretch)

    the keyboard (S or STRETCH)

To perform this type of operation, you must use the Crossing option of "Select objects." In Figure 4–89, a group of objects is selected with the crossing box. Next, a base point is identified by the endpoint at "C." Finally, a second point of displacement is identified with a polar coordinate; the Direct Distance mode could also be used. Once the objects selected in the crossing box are stretched, the objects not only move to the new location but also mend themselves.

Try It! – Open the drawing file 04_Stretch1. Use Figure 4–89 and the command sequence below to perform this task.

Command: **S** *(For STRETCH)*

Select objects to stretch by crossing-window or crossing-polygon...
Select objects: *(Pick a point at "A")*
Specify opposite corner: *(Pick a point at "B")*
Select objects: *(Press* ENTER *to continue)*
Specify base point or displacement: *(Select the endpoint of the line at "C")*
Specify second point of displacement: **@1.75<180**

 Try It! – Open the drawing file 04_Stretch2. Figure 4–90 is another example of the use of the STRETCH command. The crossing window is employed along with a base point at "C" and a polar coordinate. Use Figure 4–90 and the command sequence below to perform this task.

 Command: **S** *(For STRETCH)*

Select objects to stretch by crossing-window or crossing-polygon...
Select objects: *(Pick a point at "A")*
Specify opposite corner: *(Pick a point at "B")*
Select objects: *(Press* ENTER *to continue)*
Specify base point or displacement: *(Select the endpoint of the line at "C")*
Specify second point of displacement: **@1.00<0**

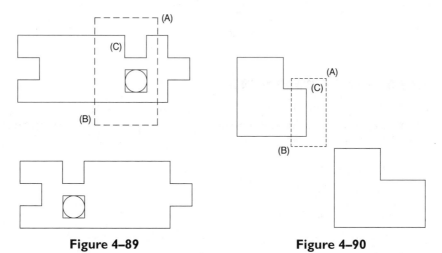

Figure 4–89 **Figure 4–90**

Applications of the STRETCH command include Figure 4–91, where a window needs to be positioned at a new location.

 Try It! – Open the drawing file 04_Stretch3. Use the following command sequence and Figure 4–91 to stretch the window at a set distance using a polar coordinate.

Command: **S** *(For STRETCH)*

Select objects to stretch by crossing-window or crossing-polygon...
Select objects: *(Pick a point at "A")*
Specify opposite corner: *(Pick a point at "B")*
Select objects: *(Press* ENTER *to continue)*
Specify base point or displacement: **Mid**
of (Select the midpoint of the line at "C")
Specify second point of displacement: **@10'6<0**

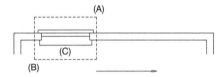

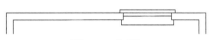

Figure 4–91

 THE TRIM COMMAND

Use the TRIM command to partially delete an object or group of objects based on a cutting edge. Choose this command from one of the following:

> the Modify toolbar
>
> the pull-down menu (Modify > Trim)
>
> the keyboard (TR or TRIM)

In Figure 4–92, the four dashed lines are selected as cutting edges. Next, segments of the circles are selected to be trimmed between the cutting edges.

 Try It! – Open the drawing file 04_Trim1. Use Figure 4–92 and the command sequence below to perform this task.

 Command: **TR** *(For TRIM)*

Current settings: Projection=UCS Edge=None
Select cutting edges ...
Select objects: *(Select the four dashed lines in Figure 4–92)*
Select objects: *(Press* ENTER *to continue)*
Select object to extend or shift-select to trim or [Project/Edge/Undo]: *(Select the circle at "A")*

Select object to extend or shift-select to trim or [Project/Edge/Undo]: *(Select the circle at "B")*

Select object to extend or shift-select to trim or [Project/Edge/Undo]: *(Select the circle at "C")*

Select object to extend or shift-select to trim or [Project/Edge/Undo]: *(Select the circle at "D")*

Select object to extend or shift-select to trim or [Project/Edge/Undo]: *(Press ENTER to exit this command)*

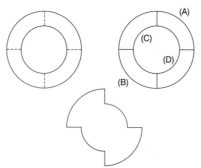

Figure 4–92

 Try It! – Open the drawing file 04_Trim2. An alternate method of selecting cutting edges is to press ENTER in response to the prompt "Select objects." This automatically creates cutting edges out of all objects in the drawing. When you use this method, the cutting edges do not highlight. In Figure 4–93, pick the lines at "A," "B," and "E" and the arc segments at "C" and "D" as the objects to trim.

Command: **TR** *(For TRIM)*

Current settings: Projection=UCS Edge=None

Select cutting edges ...

Select objects: *(Press ENTER to select all objects as cutting edges)*

Select object to extend or shift-select to trim or [Project/Edge/Undo]: *(Select the segment at "A")*

Select object to extend or shift-select to trim or [Project/Edge/Undo]: *(Select the segment at "B")*

Select object to extend or shift-select to trim or [Project/Edge/Undo]: *(Select the segment at "C")*

Select object to extend or shift-select to trim or [Project/Edge/Undo]: *(Select the segment at "D")*

Select object to extend or shift-select to trim or [Project/Edge/Undo]: *(Select the segment at "E")*

Select object to extend or shift-select to trim or [Project/Edge/Undo]: *(Press ENTER to exit this command)*

Try It! – Open the drawing file 04_Trim Fence. Yet another application of the TRIM command uses the Fence option of "Select objects." First, invoke the TRIM command and select the small circle as the cutting edge. Begin the response to the prompt of "Select object to trim" with "Fence." See Figure 4–94. Note: Turn off Running OSNAP before conducting this exercise.

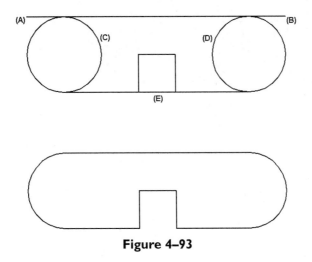

Figure 4–93

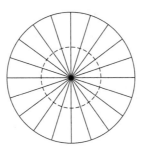

-/-- Command: **TR** *(For TRIM)*
Current settings: Projection=UCS Edge=None
Select cutting edges ...
Select objects: *(Select the small circle)*
Select objects: *(Press ENTER to continue)*
Select object to extend or shift-select to trim or [Project/Edge/Undo]: **F** *(For Fence)*

Figure 4–94

Continue with the TRIM command by identifying a Fence. This consists of a series of line segments that take on a dashed appearance. This means the fence will select any object it crosses. When you have completed the construction of the desired fence shown in Figure 4–95, press ENTER. Begin inside the smaller circle.

First fence point: *(Pick a point at "A")*

Specify endpoint of line or [Undo]: *(Pick a point at "B")*

Specify endpoint of line or [Undo]: *(Pick a point at "C")*

Specify endpoint of line or [Undo]: *(Pick a point at "D")*

Specify endpoint of line or [Undo]: *(Pick a point at "E")*

Specify endpoint of line or [Undo]: *(Pick a point at "F")*

Specify endpoint of line or [Undo]: *(Pick a point at "G")*

Specify endpoint of line or [Undo]: *(Pick a point at "H")*

Specify endpoint of line or [Undo]: *(Press* ENTER *to end the Fence and execute the trim operation)*

Select object to extend or shift-select to trim or [Project/Edge/Undo]: *(Press* ENTER *to exit this command)*

The power of the Fence option of "Select objects" is shown in Figure 4–96. Eliminating the need to select each individual line segment inside the small circle to trim, the Fence mode trims all objects it touches in relation to the cutting edge.

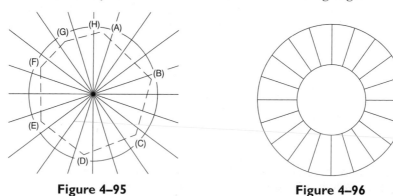

Figure 4–95   **Figure 4–96**

The TRIM command also allows you to trim to an extended cutting edge. In Extended Cutting Edge mode, an imaginary cutting edge is formed; all objects sliced along this cutting edge will be trimmed if selected individually or by the Fence mode.

 Try It! – Open the drawing file 04_Trim Edge. Study Figure 4–97 and the following prompts on this feature of the TRIM command.

Command: **TR** *(For TRIM)*

Current settings: Projection=UCS Edge=None

Select cutting edges ...

Select objects: *(Pick line "A")*

Select objects: *(Press* ENTER *to continue)*

Select object to extend or shift-select to trim or [Project/Edge/Undo]: **E** *(For Edge)*

Enter an implied edge extension mode [Extend/No extend] <No extend>: **E** *(For Extend)*

Select object to extend or shift-select to trim or [Project/Edge/Undo]: *(Pick the line at "B" along with the other segments)*

Select object to extend or shift-select to trim or [Project/Edge/Undo]: *(Press* ENTER *to exit this command)*

The Fence mode can also be used to select all line segments at once to trim.

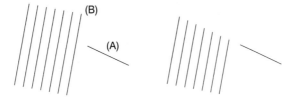

Figure 4–97

 Note: Reset the Edge setting in the TRIM command back to "No Extend" from "Extend." This can be done from the keyboard by entering the command EDGEMODE and setting it to 0 (Zero).

Command: **EDGEMODE**
Enter new value for EDGEMODE <1>: **0**

Care must be taken when it is appropriate to press ENTER and select all objects in your drawing as cutting edges with using the TRIM command. To see this in effect, try the next exercise.

 Try It! – Open the drawing file 04_Trim3. You need to remove the six vertical lines from the inside of the object. However, if you press ENTER to select all cutting edges, each individual segment would need to be trimmed, which is considered unproductive. Select lines "A" and "B" in Figure 4–98 and select the inner vertical lines. This is considered a more efficient way of using this command.

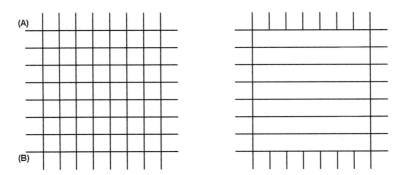

Figure 4–98

228

 Try It! – Open the drawing file 04_Trim Walls. Using Figure 4–99 as a guide, use the TRIM command to trim away the extra overshoots and complete the floor plan illustrated in the figure.

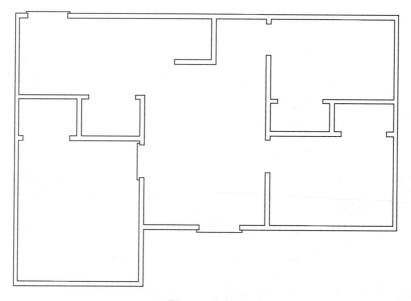

Figure 4–99

 Tip: While inside the TRIM command, you can easily toggle to the EXTEND command by holding down the SHIFT key at the following command prompt:

Select object to trim or shift-select to extend or [Project/Edge/Undo]:

THE UNDO COMMAND

The UNDO command can be used to undo the previous task or command action. Choose this command from one of the following:

the Standard toolbar

the pull-down menu (Edit > Undo)

the keyboard (U or UNDO)

For example, if you draw an arc followed by a line followed by a circle, issuing the UNDO command will undo the action caused by the most recent command; in this case, the circle would be removed from the drawing database. This represents one of the easiest ways to remove data or backtrack the design process.

Expanding the Undo list found in the Standard toolbar in Figure 4–100 allows you to undo several actions at once. From this example, notice that the Line, Trim, Fillet, and Offset actions are highlighted for removal.

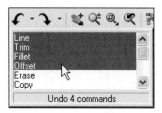

Figure 4–100

 Note: When grouping actions to be undone, you cannot, in Figure 4–100, highlight Line, skip Trim, and highlight Fillet to be removed. The groupings to undo must be strung together in this dialog box.

THE REDO COMMAND

You can also reverse the effect of the UNDO command by using REDO immediately after the UNDO operation. Choose this command from one of the following:

> the Standard toolbar
>
> the pull-down menu (Edit > Redo)
>
> the keyboard (REDO)

Clicking the REDO command button from the Standard toolbar will negate one undo operation. You can click on this button to cancel the effects of numerous UNDO operations.

As with UNDO, you can also REDO several actions at once through the Redo list shown in Figure 4–101. This list can be accessed from the Standard toolbar.

Figure 4–101

 Note: REDO will only work if you have undone a previous operation. Otherwise, REDO remains inactive.

TUTORIAL EXERCISE: CLUTCH.DWG

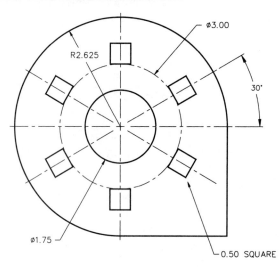

Ø3.00

R2.625

30°

Ø1.75

0.50 SQUARE

Figure 4–102

Purpose

This tutorial is designed to allow you to construct a one-view drawing of the Clutch shown in Figure 4–102 using a combination of coordinate modes and the ARRAY command.

System Settings

Use the current default settings for the units and limits of this drawing, (0,0) for the lower-left corner and (12,9) for the upper-right corner. Check to see that the following Object Snap modes are already set: Endpoint, Extension, Intersection, Center, and Quadrant.

Layers

Create the following layers with the format:

Name	Color	Linetype
Object	White	Continuous
Center	Yellow	Center

Suggested Commands

Draw the basic shape of the object using the LINE and CIRCLE commands. Lay out a center-line circle, draw one square shape, and use array to create a multiple copy of the square in a circular pattern.

Whenever possible, substitute the appropriate command alias in place of the full AutoCAD command in each tutorial step. For example, use "CP" for the COPY command, "L" for the LINE command, and so on. The complete listing of all command aliases is located in chapter 1, table 1–2.

STEP I

Check that the current layer is set to "Object." Use the Layer Control box in Figure 4–103 to accomplish this task. Begin drawing the clutch by placing a circle with the center at absolute coordinate (6.00,5.00) and radius of 2.625 units as shown in Figure 4–104. Place another circle using the same center point and a diameter of 1.75 units.

 Command: **C** *(For CIRCLE)*
Specify center point for circle or [3P/2P/ Ttr (tan tan radius)]: **6.00,5.00**
Specify radius of circle or [Diameter]: **2.625**

 Command: **C** *(For CIRCLE)*
Specify center point for circle or [3P/2P/ Ttr (tan tan radius)]: **6.00,5.00** *(or acquire the center of the first circle using Object Snap)*
Specify radius of circle or [Diameter] <2.6250>: **D** *(For Diameter)*
Specify diameter of circle <5.2500>: **1.75**

STEP 2

Use the Direct Distance mode along with Polar tracking to draw the lower-right corner of the Clutch. See Figure 4–105.

 Command: **L** *(For LINE)*
Specify first point: *(Pick the quadrant of the circle at "A")*
Specify next point or [Undo]: *(Move your cursor over the quadrant at "B" to acquire this point. Do not pick this location. Slide your cursor down until it aligns with the previous quadrant location and pick this point at "C")*
Specify next point or [Undo]: *(Select the quadrant or polar intersection at "B")*
Specify next point or [Undo]: *(Press ENTER to exit this command)*

Figure 4–103

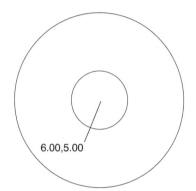

Figure 4–104

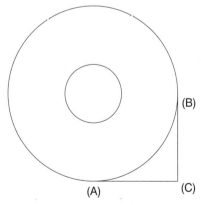

Figure 4–105

232

STEP 3

Use the TRIM command to partially delete one-fourth of the circle. Press ENTER to select the horizontal and vertical lines shown in Figure 4–106 as the cutting edges, and select the circle at "A" as the object to trim.

 Command: **TR** *(For TRIM)*

Current settings: Projection=UCS Edge=None

Select cutting edges ...

Select objects: *(Press ENTER which will select all objects cutting edges)*

Select object to extend or shift-select to trim or [Project/Edge/Undo]: *(Select the circle at "A")*

STEP 4

Make the "Center" layer current using the Layer Control box in Figure 4–107. Set the DIMCEN variable to a value of -0.09 units. Use DIMCENTER to construct a centerline. This will place a center mark at the center of the circle and extend centerlines just outside the larger circle, as shown in Figure 4–108. Erase the bottom centerline segment. This line will be placed in a later step.

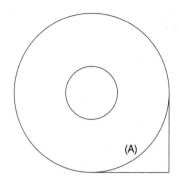

Figure 4–106

Select object to extend or shift-select to trim or [Project/Edge/Undo]: *(Press ENTER to exit this command)*

Command: **DIMCEN**

Enter new value for DIMCEN <0.0900>: **-0.09**

 Command: **DCE** *(For DIMCENTER)*

Select arc or circle: *(Select the edge of the arc at "A")*

 Command: **E** *(For ERASE)*

Select objects: *(Select the bottom centerline at "B")*

Select objects: *(Press ENTER to perform the erase operation)*

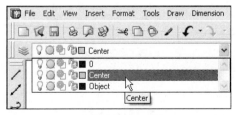

Figure 4–107

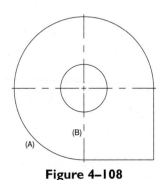

Figure 4–108

STEP 5

While in the "Center" layer, construct the 3.00 diameter centerline circle shown in Figure 4–109.

 Command: **C** *(For CIRCLE)*

Specify center point for circle or [3P/2P/ Ttr (tan tan radius)]: **6.00,5.00**

Specify radius of circle or [Diameter] <0.8750>: **D** *(For Diameter)*

Specify diameter of circle <1.7500>: **3.00**

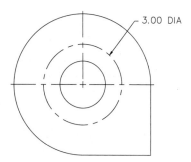

Figure 4–109

STEP 6

Set the current layer back to "Object" (see Figure 4–103). Then use the ZOOM command to magnify the upper portion of the clutch shown in Figure 4–110 for constructing a square in the next step.

 Command: **Z** *(For ZOOM)*

Specify corner of window, enter a scale factor (nX or nXP), or

[All/Center/Dynamic/Extents/Previous/ Scale/Window] <real time>: *(Pick a point at "A')*

Specify opposite corner: *(Pick a point at "B")*

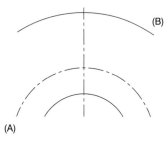

Figure 4–110

STEP 7

Construct the 0.50-unit square using the RECTANG command with the assistance of the From and Intersection OSNAP options (See Figure 4–111).

 Command: **REC** *(For RECTANG)*

Specify first corner point or [Chamfer/ Elevation/Fillet/Thickness/Width]: **From**

Base point: *(Select the intersection at "A" in Figure 4–111)*

<Offset>: **@0.25<180**

Specify other corner point: **@0.50,0.50**

Perform a ZOOM-PREVIOUS operation to return to the original display.

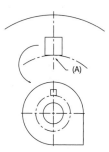

Figure 4–111

 Command: **Z** *(For ZOOM)*

Specify corner of window, enter a scale factor (nX or nXP), or

[All/Center/Dynamic/Extents/Previous/Scale/ Window] <real time>: **P** *(For Previous)*

STEP 8

Use the Array dialog box in Figure 4–112 to copy the square and vertical centerlines in a circular pattern. First, select the square and vertical centerlines at "A" in Figure 4–113. Select the intersection at "B" in Figure 4–113 as the center of the array. Enter the total number of items as 6.

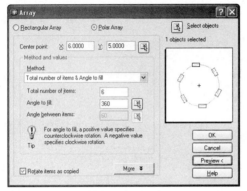

Figure 4–112

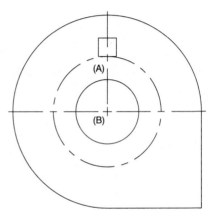

Figure 4–113

STEP 9

The completed problem is shown in Figure 4–114. Dimensions may be added at a later date.

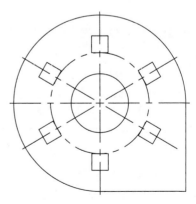

Figure 4–114

TUTORIAL EXERCISE: ANGLE.DWG

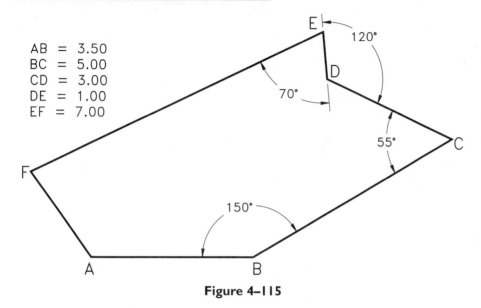

```
AB = 3.50
BC = 5.00
CD = 3.00
DE = 1.00
EF = 7.00
```

Figure 4–115

Purpose

This tutorial is designed to allow you to construct a one-view drawing of the angle shown in Figure 4–115 using the ARRAY and LENGTHEN commands.

System Settings

Use the Drawing Units dialog box and change the precision of decimal units from 4 to 2 places. Use the current default settings for the limits of this drawing, (0,0) for the lower-left corner and (12,9) for the upper-right corner. Check to see that the following Object Snap modes are already set: Endpoint, Extension, Intersection, Center.

Layers

Create the following layer with the format:

Name	Color	Linetype
Object	White	Continuous

Suggested Commands

Make the "Object" layer current. Begin this drawing by constructing line AB, which is horizontal. Use the ARRAY command to copy and rotate line "AB" at an angle of 150° in the clockwise direction. Once the line is copied, use the LENGTHEN command and modify the line to the proper length. Repeat this procedure for lines "CD," "DE," and "EF." Complete the drawing by constructing a line segment from the endpoint at vertex "F" to the endpoint at vertex "A."

Whenever possible, substitute the appropriate command alias in place of the full AutoCAD command in each tutorial step. For example, use "CP" for the COPY command, "L" for the LINE command, and so on. The complete listing of all command aliases is located in chapter 1, table 1–2.

STEP 1

Draw line "AB" using the Polar coordinate or Direct Distance mode, as shown in Figure 4–116. (Note: Line "AB" is considered a horizontal line.)

A _____ B

Figure 4–116

Command: **L** *(For LINE)*
Specify first point: **2,1**
Specify next point or [Undo]: *(Move your cursor directly to the right of the last point and enter a value of **3.50**)*
Specify next point or [Undo]: *(Press ENTER to exit this command)*

STEP 2

One technique of constructing the adjacent line at 150° from line "AB" is to use the Array dialog box in Figure 4–117 and perform a polar operation. Select line "AB" as the object to array in Figure 4–118, pick the intersection at "B" as the center of the array, and enter a value of -150° for the angle to fill. Entering a negative angle will copy the line in the clockwise direction.

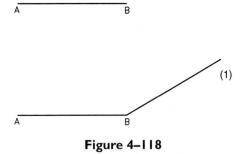

Figure 4–118

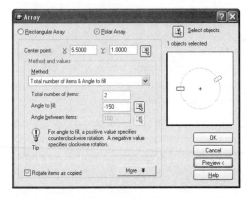

Figure 4–117

The result is shown in Figure 4–118.

STEP 3

The array operation allowed line "AB" to be rotated and copied at the correct angle, namely -150°. However, the new line is the same length as line "AB." Use the LENGTHEN command to increase the length of the new line to a distance of 5.00 units.

Command: **LEN** *(For LENGTHEN)*
Select an object or [DElta/Percent/Total/
    DYnamic]: **T** *(For Total)*
Specify total length or [Angle] <1.00)>:
    5.00
Select an object to change or [Undo]:
    (Pick the end of the line at "1")
Select an object to change or [Undo]:
    (Press ENTER to exit this command)

The result is shown in Figure 4–119.

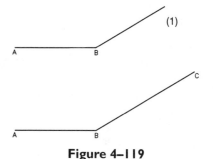

Figure 4–119

STEP 4

Use the Array dialog box shown in Figure 4–120 and perform a polar operation. Select line "BC" as the object to array in Figure 4–121, pick the intersection at "C" as the center of the array, and enter a value of -55° for the angle to fill. Entering a negative angle will copy the line in the clockwise direction.

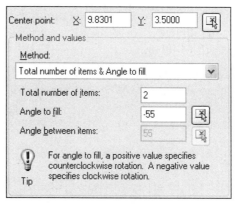

Figure 4–120

Then use the LENGTHEN command to reduce the length of the new line from 5.00 units to 3.00 units (see Figure 4–121.

Command: **LEN** (For LENGTHEN)
Select an object or [DElta/Percent/Total/
  DYnamic]: **T** (For Total)
Specify total length or [Angle] <5.00)>: **3.00**
Select an object to change or [Undo]:
  (Pick the end of the line at "1")
Select an object to change or [Undo]:
  (Press ENTER to exit this command)

The result is shown in Figure 4–121.

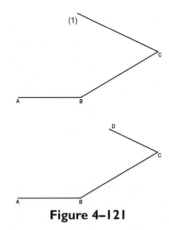

Figure 4–121

STEP 5

Use the Array dialog box shown in Figure 4–122 and perform another polar operation. Select line "CD" as the object to array in Figure 4–123, pick the intersection at "D" as the center of the array, and enter a value of 120° for the angle to fill. Entering a positive angle will copy the line in the counterclockwise direction.

Then use the LENGTHEN command to reduce the length of the new line from 3.00 units to 1.00 unit (see Figure 4–123).

Command: **LEN** (For LENGTHEN)
Select an object or [DElta/Percent/Total/
  DYnamic]: **T** (For Total)
Specify total length or [Angle] <3.00)>:
  1.00
Select an object to change or [Undo]:
  (Pick the end of the line at "1")
Select an object to change or [Undo]:
  (Press ENTER to exit this command)

The result is shown in Figure 4–123.

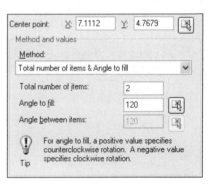

Figure 4–122

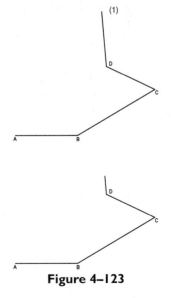

Figure 4–123

STEP 6

Use the Array dialog box shown in Figure 4–124 and perform the final polar operation. Select line "DE" as the object to array in Figure 4–125, pick the intersection at "E" as the center of the array, and enter a value of -70° for the angle to fill. Entering a negative angle will copy the line in the clockwise direction.

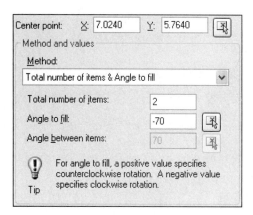

Figure 4–124

Then use the LENGTHEN command to increase the length of the new line from 1.00 unit to 7.00 units (see Figure 4–125).

Command: **LEN** *(For LENGTHEN)*
Select an object or [DElta/Percent/Total/
DYnamic]: **T** *(For Total)*
Specify total length or [Angle] <1.00)>:
7.00
Select an object to change or [Undo]:
(Pick the end of the line at "1")

Select an object to change or [Undo]:
(Press ENTER to exit this command)

The result is shown in Figure 4–125.

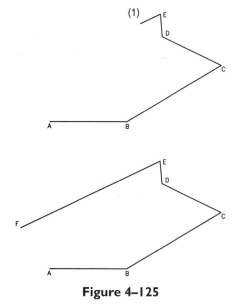

Figure 4–125

STEP 7

Connect endpoints "F" and "A" with a line as shown in Figure 4–126.

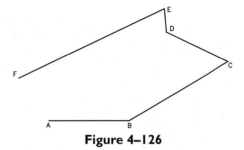

Figure 4–126

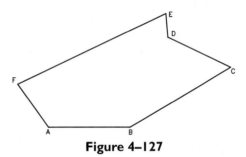

Command: **L** *(For LINE)*

Specify first point: *(Pick the endpoint of the line at "F")*

Specify next point or [Undo]: *(Pick the endpoint of the line at "A")*

Specify next point or [Undo]: *(Press* ENTER *to exit this command)*

The completed drawing is illustrated in Figure 4–127. You may add dimensions at a later date.

Figure 4–127

Open the Exercise Manual PDF file for Chapter 4 on the accompanying CD for more tutorials and exercises.

If you have the accompanying Exercise Manual, refer to Chapter 4 for more tutorials and exercises.

CHAPTER 5

Performing Geometric Constructions

In chapter 1, you were introduced to the LINE, CIRCLE, and PLINE commands. This chapter introduces the remainder of the drawing commands used for object creation. The following commands will be explained in this chapter:

ARC	DIVIDE	MLSTYLE	RECTANG
BOUNDARY	DONUT	POINT	REVCLOUD
CIRCLE-2P	ELLIPSE	MEASURE	SPLINE
CIRCLE-3P	MLEDIT	POLYGON	WIPEOUT
CIRCLE-TTR	MLINE	RAY	XLINE

Various scenarios of the CIRCLE-TTR command will be used to show the power of how AutoCAD can be used to create circles tangent to other object types. When covering the POINT command, the Point Style dialog box will also be discussed as a means of changing the appearance of points in your drawing. Multilines will also be covered, as will the ability to create a multiline style and then edit any intersections of a multiline.

METHODS OF SELECTING OTHER DRAW COMMANDS

Most drawing commands can be found on the Draw menu, shown in Figure 5–1. Arrowheads displayed to the right of the command indicate a cascading menu that holds additional options of the main command.

Figure 5–2 illustrates the Draw toolbar, which holds most drawing commands.

You can also enter most drawing commands directly from the keyboard by using their entire name, such as POINT for the POINT command. The following commands may be entered with only the first letter of the command as part of AutoCAD's command aliasing feature:

> Enter A for the ARC command
>
> Enter C for the CIRCLE command
>
> Enter L for the LINE command

See chapter 1, table 1–2 for the complete listing of all command aliases supported in AutoCAD.

Figure 5–1

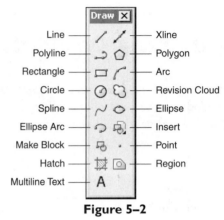

Figure 5–2

 THE ARC COMMAND

Use the ARC command to construct portions of circular shapes by radius, diameter, arc length, included angle, and direction. Choose this command from one of the following:

> the Draw toolbar
>
> the pull-down menu (Draw > Arc)
>
> the keyboard (A or ARC)

Choosing Arc from the Draw pull-down menu displays the cascading menu shown in Figure 5–3. All supported methods of constructing arcs are displayed in the list. By default, the 3 Points Arc mode supports arc constructions in the clockwise as well as the counterclockwise direction. All other arc modes support the ability to construct arcs only in the counterclockwise direction. The following pages detail most of the arc modes labeled in Figure 5–3.

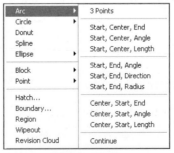

Figure 5–3

3 POINTS ARC MODE

By default, arcs are drawn with the 3 Points method. The first and third points identify the endpoints of the arc. This arc may be drawn in either the clockwise or counter-clockwise direction.

 Try It! – Create a new drawing file starting from scratch. Use the following command sequence along with Figure 5–4 to construct a 3-point arc.

 Command: **A** *(For ARC)*

Specify start point of arc or [Center]: *(Pick a point at "A")*
Specify second point of arc or [Center/End]: *(Pick a point at "B")*
Specify end point of arc: *(Pick a point at "C")*

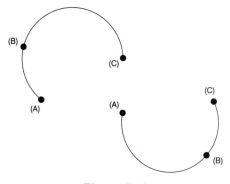

Figure 5–4

START, CENTER, END MODE

Use this ARC mode to construct an arc by defining its start point, center point, and endpoint. This arc will always be constructed in a counterclockwise direction.

 Try It! – Create a new drawing file starting from scratch. Use the following command sequence along with Figure 5–5 for constructing an arc by start, center, and endpoints.

 Command: **A** *(For ARC)*

Specify start point of arc or [Center]: *(Pick a point at "A")*
Specify second point of arc or [Center/End]: **C** *(For Center)*
Specify center point of arc: *(Pick a point at "B")*
Specify end point of arc or [Angle/chord Length]: *(Pick a point at "C")*

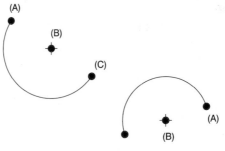

Figure 5–5

START, CENTER, ANGLE MODE

Use this mode to construct an arc by start point, center point, and included angle. If the angle is positive, the arc is drawn in the counterclockwise direction; a negative angle constructs the arc in a clockwise direction. See Figure 5–6.

 Try It! – Create a new drawing file starting from scratch. Use the following command sequence for constructing this type of arc.

Command: **A** *(For ARC)*
Specify start point of arc or [Center]: *(Pick a point at "A")*
Specify second point of arc or [Center/End]: **C** *(For Center)*
Specify center point of arc: *(Pick a point at "B")*
Specify end point of arc or [Angle/chord Length]: **A** *(For Angle)*
Specify included angle: **135**

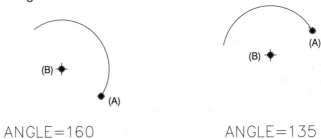

ANGLE=160   ANGLE=135

Figure 5–6

START, CENTER, LENGTH MODE

Use this mode to construct an arc by start point, center point, and length of chord. Figure 5–7 illustrates the definition of a chord.

 Try It! – Create a new drawing file starting from scratch. Use the following command sequence for constructing this type of arc.

Command: **A** *(For ARC)*

Specify start point of arc or [Center]: *(Pick a point at "A")*
Specify second point of arc or [Center/End]: **C** *(For Center)*
Specify center point of arc: *(Pick a point at "B")*
Specify end point of arc or [Angle/chord Length]: **L** *(For Length)*
Specify length of chord: **2.5**

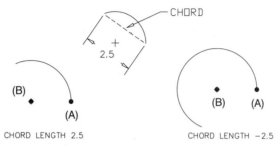

CHORD LENGTH 2.5   CHORD LENGTH −2.5

Figure 5–7

START, END, ANGLE MODE

Use this Arc mode to construct an arc by defining its starting point, endpoint, and included angle. This arc is drawn in a counterclockwise direction when a positive angle is entered; if the angle is negative, the arc is drawn clockwise.

Try It! – Create a new drawing file starting from scratch. Use the following command sequence along with Figure 5–8 for constructing an arc by start point, endpoint, and included angle.

Command: **A** *(For ARC)*

Specify start point of arc or [Center]: *(Pick a point at "A")*
Specify second point of arc or [Center/End]: **E** *(For End)*
Specify end point of arc: *(Pick a point at "B")*
Specify center point of arc or [Angle/Direction/Radius]: **A** *(For Angle)*
Specify included angle: **90**

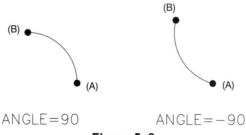

ANGLE=90   ANGLE=−90

Figure 5–8

START, END, DIRECTION MODE

Use this mode to construct an arc in a specified direction. This mode is especially helpful for drawing arcs tangent to other objects.

 Try It! – Create a new drawing file starting from scratch. Use the following prompt sequence along with Figure 5–9 for constructing an arc by direction.

 Command: **A** *(For ARC)*
Specify start point of arc or [Center]: *(Pick a point at "A")*
Specify second point of arc or [Center/End]: **E** *(For End)*
Specify end point of arc: *(Pick a point at "B")*
Specify center point of arc or [Angle/Direction/Radius]: **D** *(For Direction)*
Specify tangent direction for the start point of arc: **@1<90**

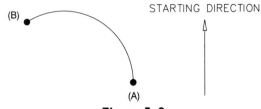

Figure 5–9

START, END, RADIUS MODE

Use this mode to construct an arc by start point, endpoint, and radius. A positive radius draws a minor arc; a negative radius draws a major arc. See Figure 5–10.

 Try It! – Create a new drawing file starting from scratch. Use the following command sequence for constructing this type of arc.

 Command: **A** *(For ARC)*
Specify start point of arc or [Center]: *(Pick a point at "A")*
Specify second point of arc or [Center/End]: **E** *(For End)*
Specify end point of arc: *(Pick a point at "B")*
Specify center point of arc or [Angle/Direction/Radius]: **R** *(For Radius)*
Specify radius of arc: **1.00**

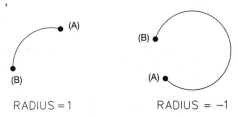

Figure 5–10

CENTER, START, END MODE

Use this mode to construct an arc by first locating the center point of the arc and then locating the start point and endpoint; this type of arc is constructed in the counter-clockwise direction. See Figure 5–11.

Try It! – Create a new drawing file starting from scratch. Use the following command sequence for constructing this type of arc.

 Command: **A** *(For ARC)*

Specify start point of arc or [Center]: **C** *(For Center)*
Specify center point of arc: *(Pick a point at "A")*
Specify start point of arc: *(Pick a point at "B")*
Specify end point of arc or [Angle/chord Length]: *(Pick a point at "C")*

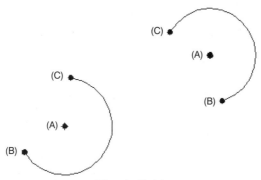

Figure 5–11

CENTER, START, ANGLE MODE

Use this mode to construct an arc by first locating the center point of the arc and then locating the start point and included angle. If the angle is positive, the arc is drawn in the counterclockwise direction; a negative angle constructs the arc in a clockwise direction. See Figure 5–12.

Try It! – Create a new drawing file starting from scratch. Use the following command sequence for constructing this type of arc.

 Command: **A** *(For ARC)*

Specify start point of arc or [Center]: **C** *(For Center)*
Specify center point of arc: *(Pick a point at "A")*
Specify start point of arc: *(Pick a point at "B")*
Specify end point of arc or [Angle/chord Length]: **A** *(For Angle)*
Specify included angle: **135**

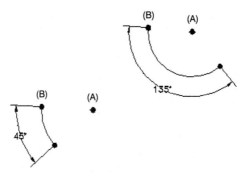

Figure 5–12

CONTINUE MODE

Use this mode to continue a previously drawn arc. All arcs drawn through Continue mode are automatically constructed tangent to the previous arc. Activate this mode from the Arc menu shown in Figure 5-3. The new arc begins at the last endpoint of the previous arc.

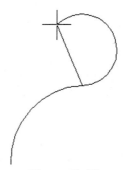

Figure 5–13

ADVANCED POLYLINE CONSTRUCTIONS – THE BOUNDARY COMMAND

The BOUNDARY command is used to create a polyline boundary around any closed shape. Choose this command from one of the following:

> the pull-down menu (Draw > Boundary)

> the keyboard (BO or BOUNDARY)

It has already been demonstrated that the Join option of the PEDIT command is used to join object segments into one continuous polyline. The BOUNDARY command automates this process even more. Start this command by choosing Boundary from the Draw pull-down menu, as shown in Figure 5–14. This activates the Boundary Creation dialog box, shown in Figure 5–15.

Figure 5–14

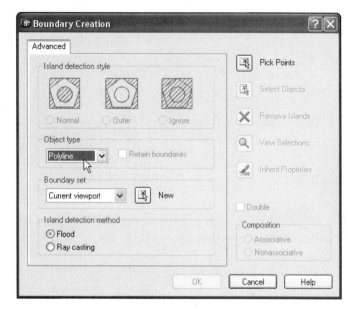

Figure 5–15

Before you use this command, it is considered good practice to create a separate layer to hold the polyline object; this layer could be called "Boundary" or "BP" for Boundary Polyline. Unlike the Join option of the PEDIT command, which converts individual objects to polyline objects, the BOUNDARY command will trace a polyline in the current layer on top of individual objects.

 Try It! – Open the drawing file 05_Boundary Extrusion. Activate the Boundary dialog box. Click the Pick Points button (see Figure 5–15). Then pick a point inside the object illustrated in Figure 5–16. Notice how the entire object is highlighted. To complete the command, press ENTER when prompted to select another internal point, and the polyline will be traced over the top of the existing objects.

Command: **BO** *(For BOUNDARY)*

(The dialog box in Figure 5–15 appears. Click the Pick Points button.)
Select internal point: *(Pick a point at "A" in Figure 5–16)*
Selecting everything...
Selecting everything visible...
Analyzing the selected data...
Analyzing internal islands...
Select internal point: *(Press ENTER to construct the boundary)*
BOUNDARY created 1 polyline

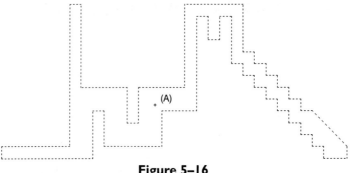

Figure 5–16

Once the boundary polyline is created, the boundary may be relocated to a different position on the screen with the MOVE command. The results are illustrated in Figure 5–17. The top object in the figure consists of the original individual objects; when the bottom object is selected, all objects highlight, signifying that it is made up of a polyline object made through the use of the BOUNDARY command.

When the BOUNDARY command is used on an object consisting of an outline and internal islands similar to the drawing in Figure 5–18, a polyline object is also traced over these internal islands.

 Try It! – Open the drawing file 05_Boundary Cover in Figure 5–18. Notice that the current layer is Boundary and the color is Magenta. Issue the BOUNDARY command and pick a point inside the middle of the object at "A" without OSNAP turned on. Notice that the polyline will be constructed on the top of all existing objects. Turn the Object layer off to display just the Boundary layer.

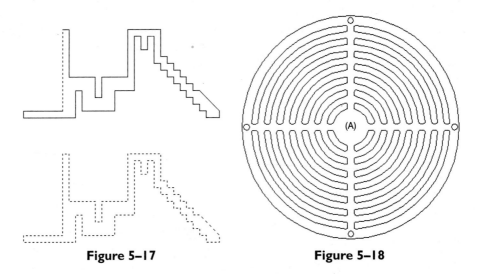

Figure 5–17   **Figure 5–18**

In addition to creating a special layer to hold the boundary polyline, another important rule to follow when using the BOUNDARY command is to be sure there are no gaps in the object. In Figure 5–19, it is acceptable to have lines cross at "A"; however, when the BOUNDARY command encounters the gap at "B," a dialog box informs you that no internal boundary could be found because the object is not completely closed. In this case, you must exit the command, close the object, and activate the BOUNDARY command again.

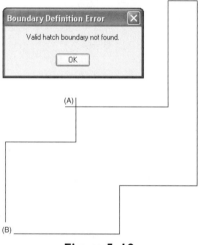

Figure 5–19

⊘ MORE OPTIONS OF THE CIRCLE COMMAND

Additional options of the CIRCLE command may be selected from the Circle pull-down menu in Figure 5–20. The 2 Points, 3 Points, Tan Tan Radius, and Tan Tan Tan modes will be explained in the next series of examples.

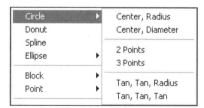

Figure 5–20

3 POINTS CIRCLE MODE

Use the CIRCLE command and the 3 Points mode to construct a circle by three points that you identify. No center point is required when you enter the 3 Points mode. Simply select three points and the circle is drawn.

252

 Try It! – Create a new drawing file starting from scratch. Study the following prompts and Figure 5–21 for constructing a circle using the 3 Points mode.

 Command: **C** *(For CIRCLE)*
Specify center point for circle or [3P/2P/Ttr (tan tan radius)]: **3P**
Specify first point on circle: *(Pick a point at "A")*
Specify second point on circle: *(Pick a point at "B")*
Specify third point on circle: *(Pick a point at "C")*

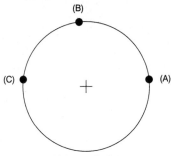

Figure 5–21

2 POINTS CIRCLE MODE

Use the CIRCLE command and the 2 Points mode to construct a circle by selecting two points. These points will form the diameter of the circle. No center point is required when you use the 2 Points mode.

 Try It! – Create a new drawing file starting from scratch. Study the following prompts and Figure 5–22 for constructing a circle using the 2 Points mode.

 Command: **C** *(For CIRCLE)*
Specify center point for circle or [3P/2P/Ttr (tan tan radius)]: **2P**
Specify first end point of circle's diameter: *(Pick a point at "A")*
Specify second end point of circle's diameter: *(Pick a point at "B")*

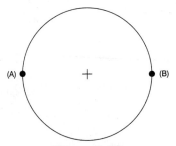

Figure 5–22

CONSTRUCTING AN ARC TANGENT TO TWO LINES USING CIRCLE-TTR

Illustrated in Figure 5–23 are two inclined lines. The purpose of this example is to connect an arc tangent to the two lines at a specified radius. The CIRCLE -TTR (Tangent-Tangent-Radius) command will be used here along with the TRIM command to clean up the excess geometry. To assist with this operation, the OSNAP-Tangent mode is automatically activated when you use the TTR option of the CIRCLE command.

First, use the CIRCLE-TTR command to construct an arc tangent to both lines, as shown in Figure 5–24.

 Command: **C** *(For CIRCLE)*
Specify center point for circle or [3P/2P/Ttr (tan tan radius)]: **T** *(For TTR)*
Specify point on object for first tangent of circle: *(Select the line at "A")*
Specify point on object for second tangent of circle: *(Select the line at "B")*
Specify radius of circle: *(Enter a radius value)*

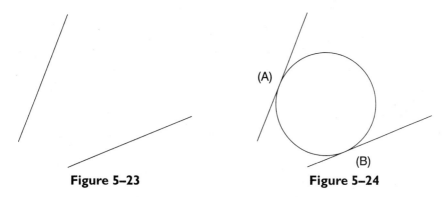

Figure 5–23   **Figure 5–24**

Use the TRIM command to clean up the lines and arc. The completed result is illustrated in Figure 5–25. The FILLET command could also have been used for this procedure. Not only will the curve be drawn, but also this command will automatically trim the lines.

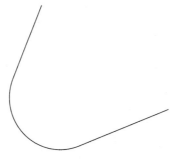

Figure 5–25

Try It! – Open the drawing file 05_TTR1. Use the CIRCLE command and the TTR option to construct a circle tangent to lines "A" and arc "B" in Figure 5–26. Use a circle radius of 0.50 units. With the circle constructed, use the TRIM command and select lines "C" and "D" as cutting edges. Trim the circle at "E", and the lines at "C" and "D." Observe the final results of this operation in Figure 5–26.

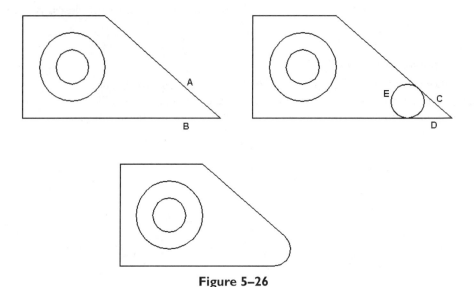

Figure 5–26

CONSTRUCTING AN ARC TANGENT TO A LINE AND ARC USING CIRCLE-TTR

Illustrated in Figure 5–27 are an arc and an inclined line. The purpose of this example is to connect an additional arc tangent to the original arc and line at a specified radius. The CIRCLE-TTR command will be used here, along with the TRIM command to clean up the excess geometry.

First, use the CIRCLE-TTR command to construct an arc tangent to the arc and inclined line as shown in Figure 5–28.

Command: **C** *(For CIRCLE)*
Specify center point for circle or [3P/2P/Ttr (tan tan radius)]: **T** *(For TTR)*
Specify point on object for first tangent of circle: *(Select the arc at "A")*
Specify point on object for second tangent of circle: *(Select the line at "B")*
Specify radius of circle: *(Enter a radius value)*

Use the TRIM command to clean up the arc and line. The completed result is illustrated in Figure 5–29.

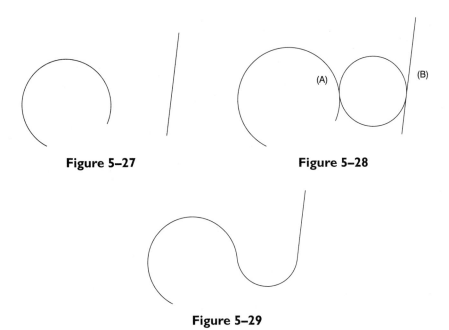

Figure 5–27   **Figure 5–28**

Figure 5–29

 Try It! – Open the drawing file 05_TTR2. Using Figure 5–30 as a guide, use the CIRCLE command and the TTR option to construct a circle tangent to the line at "A" and arc at "B." Use a circle radius value of 0.50 units. Construct a second circle tangent to the arc at "C" and line at "D" using the default circle radius value of 0.50 units. With the circles constructed, use the TRIM command, press ENTER to select all cutting edges and trim the lines at "E" and "J," the arc at "F" and "H" in addition to the circles at "G" and "K." Observe the final results of this operation in Figure 5–30.

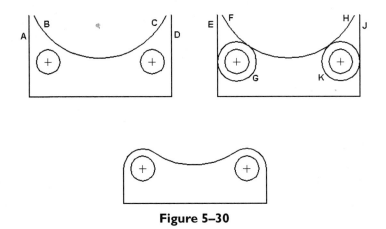

Figure 5–30

CONSTRUCTING AN ARC TANGENT TO TWO ARCS USING CIRCLE-TTR

Method #1

Illustrated in Figure 5–31 are two arcs. The purpose of this example is to connect a third arc tangent to the original two at a specified radius. The CIRCLE-TTR command will be used here along with the TRIM command to clean up the excess geometry.

Use the CIRCLE-TTR command to construct an arc tangent to the two original arcs as shown in Figure 5–32.

Command: **C** *(For CIRCLE)*
Specify center point for circle or [3P/2P/Ttr (tan tan radius)]: **T** *(For TTR)*
Specify point on object for first tangent of circle: *(Select the arc at "A")*
Specify point on object for second tangent of circle: *(Select the arc at "B")*
Specify radius of circle: *(Enter a radius value)*

Use the TRIM command to clean up the two arcs, using the circle as a cutting edge. The completed result is illustrated in Figure 5–33.

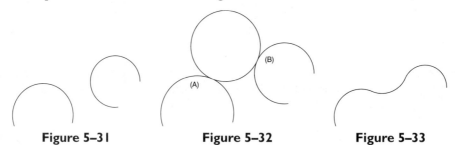

| Figure 5–31 | Figure 5–32 | Figure 5–33 |

 Try It! – Open the drawing file 05_TTR3. Use the CIRCLE command and the TTR option to construct a circle tangent to the circles at "A" and "B." Construct a second circle tangent to the circles at "C" and "D" in Figure 5–24D. Use a circle radius of 4.50 units for both circles. With the circles constructed, use the TRIM command and select both small circles as cutting edges. Trim the circles at "E" and "F". Observe the final results of this operation in Figure 5–34.

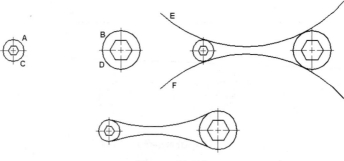

Figure 5–34

CONSTRUCTING AN ARC TANGENT TO TWO ARCS USING CIRCLE-TTR

Method #2

Illustrated in Figure 5–35 are two arcs. The purpose of this example is to connect an additional arc tangent to and enclosing both arcs at a specified radius. The CIRCLE-TTR command will be used here along with the TRIM command.

First, use the CIRCLE-TTR command to construct an arc tangent to and enclosing both arcs, using the indications in Figure 5–36.

Command: **C** *(For CIRCLE)*
Specify center point for circle or [3P/2P/Ttr (tan tan radius)]: **T** *(For TTR)*
Specify point on object for first tangent of circle: *(Select the arc at "A")*
Specify point on object for second tangent of circle: *(Select the arc at "B")*
Specify radius of circle: *(Enter a radius value)*

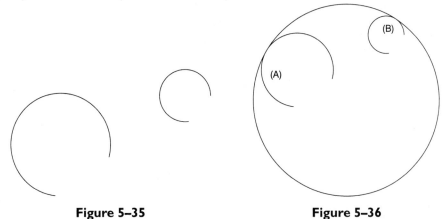

Figure 5–35   **Figure 5–36**

Use the TRIM command to clean up all arcs. The completed result is illustrated in Figure 5–37.

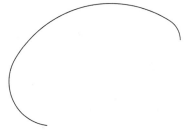

Figure 5–37

Try It! – Open the drawing file 05_TTR4. Use the CIRCLE command and the TTR option to construct a circle tangent to the circles at "A" and "B." Construct a second circle tangent to the circles at "C" and "D" in Figure 5–38. Use a circle radius of 1.50 units for both circles. With the circles constructed, use the TRIM command and trim all circles until you achieve the final results of this operation in Figure 5–38.

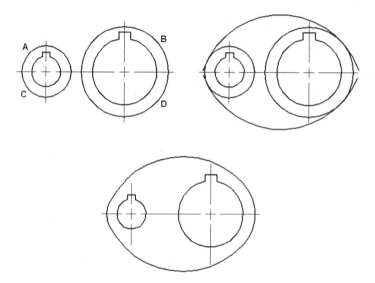

Figure 5–38

CONSTRUCTING AN ARC TANGENT TO TWO ARCS USING CIRCLE-TTR
Method #3

Illustrated in Figure 5–39 are two arcs. The purpose of this example is to connect an additional arc tangent to one arc and enclosing the other. The CIRCLE-TTR command will be used here along with the TRIM command to clean up unnecessary geometry.

First, use the CIRCLE-TTR command to construct an arc tangent to the two arcs. Study the illustration in Figure 5–40 and the following prompts to understand the proper pick points for this operation.

 Command: **C** *(For CIRCLE)*
Specify center point for circle or [3P/2P/Ttr (tan tan radius)]: **T** *(For TTR)*
Specify point on object for first tangent of circle: *(Select the arc at "A")*
Specify point on object for second tangent of circle: *(Select the arc at "B")*
Specify radius of circle: *(Enter a radius value)*

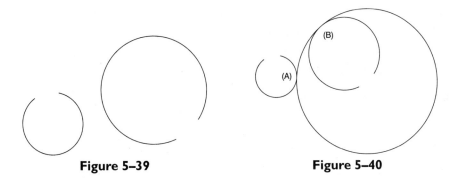

Figure 5–39   **Figure 5–40**

Use the TRIM command to clean up the arcs. The completed result is illustrated in Figure 5–41.

Figure 5–41

 Try It! – Open the drawing file 05_TTR5. Use the CIRCLE command and the TTR option to construct a circle tangent to the circles at "A" and "B." Construct a second circle tangent to the circles at "C" and "D" in Figure 5–42. Use a circle radius of 3.00 units for both circles. With the circles constructed, use the TRIM command and trim all circles until you achieve the final results of this operation in Figure 5–42.

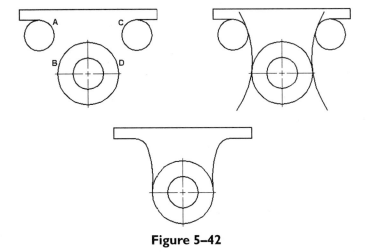

Figure 5–42

CONSTRUCTING A LINE TANGENT TO TWO ARCS OR CIRCLES

Illustrated in Figure 5–43 are two circles. The purpose of this example is to connect the two circles with two tangent lines. This can be accomplished with the LINE command and the OSNAP-Tangent option.

Use the LINE command to connect two lines tangent to the circles as shown in Figure 5–44. The following procedure is used for the first line. Use the same procedure for the second.

Command: **L** *(For LINE)*
Specify first point: **Tan**
to *(Select the circle near "A")*
Specify next point or [Undo]: **Tan**
to *(Select the circle near "B")*
Specify next point or [Undo]: *(Press ENTER to exit this command)*

When you use the Tangent option, the rubber-band cursor is not present when you draw the beginning of the line. This is due to calculations required when you identify the second point.

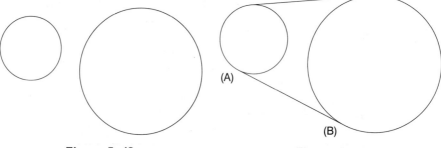

Figure 5–43   Figure 5–44

Use the TRIM command to clean up the circles so that the appearance of the object is similar to the illustration in Figure 5–45.

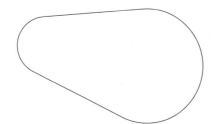

Figure 5–45

 Try It! – Open the drawing file 05_Tangent Lines. Construct a line tangent to the two circles at "A" and "B." Repeat this procedure for the circles at "C" and "D" in Figure 5–46. Use the TRIM command and trim all circles until you achieve the final results of this operation in Figure 5–46.

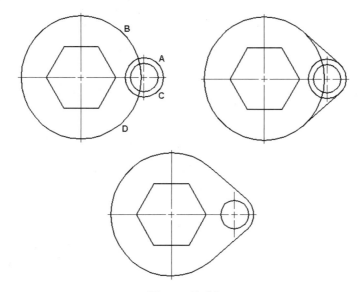

Figure 5–46

CIRCLE — TAN TAN TAN

Yet another mode of the CIRCLE command allows you to construct a circle based on three tangent points. This mode is actually a variation of the 3 Points mode together with using the OSNAP-Tangent mode three times. Clicking this mode from the Circle cascading menu shown in Figure 5–20 automates the process, enabling you to select three objects, as in the example in Figure 5–47.

Command: **C** *(For CIRCLE)*
Specify center point for circle or [3P/2P/Ttr (tan tan radius)]: **3P**
Specify first point on circle: **Tan**
to *(Select the line at "A")*
Specify second point on circle: **Tan**
to *(Select the line at "B")*
Specify third point on circle: **Tan**
to *(Select the line at "C")*

The result is illustrated in Figure 5–48, with the circle being constructed tangent to the edges of all three line segments.

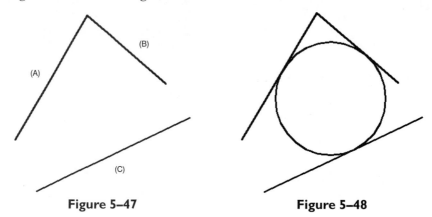

Figure 5–47   **Figure 5–48**

 Try It! – Open the drawing file 05_TTT. Use the illustration in Figure 5–47 and the previous command sequence for constructing a circle tangent to three lines.

QUADRANT VERSUS TANGENT OSNAP OPTION

Various examples have been given on previous pages concerning drawing lines tangent to two circles, two arcs, or any combination of the two. The object in Figure 5–49 illustrates the use of the OSNAP-Tangent option when used along with the LINE command.

 Command: **L** *(For LINE)*
Specify first point: **Tan**
to *(Select the arc near "A")*
Specify next point or [Undo]: **Tan**
to *(Select the arc near "B")*
Specify next point or [Undo]: *(Press ENTER to exit this command)*

Note that the angle of the line formed by points "A" and "B" is neither horizontal nor vertical. The object in Figure 5–49 is a typical example of the capabilities of the OSNAP-Tangent option.

The object illustrated in Figure 5–50 is a modification of the drawing in Figure 5–49 with the inclined tangent lines changing to horizontal and vertical tangent lines. This example is to inform you that two OSNAP options are available for performing tangencies, namely OSNAP-Tangent and OSNAP-Quadrant. However, it is up to you to evaluate under what conditions to use these OSNAP options. In Figure 5–50, you can use the OSNAP-Tangent or OSNAP-Quadrant option to draw the lines tangent to the arcs. The Quadrant option could be used only because the lines to be drawn are perfectly horizontal or vertical. Usually it is impossible to know this ahead of time, and in this case, the OSNAP-Tangent option should be used whenever possible.

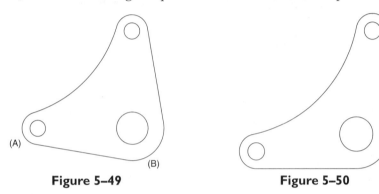

Figure 5–49   **Figure 5–50**

THE DONUT COMMAND

Use the DONUT command to construct a filled-in circle. Choose this command from one of the following:

> the pull-down menu (Draw > Donut)
>
> the keyboard (DO or DONUT)

This object belongs to the polyline family. Figure 5–51 is an example of a donut with an inside diameter of 0.50 units and an outside diameter of 1.00 units. When you place donuts in a drawing, the multiple option is automatically invoked. This means you can place as many donuts as you like until you choose another command from the menu or issue a Cancel by pressing ESC.

 Try It! – Create a new drawing file starting from scratch. Use the following command sequence for constructing this type of donut.

Command: **DO** *(For DONUT)*
Specify inside diameter of donut <0.50>: *(Press ENTER to accept the default)*
Specify outside diameter of donut <1.00>: *(Press ENTER to accept the default)*
Specify center of donut or <exit>: *(Pick a point to place the donut)*
Specify center of donut or <exit>: *(Pick a point to place another donut or press ENTER to exit this command)*

Setting the inside diameter of a donut to a value of zero (0) and an outside diameter to any other value constructs a donut representing a dot. See Figure 5–52.

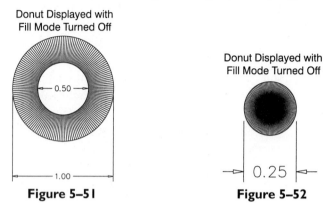

Figure 5–51

Figure 5–52

 Try It! – Create a new drawing file starting from scratch. Use the following command sequence for constructing this type of donut.

Command: **DO** *(For DONUT)*
Specify inside diameter of donut <0.50>: **0**
Specify outside diameter of donut <1.00>: **0.25**
Specify center of donut or <exit>: *(Pick a point to place the donut)*
Specify center of donut or <exit>: (Pick a point to place another donut or press ENTER to exit this command)

Figures 5–53 and 5–54 show two examples where donuts could be useful; donuts are sometimes used in place of arrows to act as terminators for dimension lines, as shown in Figure 5–53. Figure 5–54 illustrates how donuts might be used to act as connection points in an electrical schematic.

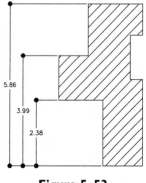

Figure 5–53

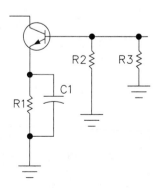

Figure 5–54

Try It! – Open the drawing file 05_Donut. Activate the DONUT command and set the inside diameter to 0 and the outside diameter to 0.10. Place four donuts at the intersections of the electrical circuit in Figure 5–54.

⬮ CONSTRUCTING ELLIPTICAL SHAPES

Use the ELLIPSE command to construct a true elliptical shape. Choose this command from one of the following:

> the Draw toolbar
>
> the pull-down menu (Draw > Ellipse) See Figure 5-55.
>
> the keyboard (EL or ELLIPSE)

Before studying the three examples for ellipse construction, see Figure 5–56 to view two important parts of any ellipse, namely its major and minor diameters.

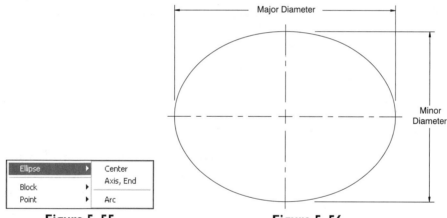

Ellipse	▶	Center
Block	▶	Axis, End
Point	▶	Arc

Figure 5–55   **Figure 5–56**

You can construct an ellipse by marking two points, which specify one of its axes (see Figure 5–57). These first two points also identify the angle with which the ellipse will be drawn. Responding to the prompt "Specify distance to other axis or [Rotation]" with another point identifies half of the other axis. The rubber-banded line is added to assist you in this ellipse construction method.

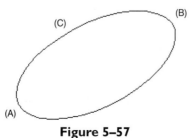

Figure 5–57

266

Try It! – Create a new drawing file starting from scratch. Use the illustration in Figure 5–57 and the command sequence below to construct an ellipse by locating three points.

 Command: **EL** *(For ELLIPSE)*
Specify axis endpoint of ellipse or [Arc/Center]: *(Pick a point at "A")*
Specify other endpoint of axis: *(Pick a point at "B")*
Specify distance to other axis or [Rotation]: *(Pick a point at "C")*

You can also construct an ellipse by first identifying its center. You can pick points to identify its axes or use polar coordinates to accurately define the major and minor diameters of the ellipse. See Figure 5–58 and the following command sequence to construct this type of ellipse. Use the Polar coordinate or Direct Distance mode for locating the two axis endpoints of the ellipse.

Try It! – Create a new drawing file starting from scratch. Use the illustration in Figure 5–58 and the command sequence below for constructing an ellipse based on a center.

 Command: **EL** *(For ELLIPSE)*
Specify axis endpoint of ellipse or [Arc/Center]: **C** *(For Center)*
Specify center of ellipse: *(Pick a point at "A")*
Specify endpoint of axis: **@2.50<0** *(To point "B")*
Specify distance to other axis or [Rotation]: **@1.50<90** *(To point "C")*

This last method, in Figure 5–59, illustrates constructing an ellipse by way of rotation. Identify the first two points for the first axis. Reply to the prompt "Specify distance to other axis or [Rotation]" with Rotation. The first axis defined is now used as an axis of rotation that rotates the ellipse into a third dimension.

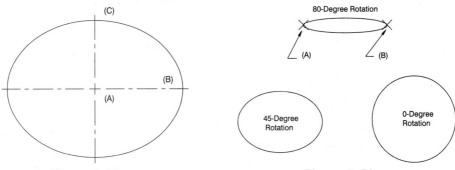

Figure 5–58   **Figure 5–59**

Try It! – Create a new drawing file starting from scratch. Use the illustration in Figure 5–59 and the command sequence below for constructing an ellipse based on a rotation around major axis.

Command: **EL** *(For ELLIPSE)*
Specify axis endpoint of ellipse or [Arc/Center]: *(Pick a point at "A")*
Specify other endpoint of axis: *(Pick a point at "B")*
Specify distance to other axis or [Rotation]: **R** *(For Rotation)*
Specify rotation around major axis: **80**

MULTILINES

A multiline consists of a series of parallel line segments. You can group as many as 16 multilines together to form a single multiline object. Multilines may take on color, may be spaced apart at different increments, and may take on different linetypes. Choose this command from one of the following:

>   the pull-down menu (Draw > Multiline)

>   the keyboard (ML or MLINE)

In Figures 5–60 through 5–62, the floor plan layout was made with the default multiline style of two parallel line segments. The spacing was changed to an offset distance of 4 units. Once the multilines are laid out as in Figure 5–60, the Multiline Edit (MLEDIT) command cleans up all corners and intersections, as displayed in Figure 5–61. Because multilines cannot be partially deleted with the BREAK command, the MLEDIT command allows you to cut between two points that you identify, as in Figure 5–62. Also, you have the option of breaking all or only one multiline segment depending on your desired results. Once a multiline segment has been cut, it can be welded back together again with the MLEDIT command.

The following pages outline the creation of a multiline style with the Multiline Style (MLSTYLE) command. Two additional dialog boxes are contained inside the Multiline Style dialog box: The Element Properties dialog box sets the spacing, color, and linetype of the individual multiline segments. The Multiline Properties dialog box provides other information on multilines, including capping modes and the ability to fill the entire multiline in a selected color.

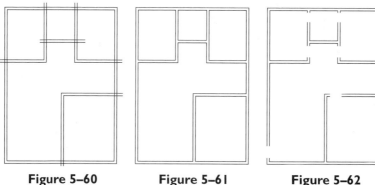

Figure 5–60   **Figure 5–61**   **Figure 5–62**

THE MLINE COMMAND

Once the element and multiline properties have been set and saved to a multiline style, loading a new current style and clicking the OK button of the Multiline Styles dialog box draws the multiline to the current multiline style configuration. Figure 5–63 is the result of setting the offset distances with the Multiline Style command. This configuration may be used for an architectural application where a block wall 8 units thick and a footing of 16 units thick is needed. Multilines are drawn with the MLINE command and the following command sequence.

Command: **ML** *(For MLINE)*
Current settings: Justification = Top, Scale = 1.00, Style = STANDARD
Specify start point or [Justification/Scale/STyle]: **J** *(For Justification)*
Enter justification type [Top/Zero/Bottom] <top>: **Z** *(For Zero)*
Current settings: Justification = Zero, Scale = 1.00, Style = STANDARD
Specify start point or [Justification/Scale/STyle]: *(Pick a point at "A")*
Specify next point: *(Pick a point at "B")*
Specify next point or [Undo]: *(Pick a point at "C")*
Specify next point or [Close/Undo]: *(Pick a point at "D")*
Specify next point or [Close/Undo]: *(Press* ENTER *to exit this command)*

See also Figure 5–65 for an example of changing the justification of multilines. Justification modes reference the top of the multiline, the bottom of the multiline, or the zero location of the multiline, which is the multiline's center.

Try It! – Open the drawing file 05_Mline Walls. A multiline style has been created consisting of block walls spaced 8" apart and a foundation wall spaced 16" apart. Use the MLINE command and the following command sequence to construct the exterior walls in Figure 5–64. Use the Direct Distance mode instead of polar coordinates to create this exterior wall.

Command: **ML** *(For MLINE)*
Current settings: Justification = Zero, Scale = 1.00, Style = FDN-WALLS
Specify start point or [Justification/Scale/STyle]: **21',10'**
Specify next point or [Close/Undo]: **16'** *(To the left)*
Specify next point or [Close/Undo]: **8'** *(Up)*
Specify next point or [Close/Undo]: **26'** *(To the left)*
Specify next point or [Close/Undo]: **28'** *(Down)*
Specify next point or [Close/Undo]: **42'** *(To the Right)*
Specify next point or [Close/Undo]: **C** *(To close the shape)*

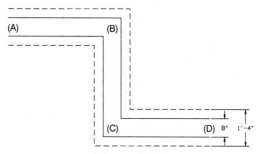

Figure 5–63

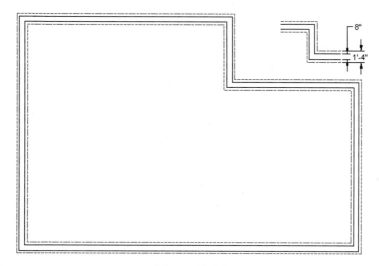

Figure 5–64

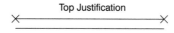

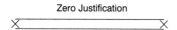

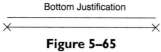

Figure 5–65

THE MULTILINE STYLE COMMAND

The Multiline Styles dialog box is used to create a new multiline style or to make an existing style current. Choose this command from one of the following:

the pull-down menu (Format > Multiline Style) See Figure 5-66.

the keyboard (MLSTYLE)

Choose Multiline Style to display the dialog box in Figure 5–67. The dialog box lists the current multiline style, which by default is named "STANDARD." To create a new multiline style, click the Element Properties button or Multiline Properties button, make the desired changes, and when finished, return to the main Multiline Styles dialog box. Change the name STANDARD to a name depicting the purpose of the new multiline style and click the Add button to add the new multiline style to the database of the current drawing.

Figure 5–66

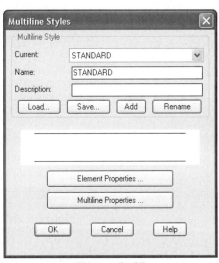

Figure 5–67

MULTILINE STYLE—ELEMENT PROPERTIES

Clicking the Element Properties button displays the Element Properties dialog box in Figure 5–68. Use this dialog box to make offset, color, and linetype assignments to the various multiline segments. Clicking the Add button adds a new multiline listing to the Elements box. Offset distances can now be changed along with Color and Linetype assignments to the selected multiline object. If a linetype is not present, a special Load button is available at the bottom of the Linetype dialog box. This in turn displays all supported linetypes. Clicking the OK button in the Element Properties dialog box returns you to the Multiline Styles dialog box, where a new multiline style name may be entered and the changes added to the new multiline style.

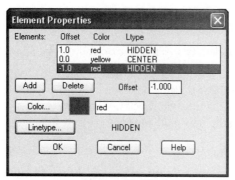

Figure 5–68

MULTILINE STYLE—MULTILINE PROPERTIES

Clicking the Multiline Properties button displays the Multiline Properties dialog box in Figure 5–69. This dialog box allows additional control of multilines: whether line segments are drawn at each multiline joint, whether the beginning and/or end of the multiline is capped with a line segment, whether the beginning and/or end of the outer parts of a multiline are capped by arcs, whether the beginning and/or end of the inner parts of a multiline are capped by arcs, or whether a user-defined angle is used to cap the start and end of the multiline. With Multiline Fill mode turned on, the entire multiline will be filled in with the selected color.

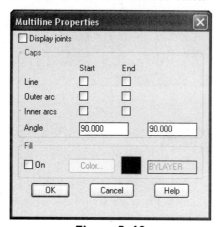

Figure 5–69

Review the steps used for creating a multiline style:

1. Open the Multiline Styles dialog box and change the name of the current style in the Name edit box.

2. After you enter a new name, click the Add button. This becomes the new multiline style.

3. Click the Element Properties button and change the characteristics such as offsets for the new multiline style.

4. Click the Multiline Properties button and change any settings in this dialog box.

5. Clicking OK at the bottom of either of these dialog boxes takes you to the main Multiline Styles dialog box. Any changes you made have already been saved to the current style. Click OK to dismiss this dialog box. You are now ready to use the new multiline style.

THE MLEDIT COMMAND

The MLEDIT command is a special feature designed for editing or cleaning up intersections of multiline objects. Choose this command from one of the following:

the pull-down menu (Modify > Object > Mline)

the keyboard (MLEDIT)

Choose Multiline from the Modify pull-down menu area, shown in Figure 5–70. This activates the Multiline Edit Tools dialog box in Figure 5–71. Clicking one of the buttons in the dialog box displays the title of the tool in the lower-left corner.

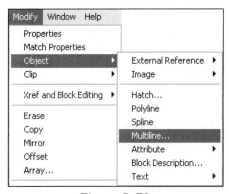

Figure 5–70

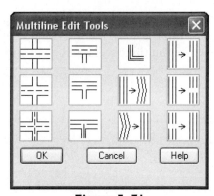

Figure 5–71

Various cleanup modes are available, such as creating a closed cross, creating an open cross, or creating a merged cross. Options are available for creating "Tee" sections in multilines. Corner joints may be created along with the adding or deleting of a vertex of a multiline. Figures 5–72, 5–73 and 5–74 depict the creation of open intersections and tees or the corner operation. In these figures, the identifying letters "A" and "B" show where to pick the multiline when using the Multiline Edit Tools dialog box.

Try It! – Open the drawing file 05_Mledit Open Cross. Activate the Multiline Edit Tools dialog box and click the Open Cross icon. Use the illustration in Figure 5–72 to complete this operation.

Try It! – Open the drawing file 05_Mledit Open Tee. Activate the Multiline Edit Tools dialog box and click the Open Tee icon. Use the illustration in Figure 5–73 to complete this operation.

 Try It! – Open the drawing file 05_Mledit Corner Joint. Activate the Multiline Edit Tools dialog box and click the Corner Joint icon. Use the illustration in Figure 5–74 to complete this operation.

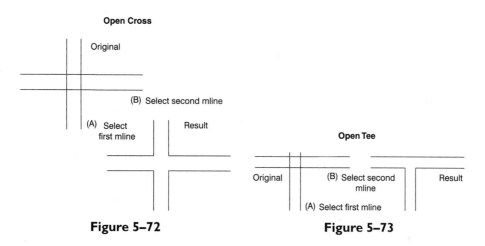

Figure 5–72

Figure 5–73

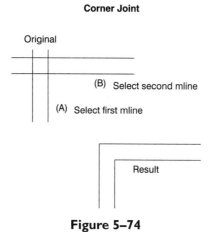

Figure 5–74

In the vertex example in Figure 5–75, deleting a vertex forces the multiline segment to straighten out. When you add a vertex as in Figure 5–76, the results may not be as evident. After the vertex was added to the straight multiline segment, the STRETCH command was used to create the shape in Figure 5–76. Multiline segments cannot be broken with the BREAK command. They must be cut with the Cut option of the Multiline Edit Tools dialog box, shown in Figure 5–71. To mend the cut, the Multiline Weld option is used. Both of these tools are located in the far right column of the Multiline Edit Tools dialog box.

 Try It! – Open the drawing file 05_Mledit Delete Vertex. Activate the Multiline Edit Tools dialog box and click the Delete Vertex icon. Use the illustration in Figure 5–75 to complete this operation.

 Try It! – Open the drawing file 05_Mledit Add Vertex. Activate the Multiline Edit Tools dialog box and click the Add Vertex icon. When prompted to select mline, pick the horizontal multiline at approximately its midpoint. To see the new vertex, press ESC and at the Command prompt, pick the horizontal multiline. Move your cursor to the middle blue box, pick it, and as you move your cursor up or down, notice how the multiline stretches based on the new vertex. Use the illustration in Figure 5–76 to complete this operation.

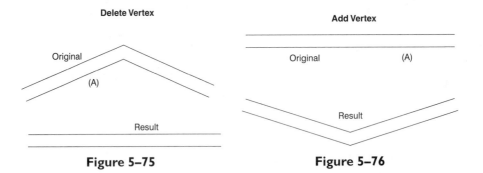

Figure 5–75   **Figure 5–76**

 Try It! – Open the drawing file 05_Mledit House. Activate the Multiline Edit Tools dialog box and click the Corner Joint icon. Use the illustration in Figure 5–77 to edit the four corners of the floor plan. Then Activate the Multiline Edit Tools dialog box and click the Open Tee icon. Edit the wall intersections using this tool. You may see different results depending upon which line you pick first. If this happens, immediately perform an undo and pick the lines in the correct order.

 Tip: Once the interior and exterior walls of a floor plan are laid out using multilines, these multilines can be exploded into individual line segments. Basic operations such as breaking, trimming, and extending can be used to add windows and doors to the floor plan.

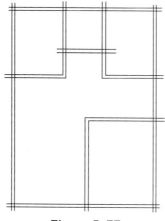

Figure 5–77

![icon] **THE POINT COMMAND**

Use the POINT command to identify the location of a point on a drawing, which may be used for reference purposes. Choose this command from one of the following:

> the Draw toolbar
>
> the pull-down menu (Draw > Point)
>
> the keyboard (PO or POINT)

Choose Point from the Draw pull-down menu shown in Figure 5–78. The OSNAP Node or Nearest option is used to snap to points. By default, a point is displayed as a dot on the screen.

![icon] Command: **PO** *(For POINT)*

Current point modes: PDMODE=0 PDSIZE=0.0000
Specify a point: *(Pick the new position of a point)*
Specify a point: *(Either pick another point location or press ESC to exit this command)*

Figure 5–78

Because the appearance of the default point as a dot may be confused with the existing grid dots already on the screen, a mechanism is available to change the appearance of the point. Choosing Point Style from the Format pull-down menu, shown in Figure 5–79, displays the Point Style dialog box in Figure 5–80. Use this icon menu to set a different point mode and point size.

Tip: Only one point style may be current in a drawing. Once a point is changed to a current style, the next drawing regeneration updates all points to this style.

Figure 5–79

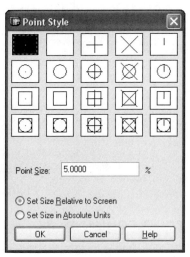

Figure 5–80

THE DIVIDE COMMAND

Illustrated in Figure 5–81 is an inclined line. The purpose of this example is to divide the line into an equal number of parts. This was a tedious task with manual drafting methods, but thanks to the DIVIDE command, this operation is much easier to perform. The DIVIDE command instructs you to supply the number of divisions and then performs the division by placing a point along the object to be divided. Choose this command from one of the following:

  the pull-down menu (Draw > Point > Divide)

  the keyboard (DIV or DIVIDE)

The Point Style dialog box shown in Figure-5–80 controls the point size and shape. Be sure the point style appearance is set to produce a visible point. Otherwise, the results of the DIVIDE command will not be obvious.

 Try It! – Open the drawing file 05_Divide. Use the DIVIDE command and select the inclined line as the object to divide, enter a value for the number of segments, and the command divides the object by a series of points as shown in Figure 5–82. This command is located on the Point cascading menu (from the Draw pull-down menu).

Command: **DIV** *(For DIVIDE)*
Select object to divide: *(Select the inclined line)*
Enter the number of segments or [Block]: **9**

Figure 5–81   Figure 5–82

One practical application of the DIVIDE command would be in drawing screw threads where a number of threads per inch are needed to form the profile of the thread. See Figure 5–83.

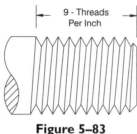

Figure 5–83

THE MEASURE COMMAND

The MEASURE command takes an object such as a line or arc and measures along it depending on the length of the segment. The MEASURE command, similar to the DIVIDE command, places a point on the object at a specified distance given in the MEASURE command. Choose this command from one of the following:

> the pull-down menu (Draw > Point > Measure)

> the keyboard (ME or MEASURE)

It is important to note that as a point is placed along an object during the measuring process, the object is not automatically broken at the location of the point. Rather, the point is commonly used to construct from, along with the OSNAP-Node option. Also, the measuring starts at the endpoint closest to the point you used to select the object (See Figure 5–84).

Choose Point from the Draw pull-down menu, and then choose the MEASURE command.

 Try It! – Open the drawing file 05_Measure. Follow the illustration in Figure 5–84 and the command sequence below for performing this operation.

Command: **ME** *(For MEASURE)*
Select object to measure: *(Select the end of the line in Figure 5–84)*
Specify length of segment or [Block]: **1.00**

The results in Figure 5–85 show various points placed at 1.00 increments. As with the DIVIDE command, the appearance of the points placed along the line is controlled through the Point Style dialog box.

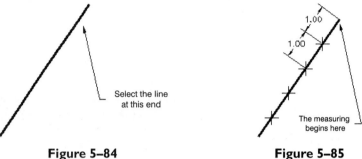

Figure 5–84   **Figure 5–85**

 THE POLYGON COMMAND

The POLYGON command is used to construct a regular polygon. Choose this command from one of the following:

    the Draw toolbar

    the pull-down menu (Draw > Polygon)

    the keyboard (POL or POLYGON)

You create polygons by identifying the number of sides for the polygon, locating a point on the screen as the center of the polygon, specifying whether the polygon is inscribed or circumscribed, and specifying a circle radius for the size of the polygon. Polygons consist of a closed polyline object with width set to zero. It is possible to create a polygon that consists of a maximum number of 1028 sides.

Try It! – Create a new drawing file starting from scratch. Use the following command sequence to construct the inscribed polygon in Figure 5–86.

 Command: **POL** *(For POLYGON)*
Enter number of sides <4>: **6**
Specify center of polygon or [Edge]: *(Mark a point at "A")*
Enter an option [Inscribed in circle/Circumscribed about circle] <I>: **I** *(For Inscribed)*
Specify radius of circle: **1.00**

Try It! – Create a new drawing file starting from scratch. Use the following command sequence to construct the circumscribed polygon in Figure 5–87.

Command: **POL** *(For POLYGON)*
Enter number of sides <4>: **7**

Specify center of polygon or [Edge]: *(Pick a point at "A")*
Enter an option [Inscribed in circle/Circumscribed about circle] <I>: **C** *(For Circumscribed)*
Specify radius of circle: **1.00**

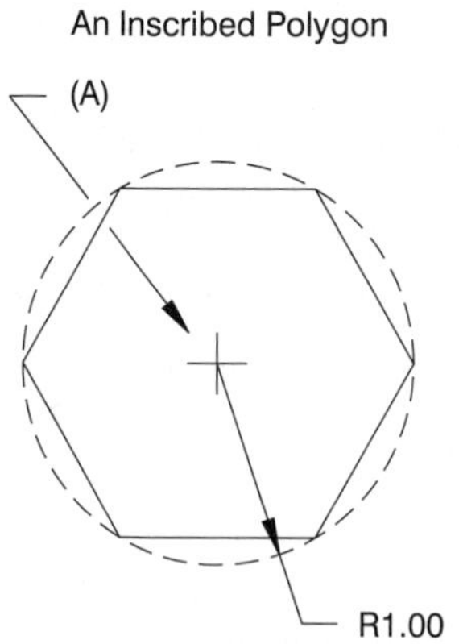

A Circumscribed Polygon

An Inscribed Polygon

(A)

R1.00

R1.00

(A)

Figure 5–86

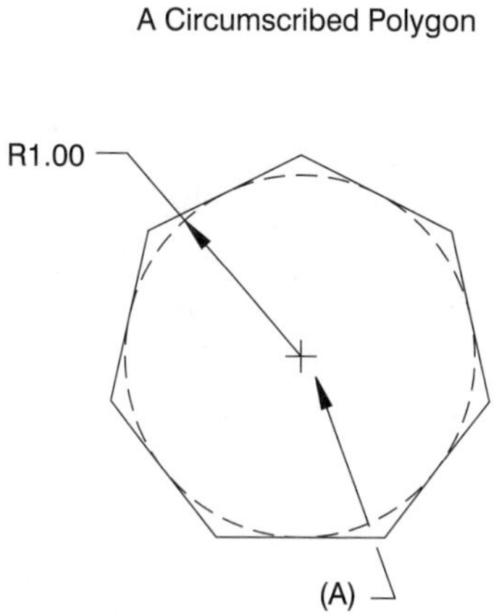

Figure 5–87

Locating the endpoints of one of its edges may specify polygons. The polygon is then drawn in a counterclockwise direction.

 Try It! – Create a new drawing file starting from scratch. Study Figure 5–88 and the following command sequence to construct a polygon by edge.

Command: **POL** *(For POLYGON)*
Enter number of sides <4>: **5**
Specify center of polygon or [Edge]: **E** *(For Edge)*
Specify first endpoint of edge: *(Mark a point at "A")*
Specify second endpoint of edge: *(Mark a point at "B")*

A Polygon by Edge

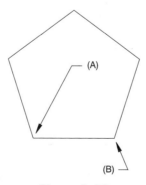

(A)

(B)

Figure 5–88

THE RAY COMMAND

A ray is a type of construction line object that begins at a user-defined point and extends to infinity only in one direction. Choose this command from one of the following:

the pull-down menu (Draw > Ray)

the keyboard (RAY)

In Figure 5–89, the quadrants of the circles identify all points where the ray objects begin and are drawn to infinity to the right. You should organize ray objects on specific layers. You should also exercise care in the editing of rays, and take special care not to leave segments of objects in the drawing database as a result of breaking ray objects. Breaking the ray object at "A" in Figure 5–89 converts one object to an individual line segment, and the other object remains as a ray. Study the following command sequence for constructing a ray.

Command: **RAY**
Specify start point: *(Pick a point on an object)*
Specify through point: *(Pick an additional point to construct the ray object)*
Specify through point: *(Pick another point to construct the ray object or press ENTER to exit this command)*

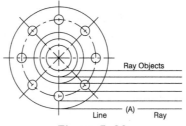

Figure 5–89

THE RECTANG COMMAND

Use the RECTANG command to construct a rectangle by defining two points. Choose this command from one of the following:

the Draw toolbar

the pull-down menu (Draw > Rectangle)

the keyboard (REC or RECTANG or RECTANGLE)

In Figure 5–90, two diagonal points are picked to define the rectangle. The rectangle is drawn as a single polyline object.

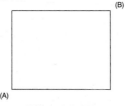

Figure 5–90

Command: **REC** *(For RECTANG)*

Specify first corner point or [Chamfer/Elevation/Fillet/Thickness/Width]: *(Pick a point at "A")*
Specify other corner point: *(Pickpoint at "B")*

Other options of the RECTANG command enable you to construct a chamfer or fillet at all corners of the rectangle, to assign a width to the rectangle, and to have the rectangle drawn at a specific elevation and at a thickness for 3D purposes. In Figure 5–91 and the following command sequence, a rectangle is constructed with a chamfer distance of 0.20 units; the width of the rectangle is also set at 0.05 units. A relative coordinate value of 1.00,2.00 will be used to construct the rectangle 1 unit in the "X" direction and 2 units in the "Y" direction. The @ symbol resets the previous point at "A" to zero.

 Try It! – Create a new AutoCAD drawing starting from scratch. Follow the command sequence below and the illustration in Figure 5–91 for constructing a wide rectangle with its corners chamfered.

 Command: **REC** *(For RECTANGLE)*

Specify first corner point or [Chamfer/Elevation/Fillet/Thickness/Width]: **C** *(For Chamfer)*
Specify first chamfer distance for rectangles <0.0000>: **0.20**
Specify second chamfer distance for rectangles <0.2000>: *(Press ENTER to accept this default value)*
Specify first corner point or [Chamfer/Elevation/Fillet/Thickness/Width]: **W** *(For Width)*
Specify line width for rectangles <0.0000>: **0.05**
Specify first corner point or [Chamfer/Elevation/Fillet/Thickness/Width]: *(Pick a point at "A")*
Specify other corner point or [Dimensions]: **@1.00,2.00** *(To identify the other corner at "B")*

 Try It! – Create a new AutoCAD drawing starting from scratch. You can also specify the dimensions (length and width) of the rectangle by following the illustration in Figure 5–92 and the command sequence below.

 Command: **REC** *(For RECTANGLE)*

Specify first corner point or [Chamfer/Elevation/Fillet/Thickness/Width]: *(Pick a point on the screen)*
Specify other corner point or [Dimensions]: **D** *(For Dimensions)*
Specify length for rectangles <0.0000>: **5.00**
Specify width for rectangles <0.0000>: **3.00**
Specify other corner point or [Dimensions]: *(Moving your cursor around positions the rectangle in four possible positions. Click the upper-right corner of your screen to anchor the upper-right corner of the rectangle)*

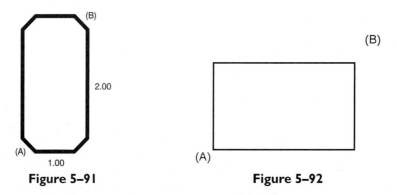

Figure 5–91   **Figure 5–92**

CREATING A REVISION CLOUD—THE REVCLOUD COMMAND

The REVCLOUD command creates a polyline object consisting of arc segments in a sequence. This feature is commonly used in drawings to identify areas where the drawing is to be changed or revised. Choose this command from one of the following:

the Draw toolbar

the pull-down menu (Draw > Revision Cloud)

the keyboard (REVCLOUD)

In Figure 5–93, an area of the house is highlighted with the revision cloud. By default, each arc segment has a minimum and maximum arc length of 0.50 units. In a drawing such as a floor plan, the arc segments would be too small to view. In cases where you wish to construct a revision cloud and the drawing is large, the arc length can be increased through a little experimentation. Once the arc segment is set, pick a point on your drawing to begin the revision cloud. Then move your cursor slowly in a counterclockwise direction. You will notice the arc segments of the revision cloud being constructed. Continue surrounding the object, while at the same time heading back to the origin of the revision cloud. Once you hover near the original start, the end of the revision cloud snaps to the start and the command ends, leaving the revision cloud as the polyline object. Study the command prompt sequence below and the illustration in Figure 5–93 for the construction of a revision cloud.

Command: **REVCLOUD**
Minimum arc length: 1/2" Maximum arc length: 1/2"
Specify start point or [Arc length/Object] <Object>: **A** *(For Arc length)*
Specify minimum length of arc <1/2">: **24**
Specify maximum length of arc <2'>: *(Press ENTER)*
Specify start point or [Object] <Object>: *(Pick a point on your screen to start the revision cloud. Surround the item with the revision cloud by moving your cursor in a counterclockwise direction. When you approach the start point, the revision cloud closes and the command exits.)*
Guide crosshairs along cloud path...
Revision cloud finished.

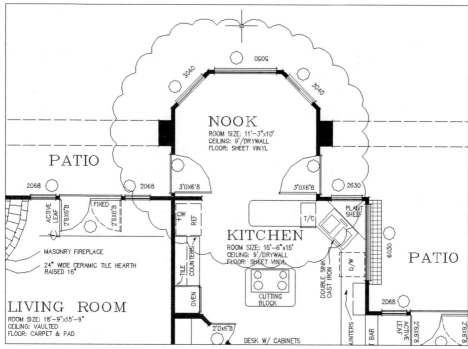

Figure 5–93

If it is difficult to construct the revision cloud using the previous command prompt sequence, you could create a closed polyline shape around the area you wish to highlight, as in the left illustration in Figure 9–94. Entering the REVCLOUD command, you will notice the presence of the Object option. Selecting Object and picking the edge of the polyline frame will convert the polyline into a revision cloud as in the illustration shown on the right in Figure 9–94.

Command: **REVCLOUD**
Minimum arc length: 2' Maximum arc length: 2'
Specify start point or [Arc length/Object] <Object>: **O** *(For Object)*
Select object: Reverse direction [Yes/No] <No>: *(Press ENTER)*
Revision cloud finished.

 Note: Once a revision cloud is created, you can easily convert it back to its original polyline frame by using the PEDIT command, selecting the edge of the revision cloud, and issuing the Decurve option.

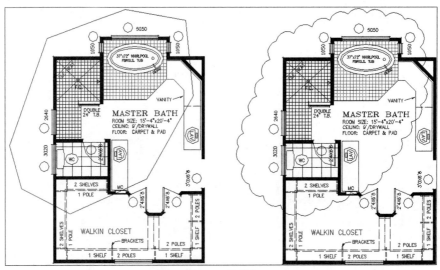

Figure 5–94

THE SPLINE COMMAND

Use the SPLINE command to construct a smooth curve given a sequence of points. Choose this command from one of the following:

the Draw toolbar

the pull-down menu (Draw > Spline)

the keyboard (SPL or SPLINE)

You have the option of changing the accuracy of the curve given a tolerance range. The basic command sequence follows, which constructs the spline segment shown in Figure 5–95.

 Command: **SPL** *(For SPLINE)*
Specify first point or [Object]: *(Pick a first point)*
Specify next point: *(Pick another point)*
Specify next point or [Close/Fit tolerance] <start tangent>: *(Pick another point)*
Specify next point or [Close/Fit tolerance] <start tangent>: *(Pick another point)*
Specify next point or [Close/Fit tolerance] <start tangent>: *(Press ENTER to continue)*
Specify start tangent: *(Press ENTER to accept)*
Specify end tangent: *(Press ENTER to accept the end tangent position, which exits the command and places the spline)*

The spline may be closed to display a continuous segment, as shown in Figure 5–96. Entering a different tangent point at the end of the command changes the shape of the curve connecting the beginning and end of the spline.

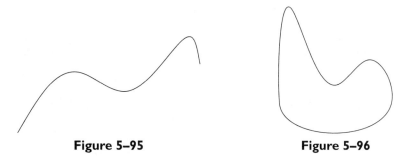

Figure 5–95   **Figure 5–96**

  Command: **SPL** *(For SPLINE)*

Specify first point or [Object]: *(Pick a first point)*
Specify next point: *(Pick another point)*
Specify next point or [Close/Fit tolerance] <start tangent>: *(Pick another point)*
Specify next point or [Close/Fit tolerance] <start tangent>: *(Pick another point)*
Specify next point or [Close/Fit tolerance] **C** *(To Close)*
Specify tangent: (Press ENTER to exit the command and place the spline)

Try It! – Open the drawing file 05_ Spline Rasp Handle (see Figure 5–97.) Turn OSNAP on and set Running OSNAP to Node. This will allow you to snap to points when drawing splines or other objects. Construct a spline by connecting all points between "A" and "B." Construct an arc with its center at "C," the start point at "D" and the ending point at "B." Connect points "D" and "E" with a line segment. Construct another spline by connecting all points between "E" and "F." Construct another arc with its center at "G," the start point at "H" and the ending point at "F." Connect points "H" and "A" with a line segment.

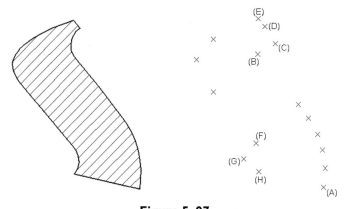

Figure 5–97

Figure 5–98 illustrates an application of using splines in the creation of a plastic bottle to hold such household products as dishwashing liquid. The bottle illustrated is in the form of a 3D wireframe model.

Once the bottle is defined with a series of splines, surfaces are applied to the spline frame forming a more realistic 3D model of the bottle in Figure 5–99. This image was created with the surface module of the Mechanical Desktop by Autodesk.

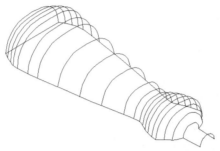

Figure 5–98

Figure 5–99

MASKING TECHNIQUES WITH THE WIPEOUT COMMAND

To mask or hide objects in a drawing without deleting them or turning off layers, the WIPEOUT command could be used. Choose this command from one of the following:

    the pull-down menu (Draw > Wipeout)

    the keyboard (WIPEOUT)

This command reads the current drawing background color and creates a mask over anything defined by a frame. In the illustration of the object on the left in Figure 5–100, a series of text objects needs to be masked over. The middle image shows a four-sided frame that was created over the text using the following command prompt sequence:

Command: **WIPEOUT**

Specify first point or [Frames/Polyline] <Polyline>: (Pick a first point)
Specify next point: (Pick a second point)
Specify next point or [Undo]: (Pick a third point)
Specify next point or [Close/Undo]: (Pick a fourth point)
Specify next point or [Close/Undo]: (Press ENTER to exit the command and create the wipeout)

As the text seems to disappear, the wipeout frame is still visible. A visible frame is important if you would like to unmask or delete the wipeout. If you want to hide all wipeout frames, use the following command sequence:

Command: **WIPEOUT**
Specify first point or [Frames/Polyline] <Polyline>: **F** *(For Frames)*
Enter mode [ON/OFF] <ON>: **Off**

Now all wipeout frames in the current drawing will be turned off as shown in the illustration on the right in Figure 5–100.

Note: You could also create a predefined polyline object and then convert it into a wipeout using the Polyline option of the WIPEOUT command.

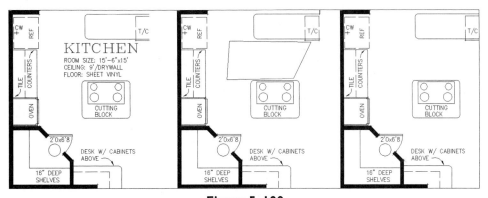

Figure 5–100

 ## CREATING CONSTRUCTION LINES WITH THE XLINE COMMAND

Xlines are construction lines drawn from a user-defined point. Choose this command from one of the following:

> the Draw toolbar
>
> the pull-down menu (Draw > Construction Line)
>
> the keyboard (XL or XLINE)

You are not prompted for any length information because the Xline extends an unlimited length, beginning at the user-defined point and going off to infinity in opposite directions from the point. Xlines can be drawn horizontal, vertical, and angular. You can bisect an angle using an Xline or offset the Xline at a specific distance. Figure 5–101, the circular view represents the Front view of a flange. To begin the creation of the Side views, lines are usually projected from key features on the adjacent view. In the case of the Front view in Figure 5–101, the key features are the top of the plate in addition to the other circular features. In this case, the Xlines were drawn with the Horizontal mode from the Quadrant of all circles. The following prompts outline the XLINE command sequence:

Command: **XL** *(For XLINE)*
Specify a point or [Hor/Ver/Ang/Bisect/Offset]: **H** *(For Horizontal)*
Specify through point: *(Pick a point on the display screen to place the first Xline)*
Specify through point: *(Pick a point on the display screen to place the second Xline)*

Since the Xlines continue to be drawn in both directions, care must be taken to manage these objects. Construction management techniques of Xlines could take the form of placing all Xlines on a specific layer to be turned off or frozen when not needed. When editing Xlines (especially with the BREAK command), you need to take special care to remove all excess objects that will still remain on the drawing screen. In Figure 5–102, breaking the Xline converts the object to a ray object. Use the ERASE command to remove any access Xlines.

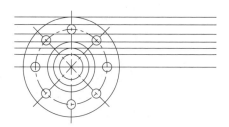

Figure 5–101

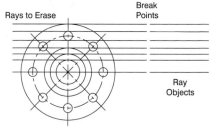

Figure 5–102

Try It! – Open the drawing file 05_Xline. Another application of Xlines is illustrated in Figure 5–103. Three horizontal and vertical Xlines are constructed. You must corner the Xlines to create the object illustrated in Figure 5–104. Follow the command prompt sequence for the FILLET to accomplish this task. Because numerous fillet operations need to be performed, use the MULTIPLE command followed by FILLET will keep you in the command. When finished, cancel the command by pressing ESC.

Command: **MULTIPLE**
Enter command name to repeat: **FILLET**
Current settings: Mode = TRIM, Radius = 0.5000
Select first object or [Polyline/Radius/Trim/mUltiple]: **R** *(For Radius)*
Specify fillet radius <0.5000>: **0**
Select first object or [Polyline/Radius/Trim/mUltiple]: *(Select the line at "A")*
Select second object: *(Select the line at "B")*
FILLET
Current settings: Mode = TRIM, Radius = 0.0000
Select first object or [Polyline/Radius/Trim/mUltiple]: *(Select the line at "B")*
Select second object: *(Select the line at "C")*

FILLET
Current settings: Mode = TRIM, Radius = 0.0000
Select first object or [Polyline/Radius/Trim/mUltiple]: *(Select the line at "C")*
Select second object: *(Select the line at "D")*
FILLET
Current settings: Mode = TRIM, Radius = 0.0000
Select first object or [Polyline/Radius/Trim/mUltiple]: *(Select the line at "D")*
Select second object: *(Select the line at "E")*
FILLET
Current settings: Mode = TRIM, Radius = 0.0000
Select first object or [Polyline/Radius/Trim/mUltiple]: *(Select the line at "E")*
Select second object: *(Select the line at "F")*
FILLET
Current settings: Mode = TRIM, Radius = 0.0000
Select first object or [Polyline/Radius/Trim/mUltiple]: *(Select the line at "F")*
Select second object: *(Select the line at "A")*
FILLET
Current settings: Mode = TRIM, Radius = 0.0000
Select first object or [Polyline/Radius/Trim/mUltiple]: *(Press the ESC key to exit this command)*

 Note: The Multiple option of the FILLET command has no effect on XLINE objects. This is the only reason for using the MULTIPLE command, which can be used to repeat any AutoCAD command.

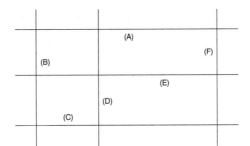

Figure 5–103

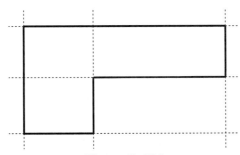

Figure 5–104

OGEE OR REVERSE CURVE CONSTRUCTION

 Try It! – Open the drawing file 05_Ogee. An ogee curve connects two parallel lines with a smooth flowing curve that reverses itself in symmetrical form.
To begin constructing an ogee curve to line segments "AB" and "CD," a line
was drawn from "B" to "C," which connects both parallel line segments. See
Figure 5–105.

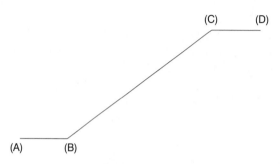

Figure 5–105

Use the DIVIDE command to divide line segment "BC" into four equal parts. Be sure
to set a new point mode by picking a new point from the Point Style dialog box.
Construct vertical lines from "B" and "C." Complete this step by constructing line
segment "XY," which is perpendicular to line "BC" as shown in Figure 5–106. Do not
worry about where line "XY" is located at this time.

Move line "XY" to the location identified by the point in Figure 5–107. Complete
this step by copying line "XY" to the location identified by point "Z" illustrated in
Figure 5–107.

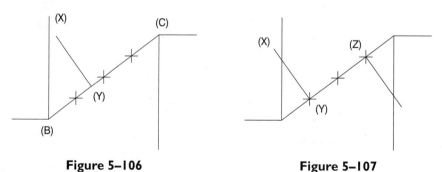

Figure 5–106   **Figure 5–107**

Construct two circles with centers located at points "X" and "Y" in Figure 5–108.
Use the OSNAP-Intersection mode to accurately locate the centers. Note: If an

intersection is not found from the previous step, use the EXTEND command to find the intersection and continue with this step. The radii of both circles are equivalent to distances "XB" and "YC."

Use the TRIM command to trim away any excess arc segments to form the ogee curve as shown in Figure 5–109. This forms the frame of the ogee for the construction of objects such as the wrench illustrated in Figure 5–110.

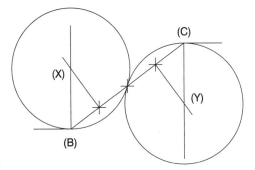

Figure 5–108

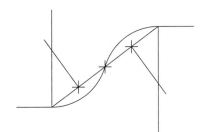

Figure 5–109

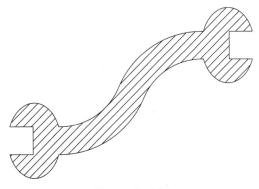

Figure 5–110

TUTORIAL EXERCISE: GEAR-ARM.DWG

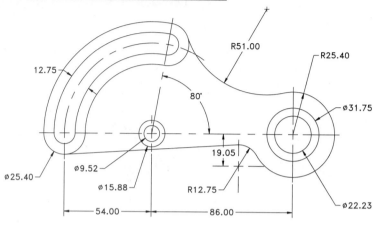

Figure 5–111

Purpose

This tutorial is designed to use geometric commands to construct a one-view drawing of the Gear-arm in metric format as illustrated in Figure 5–111.

System Settings

Use the Drawing Units dialog box to change the number of decimal places past the zero from four to two. Keep the remaining default unit values. Using the LIMITS command, keep (0,0) for the lower-left corner and change the upper-right corner from (12,9) to (265.00,200.00). Perform a ZOOM-ALL after changing the drawing limits. Since a layer called "Center" must be created to display centerlines, use the LTSCALE command and change the default value of 1.00 to 15.00. This will make the long and short dashes of the centerlines appear on the display screen. Check to see that the following Object Snap modes are already set: Endpoint, Extension, Intersection, Center.

Layers

Create the following layers with the format:

Name	Color	Linetype
Object	White	Continuous
Center	Yellow	Center
Dimension	Yellow	Continuous

Suggested Commands

Begin a new drawing called "Gear-arm." The object consists of a combination of circles and arcs along with tangent lines and arcs. Before beginning, be sure to set "Object" as the new current layer. Use the POINT command to identify and lay out the centers of all circles for construction purposes. Use the ARC command to construct a series of arcs for the left side of the Gear-arm. The TRIM command will be used to trim circles, lines, and arcs to form the basic shape. Also, use the CIRCLE-TTR command for tangent arcs to existing geometry.

Whenever possible, substitute the appropriate command alias in place of the full AutoCAD command in each tutorial step. For example, use "CP" for the COPY command, "L" for the LINE command, and so on. The complete listing of all command aliases is located in chapter 1, table 1–2.

STEP I

Check that the current layer is set to "Object." Use the Layer Control box in Figure 5–112 to accomplish this task.

Begin the gear-arm by drawing two circles of diameters 9.52 and 15.88 using the CIRCLE command and coordinate 112.00,90.00 as the center of both circles. See Figure 5–113.

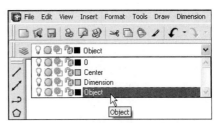

Figure 5–112

 Command: **C** *(For CIRCLE)*

Specify center point for circle or [3P/2P/ Ttr (tan tan radius)]: **112.00,90.00** *(At Point "A")*

Specify radius of circle or [Diameter]: **D** *(For Diameter)*

Specify diameter of circle: **9.52**

 Command: **C** *(For CIRCLE)*

Specify center point for circle or [3P/2P/Ttr (tan tan radius)]: **@** *(To identify the last known point as the center of the circle)*

Specify radius of circle or [Diameter] <4.76>: **D** *(For Diameter)*

Specify diameter of circle <9.52>: **15.88**

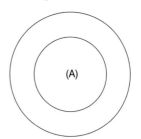

Figure 5–113

STEP 2

Change the default point appearance from a "dot" to the "plus" through the Point Style dialog box (see Figure 5–114), which you can open by choosing Point Style from the Format pull-down menu. While inside this dialog box, change the Point Size: to 3.00 units because this is a Metric drawing. Then use the POINT command to place a point at the center of the two circles. See Figure 5–115.

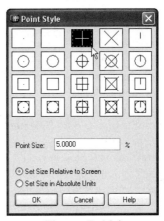

Figure 5–114

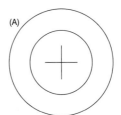 Command: **PO** *(For POINT)*

Current point modes: PDMODE=2 PDSIZE=3.00

Specify a point: *(Select the edge of the circle at "A" in Figure 5–115)*

Figure 5–115

STEP 3

Use the COPY command to duplicate the point using a polar coordinate distance of 86 units in the 0° direction. The Direct Distance mode may also be used for this operation. See Figure 5–116.

 Command: **CP** *(For COPY)*

Select objects: **L** *(For Last to select the last point)*

Select objects: *(Press* ENTER *to continue)*
Specify base point or displacement, or
  [Multiple]: *(Select the edge of the circle at "A" to identify its center)*
Specify second point of displacement
  or <use first point as displacement>:
  @86<0

(A)

Figure 5–116

STEP 4

Use the CIRCLE command to place three circles of different sizes from the same point at "A." See Figure 5–117.

Command: **C** *(For CIRCLE)*
Specify center point for circle or [3P/2P/
  Ttr (tan tan radius)]: **Nod**
of *(Select the point at "A")*
Specify radius of circle or [Diameter]
  <7.94>: **25.40**
Command: **C** *(For CIRCLE)*
Specify center point for circle or [3P/2P/Ttr
  (tan tan radius)]: **@** *(To identify the last known point as the center of the circle)*

Specify radius of circle or [Diameter]
  <25.40>: **D** *(For Diameter)*
Specify diameter of circle <50.80>: **31.75**

Command: **C** *(For CIRCLE)*
Specify center point for circle or [3P/2P/
  Ttr (tan tan radius)]: **@** *(To identify the last known point as the center of the circle)*
Specify radius of circle or [Diameter]
  <15.88>: **D** *(For Diameter)*
Specify diameter of circle <31.75>: **22.23**

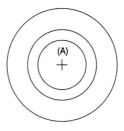

(A)

Figure 5–117

STEP 5

Use the COPY command to duplicate point "A" using a polar coordinate distance of 54 units in the 180° direction. The Direct Distance mode could also be used for this application. See Figure 5–118.

 Command: **CP** *(For COPY)*
Select objects: *(Select the point at "A")*
Select objects: *(Press ENTER to continue)*

Specify base point or displacement, or [Multiple]: *(Select the edge of the large circle at "A" to identify its center)*
Specify second point of displacement or <use first point as displacement>: **@54<180**

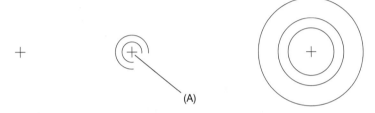

Figure 5–118

STEP 6

Use the CIRCLE command to place two circles of diameters 25.40 and 12.75 using point "A" in Figure 5–119 as the center. These circles will be converted to arcs in later steps.

 Command: **C** *(For CIRCLE)*
Specify center point for circle or [3P/2P/ Ttr (tan tan radius)]: **Nod**
of *(Select the point at "A")*
Specify radius of circle or [Diameter] <11.12>: **D** *(For Diameter)*
Specify diameter of circle <22.23>: **25.40**

Command: **C** *(For CIRCLE)*
Specify center point for circle or [3P/2P/ Ttr (tan tan radius)]: **@** *(To identify the last known point as the center of the circle)*
Specify radius of circle or [Diameter] <12.70>: **D** *(For Diameter)*
Specify diameter of circle <25.40>: **12.75**

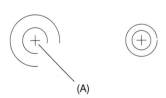

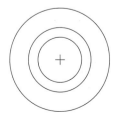

Figure 5–119

STEP 7

Use the COPY command to duplicate point "A" using a polar coordinate distance of 54 units in the 80° direction. See Figure 5–120.

 Command: **CP** *(For COPY)*

Select objects: *(Select the point at "A")*
Select objects: *(Press ENTER to continue)*

Specify base point or displacement, or [Multiple]: *(Select the edge of the large circle at "A" to identify its center)*
Specify second point of displacement or <use first point as displacement>: **@54<80**

Figure 5–120

STEP 8

Use the LINE command to draw a line using a polar coordinate distance of 70 and an 80° direction. Start the line at point "A." Then use the CIRCLE command to place two circles of diameters 25.40 and 12.75 at point "B" as in Figure 5–121. These circles will be converted to arcs in later steps.

 Command: **L** *(For LINE)*

Specify first point: *(Select the edge of the large circle at "A" to identify its center)*
Specify next point or [Undo]: **@70<80**
Specify next point or [Undo]: *(Press ENTER to exit this command)*

 Command: **C** *(For CIRCLE)*

Specify center point for circle or [3P/2P/Ttr (tan tan radius)]: **Nod**
of *(Select the point at "B")*
Specify radius of circle or [Diameter] <6.38>: **D** *(For Diameter)*
Specify diameter of circle <12.75>: **25.40**

Command: **C** *(For CIRCLE)*

Specify center point for circle or [3P/2P/Ttr (tan tan radius)]: **@** *(To identify the last known point as the center of the circle)*
Specify radius of circle or [Diameter] <12.70>: **D** *(For Diameter)*
Specify diameter of circle <25.40>: **12.75**

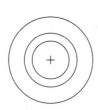

Figure 5–121

STEP 9

Use the ARC command and draw an arc using point "A" as the center, point "B" as the start point, and point "C" as the endpoint, as shown in Figure 5–122.

 Command: **A** *(For ARC)*

Specify start point of arc or [Center]: **C** *(For Center)*

Specify center point of arc: *(Select the edge of the circle at "A" to identify its center)*
Specify start point of arc: *(Select the intersection of the line and circle at "B")*
Specify end point of arc or [Angle/chord Length]: *(Using polar tracking, select the polar intersection at "C")*

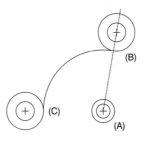

Figure 5–122

STEP 10

Use the ARC command and draw an arc using point "A" as the center, point "B" as the start point, and point "C" as the endpoint, as shown in Figure 5–123.

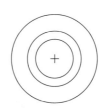 Command: **A** *(For ARC)*

Specify start point of arc or [Center]: **C** *(For Center)*

Specify center point of arc: *(Select the edge of the circle at "A")*
Specify start point of arc: *(Select the intersection of the line and circle at "B")*
Specify end point of arc or [Angle/chord Length]: *(Using polar tracking, select the polar intersection at "C")*

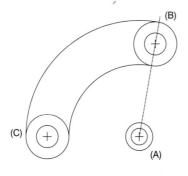

Figure 5–123

STEP 11

Use the TRIM command, select the two dashed arcs in Figure 5–124 as cutting edges, and trim the two circles at points "C" and "D."

⊟ Command: **TR** *(For TRIM)*

Current settings: Projection=UCS
   Edge=None
Select cutting edges ...
Select objects: *(Select the two dashed arcs "A" and "B")*
Select objects: *(Press ENTER to continue)*

Select object to trim or shift-select to extend or [Project/Edge/Undo]: *(Select the circle at "C")*
Select object to trim or shift-select to extend or [Project/Edge/Undo]: *(Select the circle at "D")*
Select object to trim or shift-select to extend or [Project/Edge/Undo]: *(Press ENTER to exit this command)*

If, during the trimming process, you trim the wrong segment, you can use a built-in Undo to trim the correct item.

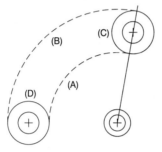
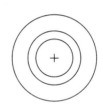

Figure 5–124

STEP 12

Your drawing should be similar to the illustration in Figure 5–125.

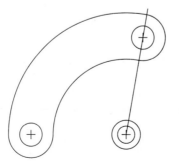
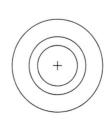

Figure 5–125

STEP 13

Use the ARC command and draw an arc using point "A" as the center, point "B" as the start point, and point "C" as the endpoint, as shown in Figure 5–126.

 Command: **A** *(For ARC)*

Specify start point of arc or [Center]: **C** *(For Center)*

Specify center point of arc: *(Select the edge of the circle at "A")*

Specify start point of arc: *(Select the intersection of the line and circle at "B")*

Specify end point of arc or [Angle/chord Length]: *(Using polar tracking, select the polar intersection at "C")*

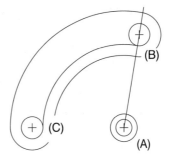

Figure 5–126

STEP 14

Use the ARC command and draw an arc using point "A" as the center, point "B" as the start point, and point "C" as the endpoint as shown in Figure 5–127.

 Command: **A** *(For ARC)*

Specify start point of arc or [Center]: **C** *(For Center)*

Specify center point of arc: *(Select the edge of the circle at "A")*

Specify start point of arc: *(Select the intersection of the line and circle at "B")*

Specify end point of arc or [Angle/chord Length]: *(Using polar tracking, select the polar intersection at "C")*

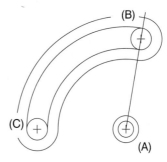

Figure 5–127

STEP 15

Use the TRIM command, select the two
dashed arcs at the right as cutting edges,
and trim the two circles at points "C" and
"D," as shown in Figure 5–128.

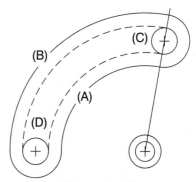

Figure 5–128

⊥⋯ Command: **TR** *(For TRIM)*
Current settings: Projection=UCS
 Edge=None
Select cutting edges ...
Select objects: *(Select the two dashed arcs
 "A" and "B")*
Select objects: *(Press* ENTER *to continue)*
Select object to trim or shift-select to
 extend or [Project/Edge/Undo]: *(Select
 the circle at "C")*
Select object to trim or shift-select to
 extend or [Project/Edge/Undo]: *(Select
 the circle at "D")*

Select object to trim or shift-select to
 extend or [Project/Edge/Undo]: *(Press
 ENTER to exit this command)*

STEP 16

Your drawing should be similar to the
illustration in Figure 5–129. Always perform periodic screen redraws using the
REDRAW command to clean up the display
screen if necessary.

Command: **R** *(For REDRAW)*

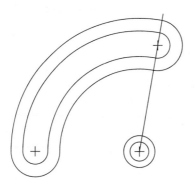
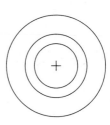

Figure 5–129

STEP 17

Use the LINE command and draw a line from the quadrant of the small circle to the quadrant of the large circle in Figure 5–130. This line will be used only for construction purposes.

 Command: **L** *(For LINE)*

Specify first point: **Qua**
of *(Select the circle at "A")*
Specify next point or [Undo]: *(Use polar tracking to select the polar intersection at "B")*
Specify next point or [Undo]: *(Press ENTER to exit this command)*

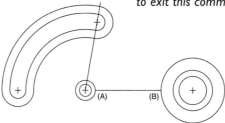

Figure 5–130

STEP 18

Use the MOVE command to move the dashed line down the distance of 19.05 units using the Polar coordinate or Direct Distance mode. Then use the OFFSET command to offset the dashed circle the distance 12.75 units. The intersection of these two objects will be used to draw a 12.75-radius circle, as shown in Figure 5–131.

 Command: **M** *(For MOVE)*

Select objects: *(Select the dashed line in Figure 5–131)*
Select objects: *(Press ENTER to continue)*
Specify base point or displacement: *(Select the endpoint of the line at "A")*

Specify second point of displacement or <use first point as displacement>: *(Move your cursor straight down and enter a value of 19.05)*

 Command: **O** *(For OFFSET)*

Specify offset distance or [Through] <1.00>: **12.75**
Select object to offset or <exit>: *(Select the dashed circle at "B")*
Specify point on side to offset: *(Pick a blank part of the screen at "C")*
Select object to offset or <exit>: *(Press ENTER to exit this command)*

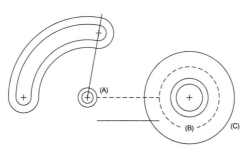

Figure 5–131

STEP 19

Use the CIRCLE command to draw a circle with a radius of 12.75. Use the center of the circle as the intersection of the large dashed circle and the dashed horizontal line illustrated in Figure 5–132. Use the ERASE command to delete the dashed circle and dashed line.

 Command: **C** *(For CIRCLE)*

Specify center point for circle or [3P/ 2P/Ttr (tan tan radius)]: *(Select the intersection of the line and circle at "A")*

Specify radius of circle or [Diameter] <6.38>: **12.75**

Command: **E** *(For ERASE)*

Select objects: *(Select the dashed circle and dashed line in Figure 5–132)*

Select objects: *(Press ENTER to execute this command)*

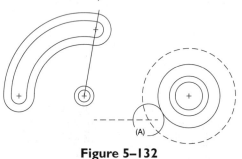

Figure 5–132

STEP 20

Use the LINE command to draw a line from a point tangent to the arc at "A" to a point tangent to the circle at "B," as shown in Figure 5–133. Use the OSNAP-Tangent option to accomplish this.

 Command: **L** *(For LINE)*

Specify first point: **Tan**
to *(Select the arc at "A")*

Specify next point or [Undo]: **Tan**
to *(Select the arc at "B")*

Specify next point or [Undo]: *(Press ENTER to exit this command)*

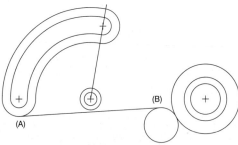

Figure 5–133

STEP 21

Use the TRIM command, select the dashed line and dashed circle illustrated in Figure 5–134 as cutting edges, and trim the circle at "A."

 Command: **TR** *(For TRIM)*

Current settings: Projection=UCS
    Edge=None
Select cutting edges ...
Select objects: *(Select the dashed line and the dashed circle in Figure 5–134)*

Select objects: *(Press ENTER to continue)*
Select object to trim or shift-select to
    extend or [Project/Edge/Undo]: *(Select the circle at "A")*
Select object to trim or shift-select to
    extend or [Project/Edge/Undo]: *(Press ENTER to exit this command)*

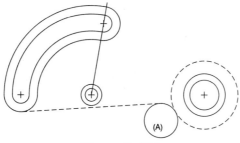

Figure 5–134

STEP 22

Use the CIRCLE-TTR command to draw a circle tangent to the arc at "A" and tangent to the circle at "B" with a radius of 51 as in Figure 5–135. When you are using the TTR option of the CIRCLE command, the OSNAP-Tangent option is automatically invoked.

 Command: **C** *(For CIRCLE)*

Specify center point for circle or [3P/2P/
    Ttr (tan tan radius)]: **T** *(For TTR)*

Specify point on object for first tangent
    of circle: *(Select the arc at "A")*
Specify point on object for second
    tangent of circle: *(Select the arc at "B")*
Specify radius of circle <12.75>: **51**

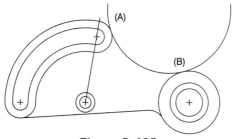

Figure 5–135

STEP 23

Use the TRIM command, select the dashed arc and dashed circle illustrated in Figure 5–136 as cutting edges, and trim the large 51-radius circle at "A."

 Command: **TR** *(For TRIM)*

Current settings: Projection=UCS Edge=None

Select cutting edges ...

Select objects: *(Select the dashed arc and dashed circle in Figure 5–136)*

Select objects: *(Press ENTER to continue)*

Select object to trim or shift-select to extend or [Project/Edge/Undo]: *(Select the circle at "A")*

Select object to trim or shift-select to extend or [Project/Edge/Undo]: *(Press ENTER to exit this command)*

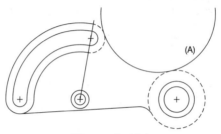

Figure 5–136

STEP 24

Use the TRIM command, select the two dashed arcs illustrated in Figure 5–137 as cutting edges, and trim the circle at "C." Use the ERASE command to delete all four points used to construct the circles.

 Command: **TR** *(For TRIM)*

Current settings: Projection=UCS Edge=None

Select cutting edges ...

Select objects: *(Select the dashed arcs "A" and "B" in Figure 5–137)*

Select objects: *(Press ENTER to continue)*

Select object to trim or shift-select to extend or [Project/Edge/Undo]: *(Select the circle at "C")*

Select object to trim or shift-select to extend or [Project/Edge/Undo]: *(Press ENTER to exit this command)*

Command: **E** *(For ERASE)*

Select objects: *(Select the four points in Figure 5–137)*

Select objects: *(Press ENTER to execute this command)*

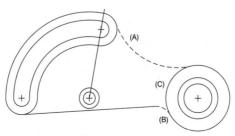

Figure 5–137

STEP 25

Your display should appear similar to the illustration in Figure 5–138. Notice the absence of the points that were erased in the previous step. Standard centerlines will be placed to mark the center of all circles in the next series of steps. Use the Layer Control box to set the current layer to "Center" in Figure 5–139. If you have not already done so, use the LTSCALE command to change the linetype scale from 1.00 to a new value of 15.00.

Command: **LTS** *(For LTSCALE)*
New scale factor <1.00>: **15.00**

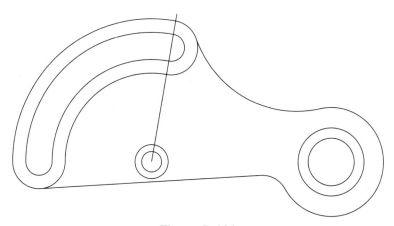

Figure 5–138

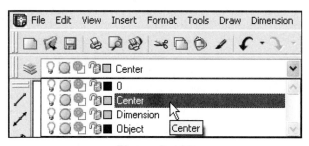

Figure 5–139

STEP 26

To place centerlines for all circles and arcs, two dimension commands or variables need to be changed. The dimension variable DIMSCALE is set to a value of 1. Because this is a metric drawing, the DIMSCALE value needs to be changed to 25.4. This will increase all variables by this value, which is necessary because you are drawing in metric units. Also, the DIMCEN variable needs to be changed from a value of 0.09 to -0.09. The negative value will extend the centerlines past the edge of the circle when you use the DIMCENTER command. After all centerlines are placed, erase the top of the centerline at "E." See Figure 5–140.

Command: **DIMSCALE**

Enter new value for DIMSCALE <1.00>: **25.4**

Command: **DIMCEN**

Enter new value for DIMCEN <0.09>: **-0.09**

 Command: **DCE** (For DIMCENTER)

Select arc or circle: (Select the arc at "A")

Repeat the preceding command to place a center marker at arcs "B," "C," and "D."

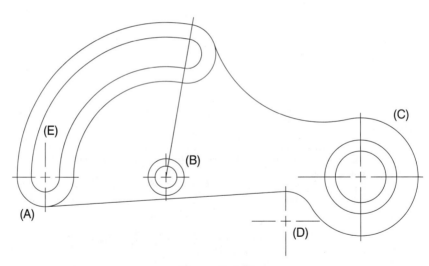 Command: **E** (For ERASE)

Select objects: (Select the top of the line at "E")

Select objects: (Press ENTER to execute this command)

Figure 5–140

STEP 27

Use the OFFSET command to offset the inside arc at "A" the distance 6.375 units. Indicate a point in the vicinity of "B" for the side to perform the offset as shown in Figure 5–141.

Select object to offset or <exit>: *(Select the arc at "A")*
Specify point on side to offset: *(Pick a point in the vicinity of "B")*
Select object to offset or <exit>: *(Press* ENTER *to exit this command)*

 Command: **O** *(For OFFSET)*
Specify offset distance or [Through] <12.75>: **6.375**

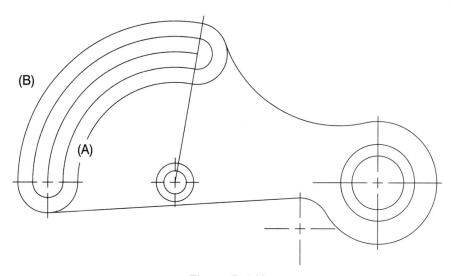

Figure 5–141

STEP 28

Change the middle arc at "A" and the line at "B" in Figure 5–142 to the "Center" layer. First select the lines. Then activate Layer Control in Figure 5–143 and pick the "Center" layer.

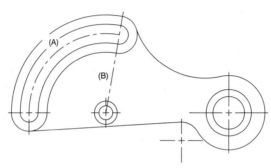

Figure 5–142

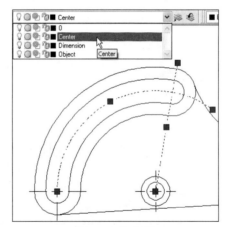

Figure 5–143

STEP 29

Use the LENGTHEN command to lengthen the centerline arc "A" in Figure 5–144. Then use the DIMCEN variable and change the default value from -0.09 to 0.09. This will change the center point to a "plus" for arc "B" without the centerlines extending beyond the arc when you use the DIMCEN-TER command.

Command: **LEN** (For LENGTHEN)
Select an object or [DElta/Percent/Total/
    DYnamic]: (Select the arc at "A")

Current length: 94.25, included angle: 100
Select an object or [DElta/Percent/Total/
   DYnamic]: **T** *(For Total)*
Specify total length or [Angle] <1.00)>:
   120
Select an object to change or [Undo]:
   (Select the arc at "A")

Select an object to change or [Undo]:
   (Press ENTER *to exit this command)*
Command: **DIMCEN**
Enter new value for DIMCEN <-0.09>:
   0.09

 Command: **DCE** *(For DIMCENTER)*
Select arc or circle: *(Select the arc at "B")*

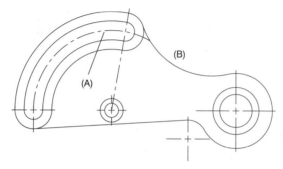

Figure 5–144

STEP 30

The finished object may be dimensioned as an optional step, as illustrated in Figure 5–145.

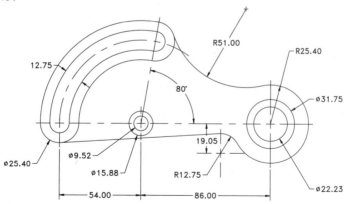

Figure 5–145

 Open the Exercise Manual PDF file for Chapter 5 on the accompanying CD for more tutorials and exercises.

 If you have the accompanying Exercise Manual, refer to Chapter 5 for more tutorials and exercises.

CHAPTER 6

Adding Text to Your Drawing

AutoCAD provides a robust set of text commands that allow you to place different text objects, edit those text objects, and even globally change the text height without affecting the justification. The heart of placing text in a drawing is through the MTEXT command. The chapter continues with a discussion on the uses of the DTEXT command, which stands for Dynamic Text or Single Line Text. You will also be given information on how to edit text created with the MTEXT and DTEXT commands. The SCALETEXT and JUSTIFYTEXT commands will be explained in detail, as will the ability to create custom text styles and assign different text fonts to these styles.

AUTOCAD TEXT COMMANDS

All text commands can be located easily by launching the Text toolbar illustrated in Figure 6–1.

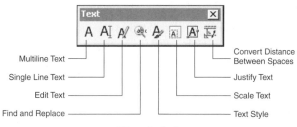

Multiline Text

Single Line Text

Edit Text

Find and Replace

Convert Distance Between Spaces

Justify Text

Scale Text

Text Style

Figure 6–1

Text can be added to your AutoCAD drawing file through two methods: Multiline Text (the MTEXT command) and Single Line Text (the DTEXT command). In addition to being able to select these commands from the Text toolbar, you can also find them by choosing the Text cascading menu from the Draw pull-down menu, as shown in Figure 6–2.

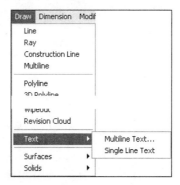

Figure 6–2

[A] THE MTEXT COMMAND

The MTEXT command allows for the placement of text in multiple lines. Entering the MTEXT command from the command prompt displays the following prompts:

[A] Command: **MT** *(For MTEXT)*
Current text style: "GENERAL NOTES" Text height: 0.2000
Specify first corner: *(Pick a point at "A" to identify one corner of the Mtext box)*
Specify opposite corner or [Height/Justify/Line spacing/Rotation/Style/Width]: *(Pick another corner forming a box).*

Picking a first corner displays a user-defined box with an arrow at the bottom similar to that shown in Figure 6–3. This box will define the area where the multiline text will be placed. If the text cannot fit on one line, it will wrap to the next line automatically. After you click on a second point marking the other corner of the insertion box, a Text Formatting toolbar will appear, along with the Multiline Text Editor. The parts of both dialog boxes are shown in Figure 6–4. One of the biggest advantages of this tool is the transparency associated with the Multiline Text Editor. As you begin entering text, you can still see your drawing in the background.

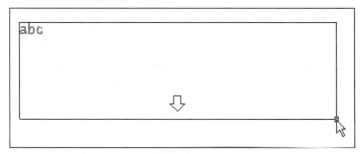

Figure 6–3

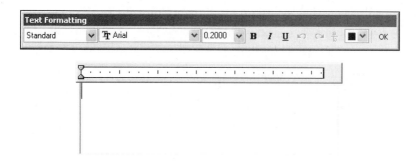

Figure 6–4

As you begin to type, your text will appear in the box shown in Figure 6–5. As the multiline text is entered, it will automatically wrap to the next line depending on the size of the initial bounding box. The current text style and font are displayed, in addition to the text height. You can make the text bold, italicized, or underlined by using the B, I, and U buttons present in the toolbar. When you have entered the text, click on the OK button to dismiss the toolbar and text editor and place the text in the drawing, as shown in Figure 6–6.

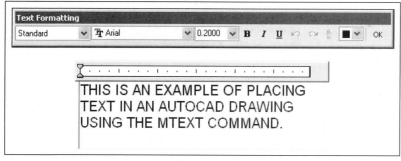

Figure 6–5

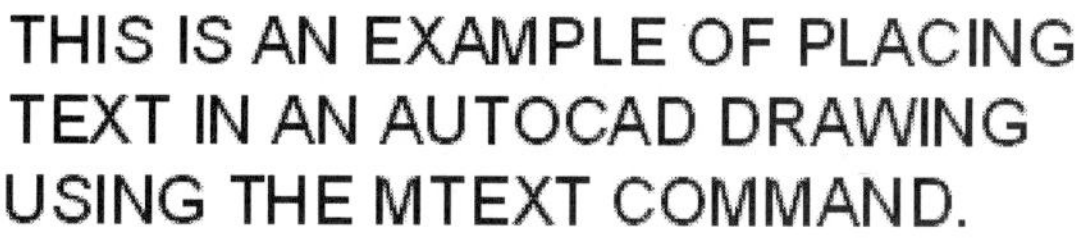

THIS IS AN EXAMPLE OF PLACING
TEXT IN AN AUTOCAD DRAWING
USING THE MTEXT COMMAND.

Figure 6–6

USING SPECIAL TEXT CHARACTERS

Special text characters and symbols are available for you to add to your text. While your cursor is positioned inside of the Multiline Text Editor, a tooltip appears, as shown in Figure 6–7, alerting you to edit the text or right-click to display a menu. Right-clicking inside of this area displays the special menu illustrated in Figure 6–8, designed exclusively for multiline text. Regarding special symbols—a degrees, plus/minus, and diameter symbol are available immediately for your use.

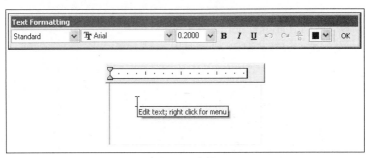

Figure 6–7

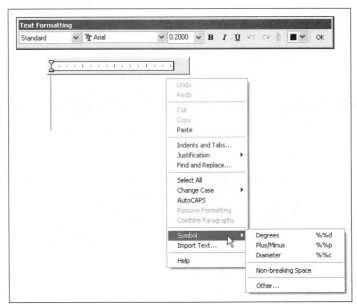

Figure 6–8

Clicking on Other… displays the Character Map dialog box shown in Figure 6–9. This dialog box contains an entire set of special characters based on the current font. Changing to a different font gives you a completely different set of character symbols. To insert a character from this dialog box, select the desired character and click Select. Then click on the Copy button. Once back inside of the Multiline Text Editor, right-click and pick Paste on the menu.

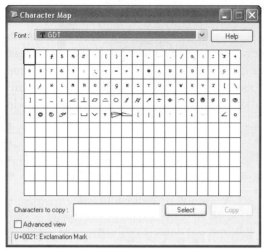

Figure 6–9

 Try It! – Create a new drawing file starting from scratch. Enter in the MTEXT command from the keyboard and construct a rectangular boundary to display the Text Formatting toolbar and Multiline Text Editor. Right-click inside of the Multiline Text Editor and pick Symbol, as shown in Figure 6–8. In the case of the diameter symbol, notice the appearance of the special code "%%c" in Figure 6–10. This is used to place the diameter symbol in the Mtext object. Clicking on the OK button exits the Multiline Text Editor and displays the text in Figure 6–11.

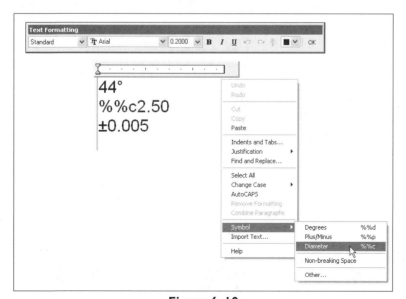

Figure 6–10

$$44°$$

$$\varnothing 2.50$$

$$\pm 0.005$$

Figure 6–11

FRACTIONAL TEXT

Fractional text can be reformatted through the Multiline Text Editor. Entering a space after the fraction value in Figure 6–12 displays the AutoStack Properties dialog box in Figure 6–13. Use this dialog box to enable the AutoStacking of fractions. You can also remove the leading space in between the whole number and the fraction. Dismissing this dialog box and completing the fractional text expression displays the results in Figure 6–14.

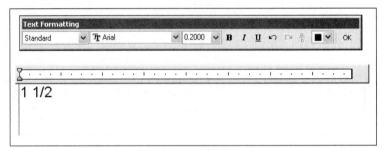

Figure 6–12

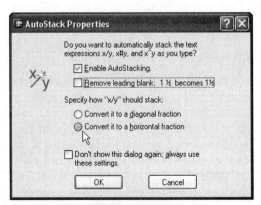

Figure 6–13

$1 \frac{1}{2}$ O.D. PIPE DIAMETER

Figure 6–14

CHANGING THE JUSTIFICATION OF MULTILINE TEXT

Right-clicking in the Multiline Text Editor and picking Justification displays information illustrated in Figure 6–15. Various justification modes are available to change the appearance of the text in your drawing. To change the justification of existing text, first highlight the text for the changes to take place. Figure 6–16 is an example of left-justified versus right-justified text placed with the MTEXT command.

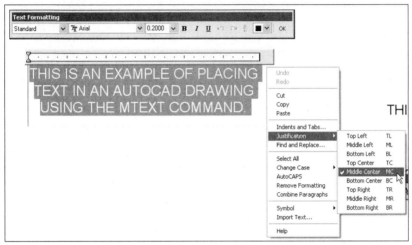

Figure 6–15

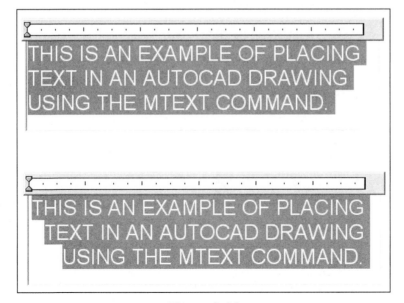

Figure 6–16

CONTROLLING THE LINE SPACING OF MULTILINE TEXT

Spacing in between lines of multiline text can be accomplished by activating the Line Spacing option in the MTEXT command prompt below.

[A] Command: **MT** *(For MTEXT)*
Current text style: "Standard" Text height: 0.2000
Specify first corner: *(Pick a first corner)*
Specify opposite corner or [Height/Justify/Line spacing/Rotation/Style/Width]: **L** *(For Line Spacing)*
Enter line spacing type [At least/Exactly] <At least>: *(Press ENTER to accept)*
Enter line spacing factor or distance <4x>: **2x**
Specify opposite corner or [Height/Justify/Line spacing/Rotation/Style/Width]: *(Pick a second corner)*

In this example, a line spacing of 2x will be applied. This will have the same effect as double spacing the lines. When you enter the desired text in the Multiline Text Editor in Figure 6–17, the text will not appear to have the new spacing. Clicking OK will place the text and display the new line spacing (see Figure 6–18).

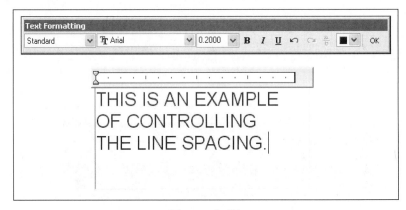

Figure 6–17

THIS IS AN EXAMPLE

OF CONTROLLING THE

LINE SPACING.

Figure 6–18

 Note: To change the line spacing of an existing mtext object, use the Properties Palette, which will be described in detail in chapter 7.

FINDING AND REPLACING MULTILINE TEXT

In the event that you need to replace text in a drawing, this can be accomplished easily through the Find and Replace feature of multiline text. In the example shown in Figure 6–19, the numbers "2004" need to be added after the word "AutoCAD". Right-click in the Multiline Text Editor and pick Find and Replace.

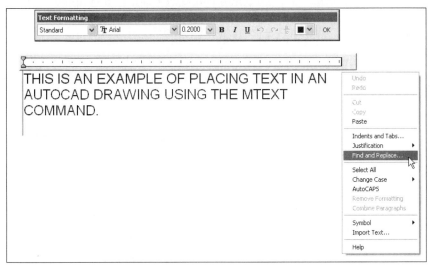

Figure 6–19

This activates the Replace dialog box shown in Figure 6–20. Enter the word you wish to find and the replacement word or words. Clicking on replace will search the Multiline Text Editor for any occurrences of this word. If a match is found, the word is replaced.

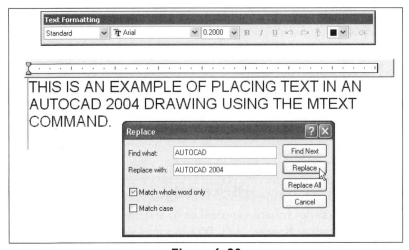

Figure 6–20

CREATING INDENTED TEXT

Multiline text can be indented in order to align text objects to form a list or table. A ruler illustrated in Figure 6–21 consisting of short and longer tick marks displays the current paragraph settings.

The tabs and indents that you set before you start to enter text apply to the whole multiline text object. If you want to set tabs and indents to individual paragraphs, click in a paragraph or select multiple paragraphs to apply the indentations.

Two sliders are available on the left side of the ruler to show the amount of indentation applied to the various mtext parts. In Figure 6–21, the top slider is used to indent the first line of a paragraph. A tooltip is available to remind you of this function.

The long tick marks on the ruler identify the default tab stops. When you click in the ruler to create your own tabs, a small L-shaped marker displays identifying it as a custom tab stop. In the event you need to delete one of these custom tab stops, pick and drag the marker off of the ruler.

Figure 6–21

In Figure 6–22, a bottom slider is also available. This slider controls the amount of indentation applied to the other lines of the paragraph. You can see the results of both sliders in the figures.

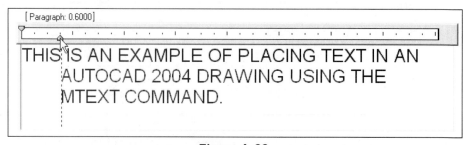

Figure 6–22

 Note: If you set tabs and indents prior to starting to enter multiline text, the tabs are applied to the whole multiline object.

You can also set indentations and tabs through the Indents and Tabs dialog box illustrated in Figure 6–23. Right-clicking in the Multiline Text Editor displays the shortcut menu. Pick on Indents and Tabs to activate the dialog box. This dialog box can also be displayed by right-clicking in the ruler in Figure 6–24 and picking Indents and Tabs.

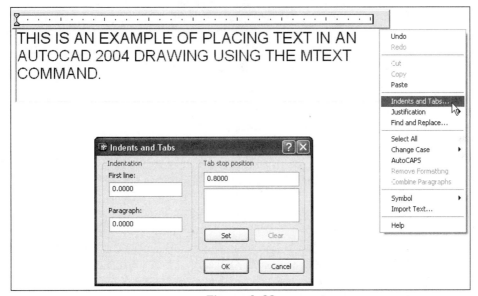

Figure 6–23

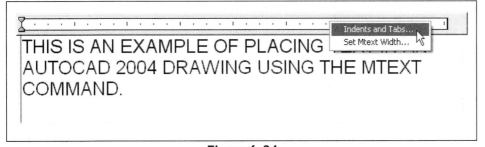

Figure 6–24

CHANGING THE MTEXT WIDTH

When you first construct a rectangle that defines the boundary of the mtext object, you are not locked into this boundary. Right-clicking inside of the tab marker area illustrated in Figure 6–25 activates the Set Mtext Width dialog box. The current value in the Width field is the current width of the rectangle used to construct the mtext object. Enter a larger or smaller value to expand or contract the mtext object.

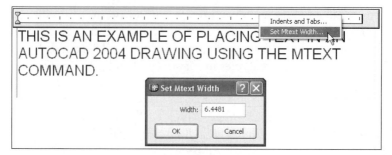

Figure 6–25

IMPORTING TEXT INTO YOUR DRAWING

If you have existing files created in a word processor, you can import these files directly into AutoCAD and have them converted into mtext objects. In Figure 6–26, right-click in the Multiline Text Editor to display the menu; then pick Import Text. An open file dialog box will display allowing you to pick the desired file to import. The Import Text feature of the Multiline Text supports two file types; those ending in a TXT extension and the other in the RTF (Rich Text Format) extension.

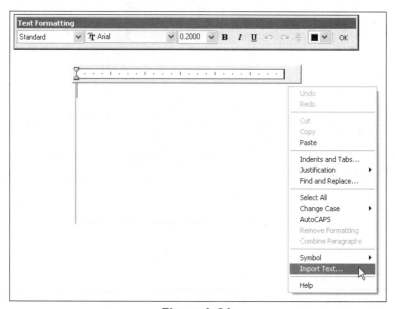

Figure 6–26

The finished result of this text importing operation is illustrated in Figure 6–27.

Note: When importing text from an RTF file, the original font type and text format are retained. Also, imported TXT or RTF files are limited to 32KB in size.

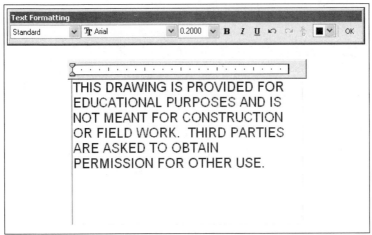

Figure 6–27

[AI] CREATING SINGLE LINE TEXT

Single line text is another method of adding text to your drawing. The actual command used to perform this operation is DTEXT, which stands for Dynamic Text mode, and allows you to place a single line of text in a drawing and view the text as you type it. Choose the Draw menu and then choose Single Line Text as shown in Figure 6–28. All of the following text examples are displayed in a font called RomanS even though the default text font is called Txt. Information about how to control text fonts will be discussed with the STYLE command.

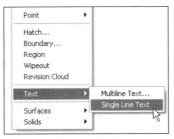

Figure 6–28

When using the DTEXT command, you are prompted to specify a start point, height, and rotation angle. You are then prompted to enter the actual text. As you do this, each letter displays on the screen. When you are finished with one line of text, pressing ENTER drops the Insert bar to the next line where you can enter more text. Again pressing ENTER drops the Insert bar down to yet another line of text (see Figure 6–29). Pressing ENTER at the "Enter text" prompt exits the DTEXT command and permanently adds the text to the database of the drawing (see Figure 6–30).

[AI] Command: **DT** *(For DTEXT)*
Current text style: "Standard" Text height: 0.2000
Specify start point of text or [Justify/Style]: *(Pick a point at "A")*
Specify height <0.2000>: **0.50**
Specify rotation angle of text <0>: *(Press* ENTER *to accept this default value)*
Enter text: **AutoCAD** *(After this text is entered, press* ENTER *to drop to the next line of text)*
Enter text: **2004** *(After this text is entered, press* ENTER *to drop to the next line of text)*
Enter text: *(Either add more text or press* ENTER *to exit this command and place the text)*

AutoCAD
2004

Figure 6–29

AutoCAD
2004

Figure 6–30

By default, the justification mode used by the DTEXT command is left justified. Study Figure 6–31 and the following command sequence to place the text string "MECHANICAL."

[AI] Command: **DT** *(For DTEXT)*
Current text style: "Standard" Text height: 0.2000
Specify start point of text or [Justify/Style]: *(Pick a point at "A")*
Specify height <0.2000>: **0.50**
Specify rotation angle of text <0>: *(Press* ENTER *to accept this default value)*
Enter text: **MECHANICAL**
Enter text: *(Press* ENTER *twice to place the text and exit this command)*

Figure 6–32 and the following command sequence demonstrate justifying text by a center point.

[AI] Command: **DT** *(For DTEXT)*
Current text style: "Standard" Text height: 0.2000
Specify start point of text or [Justify/Style]: **C** *(For Center)*
Specify center point of text: *(Pick a point at "A")*
Specify height <0.2000>: **0.50**
Specify rotation angle of text <0>: *(Press* ENTER *to accept this default value)*
Enter text: **CIVIL ENGINEERING**
Enter text: *(Press* ENTER *twice to place the text and exit this command)*

MECHANICAL
(A)

Figure 6–31

CIVIL ENGINEERING
(A)

Figure 6–32

Figure 6–33 and the following command sequence demonstrate justifying text by a middle point.

A͟ Command: **DT** *(For DTEXT)*
Current text style: "Standard" Text height: 0.2000
Specify start point of text or [Justify/Style]: **M** *(For Middle)*
Specify middle point of text: *(Pick a point at "A")*
Specify height <0.2000>: **0.50**
Specify rotation angle of text <0>: *(Press* ENTER *to accept this default value)*
Enter text: **CIVIL ENGINEERING**
Enter text: *(Press* ENTER *twice to place the text and exit this command)*

Figure 6–34 and the following command sequence demonstrate justifying text by aligning the text between two points. The text height is automatically scaled depending on the length of the points and the number of letters that make up the text.

A͟ Command: **DT** *(For DTEXT)*
Current text style: "Standard" Text height: 0.5000
Specify start point of text or [Justify/Style]: **A** *(For Aligned)*
Specify first endpoint of text baseline: *(Pick a point at "A")*
Specify second endpoint of text baseline: *(Pick a point at "B")*
Enter text: **MECHANICAL**
Enter text: *(Press* ENTER *twice to place the text and exit this command)*

CIVIL ENGINEERING
(A)

Figure 6–33

MECHANICAL
(A)   (B)

Figure 6–34

Figure 6–35 and the following command sequence demonstrate justifying text by fitting the text in between two points and specifying the text height. Notice how the text appears compressed due to the large text height and short distance of the text line.

A͟ Command: **DT** *(For DTEXT)*
Current text style: "Standard" Text height: 0.2000
Specify start point of text or [Justify/Style]: **F** *(For Fit)*
Specify first endpoint of text baseline: *(Pick a point at "A")*
Specify second endpoint of text baseline: *(Pick a point at "B")*
Specify height <0.2000>: **0.50**
Enter text: **MECHANICAL**
Enter text: *(Press* ENTER *twice to place the text and exit this command)*

Figure 6–36 and the following command sequence demonstrate justifying text by a point at the right.

[A] Command: **DT** *(For DTEXT)*
Current text style: "Standard" Text height: 0.2000
Specify start point of text or [Justify/Style]: **R** *(For Right)*
Specify right endpoint of text baseline: *(Pick a point at "A")*
Specify height <0.2000>: **0.50**
Specify rotation angle of text <0>: *(Press* ENTER *to accept this default value)*
Enter text: **MECHANICAL**
Enter text: *(Press* ENTER *twice to place the text and exit this command)*

MECHANICAL
(A)   (B)

Figure 6–35

MECHANICAL
(A)

Figure 6–36

Figure 6–37 and the following command sequence demonstrate rotating text along a user-specified angle.

[A] Command: **DT** *(For DTEXT)*
Current text style: "Standard" Text height: 0.2000
Specify start point of text or [Justify/Style]: *(Pick a point at "A")*
Specify height <0.2000>: **0.50**
Specify rotation angle of text <0>: **45**
Enter text: **ENGINEERING**
Enter text: **DESIGN**
Enter text: **GRAPHICS**
Enter text: *(Press* ENTER *twice to place the text and exit this command)*

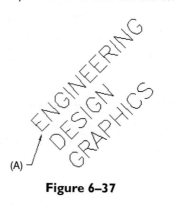
(A)

Figure 6–37

TEXT JUSTIFICATION MODES

Figure 6–38 illustrates a sample text item and the various locations by which the text can be justified. By default, when you place text using the DTEXT command, it is left justified. Enter one of the sets of letters shown in Figure 6–39 to justify text in a different location when prompted to "Specify start point of text or [Justify/Style]." If you cannot remember the letter designations, type "J" at the prompt and ENTER. This will cause the various letter designations to choose from to appear at the command line.

Figure 6–38

LEFT	Align Left (Default)
C	Center
M	Middle
R	Right
TL	Top/Left
TC	Top/Center
TR	Top/Right
ML	Middle/Left
MC	Middle/Center
MR	Middle/Right
BL	Bottom/Left
BC	Bottom/Center
BR	Bottom/Right
A	Align
F	Fit

Figure 6–39

SPECIAL TEXT CHARACTERS USED WITH DTEXT

Special text characters called control codes enable you to apply certain symbols and styles to text objects. All control codes begin with the double percent sign (%%) followed by the special character that invokes the symbol. These special text characters are explained in the following paragraphs.

UNDERSCORE (%%U)

Use the double percent signs followed by the letter "U" to underscore a particular text item. Figure 6–40 shows the word "MECHANICAL," which is underscored. When you are prompted to enter the text, type the following:

Enter text: **%%UMECHANICAL**

MECHANICAL

Figure 6–40

DIAMETER SYMBOL (%%C)

The double percent signs followed by the letter "C" create the diameter symbol shown in Figure 6–41. When you are prompted to enter text, the diameter symbol is displayed if you type the following:

Enter text: **%%C0.375**

∅0.375

Figure 6–41

PLUS/MINUS SYMBOL (%%P)

The double percent signs followed by the letter "P" create the plus/minus symbol shown in Figure 6–42. When you are prompted to enter text, the plus/minus symbol is displayed if you type the following:

Enter text: **%%P0.005**

±0.005

Figure 6–42

DEGREE SYMBOL (%%D)

The double percent signs followed by the letter "D" create the degree symbol shown in Figure 6–43. When you are prompted to enter text, the degree symbol is displayed if you type the following:

Enter text: **37%%D**

37°

Figure 6–43

OVERSCORE (%%O)

Similar to the underscore, the double percent signs followed by the letter "O" over-score a text object as shown in Figure 6–44. When you are prompted to enter text, the overscore is displayed if you type the following:

Enter text: **%%OMECHANICAL**

MECHANICAL

Figure 6–44

COMBINING SPECIAL TEXT CHARACTER MODES

Figure 6–45 shows how the control codes are used to toggle on or off the special text characters. Enter the following at the text prompt:

Enter text: **%%UTEMPERATURE%%U 29%%D F**

The first "%%U" toggles underscore mode on and the second "%%U" turns underscore off.

TEMPERATURE 29° F
Figure 6–45

ADDITIONAL DYNAMIC TEXT APPLICATIONS

You place a line of text using the DTEXT command and press ENTER. This drops the Insert bar down to the next line of text. This continues until you press ENTER at the "Enter text" prompt, which exits the command and places the text permanently in the drawing. You can also control the placement of the Insert bar by clicking a new location in response to the "Enter text" prompt. In Figure 6–46, various labels need to be placed in the pulley assembly. The first label, "BASE PLATE," is placed with the DTEXT command. After pressing ENTER, pick a new location at "B" and place the text "SUPPORT." Continue this process with the other labels. The Insert bar at "E" denotes the last label that needs to be placed. When you perform this operation, pressing ENTER one last time at the "Enter text" prompt places the text and exits the command. Follow the command sequence below for a better idea on how to perform this operation.

Try It! – Open the drawing file 06_Pulley Text. Use Figure 6–46 and the following command sequence for performing this task.

[A] Command: **DT** *(For DTEXT)*
Current text style: "STANDARD" Text height: 0.3000
Specify start point of text or [Justify/Style]: *(Pick a point at "A")*
Specify height <0.3000>: **0.25**
Specify rotation angle of text <0>: *(Press ENTER to accept this default value)*
Enter text: **BASE PLATE** *(After this text is entered, pick approximately at "B")*
Enter text: **SUPPORT** *(After this text is entered, pick approximately at "C")*
Enter text: **SHAFT** *(After this text is entered, pick approximately at "D")*
Enter text: **PULLEY** *(After this text is entered, pick approximately at "E")*
Enter text: *(You would enter **BUSHING** for this part and then press ENTER)*
Enter text: *(Press ENTER to exit this command)*

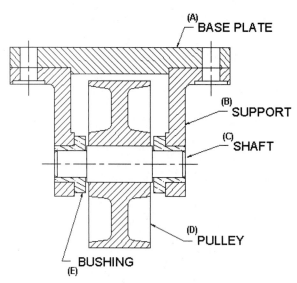

Figure 6–46

 EDITING TEXT

Text constructed with the MTEXT and DTEXT commands is easily modified with the DDEDIT command. Start this command by choosing Edit… which is found in the Text category of the Modify pull-down menu, shown in Figure 6–47. Selecting the text object in Figure 6–48 displays the Multiline Text Editor in Figure 6–49. This is the same dialog box as that used to initially create text through the MTEXT command. Use this dialog box to change the text height, font, color, and justification.

 Try It! – Open the drawing file 06_Edit Text1. Find the Edit text command in the Modify pull-down menu or use the following command sequence to display the Multiline Text Editor.

Command: **ED** *(For DDEDIT)*

Select an annotation object or [Undo]: *(Select the Mtext object and the Multiline Text Editor dialog box will appear. Perform any text editing task and click the OK button.)*
Select an annotation object or [Undo]: *(Pick another Mtext object to edit or press* ENTER *to exit this command)*

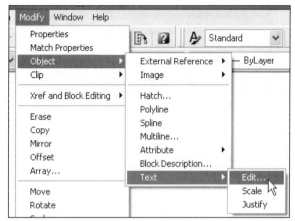

Figure 6–47

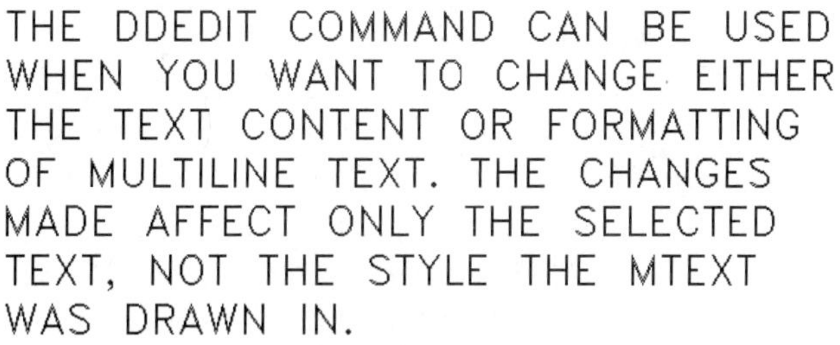

THE DDEDIT COMMAND CAN BE USED
WHEN YOU WANT TO CHANGE EITHER
THE TEXT CONTENT OR FORMATTING
OF MULTILINE TEXT. THE CHANGES
MADE AFFECT ONLY THE SELECTED
TEXT, NOT THE STYLE THE MTEXT
WAS DRAWN IN.

Figure 6–48

Examples of editing Mtext objects while inside the Multiline Text Editor dialog box are illustrated in Figure 6–49.

 Try It! – Open the drawing file 06_Edit Text2. Activate the Multiline Text Editor dialog box by entering the ED command and picking any text object. In this dialog box, the text is currently drawn in the RomanD font and needs to be changed to Arial. To accomplish this, first highlight all of the text. Next, change to the desired font. Since all text was highlighted, changing the font to Arial updates all text in Figure 6–50. Clicking on the OK button dismisses the dialog box and changes the text font in the drawing shown in Figure 6–51.

When editing Mtext objects, you can selectively edit only certain words that are part of a multiline text string.

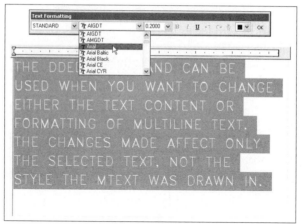

Figure 6–49

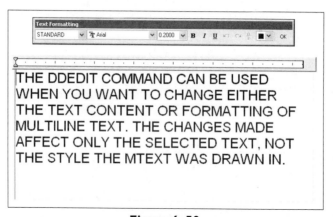

Figure 6–50

THE DDEDIT COMMAND CAN BE USED
WHEN YOU WANT TO CHANGE EITHER
THE TEXT CONTENT OR FORMATTING OF
MULTILINE TEXT. THE CHANGES MADE
AFFECT ONLY THE SELECTED TEXT, NOT
THE STYLE THE MTEXT WAS DRAWN IN.

Figure 6–51

Try It! – Open the drawing file 06_Edit Text3. Activate the Multiline Text Editor dialog box by entering the ED command and picking any text object. In Figure 6–52, the text "PLACING" is underscored; the text "AutoCAD" is changed to a Swiss font; the text "MTEXT" is increased to a text height of 0.30 units. When performing this type of operation, you only need to highlight the text object you want to change. Clicking on the OK button dismisses the dialog box and changes the individual text objects in the drawing, as in Figure 6–53.

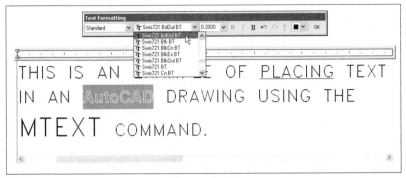

Figure 6–52

THIS IS AN EXAMPLE OF PLACING TEXT
IN AN AutoCAD DRAWING USING THE
MTEXT COMMAND.

Figure 6–53

Using the DDEDIT command on a text object created with the DTEXT command, as shown in Figure 6–54, displays the Edit Text dialog box shown in Figure 6–55. Use this to change text in the edit box provided. Font, justification, and text height are not supported in this dialog box.

MECHANICAL ENGINEERING

Figure 6–54

Figure 6–55

MODIFYING THE HEIGHT OF TEXT

The height of a text object can be easily changed through a command found in the Modify pull-down menu. Click on Object followed by Text and then Height. This command allows you to pick a text object and change its text height.

Try It! – Open the drawing file 06_Edit Text Height. All text in this drawing is considered a single mtext object. Use Figure 6–56 and the following command sequence for performing this task.

Command: **SCALETEXT**

Select objects: *(Pick the text object)*
Select objects: *(Press* ENTER *to continue)*
Enter a base point option for scaling
[Existing/Left/Center/Middle/Right/TL/TC/TR/ML/MC/MR/BL/BC/BR] <Existing>: *(Press* ENTER*)*
Specify new height or [Match object/Scale factor] <0.2000>: **0.30**

THE CUTTING PLANE LINE USED IN THE
CREATION OF A SECTION VIEW CONSISTS OF A
VERY THICK LINE SEGMENT WITH A SERIES OF
DASHES APPROXIMATELY 0.25" IN LENGTH.

Figure 6–56

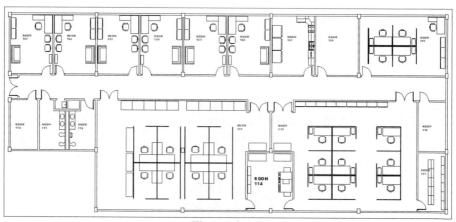

Figure 6–57

 Try It! – Open the drawing file 06_Scale Text illustrated in Figure 6–57. All offices in this facilities plan have been assigned room numbers. One of the room numbers (ROOM 114) has a text height of 12 inches, while all other room numbers are 8 inches in height. You could edit each individual room number until all match the height of ROOM 114; or you could use the SCALE-TEXT command by using the following command sequence:

 Command: **SCALETEXT**

Select objects: *(Pick all 16 mtext objects except for ROOM 114)*
Select objects: *(Press ENTER to continue)*
Enter a base point option for scaling
[Existing/Left/Center/Middle/Right/TL/TC/TR/ML/MC/MR/BL/BC/BR] <Existing>: *(Press ENTER to accept this value)*
Specify new height or [Match object/Scale factor] <8.0000>: **M** *(For Match object)*
Select a text object with the desired height: *(Pick the text identified by ROOM 114)*
Height=12.0000

The results are displayed in Figure 6–58. All the text was properly scaled without affecting the justification points.

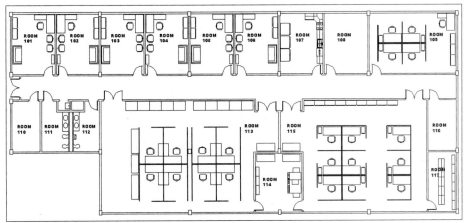

Figure 6–58

Note: The Quick Select dialog box could be used to select all mtext objects.

[A] MODIFYING THE JUSTIFICATION OF TEXT

Yet another text editing mode can be found in the Modify pull-down menu. Click on Object followed by Text and then Justify. This command will allow you to pick a text item and change its current justification. If you pick a multiline text object, all text in this mtext object will have its justification changed. In the case of regular text placed with the DTEXT command, you would have to select each individual line of text for it to be justified.

Try It! – Open the drawing file 06_Edit Text Justification. All text in this drawing is considered a single mtext object. Use Figure 6–59 and the following command sequence for performing this task.

[A] Command: **JUSTIFYTEXT**
Select objects: *(Pick the text object)*
Select objects: *(Press* ENTER *to continue)*
Enter a justification option
[Left/Align/Fit/Center/Middle/Right/TL/TC/TR/ML/MC/MR/BL/BC/BR] <Left>: **C**
   (For Center)

```
THE ALPHABET OF LINES       THE ALPHABET OF LINES
OBJECT LINE                       OBJECT LINE
CUTTING PLANE LINE            CUTTING PLANE LINE
HIDDEN LINE                       HIDDEN LINE
CENTER LINE                       CENTER LINE
DIMENSION LINE                  DIMENSION LINE
EXTENSION LINE                   EXTENSION LINE
SECTION LINE                      SECTION LINE
PHANTOM LINE                    PHANTOM LINE
BREAK LINE                          BREAK LINE
```

Figure 6–59

 Try It! – Open the drawing file 06_Justify Text illustrated in Figure 6–60. All of the text justification points in this facilities drawing need to be changed from left justified to top center justified. Use the following command prompt and illustration in Figure 6–60 to accomplish this task.

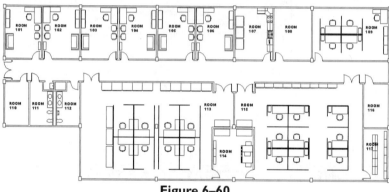

Figure 6–60

Command: **JUSTIFYTEXT**

Select objects: (Select all 17 text objects)
Select objects: (Press ENTER to continue)
Enter a justification option
[Left/Align/Fit/Center/Middle/Right/TL/TC/TR/ML/MC/MR/BL/BC/BR] <Left>: **TC** (For Top Center)

The results are illustrated in Figure 6–61. All text objects have been globally changed from left justified to top center justified.

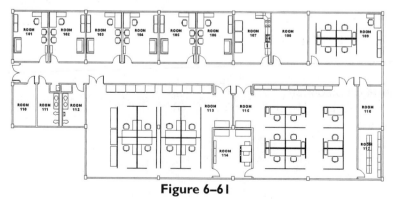

Figure 6–61

THE SPELL COMMAND

Issuing the SPELL command and selecting the multiline text object displays the Check Spelling dialog box shown in Figure 6–62, with the following components: the current dictionary and the Current word identified by the spell checker as being misspelled. The presence of the word does not necessarily mean that the word is misspelled, such as "AUTOCAD." The word is not part of the current dictionary.

The Suggestions area displays all possible alternatives to the word identified as being misspelled. The Ignore button allows you to skip the current word; this would be applicable especially in the case of acronyms such as CAD and GDT. Clicking on Ignore All skips all remaining words that match the current word. If the word "AUTOCAD" keeps coming up as a misspelled word, instead of constantly choosing the Ignore button, use Ignore All to skip the word "AUTOCAD" in any future instances. The Change button replaces the word in the Current word box with a word in the Suggestions box. The Add button adds the current word to the current dictionary. The Lookup button checks the spelling of a selected word found in the Suggestions box. The Change Dictionaries button allows you to change to a different dictionary containing other types of words. The Context area displays a phrase showing how the current word is being used in a sentence. Use this area to check proper sentence structure. In Figure 6–62, the word "AUTOCAD" was identified as being misspelled. Clicking on the Ignore button continues with the spell checking operation until completed. After undergoing spell checking, the Mtext object is displayed in Figure 6–63.

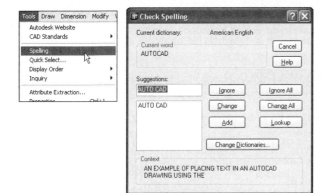

THIS IS AN EXAMPLE OF PLACING TEXT IN AN AUTOCAD DRAWING USING THE MTEXT COMMAND

Figure 6–62   **Figure 6–63**

Try It! – Open the drawing 06_Spell Check1 and perform a spell check operation on this mtext object.

Another feature of the spell checker is the ability to correct the spelling of special text called an attribute that is embedded into a symbol object called a block. Blocks are covered in chapter 16 and Attributes in chapter 17.

 Try It! – Open the drawing file 06_Spell Check Attrib. The partial floor plan displays in Figure 6–64. Using the SPELL command and picking all four groups of text will display the Spell Checking dialog box in Figure 6–65. Cycle through all text items until you have successfully examined all text items for their correctness in spelling.

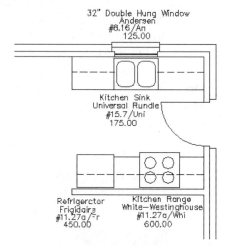

Figure 6–64

Figure 6–65

FINDING AND REPLACING TEXT

AutoCAD provides more capabilities for editing text with a Find and Replace tool. In the multiline text object in Figure 6–66, all words that contain "entity" must be replaced with the new word "object."

 Try It! – Open the drawing file 06_Text Find. Begin this process by starting the Find and Replace tool, which can be selected by right-clicking anywhere on your screen to display the menu shown in Figure 6–67. This activates the Find and Replace dialog box in Figure 6–68. While inside the dialog box, enter the text to find (entity) and the text to replace with (object). Clicking on the Find button performs a search through the entire drawing and displays the text objects in the Context edit box. AutoCAD finds the search text and highlights the item. Click on the Replace button to replace this text item. Click on the Replace All box to automatically replace all text with the new text "object."

Part of an entity can be removed using the Break command. You can break lines, circles, arcs, polylines, ellipses, splines, xlines, and rays. When breaking an entity, you can either select the entity at the first break point and then specify a second break point, or you can select the entire entity and then specify the two break points.

Figure 6–66

Figure 6–67

Clicking on the Options… button displays the Find and Replace Options dialog box in Figure 6–69. You can control the type of text AutoCAD will search for through the check boxes in the dialog box. By default, all text is searched for. To concentrate your search on all text and Mtext objects, be sure that the box adjacent to Text (Mtext, Dtext, Text) is the only one checked.

Once the text "entity" is replaced with "object," click the Close button of the Find and Replace text dialog box in Figure 6–68. This will display the text paragraph in Figure 6–70.

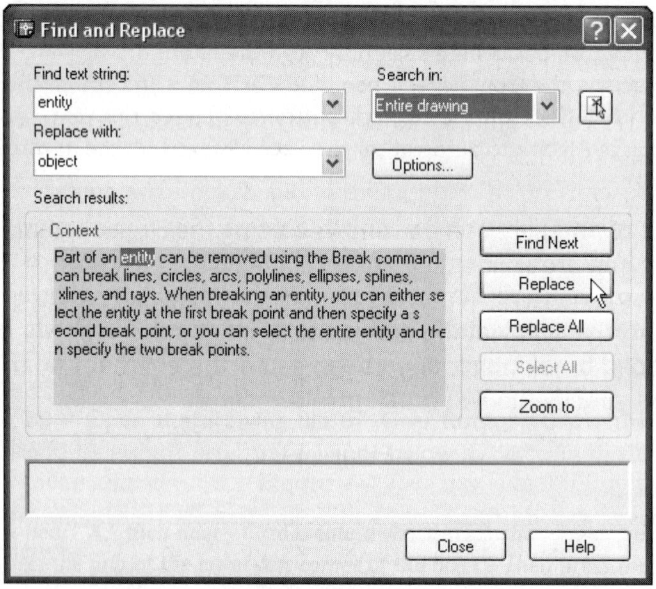

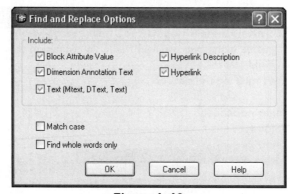

Figure 6–68

Figure 6–69

Part of an object can be removed using the Break command. You can break lines, circles, arcs, polylines, ellipses, splines, xlines, and rays. When breaking an object, you can either select the object at the first break point and then specify a second break point, or you can select the entire object and then specify the two break points.

Figure 6–70

The previous example of using the Find and Replace dialog box illustrated how you can find all occurrences of a word and replace these occurrences with a new word. Another very productive feature of this dialog box is the ability to find not only an item of text, but to zoom into your drawing where this text is located. Figure 6–71 illustrates a base map consisting of various buildings labeled from 100 through 940.

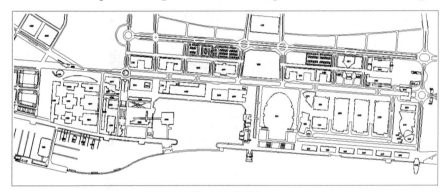

Figure 6–71

Activate the Find and Replace dialog box. Suppose you wish to locate Building 540. In the Find text string edit box shown in Figure 6–72, enter « 540 » and press ENTER. Doing so will add this number to the Context list (see Figure 6–72). Then press the Zoom to button.

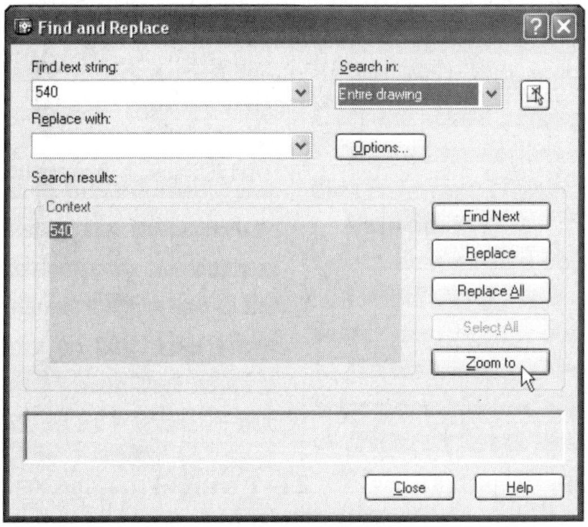

Figure 6–72

The results are illustrated in Figure 6–73. Your drawing will be magnified in relation to the text you are searching for, in this case, « 540 ». At first, however, you may not see the text. This is due to the fact that the Find and Replace dialog box positions itself directly on top of the text you are searching for. Simply move the dialog box away to view the text.

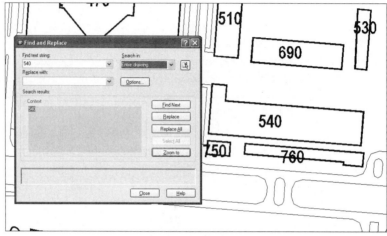

Figure 6–73

CREATING DIFFERENT TEXT STYLES

A text style is a collection of settings that are applied to text placed with the DTEXT or MTEXT command. These settings could include pre-setting the text height and font, in addition to providing special effects, such as an oblique angle for inclined text. Choose Text Style... from the Format pull-down menu or from the Text Style toolbar, shown in Figure 6–74, to activate the Text Style dialog box, shown in Figure 6–75.

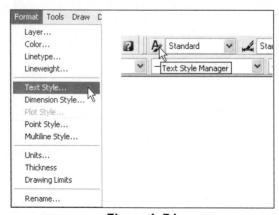

Figure 6–74

Use the Text Style dialog box in Figure 6–75 to create new text styles. As numerous styles are created, this dialog can also be used to make an existing style current in the drawing. By default, when you first begin a drawing, the current Style Name is STANDARD. It is considered good practice to create your own text style and not rely on STANDARD. Once a new style is created, a font name is matched with the style. Clicking in the edit box for Font Name displays a list of all text fonts supported by the operating system. These fonts have different extensions, such as SHX and TTF. TTF or TrueType Fonts are especially helpful in AutoCAD because these fonts display in the drawing in their true form. If the font is bold and filled-in, the font in the drawing file displays as bold and filled-in.

When a Font Name is selected, it displays in the Preview area located in the lower-right corner of the dialog box. The Effects area allows you to display the text upside down, backwards, or vertically. Other effects include a width factor, explained later, and the oblique angle for text displayed at a slant.

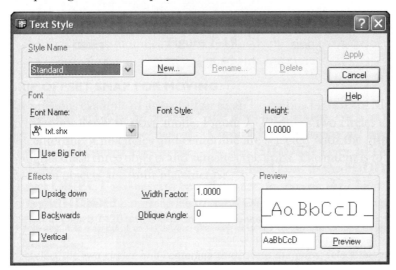

Figure 6–75

Clicking on the New... button of the Text Style dialog box displays the New Text Style dialog box shown in Figure 6–76. A new style is created called GENERAL NOTES. Clicking on the OK button returns you to the Text Style dialog box shown in Figure 6–77. Clicking in the Font Name edit box displays all supported fonts. Clicking on Romand.shx assigns the font to the style name GENERAL NOTES. Clicking on the Apply button saves the font to the database of the current drawing file.

In the Text Style dialog box shown in Figure 6–77, an edit box is present that deals with the text height. By default, the value is set to 0.0000 units. This allows the DTEXT or MTEXT command to control the text height. Entering a value for Height in the edit box places all text under this style at that height. When you use the DTEXT or MTEXT commands, the text height is already set to a value defined in the Text Style dialog box.

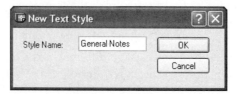

Figure 6–76

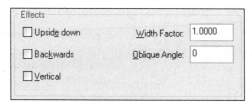

Figure 6–77

The Effects area of the Text Style dialog box, shown in Figure 6–78, allows for other settings to be applied to the text style being created. A few of the more unusual settings are "Upside down" and "Backwards," illustrated in Figure 6–79. Other more useful settings include the ability to have text displayed vertically, adding a Width Factor to condense or expand text, and an Obliquing Angle used to provide a slant or incline in the text (see Figure 6–79).

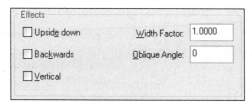

Effects	
☐ Upside down	Width Factor: 1.0000
☐ Backwards	Oblique Angle: 0
☐ Vertical	

Figure 6–78

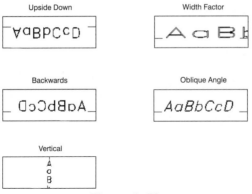

Figure 6–79

All of the images in Figure 6–79 were taken from the Preview area located in the lower-right corner of the Text Style dialog box. All changes to fonts and effects will be updated in this area. Changing the text height will not be reflected in the Preview area. By default, the letters "AaBbCcDd" preview. You can enter different letters and have them preview to have an idea of how these letters or numbers will appear based on the selected font. In Figure 6–80, the characters "1 Q X 9" are entered in the box provided. Clicking on the Preview button displays these characters in the Preview area.

Figure 6–80

THE TEXT STYLE CONTROL BOX

A special Text Style toolbar is available next to the Standard toolbar on your display screen. Use the Text Style Control box illustrated in Figure 6–81 to make an existing text style current. This control box allows you to change from one text style to another easily.

 Note: You can also highlight a text object and use the Text Style Control box to change the selected text to a different text style. This action is similar to changing an object from one layer to another through the Layer Control box.

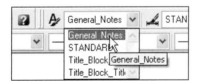

Figure 6–81

346

SUPPORTED TEXT FONTS

Two types of text fonts can be used inside AutoCAD drawings. One type of font has the extension .SHX. This is a native AutoCAD font, which has been compiled after the font was defined through a series of pen motions that created the shape of a letter. The other type of font has an extension .TTF, which stands for true type font. This font consists of high quality text. Also, a variety of true type fonts are available to add better contrast to your drawing. Figure 6–82 shows the names of a few of these native AutoCAD and true type fonts. The AutoCAD-defined fonts have a Caliper icon adjacent to the name of the font. The overlapping letters TT, also adjacent to the name of the font, identify the true type font.

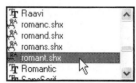

Figure 6–82

To have an idea of how the various fonts would appear using the same text, see Figure 6–83. The text "AutoCAD 2004" is displayed in the Txt, Swiss Block Outline Bold, City Blueprint, Roman Simplex, and Roman Duplex fonts. By default, a new drawing is automatically assigned to construct text in the Txt font. Use the Text Style dialog box to assign new text fonts. While the true type fonts appear inviting, care needs to be taken with using these fonts in your drawing. Having a large number of text items and notes in a true type font will cause your drawing to slow down considerably compared to native AutoCAD font use. In Figure 6–83, two very popular native AutoCAD fonts are displayed. The Roman Simplex font is similar to the standard block lettering style used in early engineering drawings. To make the text stand out, Roman Duplex is also available. This is similar to the Roman Simplex font, with the exception that two sets of lines are present to describe each letter and number.

AUTOCAD 2004

AUTOCAD 2004

AUTOCAD 2004

AUTOCAD 2004

AUTOCAD 2004
Figure 6–83

CHARACTER MAPPING FOR SYMBOL FONTS

A few of the native AutoCAD fonts actually consist of symbols, which can be displayed in your drawing through the DTEXT command. These symbols include mathematical, astronomical, music, and even mapping symbols. In order for these symbols to display, each symbol must be mapped to a particular letter of the alphabet. The image in Figure 6–84 displays all symbols and the corresponding letter used to bring up the symbol.

	A	B	C	D	E	F	G	H	I	J	K	L	M	N	O	P	Q	R	S	T	U	V	W	X	Y	Z
GREEKC	Α	Β	Χ	Δ	Ε	Φ	Γ	Η	Ι	ϑ	Κ	Λ	Μ	Ν	Ο	Π	Θ	Ρ	Σ	Τ	Υ	∇	Ω	Ξ	Ψ	Ζ
GREEKS	Α	Β	Χ	Δ	Ε	Φ	Γ	Η	Ι	ϑ	Κ	Λ	Μ	Ν	Ο	Π	Θ	Ρ	Σ	Τ	Υ	∇	Ω	Ξ	Ψ	Ζ
SYASTRO	⊙	☿	♀	⊕	♂	♃	♄		Ψ	♇	☾		✳		♈	♉	♊	♋		♌	♍	♎	♏	♐	♑	♒
SYMAP	○	□	△	◇	☆	+	×	∗	•	■	▲	◄	▼	►	★	↑	↓	→	×	♠	♣	♥	♦	○	✿	△
SYMATH	ℵ	′	|	‖	±	∓	×	·	÷	=	≠	≡	<	>	≤	≥	∝	~	√	⊂	∪	⊃	∩	∈	→	↑
SYMETEO	·	·	·	▲	■	⬝	∧	∩	∩	∪	∪	,	,	S	∼	∞	℞	ϱ	—	∕	|	╲	―	╱	∕
SYMUSIC	·	♪	♩	○	○	●	♯	♮	♭	━	-	×	♩	𝄞	𝄢	𝄡	·	,	⁝	⌐	∧	≂	▽			

	a	b	c	d	e	f	g	h	i	j	k	l	m	n	o	p	q	r	s	t	u	v	w	x	y	z
GREEKC	α	β	χ	δ	ε	φ	γ	η	ι	∂	κ	λ	μ	ν	ο	π	ϑ	ρ	σ	τ	υ	ε	ω	ξ	ψ	ζ
GREEKS	α	β	χ	δ	ε	φ	γ	η	ι	∂	κ	λ	μ	ν	ο	π	ϑ	ρ	σ	τ	υ	ε	ω	ξ	ψ	ζ
SYASTRO	✳	′	′	⊂	∪	⊃	∩	∈	→	↑	←	↓	∂	∇	ˆ	ˇ	`	˘	ℵ	§	†	‡	∃	ℒ	®	©
SYMAP	♁	♆	⚘	♀	ˮ	·	·	∘	∘	○	○	○	◯	◌	◐	♨	‖	⊥	∠	∴	♤	♡	◇	♣	♠	♣
SYMATH	←	↓	∂	∇	√	∫	∮	∞	§	†	‡	∃	Π	Σ	(	)	[	]	{	}	⟨	⟩	√	∫	≈	≅
SYMETEO	|	╲	╲	―	╱	|	╲	⌐	⌐	⌐	⌣	(	)	⌢	∠	∧	⊓	♌	∝	♉	♭	♉	♉	·		
SYMUSIC	·	♪	♩	○	○	●	♯	♮	♭	━	-	♩	♩	𝄞	𝄢	𝄐	⊙	☿	♀	⊕	♂	♃	♄		Ψ	♇

Figure 6–84

 Try It! – Create a new drawing file starting from scratch. Activate the Text Style dialog box in Figure 6–85 by entering the STYLE command in from the keyboard. Click the New button and create a new text style called Map Symbols. Next, assign the font called "Symap.SHX," which contains all mapping symbols. Notice the appearance of a few of these symbols in the Preview area located in the lower-right corner of the dialog box.

348

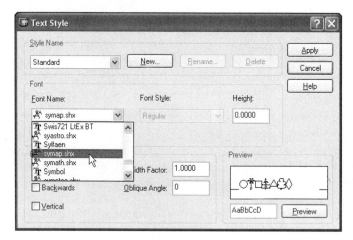

Figure 6–85

Next, use the chart in Figure 6–84 along with the DTEXT command to place a few of the symbols in the drawing. Figure 6–86 shows the effects of the DTEXT command when the letters "o," "p," "Q," "R," and "S" are entered. It is important to note that a different symbol is mapped to each uppercase and lowercase letter.

[AI] Command: **DT** *(For DTEXT)*
Current text style: "Map_Symbols" Text height: 0.5000
Specify start point of text or [Justify/Style]:
Specify height <0.2000>: **0.50**
Specify rotation angle of text <0>: *(Press ENTER to accept this default value)*
Enter text: **o**
Enter text: **p**
Enter text: **Q**
Enter text: **R**
Enter text: **S**
Enter text: *(Press ENTER twice to place the map symbols and exit this command)*

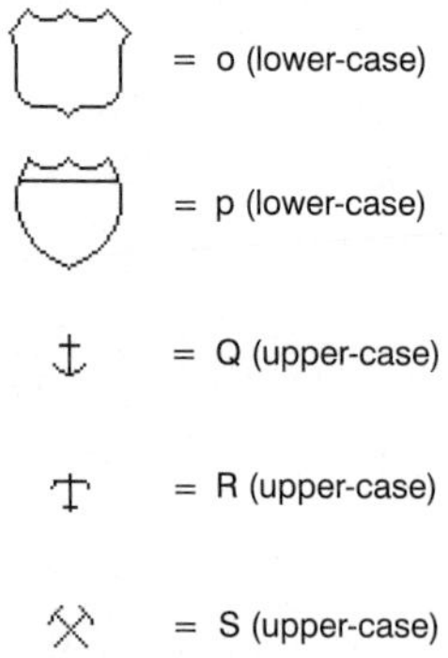

= o (lower-case)

= p (lower-case)

= Q (upper-case)

= R (upper-case)

= S (upper-case)

Figure 6–86

TUTORIAL EXERCISE: 06_PUMP_ASSEMBLY.DWG

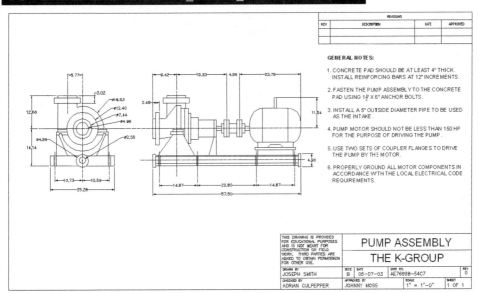

Figure 6–87

Purpose

This tutorial is designed to create numerous text styles and add different types of text objects to the title block illustrated in Figure 6–87.

System Settings

Since this drawing is already provided on the CD, open an existing drawing file called "06_Pump_Assembly." Follow the steps in this tutorial for creating a number of text styles and then placing text in the title block area. Check to see that the following Object Snap modes are already set: Endpoint, Extension, Intersection, Center.

Layers

Various layers have already been created for this drawing. Since this tutorial covers the topic of text, the current layer is "Text."

Suggested Commands:

Open the drawing called "06_Pump_Assembly." Create four text styles called "Company Drawing," "Title Block Text," "Disclaimer," and "General Notes." The text style "Company Drawing" will be applied to one part of the title block. "Title Block Text" will be applied to a majority of the title block where questions such as Drawn By, Date, and Scale are asked. A disclaimer will be imported into the title block on the "Disclaimer" text style. Finally a series of notes will be imported in the "General Notes" text style. Once these general notes are imported, a spell check operation will be performed on the notes.

Whenever possible, substitute the appropriate command alias in place of the full AutoCAD command in each tutorial step. For example, use "CP" for the COPY command, "L" for the LINE command, and so on. The complete listing of all command aliases is located in chapter 1, table 1–2.

STEP I

Opening the drawing file "06_Pump _Assembly" displays the image similar to Figure 6–88. The purpose of this tutorial is to fill in the title block area with a series of text and Mtext objects. Also, a series of general notes will be placed to the right of the pump assembly. Before you place the text, four text styles will be created to assist with the text creation. The four text styles are also identified in Figure 6–88. The text style "Company Drawing" will be used to place the name of the company and drawing title. Information in the form of scale, date, and who performed the drawing will be handled by the text style "Title Block

Text." A disclaimer will be imported from an existing .TXT file available on the CD; this will be accomplished in the "Disclaimer" text style. A listing of six general notes is also available on the CD. It is in .RTF format, originally created in Microsoft Word. It will be imported into the drawing and placed in the "General Notes" text style.

Since you will be locating various justification points for the placement of text, turn OSNAP OFF by single-clicking on OSNAP in the Status bar at the bottom of the display screen.

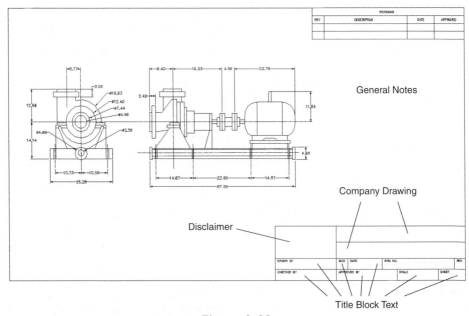

Figure 6–88

STEP 2

Before creating the first text style, first see what styles are already defined in the drawing by choosing Text Style... from the Format pull-down menu in Figure 6–89. This displays the Text Style dialog box in Figure 6–90. Click in the Style Name edit box to display the current text styles defined in the drawing. Every Auto-

CAD drawing contains the STANDARD text style. This is created by default and cannot be deleted. Also, the "Title_ Block_Headings" text style is present. This text style is used to create the headings in each title block box such as "Scale" and "Date."

Figure 6–89

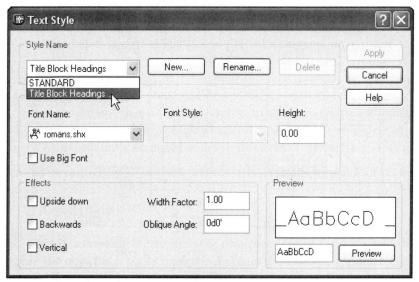

Figure 6–90

STEP 3

Create the first text style by clicking on the New... button in Figure 6–90. This will activate the New Text Style dialog box in Figure 6–91. For Style Name, enter "Company Drawing." When finished, click the OK button. This will take you back to the Text Style dialog box in Figure 6–92. In the Font area, change the name of the font to Arial. Notice the font appearing in the Preview area.

All company and drawing names will be drawn to a height of 0.25 units. Rather than enter the text height in the DTEXT command, change the text height in the Height edit box from 0.00 to 0.25. This locks the text at 0.25 units whenever you place text in the "Company Drawing" text style. When finished, click the Apply button to complete the text style creation process.

Figure 6–91

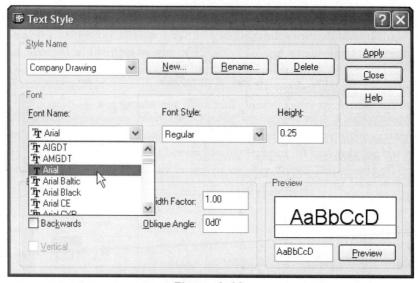

Figure 6–92

STEP 4

While in the Text Style dialog box, create the next text style by clicking on the New... button and entering "General Notes" as the new text style name. When finished, click the OK button. In the Font area of the Text Style dialog box, keep the name of the font as Arial. This font will appear in the Preview area. All general drawing notes will be drawn to a height of 0.12 units. Change the text height in the Height: edit box from 0.25 to 0.12. This locks the text at 0.12 units whenever you place text in the "General Notes" text style. Your dialog box should appear similar to Figure 6–93. When finished, click the Apply button to complete the text style creation process.

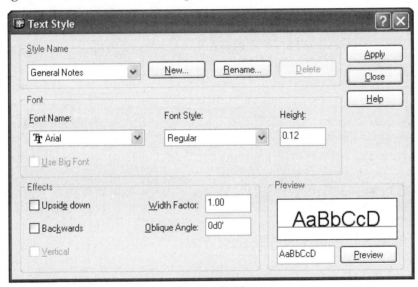

Figure 6–93

STEP 5

While in the Text Style dialog box, create the next text style by clicking on the New... button and entering "Title Block Text" as the new text style name. When finished, click the OK button. In the Font area of the Text Style dialog box, change the name of the font to RomanS. This font will appear in the Preview area. All title block text information such as date and scale will be drawn to a height of 0.10 units. Change the text height in the Height edit box from 0.12 to 0.10. This locks the text at 0.10 units whenever you place text in the "Title Block Text" text style. Your dialog box should appear similar to Figure 6–94. When finished, click the Apply button to complete the text style creation process.

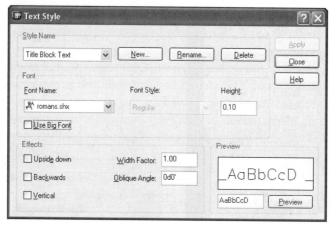

Figure 6–94

STEP 6

While in the Text Style dialog box, create the final text style by clicking on the New… button and entering "Disclaimer" as the new text style name. When finished, click the OK button. In the Font area of the Text Style dialog box, ensure that Romans is still the font. This font will appear in the Preview area. A disclaimer will be imported into an area of the title block and will be drawn to a height of 0.08 units. Change the text height in the Height edit box from 0.10 to 0.08. This locks the text at 0.08 units whenever you place text in the "Disclaimer" text style. Your dialog box should appear similar to Figure 6–95. When finished, click the Apply button to complete the text style creation process.

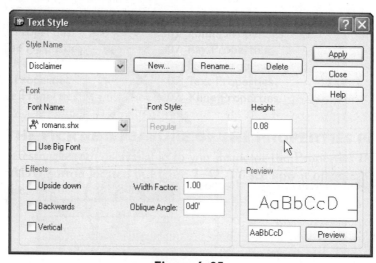

Figure 6–95

STEP 7

Double check that all text styles have been created by clicking in the dropdown box in the Style Name area of the Text Style dialog box. All text styles should be created and appear similar to Figure 6–96.

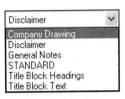

Figure 6–96

STEP 8

It is now time to begin placing the first series of text objects. We will start with the drawing and company titles. First, while in the Text Style dialog box, click on the "Company Drawing" text style in Figure 6–97. Click the Close button when finished to return to the drawing. Zoom in to the title block in Figure 6–98. To better assist in picking the text location, check to see that OSNAP is turned OFF by clicking in the appropriate button located in the Status bar at the bottom of the screen. Use the DTEXT command to place the two text objects in Figure 6–98. Both these text objects will be middle justified. The height of the text is already set through the current text style.

 Command: **DT** *(For DTEXT)*

Current text style: "Company Drawing"
    Text height: 0.25
Specify start point of text or [Justify/
    Style]: **M** *(For Middle)*
Specify middle point of text: *(Pick a point
    near the middle of the Title Block area
    represented by "A" in Figure 6–98)*

Specify rotation angle of text <0d0'>:
    (Press ENTER to accept this default value)
Enter text: **PUMP ASSEMBLY** *(After
    this text is entered, press ENTER)*
Enter text: *(Press ENTER to exit this command)*

 Command: **DT** *(For DTEXT)*

Current text style: "Company Drawing"
    Text height: 0.25
Specify start point of text or [Justify/
    Style]: **M** *(For Middle)*
Specify middle point of text: *(Pick a point
    near the middle of the Title Block area
    represented by "B" in Figure 6–98)*
Specify rotation angle of text <0d0'>:
    (Press ENTER to accept this default value)
Enter text: **THE K-GROUP** *(After this
    text is entered, press ENTER)*
Enter text: *(Press ENTER to exit this command)*

Figure 6–97

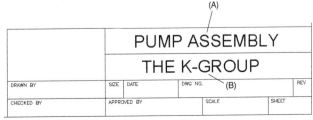

Figure 6–98

356

STEP 9

Open the Text Style dialog box and click on the "Title Block Text" text style in Figure 6–99. Click the Close button when finished to return to the drawing. While still zoomed in to the title block, use the DTEXT command to place the nine text objects in Figure 6–100. All of these text objects will be left justified. The height of the text is already set through the current text style.

[A] Command: **DT** *(For DTEXT)*

Current text style: "Title Block Text" Text height: 0.12

Specify start point of text or [Justify/ Style]: *(Pick a point approximately at "A")*

Specify rotation angle of text <0d0'>: *(Press ENTER to accept this default value)*

Enter text: **JOSEPH SMITH** *(After this text is entered, pick approximately at "B")*

Enter text: **ADRIAN CULPEPPER** *(After this text is entered, pick approximately at "C")*

Enter text: **B** *(After this text is entered, pick approximately at "D")*

Enter text: **JOHNNY MOSS** *(After this text is entered, pick approximately at "E")*

Enter text: **05-07-03** *(After this text is entered, pick approximately at "F")*

Enter text: **AE76998-54C7** *(After this text is entered, pick approximately at "G")*

Enter text: **1" = 1'-0"** *(After this text is entered, pick approximately at "H")*

Enter text: **0** *(After this text is entered, pick approximately at "I")*

Enter text: **1 OF 1** *(After this text is entered, press ENTER)*

Enter text: *(Press ENTER to exit this command)*

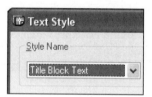

Figure 6–99

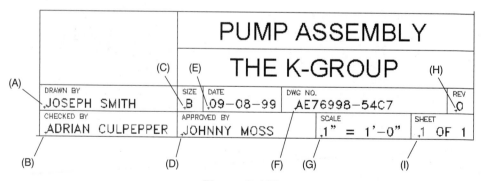

Figure 6–100

STEP 10

Open the Text Style dialog box and click on the "Disclaimer" text style in Figure 6–101. Click the Close button when finished to return to the drawing. While still zoomed in to the title block, use the MTEXT command to create a rectangle, as in Figure 6–102. This will be used to hold the multiline text.

A Command: **MT** *(For MTEXT)*
Current text style: "Disclaimer" Text height: 0.08
Specify first corner: *(Pick a point at "A")*

Specify opposite corner or [Height/ Justify/Line spacing/Rotation/Style/ Width]: *(Pick a point at "B")*

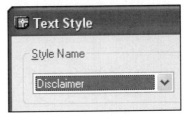

Figure 6–101

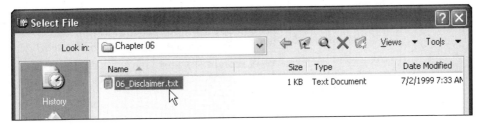

(A)		PUMP ASSEMBLY			
	(B)	THE K-GROUP			
DRAWN BY JOSEPH SMITH	SIZE B	DATE 09–08–99	DWG NO. AE76998–54C7		REV 0
CHECKED BY ADRIAN CULPEPPER	APPROVED BY JOHNNY MOSS		SCALE 1" = 1'–0"	SHEET 1 OF 1	

Figure 6–102

When the Multiline Text Editor dialog box appears, click on the Import Text... button. Then click on the location of your CD and pick the file 06_Disclaimer.txt, as in Figure 6–103. Since this information was already created in an application outside AutoCAD, it will be imported into

the Multiline Text Editor dialog box in Figure 6–104. The text height and font are already set through the current text style. Click the OK button to place the text in the title block.

The results of this operation are displayed in Figure 6–105.

Select File						? X
Look in:	Chapter 06				Views ▼	Tools ▼
	Name ▲		Size	Type	Date Modified	
History	06_Disclaimer.txt		1 KB	Text Document	7/2/1999 7:33 AM	

Figure 6–103

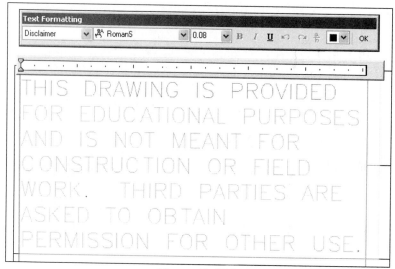

Figure 6–104

THIS DRAWING IS PROVIDED FOR EDUCATIONAL PURPOSES AND IS NOT MEANT FOR CONSTRUCTION OR FIELD WORK. THIRD PARTYS ARE ASKED TO OBTAIN PERMISSION FOR OTHER USE.	PUMP ASSEMBLY				
	THE K-GROUP				
DRAWN BY JOSEPH SMITH	SIZE B	DATE 09-08-99	DWG NO. AE76998-54C7		REV O
CHECKED BY ADRIAN CULPEPPER	APPROVED BY JOHNNY MOSS		SCALE 1" = 1'-0"	SHEET 1 OF 1	

Figure 6–105

STEP 11

Before placing the last text object, zoom to the extents of the drawing. Then click on the "General Notes" text style located in the Text Style Control Box in Figure 6–106. Use the MTEXT command to create a rectangle to the right of the Pump Assembly in Figure 6–107. This will be used to hold a series of general notes in multiline text format.

A Command: **MT** *(For MTEXT)*
Current text style: "General Notes" Text height: 0.12
Specify first corner: *(Pick a point at "A")*
Specify opposite corner or [Height/Justify/Line spacing/Rotation/Style/Width]: *(Pick a point at "B")*

Figure 6–106

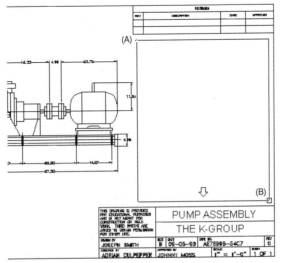

Figure 6–107

When the Multiline Text Editor dialog box appears, click on the Import Text... button. Then click on the location of your CD and select the file 06_General_Notes.rtf, as in Figure 6–108. This rtf file (Rich Text File) was created outside AutoCAD in Microsoft Word. Notice that this file is imported into the Multiline Text Editor dialog box, as in Figure 6–109. Click the OK button to place the text in the title block.

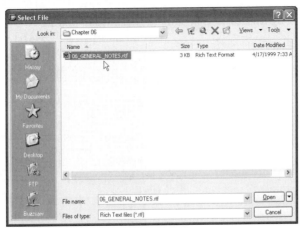

Figure 6–108

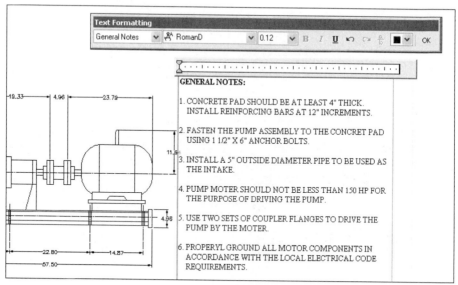

Figure 6–109

The results of this operation are displayed in Figure 6–110. However, the text appears to be displayed in a different font. This is due to the fact that when files are imported in RTF format, the original format of the text is kept. This means that the Times New Roman font was used even though the current text style uses the Arial font.

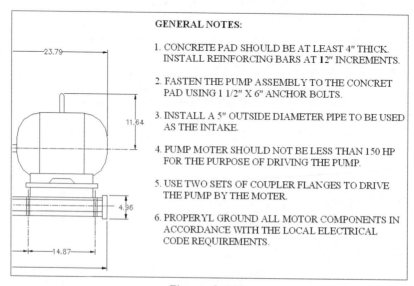

Figure 6–110

To remedy this, use the DDEDIT command and select one of the general notes. When the Multiline Text Editor dialog box displays, highlight all of the general notes. Change the font to Arial, as in Figure 6–111.

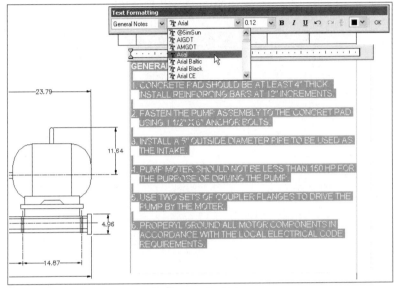

Figure 6–111

Clicking the OK button in the Multiline Text Editor dialog box displays the text that has been edited. Notice the text font is Arial (see Figure 6–112). Your results may be somewhat different depending on the size of the mtext box.

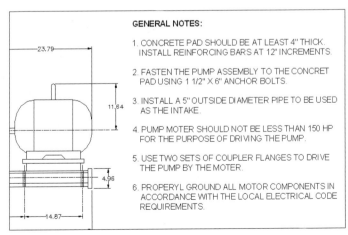

Figure 6–112

STEP 12

Use the SPELL command and click on the multiline text object that holds the six general notes. When the Check Spell-ing dialog box appears in Figure 6–113, make the corrections to the words CON-CRETE, MOTOR, and PROPERLY.

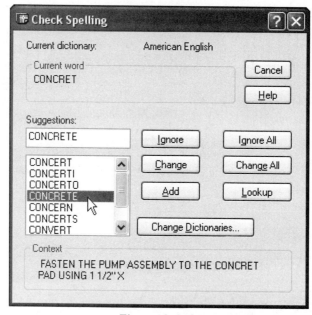

Figure 6–113

STEP 13

Use the DDEDIT command one more time on the general notes. Selecting the general notes displays the Multiline Text Editor dialog box in Figure 6–114. In the second general note, highlight the fraction "1/2" in Figure 6–114. Then click on the Stack/Unstack button.

This stacks the fraction in Figure 6–115.

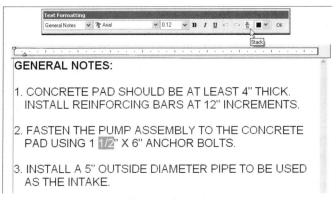

Text Formatting
General Notes
Arial
0.12
B I U
OK
Stack
GENERAL NOTES:
1. CONCRETE PAD SHOULD BE AT LEAST 4" THICK.
INSTALL REINFORCING BARS AT 12" INCREMENTS.
2. FASTEN THE PUMP ASSEMBLY TO THE CONCRETE
PAD USING 1 1/2" X 6" ANCHOR BOLTS.
3. INSTALL A 5" OUTSIDE DIAMETER PIPE TO BE USED
AS THE INTAKE.

Figure 6–114

Text Formatting
General Notes
Arial
0.12
B I U
OK
GENERAL NOTES:
1. CONCRETE PAD SHOULD BE AT LEAST 4" THICK.
INSTALL REINFORCING BARS AT 12" INCREMENTS.
2. FASTEN THE PUMP ASSEMBLY TO THE CONCRETE
PAD USING 1½" X 6" ANCHOR BOLTS.
3. INSTALL A 5" OUTSIDE DIAMETER PIPE TO BE USED
AS THE INTAKE.

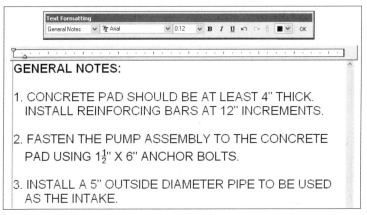

Figure 6–115

Clicking the OK button returns to the drawing editor with all changes made to the general notes in Figure 6–116.

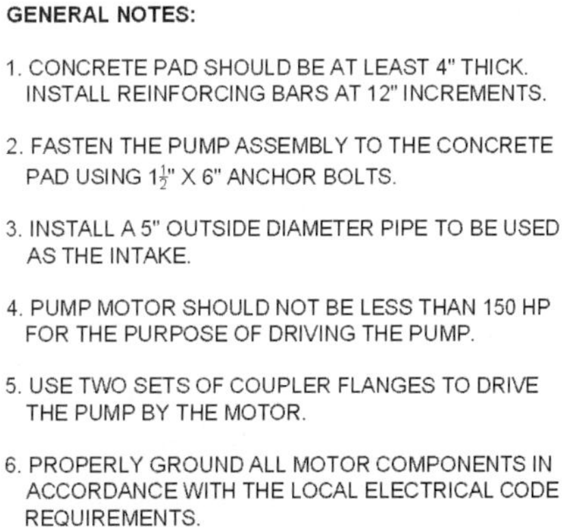

GENERAL NOTES:
1. CONCRETE PAD SHOULD BE AT LEAST 4" THICK.
INSTALL REINFORCING BARS AT 12" INCREMENTS.
2. FASTEN THE PUMP ASSEMBLY TO THE CONCRETE
PAD USING 1½" X 6" ANCHOR BOLTS.
3. INSTALL A 5" OUTSIDE DIAMETER PIPE TO BE USED
AS THE INTAKE.
4. PUMP MOTOR SHOULD NOT BE LESS THAN 150 HP
FOR THE PURPOSE OF DRIVING THE PUMP.
5. USE TWO SETS OF COUPLER FLANGES TO DRIVE
THE PUMP BY THE MOTOR.
6. PROPERLY GROUND ALL MOTOR COMPONENTS IN
ACCORDANCE WITH THE LOCAL ELECTRICAL CODE
REQUIREMENTS.

Figure 6–116

STEP 14

The completed title block is displayed in
Figure 6–117.

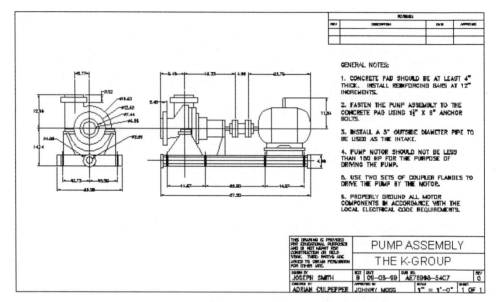

Figure 6–117

 Open the Exercise Manual PDF file for Chapter 6 on the accompanying CD for more
tutorials and exercises.

 If you have the accompanying Exercise Manual, refer to Chapter 6 for more tutorials and
exercises.

CHAPTER 7

Object Grips and Changing the Properties of Objects

This chapter begins with a discussion of what grips are and how they are used to edit portions of your drawing. Various Try It! exercises are available to practice using grips. This chapter continues by examining a number of methods used to modify objects. These methods are different from the editing commands learned in chapter 4. Sometimes you will want to change the properties of objects such as layer, color, and even linetype. This is easily accomplished through the Properties Palette. The Match Properties command will also be introduced in this chapter. This powerful command allows you to select a source object and have the properties of the source transferred to other objects that you pick.

USING OBJECT GRIPS

An alternate method of editing is to use object grips. A grip is a small box appearing at key object locations such as the endpoints and midpoints of lines and arcs or the center and quadrants of circles. Once grips are selected, the object may be stretched, moved, rotated, scaled, or mirrored. Grips are at times referred to as visual Object Snaps because your cursor automatically snaps to all grips displayed along an object.

 Try It! – Open the drawing file called 07_Grip Objects. While in the Command: prompt, click on each object type displayed in Figure 7–1 to activate the grips. Examine the grip locations on each object.

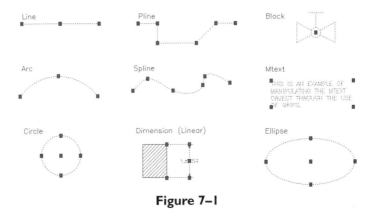

Figure 7–1

The Grips dialog box is available for changing settings, color, and grip size. Choose Options... from the Tools pull-down menu, as shown in Figure 7–2, which will display the main Options dialog box. Click on the Selection tab to display the grip settings (see Figure 7–3). By default, grips are enabled; a check in the Enable grips box means that grips will display when you select an object. Also by default, a grip is placed at the insertion point when a block is selected. Check the Enable grips within blocks box if you want grips to be displayed along with all individual objects that makeup the block. Color is applied to selected and unselected grips. Selecting the arrow in the Unselected grip color box or Selected grip color box displays a color dialog box used to change the color of selected or unselected grips. It is considered good practice to change the Unselected grip color to a light color if you are using a black screen background. When you hover your cursor over a grip, a green color appears below your cursor. This is called a Hover grip color. Use the Grip Size area to move a slider bar left to make the grip smaller or to the right to make the grip larger.

Figure 7–2

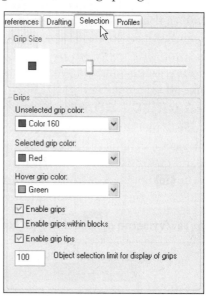

Figure 7–3

OBJECT GRIP MODES

Figure 7–4 shows the main types of grips. When an object is first selected with the grip pickbox located at the intersection of the crosshairs, the object highlights and the square grips are displayed in the default color of blue. The entire object is subject to the many grip edit commands. When one of the grips is selected, it turns red by default. Once a grip is selected, the following prompts appear in the command prompt area:

** STRETCH **
Specify stretch point or [Base point/Copy/Undo/eXit]: *(Press the Spacebar)*
** MOVE **
Specify move point or [Base point/Copy/Undo/eXit]: *(Press the Spacebar)*
** ROTATE **
Specify rotation angle or [Base point/Copy/Undo/Reference/eXit]: *(Press the Spacebar)*
** SCALE **
Specify scale factor or [Base point/Copy/Undo/Reference/eXit]: *(Press the Spacebar)*
** MIRROR **
Specify second point or [Base point/Copy/Undo/eXit]: *(Press the Spacebar to begin STRETCH mode again or enter X to exit grip mode)*

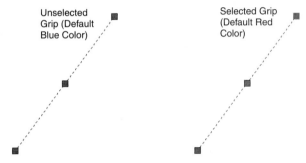

Figure 7–4

 Note: When you hover your cursor over an unselected grip, the grip turns green underneath your cursor.

To move from one edit command mode to another, press the Spacebar. Once an editing operation is completed, pressing ESC removes the highlight and removes the grips from the object. Figure 7–5 shows various examples of each editing mode.

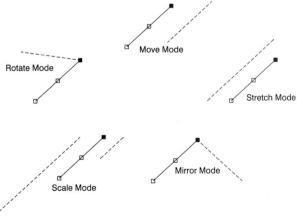

Figure 7–5

ACTIVATING THE GRIP SHORTCUT MENU

In Figure 7–6, a horizontal line has been selected and grips appear. The rightmost end-point grip has been selected. Rather than use the Spacebar to scroll through the various grip modes, click the right mouse button. Notice that a shortcut menu on grips appears. This provides an easier way of navigating from one grip mode to another.

 Try It! – Open the drawing file 07_Grip Shortcut. Click on any object, pick on a grip, and then press the right mouse button to activate the grip shortcut menu.

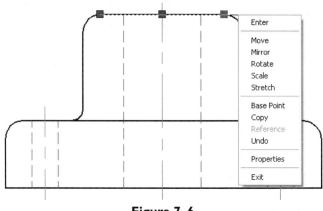

Figure 7–6

USING THE GRIP—STRETCH MODE

The STRETCH mode of grips operates similarly to the normal STRETCH command. Use STRETCH mode to move an object or group of objects and have the results mend themselves similar to Figure 7–7. The line segments "A" and "B" are both too long by two units. To decrease these line segments by two units, use the STRETCH mode by selecting lines "A," "B," and "C" with the grip pickbox at the command prompt. Next, while holding down SHIFT, select the grips "D," "E," and "F." This will turn the grips red and ready for the stretch operation. Release SHIFT and pick the grip at "E" again. The STRETCH mode appears in the command prompt area. Since the last selected grip is considered the base point, entering a polar coordinate value of @2.00<180 will stretch the three highlighted grip objects to the left a distance of two units. The Direct Distance mode could also be used with grips. To remove the object highlight and grips, press ESC at the command prompt.

 Try It! – Open the drawing file 07_Grip Stretch. Follow the illustration in Figure 7–7 and the prompt sequence below to perform this task.

Command: (Select the three dashed lines shown in Figure 7–7. Then, while holding down SHIFT, select the grips at "D," "E," and "F". Release the SHIFT key and pick the grip at "E" again)

STRETCH
Specify stretch point or [Base point/Copy/Undo/eXit]: **@2<180**
Command: (*Press* ESC *to remove the object highlight and grips*)

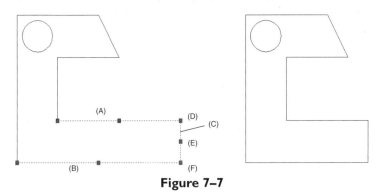

Figure 7–7

USING THE GRIP—SCALE MODE

Using the SCALE mode of object grips allows an object to be uniformly scaled in both the X and Y directions. This means that a circle, such as the one shown in Figure 7–8, cannot be converted to an ellipse through different X and Y values. As the grip is selected, any cursor movement will drag the scale factor until a point is marked where the object will be scaled to that factor. Figure 7–8 and the following prompt show the use of an absolute value to perform the scaling operation of half the circle's normal size.

 Try It! – Open the drawing file 07_Grip Scale. Follow the illustration in Figure 7–8 and the prompt sequence below to perform this task.

Command: (*Select the circle to enable grips, then select the grip at the center of the circle. Press the Spacebar until the SCALE mode appears at the bottom of the prompt line or press the right mouse button to activate the grip cursor menu to choose Scale*)

SCALE
Specify scale factor or [Base point/Copy/Undo/Reference/eXit]: **0.50**
Command: (*Press* ESC *to remove the object highlight and grips*)

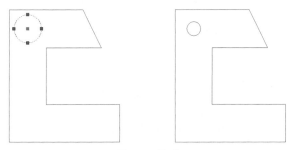

Figure 7–8

USING THE GRIP—MOVE/COPY MODE

The Multiple Copy option of the MOVE mode is demonstrated with the circle shown in Figure 7–9 being copied with polar coordinates at distances 2.50 and 5.00, both in the 270° direction. The Direct Distance mode could also be used to accomplish this task. This Multiple Copy option is actually disguised under the command options of object grips.

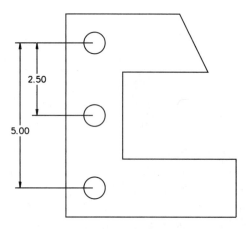

Figure 7–9

Select the circle, and then select the center grip of the circle. Use the Spacebar to scroll past the STRETCH mode to the MOVE mode. Issue a Copy within the MOVE mode to be placed in Multiple MOVE mode. See Figures 7–10 and 7–11.

 Try It! – Open the drawing file 07_Grip Copy. Follow the illustration in Figure 7–10 and the prompt sequence below to perform this task.

Command: (Select the circle to activate the grips at the center and quadrants; select the grip at the center of the circle. Then press the Spacebar until the MOVE mode appears at the bottom of the prompt line or press the right mouse button to activate the grip cursor menu to choose Move)

MOVE
Specify move point or [Base point/Copy/Undo/eXit]: **C** (For Copy)
MOVE (multiple)
Specify move point or [Base point/Copy/Undo/eXit]: **@2.50<270**
MOVE (multiple)
Specify move point or [Base point/Copy/Undo/eXit]: **@5.00<270**
MOVE (multiple)
Specify move point or [Base point/Copy/Undo/eXit]: **X** (For exit)
Command: (Press ESC to remove the object highlight and grips)

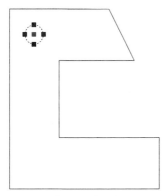

Figure 7–10

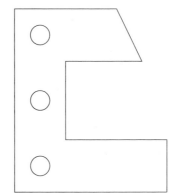

Figure 7–11

USING THE GRIP—MIRROR MODE

Use the grip MIRROR mode to flip an object along a mirror line similar to the one used in the regular MIRROR command. Follow the prompts for the MIRROR option if an object needs to be mirrored but the original does not need to be saved. This performs the mirror operation but does not produce a copy of the original. If the original object needs to be saved during the mirror operation, use the Copy option of MIRROR mode. This places you in multiple MIRROR mode. Locate a base point and a second point to perform the mirror operation. See Figure 7–12.

 Try It! – Open the drawing file 07_Grip Mirror. Follow the illustration in Figure 7–12 and the prompt sequence below to perform this task.

Command: *(Select the circle in Figure 7–12 to enable grips, then select the grip at the center of the circle. Press the Spacebar until the MIRROR mode appears at the bottom of the prompt line or press the right mouse button to activate the grip cursor menu to choose Mirror)*

MIRROR
Specify second point or [Base point/Copy/Undo/eXit]: **C** *(For Copy)*
MIRROR (multiple)
Specify second point or [Base point/Copy/Undo/eXit]: **B** *(For Base Point)*
Base point: **Mid**
of *(Pick the midpoint at "A")*
MIRROR (multiple)
Specify second point or [Base point/Copy/Undo/eXit]: **@1<90**
MIRROR (multiple)
Specify second point or [Base point/Copy/Undo/eXit]: **X** *(For Exit)*
Command: *(Press ESC to remove the object highlight and grips)*

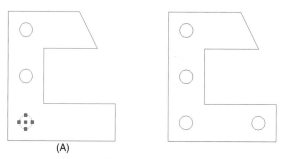

(A)

Figure 7–12

USING THE GRIP—ROTATE MODE

Numerous grips may be selected with window or crossing boxes. At the command prompt, pick a blank part of the screen; this should place you in Window/Crossing selection mode. Picking up or below and to the right of the previous point places you in Window selection mode; picking up or below and to the left of the previous point places you in Crossing selection mode. This method is used on all objects shown in Figure 7–13. Selecting the lower-left grip and using the Spacebar to advance to the ROTATE option allows all objects to be rotated at a defined angle in relation to the previously selected grip.

 Try It! – Open the drawing file 07_Grip Rotate. Follow the illustration in Figure 7–13 and the prompt sequence below to perform this task.

Command: *(Pick near "X," then near "Y" to create a window selection set and enable all grips in all objects. Select the grip at the lower-left corner of the object. Then press the Spacebar until the ROTATE mode appears at the bottom of the prompt line or click the right mouse button to activate the grip cursor menu to choose Rotate)*

ROTATE

Specify rotation angle or [Base point/Copy/Undo/Reference/eXit]: **30**

Command: *(Press ESC to remove the object highlight and grips)*

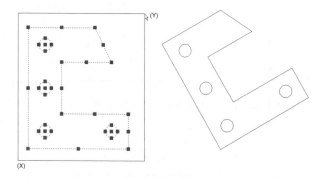

Figure 7–13

USING THE GRIP—MULTIPLE ROTATE MODE

One problem of the regular ROTATE command is that although an object can be rotated, the original location cannot be saved. You have to use points to mark the original location of an object before it is rotated. The ARRAY command has been used as a substitute to rotate and copy an object. Now with object grips, you may use ROTATE mode to rotate and copy an object without the use of reference points or the ARRAY command. Figure 7–14 is a line that needs to be rotated and copied at a 40° angle. With a positive angle, the direction of the rotation will be counterclockwise.

Selecting the line in Figure 7–14 enables grips located at the endpoints and midpoint of the line. Select the grip at "B" in Figure 7–15. This grip also locates the vertex of the required angle. Press the Spacebar until the ROTATE mode is reached. Enter Multiple ROTATE mode by entering C for Copy when you are prompted in the following command sequence. Finally, enter a rotation angle of 40 to produce a copy of the original line segment at a 40° angle in the counterclockwise direction. (See Figure 7–15.) The result is illustrated in Figure 7–16.

 Try It! – Open the drawing file 07_Grip Multiple Rotate. Follow the illustration in Figure 7–14 and the prompt sequence below to perform this task.

Command: *(Select line segment "A"; then select the grip at "B". Press the Spacebar until the ROTATE mode appears at the bottom of the prompt line or click the right mouse button to activate the grip cursor menu to choose Rotate)*
ROTATE
Specify rotation angle or [Base point/Copy/Undo/Reference/eXit]: **C** *(For Copy)*
ROTATE (multiple)
Specify rotation angle or [Base point/Copy/Undo/Reference/eXit]: **40**
Specify rotation angle or [Base point/Copy/Undo/Reference/eXit]: **X** *(For Exit)*
Command: *(Press* ESC *to remove the object highlight and grips)*

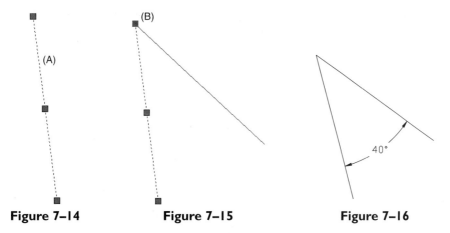

Figure 7–14　　　　**Figure 7–15**　　　　**Figure 7–16**

USING GRIP OFFSET SNAP FOR ROTATIONS

All Multiple Copy modes within grips may be operated in a snap location mode while you hold down SHIFT. Here's how it works. In Figure 7–17, the vertical centerline and circle are selected with the grip pickbox. The objects highlight and the grips appear. A multiple copy of the selected objects needs to be made at an angle of 45°. The grip ROTATE option is used in Multiple Copy mode.

 Try It! – Open the drawing file 07_Grip Offset Snap Rotate. Follow the illustration in Figure 7–17 and the prompt sequence below to perform this task.

Command: *(Select centerline segment "A" and circle "B"; and then select the grip near the center of circle "C". Press the Spacebar until the ROTATE mode appears at the bottom of the prompt line)*
ROTATE
Specify rotation angle or [Base point/Copy/Undo/Reference/eXit]: **C** *(For Copy)*
ROTATE (multiple)
Specify rotation angle or [Base point/Copy/Undo/Reference/eXit]: **B** *(For Base Point)*
Base point: **Cen**
of *(Select the circle at "C" to snap to the center of the circle)*
ROTATE (multiple)
Specify rotation angle or [Base point/Copy/Undo/Reference/eXit]: **45**

Rather than enter another angle to rotate and copy the same objects, you hold down SHIFT, which places you in offset snap location mode. Moving the cursor snaps the selected objects to the angle just centered, namely 45° (see Figure 7–18).

ROTATE (multiple)
Specify rotation angle or [Base point/Copy/Undo/Reference/eXit]: *(Hold down the SHIFT key and move the circle and centerline until it snaps to the next 45° position shown in Figure 7–18)*
ROTATE (multiple)
Specify rotation angle or [Base point/Copy/Undo/Reference/eXit]: **X** *(For Exit)*
Command: *(Press ESC to remove the object highlight and grips)*

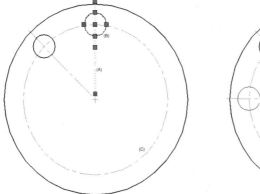

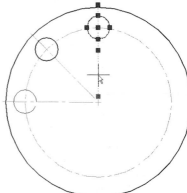

Figure 7–17   **Figure 7–18**

The Rotate-Copy-Snap Location mode could allow you to create Figure 7–19 without the aid of the ARRAY command. Since all angle values are 45°, continue holding down SHIFT to snap to the next 45° location, and mark a point to place the next group of selected objects.

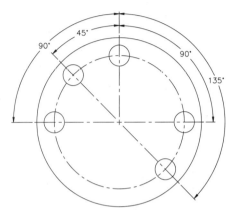

Figure 7–19

USING GRIP OFFSET SNAP FOR MOVING

As with the previous example of using Offset Snap Locations for ROTATE Mode, these same snap locations apply to MOVE mode. Figure 7–20 shows two circles along with a common centerline. The circles and centerline are selected with the grip pickbox, which highlights these three objects and activates the grips. The intent is to move and copy the selected objects at 2-unit increments.

 Try It! – Open the drawing file 07_Grip Offset Snap Move. Follow the illustration in Figure 7–20 and the prompt sequence below to perform this task.

Command: *(Select the two circles and centerline to activate the grips; select the grip at the midpoint of the centerline. Then press the Spacebar until the MOVE mode appears at the bottom of the prompt line or click the right mouse button to activate the grip cursor menu to choose Move)*
MOVE
Specify move point or [Base point/Copy/Undo/eXit]: **C** *(For Copy)*
MOVE (multiple)
Specify move point or [Base point/Copy/Undo/eXit]: **@2.00<270**

In Figure 7–21, instead of remembering the previous distance and entering it to create another copy of the circles and centerline, hold down SHIFT and move the cursor down to see the selected objects snap to the distance already specified.

MOVE (multiple)

Specify move point or [Base point/Copy/Undo/eXit]: *(Hold down the* SHIFT *key and move the cursor down to have the selected objects snap to another 2.00-unit distance)*
MOVE (multiple)

Specify move point or [Base point/Copy/Undo/eXit]: **X** *(To exit)*
Command: *(Press* ESC *to remove the object highlight and grips)*

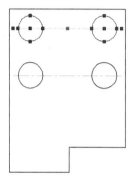

Figure 7–20

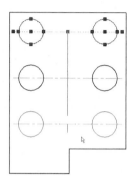

Figure 7–21

Figure 7–22 shows the completed hole layout, the result of using the Offset Snap Location method of object grips.

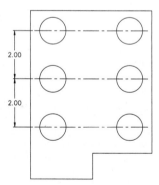

Figure 7–22

☒ MODIFYING THE PROPERTIES OF OBJECTS

At times, objects are drawn on the wrong layer, color, or even in the wrong linetype. The lengths of line segments are incorrect, or the radius values of circles and arcs are incorrect. Eliminating the need to erase these objects and reconstruct them to their correct specifications, a series of tools are available to modify the properties of these objects. Illustrated in Figure 7–23 are three line segments. One of the line segments has been pre-selected, as shown by the highlighted appearance and the presence of grips.

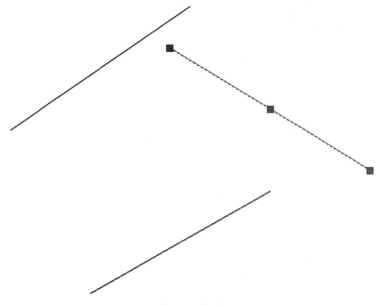

Figure 7–23

Clicking on the Properties button on the Standard toolbar in Figure 7–24 displays the Properties Palette in Figure 7–25. This dialog box displays information about the object already selected; in this case the information is about the line segment, which is identified at the top of the dialog box.

Figure 7–24

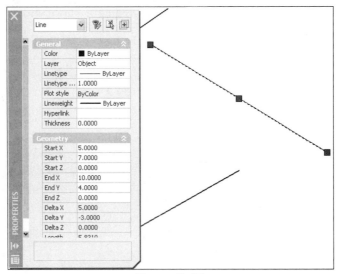

Figure 7–25

When changing a selected object to a different layer, click in the layer field in Figure 7–26 and the current layer displays (in this example, Layer Object). Clicking the down arrow displays all layers defined in the drawing. Clicking on one of these layers changes the selected object to a different layer.

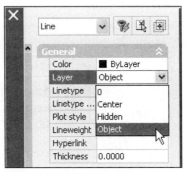

Figure 7–26

A number of options that control the Properties Palette are illustrated in Figure 7–27. You can elect to move, size, or close the Properties Palette from this menu. You can allow or prevent the Properties Palette from docking. Auto-hiding displays the Properties Palette when your cursor lies anywhere inside of the window. When your cursor moves off of the Properties Palette, the window collapses to display only the blue side strip. Checking Description controls the display of a description area located at the bottom of the palette.

Figure 7–27

The same three line segments are illustrated in Figure 7–28. This time, all three segments have been pre-selected, as shown by their highlighted appearance and the presence of grips.

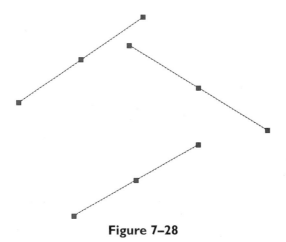

Figure 7–28

Clicking on the Properties button on the Standard toolbar displays the Properties Palette, as shown in Figure 7–29. At the top of the dialog box, the three lines are identified. You can change the color, layer, linetype, and other general properties of all three lines. However, you are unable to enter the Start and End X, Y, and Z values, and there is no length or angle information. Whenever you select more than one of the same object, you can change the general properties of the object but not individual values that deal with the object's geometry.

 Try It! – Open the drawing file 07_Line Properties. Pre-select all three lines and click on the Properties button. The Properties Palette will display in Figure 7–29 with general properties about all three lines.

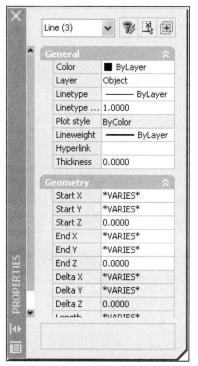

Figure 7–29

In Figure 7–30, an arc, circle, and line are pre-selected along with the display of grips. Activating the Properties Palette in Figure 7–31 displays a number of object types at the top of the dialog box. You can click which object or group of objects to modify. With "All (3)" highlighted, you can change the general properties, such as layer and linetype, but not any geometry settings.

Try It! – Open the drawing file 07_Clutch Properties. What if you need to increase the radius of the circle to 1.25 units? Click on the inner circle and then select the Properties button. Click on the Circle object type at the top of the Properties Palette. Since only one object was selected, the full complement of general and geometry settings is present for you to modify. Click in the Radius field and change the current value to 1.25 units. Pressing ENTER automatically updates the other geometry settings in addition to the actual object in the drawing (see Figure 7–32). When finished, dismiss this dialog box to return to the drawing.

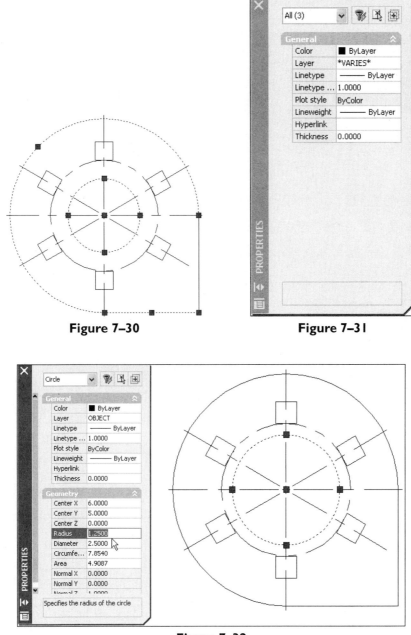

All (3)	

General
Color	■ ByLayer
Layer	*VARIES*
Linetype	——— ByLayer
Linetype ...	1.0000
Plot style	ByColor
Lineweight	——— ByLayer
Hyperlink	
Thickness	0.0000

Figure 7–30   **Figure 7–31**

Circle	

General
Color	■ ByLayer
Layer	OBJECT
Linetype	——— ByLayer
Linetype ...	1.0000
Plot style	ByColor
Lineweight	——— ByLayer
Hyperlink	
Thickness	0.0000

Geometry
Center X	6.0000
Center Y	5.0000
Center Z	0.0000
Radius	1.2500
Diameter	2.5000
Circumfe...	7.8540
Area	4.9087
Normal X	0.0000
Normal Y	0.0000
Normal Z	1.0000

Specifies the radius of the circle

Figure 7–32

You can also open the Properties Palette by choosing Properties from the Modify pull-down menu illustrated in Figure 7–33. This displays the Properties Palette shown in Figure 7–34. Notice that "No selection" is listed at the top of the dialog box, meaning that no objects have been selected to modify. Even though nothing is selected, you can still change the current color, linetype, and even make a layer current.

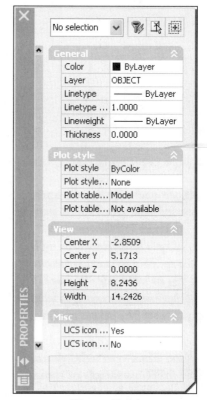

Figure 7–33   Figure 7–34

Clicking on the Quick Select button in Figure 7–34 displays the Quick Select dialog box in Figure 7–35. Use this dialog box to build a selection set to modify its object properties.

Try It! – Open the drawing file 07_Change Text Height. In Figure 7–36, the room numbers in the rectangular boxes are currently set to a height of 18". All text items need to be changed to a new height of 12". Rather than individually change each text item, the Quick Select dialog box will be used to assist with this operation.

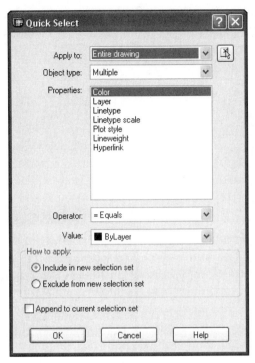

Figure 7–35

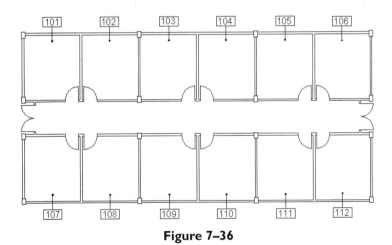

Figure 7–36

First, activate the Properties Palette and click on the Quick Select button in Figure 7–37.

Once inside of the Quick Select dialog box, click in the Object type window and select "Text" in Figure 7–38. Clicking the OK button will return you back to your drawing. Notice all text is highlighted.

Figure 7–37

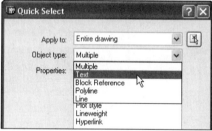

Figure 7–38

Notice the text height listing of 18.0000. Change this value to 12 in Figure 7–39. Then press the ENTER key followed by the ESC key and notice all text in Figure 7–40 has been changed to the new height of 12".

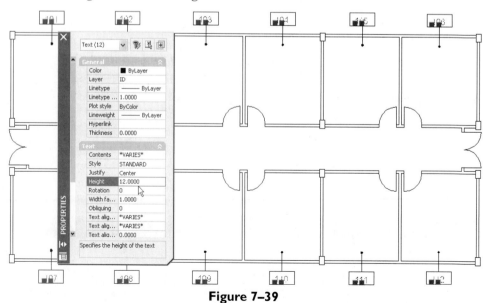

Figure 7–39

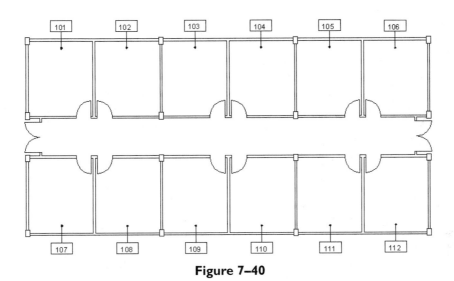

Figure 7–40

MODIFYING THE PROPERTIES OF INDIVIDUAL OBJECTS

This segment of the chapter begins a study of the Properties Palette and all of the properties displayed on the particular object selected. In each example, the object is already pre-selected. Grips are also present at key locations of each object. You can modify the following General properties of all objects through this dialog box: Color, Layer, Linetype, Linetype scale, Plot style, Lineweight, Hyperlink, and Thickness. The remainder of this segment will concentrate on the individual geometric properties that can be modified through the Properties Palette, which include:

- Name of the object being listed (Arc) at the top of the dialog box
- Starting and Ending X,Y, and Z values of the arc (grayed out)
- X,Y, and Z values of the center of the arc
- Radius of the arc
- Start and End angles of the arc
- Total angle (grayed out)
- Total arc length (grayed out)
- Length of the arc (grayed out)
- Normal X,Y, and Z values (grayed out)

 Try It! – Open the drawing file 07_Arc Properties. Pick the arc object and then click the Properties button. The Properties Palette for an arc provides you with the following information, which is displayed in Figure 7–41. Values in bold can be modified; items grayed out cannot.

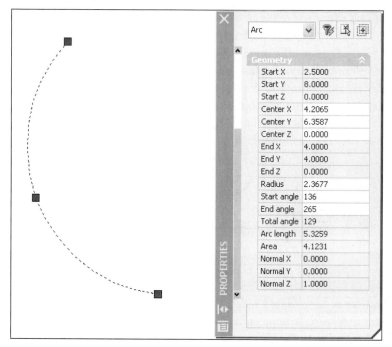

Figure 7–41

 Note: The following drawing files are provided on CD. Open each one individually, select the object, click on the Properties button, and observe the results displayed in the Properties Palette.

07_Block Properties
07_Circle Properties
07_Dim Properties
07_Ellipse Properties
07_Hatch Properties
07_Line Properties
07_Mline Properties

07_Mtext Properties
07_Polyline Properties
07_Points Properties
07_Ray Properties
07_Spline Properties
07_Text Properties
07_Xline Properties

USING THE PICKADD FEATURE OF THE PROPERTIES PALETTE

A very interesting feature is available to you inside of the Properties Palette; it is the PICKADD button located in Figure 7–42. To see how it operates, follow the next exercise.

Try It! – Open the drawing file 07_PickAdd. A number of object types ranging from lines to text to multiline text and polylines with dimensions are displayed. Activate the Properties Palette and pick on one object. Notice that it highlights, the grips appear, and information about the object is displayed in the Properties Palette. Suppose however that you want to list information about another object. You must first press ESC to deselect the original object. Now select another different object and this information is displayed in the Properties Palette. This time, select the PICKADD button in Figure 7–42. Notice the button changes in appearance. Instead of the "plus" sign, a "1" is displayed in Figure 7–43. Click on any object and notice the information displayed in the Properties Palette. Without pressing ESC, click on another object. The original object deselects and the new object highlights with its information displayed in the Properties Palette. Very simply, the PICKADD button eliminates the need for the ESC key when displaying information about individual objects.

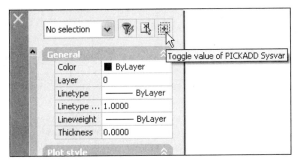

Figure 7–42

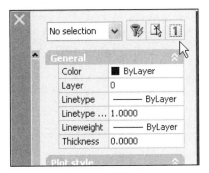

Figure 7–43

Note: Changing the PICKADD button will affect all drawings. Since this change is global, click the PICKADD button to set it back to the "plus" sign before continuing on.

388

USING THE LAYER CONTROL BOX TO MODIFY OBJECT PROPERTIES

If all you need to do is to change an object or a group of objects from one layer to another, the Layer Control box can easily perform this operation.

 Try It! – Open the drawing file 07_Change Layer. In Figure 7–44, select the arc and two line segments. Notice that the current layer is 0 in the Layer Control box. These objects need to be on the OBJECT layer.

Click in the Layer Control box to display all layers defined in the drawing. Then click on the desired layer for all highlighted objects (in this case, the OBJECT layer in Figure 7–45).

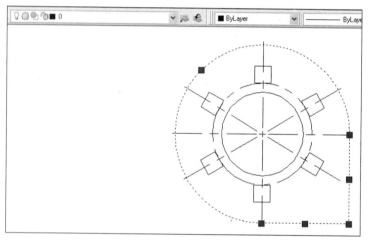

Figure 7–44

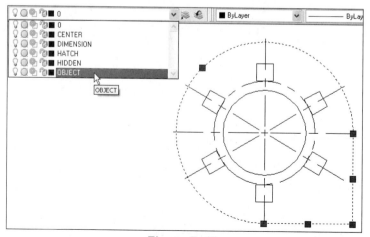

Figure 7–45

Notice that in Figure 7–46, with the objects still highlighted, the layer listed is OBJECT. This is one of the quickest and most productive ways of changing an object from one layer to another.

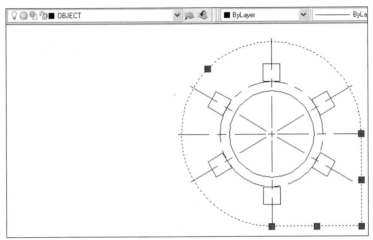

Figure 7–46

 Try It! – Open the drawing file 07_Mosaic. In Figure 7–47, this drawing consists of 5 different types of objects all drawn on Layer 0; circles, squares drawn as closed polylines, lines, text (the letter X) and text (the letter Y). The Quick Select dialog box will be used to select these object types individually.

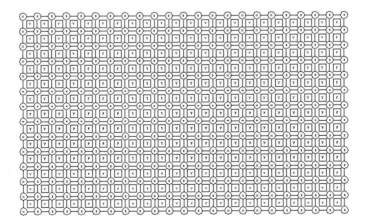

Figure 7–47

Once selected, you change the objects to the correct layers, which are also supplied with this drawing and identified in Figure 7–48.

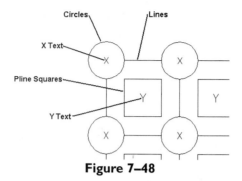

Figure 7–48

First, activate the Quick Select dialog box with the QSELECT command from the Tools pull-down menu shown in Figure 7–49. In the Object type box, select Circle, as you see in Figure 7–49, and click the OK button.

With all circles selected, click the Layer Control box and pick the Circles layer in Figure 7–50. All circles in the mosaic pattern should turn red. Press ESC to remove the object highlight and grips.

Figure 7–49

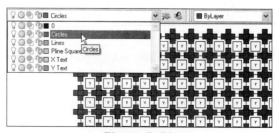

Figure 7–50

Next, you need to select all of the text with the letter "X" and change this text to the layer called X Text. Activate the Quick Select dialog box and make the following changes: Change the Object type to Text, set the Properties to Contents, and enter "X" as the value. Your display should be similar to Figure 7–51. Click the OK button to dismiss the Quick Select dialog box and select all text with the letter "X".

Change all letters to the X Text layer in the Layer Control box in Figure 7–52. Press ESC to remove the object highlight and grips. Follow the same procedures for changing all line segments to the Lines layer, all polylines to the Pline Squares layer, and all "Y" text to the Y Text layer.

Figure 7–51

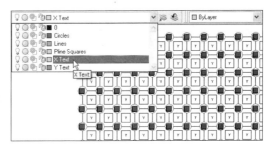

Figure 7–52

 Try It! – Open the drawing file 07_Qselect Duplex. Use the illustration in Figure 7–53 and the following information to change the items to their proper layers.

Change all lines to the Walls layer.
Change all text to the Room Labels layer
Change the blocks "Door" and "Louver" to the Door layer
Change the block "Window" to the Window layer
Change the blocks "Countertop", "Range", "Sink", and "Refrigerator" to the Countertop layer

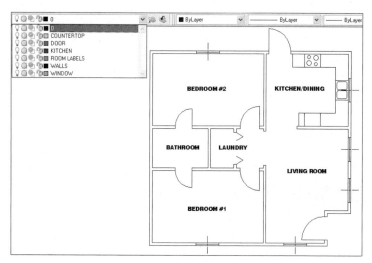

Figure 7–53

Tip: For the blocks, see the illustration in Figure 7–54 for supplying the correct information in the Quick Select dialog box; change the Object type to Block Reference. Change the Properties to Name. Change the Value to Louver (the name of the block). Then, after clicking the OK button and all louver doors are selected, change the highlighted objects to the Door layer. Follow this procedure for selecting the Window, Countertop, Range, Sink, Door, and Refrigerator blocks.

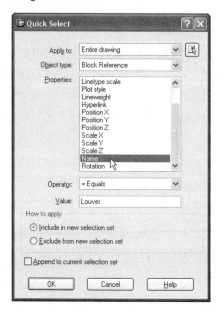

Figure 7–54

DOUBLE-CLICK EDIT ON ANY OBJECT

Double-clicking on any object provides you with a quick way of launching the Properties Palette or other related dialog boxes depending on the object type. For instance, double-clicking on a line segment launches the Properties Palette, which displays information about the line. Double clicking on a text object will display the Edit Text dialog box allowing you to enter or delete words. Whenever the Properties Palette is launched and you want to modify a different object, first press ESC to deselect the current object. You may also want to dismiss the Properties Palette.

Try It! – Open the drawing file 07_Double Click Edit. In Figure 7–55, double-click on the magenta center line and the Properties Palette launches with information about the line. Press ESC and dismiss the Properties Palette. Double-click on the word "BLOCK" to launch the Edit Text dialog box. Press ESC and dismiss the dialog box when finished. Double click on the sentence "THE OBJECT SHOWN ABOVE IS A WINDOW SYMBOL." to launch the Multiline Text Editor dialog box. Press ESC and dismiss the dialog box when finished. Double-click on the hatch pattern and the Hatch Edit dialog box will launch. Press ESC and dismiss this dialog box when finished. Continue by double-clicking on the circle, dimension, and rectangle and observe the type of dialog box launched through this method. Press ESC and dismiss the dialog boxes when finished with each operation.

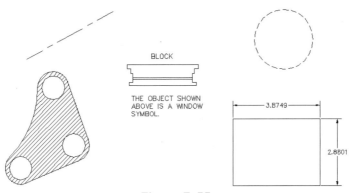

Figure 7–55

MATCHING THE PROPERTIES OF OBJECTS

At times objects are drawn on the wrong layers or the wrong color scheme is applied to a group of objects. Text objects are sometimes drawn with an incorrect text style. You have just seen how the Properties Palette and the Layer Control box provide quick ways to fix such problems. Yet another tool is available for changing the properties of objects—the MATCHPROP command. Choose this command from the Standard toolbar in Figure 7–56. You could also select this command from the Modify pull-down menu shown in Figure 7–57.

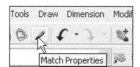

Figure 7–56

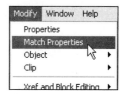

Figure 7–57

When you start the command, a source object is required. This source object transfers all of its current properties to other objects designated as "Destination Objects." In Figure 7–58, the flange requires the object lines located at "B," "C," "D," and "E" to be converted to hidden lines. Using the MATCHPROP command, select the existing hidden line "A" as the source object. Notice the appearance of the Match Properties icon. Select lines "B" through "E" as the destination objects using this icon.

 Try It! – Open the drawing file 07_Matchprop Flange. Use the illustration in Figure 7–58 and the command sequence below for performing this task.

Command: **MA** *(For MATCHPROP)*
Select source object: *(Select the hidden line at "A")*
Current active settings: Color Layer Ltype Ltscale Lineweight Thickness PlotStyle Text Dim Hatch Polyline Viewport
Select destination object(s) or [Settings]: *(Select line "B")*
Select destination object(s) or [Settings]: *(Select line "C")*
Select destination object(s) or [Settings]: *(Select line "D")*
Select destination object(s) or [Settings]: *(Select line "E")*
Select destination object(s) or [Settings]: *(Press ENTER to exit this command)*

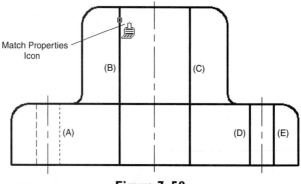

Figure 7–58

The results appear in the flange illustrated in Figure 7–59, where the continuous object lines were converted to hidden lines. Not only did the linetype property get transferred, but also the color, layer, lineweight, and linetype scale information.

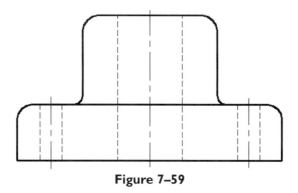

Figure 7–59

To get a better idea of what object properties are affected by the MATCHPROP command, re-enter the command, pick a source object, and instead of picking a destination object immediately, enter s for settings. This will display the Property Settings dialog box in Figure 7–60.

 Command: **MA** *(For MATCHPROP)*

Select source object: *(Select the hidden line at "A" in Figure 7–58)*
Current active settings: Color Layer Ltype Ltscale Lineweight Thickness
PlotStyle Text Dim Hatch Polyline Viewport
Select destination object(s) or [Settings]: **S** *(For Settings; this displays the Property Settings dialog box in Figure 7–60)*

Any box with a check displayed in it will transfer that property from the source object to all destination objects. If you need to transfer only the layer information and not the color and linetype properties of the source object, remove the checks from the Color and Linetype properties before you select the destination objects, which prevents these properties from being transferred to any destination objects.

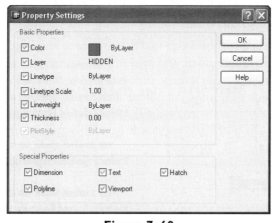

Figure 7–60

MATCHING DIMENSION PROPERTIES

The MATCHPROP command can control special properties of dimensions, text, hatch patterns and polylines. The Dimension Special Property will be featured next (see Figure 7–61).

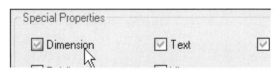

Figure 7–61

Try It! – Open the drawing file 07_Matchprop Dim. Figure 7–62 shows two blocks: the block assigned a dimension value of 46.6084 was dimensioned with the METRIC dimension style with the RomanD font applied. The block assigned a dimension value of 2.3872 was dimensioned with the STANDARD dimension style with the TXT font applied. Both blocks need to be dimensioned with the METRIC dimension style. Issue the MATCHPROP command and select the 46.6084 dimension as the source object and then select the 2.3872 dimension as the destination object.

Command: **MA** *(For MATCHPROP)*
Select source object: *(Select the dimension at "A" in Figure 7–62)*
Current active settings: Color Layer Ltype Ltscale Lineweight Thickness
PlotStyle Text Dim Hatch Polyline Viewport
Select destination object(s) or [Settings]: *(Select the dimension at "B")*
Select destination object(s) or [Settings]: *(Press ENTER to exit this command)*

The results are shown in Figure 7–63, with the METRIC dimension style applied to the STANDARD dimension style through the use of the MATCHPROP command. Because the text font was associated with the dimension style, it also changed in the destination object.

Dimension Styles

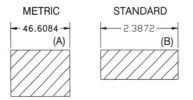

Figure 7–62

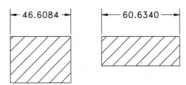

Figure 7–63

MATCHING TEXT PROPERTIES

The following is an example of how the MATCHPROP command affects a text object with the Text Special Property shown in Figure 7–64.

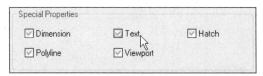

Figure 7–64

 Try It! – Open the drawing file 07_Matchprop Text. Figure 7–65 shows two text items displayed in different fonts. The text "Coarse Knurl" at "A" was constructed with a text style called RomanD. The text "Medium Knurl" at "B" was constructed with the default text style called STANDARD. Use the following command sequence to match the STANDARD text style with the RomanD text style using the MATCHPROP command.

 Command: **MA** *(For MATCHPROP)*
Select source object: *(Select the text at "A")*
Current active settings: Color Layer Ltype Ltscale Lineweight Thickness
PlotStyle Text Dim Hatch Polyline Viewport
Select destination object(s) or [Settings]: *(Select the text at "B")*
Select destination object(s) or [Settings]: *(Press ENTER to exit this command)*

The result is shown in Figure 7–66. Both text items now share the same text style. Notice that the text string stays intact when text properties are matched. Only the text style of the source object is applied to the destination object.

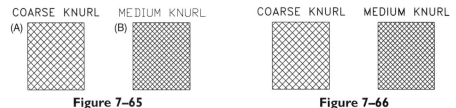

Figure 7–65   **Figure 7–66**

MATCHING HATCH PROPERTIES

A source hatch object can also be matched to a destination pattern with the MATCHPROP command and the Hatch Special Property (see Figure 7–67).

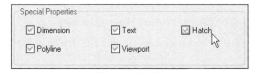

Figure 7–67

398

 Try It! – Open the drawing file 07_Matchprop Hatch. In Figure 7–68, the crosshatch patterns at "B" and "C" are at the wrong angle and scale. They should reflect the pattern at "A" because it is the same part. Use the MATCH-PROP command, select the hatch pattern at "A" as the source object, and select the patterns at "B" and "C" as the destination objects.

Command: **MA** *(For MATCHPROP)*
Select source object: *(Select the hatch pattern at "A")*
Current active settings: Color Layer Ltype Ltscale Lineweight Thickness
PlotStyle Text Dim Hatch Polyline Viewport
Select destination object(s) or [Settings]: *(Select the hatch pattern at "B")*
Select destination object(s) or [Settings]: *(Press ENTER to exit this command*

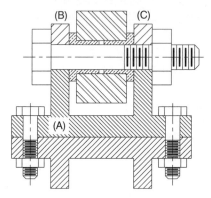

Figure 7–68

The results appear in Figure 7–69, where the source hatch pattern property was applied to all destination hatch patterns.

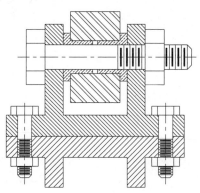

Figure 7–69

MATCHING POLYLINE PROPERTIES

A source polyline object can also be matched to a destination pattern with the MATCH-PROP command and the Polyline Special Property (see Figure 7–70).

Figure 7–70

 Try It! – Open the drawing file 07_Matchprop Pline. In Figure 7–71, the polylines at "B" and "C" are at the wrong width. They should match the width of the polyline at "A". Using the MATCHPROP command, select the polyline at "A" as the source object, and select the polylines at "B" and "C" as the destination objects.

 Command: **MA** *(For MATCHPROP)*
Select source object: *(Select the hatch pattern at "A")*
Current active settings: Color Layer Ltype Ltscale Lineweight Thickness
PlotStyle Text Dim Hatch Polyline Viewport
Select destination object(s) or [Settings]: *(Select the hatch pattern at "B")*
Select destination object(s) or [Settings]: *(Select the hatch pattern at "C")*
Select destination object(s) or [Settings]: *(Press ENTER to exit this command)*

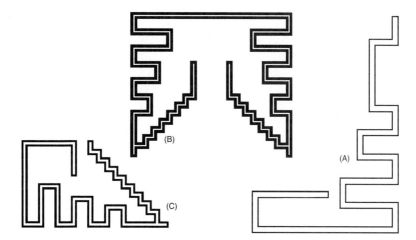

Figure 7–71

The results appear in Figure 7–72, where the source polyline property was applied to all destination polylines.

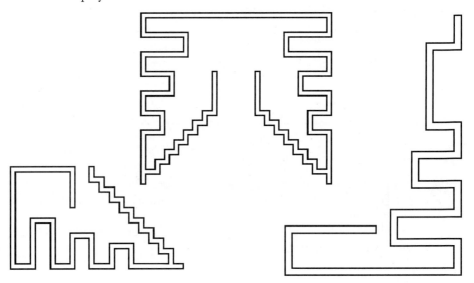

Figure 7–72

TUTORIAL EXERCISE: 07_MODIFY-EX.DWG

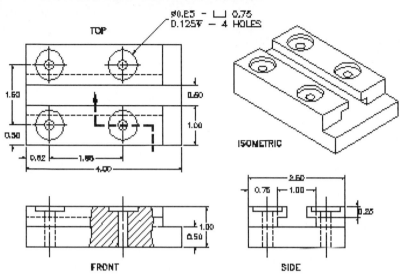

Figure 7–73

Purpose

This tutorial exercise is designed to change the properties of existing objects displayed in Figure 7–73.

System Settings

Since this drawing is provided on the CD, open an existing drawing file called "07_Modify-Ex." Follow the steps in this tutorial for changing various objects to the correct layer, text style, and dimension style.

Layers

Layers have already been created in this drawing.

Suggested Commands

Begin this tutorial by using the Properties Palette to change the isometric object to a different layer. Continue by changing the text height and layer of the view identifiers (FRONT, TOP, SIDE, ISOMETRIC). The MATCH-PROP command will be used to transfer the properties from one dimension to another, one text style to another, and one hatch pattern to another. The Layer Control box will be used to change the layer of various objects located in the Front and Top views.

Whenever possible, substitute the appropriate command alias in place of the full AutoCAD command in each tutorial step. For example, use "CP" for the COPY command, "L" for the LINE command, and so on. The complete listing of all command aliases is located in chapter 1, table 1–2.

STEP 1

Loading this drawing displays the objects in a page layout called "Orthographic Views." Page layout is where the drawing will be plotted out in the future. This name is present next to the Model tab in the bottom portion of the drawing screen. Since a majority of changes will be made in Model mode, click on the Model tab. Your image will appear similar to Figure 7–74.

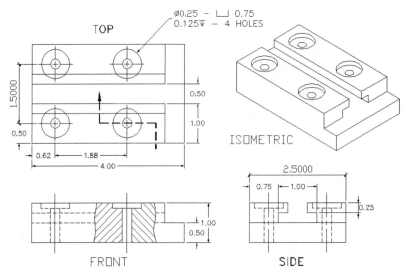

Figure 7–74

STEP 2

While in the Model environment, select all lines that make up the isometric view in Figure 7–75. You can accomplish this by using the Window mode from the Command prompt. If you accidentally select the word ISOMETRIC, deselect this word. Activate the Properties Palette and click in the Layer field. This will display the current layer the objects are drawn on (DIM) in Figure 7–75. Click the down arrow to display the other layers and pick the OBJECT layer. This will change all selected objects that make up the isometric view to the OBJECT layer. When finished, dismiss the Properties Palette. Press ESC to remove the object highlight and the grips from the drawing.

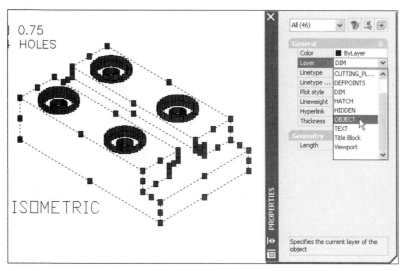

Figure 7–75

STEP 3

Select the view titles (FRONT, TOP, SIDE, ISOMETRIC). These text items need to be changed to a height of 0.15 and the TEXT layer. With all four text objects highlighted, activate the Properties Palette, click in the Layer field, and click the down arrow to change the selected objects to the TEXT layer in Figure 7–76.

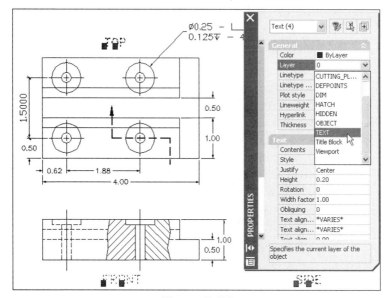

Figure 7–76

With the text still highlighted, change the text height in the Properties Palette from 0.20 to a new value of 0.15. Pressing ENTER after making this change automatically updates all selected text objects to this new height in Figure 7–77. Dismiss this dialog box when finished. Press ESC to remove the object highlight and the grips from the drawing.

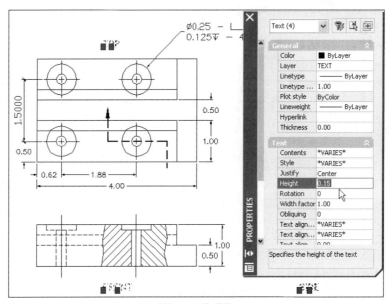

Figure 7–77

STEP 4

When you examine the view titles, TOP and SIDE are in one text style while FRONT and ISOMETRIC are in another. To remain consistent in the design process, you should make sure all text identifying the view titles has the same text style as TOP. Activate the MATCHPROP command by clicking on the button in Figure 7–78. Click the text object TOP as the source object. When the Match Property icon appears in Figure 7–79, select ISOMETRIC and FRONT as the destination objects. All properties associated with the TOP text object will be transferred to ISOMETRIC and FRONT, including the text style.

Command: **MA** *(for MATCHPROP)*

Select source object: *(Select the text object "TOP" which should highlight in Figure 7–79)*

Current active settings: Color Layer Ltype Ltscale Lineweight Thickness PlotStyle Text Dim Hatch Polyline Viewport

Select destination object(s) or [Settings]: *(Select the text object "ISOMETRIC")*

Select destination object(s) or [Settings]: *(Select the text object "FRONT")*

Select destination object(s) or [Settings]: *(Press enter to exit this command)*

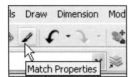

Figure 7–78

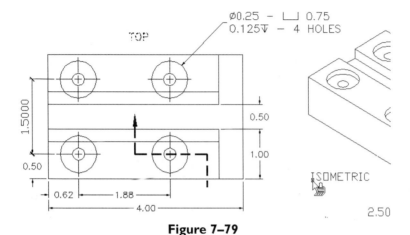

Figure 7–79

STEP 5

Notice the dimensions in this drawing. Two dimensions stand out above the rest (the 1.5000 vertical dimension in the Top view and the 2.5000 horizontal dimension in the Side view). Again to remain consistent in the design process, you should make sure all dimensions have the same appearance (dimension text height, number of decimal places, broken inside instead of placed above the dimension line). Activate the MATCHPROP command again by clicking on the button in the Standard Toolbar. Click the 4.00 horizontal dimension in the Top view as the source object. When the Match Property icon appears as in Figure 7–80, select the 1.5000 and 2.5000 vertical dimensions as the destination objects.

All dimension properties associated with the 4.00 dimension will be transferred to the 1.5000 and 2.5000 dimensions.

Command: **MA** *(for MATCHPROP)*
Select source object: *(Select the 4.00 dimension, which should highlight)*
Current active settings: Color Layer Ltype Ltscale Lineweight Thickness PlotStyle Text Dim Hatch Polyline Viewport
Select destination object(s) or [Settings]: *(Select the 2.5000 dimension at "A")*
Select destination object(s) or [Settings]: *(Select the 1.5000 dimension at "B")*
Select destination object(s) or [Settings]: *(Press ENTER to exit this command)*

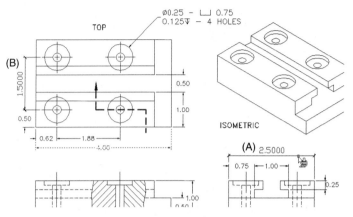

Figure 7–80

STEP 6

In the Front view, an area is crosshatched. However, both sets of crosshatching lines need to be drawn in the same direction, rather than opposing each other. Activate the MATCHPROP command again by clicking on the button in the Standard Toolbar. Click the left hatch pattern as the source object. When the Match Property icon appears, as in Figure 7–81, select the right hatch pattern as the destination object. All hatch properties associated with the left hatch pattern will be transferred to the right hatch pattern.

Command: **MA** *(for MATCHPROP)*
Select source object: *(Select the left hatch pattern, which should highlight)*
Current active settings: Color Layer Ltype Ltscale Lineweight Thickness
PlotStyle Text Dim Hatch Polyline Viewport
Select destination object(s) or [Settings]: *(Pick the right hatch pattern)*
Select destination object(s) or [Settings]: *(Press ENTER to exit this command.)*

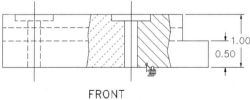

FRONT
Figure 7–81

STEP 7

Your display should appear similar to Figure 7–82. All view titles (FRONT, TOP, SIDE, ISOMETRIC) share the same text style and are at the same height. All dimensions share the same parameters and text orientation. Both crosshatch patterns are drawn in the same direction.

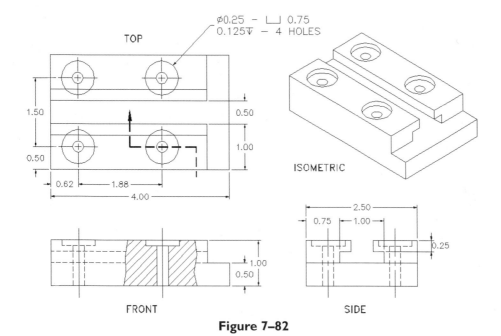

Figure 7–82

STEP 8

A different method will now be used to change the specific layer properties of objects. The two highlighted lines in the Top view in Figure 7–83 were accidentally drawn on Layer OBJECT and need to be transferred to the HIDDEN layer. With the lines highlighted, click in the Layer Control box to display all layers. Click on the HIDDEN layer to change the highlighted lines to the HIDDEN layer. Press ESC to remove the object highlight and the grips from the drawing.

 Note: You may need to REGEN the drawing in order to get the dashes to appear in the hidden lines.

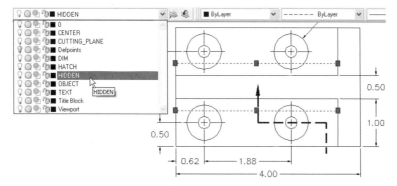

Figure 7–83

STEP 9

The two highlighted lines in the Front view in Figure 7–84 were accidentally drawn on the TEXT Layer and need to be transferred to the CENTER layer. With the lines highlighted, click in the Layer Control box to display all layers. Click on the CENTER layer to change the highlighted lines to the CENTER layer. Press ESC to remove the object highlight and the grips from the drawing.

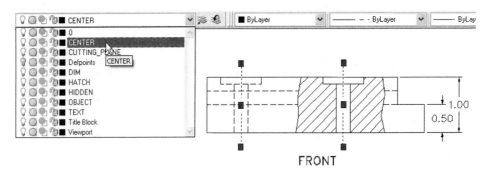

Figure 7–84

STEP 10

Your display should appear similar to Figure 7–85. Notice the hidden lines in the Top view and the centerlines in the Front view.

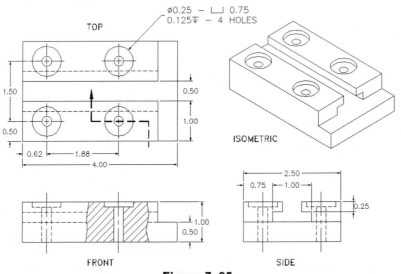

Figure 7–85

STEP 11

Click on the Orthographic Views Tab. Select the rectangular viewport and change this object's layer to "Viewport" in Figure 7–86.

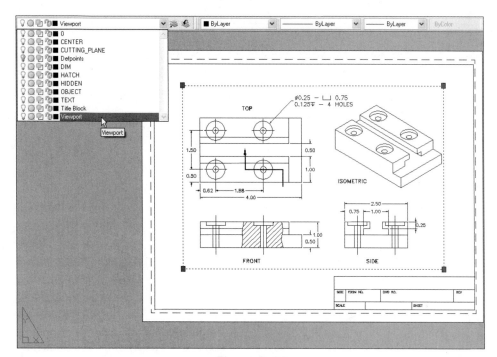

Figure 7–86

STEP 12

Turn off the Viewport layer. Your display should appear similar to Figure 7–87. This completes this tutorial exercise.

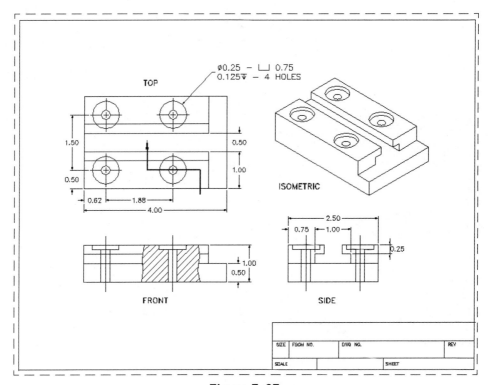

Figure 7–87

 Open the Exercise Manual PDF file for Chapter 7 on the accompanying CD for more tutorials and exercises.

 If you have the accompanying Exercise Manual, refer to Chapter 7 for more tutorials and exercises.

8

Shape Description/ Multiview Projection

Before any object is made in production, some type of drawing needs to be created. This is not just any drawing, but rather an engineering drawing consisting of overall object sizes with various views of the object organized on the computer screen. This chapter introduces the topic of shape description or how many views are really needed to describe an object. The art of multiview projection includes methods of constructing one-view, two-view, and three-view drawings using AutoCAD commands.

SHAPE DESCRIPTION

Begin constructing an engineering drawing by first analyzing the object being drawn. To accomplish this, describe the object by views or how an observer looks at the object. Figure 8–1 shows a simple wedge; this object can be viewed at almost any angle to get a better idea of its basic shape. However, to standardize how all objects are to be viewed, and to limit the confusion usually associated with complex multiview drawings, some standard method of determining how and where to view the object must be exercised.

Even though the simple wedge is easy to understand because it is currently being displayed in picture or isometric form, it would be difficult to produce this object because it is unclear what the sizes of the front and top faces are. Illustrated in Figure 8–2 are six primary ways or directions in which to view an object. The Front view begins the shape description and is followed by the Top view and Right Side view. Continuing on, the Left Side view, Back view, and Bottom view complete the primary ways to view an object.

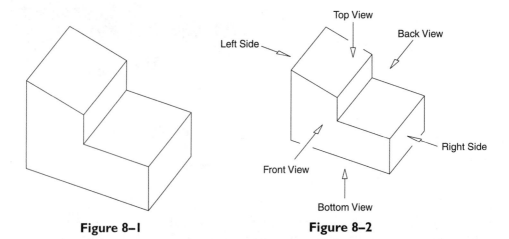

Figure 8–1   **Figure 8–2**

Now that the primary ways of viewing an object have been established, the views need to be organized to promote clarity and have the views reference one another. Imagine the simple wedge positioned in a transparent glass box similar to the illustration in Figure 8–3. With the entire object at the center of the box, the sides of the box represent the ways to view the object. Images of the simple wedge are projected onto the glass surfaces of the box.

With the views projected onto the sides of the glass box, we must now prepare the views to be placed on a two-dimensional (2D) drawing screen. For this to be accomplished, the glass box, which is hinged, is unfolded as in Figure 8–4. All folds occur from the Front view, which remains stationary.

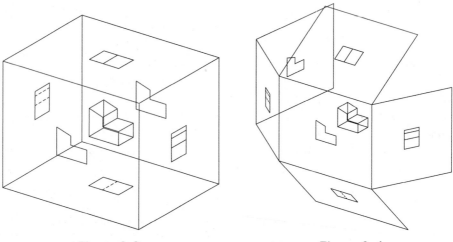

Figure 8–3   **Figure 8–4**

Figure 8–5 shows the views in their proper alignment to one another. However, this illustration is still in a pictorial view. These views need to be placed flat before continuing.

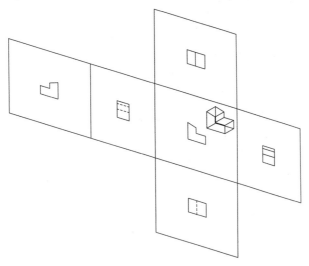

Figure 8–5

As the glass box is completely unfolded and laid flat, the result is illustrated in Figure 8–6. The Front view becomes the main view, with other views placed in relation to it. Above the Front view is the Top view. To the right of the Front view is the Right Side view. To the left of the Front view is the Left Side view, followed by the Back view. Underneath the Front view is the Bottom view. This becomes the standard method of laying out the views needed to describe an object. But are all the views necessary? Upon closer inspection we find that, except for being a mirror image, the Front and Back views are identical. The Top and Bottom views appear similar, as do the Right and Left Side views. One very important rule to follow in multiview objects is to select only those views that accurately describe the object and discard the remaining views.

The complete multiview drawing of the simple wedge is illustrated in Figure 8–7. Only the Front, Top, and Right Side views are needed to describe this object. When laying out views, remember that the Front view is usually the most important—it holds the basic shape of the object being described. Directly above the Front view is the Top view, and to the right of the Front view is the Right Side view. All three views are separated by a space of varying size. This space is commonly called a dimension space because it is a good area in which to place dimensions describing the size of the object. The space also acts as a separator between views; without it, the views would touch one another, making them difficult to read and interpret. The minimum distance of this space is usually 1.00 unit.

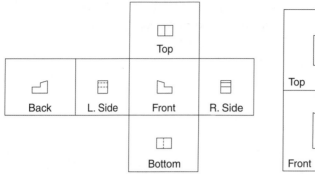

Figure 8–6

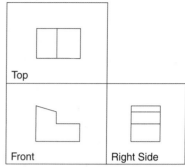

Figure 8–7

RELATIONSHIPS BETWEEN VIEWS

Some very interesting and important relationships are set up when the three views are placed in the configuration illustrated in Figure 8–8. Notice that because the Top view is directly above the Front view, both share the same width dimension. The Front and Right Side views share the same height. The relationship of depth on the Top and Right Side views can be explained with the construction of a 45°-projection line at "A" and the projection of the lines over and down or vice versa to get the depth. Another principle illustrated by this example is that of projecting lines up, over, across, and down to create views. Editing commands such as ERASE and TRIM are then used to clean up unnecessary lines.

With the three views identified in Figure 8–9, the only other step—and it is an important one—is to annotate the drawing, or add dimensions to the views. With dimensions, the object drawn in multiview projection can now be produced. Even though this has been a simple example, the methods of multiview projection work even for the most difficult and complex of objects.

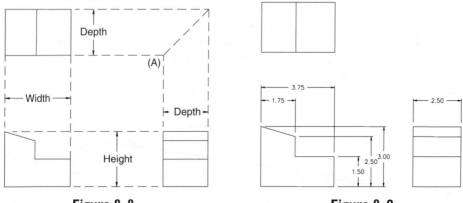

Figure 8–8   **Figure 8–9**

REVIEW OF LINETYPES AND CONVENTIONS

At the heart of any engineering drawing is the ability to assign different types of lines to convey meaning to the drawing. When plotted, all lines of a drawing are dark; border and title block lines are the thickest. Object lines outline visible features of a drawing and are made thick and dark, but not as thick as a border line. Features that are behind a surface but important to the description of the object are identified by a hidden line. This line is a series of dashes 0.12 units in length with a spacing of 0.06 units. Centerlines identify the centers of circular features such as holes or show that a hidden feature in one view is circular in another. The centerline consists of a series of long and short dashes. The short dash measures approximately 0.12 units, whereas the long dash may vary from 0.75 to 1.50 units. A gap of 0.06 separates dashes. Study the examples of these lines in Figure 8–10.

The object shown in Figure 8–11 is a good illustration of the use of phantom lines in a drawing. Phantom lines are especially useful where motion is applied. The arm on the right is shown by standard object lines consisting of the continuous linetype. To show that the arm rotates about a center pivot, the identical arm is duplicated through the ARRAY command, and all lines of this new element are converted to phantom lines through the Properties dialog box, MATCHPROP command, or Layer Control box. Notice that the smaller segments are not shown as phantom lines due to their short lengths.

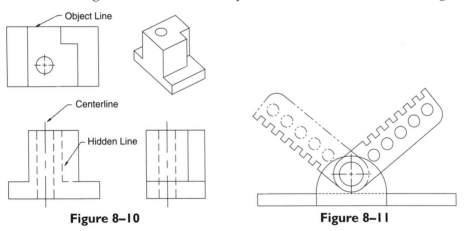

Figure 8–10   **Figure 8–11**

Use of linetypes in a drawing is crucial to the interpretation of the views and the final design before the object is actually made. Sometimes the linetype appears too long; in other cases, the linetype does not appear at all, even though using the LIST command on the object will show the proper layer and linetype. The LTSCALE or Linetype Scale command is used to manipulate the size of all linetypes loaded into a drawing. By default, all linetypes are assigned a scale factor of 1.00. This means that the actual dashes and/or spaces of the linetype are multiplied by this factor. The views illustrated in Figure 8–12 show linetypes that use the default value of 1.00 from the LTSCALE command.

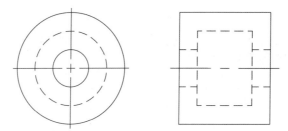

Figure 8–12

If a linetype dash appears too long, use the LTSCALE command and set a new value to less than 1.00. If a linetype appears too short, use the LTSCALE command and set a new value to greater than 1.00. The same views illustrated in Figure 8–13 show the effects of the LTSCALE command set to a new value of 0.75. Notice that the center in the Right Side view has one more series of dashes than the same object illustrated previously. The 0.75-multiplier affects all dashes and spaces defined in the linetype.

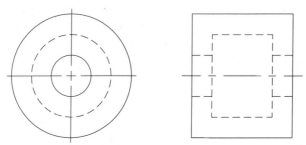

Figure 8–13

Figure 8–14 illustrates how using the LTSCALE command with a new value of 0.50 shortens the linetype dashes and spaces even more. Now even the center marks identifying the circles have been changed to centerlines. When you use the LTSCALE command, the new value, whether larger or smaller than 1.00, affects all linetypes visible on the display screen.

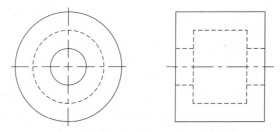

Figure 8–14

ONE-VIEW DRAWINGS

An important rule to remember concerning multiview drawings is to draw only enough views to accurately describe the object. In the drawing of the gasket in Figure 8–15, Front and Side views are shown. However, the Side view is so narrow that it is difficult to interpret the hidden lines drawn inside. A better approach would be to leave out the Side view and construct a one-view drawing consisting of just the Front view.

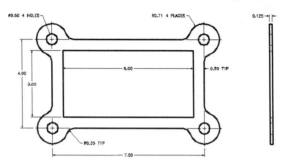

Figure 8–15

Begin the one-view drawing of the gasket by first laying out centerlines marking the centers of all circles and arcs, as in Figure 8–16. A layer containing centerlines could be used to show all lines as centerlines.

Use the CIRCLE command to layout all circles representing the bolt holes of the gasket shown in Figure 8–17. A layer containing continuous object lines could be used for these circles. You could use the OFFSET command to form the large rectangle on the inside of the gasket. If lines of the rectangle extend past each other, use the FILLET command set to a value of 0. Selecting two lines of the rectangle will form a corner. Repeat this procedure for any other lines that do not form exact corners.

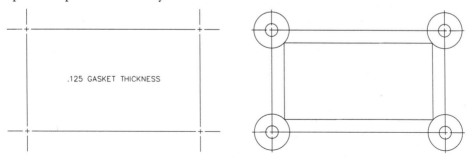

Figure 8–16   **Figure 8–17**

Use the TRIM command to begin forming the outside arcs of the gasket shown in Figure 8–18.

Use the FILLET command set to the desired radius to form a smooth transition from the arcs to the outer rectangle shown in Figure 8–19.

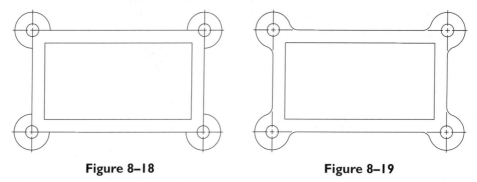

| Figure 8–18 | Figure 8–19 |

TWO-VIEW DRAWINGS

Before attempting any drawing, determine how many views need to be drawn. A minimum number of views are needed to describe an object. Drawing extra views is not only time-consuming, but it may result in two identical views with mistakes in each view. You must interpret which is the correct set of views. The illustration in Figure 8–20 is a three-view multiview drawing of a coupler. The circles and circular hidden circle identify the Front view. Except for their rotation angles, the Top and Right Side views are identical. In this example or for other symmetrical objects, only two views are needed to accurately describe the object being drawn. The Top view has been deleted to leave the Front and Right Side views. The Side view could have easily been deleted in favor of leaving the Front and Top views. This decision is up to the designer, depending on sheet size and which views are best suited for the particular application.

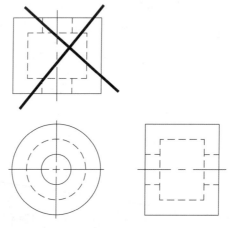

Figure 8–20

To illustrate how AutoCAD is used as the vehicle for creating a two-view engineering drawing, study the pictorial drawing illustrated in Figure 8–21 to get an idea of how the drawing will appear. Begin the two-view drawing by using the LINE command to lay out the Front and Side views. You can find the width of the Top view by projecting lines up from the front because both views share the same width. Provide a space of 1.50 units between views to act as a separator and allow for dimensions at a later time.

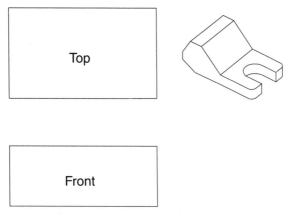

Figure 8–21

Begin adding visible details to the views, such as circles, filleted corners, and angles, as shown in Figure 8–22. Use various editing commands such as TRIM, EXTEND, and OFFSET to clean up unnecessary geometry.

From the Front view, project corners up to the Top view. These corners will form visible edges in the Top view. Use the same projection technique to project features from the Top view to the Front view (see Figure 8–23).

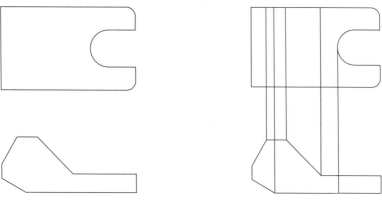

Figure 8–22   **Figure 8–23**

Use the TRIM command to delete any geometry that appears in the 1.50 dimension space. The views now must conform to engineering standards by showing which lines are visible and which are invisible, as shown in Figure 8–24. The corner at "A" represents an area hidden in the Top view. Use the Property dialog box or MATCHPROP command to convert the line in the Top view from the continuous linetype to the hidden linetype. In the same manner, the slot visible in the Top view is hidden in the Front view. Again convert the continuous line in the Front view to the hidden linetype, using one of the methods already described. Since the slot in the Top view represents a circular feature, use the DIMCENTER command to place a center marker at the center of the semicircle. To show in the Front view that the hidden line represents a circular feature, add one centerline consisting of one short dash and two short dashes. If the slot in the Top view were square instead of circular, centerlines would not be necessary.

Use the spaces in-between views to properly add dimensions to the drawing, as shown in Figure 8–25. As these spaces fill up with dimensions, use outside areas to place other dimensions.

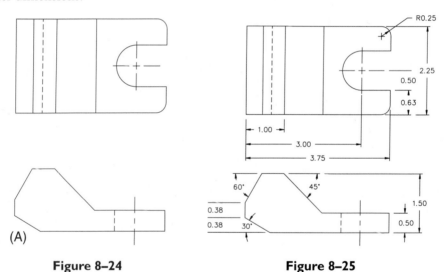

(A)

Figure 8–24   **Figure 8–25**

THREE-VIEW DRAWINGS

If two views are not enough to describe an object, draw three views. This consists of Front, Top, and Right Side views. A three-view drawing of the guide block, as illustrated in pictorial format in Figure 8–26, will be the focus of this segment. Notice the broken section exposing the Spotface operation above a drill hole. Begin this drawing by laying out all views using overall dimensions of width, depth, and height. The LINE and OFFSET commands are popular commands used to accomplish this. Provide a space between views to accommodate dimensions at a later time.

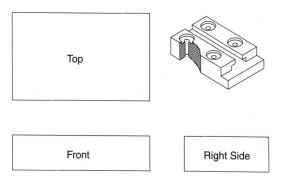

Figure 8–26

Begin drawing features in the views where they are visible, as illustrated in Figure 8–27. Since the Spotface holes appear above, draw these in the Top view. The notch appears in the Front view; draw it there. A slot is visible in the Right Side view and is drawn there.

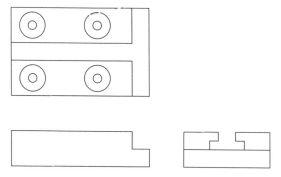

Figure 8–27

As in two-view drawings, all features are projected down from the Top to the Front view. To project depth measurements from the Top to the Right Side view, construct a 45° line at "A." See Figure 8–28.

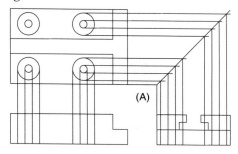

Figure 8–28

Use the 45° line to project the slot from the Right Side view to the Top view as shown in Figure 8–29. Project the height of the slot from the Right Side view to the Front view. Convert the continuous lines to hidden lines where features appear invisible, such as the holes in the Front and Right Side views.

Change the remaining lines from continuous to hidden. Erase any construction lines, including the 45°-projection line (see Figure 8–30).

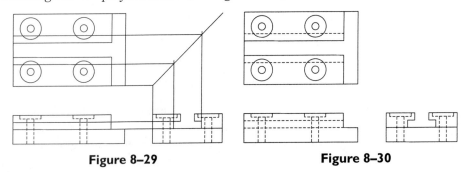

Figure 8–29   **Figure 8–30**

Begin adding centerlines to label circular features as shown in Figure 8–31. The DIM-CENTER command is used where the circles are visible. Where features are hidden but represent circular features, the single centerline consisting of one short dash and two long dashes is used. In Figure 8–32, dimensions remain the final step in completing the engineering drawing before it is checked and shipped off for production.

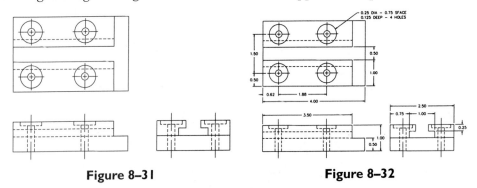

Figure 8–31   **Figure 8–32**

RUNOUTS

Where flat surfaces become tangent to cylinders, there must be some method of accurately representing this using fillets. In the object illustrated in Figure 8–33, the Front view shows two cylinders connected to each other by a tangent slab. The Top view is complete; the Front view has all the geometry necessary to describe the object with the exception of the exact intersection of the slab with the cylinder.

Figure 8–34 displays the correct method for finding intersections or runouts—areas where surfaces intersect others and blend in, disappear, or simply run out. A point of intersection is found at "A" in the Top view with the intersecting slab and the cylinder. This actually forms a 90° angle with the line projected from the center of the cylinder and the angle made by the slab. A line is projected from "A" in the Top view to intersect with the slab found in the Front view.

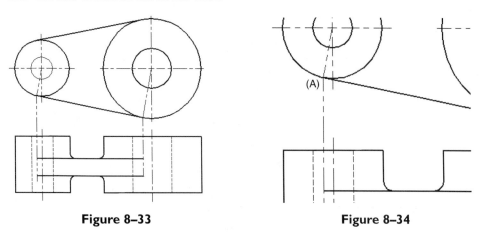

Figure 8–33   **Figure 8–34**

In Figure 8–35, fillets are copied to represent the slab and cylinder intersections. This forms the runout.

The resulting two-view drawing, complete with runouts, is illustrated in Figure 8–36.

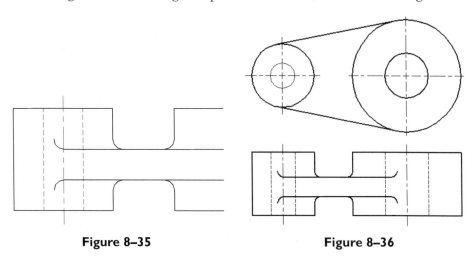

Figure 8–35   **Figure 8–36**

TUTORIAL EXERCISE: SHIFTER.DWG

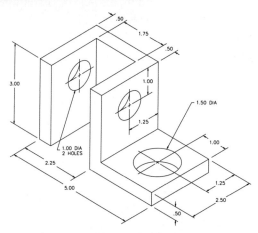

Figure 8–37

Purpose

This tutorial is designed to allow the user to construct a three-view drawing of the Shifter as shown in Figure 8–37.

System Settings

Use the Drawing Units dialog box and change the number of decimal places from four to two. Set linetype scale to 0.50 using the LTSCALE (LTS) command. Keep the remaining default settings. Use the default settings for the screen limits: (0,0) for the lower-left corner and (12.00,9.00) for the upper-right corner. The grid and snap do not need to be set to any certain values. Check to see that the following Object Snap modes are already set: Endpoint, Extension, Intersection, Center.

Layers

Create the following layers with the format:

Name	Color	Linetype
Object	Green	Continuous
Hidden	Red	Hidden
Center	Yellow	Center
Dimension	Yellow	Continuous
Projection	Cyan	Continuous

Suggested Commands

The primary commands used during this tutorial are OFFSET and TRIM. The OFFSET command is used for laying out all views before the TRIM command is used to clean up excess lines. Since different linetypes represent certain features of a drawing, the Layer Control box is used to convert to the desired linetype needed as set in the Layer Properties Manager dialog box. Once all visible details are identified in the primary views, project the visible features to the other views using the LINE command. A 45° inclined line is constructed to project lines from the Top view to the Right Side view and vice versa.

Whenever possible, substitute the appropriate command alias in place of the full AutoCAD command in each tutorial step; for example, use "CP" for the COPY command, "L" for the LINE command, and so on. The complete listing of all command aliases is located in chapter 1, table 1–2.

STEP 1

Make the "Object" layer current in Figure 8–38. Begin the orthographic drawing of the Shifter by constructing a right angle consisting of one horizontal and one vertical line. The corner formed by the two lines will be used to orient the Front view (see Figure 8–39).

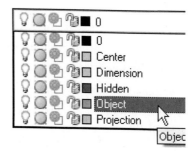

Figure 8–38

 Command: **L** *(For LINE)*

Specify first point: **1,1**
Specify next point or [Undo]: **@11,0**
Specify next point or [Undo]: *(Press* ENTER *to exit this command)*

 Command: **L** *(For LINE)*

Specify first point: **1,1**
Specify next point or [Undo]: **@8<90**
Specify next point or [Undo]: *(Press* ENTER *to exit this command)*

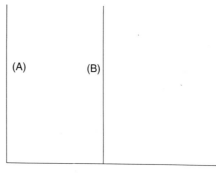

Figure 8–39

STEP 2

Begin the layout of the primary views by using the OFFSET command to offset the vertical line at "A" a distance of 5.00 units, which represents the length of the Shifter (see Figure 8–40).

 Command: **O** *(For OFFSET)*

Specify offset distance or [Through] <1.0000>: **5.00**
Select object to offset or <exit>: *(Select the vertical line at "A")*
Specify point on side to offset: *(Pick a point anywhere near "B")*
Select object to offset or <exit>: *(Press* ENTER *to exit this command)*

Figure 8–40

STEP 3

Use the OFFSET command to offset the horizontal line at "A" a distance of 3.00 units, which represents the height of the Shifter (see Figure 8–41).

 Command: **O** *(For OFFSET)*

Specify offset distance or [Through]
   <5.00>: **3.00**
Select object to offset or <exit>: *(Select the horizontal line at "A")*
Specify point on side to offset: *(Pick a point anywhere near "B")*
Select object to offset or <exit>: *(Press ENTER to exit this command)*

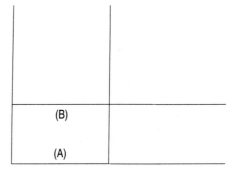

Figure 8–41

STEP 4

Begin laying out dimension spaces that will act as separators between views and allow for the placement of dimensions once the Shifter is completed (see Figure 8–42). A spacing of 1.50 units will be more than adequate for this purpose. Again, use the OFFSET command to accomplish this.

 Command: **O** *(For OFFSET)*

Specify offset distance or [Through]
   <3.00>: **1.50**
Select object to offset or <exit>: *(Select the vertical line at "A")*
Specify point on side to offset: *(Pick a point anywhere near "B")*
Select object to offset or <exit>: *(Select the horizontal line at "C")*
Specify point on side to offset: *(Pick a point anywhere near "D")*
Select object to offset or <exit>: *(Press ENTER to exit this command)*

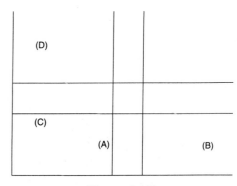

Figure 8–42

STEP 5

Use the OFFSET command to lay out the depth of the Shifter at a distance of 2.50 units (see Figure 8–43).

 Command: **O** *(For OFFSET)*

Specify offset distance or [Through]
    <1.50>: **2.50**
Select object to offset or <exit>: *(Select the vertical line at "A")*
Specify point on side to offset: *(Pick a point anywhere near "B")*
Select object to offset or <exit>: *(Select the horizontal line at "C")*
Specify point on side to offset: *(Pick a point anywhere near "D")*
Select object to offset or <exit>: *(Press ENTER to exit this command)*

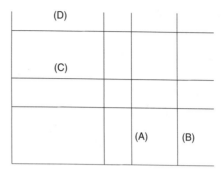

Figure 8–43

STEP 6

Use the TRIM command to trim away the excess construction lines you used when laying out the primary views of the Shifter (see Figure 8–44).

 Command: **TR** *(For TRIM)*

Current settings: Projection=UCS
    Edge=None
Select cutting edges ...
Select objects: *(Select lines "A" and "B")*
Select objects: *(Press ENTER to continue)*
Select object to trim or shift-select to
    extend or [Project/Edge/Undo]: *(Select the line at "C")*
Select object to trim or shift-select to
    extend or [Project/Edge/Undo]: *(Select the line at "D")*
Select object to trim or shift-select to
    extend or [Project/Edge/Undo]: *(Select the line at "E")*
Select object to trim or shift-select to
    extend or [Project/Edge/Undo]: *(Select the line at "F")*
Select object to trim or shift-select to
    extend or [Project/Edge/Undo]: *(Press ENTER to exit this command)*

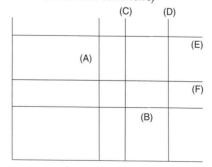

Figure 8–44

STEP 7

Use the TRIM command again to complete trimming away the excess construction lines you used when laying out the primary views of the Shifter (see Figure 8–45).

Command: **TR** *(For TRIM)*

Current settings: Projection=UCS
Edge=None
Select cutting edges ...
Select objects: *(Press ENTER to accept all objects as cutting edges)*
Select object to trim or shift-select to extend or [Project/Edge/Undo]: *(Select the lines at "A" through "D" to trim)*
Select object to trim or shift-select to extend or [Project/Edge/Undo]: *(Press ENTER to exit this command)*

Figure 8–45

STEP 8

Your display should appear similar to the illustration in Figure 8–46 with the layout of the Front, Top, and Right Side views. Begin adding details to all views through methods of projection. Use the OFFSET command to offset lines "A," "C," and "E" a distance of 0.50.

Command: **O** *(For OFFSET)*

Specify offset distance or [Through] <2.50>: **0.50**
Select object to offset or <exit>: *(Select the vertical line at "A")*
Specify point on side to offset: *(Pick a point anywhere near "B")*
Select object to offset or <exit>: *(Select the horizontal line at "C")*
Specify point on side to offset: *(Pick a point anywhere near "D")*
Select object to offset or <exit>: *(Select the horizontal line at "E")*

Specify point on side to offset: *(Pick a point anywhere near "F")*
Select object to offset or <exit>: *(Press ENTER to exit this command)*

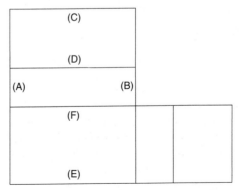

Figure 8–46

STEP 9

Use the OFFSET command and offset the vertical line at "A" a distance of 1.75 (see Figure 8–47).

 Command: **O** *(For OFFSET)*

Specify offset distance or [Through] <0.50>: **1.75**

Select object to offset or <exit>: *(Select the vertical line at "A")*

Specify point on side to offset: *(Pick a point anywhere near "B")*

Select object to offset or <exit>: *(Press ENTER to exit this command)*

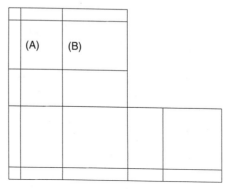

Figure 8–47

STEP 10

Use the OFFSET command to offset the vertical line at "A" a distance of 0.50 (see Figure 8–48).

 Command: **O** *(For OFFSET)*

Specify offset distance or [Through] <1.75>: **0.50**

Select object to offset or <exit>: *(Select the vertical line at "A")*

Specify point on side to offset: *(Pick a point anywhere near "B")*

Select object to offset or <exit>: *(Press ENTER to exit this command)*

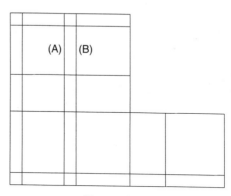

Figure 8–48

STEP 11

Your display should appear similar to the illustration in Figure 8–49. Next, use the TRIM command to partially delete the line segments located in the spaces between views.

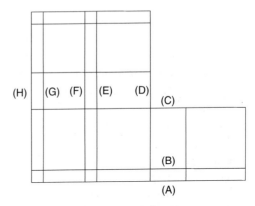

Figure 8–49

 Command: **TR** *(For TRIM)*

Current settings: Projection=UCS
    Edge=None
Select cutting edges ...
Select objects: *(Press ENTER to accept all objects as cutting edges)*
Select object to trim or shift-select to extend or [Project/Edge/Undo]: *(Select the lines at "A" through "H" to trim)*
Select object to trim or shift-select to extend or [Project/Edge/Undo]: *(Press ENTER to exit this command)*

STEP 12

Use the ZOOM command and the Window option to magnify the Top view similar to the illustration in Figure 8–50. Then use the TRIM command to trim the line labeled "B."

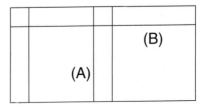

Figure 8–50

 Command: **TR** *(For TRIM)*

Current settings: Projection=UCS
    Edge=None
Select cutting edges ...
Select objects: *(Select the vertical line at "A")*
Select objects: *(Press ENTER to continue)*
Select object to trim or shift-select to extend or [Project/Edge/Undo]: *(Select the line at "B")*
Select object to trim or shift-select to extend or [Project/Edge/Undo]: *(Press ENTER to exit this command)*

STEP 13

Zoom back to the original display using the ZOOM command and the Previous option. Use the ZOOM command and the Window option to magnify the display to show the Front view illustrated in Figure 8–51. Use the TRIM command to clean up the excess lines in the Front view, using the illustration in Figure 8–51 as a guide.

 Command: **TR** *(For TRIM)*

Current settings: Projection=UCS
   Edge=None
Select cutting edges ...
Select objects: *(Select lines "A" and "B")*
Select objects: *(Press ENTER to continue)*
Select object to trim or shift-select to
   extend or [Project/Edge/Undo]: *(Select
   lines "C" through "F" to trim)*

Select object to trim or shift-select to extend or [Project/Edge/Undo]: *(Press ENTER to exit this command)*

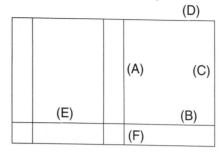

Figure 8–51

STEP 14

Use the ZOOM command and the Previous option to zoom back to the previous screen display containing the three views. Use the TRIM command to partially delete the lines at "A" through "D" in the top view, as shown in Figure 8–52.

 Command: **TR** *(For TRIM)*

Current settings: Projection=UCS
   Edge=None
Select cutting edges ...
Select objects: *(Press ENTER to accept all
   objects as cutting edges)*
Select object to trim or shift-select to
   extend or [Project/Edge/Undo]: *(Select
   the lines at "A" through "D" to trim)*
Select object to trim or shift-select to
   extend or [Project/Edge/Undo]: *(Press
   ENTER to exit this command)*

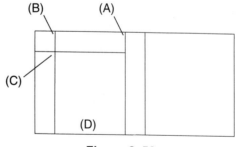

Figure 8–52

STEP 15

Begin placing a circle in the Top view representing the 1.50 diameter drill hole in Figure 8–53. Use the CIRCLE command and the OSNAP-From and OSNAP-Intersect modes to set up a temporary point of reference as illustrated in Figure 8–53.

 Command: **C** *(For CIRCLE)*

Specify center point for circle or [3P/2P/
    Ttr (tan tan radius)]: **From**

Base point: **End**
of *(Pick the endpoint at "A")*
<Offset>: **@-1,-1.25**

Specify radius of circle or [Diameter]: **D**
    (For Diameter)

Specify diameter of circle: **1.50**

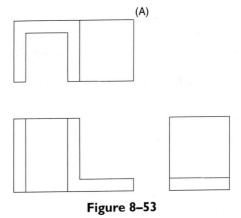
(A)

Figure 8–53

STEP 16

Place a circle in the Right Side view representing the 1.00-diameter drill hole. Use the CIRCLE command and the OSNAP-From and OSNAP-Intersect modes to set up a temporary point of reference as illustrated in Figure 8–54 at "A."

 Command: **C** *(For CIRCLE)*

Specify center point for circle or [3P/2P/
    Ttr (tan tan radius)]: **From**

Base point: **End**
of *(Pick the endpoint at "A")*
<Offset>: **@-1.25,-1.00**

Specify radius of circle or [Diameter]
    <0.75>: **D** *(For Diameter)*

Specify diameter of circle <1.50>: **1.00**

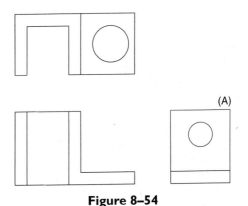
(A)

Figure 8–54

STEP 17

Before continuing with this step, first make the "Projection" layer current in Figure 8–55. Use the LINE command to draw projection lines from both circles to the Front view as shown in Figure 8–56. Use the OSNAP-Quadrant mode along with the Polar Intersection mode to accomplish this.

 Command: **L** *(For LINE)*

Specify first point: **Qua**
of *(Select the quadrant of the circle at "A")*
Specify next point or [Undo]: *(Select the polar intersection at "B")*
Specify next point or [Undo]: *(Press ENTER to exit this command)*

Repeat the above procedure for the circle locations at "C," "D," and "E" using quadrants and polar intersections.

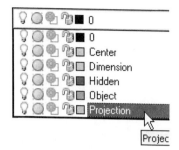

Figure 8–55

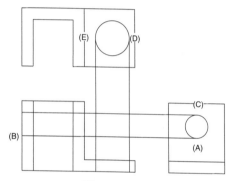

Figure 8–56

STEP 18

Before continuing, first make the "Center" layer current in as shown in Figure 8–57. Then place center marks at the centers of both circles as illustrated in Figure 8–58. Before you proceed with this operation, you need to set a system variable to a certain value to achieve the desired results. Set the DIMCEN variable to a value of -0.12. This will not only place the center mark when the DIMCENTER command is used, but will also extend the centerline a short distance outside both circles.

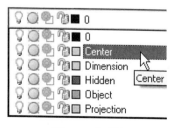

Figure 8–57

Command: **DIMCEN**

Enter new value for DIMCEN <0.09>:
   -0.12

Command: **DCE** *(For DIMCENTER)*

Select arc or circle: *(Select the circle at "A")*

Command: **DCE** *(For DIMCENTER)*

Select arc or circle: *(Select the circle at "B")*

When you're finished with these operations, make the "Projection" layer current again.

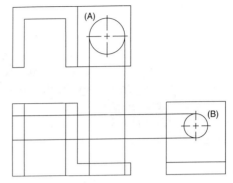

Figure 8–58

STEP 19

With the "Projection" layer current, construct two lines from the endpoints of both center marks using the OSNAP-Endpoint mode. Turn Ortho mode on to assist with this operation. The lines will be converted to centerlines at a later step. See Figure 8–59.

 Command: **L** *(For LINE)*

Specify first point: *(Select the endpoint of the center mark in the Top view at "A")*

Specify next point or [Undo]: *(Pick a point just below the Front view at "B")*

Specify next point or [Undo]: *(Press ENTER to exit this command)*

 Command: **L** *(For LINE)*

Specify first point: *(Select the endpoint of the center mark in the Side view at "C")*

Specify next point or [Undo]: *(Pick a point to the left of the Front view at "D")*

Specify next point or [Undo]: *(Press ENTER to exit this command)*

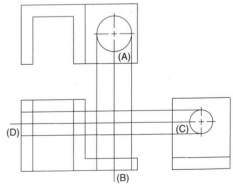

Figure 8–59

STEP 20

Use the TRIM command and the illustration in Figure 8–60 to trim away excess lines.

🔲 Command: **TR** *(For TRIM)*

Current settings: Projection=UCS
   Edge=None
Select cutting edges ...
Select objects: *(Select the lines at "A," "B," "C," and "D")*
Select objects: *(Press ENTR to continue)*
Select object to trim or shift-select to extend or [Project/Edge/Undo]: *(Select the lines at "E" through "J" to trim)*
Select object to trim or shift-select to extend or [Project/Edge/Undo]: *(Press ENTER to exit this command)*

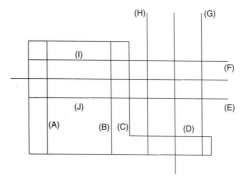

Figure 8–60

STEP 21

At the Command prompt, select the six lines in Figure 8–61. Use the Layer Control box and change all highlighted lines to the "Hidden" layer. When finished, press ESC to remove the object highlight and the grips. If the hidden line dashes look too large, change the value in the LTSCALE command from 1.0000 to 0.5000 units.

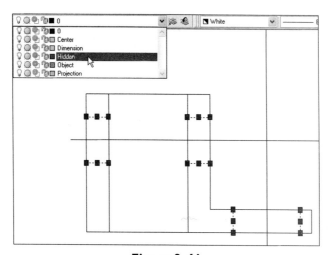

Figure 8–61

436

STEP 22

Before continuing with this step, turn off OSNAP mode. Then use the BREAK command to partially delete the lines illustrated in Figure 8–62 before converting them to center lines. Remember, centerlines extend past the object lines when identifying hidden drill holes; it would be inappropriate to use the TRIM command for this step.

Command: <Osnap off>
Command: **BR** *(For BREAK)*
Select object: *(Select the horizontal line approximately at "A")*
Specify second break point or [First point]: *(Select the line approximately at "B")*
Command: **BR** *(For BREAK)*
Select object: *(Select the vertical line approximately at "C")*
Specify second break point or [First point]: *(Select the line approximately at "D")*

Command: **BR** *(For BREAK)*
Select object: *(Select the horizontal line approximately at "E")*
Specify second break point or [First point]: **@** *(To break the line at the last point)*

Using the @ symbol breaks the object at the exact location identified by the first point selected. Remember, the @ symbol means "last point."

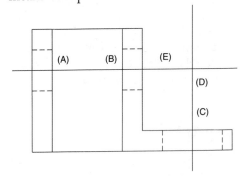

Figure 8–62

STEP 23

The purpose of the @ symbol in the previous step is to break a line into two segments without showing the break. The @ symbol means "the last known point," which completes the BREAK command by satisfying the "Second point" prompt. To prove this, use the ERASE command to delete the segments no longer needed (see Figure 8–63).

Command: **E** *(For ERASE)*
Select objects: *(Carefully select the lines at "A" and "B")*
Select objects: *(Press ENTER to perform the erase operation)*

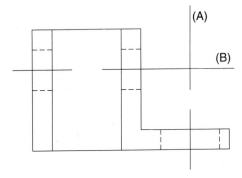

Figure 8–63

STEP 24

At the Command prompt, select the three lines in Figure 8–64. Then use the Layer Control box change all highlighted lines to the "Center" layer. When finished, press ESC to remove the object highlight and the grips.

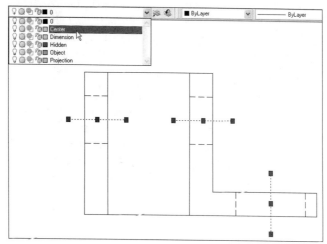

Figure 8–64

STEP 25

Turn OSNAP back on. You need to construct a 45° angle in order to begin projecting features from the Top view to the Right Side view and then back again. You form this angle by extending the bottom edge of the Top view to intersect with the left edge of the Side view, as shown in Figure 8–65. Use the FILLET command with the Radius set to 0 to accomplish this. Then draw the 45° line; the length of this line is not important. Turn Ortho mode off for this step.

Command: **F** *(For FILLET)*

Current settings: Mode = TRIM, Radius = 0.50

Select first object or [Polyline/Radius/Trim]: **R** *(For Radius)*

Specify fillet radius <0.50>: **0**

Select first object or [Polyline/Radius/Trim]: *(Select line "A")*

Select second object: *(Select line "B")*

Command: **L** *(For LINE)*

Specify first point: *(Select the endpoint at "C")*

Specify next point or [Undo]: **@4<45**

Specify next point or [Undo]: *(Press ENTER to exit this command)*

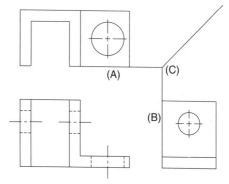

Figure 8–65

STEP 26

The "Projector" layer should still be current. Draw lines from points "A," "B," and "C" to intersect with the 45°-projection line as shown in Figure 8–66. Be sure Ortho mode is on to draw horizontal lines. Use the OSNAP-Intersect, Endpoint, and Quadrant modes to assist in this operation.

Command: **L** *(For LINE)*

Specify first point: *(Select the intersection of the corner at "A")*

Specify next point or [Undo]: *(Draw a horizontal line just past the 45° angle)*

Specify next point or [Undo]: *(Press ENTER to exit this command)*

Command: **L** *(For LINE)*

Specify first point: *(Select the endpoint of the centerline at "B")*

Specify next point or [Undo]: *(Draw a horizontal line just past the 45° angle)*

Specify next point or [Undo]: *(Press ENTER to exit this command)*

Command: **L** *(For LINE)*

Specify first point: **QUA**

of *(Select the quadrant of the circle at "C")*

Specify next point or [Undo]: *(Draw a horizontal line just past the 45° angle)*

Specify next point or [Undo]: *(Press ENTER to exit this command)*

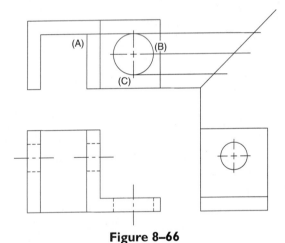

Figure 8–66

STEP 27

Make sure Polar Tracking is turned on. Draw lines from the intersection of the 45°-projection line to points in the Right Side view as shown in Figure 8–67. Use the following commands to accomplish this.

Command: **L** *(For LINE)*

Specify first point: *(Select the intersection of the angle at "A")*

Specify next point or [Undo]: *(Select the polar intersection at "B")*

Specify next point or [Undo]: *(Press ENTER to exit this command)*

Command: **L** *(For LINE)*

Specify first point: *(Select the intersection of the angle at "C")*

Specify next point or [Undo]: *(Select the polar intersection at "D")*

Specify next point or [Undo]: *(Press* ENTER *to exit this command)*

 Command: **L** *(For LINE)*

Specify first point: *(Select the intersection of the angle at "E")*

Specify next point or [Undo]: *(Pick a point below the bottom of the Side view at "F")*

Specify next point or [Undo]: *(Press* ENTER *to exit this command)*

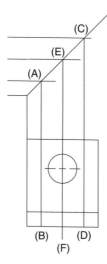

Figure 8–67

STEP 28

Delete the three projection lines from the Top view using the ERASE command as shown in Figure 8–68.

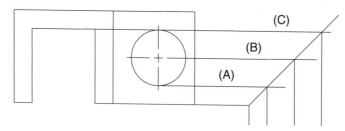 Command: **E** *(For ERASE)*

Select objects: *(Select lines "A," "B," and "C")*

Select objects: *(Press* ENTER *to perform this command)*

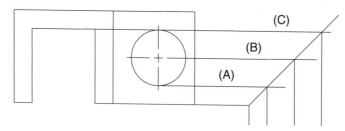

Figure 8–68

STEP 29

Use the TRIM command to trim away unnecessary geometry using the illustration in Figure 8–69 as a guide. The hidden hole and slot will be formed in the Side view through this operation.

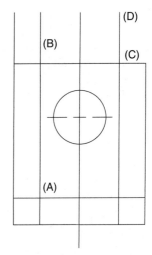

Figure 8–69

⊢ Command: **TR** (For TRIM)

Current settings: Projection=UCS
  Edge=None
Select cutting edges ...
Select objects: (Select the horizontal line at
  "A")
Select objects: (Press ENTER to continue)
Select object to trim or shift-select to
  extend or [Project/Edge/Undo]: (Select
  the line at "B")
Select object to trim or shift-select to
  extend or [Project/Edge/Undo]: (Press
  ENTER to exit this command)

Repeat this procedure for the line illustrated in Figure 8–69 using the horizontal line "C" as the cutting edge and the vertical line "D" as the line to trim.

STEP 30

Use the BREAK command to split the vertical line at "A" into two separate objects as shown in Figure 8–70. Accomplish this by typing @ in response to the prompt "Enter second point." This will split the line in two without the break being noticeable.

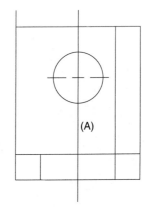

Figure 8–70

🔲 Command: **BR** (For BREAK)

Select object: (Select the vertical line at
  "A")
Specify second break point or [First
  point]: @ (To select the last point)

STEP 31

Use the ERASE command to delete the top half of the line broken in the previous step. This will leave a short line segment that will be converted to a centerline marking the center of the hidden hole (see Figure 8–71).

 Command: **E** *(For ERASE)*

Select objects: *(Select the vertical line at "A")*

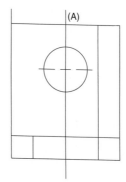

Figure 8–71

STEP 32

Change the two vertical lines labeled "A" and "B" in Figure 8–72 to the "Hidden" layer. First select the lines, activate the Layer Control box in Figure 8–72, and pick the "Hidden" layer. When finished, press ESC to remove the object highlight and the grips.

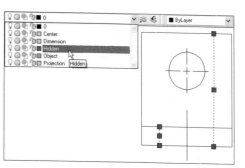

Figure 8–72

Change vertical line "C" in Figure 8–73 to the "Center" layer. First select the line, activate the Layer dropdown list in Figure 8–73, and pick the "Center" layer. When finished, press ESC to remove the object highlight and the grips.

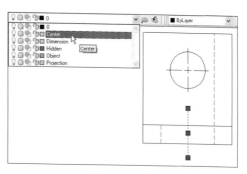

Figure 8–73

STEP 33

Draw three lines from key features on the Side view to intersect with the 45°- projection line as shown in Figure 8–74. Use Object Snap options whenever possible. Ortho mode must be on.

✏ Command: **L** *(For LINE)*
Specify first point: **Qua**
of *(Select the quadrant of the circle at "A")*
Specify next point or [Undo]: *(Pick a point just past the 45° angle)*
Specify next point or [Undo]: *(Press ENTER to exit this command)*

Repeat the same procedure for the other two projection lines. Use the OSNAP Endpoint option and begin the second projection line from the endpoint of the center marker at "B." Begin the third projection line from the quadrant of the circle at "C."

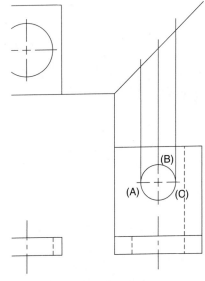

Figure 8–74

STEP 34

Draw three lines from the intersection at the 45°-projection line across to the Top view as shown in Figure 8–75. Draw the middle line longer because it will be converted to a centerline at a later step.

✏ Command: **L** *(For LINE)*
Specify first point: *(Select the intersection at "A")*
Specify next point or [Undo]: *(Snap to the polar intersection at "B")*
Specify next point or [Undo]: *(Press ENTER to exit this command)*

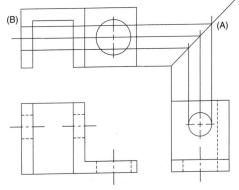

Figure 8–75

Repeat the same procedure for the other two lines. Do not use an Object Snap option for the opposite end of the middle line but rather identify a point just to the left of the vertical line at "B."

STEP 35

Use the ERASE command to erase the three vertical projection lines from the Side view as shown in Figure 8–76.

 Command: **E** *(For ERASE)*

Select objects: *(Select the vertical lines labeled "A," "B," and "C")*

Select objects: *(Press ENTER to execute this command)*

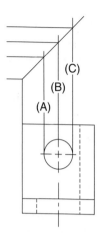

Figure 8–76

STEP 36

Use the TRIM command to trim away excess lines in the Top view as illustrated in Figure 8–77.

 Command: **TR** *(For TRIM)*

Current settings: Projection=UCS Edge=None

Select cutting edges ...

Select objects: *(Select the three vertical lines labeled "A," "B," and "C")*

Select objects: *(Press ENTER to continue)*

Select object to trim or shift-select to extend or [Project/Edge/Undo]: *(Select lines "D" through "G" to trim)*

Select object to trim or shift-select to extend or [Project/Edge/Undo]: *(Press ENTER to exit this command)*

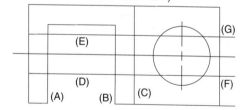

Figure 8–77

STEP 37

Before continuing, first turn OSNAP off. Then use the BREAK command to partially delete the horizontal line segment from "A" to "B" as shown in Figure 8–78. Use the BREAK command with the @ option to break the line into two segments at "C." Use the ERASE command to delete the trailing line segment at "D."

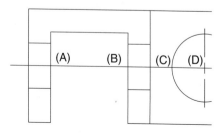

Figure 8–78

Command: **BR** (For BREAK)

Select object: (Select the horizontal line at "A")
Specify second break point or [First point]: (Select the line at "B")

Command: **BR** (For BREAK)

Select object: (Select the horizontal line at "C")
Specify second break point or [First point]: @ (To perform the break at the last point)

Command: **E** (For ERASE)

Select objects: (Select the horizontal line segment at "D")
Select objects: (Press ENTER to execute this command)

Turn OSNAP back on.

STEP 38

Change the four horizontal lines labeled "A" through "D" in Figure 8–79 to the "Hidden" layer. First select the lines, activate the Layer Control box in Figure 8–79, and pick the "Hidden" layer. When finished, press ESC to remove the object highlight and the grips.

Change the two horizontal lines labeled "E" and "F" in Figure 8–80 to the "Center" layer. First select the lines and then activate the Layer Control box in Figure 8–80 and pick the "Center" layer. When finished, press ESC to remove the object highlight and the grips.

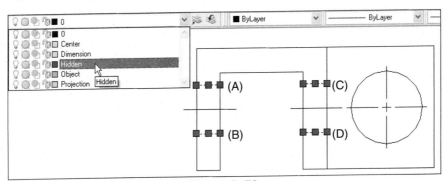

Figure 8–79

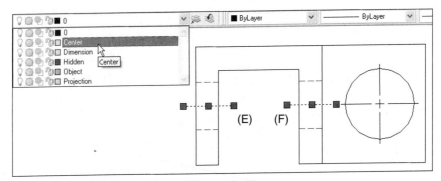

Figure 8–80

STEP 39

Use the ERASE command to delete the 45°
line. Use the FILLET command to create
corners in the Top view and Right Side
view as shown in Figure 8–81. Check that
the fillet radius is currently set to 0.00.

 Command: **E** *(For ERASE)*

Select objects: *(Select the inclined line at
"A")*

Select objects: *(Press ENTER to execute this
command)*

 Command: **F** *(For FILLET)*

Current settings: Mode = TRIM, Radius =
0.00
Select first object or [Polyline/Radius/
Trim]: *(Select line "B")*
Select second object: *(Select line "C")*

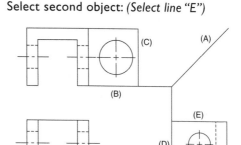 Command: **F** *(For FILLET)*

Current settings: Mode = TRIM, Radius =
0.00
Select first object or [Polyline/Radius/
Trim]: *(Select line "D")*
Select second object: *(Select line "E")*

Figure 8–81

STEP 40

Perform a ZOOM-Extents to view the
entire drawing (see Figure 8–82). The
final process in completing a multiview
drawing is to place dimensions to define
size and locate features. This topic will be
discussed in a later chapter.

 Command: **Z** *(For ZOOM)*

Specify corner of window, enter a scale
factor (nX or nXP), or [All/Center/
Dynamic/Extents/Previous/Scale/
Window] <real time>: **E** *(For Extents)*

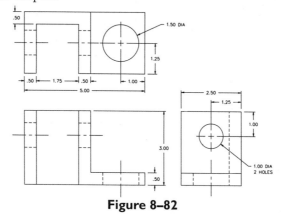

Figure 8–82

 Open the Exercise Manual PDF file for Chapter 8 on the accompanying CD for more
tutorials and exercises.

 If you have the accompanying Exercise Manual, refer to Chapter 8 for more tutorials and
exercises.

CHAPTER 9

An Introduction to
Drawing Layouts

INTRODUCING DRAWING LAYOUTS

Up to this point in the book, all drawings have been constructed in Model Space. This consisted of part geometry such as arcs, circles, lines, and even dimensions. You can even plot your drawing from Model Space; this gets a little tricky, especially when you attempt to plot the drawing at a known scale, or even trickier if multiple details need to be arranged on a single sheet at different scales.

You can elect to work in two separate spaces on your drawing; namely Model Space and Layout mode, sometimes referred to as Paper Space. Part of the geometry and dimensions is still drawn in Model Space. However, items such as notes, annotations and title blocks are laid out separately in a drawing layout, which is designed to simulate an actual sheet of paper. To arrange a single view of a drawing or multiple views of different drawings, you arrange a series of viewports in the drawing layout to view the images. These viewports are mainly rectangular in shape and can be made any size. You can even create circular and polygonal viewports. An option of the ZOOM command allows you to scale the image inside the viewport. In this way, a series of images may be scaled differently in order for the drawing to be plotted out at a scale of 1:1. A dedicated toolbar containing standard scales is also available to perform this task

This chapter introduces you to the controls used in a drawing layout to manage information contained in a viewport. A tutorial exercise is provided to help you practice creating drawing layouts.

MODEL SPACE

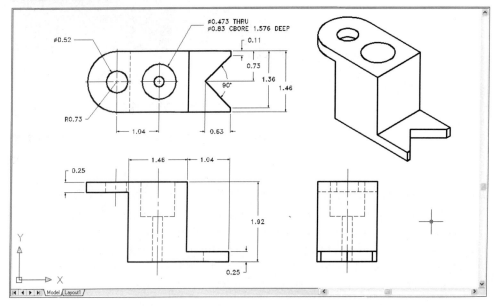

Figure 9–1

Before starting on the topic of Paper Space, you must first understand the current environment in which all drawings are originally constructed, namely Model Space. It is here that the drawing is drawn full size or in real world units. Model Space is easily identified by the appearance of the User Coordinate System icon located in the lower-left corner of the active drawing area in Figure 9–1. This icon is associated with the Model Space environment. Past and present users of AutoCAD have felt that it clutters the display screen and gets in the way of geometry, and they have used the UCSICON command to turn the icon off. While these points may seem true, the icon should be present at all times to identify the space you are working in. Another indicator that you are in Model Space is in the presence of the Model tab located just below the User Coordinate System icon, also in Figure 9–1.

TILED VIEWPORTS IN MODEL SPACE

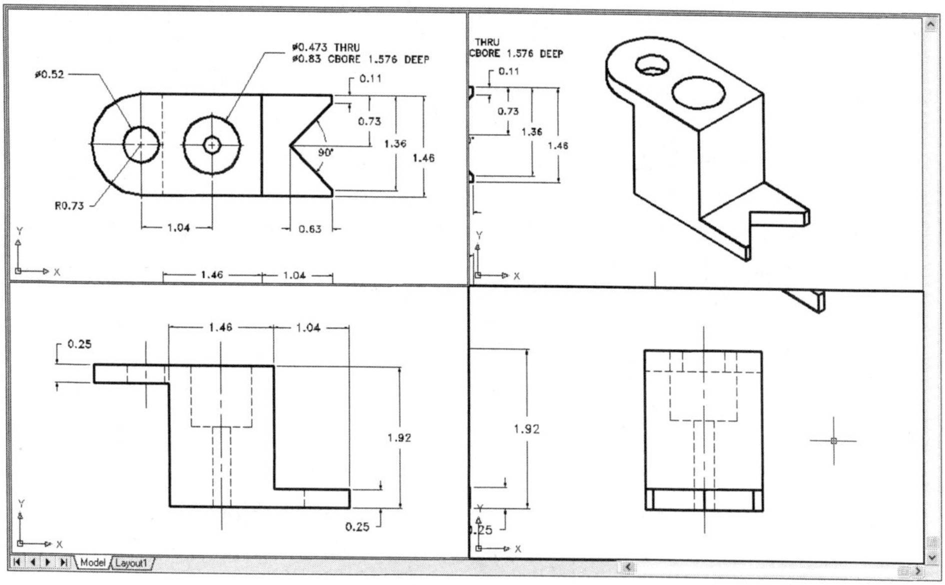

Figure 9–2

An advantage of constructing in the Model Space environment is the ability to divide the display screen into a number of smaller screens, as illustrated in Figure 9–2. **(Drawing file 09_V-Step is provided and can be opened to practice creating viewports in Model Space).** In the illustration, the display screen is divided into four smaller screens or viewports. These viewports are further arranged in a certain pattern or configuration. You do not have much control over this viewport arrangement. For this reason these are called tiled viewports, because they resemble the tile pattern commonly used in floors. Do not be confused that you have four separate files open in the illustration. Rather, it is the same drawing file, which enables you to view separate parts of the drawing in each viewport. Again, notice the appearance of the familiar User Coordinate System icon present in the lower left corner of each viewport. This signifies you are in Model Space in each viewport. Also in the illustration, notice how the lower-right viewport has a thick border around it along with the presence of the AutoCAD cursor. This happens to be the active viewport. Only one viewport can be active at one time. If a viewport is inactive, the appearance of an arrow instead of a cursor is present. To activate another viewport, simply move the arrow to the viewport and click inside the viewport; this activates a new current viewport.

THE VPORTS COMMAND

Viewports are created through the VPORTS command. You can access this command by choosing Viewports from the View pull-down menu. Choosing New Viewports… as shown in Figure 9–3, activates the Viewports dialog box in Figure 9–4. A toolbar dedicated to viewports is also illustrated in Figure 9–3. Numerous layout examples and configurations for viewports are available. Clicking on the desired viewport in the Standard viewports area previews the viewport layout at the right of the dialog box. These changes will be made to the display after you click the OK button at the bottom of the dialog box.

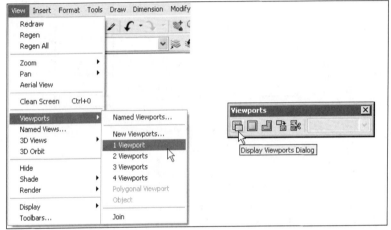

Figure 9–3

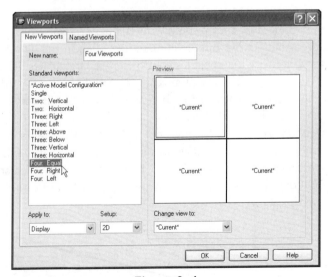

Figure 9–4

MODEL SPACE AND PAPER SPACE

You have already seen that Model Space is the environment set aside for constructing your drawing in real world units or full size. Paper Space is considered an area used to lay out your drawing before it is plotted. Paper Space is also used to place title block information and notes associated with the drawing. The drawing illustrated in Figure 9–5 has been laid out in Paper Space and shows a sheet of paper with the drawing surrounded by a viewport. The dashed lines along the outer perimeter of the sheet are referred to as margins. Anything inside the margins will plot; for this reason, this is called the printable area. Notice also at the bottom of the screen the Layout1 tab is activated. Both tabs at the bottom of the screen can be used to easily display your drawing in either Model Space or Paper Space. Also notice the icon in the lower-left corner of the illustration; this icon is in the form of a triangle and is used to quickly identify the Paper Space environment.

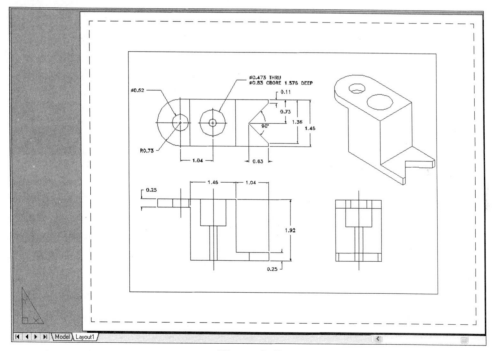

Figure 9–5

PAPER SPACE FEATURES

When you draw and plot from Model Space, the limits of the drawing have to be properly set depending on the scale of the drawing and the sheet of paper the drawing is planned for. Also, title block information has to be scaled depending on the scale of the drawing. The text height that made up the notes in a drawing also has to be scaled depending on the scale of the drawing. Finally, the scale of the drawing is applied to Plot Settings tab of the Plot dialog box, which scales the drawing for placement on a certain sheet of paper. There can be many problems when you use this method. For instance, a certain design office assigned an individual to plot all drawings produced by the engineers and designers. It was this individual's responsibility to plot the drawings at the proper scale. However, with numerous drawings to plot, it was easy to get the plot scales confused. As a result, drawings were plotted at the wrong scale. This not only cost paper but valuable time. This scenario is the reason for Paper Space; All drawings, no matter how many details are arranged, can be plotted at 1=1 with little or no error encountered, through the use of Paper Space. Plotting becomes the fundamental reason for using the Paper Space environment. Here are a few other reasons for arranging a drawing in the Paper Space:

- Paper Space is based on the actual sheet size. If you are plotting a drawing on a "D" size sheet of paper, you use the actual size (36 x 24) in Paper Space.

- Title blocks and the text used for notes do not have to be scaled as in the past with Model Space.

- Viewports created in Paper Space are user-defined. Viewports created in Model Space are dependent on a configuration set by AutoCAD. In Paper Space, the viewports can be different sizes and shapes, depending on the information contained in the viewport.

- Multiple viewports can be created in Paper Space, as in Model space. However, the images assigned to Paper Space viewports can be at different scales. Also, the control of layers in Paper Space is viewport-dependent. In other words, layers turned on in one viewport can be turned off or frozen in another viewport.

- All drawings, no matter how many viewports are created or details arranged, are plotted out at a scale of 1=1.

FLOATING MODEL SPACE

When a viewport is created in paper space, the drawing image automatically fills up the viewport as shown in Figure 9–6. Notice also the appearance of the drafting triangle icon in the lower-left corner of the display screen. This signifies that you are presently in the paper space environment.

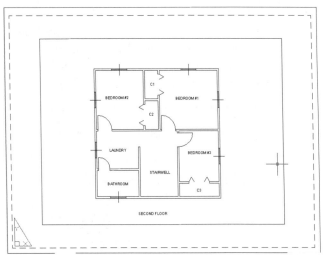

Figure 9–6

While in paper space, operations such as adding a border, title block, and general notes are usually performed. Paper space is also the area in which you plot your drawing.

Other operations need to be performed in model space without leaving the layout environment. This is referred to as floating model space. To activate floating model space, double-click inside of a viewport. The UCS icon normally will display (see Figure 9-7).

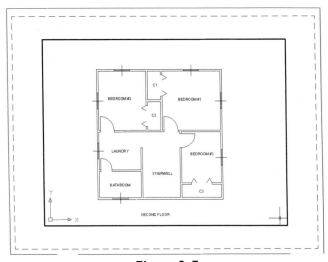

Figure 9–7

     Note: If you need to switch back to the layout mode, double-click outside of the viewport. The paper space drafting triangle will reappear.

SCALING VIEWPORT IMAGES

One of the operations performed while in a floating model space viewport is the scaling of the image to paper space. It was previously mentioned that by default, images that are brought into viewports display in a zoomed-extents appearance. To scale an image to a viewport, use the following steps and refer to Figure 9–8:

- Have the Viewports toolbar present somewhere on your screen.
- Activate floating model space by double-clicking inside of the viewport to scale.
- In the Viewports toolbar, click in the scale box to display other scales.
- Select various scales until the image fits in the viewport.

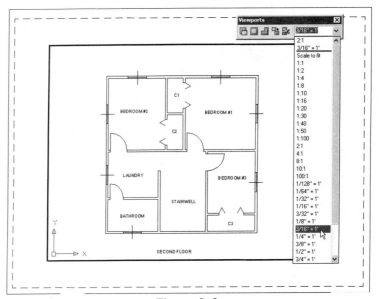

Figure 9–8

If you find that the image is very close to fitting but is cut off by one edge of a viewport, there are two options to resolve the problem:

- Stretch the viewport (using grips or the STRETCH command)
- Pan the image into position.

A combination of both methods usually proves to be the most successful method.

To pan the image, double--click inside the viewport to start floating model space. You will notice the thick borders on the viewport. Using the PAN command or holding down the wheel on a wheel mouse, move the image to fit the viewport. Once the image is panned into position, double-click outside of the viewport on the drawing surface (usually the area surrounding the drawing sheet) to return from the floating model space back to the layout space.

USING THE ZOOM XP OPTION

The ZOOM command can also be used to scale images inside of viewports to paper space. This is accomplished using the XP option, which stands for "Times Paper". For example, to size a floating model space image to the scale of 1/8":1'-0", you would enter the following at the ZOOM command:

Command: **Z** *(For ZOOM)*
Specify corner of window, enter a scale factor (nX or nXP), or
[All/Center/Dynamic/Extents/Previous/Scale/Window] <real time>: 1/96XP

LOCKING VIEWPORTS

When you have scaled the contents of a floating model space viewport, it is very easy to accidentally change this scale by rolling the wheel of a mouse or performing another ZOOM operation. To prevent accidental panning and zooming of a viewport image once it has been scaled, a popular technique is to lock the viewport. To perform this task, use the following steps and refer to Figure 9–9:

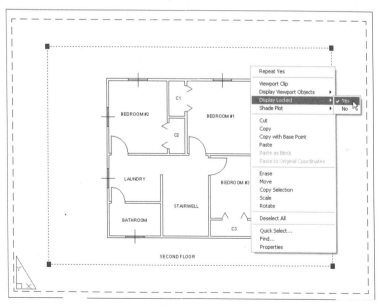

Figure 9–9

- Be sure you are in layout mode (look for the drafting triangle in the lower-left corner)
- Select the viewport to lock. It will highlight and grips will appear.
- Right-click on your screen. A menu will appear.
- Select Display Locked followed by "Yes".

Note: When locking a viewport, your scale will be grayed out inside of the Viewports toolbar. If you need to change the scale of an image inside of a viewport, you must first unlock the viewport.

CREATING A PAPER SPACE LAYOUT

Try It! – Open the drawing file 09_Gasket1. Figure 9–10 shows the drawing originally created in Model Space. You decide to lay this drawing out in Paper Space.

As you click on the Layout1 tab at the bottom of the display screen, you activate the Page Setup dialog box in Figure 9–11. While in the Plot Device tab, click on the plotter configuration desired. Plotter configurations list all plotters that can be used by your drawing. The HP DesignJet 750C Plus will be used throughout this discussion of layouts in Paper Space. You can configure this plotter using the first tutorial exercise in chapter 10, "Plotting Your Drawings." The concepts described here do not change for different plotting devices. Notice the name at the top of this dialog box, change this to One View Drawing. It is considered good practice to assign a meaningful name to your layout.

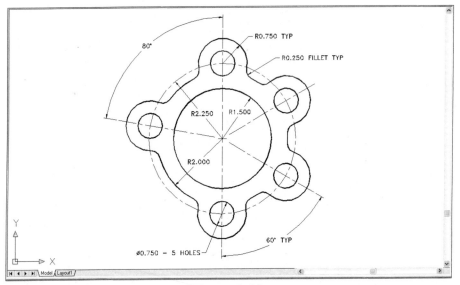

Figure 9–10

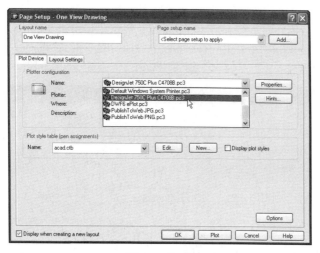

Figure 9–11

Clicking on the Layout Settings tab of the Page Setup dialog box displays additional settings, shown in Figure 9–12, for controlling a plot. Since the drawing will be laid out in Paper Space, the plot scale reflects 1:1. You can also control the Plot offset (where the plot begins), the Plot area, and additional Plot options. Clicking in the Paper size edit box will display all supported paper sizes for performing the plot. With the plot device currently set to the DesignJet 750C Plus, numerous paper sizes can be used, as shown in Figure 9–13. For this example, the expanded B (17.00 x 11.00) is being used as the layout sheet.

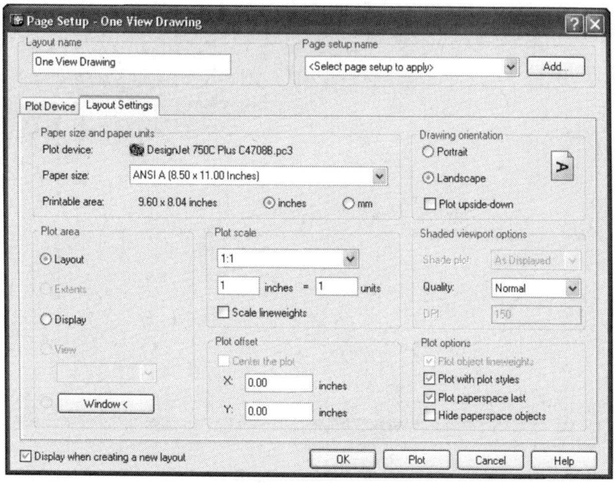

Figure 9–12

Clicking the OK button displays the image in Figure 9–14. The expanded B size draw-ing sheet is displayed in this figure along with a viewport that contains the drawing image brought in from Model Space. Notice also the title of the layout is One View Drawing, located at the bottom of the screen next to the Model tab.

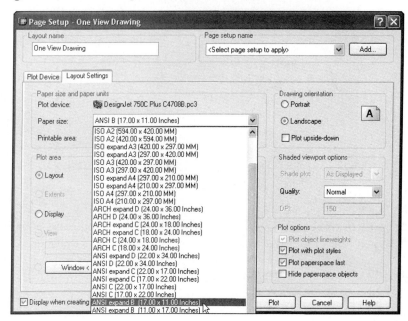

Figure 9–13

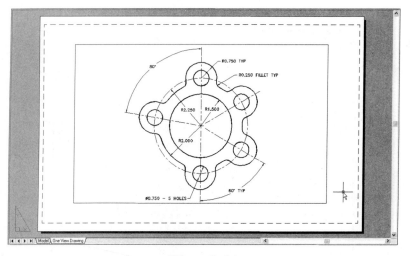

Figure 9–14

In the previous figure, the viewport was automatically created and the image displayed inside the viewport. However, a scale factor has not yet been applied to the image. Before applying a scale factor, you must first be placed in floating Model Space. This is signified by the sheet layout in Figure 9–15 along with the presence of the User Coordinate System icon inside the viewport. To accomplish this task, double-click anywhere inside the viewport. This places you into floating Model Space.

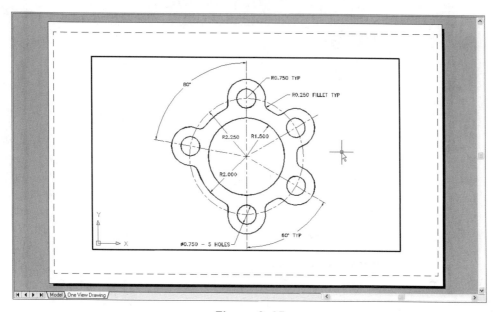

Figure 9–15

From inside floating Model Space, the image needs to be scaled to the Paper Space viewport. A special toolbar is available to control viewports. To display it, first right-click any button on your display screen. This displays the toolbar pop-up menu in Figure 9–16. Clicking on Viewports displays the Viewports toolbar also illustrated in Figure 9–16. The number inside the edit box in Figure 9–17 is the current scale of the image inside the viewport. Whenever you create a drawing layout, the image is automatically zoomed to the extents of the viewport. For this reason, the desired scale will never be correctly displayed in this area.

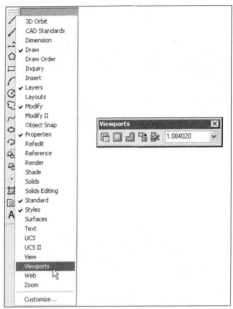

Figure 9–16

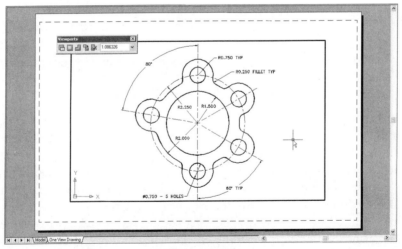

Figure 9–17

To have the image properly scaled inside the viewport, click the down arrow in Figure 9–18. This displays the drop-down list of all standard scales. These include standard engineering and architectural scales. The image is assigned a scale of 1:1, which is applied to the viewport. You can see in the figure that the image has shrunk inside the viewport. However, it is properly scaled.

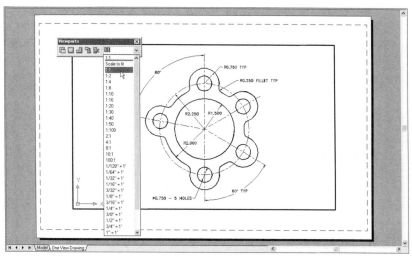

Figure 9–18

With the image scaled, double-click anywhere outside the viewport; this returns you to the Paper Space environment. Clicking once on the viewport displays grips located at the four corners as in Figure 9–19. Use these grips to size the viewport to the image. Accomplish this by clicking on one of the grips in the corner and stretching the viewport to a new location.

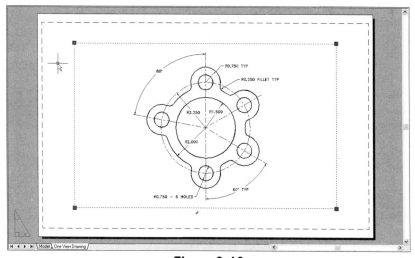

Figure 9–19

The results of this operation are displayed in Figure 9–20. Press ESC to remove the object highlight and the grips.

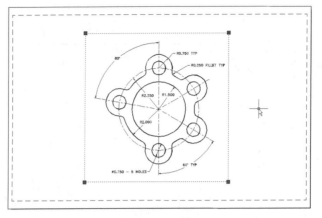

Figure 9–20

Before the drawing is plotted, a title block containing information such as drawing scale, date, title, company name, and designer is inserted in the Paper Space environment. Accomplish this by first choosing Block… from the Insert pull-down menu as shown in Figure 9–21. This activates the Insert dialog box shown in Figure 9–22. This feature of AutoCAD will be covered in greater detail in chapter 16. Be sure the block "ANSI B title block" is listed in the Name: edit box.

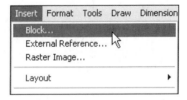

Figure 9–21

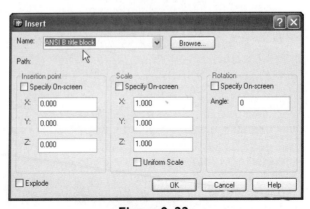

Figure 9–22

Click the OK button to place the title block in Figure 9–23. Be sure the title block displays completely inside the paper margins; otherwise part of it will not plot.

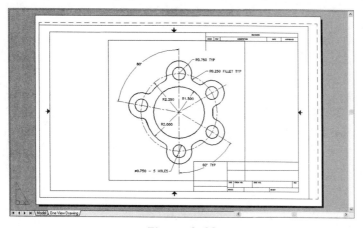

Figure 9–23

At this point, the MOVE command is used to move the viewport containing the drawing image to a better location based on the title block. Unfortunately, if the drawing were to be plotted out at this point, the viewport would also plot. It is considered good practice to assign a layer to the viewport and then turn it off. In Figure 9–24, the viewport is first selected and grips appear. Click in the Layer Control box and click in the Viewport layer. Then click the light bulb icon in the Viewport layer row to turn it off.

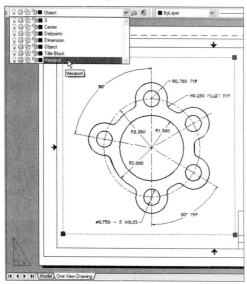

Figure 9–24

The drawing display will appear similar to Figure 9–25, with the drawing laid out and properly scaled in Paper Space.

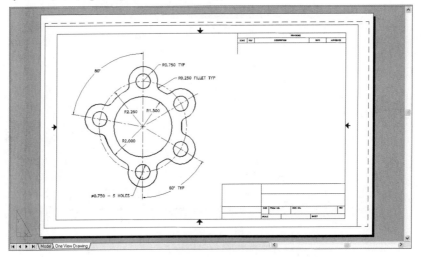

Figure 9–25

CREATING VIEWPORTS WITH THE MVIEW COMMAND

Open the drawing file 09_Mview. In the unlikely event that you delete a Paper Space viewport, the image tied to that viewport is also deleted (see Figure 9–26). Not to worry—you still have all drawing information contained in the Model tab. You need to re-create the viewport in Paper Space, which will retrieve the information from the Model tab.

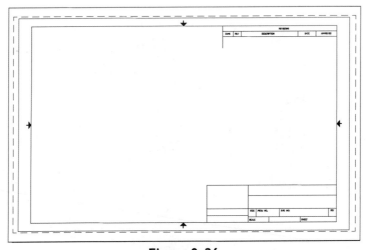

Figure 9–26

To do this, choose Viewports from the View pull-down menu or activate the Viewports toolbar. Numerous viewport options are displayed in Figure 9–27.

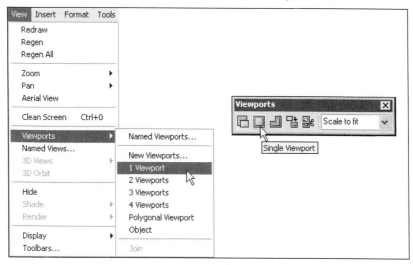

Figure 9–27

Clicking on 1 Viewport will prompt you to pick a first corner and other corner, which will create the viewport in Figure 9–28. Whenever a new viewport is created, the image automatically zooms to the extents of the viewport no matter how large or small the viewport is. Notice also in Figure 9–28 the appearance of the Status bar. Clicking on the Paper switches you to floating Model Space; the button changes to MODEL.

Remember that you scale the image inside floating Model Space. The Viewports toolbar was used earlier to accomplish this task. You could also use the ZOOM command along with the XP option; it is this special XP option that scales the image inside the viewport to Paper Space units. Follow the command prompt sequence below to perform this task:

Command: **Z** *(For ZOOM)*
Specify corner of window, enter a scale factor (nX or nXP), or
[All/Center/Dynamic/Extents/Previous/Scale/Window] <real time>: **I XP**

Complete the layout by returning to Paper Space, changing the viewport to the proper layer and then turning it off, and plotting the drawing.

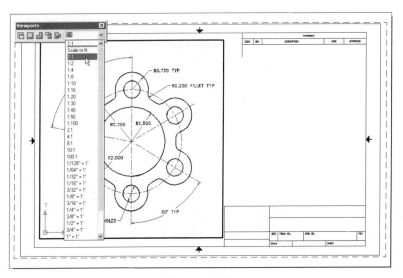

Figure 9–28

USING A WIZARD TO CREATE A LAYOUT

The AutoCAD Create Layout Wizard can be especially helpful in laying out a drawing in the Paper Space environment.

 Try It! – Open the drawing file 09_Lifter illustrated in Figure 9–29. Choosing Wizards from the Tools pull-down menu displays additional wizards, as shown in Figure 9–30.

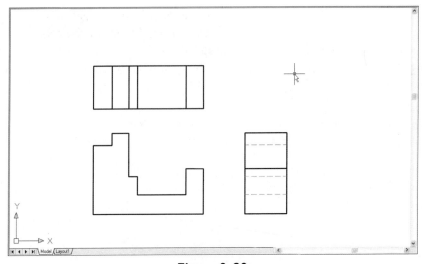

Figure 9–29

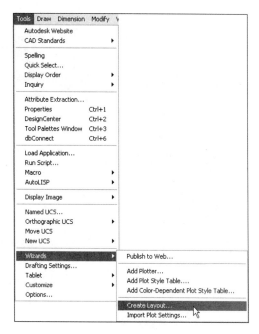

Figure 9–30

Clicking on Create Layout… displays the Create Layout-Begin dialog box in Figure 9–31. You cycle through the different categories and when finished, the drawing layout is displayed, complete with title block and viewport. Figure 9–31 displays the first category, namely Begin. You have the opportunity to name the layout of your choice in this dialog box. Notice the name "Three Views" will be used for this segment. Click the Next> button to continue to the Printer category.

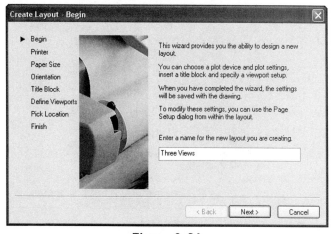

Figure 9–31

Use the Printer category to select a configured plotter from the list provided as in Figure 9–32. The HP DesignJet 750C Plus will be selected for use during this segment. Click the Next> button to continue to the Paper Size category.

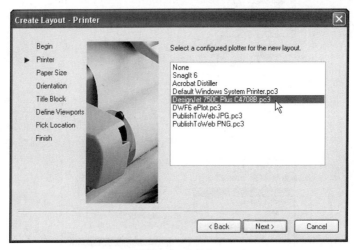

Figure 9–32

The Paper Size category displays all available paper sizes supported by the currently configured plotter. The size ANSI expand C (22.00 x 17.00 Inches) will be used during this segment. Selecting a different printer or plotter could change the number of selections. You can also select whether your paper size is in Millimeter (Metric) or Inches (English) units (see Figure 9–33). Click the Next> button to continue to the Orientation category.

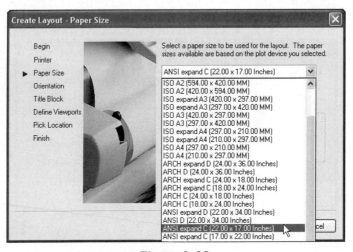

Figure 9–33

Use the Orientation category to designate whether to plot the drawing in Landscape or Portrait mode in Figure 9–34. Click the Next> button to continue to the Title Block category.

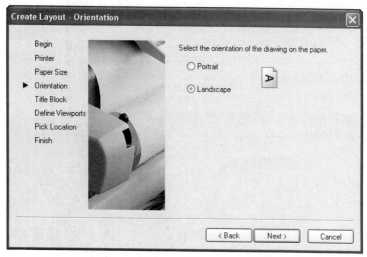

Figure 9–34

Depending on the paper size, choose a corresponding title block to automatically be inserted in the layout sheet. When you pick the title block as shown in Figure 9–35, a preview of the title is displayed at the right of the dialog box. Click the Next> button to continue to the Define Viewports category.

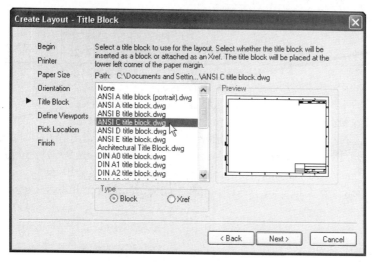

Figure 9–35

The Define Viewports category is used to either create a viewport or leave the drawing layout empty of viewports. If you select the None option, you will have to use the MVIEW command to create a user-defined viewport. Other choices include creating a single viewport, a series of standard engineering views (Front, Top, Side, and Isometric), or an array of viewports in a certain pattern (see Figure 9–36). You can also set the scale for the image inside the viewport. The scale of 1:1 shown in Figure 9–36 will automatically be applied to the image; you will not have to use the Viewports toolbar to set this value. Click the Next> button to continue to the Pick Location category.

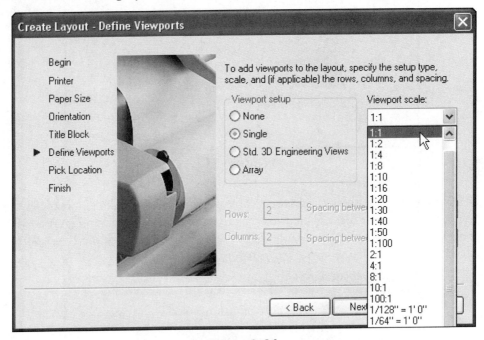

Figure 9–36

The Pick Location category creates the viewport to hold the image in Paper Space. If you click the Next> button, the viewport will be constructed to match the margins of the paper size. If you click the Select location< button in Figure 9–37, you return to the drawing and pick two diagonal points to define the viewport. In this way, you control the size of the viewport. Click the Next> button to continue to the Finish category.

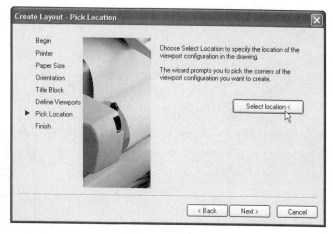

Figure 9–37

The Finish category alerts you to a new layout named "Three Views" that will be created. Once it is created, modifications can be made through the Page Setup dialog box. Click the Finish button in Figure 9–38 to dismiss the Create Layout wizard and create the drawing layout.

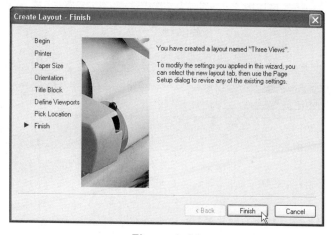

Figure 9–38

The completed layout is illustrated in Figure 9–39. Notice the new layout tab called Three Views at the bottom of the screen next to the Layout1 tab. Also, notice the lack of dashed margins to identify the edges of the plotted area. This is because a viewport the size of the paper margins was created as a result of the Pick Location category of the Create Layout wizard. Since this viewport is quite large compared to the drawing image, picking the edge of the viewport highlights the object and displays grips at its four corners (see Figure 9–39). Click on the corners of the grips to size the viewport to match the image.

Figure 9–40 displays an image of a viewport that has been sized from the previous operation. Double-clicking inside the viewport switches you to floating Model Space. Here, the Viewports toolbar verifies the scale of the image that occupies the viewport as 1:1 (set in the Define Viewports category of the Create Layout wizard).

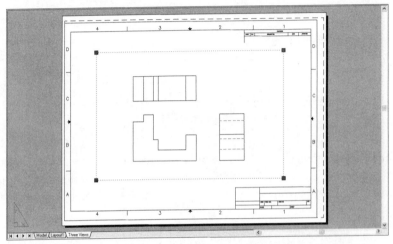

Figure 9–39

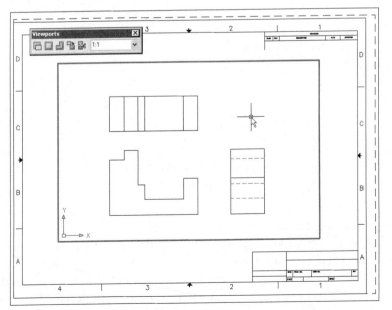

Figure 9–40

Double-clicking outside the viewport returns you to Paper Space. Selecting the viewport and then right clicking displays the cursor menu in Figure 9–41. The upper part of this cursor menu holds special controls for viewports. One such control is the ability to lock a viewport. This will prevent any accidental zoom operations from being performed in floating Model Space.

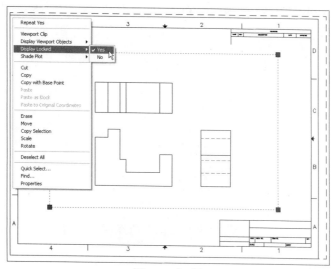

Figure 9–41

Changing the viewport to a layer and then turning the layer off displays the image as in Figure 9–42.

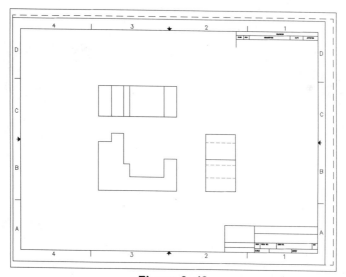

Figure 9–42

One more operation needs to be performed before you save the drawing. When you use the Create Layout wizard, the new layout tab (Three Views) was added to an existing layout (Layout1). Right-clicking on the Layout1 tab displays the cursor menu in Figure 9–43. Clicking on Delete displays the AutoCAD alert box in Figure 9–44. Click the OK button to remove the tab. (Note: You cannot remove the Model tab.) The completed drawing layout, ready for plotting, is displayed in Figure 9–45.

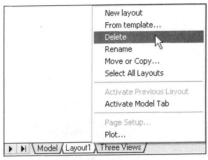

Figure 9–43

Figure 9–44

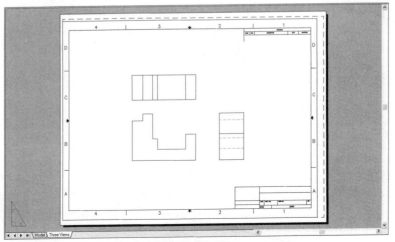

Figure 9–45

ARRANGING ARCHITECTURAL DRAWINGS IN PAPER SPACE

Architectural drawings pose certain challenges when you lay them out in Paper Space based on the scale of the drawing. For example, Figure 9–46 shows a drawing of a stair detail originally created in Model Space. The scale of this drawing is 3/8" = 1'-0". This scale will be referred to later on in this segment on laying out architectural drawings in Paper Space.

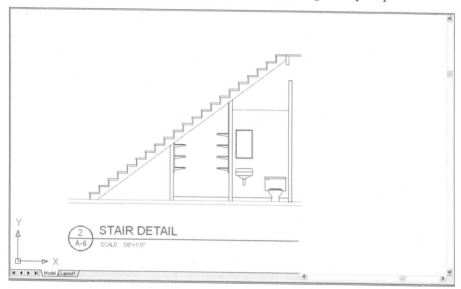

Figure 9–46

Try It! – Open the drawing file 09_Stair Detail. Click on the Layout1 tab or use the Create Layout wizard to arrange the stair detail in Paper Space so that your drawing appears similar to Figure 9–47. In the figure, we see the drawing sheet complete with viewport that holds the image and a title block that has been inserted. Use ARCH expand C as the paper size and insert the ANSI C title block. Because the image in Figure 9–47 is now visible in Paper Space, it is not necessarily scaled to 3/8" = 1'-0". Double-clicking inside the viewport places you in floating Model Space (notice the appearance of the User Coordinate System icon only in the viewport). Activate the Viewports toolbar and scale the image in floating Model Space to the proper scale. Notice, in the figure, a complete listing of the more commonly used Architectural scales. You could also have used the ZOOM command and entered a XP value of 1/32 (found by dividing 1' [12"] by 3/8").

Command: **Z** *(For ZOOM)*
Specify corner of window, enter a scale factor (nX or nXP), or
[All/Center/Dynamic/Extents/Previous/Scale/Window] <real time>: 1/**32XP**

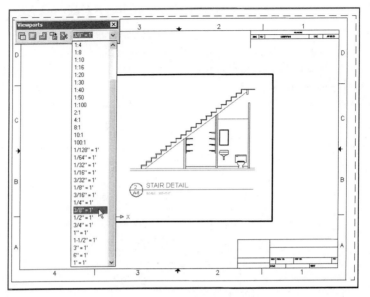

Figure 9–47

Changing the viewport to a different layer and then turning that layer off displays the completed layout of the stair detail in Figure 9–48. Since the purpose of Paper Space is to lay out a drawing and scale the image inside the viewport, this drawing will be plotted at a scale of 1:1.

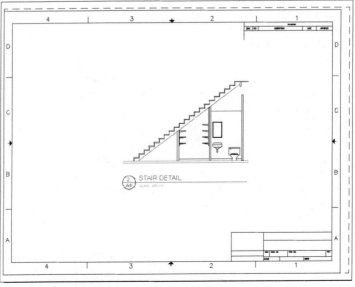

Figure 9–48

TYPICAL ARCHITECTURAL DRAWING SCALES

When the ZOOM command is used along with the XP option to scale architectural drawings drawn full size in Model Space, below are a few typical architectural scales and the XP scaling factors to use inside the floating Model Space viewport:

1/8" = 1'-0"	Zoom 1/96XP
1/4" = 1'-0"	Zoom 1/48XP
3/8" = 1'-0"	Zoom 1/32XP
1/2" = 1'-0"	Zoom 1/24XP
3/4" = 1'-0"	Zoom 1/16XP
1" = 1'-0"	Zoom 1/12XP
1 1/2" = 1'-0"	Zoom 1/8XP

All of the XP values are arrived at by the division of the value located to the right of the "=" by the value located to the left of the "=." In other words, to arrive at the XP value for the scale 1/2" = 1'-0", divide 1' (12") by 1/2 or 0.50; the result is 24. The value to enter when using the ZOOM command is 1/24XP.

ARRANGING METRIC DRAWINGS IN PAPER SPACE

The methods used for arranging metric drawings in Paper Space are no different from what has been already discussed for architectural drawings. For example, Figure 9–49 shows a drawing of an object in decimal units, originally created in Model Space. It just so happens that decimal units can also support metric drawings.

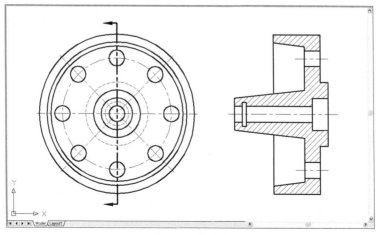

Figure 9–49

478

 Try It! – Open the drawing file 09_Metric Bearing. If the lineweights looks too thick, open the Lineweight Settings dialog box found under the Format pull-down menu and adjust the Adjust Display Scale control down. Click on the Layout1 tab or use the Create Layout wizard to arrange the metric drawing in Paper Space so that your drawing appears similar to the illustration in Figure 9–50. In the figure, we see the drawing sheet complete with viewport that holds the image and a metric title block that has been inserted. Use ISO expand A3 as the paper size and insert the ISO A3 title block. Because the image in Figure 9–50 is now visible in Paper Space, it is not necessarily scaled at 1:1. Double-clicking inside the viewport places you in floating Model Space (notice the appearance of the User Coordinate System icon only in the viewport). Activate the Viewports toolbar and scale the metric image in floating Model Space to the 1:1 scale.

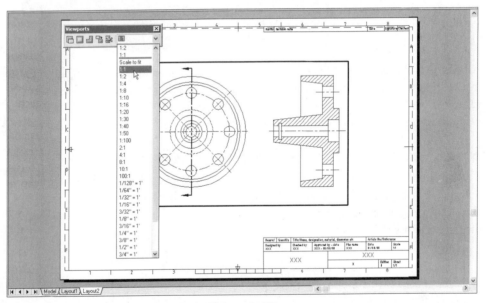

Figure 9–50

Changing the viewport to a different layer and then turning that layer off displays the completed layout of the object in Figure 9–51. Since the purpose of Paper Space is to lay out a drawing and scale the image inside the viewport, this drawing will be plotted out at a scale of 1:1.

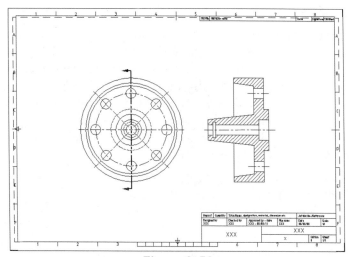

Figure 9–51

CREATING MULTIPLE DRAWING LAYOUTS

The methods explained so far have dealt with the arrangement of a single layout in Paper Space. AutoCAD provides for greater flexibility with working in Paper Space by enabling you to create multiple layouts of the same drawing. This will be explained through the illustration in Figure 9–52, which depicts a floor plan complete with dimensions. The drawing appears very busy, with a foundation plan with its dimensions and an electrical plan with its notes all overlaid on the floor plan. Because it is difficult to read each plan, individual layouts will be created to display separate images of the floor, foundation, and electrical plans. Because multiple layouts are used, a special feature of layers will allow certain layers frozen in one layout to be visible in another layout.

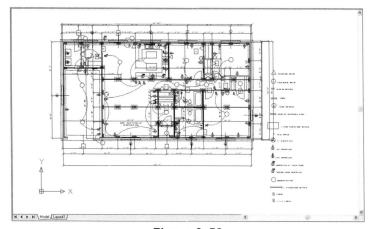

Figure 9–52

Begin the multiple layout process by clicking on the Layout1 tab. Make the proper plotter and sheet size settings to arrange the floor plan in Paper Space so that your drawing appears similar to the illustration in Figure 9–53. In the figure, we see the drawing sheet complete with viewport that holds the image and an architectural title block that has been inserted. Because the image in Figure 9–53 is now visible in Paper Space, it is not necessarily scaled to 1/4" = 1'-0". Double-clicking inside the viewport places you in floating Model Space (notice the appearance of the User Coordinate System icon only in the viewport). Activate the Viewports toolbar and scale the image in floating Model Space to the proper scale.

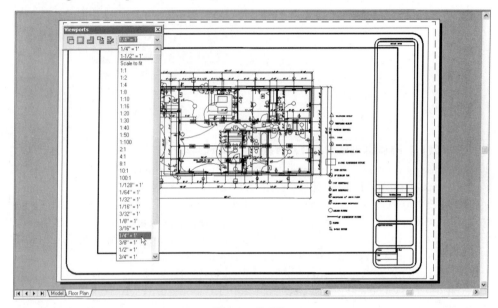

Figure 9–53

You want to see only the floor plan and its dimensions inside this viewport. Activating the Layer Properties Manager dialog box displays the layer information in Figure 9–54. When you create a new drawing layout and use this dialog box, two additional layer modes are added. These are the ability to freeze layers only in the active viewport and the ability to freeze layers in new viewports. Freezing layers in all viewports is not an effective means of controlling layers when you create multiple layouts. You need to be very familiar with the layers created in order to perform this task. To display only the floor plan information, notice that the following layers have been frozen under the Current... heading in Figure 9–54:

Construction

Electrical_Symbols

Electrical_Wiring

Foundation

Foundation_Centers

Foundation_Dimensions

Piers

All other layers such as Appliances, First_Floor, Floor_Dimensions, Title Block, Title Text, Viewport, and Windows remain visible in this viewport.

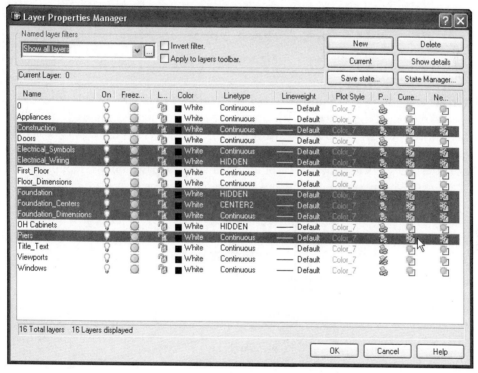

Figure 9–54

The resulting image is displayed in Figure 9–55 with only the floor plan information visible in the viewport. The next series of steps involve the creation of additional layouts to display the foundation and electrical plans.

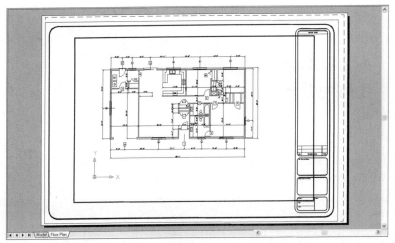

Figure 9–55

To create an additional layout, right-click on the Floor Plan tab at the bottom of the display screen as in Figure 9–55. This displays the cursor menu in Figure 9–56. Clicking on Move or Copy... displays the Move or Copy dialog box in Figure 9–57. You have the opportunity to shuffle drawing layouts around to change their order. You can also click in the check box to create a copy of the existing layout based on the floor plan. This new layout will be moved to the end; it will be placed after the floor plan.

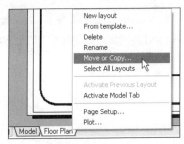

Figure 9–56

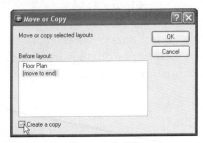

Figure 9–57

Clicking the OK button displays the new layout in Figure 9–58. When you copy a layout, all information such as title block, viewport, and information inside the viewport is copied. Begin the conversion of the new layout to the foundation plan by right-clicking on the Floor Plan (2) tab at the bottom of the display screen. This displays the cursor menu in Figure 9–59. Click on the Rename option to activate the Rename Layout dialog box in Figure 9–60. Rename this layout to Foundation Plan and click the OK button.

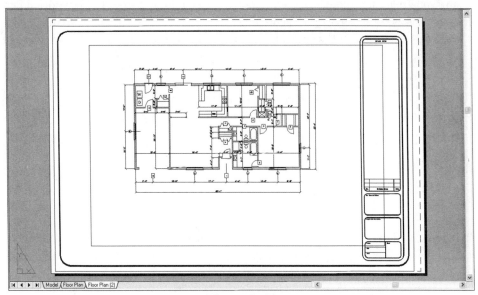

Figure 9–58

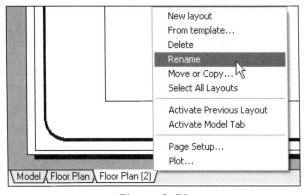

Figure 9–59

Figure 9–60

Be sure you are in floating Model Space by double-clicking inside the viewport. Then activate the Layer Properties Manager dialog box in Figure 9–61. Because this layout will hold all information that pertains to the Foundation Plan, the following layers are frozen in the current viewport:

- Appliances
- Construction
- Electrical_Symbols
- Electrical_Wiring
- Floor_Dimensions

The remaining layers, thawed in the current viewport, will display the foundation plan while freezing all other layers in this viewport.

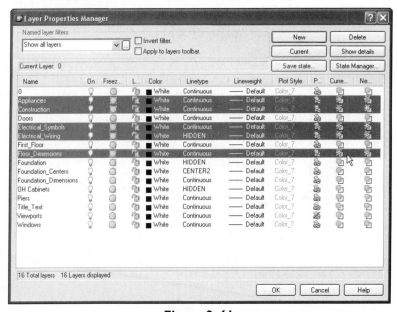

Figure 9–61

The resulting image is displayed in Figure 9–62, which shows the information needed to describe the foundation plan. Notice also at the bottom of the display screen that the new layout tab is renamed to reflect the Foundation Plan.

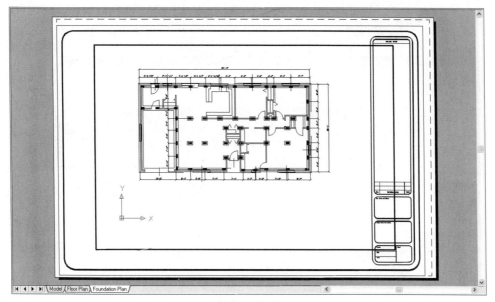

Figure 9–62

Follow the previous steps to create a new layout and rename the layout to Electrical Plan. Then activate the Layer Properties Manager dialog box in Figure 9–63 and freeze the following layers in the current viewport:

> Construction
>
> Floor_Dimensions
>
> Foundation
>
> Foundation_Centers
>
> Foundation_Dimensions
>
> Piers

The remaining layers, thawed in the current viewport, will display the electrical plan in Figure 9–64.

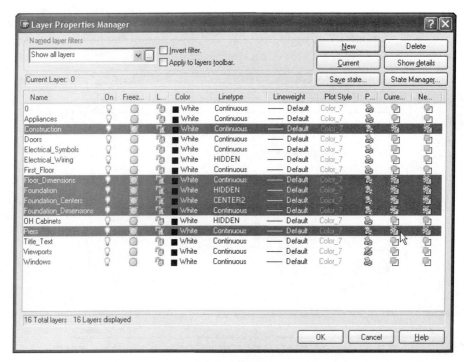

Figure 9–63

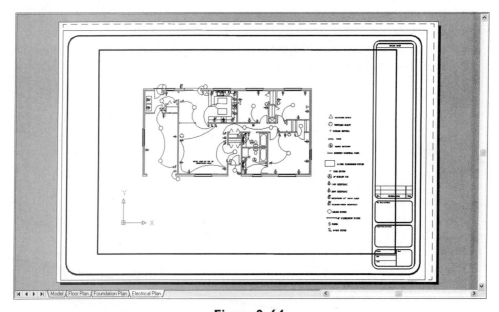

Figure 9–64

Complete the multiple drawing layout in Figure 9–65 by assigning the viewports to a layer (called Viewports) and turn this layer off. Add text to each layout to identify the purpose and scale of each layout (Floor Plan, Foundation Plan, and Electrical Plan).

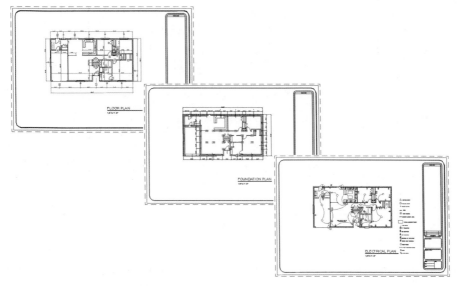

Figure 9–65

TRANSLATING DISTANCES BETWEEN SPACES

Suppose you want to label the rooms located in the floor plan illustrated in Figure 9–66. It is considered good practice to add this type of text while still inside of model space. This will allow the text to be selected with the floor plan if the drawing needs to move to a new location. Unfortunately, you must first determine the plotted text size. You then multiply plotted text height by the plotted scale of the drawing. This will give you the text height while inside of the model.

A better way to accomplish this task is to use the "Convert distance between spaces" tool located in the Text toolbar illustrated in Figure 9–66. The actual command name is SPACETRANS or "translate between model space and paper space".

 Try It! – Open the drawing 09_SpaceTrans as shown in Figure 9–66. The floor plan is currently displayed in floating model space. Also, the scale of the floor plan inside of the viewport is ¼" = 1" (The scale is grayed out in the figure because the viewport is locked.) You want the text height of all rooms to plot out at a height of 0.25 units. Instead of calculating this height manually based on the viewport scale, the spacetrans command will be used instead

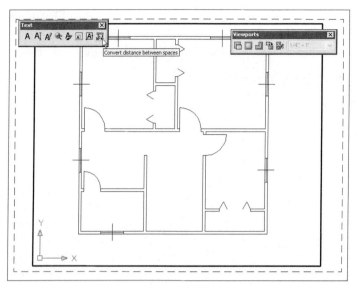

Figure 9–66

Activate the MTEXT command and pick a point to specify the first corner of the mtext rectangle. However, do not pick the second rectangle point. You will be prompted to enter a number of text properties; one property is Height. After you enter Height from the keyboard, click on the "Convert distance between spaces" button located in Figure 9–66. Follow the prompts and enter the desired paper space distance of 0.25 (remember, this is the plotted text height). The MTEXT command resumes and a new text height of 12 is applied. This value was automatically calculated by the SPACETRANS command by multiplying 0.25 by 48 (the viewport scale). Picking the opposite corner of the mtext rectangle in Figure 9–67 and entering the text "SECOND FLOOR" completes the command sequence.

[A] Command: **MT** *(For MTEXT)*
Current text style: "STANDARD" Text height: 7 3/16"
Specify first corner: *(Pick a first corner on your screen...do not pick the opposite corner yet)*
Specify opposite corner or [Height/Justify/Line spacing/Rotation/Style/Width]: **H** *(For Height)*
Specify height <7 3/16">: **'_spacetrans** *(Pick the SpaceTrans button from the Text toolbar)*
>>Specify paper space distance <1">: **0.25** *(This represents the plotted text height)*
Resuming MTEXT command.
Specify height <7 3/16">: **12.00000000000000** *(This is the text height calculated by the SPACETRANS command based on the viewport scale of ¼ " = 1"-0")*

Specify opposite corner or [Height/Justify/Line spacing/Rotation/Style/Width]: *(Now pick the opposite corner to define the width of the multiline text as shown in Figure 9–67. When the Multiline Text Editor appears, enter the text SECOND FLOOR).*

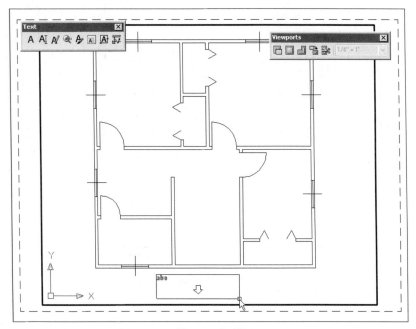

Figure 9–67

Now try the SPACETRANS command again by placing the text BEDROOM 1 as shown in Figure 9–68. The plotted text height of the room names will be 0.15 units. Follow the MTEXT prompt for completing this task using the SPACETRANS command.

A Command: **MT** *(For MTEXT)*
Current text style: "STANDARD" Text height: 1'
Specify first corner: *(Pick a first corner on your screen…do not pick the opposite corner yet)*
Specify opposite corner or [Height/Justify/Line spacing/Rotation/Style/Width]: **H** *(For Height)*
Specify height <1'>: **'spacetrans** *(Pick the SpaceTrans button from the Text toolbar)*
>>Specify paper space distance <1/4">: **0.15** *(This represents the plotted text height)*
Resuming MTEXT command.
Specify height <1'>: **7.200000000000001** *(This is the new text height calculated by the SPACETRANS command based on the viewport scale of ¼" = 1'-0")*
Specify opposite corner or [Height/Justify/Line spacing/Rotation/Style/Width]: *(Now pick the opposite corner to define the width of the multiline text. When the Multiline Text Editor appears, enter the text BEDROOM 1).*

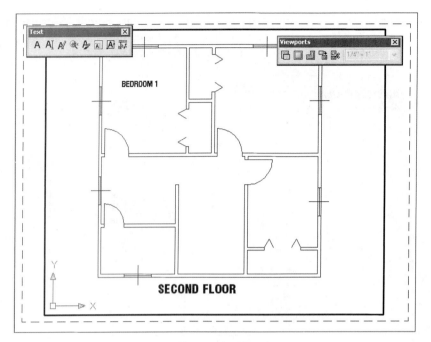

Figure 9–68

USING THE OPTIONS DIALOG BOX TO CONTROL LAYOUTS

Moving your cursor into the command prompt area at the bottom of your screen and right-clicking the mouse button activates the cursor menu in Figure 9–69. Click on Options to activate the Options dialog box. (You can also activate this dialog box by choosing Options from the Tools pull-down menus.) Clicking on the Display tab displays various controls that affect the display of AutoCAD. One of these controls focuses on Layout Elements in Figure 9–70. By default, all boxes are checked, which means the individual layout settings are all turned on. Removing the check turns the setting off.

Figure 9–69

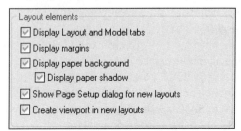

Figure 9–70

All of these settings are labeled in Figure 9–71. Turning off the Display Layout and Model tabs setting removes the Model and Layout1 tabs from the screen. Previously created layouts are still present; you just can't access them by clicking on the tab. Turning off the Display margins setting turns off the dashed margin that identifies the printable area of the paper. Turning off the Display paper background setting turns the background from gray to white. At the right and bottom edges of the paper, two shadows are created to give the appearance of a 3D paper sheet. Turning off the Display paper shadow setting displays a paper border similar to the top and left edges. If the Create viewport in new layouts setting is turned off, a blank sheet empty of any viewports is displayed only for new layouts. You would have to use the MVIEW command to create your own viewport.

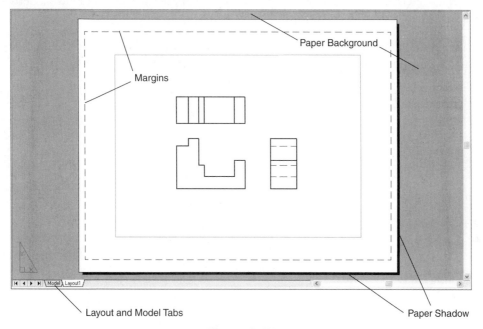

Figure 9–71

NOTES TO THE STUDENT AND INSTRUCTOR

A tutorial exercises has been designed around this introductory topic of laying out drawings in Paper Space. As with all tutorials, you will tend to follow the steps very closely, taking care not to make a mistake. This is to be expected. However, most individuals rush through tutorials to get the correct solution, only to forget the steps used to complete the tutorial. This is also to be expected.

Of all tutorial exercises in this book, the Drawing Layout tutorials will probably pose the greatest challenge. Certain operations have to be performed in Paper Space while other operations require floating Model Space. For example, it was already illustrated that user-defined viewports are constructed in Paper Space; scaling an image to Paper Space units through the Viewports toolbar or through the XP option of the ZOOM command must be accomplished in floating Model Space, and so on.

It is recommended to both student and instructor that all tutorial exercises, and especially this tutorial that deals with Paper Space, be performed two or even three times. Completing the tutorial the first time will give you the confidence that it can be done; however, you may not understand all of the steps involved. Completing the tutorial a second time will allow you to focus on where certain operations are performed and why things behave the way they do. This still may not be enough. Completing the tutorial exercise a third time will allow you to anticipate each step and have a better idea of which operation to perform in each step.

This recommendation should be exercised with all tutorials; however, it should especially be practiced when you work on the following Paper Space tutorial.

TUTORIAL EXERCISE: 09_HVAC.DWG

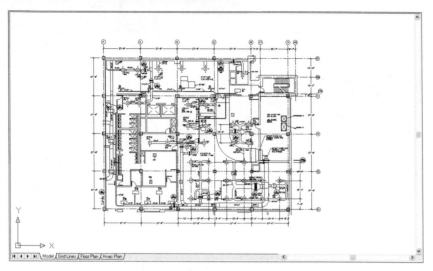

Figure 9–72

Purpose

This tutorial is designed to create multiple layouts of the HVAC drawing in Figure 9–72 in Paper Space.

System Settings

All unit and limit values have been pre-set. In order for this tutorial to function properly, the HP DesignJet 750C Plus C4708B must be configured. If this not the case, please perform the first tutorial exercise in chapter 10 to configure this plotter.

Layers

Layers have already been created for this exercise.

Suggested Commands

Begin by opening the drawing file 09_ Hvac.Dwg. Click on Layout1 to begin the first drawing layout that consists of the grid lines. When the layout is created, enter floating Model space and scale the image to 1/8"=1'-0". Also, freeze the layers that pertain to the floor and HVAC plans only in the current viewport. Press ENTER to return to Paper Space and insert an architectural title block and change the viewport to the Viewports layer. From this layout, create another layout called Floor Plan. While in floating Model Space, freeze the layers that pertain to the grid lines and HVAC plans. From this layout, create another layout called Hvac Plan. While in floating Model Space, freeze the layers that pertain to the grid lines. Turn off all viewports and add drawing titles to each layout.

Whenever possible, substitute the appropriate command alias in place of the full AutoCAD command in each tutorial step; for example, use "CP" for the COPY command, "L" for the LINE command, and so on. The complete listing of all command aliases is located in chapter 1, table 1–2.

STEP 1

Illustrated in Figure 9–73 is a drawing of an HVAC plan, which includes grid lines used for the layout of columns, a floor plan, and the actual HVAC plan, which consists of ductwork. The drawing was created in Model Space. This is easily identified by the appearance of the User Coordinate System icon displayed in the lower-left corner. Three separate layouts will be created. The first will consist of the grid lines; the second will be the floor plan information; the third will consist of the floor plan along with HVAC plan. First, click on the Layout1 tab at the bottom of the screen.

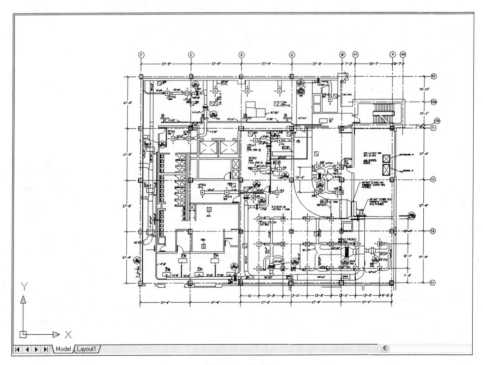

Figure 9–73

STEP 2

When the Page Setup dialog box appears in Figure 9–74, click on the Plot Device tab and change the plotter to the Design-Jet 750C Plus C4708B.pc3. (Note: If this printer device is not listed, stop this tutorial, go to the first tutorial exercise in chapter 10 and follow the steps for configuring this device; begin this exercise

again.) Also, in the upper-left corner of the dialog box, change the Layout name from "Layout1" to "Grid Lines." Then click on the Layout Settings tab in Figure 9–75 and change the Paper size to ARCH expand D (24.00 x 36.00 Inches). Click the OK button to create the layout.

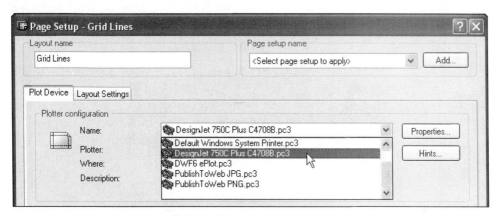

Figure 9–74

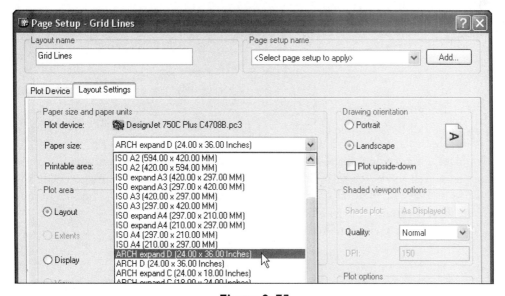

Figure 9–75

STEP 3

Your display should be similar to Figure 9–76. Notice that the name of the layout next to the Model tab is "Grid Lines." Click on the rectangular viewport at "A." The viewport should highlight and grips will appear at the corners. Then click in the Layer Control box to change the highlighted viewport to the VIEW-PORTS layer. Press ESC to remove the object highlight and grips.

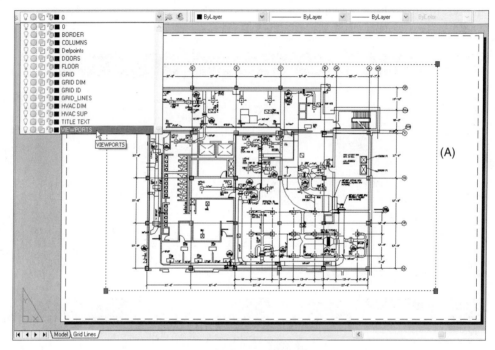

Figure 9–76

STEP 4

Double-click anywhere inside the viewport to switch to floating Model Space. Notice the re-appearance of the User Coordinate System icon in the lower-left corner of the viewport. Activate the Viewports dialog box in Figure 9–77 and change the scale of the image inside the viewport to 1/8"=1'-0". The image may get smaller.

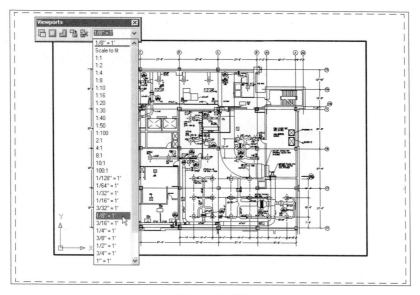

Figure 9–77

STEP 5

Double-click anywhere outside the viewport to return to Paper Space in Figure 9–78. Click on the edge of the viewport and use the corner grips to size the viewport to the image. When finished, press ESC to remove the object highlight and grips.

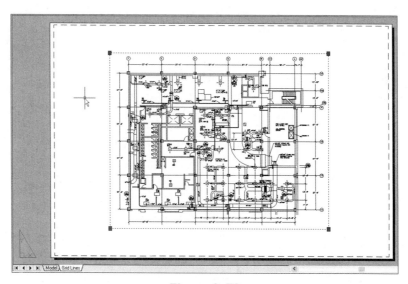

Figure 9–78

STEP 6

Choosing Block... from the Insert pull-down menu activates the Insert dialog box in Figure 9–79. Search in the Name box for a block called Architectural Title Block (this block has been pre-defined in the drawing). Click the OK button to return to the drawing.

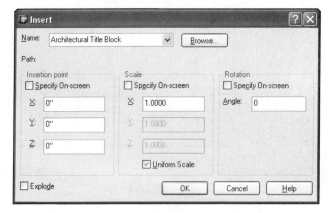

Figure 9–79

STEP 7

Position the title block in Figure 9–80. Then click on the title block to have the highlight appear along with the grips. Activate the Layer Control box and change the highlighted title block to the BORDER layer. When finished, press ESC to remove the object highlight and grips. The title block may change its color to red.

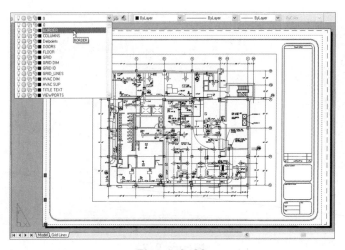

Figure 9–80

STEP 8

It will be necessary to center the viewport containing the image, moving it to a better location. Use the Insert dialog box again to insert a block called "Title." This block contains the title and scale of the drawing and is also predefined in the drawing.

After inserting this in your drawing, use the EXPLODE command to break this block up into individual objects. When finished, your drawing should appear similar to Figure 9–81.

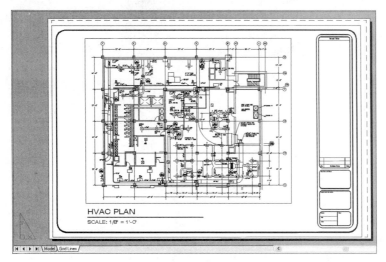

Figure 9–81

STEP 9

Click on the drawing title (HVAC PLAN); the title should highlight and grips will appear. Then right-click the mouse to activate the Properties palette. Change HVAC PLAN to GRID LINES

PLAN as shown in Figure 9–82; pressing ENTER automatically updates the text to the new value. Dismiss this dialog box when finished. Press ESC to remove the object highlight and grips.

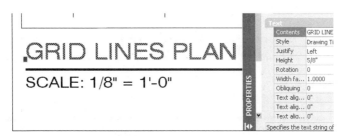

Figure 9–82

STEP 10

Prepare to create the second layout by right-clicking on the Grid Lines tab to display the cursor menu in Figure 9–83. Click on the Move or Copy... option to make a copy of the current layout. With the Move or Copy dialog box displayed in Figure 9–84, place a check in the box to make a copy. Also highlight the area to move the new layout to the end. When finished, click the OK button to create the new layout.

Figure 9–83

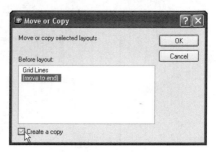

Figure 9–84

STEP 11

Before continuing, right-click on the new layout "Grid Lines (2)" and click on the Rename option in the cursor menu in Figure 9–85. In the Rename Layout dialog box shown in Figure 9–86, change the current layout name "Grid Lines (2)" to the new name of "Floor Plan." Click the OK button to dismiss the dialog box and make the change.

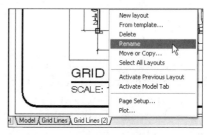

Figure 9–85

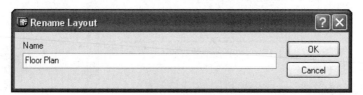

Figure 9–86

STEP 12

Your display should appear similar to Figure 9–87. Not only has the image been copied in the new layout, but the viewport, border, and drawing title have been copied as well.

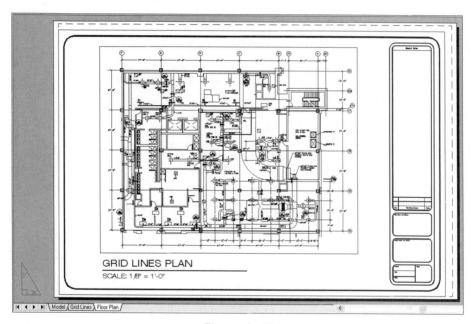

Figure 9–87

STEP 13

While in this new layout, right-click on the Floor Plan tab to display the cursor menu in Figure 9–88. Click on the Move or Copy... mode to make one more copy of the current layout. With the "Move or Copy" dialog box displayed in Figure 9–89, place a check in the box to create a copy. Also, highlight the area to move the new layout to the end. When finished, click the OK button to create the new layout.

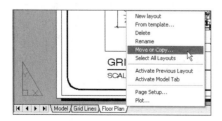

Figure 9–88

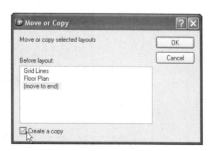

Figure 9–89

STEP 14

Before continuing, right-click on the new layout "Floor Plan (2)" and click on the Rename option in the cursor menu in Figure 9–90. Change the current layout name "Floor Plan (2)" to the new name of "Hvac Plan" in Figure 9–91. Click the OK button to dismiss the dialog box and make the change.

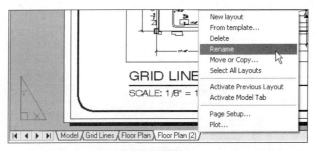

Figure 9–90

Figure 9–91

STEP 15

Your display should appear similar to Figure 9–92.

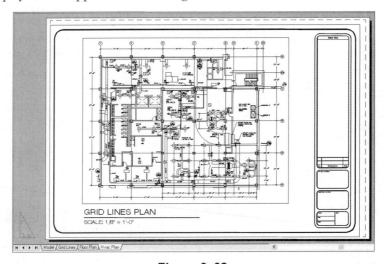

Figure 9–92

STEP 16

Click on the Grid Lines tab. Double-click anywhere inside the viewport in Figure 9–93; this places you in floating Model Space. You will need to turn off all layers that deal with the floor plan and HVAC, which leave only the layers with the grid lines visible.

Activate the Layer Properties Manager dialog box in Figure 9–94 and freeze the highlighted layers in the current viewport that pertain to the floor plan and HVAC plan. Pick one of the layers and hold down the CTRL key as you pick the other three. This allows you to freeze all four layers at once.

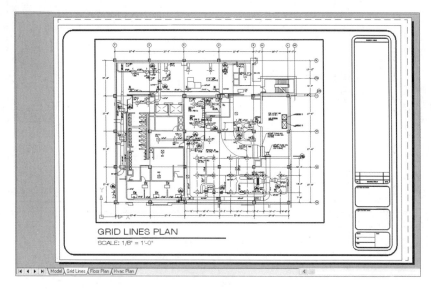

Figure 9–93

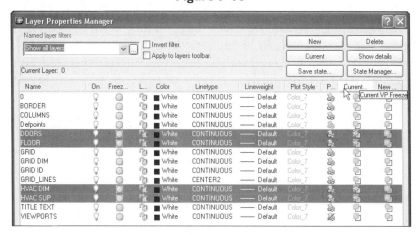

Figure 9–94

Clicking the OK button returns to the drawing editor. Double-click anywhere outside the viewport to return to Paper Space. Your drawing should appear similar to Figure 9–95.

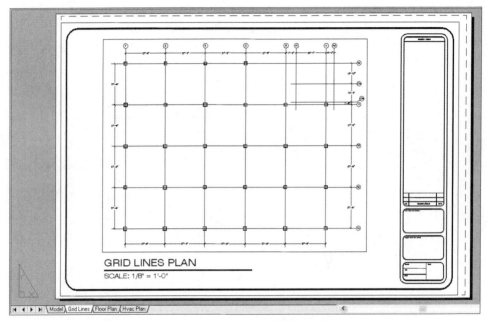

GRID LINES PLAN
SCALE: 1/8" = 1'-0"

Figure 9–95

STEP 17

Click on the Floor Plan tab. Double-click anywhere inside the viewport in Figure 9–96; this places you in floating Model Space. You will need to turn off all layers that deal with the grid lines and HVAC.

Activate the Layer Properties Manager dialog box in Figure 9–97 and freeze the layers in the current viewport that pertain to the grid lines plan and HVAC plan. Pick the GRID layer and hold down the SHIFT key as you pick HVAC SUP. This allows you to freeze all six layers at once.

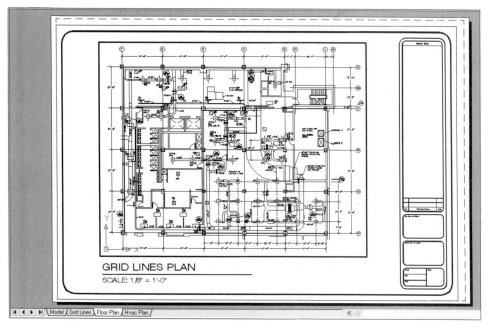

Figure 9–96

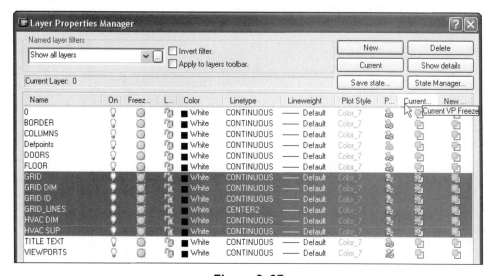

Figure 9–97

Clicking the OK button returns you to the drawing editor. Double-click anywhere outside the viewport to return to Paper Space. Use the Properties Palette to change the drawing title from GRID LINES PLAN, as shown in Figure 9–98, to FLOOR PLAN. Do this by first selecting GRID LINES PLAN. With this object highlighted, right click to activate the Properties Palette and change the text. When finished, press ESC to remove the object highlight and grips.

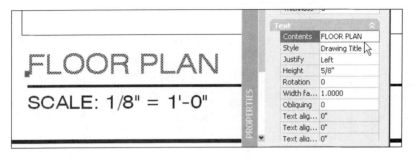

Figure 9–98

Your drawing should appear similar to Figure 9–99.

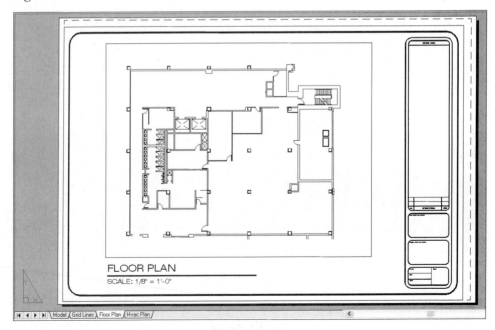

Figure 9–99

STEP 18

Click on the Hvac Plan tab. Double-click anywhere inside the viewport in Figure 9–100; this places you in floating Model Space. You will need to turn off all layers that deal with the grid lines plan.

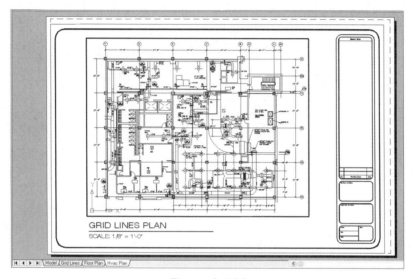

Figure 9–100

Activate the Layer Properties Manager dialog box in Figure 9–101 and freeze the layers in the current viewport that pertain to the grid lines plan.

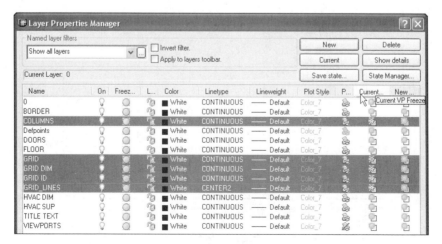

Figure 9–101

Clicking the OK button returns to the drawing editor. Double-click anywhere outside the viewport to return to Paper Space. Use the Properties Palette to change the drawing title from GRID LINES PLAN, as shown in Figure 9–102

to FLOOR PLAN. Do this by first selecting GRID LINES PLAN. With this object highlighted, right-click to activate the Properties Palette and change the text. When finished, press ESC to remove the object highlight and grips.

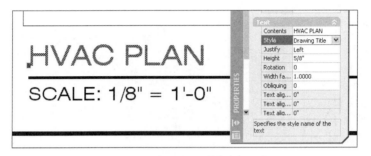

Figure 9–102

Your drawing should appear similar to Figure 9–103.

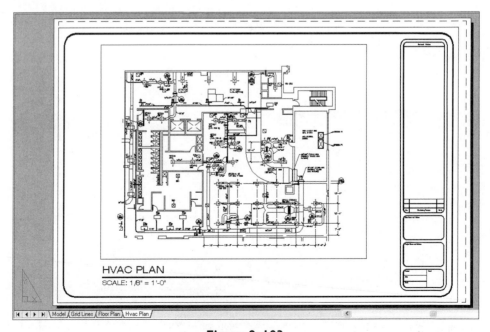

Figure 9–103

STEP 19

Turn off the Viewports layer. The completed drawing is displayed in Figure 9–104.

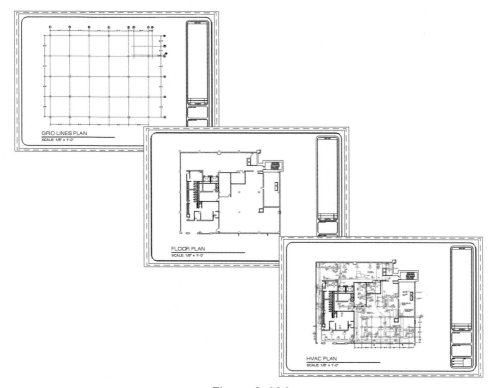

Figure 9–104

 Open the Exercise Manual PDF file for Chapter 9 on the accompanying CD for more tutorials and exercises.

 If you have the accompanying Exercise Manual, refer to Chapter 9 for more tutorials and exercises.

CHAPTER 10

Plotting Your Drawings

This chapter discusses plotting through a series of tutorial exercises designed to perform the following tasks:

- Configure a new plotter
- Plot from a drawing layout (Paper Space)
- Plot from Model Space
- Control lineweights
- Use the Plotter Configuration Editor to filter available paper sizes
- Create a color-dependent plot style table
- Create a web page consisting of various drawing layouts for viewing over the Internet

CONFIGURING A PLOTTER

Before plotting, you must first establish communication between AutoCAD and the plotter. This is called configuring. From a list of supported plotting devices, you choose the device that matches the model of plotter you own. This plotter becomes part of the software database, which allows you to choose this plotter many times. If you have more than one output device, each device must be configured before being used. This section discusses the configuration process used in AutoCAD.

STEP I

Begin the plotter configuration process by choosing Plotter Manager from the File pull-down menu as in Figure 10–1. This activates the Plotters program group in Figure 10–2, which lists all valid plotters that are currently configured. The listing in Figure 10–2 displays the default plotters configured after the software is loaded. Except for the DWF devices, which allow you to publish a drawing for viewing over the Internet, a plotter has not yet been configured. Double-click on the Add-A-Plotter Wizard icon to continue with the configuration process.

Figure 10–1   **Figure 10–2**

STEP 2

Double-clicking on the Add-A-Plotter Wizard icon displays the Add Plotter-Intro-duction Page in Figure 10–3. This dialog box states you are about to configure a Windows or non-Windows system plotter. This configuration information will be saved in a file with the extension .PC3. Click the Next> button to continue on to the next dialog box.

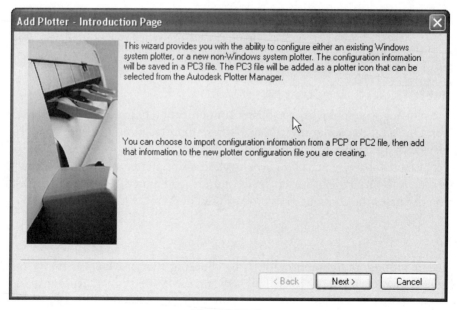

Figure 10–3

STEP 3

In the Add Plotter-Begin dialog box in Figure 10–4, decide how the plotter will be controlled: by the computer you are currently using, by a network plot server, or by an existing system printer where changes can be made specifically for AutoCAD. Click on the radio button next to My Computer. Then click on the Next> button to continue on to the next dialog box.

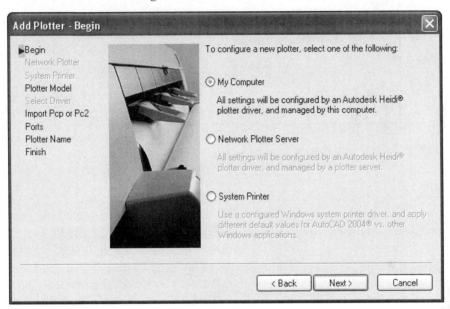

Figure 10–4

STEP 4

Use the Add Plotter-Plotter Model dialog box in Figure 10–5 to associate your model plotter with AutoCAD. You would first choose the appropriate plotter manufacturer from the list provided. Once this is done, all models supported by the manufacturer appear to the right. If your plotter model is not listed, you are told to consult the plotter documentation for a compatible plotter. If you have an installation disk containing an HDI driver, you could click on the Have Disk... button to copy this driver from disk to your computer. For the purposes of this tutorial, click on Hewlett-Packard in the list of Manufacturers. Click on the DesignJet 750C Plus C4708B for the plotter model. A Driver Info dialog box may appear giving you more directions regarding the type of HP DeskJet plotter selected. Click the Continue button to move on to the next dialog box used in the plotter configuration process.

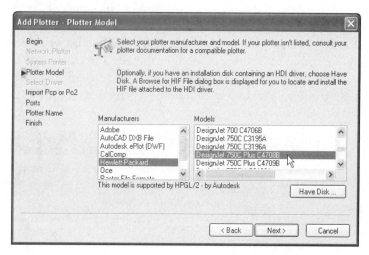

Figure 10–5

STEP 5

PCP and PC2 files have been in existence for many years. They are designed to hold plotting information such as pen assignments. In this way, you use the PCP or PC2 files to control pen settings instead of constantly making pen assignments every time you perform a plot. The Add Plotter-Import Pcp or Pc2 dialog box in Figure 10–6 allows you to import those files for use in AutoCAD in a PC3 format. If you will not be using any PCP or PC2 files from previous versions of AutoCAD, click on the Next> button to move on to the next dialog box.

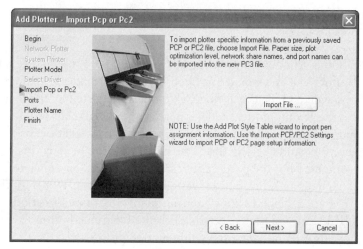

Figure 10–6

STEP 6

In the Add Plotter-Ports dialog box, shown in Figure 10–7, click on the port used for communication between your computer and the plotter. The LPT1 port will be used for the purposes of this tutorial. Place a check in its box and then click on the Next> button to continue on to the next dialog box.

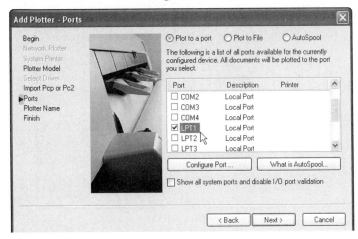

Figure 10–7

STEP 7

In the Add Plotter-Plotter Name dialog box in Figure 10–8, you have the option of giving the plotter a name other than the name displayed in the dialog box. For the purpose of this tutorial, accept the name that is given. This name will be displayed whenever you use the Page Setup and Plot dialog boxes.

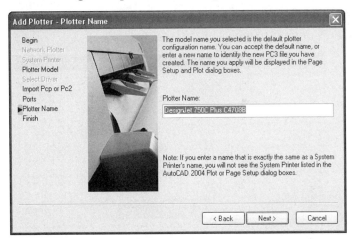

Figure 10–8

STEP 8

The last dialog box is displayed in Figure 10–9. In the Add Plotter-Finish dialog box, you can modify the default settings of the plotter you just configured. You can also test and calibrate the plotter if desired. Click the Finish button to dismiss the Add Plotter dialog box.

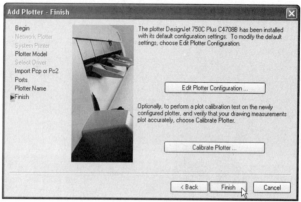

Figure 10–9

STEP 9

Exiting the Add Plotters dialog box returns you to the Plotters program group shown in Figure 10–10. Notice that the additional icon for the DesignJet 750C Plus plotter has been added to this list. In the View pull-down menu, you have additional controls used to display the plotters. By default, the Large Icons mode is selected. Clicking on Small Icons displays the plotter list in Figure 10–11. Clicking on Details displays additional information on your plotters in Figure 10–12. This completes the steps used to configure the DesignJet 750C Plus plotter. Follow these same steps if you need to configure another plotter.

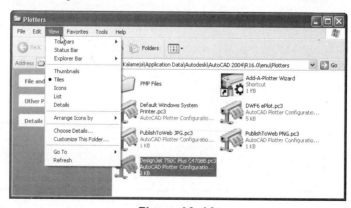

Figure 10–10

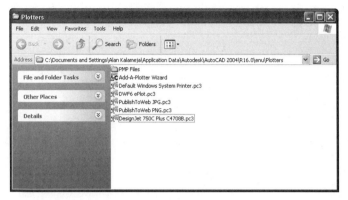

Figure 10–11

Figure 10–12

PLOTTING FROM PAPER SPACE

Open the drawing 10-Center Guide to step through the process of plotting the drawing from layout mode, or Paper Space.

STEP I

When you open the drawing 10_Center_Guide.Dwg, your drawing should appear similar to Figure 10–13. This drawing should be laid out in Paper Space. A layout called "Four Views" should be present at the bottom of the screen next to the Model tab.

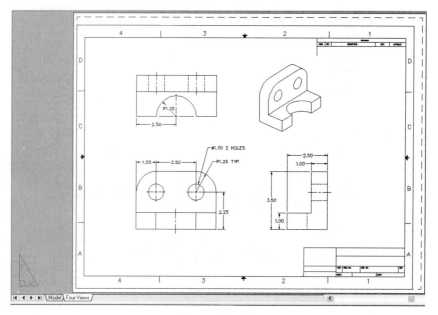

Figure 10–13

STEP 2

Begin the process of plotting this drawing by choosing Plot... from the File pull-down menu as in Figure 10–14. This activates the Plot dialog box in Figure 10–15. In the Plot Device tab, verify the correct plotter configuration. You could click the down arrow for the current list of all configured plotters and then pick the desired plotter from this list. For the purposes of this tutorial, the DesignJet 750C Plus is being used as the output device. Although your plotter may be different, the steps used in performing the plot are the same for any device. (The Properties... button adjacent to the plotter name will be explained in greater detail later in this chapter, as will the Plot style table area located in the middle of the dialog box.) By default, the Current tab radio button is clicked in the What to plot area. If numerous layouts were created, you could plot all layouts at once by choosing this radio button. Also, you can opt to create additional copies of your plot; by default this value is set to 1. In the Plot to file area in the lower-right corner of the dialog box, you could output your plot to a file instead of to paper (hard copy). AutoCAD drawings can be merged into other Windows applications; however, the file must be converted to a different format. Plotting the drawing to a file creates a file with the PLT extension. This will always be the extension if you configured any type of plotter manufactured by Hewlett-Packard.

Figure 10–14

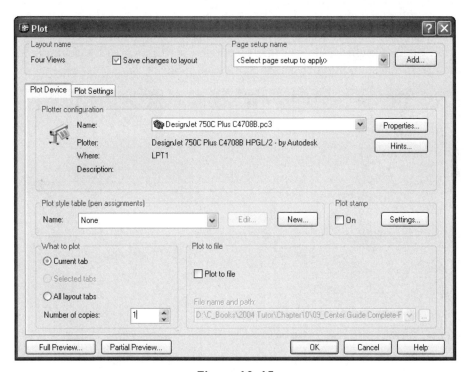

Figure 10–15

Click on the Plot Settings tab of the Plot dialog box in Figure 10–16 and verify the following areas are properly set. Make sure the paper size is currently set to ANSI expand C (22.00 x 17.00 Inches). In the Drawing orientation area, make sure that the radio button adjacent to Landscape is selected. You could also plot the drawing out in Portrait mode, where the short edge of the paper is the top of the page. For special plots, you could plot the drawing upside-down.

In the Plot area, the Layout radio button is selected. Since you created a layout, this is the obvious choice. The Extents mode allows you to plot the drawing based on all objects that make up the drawing. Plotting the Display will plot your current drawing view, but be careful. If you are currently zoomed in to your drawing, plotting the Display will only plot this view. In this case, it would be more predictable to use Layout or Extents to plot. When you plot a Layout in Paper Space, the Plot Scale will be set to 1:1. Since you pre-scaled the drawing to the Paper Space viewport using the Viewports toolbar, all drawings in Paper Space are designed to be plotted at this scale. The Plot Offset is designed to move or shift the location of your plot on the paper if it appears off center. In the Plot options area, you have more control over plots by applying lineweights, using existing plot styles, plotting Paper Space last (Model Space first), or even hiding objects, on a 3D solid model.

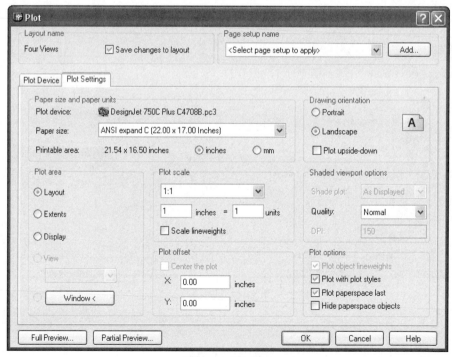

Figure 10–16

STEP 3

One of the more efficient features of plotting is that you can preview your plot before sending the plot information to the plotter. In this way, you can determine if the entire drawing will plot based on the sheet size (this includes the border and title block). Clicking on the Full Preview… button activates the image in Figure 10–17. The sheet size is present along with the border and four-view drawing. Right-clicking anywhere on this preview image displays the cursor menu, allowing you to perform various display functions such as ZOOM and PAN to assist with the verification process. If everything appears satisfactory, click on the Plot option to send the drawing information to the plotter. Clicking on the Exit option returns you to the Plot dialog box, where you can make changes in the Plot Device or Plot Settings tabs.

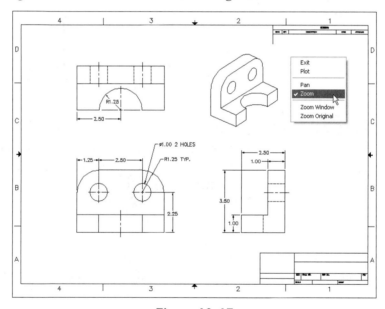

Figure 10–17

STEP 4

A faster way of previewing a plot is through the Partial Preview… button. Clicking on this button displays the Partial Plot Preview dialog box in Figure 10–18. This method of previewing a plot shows a quick representation of the effective plot area relative to the paper size and printable area. You will also be given warnings in the event that the drawing is outside the printable area. This usually affects the border and title block.

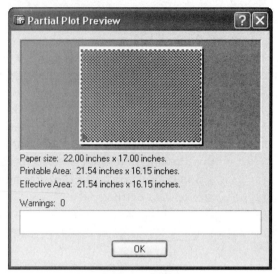

Paper size: 22.00 inches x 17.00 inches.
Printable Area: 21.54 inches x 16.15 inches.
Effective Area: 21.54 inches x 16.15 inches.

Warnings: 0

[OK]

Figure 10–18

PLOTTING FROM MODEL SPACE

You can also plot from Model Space. However, there is more involved in bringing in such basic items as borders, title blocks or notes into the drawing. Open the drawing called 10_Roof_Plan.Dwg and follow the next series of steps, designed to plot from Model Space.

STEP 1

As with Paper Space, you must determine the scale at which you wish to plot. The drawing in Figure 10–19 illustrates a Roof Plan. Verify the area occupied by the drawing using the LIMITS command.

Command: **LIMITS**
Reset Model space limits:
Specify lower left corner or [ON/OFF] <0'-0",0'-0">: *(Press* ENTER*)*
Specify upper right corner <276'-0",184'-0">: *(Press* ENTER*)*

From the information in the LIMITS command, this drawing is currently occupying a sheet of paper 276' by 184'. This makes sense because Model Space is where the drawing is constructed full size or in real world units. So you need a very large size sheet of paper to draw on. Notice also the scale of the drawing under the text "ROOF PLAN"; it reads 1/8" = 1'-0". This becomes the plot scale, which you will use in a later step.

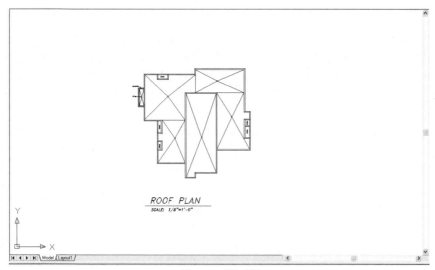

Figure 10–19

STEP 2

Since you will not be using the Layout1 tab, you must insert a title block in this drawing. The Architectural Title Block will be used in this example. This title block normally measures 34.5" x 23". Remember, however, that the limits of the drawing used to define the sheet measure 276' x 184'. This means the title block must be blown up or enlarged when you bring it into the drawing of the Roof Plan. Also, you must enlarge it based on the scale of the drawing. Since the drawing will be plotted out at 1/8" = 1'-0", you need to enlarge the title block 96 times its normal size. The value 96 is found by the division of 1' or 12" by 1/8 or 0.125 units. Activate the Insert dialog box in Figure 10–20 and make the changes in the dialog box such as the insertion point, scale, and rotation angle. The title block will be inserted at 0,0 and have a scale value of 96, as previously explained. A rotation angle of 0 will be used.

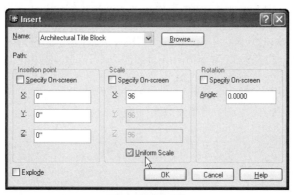

Figure 10–20

STEP 3

Your drawing should appear similar to Figure 10–21. Since the title block had to be enlarged, adding additional support items to the drawing such as dimensions and notes must also include the enlargement factor of 96 when the note height and overall dimension scale factor are calculated. This makes this process more involved and time-consuming.

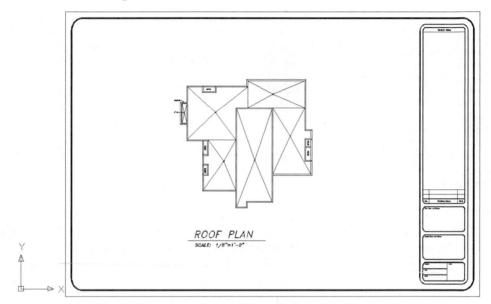

Figure 10–21

STEP 4

Activate the Plot dialog box and make sure you are in the Plot Settings tab in Figure 10–22. Verify that the current plot device is the HP DesignJet 750C Plus. If this is not the device, click on the Plot Device tab to activate this plotter. In addition, Landscape mode should be selected under the Drawing orientation area. Continue on to the next step for making additional changes in this dialog box.

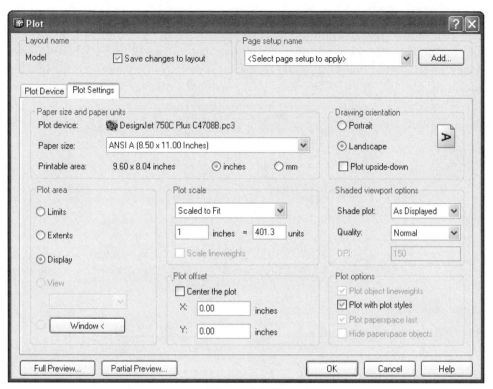

Figure 10–22

STEP 5

In the Plot area portion of the Plot dialog box in Figure 10–23, you are required to click on the radio button of your choice to determine the type of area you desire to plot the drawing by. These modes will now be explained so that you can see their results. The image of the Roof Plan without the title block will be used to explain each mode. Clicking on the Limits radio button plots the Roof Plan based on the original sheet size created through the LIMITS command. The results are illustrated in Figure 10–24 (The title block has been temporarily turned off in this figure.) Clicking on the Extents radio button calculates the plot based on the objects placed in the drawing. This results in a plot similar to Figure 10–25 (Again, the title block has been temporarily turned off in this figure.) Clicking on the Display radio button could result in a plot based on the image in Figure 10–24. It could also result in the plot appearing similar to Figure 10–26. Here, the drawing was magnified through the ZOOM command; however, the entire drawing needs to be plotted out. If you forget to perform a ZOOM-ALL when using the plot Display mode, the plot displays as in Figure 10–26, which is a common occurrence. The other two plot area modes are

View and Window. It you have previously created a named view of your drawing, you can retrieve this view and plot it out. You can also define a plot area by a window by clicking on the Window< button. This returns you to your drawing, where you can define a rectangular box as the plot area. Continue on to the next step for making additional changes in this dialog box.

Figure 10–23

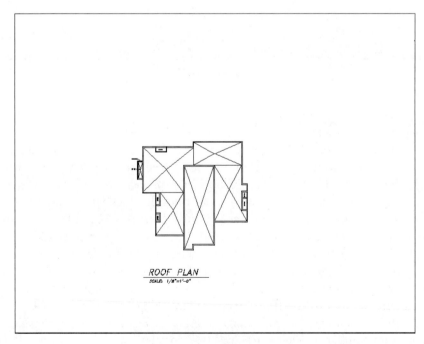

Figure 10–24

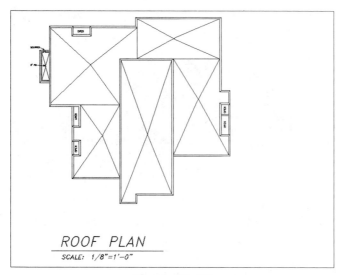

Figure 10–25

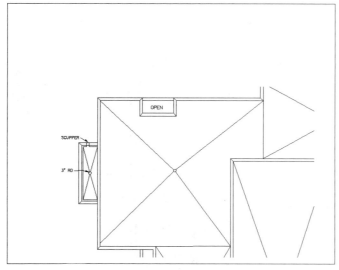

Figure 10–26

STEP 6

Next, click in the Paper size and paper units area of the Plot dialog box. For this drawing of the Roof Plan, a D-size sheet of paper will be used. Clicking on the ARCH expand D (24.00 x 36.00 Inches) sheet size in Figure 10–27 performs this task. Continue on to the next step for making additional changes in this dialog box.

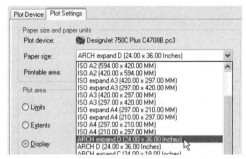

Figure 10–27

STEP 7

In the previous topic on plotting using a layout, a plot scale of 1:1 is used for all plotting from Paper Space. When plotting from Model Space, you must enter the scale of the drawing as the plot scale. In Figure 10–28, click on the scale 1/8"=1'0" in the Plot scale area of the Plot dialog box. Continue on to the next step for making additional changes in this dialog box.

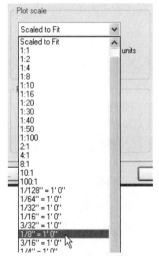

Figure 10–28

STEP 8

The Plot dialog box should appear similar to Figure 10–29. Notice, in the Plot scale area, the Custom mode displays 1 inches = 96 drawing units. This means that for every 96 units in the drawing (since it is full size), the plotter plots out 1 unit. Also, in the lower-right corner under Plot options, you could check the box labeled Plot object lineweights to control the line quality of your drawing. This will be discussed in the next section of this chapter. Before plotting your drawing, it is considered good practice to click on the Full Preview… button to preview the appearance of your plot.

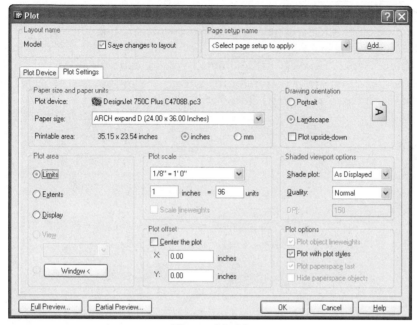

Figure 10–29

STEP 9

The previewed plot is displayed in Figure 10–30. If the results are acceptable, click Plot in the cursor menu to plot the drawing. If you want to make adjustments or if you choose not to plot at this time, click Exit to return to the Plot dialog box.

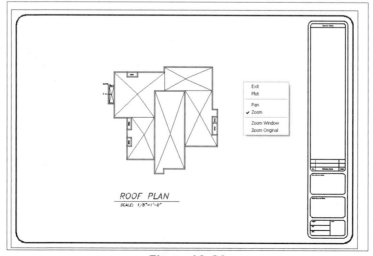

Figure 10–30

ENHANCING YOUR PLOTS WITH LINEWEIGHTS

This section on plotting will describe the process of assigning lineweights to objects and then having the lineweights appear in the finished plot. Open the drawing called 10_V_Step.Dwg in Figure 10–31 and notice you are currently in Model Space (the Model tab is current at the bottom of the screen). Follow the next series of steps to assign lineweights to a drawing before it is plotted.

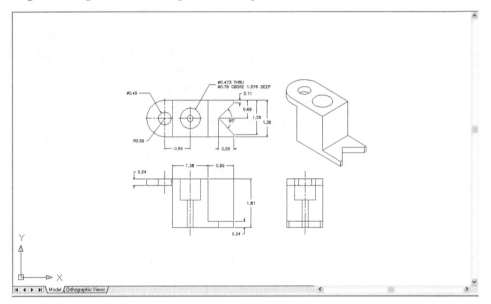

Figure 10–31

STEP 1

From the image of the drawing in Figure 10–31, all lines on the Object layer need to be assigned a lineweight of 0.60 mm. All objects on the Hidden layer need to be assigned a lineweight of 0.30 mm. There is also a border and title block that will be used with this drawing. The Title Block layer needs a lineweight assignment of 0.80 mm. Click in the Layer Properties Manager dialog box in Figure 10–32 and make these lineweight assignments.

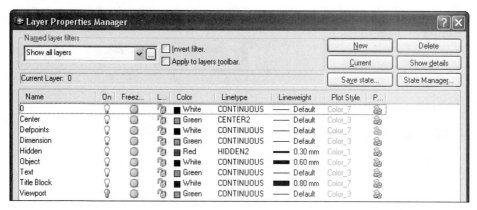

Figure 10–32

STEP 2

Click on the LWT button in the Status bar to display the lineweights in Figure 10–33. Notice however that all lineweights appear extremely thick. You can control the lineweight scale to have the linetypes give off a more pleasing appearance in your drawing.

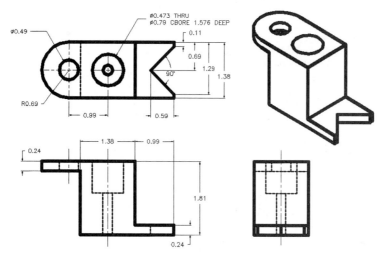

Figure 10–33

Clicking on Lineweight in the Format pull-down menu in Figure 10–34 displays the Lineweight Settings dialog box in Figure 10–35. Notice the slider bar in the area called Adjust Display Scale. Use this slider bar to reduce or increase the scale of the lineweights when being viewed in your drawing. In Figure 10–35, the slider bar has been adjusted towards the minimum side of the scale. This should make the lineweights better to read in the drawing. Click the OK button to return to the drawing and observe the results.

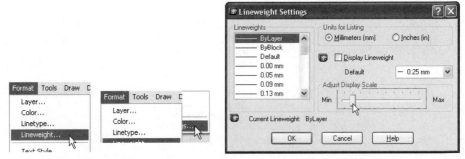

Figure 10–34   **Figure 10–35**

All hidden and object lines have been reduced in scale in Figure 10–36. This action adds to the clarity and appearance of the drawing. If after making changes to the lineweights you decide to increase the scale, return to the Lineweight Settings dialog box and readjust the slider bar until you achieve the desired results.

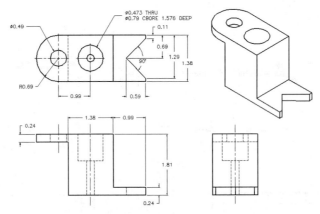

Figure 10–36

STEP 3

Clicking on the Orthographic Views tab switches you to Paper Space in Figure 10–37. The lineweights will not appear in Paper Space at first glance. Zooming in to the drawing will display all lineweights at their proper widths.

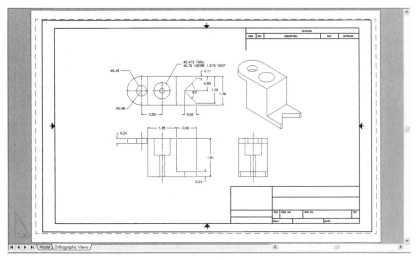

Figure 10–37

STEP 4

Activate the Plot dialog box and make sure you are in the Plot Settings tab as shown in Figure 10–38. Verify that the current plot device is the HP DesignJet 750C Plus. If this is not the device, click on the Plot Device tab to activate this plotter. In the Plot options area of the main Plot dialog box, be sure to check the Plot object lineweights box. This will ensure that the linetypes will plot. If this box appears to be inactive, remove the check from the Plot with plot styles area.

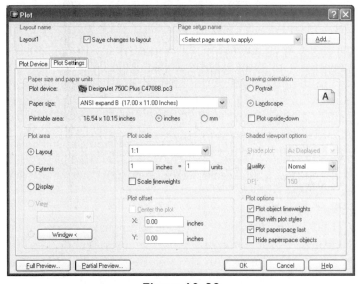

Figure 10–38

STEP 5

Click on the Full Preview… button; your display should appear similar to Figure 10–39. Notice that the viewport is not present in the plot preview. The Viewports layer was either turned off or set to a non-plot state inside the Layer Properties Manager dialog box. To view the lineweights in Preview mode, zoom in to segments of your drawing.

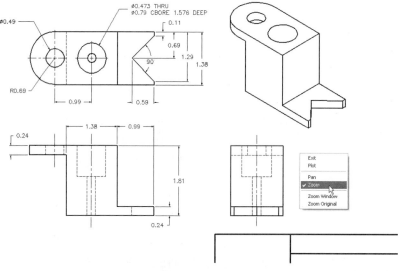

Figure 10–39

MODIFYING THE PROPERTIES OF A PLOTTER CONFIGURATION–FILTERING AVAILABLE PAPER SIZES

You have seen that a large number of sheet sizes are available to you depending on the type of plotter you are using. For the DesignJet 750C Plus C4708B plotter that you configured in the first pages of this chapter, you have the choice of sheet sizes that range from "A" through "D" in the ANSI standard. Notice you also have access to expanded sheet sizes that deal with a larger plottable area in addition to ARCH and ISO standard sheet sizes. What if you wanted to display only those sheet sizes that you use all of the time and not have to sort through all sheet sizes supported by the default plot device? Follow the following steps to accomplish this task.

STEP 1

First activate either the Plot dialog box or Page Setup dialog box in Figure 10–40 (the figure is taken from the Page Setup dialog box), and verify under the Plot Device tab that the plotter is the DesignJet 750C Plus plotter (this plotter has been used in all plotting tutorials throughout this chapter). Then click on the Properties button.

Figure 10–40

STEP 2

This activates the Plotter Configuration Editor for the current plot device (see Figure 10–41). Find the item in the display box that deals with the User-defined Paper Sizes & Calibration and click on the Filter Paper Sizes section. Then find the section on Filter Paper Sizes. Clicking on this section displays all standard paper sizes supported by the selected plotter.

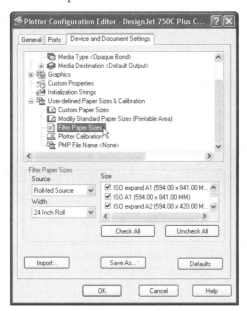

Figure 10–41

Suppose you commonly use the ANSI expanded C and D sheets in addition to the ARCH expanded C and D paper sizes. First click on the Uncheck All button in Figure 10–41 to deselect all sheet sizes. A file error dialog box in Figure 10–42 may appear directing you to first save the action you are about to perform with a PMP extension. Click the OK button and then proceed through the list of paper sizes and place checks next to all ANSI and ARCH expanded C and D paper sizes in Figure 10–43. You will notice two types of ARCH and ANSI expanded C sheets. These differ only on how the paper is loaded into the plotter. When finished, click the OK button.

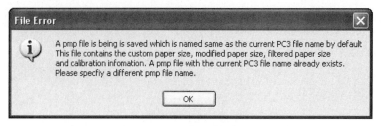

Figure 10–42

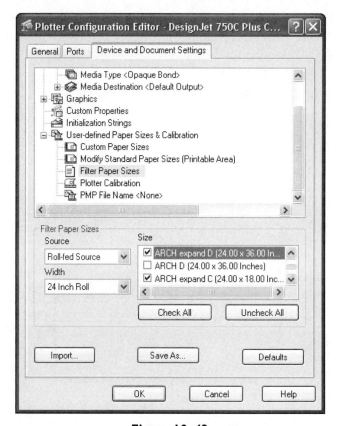

Figure 10–43

STEP 3

Clicking the OK button in Figure 10–43 displays the Changes to a Printer Configuration File dialog box in Figure 10–44. You have the opportunity to use the filtered paper sizes for only the current drawing or you can save these changes to the original PC3 file. This will allow only the four sheet sizes to appear for all plots with using this plotter. Click OK to dismiss this dialog box.

Once back in the main Plot dialog box, click on the Plot Settings tab and notice the available sheet sizes in Figure 10–45. The filtered sizes reflect ANSI and ARCH expanded C and D sheet sizes.

Figure 10–44

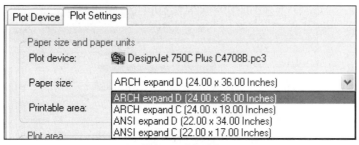

Figure 10–45

STEP 4

To display all available paper sizes, return back to the Plotter Configuration Editor and click on the PMP File Name listing in Figure 10–46. Click on the Detach button and then click OK. You must once again save these changes to the current PC3 file in Figure 10–47. Click the OK button to detach the PMP file and have all paper sizes return to your list.

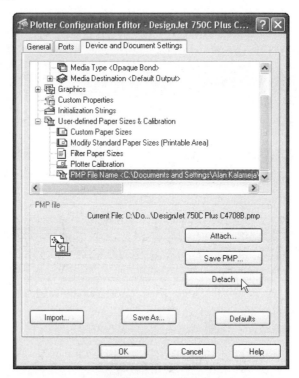

Figure 10–46

Figure 10–47

THE OPTIONS DIALOG BOX–USING THE PLOTTING TAB

Additional plotting controls can be set through the Plotting tab of the Options dialog box. To call up this dialog box, first move your cursor into the Command line area and right-click your mouse button. This displays the cursor menu in Figure 10–48. Clicking on Options… displays the Options dialog box in Figure 10–49. Click on the Plotting tab because we will be concentrating on this area at this time. (You can also launch the Options dialog box by choosing Options from the Tools pull-down menu.)

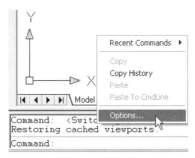

Figure 10–48

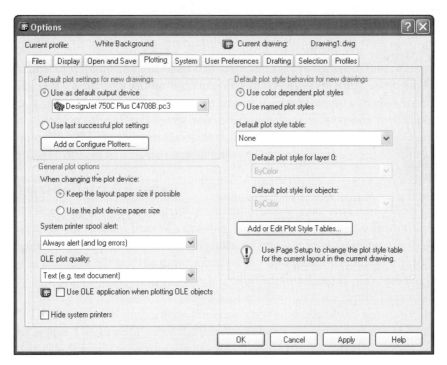

Figure 10–49

The Default plot settings for new drawings area of the Plotting tab, shown in Figure 10–50, allows you to choose a plotter and have it serve as the default output device whenever you set up a page layout or activate the Plot dialog box. A button is also available to allow you to conveniently add or configure additional plotters.

Figure 10–50

The Default plot style behavior for new drawings area of the Plotting tab, shown in Figure 10–51, displays two types of plot styles to use with your drawings and lists available plot style table, which are device-independent files that control the appearance of color, linetypes, and lineweights. You can also apply additional controls such as screening, grayscale, dithering, end and joint styles, and fill patterns to your plots. Plot styles are attached to your drawing. You can also apply a different plot style to individual page layouts. In this way, you can produce different plots of the same drawing.

Two types of plot styles are listed in Figure 10–51; the first is a Color-Dependent plot style. This type of plot style determines how your drawing plots based on color. You have 255 colors to work with when using this type of table. Plot characteristics such as linetype and lineweight are assigned to color, which determines how the drawing will plot. Color-dependent plot styles have the file extension .CTB. The plot styles listed in the figure are the default plot style tables that come with AutoCAD 2004.

The second type of plot style table is illustrated in Figure 10–52; this is the Named plot style table. This type of plot style does not depend on color. Rather, it is a list of names that you create. Inside each name is a list of plot characteristics such as color, linetype, and lineweight. The name is commonly referred to as a plot style. What makes this type of plot style different from the color-dependent variety is your ability to assign different plot style names to objects with the same colors. This provides a very powerful way of plotting your drawings. Named plot styles have the file extension .STB.

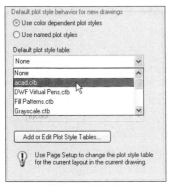

Figure 10–51

Figure 10–52

Figure 10–53 displays the upper corner of your display screen. This is the Plot Style Control box used for changing named plot styles. However, if you are currently set to a color-dependent plot style, this area is grayed out in the figure.

Clicking on the Use named plot styles radio button back in the Plotting tab of the Options dialog box in Figure 10–49 activates the Plot Style Control in Figure 10–54. This area will activate only for new drawing files.

Figure 10–53

Figure 10–54

Displaying the Layer Properties Manager dialog box in Figure 10–55 also shows the Plot Style area active when you use named plot styles. Color-dependent plot styles will have this layer area grayed out.

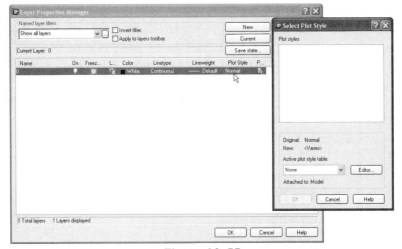

Figure 10–55

CREATING A COLOR-DEPENDENT PLOT STYLE TABLE

This section of the chapter is devoted to the creation of a color-dependent plot style table. Once the table is created, it will be applied to a drawing. From there, the drawing will be previewed to see how this type of plot style table affects the final plot. Open the drawing 10_Color_R-Guide.Dwg. Your display should appear similar to the image in Figure 10–56. A two-view drawing together with an isometric view is arranged in a layout called "Orthographic Views." The drawing is also organized by layer names and color assignments. The object is to create a color-dependent plot style table where all layers will plot out black. Also, through the color-dependent plot style table, the hidden lines will be assigned a lineweight of 0.30 mm, object lines 0.70 mm, and the title block 0.80 mm. Follow the next series of steps to perform this task

Color	Layer
Red	Hidden
Yellow	Center
Green	Viewports
Cyan	Text
Blue	Title Block
Magenta	Dimensions
Black	Object

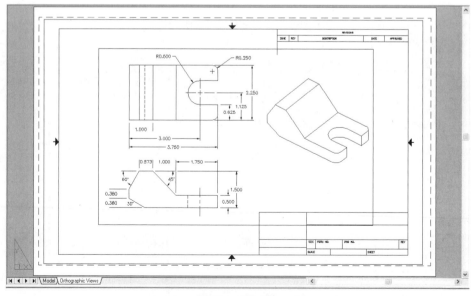

Figure 10–56

STEP 1

Begin the process of creating a color-dependent plot style table by choosing *Plot Style Manager* from the File pull-down menu as in Figure 10–57. This activates the Plot Styles dialog box in Figure 10–58. Various color-dependent and named plot styles already exist in this dialog box. To create a new plot style, double-click on the Add-A-Plot Style Table Wizard.

Figure 10–57

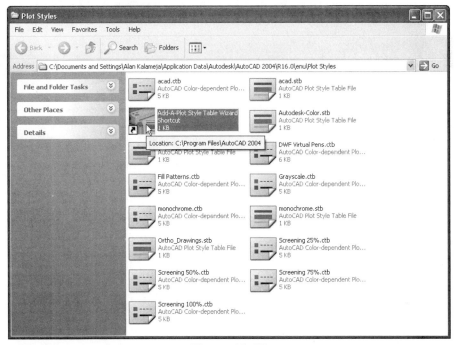

Figure 10–58

STEP 2

The Add Plot Style Table dialog box appears as in Figure 10–59 and introduces you to the process of creating plot style tables. Plot styles contain plot definitions for color, lineweight, linetype, end capping, fill patterns, and screening. You will be presented with various choices in creating a plot style from scratch, using the parameters in an existing plot style, or importing pen assignments information from a PCP, PC2, or CFG file. You will also have the choice of saving this plot style information in a CTB (color-dependent) or STB (named) plot style. Click the Next> button to display the next dialog box.

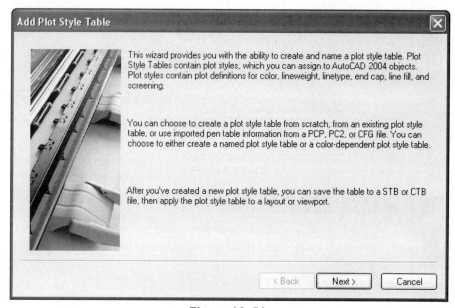

Figure 10–59

STEP 3

In the Add Plot Style Table – Begin dialog box in Figure 10–60, four options are available for you to choose, depending on how you want to create the plot style table. Click on the Start from scratch radio button to create this plot style from scratch. If you have made pen assignments from previous releases of AutoCAD, you can import them through this dialog box. Click the Next> button to display the next dialog box.

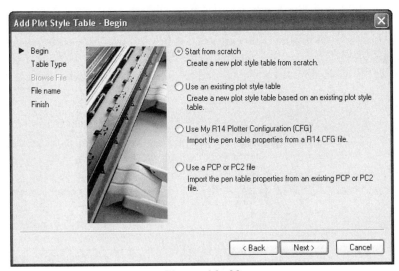

Figure 10–60

STEP 4

In the Add Plot Style Table – Pick Plot Style Table dialog box in Figure 10–61, click on the Color-Dependent Plot Style Table radio button to make this the type of plot style you will create. Click the Next> button to display the next dialog box.

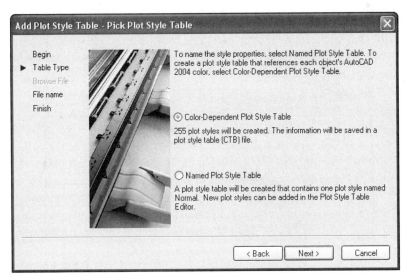

Figure 10–61

STEP 5

Use the Add Plot Style Table – File name dialog box in Figure 10–62 to assign a name to the plot style table. Enter the name Ortho_Drawings in the File name area. The extension CTB will automatically be added to this file name. Click the Next> button to display the next dialog box.

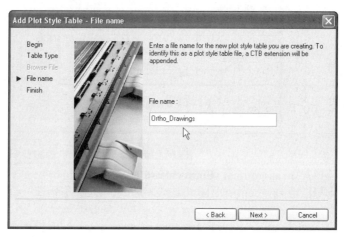

Figure 10–62

STEP 6

The Finish dialog box in Figure 10–63 alerts you that a plot style called Ortho_ Drawings.ctb has been created. However, you want to have all colors plot out black and you need to assign different lineweights to a few of the layers. To accomplish this, click on the Plot Style Table Editor button.

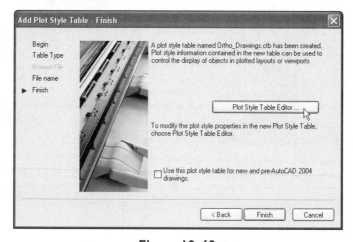

Figure 10–63

STEP 7

Clicking on the Plot Style Table Editor button displays the Plot Style Table Editor dialog box in Figure 10–64. Notice the name of the plot style table present at the top of the dialog box. Also, three tabs are available for making changes to the current plot style table (Ortho-Drawings.ctb). The first tab is General and displays file information about the current plot style table being edited. It is considered good practice to add a description to further document the purpose of this plot style table. It must be pointed out at this time that this plot style table is not going to be used only on the current drawing. Rather, if layers are standard across projects, the same plot style dialog box can be used. This is typical information that can be entered in the Description area.

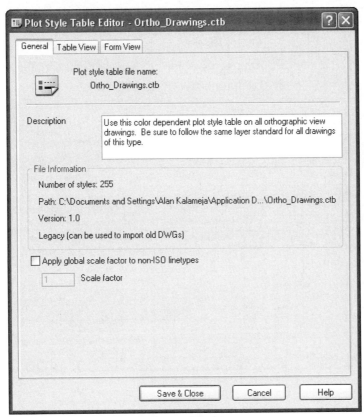

Figure 10–64

STEP 8

Clicking on the Table View tab activates the image in Figure 10–65. You can use the horizontal scroll bar to get a listing of all 255 colors along with special properties that can be changed. This information is presented in a spreadsheet format. You can click in any of the categories under a specific color and make changes, which will be applied to the current plot style table. The color and lineweight changes will be made through the next tab.

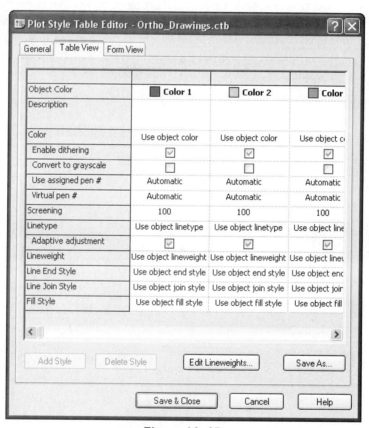

Figure 10–65

STEP 9

Clicking on the Form View tab displays the image in Figure 10–66. Here the colors are arranged vertically with the properties displayed on the right. Click on Color 1 (Red) and change the Color property to Black. Whatever is red in your drawing will plot out in the color black. Since red is used to identify hidden lines, click in the Lineweight area and set the lineweight for all red lines to 0.30 mm.

For Color 2 (Yellow), Color 3 (Green), Color 4 (Cyan), and Color 6 (Magenta), change the color to Black in the Properties area. All colors can be selected at one time by holding down CTRL while picking each of them. Changes can then be made to all colors simultaneously. Whatever is yellow, green, cyan, and magenta will plot out in the color black. No other changes need to be made in the dialog box for these colors.

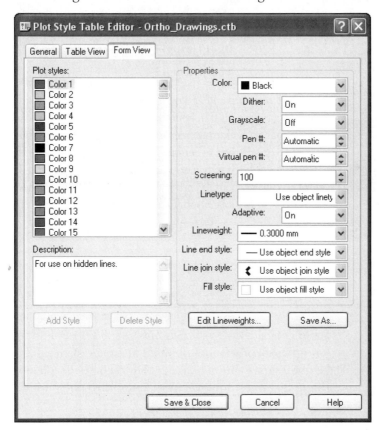

Figure 10–66

STEP 10

Click on Color 5 (Blue) in Figure 10–67 and change the Color property to Black. Whatever is blue in your drawing will plot out in the color black. Since blue is used to identify the title block lines, click in the Lineweight area and set the lineweight for all blue lines to 0.80 mm.

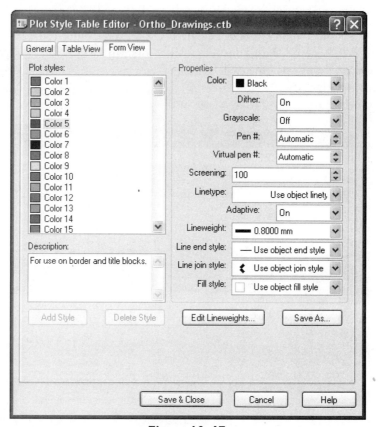

Figure 10–67

STEP 11

Click on Color 7 (Black) in Figure 10–68 and change the Color property to Black. Since black is used to identify the object lines, click in the Lineweight area and set the lineweight for all object lines to 0.70 mm. This completes the editing process of the current plot style table. Click on the Save & Close button; this returns you to the Finish dialog box. Clicking on the Finish button will return you to the Plot Styles dialog box. Close this box to display your drawing.

Figure 10–68

STEP 12

Activate the Plot dialog box in the Plot device tab, verify that the plotter is the Design-Jet 750C Plus plotter (this plotter has been used in all plotting tutorials throughout this chapter). In the Plot style table (pen assignments) area (see Figure 10–69), make the current plot style Ortho_Drawings.ctb. Notice at the top of the dialog box that the plot style table will be saved to this layout. This means if you need to plot this drawing again in the future, you will not have to look for the desired plot style table.

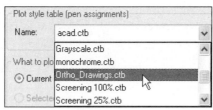

Figure 10–69

STEP 13

Click on the Plot Settings tab in the Plot dialog box and verify in the lower-right corner of the dialog box under Plot options (see Figure 10–70) that you will be plotting with plot styles. Click on the Full Preview button to display the results. Unless a plotter is attached to your computer, it is not necessary to click the OK button.

Figure 10–70

STEP 14

The results of performing a full preview are illustrated in Figure 10–71. Notice that all lines are black even though they appear in color in the drawing file. Notice that when you zoom in on a part of the preview different line weights appear in Figure 10–72 even though they all appear the same in the drawing file. This is the result of using a color-dependent plot style table on this drawing. This file can also be attached to other drawings that share the same layer names and colors.

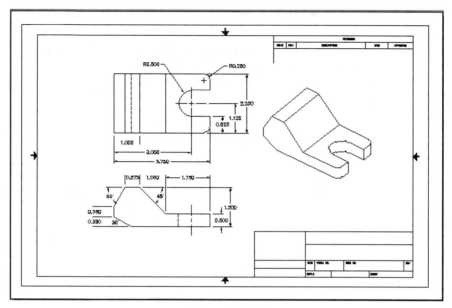

Figure 10–71

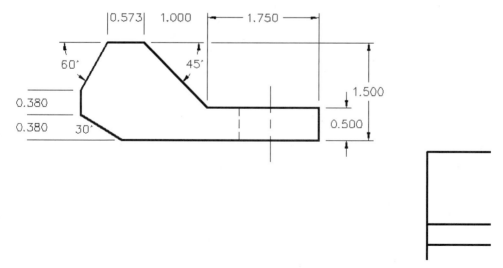

Figure 10–72

CONVERTING DRAWINGS TO DIFFERENT PLOT STYLES

In the event that you encounter a drawing that has a Named plot style (STB) attached, you can easily convert this drawing to a Color Dependent plot style by using the CONVERTPSTYLES command. This command must be entered from the keyboard at the Command: prompt. Follow the next series of steps that illustrate this important process.

STEP 1

Open the drawing file 10_Convert from STB in Figure 10–73.

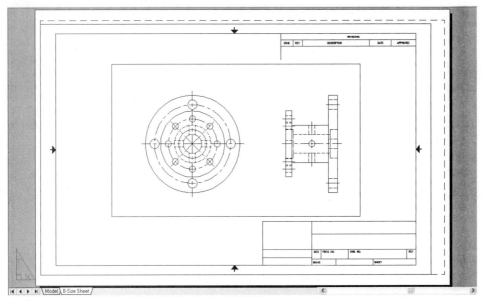

Figure 10–73

STEP 2

Activate the Page Setup dialog box illustrated in Figure 10–74. Notice that the current plot style is set to ACAD-Color.STB. This means that you are not plotting by color, which is the plotting method preferred by most AutoCAD users. Unfortunately, there is no control in the Page Setup dialog box that allows you to easily switch to the Color Dependent plot style (CTB). Click OK to proceed to the next step.

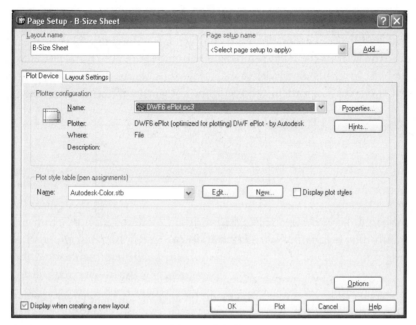

Figure 10–74

STEP 3

Begin the conversion process from an STB to a CTB by entering the following command at the prompt below:

Command: **CONVERTPSTYLES**

Pressing ENTER after this command will display the AutoCAD alert box illustrated in Figure 10–75. The alert box informs you that all plot style names will be removed from objects and that any Named plot style tables will be detached from the drawing. Click OK to initiate the conversion.

Figure 10–75

STEP 4

Notice that after the conversion process is complete, your drawing will not appear to have changed. Again activate the Page Setup dialog box in Figure 10–76 and notice by searching through the available plot styles that all file extensions are CTB. You can now make lineweight and color assignments based on the colors that make up your drawing.

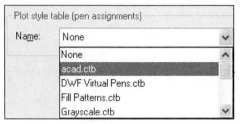

Figure 10–76

 Note: A drawing that has a Color Dependent plot style (CTB) attached can be changed to the Named plot style (STB) by using the CONVERTCTB command. A dialog box will display allowing you to select the Color Dependent plot style to convert. Then another dialog box will appear allowing you to select an existing Named plot style from the list or create a new Named plot style.

PUBLISHING TO THE WEB

Yet another way of applying electronic plots is through the Publish to the Web utility that is provided with AutoCAD. This feature creates a project that consists of a formatted HTML page and your drawing content in either DWF, JPG, or PNG image formats. Through the Publish to Web Wizard, you can select how your layout will look from a number of preformatted designs. Once the HTML page is created, you can post the page to an Internet location through the wizard. Follow the next series of steps that demonstrate how easy it is to publish drawings to the web.

STEP 1

Open the drawing file 10_Publish to Web. In Figure 10–77, notice that four layouts exist, namely Four Views, Front View, Top View, and Right Side View. All four of these layouts will be arranged in a single web page to demonstrate how easy it is to perform this task. This example demonstrates how you can publish various layouts of the same drawing to the web; you can also publish different drawings as well.

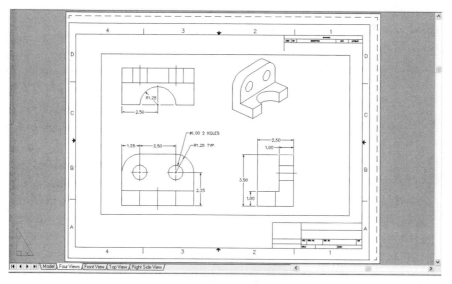

Figure 10–77

STEP 2

Begin the process of publishing to the web by clicking on the File pull-down followed by Publish to Web... in Figure 10–78. This will launch the Publish to Web wizard in Figure 10–79. In this first dialog box (called Begin), be sure the radio button next to "Create New Web Page" is selected. If you already have an existing web page that needs updating or editing, you could click the radio button next to "Edit Existing Web Page". When finished, click the Next> button and continue on to the next step of this wizard.

Figure 10–78

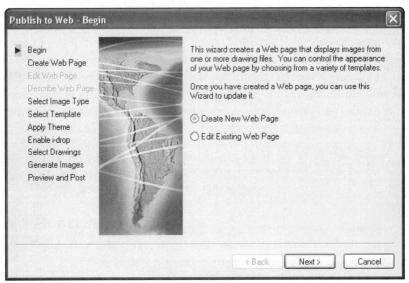

Figure 10–79

STEP 3

In the Create Web Page dialog box in Figure 10–80, add the name of the web page as "Four Views with Details"; then specify the location of this web page. It is also considered good practice to add a description of the web page in the event that others will be manipulating your web page. After completing the description, click the Next> button to go to the next step.

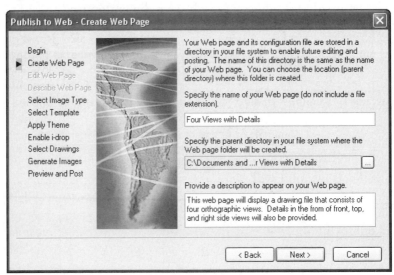

Figure 10–80

STEP 4

Figure 10–81 displays the Select Image Type dialog box. By default, your web page will be created in DWF format. These are vector-based images of your drawing that are viewed in Autodesk Express Viewer, which is a free viewer. Another image type option is illustrated in Figure 10–82 where the AutoCAD drawings consist of JPG images. JPG files consist of raster images that do not perform very well if your drawing contains a lot of text. However, for this tutorial exercise, JPG will be the image format used. Yet another image format is displayed in Figure 10–83. The PNG formats are also raster-based representations of your AutoCAD drawing. They produce high-quality images when compared with JPG files. Be sure that the image type reads JPG and click the Next> button when finished.

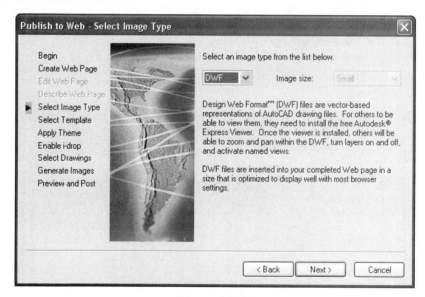

Figure 10–81

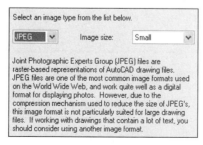

Figure 10–82

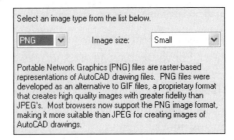

Figure 10–83

STEP 5

The next step in the process of publishing to the web is to select a web template. The Select Template dialog box in Figure 10–84 displays four different templates for you to choose from. Click on each one to preview each template at the right of the dialog box. For this tutorial, click on "Array of Thumbnails" and click the Next> button when finished.

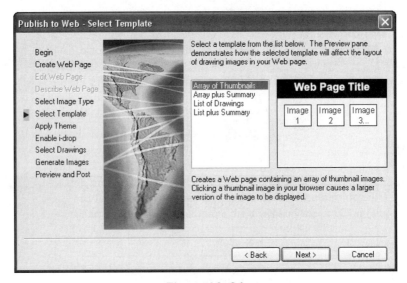

Figure 10–84

STEP 6

The next step when publishing to the web is the selection of a theme. Figure 10–85 shows seven possible themes that you can apply. Click each one to preview the contents of each. For this tutorial, the Classic theme will be used which is illustrated in Figure 10–86. When finished, click the Next> button to continue.

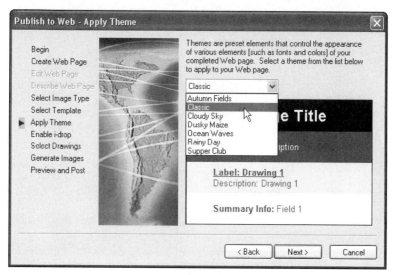

Figure 10–85

Figure 10–86

STEP 7

The next step displays the Enable I-drop dialog box in Figure 10–87. Placing a check in the Enable I-drop box allows those who visit your web page to drag and drop drawing files into their session of AutoCAD. While this is a very powerful feature, we will not be demonstrating it during this tutorial. Click the Next> button to continue.

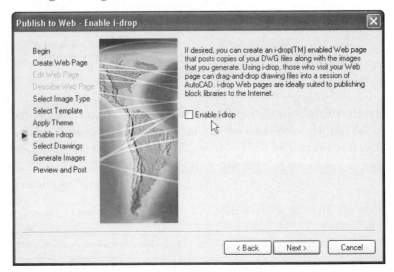

Figure 10–87

STEP 8

The next dialog box allows you to add the drawings that will make up the web page. In Figure 10–88, the drawing name is listed along with layout and label information. The label can be any name or series of names that you wish to appear in your web page. After filling in a description of this drawing, click the Add-> button to add the label to the image list. Under the Layout heading in Figure 10–89, click on Front View as the new layout, add a description, and click the Add-> button to add this label to the image list. Do the same for the Right Side View and Top View. After adding the four layouts to the image list, your dialog should appear similar to Figure 10–90. Click the Next> button to continue.

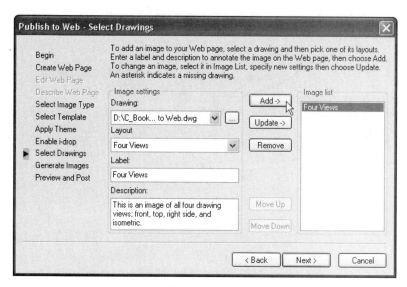

Figure 10–88

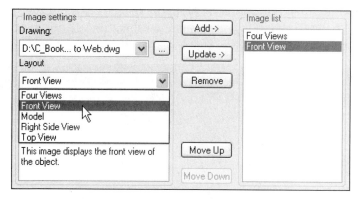

Figure 10–89

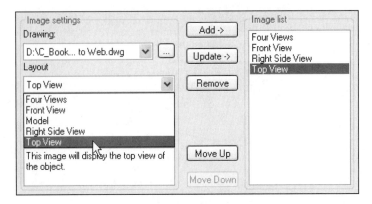

Figure 10–90

STEP 9

The Generate Images dialog box in Figure 10–91 will create the web page in the folder that you specified earlier. Clicking on the Next> button of this dialog box will pause your system while all of the layouts are plotted. This may take time depending on the number and complexity of drawings being published. Be sure the radio button next to "Regenerate images for drawings that have changed" is selected.

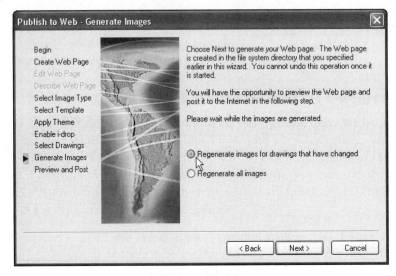

Figure 10–91

STEP 10

After the regeneration of all images, the Preview and Post dialog box appears in Figure 10–92. You can either preview your results or post the web page to a web site at a later time. Click the Preview button.

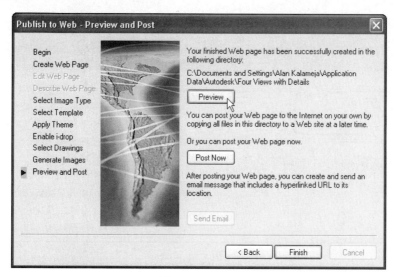

Figure 10–92

STEP 11

Clicking on the Preview button in the previous step will launch the Microsoft Internet Explorer and display your web page in Figure 10–93. To view each image separately, click on the thumbnail to enlarge the image and show more detail (see the image of the Right Side View in Figure 10–94). Dismissing Internet Explorer takes you back to the Preview and Post dialog box. Pick the Finish button to end this task.

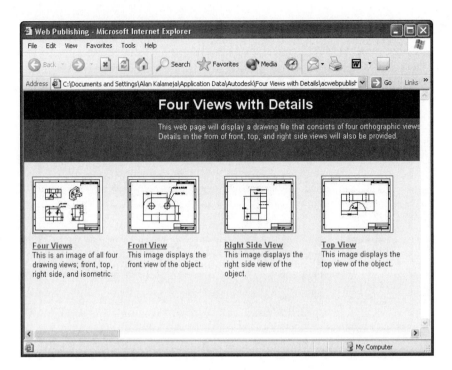

Figure 10–93

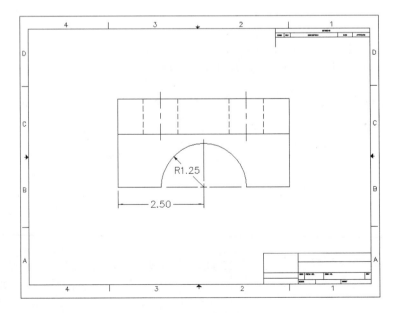

Figure 10–94

THE DESIGN PUBLISHER

The Design Publisher in AutoCAD allows you to arrange a set of drawings that can then be plotted directly to paper or published to a Design Web Format file (DWF). Drawing sets can be published as either single- or multiple-sheet DWF6 files. Once the DWF files are created, you can share these files with other users who do not even own AutoCAD. These recipients of the DWF files can view and plot the files through the Autodesk Express Viewer, which is free and can be found on the main Autodesk Web site:

www.autodesk.com

DWF files differ from DWG files in that DWF files cannot be changed by a recipient; they can only be viewed and plotted.

You access the Design Publisher by clicking on Publish located in the File pull-down menu as shown in Figure 10-95. You could also enter PUBLISH on the command line. Either one of these operations will launch the Publish Drawing Sheets dialog box illustrated in Figure 10-96.

Figure 10–95

The Publish Drawing Sheets dialog box illustrated in Figure 10-96 allows you to arrange, reorder, rename, and save a number of drawings for publishing. You can elect to publish the drawings to a DWF file, send the drawings to a plotter for hardcopy, or save the drawing set as a plot file.

The areas of the Publish Drawing Sheets dialog box are described in the following sections.

LIST OF DRAWING SHEETS

This is the area where you assemble the drawings to be published. The Sheet Name column is a combination of the drawing name and the name of the layout separated by a dash (-). The Drawing Name column displays the name of the drawing file. The Layout Name column displays the selected drawing layout name. The Page Setup column displays the page setup name for the layout. The Status column will display a message regarding the status of each drawing sheet.

You control the contents of the Publish Drawing Sheets dialog box with the following buttons and areas:

- Add Sheets—Clicking on this button will launch the standard File Selection dialog box. Here you can add sheets to the list of drawings. If you add a sheet with a name that already exists, you will have to enter a new drawing sheet name.

- Load List—Clicking on this button will display the standard File Selection dialog box if the contents of the List of Drawing Sheets is empty. Here you can select a Drawing Set Description file (DSD) or a Batch Plot file (BP3) to load.

- Save List—Clicking on this button will display the Save List dialog box. It is here that you save the current list of drawings as a Drawing Set Descriptions file (DSD).

- Remove—Clicking on this button will allow you to remove or delete the currently selected drawing sheet from the List of Drawing Sheets list.

- Remove All—Clicking on this button will delete all drawing sheets only if the current list of drawings has been saved. If the current drawing sheet list has not been saved, you will be prompted to first save the new list before the drawings are removed.

- Move Up—Allows you to move a selected drawing sheet up one position in the drawing list.

- Move Down—Allows you to move a selected drawing sheet down one position in the drawing list.

- Publish To—This area allows you to choose the method of publishing the list of drawings. You can choose to publish to a multi-sheet DWF file or to produce paper plots.

- Multi-Sheet DWF File—Clicking on this radio button will generate a single multi-sheet DWF file. You can also elect to enter a password as a means of protecting your date.

- Plotters Named in Page Setups—Clicking in this radio button will indicate that the output devices used for each drawing sheet in a page setup will be used.

- Publish—Clicking on this button will begin the publishing process. One of the following outputs will be produced:

  - A single-sheet or multi-sheet DWF file will be created.

  - A plot will be produced to a device or file depending on which radio button was selected in the Publish to area.

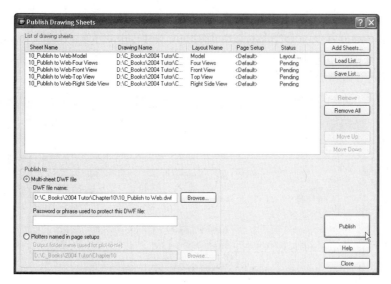

Figure 10–96

The result of publishing the set of drawings in this example will be a DWF file, as shown in Figure 10–96. Locating and double-clicking on this file in the Windows Explorer will launch the Autodesk Express Viewer as shown in Figure 10-97. In this figure, all sheets generated are present in this multi-sheet DWF file. To view the other sheets, click the down arrow to expand the available sheets.

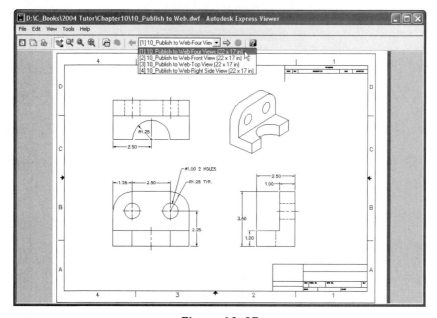

Figure 10–97

While inside a drawing sheet, you can also right-click in the sheet to display the shortcut menu as shown in Figure 10-98. Here you can perform basic viewing functions such as panning and zooming. You can also display named views and turn layers on and off and even plot the DWF file out.

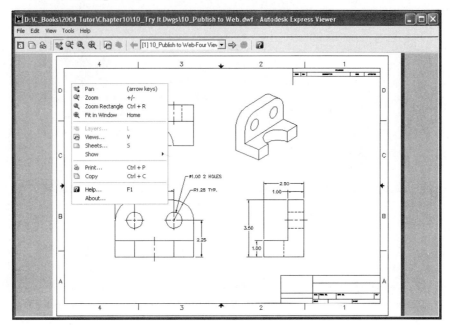

Figure 10–98

 Open the Exercise Manual PDF file for Chapter 10 on the accompanying CD for more tutorials and exercises.

 If you have the accompanying Exercise Manual, refer to Chapter 10 for more tutorials and exercises.

CHAPTER 11

Dimensioning Basics

Once orthographic views have been laid out, a design is not ready for the production line until dimensions describing the width, height, and depth of the object are added to the drawing. However, these numbers must be added in a certain organized fashion; otherwise, the drawing becomes difficult to read. That may lead to confusion and the possible production of a part that is incorrect according to the original design. This chapter begins with an introduction to the proper placement of dimensions in a drawing. The chapter continues with a discussion of the commands used in Auto-CAD for creating dimensions. These commands include Linear, Aligned, Ordinate, Radius, Diameter, Angular, Quick, Baseline, and Continue dimensions. A section in this chapter deals with the placement of Geometric Dimensioning and Tolerancing symbols. Also, a section on the meaning of various dimensioning symbols (Counterbore, Countersink, Depth, Slope, and so on) is illustrated by various examples. A short segment on the effects grips have on dimensions is included in this chapter. The topic of associative dimensioning and the ability to add dimensions in a drawing layout (Paper Space) will be demonstrated.

DIMENSIONING BASICS

Before discussing the components of a dimension, remember that object lines (at "A") continue to be the thickest lines of a drawing, with the exception of border or title blocks (see Figure 11–1). To promote contrasting lines, dimensions become visible, yet thin, lines. The heart of a dimension is the dimension line (at "B"), which is easily identified by the location of arrow terminators at both ends (at "D"). In mechanical cases, the dimension line is broken in the middle, which provides an excellent location for the dimension text (at "E"). For architectural applications, dimension text is usually placed above an unbroken dimension line. The extremities of the dimension lines are limited by the placement of lines that act as stops for the arrow terminators. These lines, called extension lines (at "C"), begin close to the object without touching the object. Extension lines will be discussed in greater detail in the pages that follow. For placing diameter and radius dimensions, a leader line consisting of an inclined line with a short horizontal shoulder is used (at "F"). Other applications of leader lines are for adding notes to drawings.

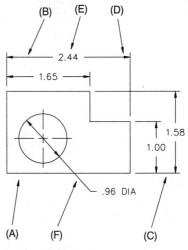

Figure 11–1

When you place dimensions in a drawing, provide a spacing of at least 0.38 units between the first dimension line and the object being dimensioned (at "A" in Figure 11–2). If you're placing stacked or baseline dimensions, provide a minimum spacing of at least 0.25 units between the first and second dimension line at "B" or any other dimension lines placed thereafter. This will prevent dimensions from being placed too close to each other.

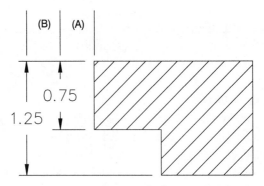

Figure 11–2

It is recommended that extensions never touch the object being dimensioned and that they begin approximately between 0.03 and 0.06 units away from the object at "A" (see Figure 11–3). As dimension lines are added, extension lines should extend no further than 0.12 beyond the arrow or any other terminator (at "B"). The height of dimension text is usually 0.125 units (at "C"). This value also applies to notes placed on objects with leader lines. Certain standards may require a taller lettering height. Become familiar with your office's standard practices, which may deviate from these recommended values.

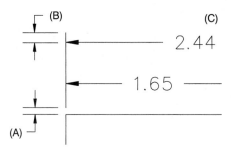

Figure 11–3

PLACEMENT OF DIMENSIONS

When placing multiple dimensions on one side of an object, place the shorter dimension closest to the object followed by the next larger dimension, as shown in Figure 11–4. When you are placing multiple horizontal and vertical dimensions involving extension lines that cross other extension lines, do not place gaps in the extension lines at their intersection points.

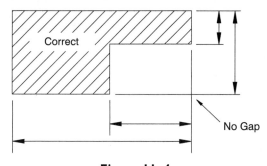

Figure 11–4

It is acceptable for extension lines to cross each other, but it is considered unacceptable practice for extension lines to cross dimension lines, as in the example in Figure 11–5. The shorter dimension is placed closest to the object, followed by the next larger dimension.

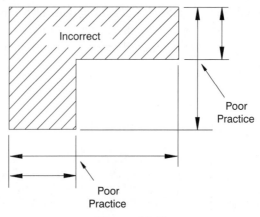

Figure 11–5

It is considered poor practice to dimension inside an object when there is sufficient room to place dimensions on the outside. There may be exceptions to this rule, however. It is also considered poor practice to cross dimension lines, because this may render the drawing confusing and possibly result in the inaccurate interpretation of the drawing. It is also considered poor practice to superimpose extension lines over object lines. See Figure 11–6.

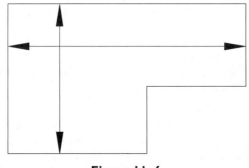

Figure 11–6

PLACING EXTENSION LINES

Because two extension lines may intersect without providing a gap, so also may extension lines and object lines intersect with each other without the need for a gap between them. The practice is the same when centerlines extend beyond the object without gapping, as illustrated in Figure 11–7.

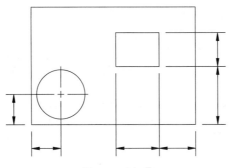

Figure 11–7

In Figure 11–8, the gaps in the extension lines may appear acceptable; however, in a very complex drawing, gaps in extension lines would render a drawing confusing. Draw extension lines as continuous lines without providing breaks in the lines.

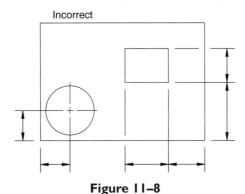

Figure 11–8

In the same manner, when centerlines are used as extension lines for dimension purposes, no gap is provided at the intersection of the centerline and the object (as in Figure 11–9). As with extension lines, the centerline should extend no further than 0.125 units past the arrow terminator.

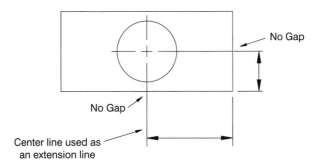

Figure 11–9

GROUPING DIMENSIONS

To promote ease of reading and interpretation, group dimensions whenever possible, as shown in Figure 11–10. In addition to making the drawing and dimensions easier to read, this promotes good organizational skills and techniques. As in previous examples, always place the shorter dimensions closest to the drawing followed by any larger or overall dimensions.

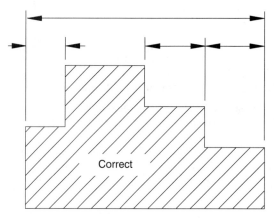

Figure 11–10

Avoid placing dimensions to an object line substituting for an extension line, as in Figure 11–11. The drawing is more difficult to follow, with the dimensions placed at different levels instead of being grouped. In some cases, however, this practice of dimensioning is unavoidable.

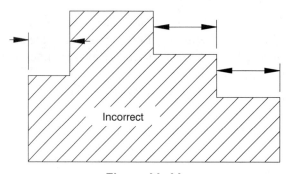

Figure 11–11

For tight spaces, arrange dimensions as in the illustration in Figure 11–12. Take extra care to follow proper dimension rules without sacrificing clarity.

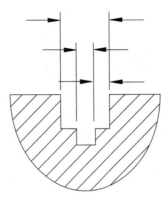

Figure 11–12

DIMENSIONING TO VISIBLE FEATURES

In Figure 11–13, the object is dimensioned correctly; however, the problem is that there are dimensions to hidden lines. Although there are always exceptions, try to avoid dimensioning to any hidden surfaces or features.

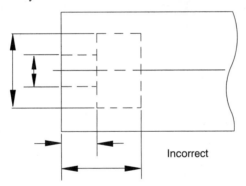

Incorrect

Figure 11–13

The object illustrated in Figure 11–14 is almost identical to that in Figure 11–13, except that it has been converted to a full section. Surfaces that were previously hidden are now exposed. This example illustrates a better way to dimension details that were previously hidden.

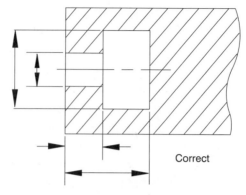

Figure 11–14

DIMENSIONING TO CENTERLINES

Centerlines are used to identify the center of circular features as in "A" in Figure 11–15. You can also use centerlines to indicate an axis of symmetry as in "B." Here, the centerline consisting of a short dash flanked by two long dashes signifies that the feature is circular in shape and form. Centerlines may take the place of extension lines when dimensions are placed in drawings.

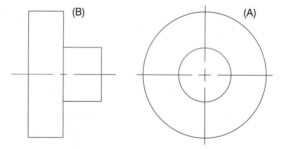

Figure 11–15

Figure 11–16 represents the top view of a U-shaped object with two holes placed at the base of the U. It also represents the correct way of utilizing centerlines as extension lines when dimensioning to holes. What makes this example correct is the rule of always dimensioning to visible features. The example in Figure 11–16 uses centerlines to dimension to holes that appear as circles. This is in direct contrast to Figure 11–17.

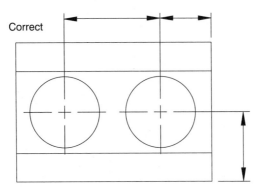

Correct

Figure 11–16

Figure 11–17 represents the front view of the U-shaped object. The hidden lines display the circular holes passing through the object along with centerlines. Centerlines are being used as extension lines for dimensioning purposes; however, it is considered poor practice to dimension to hidden features or surfaces. Always attempt to dimension to a view where the features are visible before dimensioning to hidden areas.

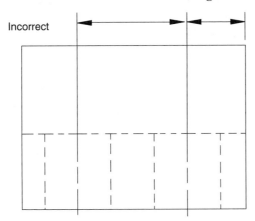

Incorrect

Figure 11–17

ARROWHEADS

Arrowheads are generally made three times as long as they are wide, or very long and narrow, as shown in Figure 11–18. The actual size of an arrowhead would measure approximately 0.125 units in length.

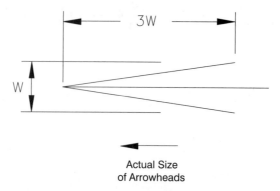

Actual Size
of Arrowheads

Figure 11–18

Dimension line terminators may take the form of shapes other than filled-in arrowheads (see Figure 11–19). The 45° slash or "tick" is a favorite dimension line terminator used by architects, although they are sometimes seen in mechanical applications.

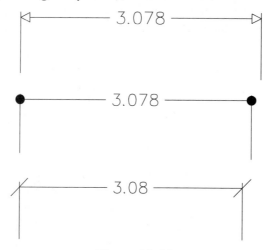

Figure 11–19

DIMENSIONING SYSTEMS

THE UNIDIRECTIONAL SYSTEM

A typical example of placing dimensions using AutoCAD is illustrated in Figure 11–20. Here, all text items are aligned horizontally. This applies to all vertical, aligned, angular, and diameter dimensions. When all dimension text can be read from one direction, this is the Unidirectional Dimensioning System. By default, all AutoCAD settings are set to dimension in the Unidirectional System.

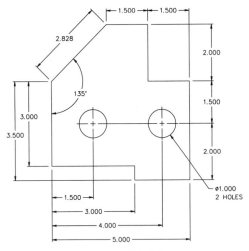

Figure 11–20

THE ALIGNED SYSTEM

The same object from the previous example is illustrated in Figure 11–21. Notice that all horizontal dimensions have the text positioned in the horizontal direction as in Figure 11–20. However, vertical and aligned dimension text is rotated or aligned with the direction being dimensioned. This is the most notable feature of the Aligned Dimensioning System. Text along vertical dimensions is rotated in such a way that the drawing must be read from the right. Angular dimensions remain unaffected in the Aligned System; however, aligned dimension text is rotated parallel with the feature being dimensioned.

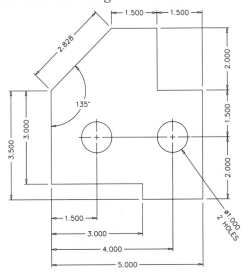

Figure 11–21

ORDINATE DIMENSIONS

A different type of dimensioning system is illustrated in Figure 11–22. Notice how all hole locations are designated by two numbers. Actually these sets of numbers are coordinates. One of the numbers measures a distance in the X-direction, while the other number measures a distance in the Y-direction. No arrows or dimension lines are used. This type of dimensioning technique is referred to as Ordinate dimensioning. All hole location coordinates reference a known 0,0 location in the drawing. As shown in Figure 11–22, this type of dimensioning is well suited for locating hole positions.

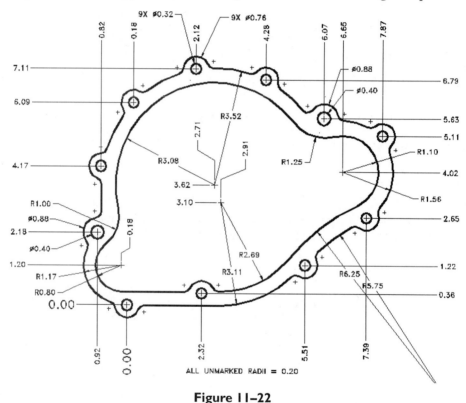

Figure 11–22

ARCHITECTURAL DIMENSIONS

Figure 11–23 illustrates another type of dimensioning technique. Architectural dimensions usually have their text drawn directly above the dimension line. Instead of an arrow, architectural dimensions are terminated with a tick mark. Architectural dimension text for vertical dimensions is read from the left side; this makes all vertical architectural dimensions aligned or parallel to the dimension line.

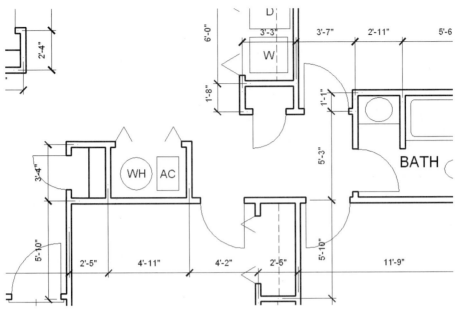

Figure 11–23

REPETITIVE DIMENSIONS

Throughout this chapter, numerous methods of dimensioning are discussed, supported by many examples. Just as it was important in multiview projection to draw only those views that accurately describe the object, so also is it important to dimension these views. In this manner, the actual production of the part may start. However, take care when placing dimensions. It takes planning to better place a dimension. The problem with the illustration in Figure 11–24 is that, even though the views are correct and the dimensions call out the overall sizes of the object, there are too many cases where dimensions are duplicated. Once a feature has been dimensioned, such as the overall width of 3.75 units shown between the Front and Top view, this dimension does not need to be repeated in the Top view. This is the purpose of understanding the relationship between views and what dimensions the views have in common with each other. Adding unnecessary dimensions also makes the drawing very busy and cluttered in addition to being very confusing to read. Also if a size changes in production and a dimension needs to change on the drawings, more instances of the dimension must be found and corrected. Compare Figure 11–24 with Figure 11–25, which shows only those dimensions needed to describe the size of the object. Do not be concerned that the Top view contains no dimensions; the designer should interpret the width of the top as 3.75 units from the Front view and the depth as 2.50 from the Right Side view.

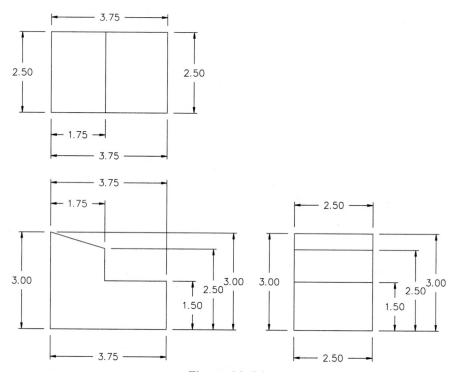

Figure 11–24

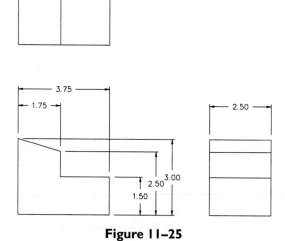

Figure 11–25

METHODS OF CHOOSING DIMENSION COMMANDS

THE DIMENSION TOOLBAR

The Dimension toolbar, illustrated in Figure 11–26, makes the selection of dimensioning commands easier. Study this image and the corresponding commands associated with each button.

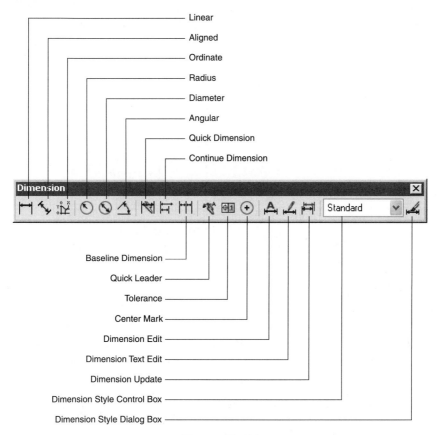

Figure 11–26

THE DIMENSION PULL-DOWN MENU

The Dimension pull-down menu, illustrated in Figure 11–27, is an additional area used for selecting the more commonly used dimension commands. Another way of activating dimension commands is through the keyboard. These commands tend to get long; the DIM-LINEAR command is one example. To spare you the effort of entering the entire command, all dimension commands have been abbreviated to three letters. These command aliases are listed in Table 1–2, Chapter 1. For example, DLI is the alias for the DIMLINEAR command.

Figure 11–27

BASIC DIMENSION COMMANDS

LINEAR DIMENSIONS

The Linear Dimensioning mode generates either a horizontal or vertical dimension depending on the location of the dimension. The following prompts illustrate the generation of a horizontal dimension with the DIMLINEAR command. Notice that identifying the dimension line location at "C" in Figure 11–28 automatically generates a horizontal dimension.

 Try It! - Open the drawing file 11_Dim Linear1. Verify that OSNAP is on and set to Endpoint. Follow the command sequence below and Figure 11–28 for performing this dimensioning task.

 Command: **DLI** *(For DIMLINEAR)*

Specify first extension line origin or <select object>: *(Select the endpoint of the line at "A")*
Specify second extension line origin: *(Select the other endpoint of the line at "B")*
Specify dimension line location or [Mtext/Text/Angle/Horizontal/Vertical/Rotated]:
  (Pick a point near "C" to locate the dimension)
Dimension text = 2.00

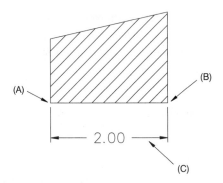

Figure 11–28

The linear dimensioning command is also used to generate vertical dimensions. The following prompts illustrate the generation of a vertical dimension with the DIMLINEAR command. Notice that identifying the dimension line location at "C" in Figure 11–29 automatically generates a vertical dimension.

 Try It! - Open the drawing file 11_Dim Linear2. Verify that OSNAP is on and set to Endpoint. Follow the command sequence below and Figure 11–29 for performing this dimensioning task.

 Command: **DLI** *(For DIMLINEAR)*

Specify first extension line origin or <select object>: *(Select the endpoint of the line at "A")*
Specify second extension line origin: *(Select the endpoint of the line at "B")*
Specify dimension line location or [Mtext/Text/Angle/Horizontal/Vertical/Rotated]:
   (Pick a point near "C" to locate the dimension)
Dimension text = 2.10

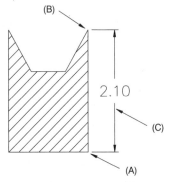

Figure 11–29

588

Rather than select two separate endpoints to dimension to, certain situations allow you to press ENTER and select the object (in this case, the line). This selects the two endpoints and prompts you for the dimension location. The completed dimension is illustrated in Figure 11–30.

Try It! - Open the drawing file 11_Dim Linear3. Follow the command sequence below and Figure 11–30 for performing this dimensioning task.

 Command: **DLI** *(For DIMLINEAR)*
Specify first extension line origin or <select object>: *(Press ENTER to select an object)*
Select object to dimension: *(Select the line at "A")*
Specify dimension line location or [Mtext/Text/Angle/Horizontal/Vertical/Rotated]:
    (Pick a point near "B" to locate the dimension)
Dimension text = 5.5523

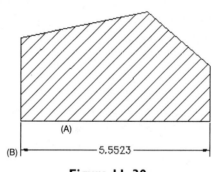

Figure 11–30

 ALIGNED DIMENSIONS

The Aligned Dimensioning mode generates a dimension line parallel to the distance specified by the location of two extension line origins as shown in Figure 11–31.

Try It! - Open the drawing file 11_Dim Aligned. Verify that OSNAP is on and set to Endpoint. The following prompts and Figure 11–31 illustrate the generation of an aligned dimension with the DIMALIGNED command.

 Command: **DAL** *(For DIMALIGNED)*
Specify first extension line origin or <select object>: *(Select the endpoint of the line at "A")*
Specify second extension line origin: *(Select the endpoint of the line at "B")*
Specify dimension line location or [Mtext/Text/Angle]: *(Pick a point at "C" to locate the dimension)*
Dimension text = 5.6525

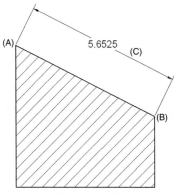

Figure 11–31

ROTATED LINEAR DIMENSIONS

This Linear Dimensioning mode generates a dimension line, which is rotated at a specific angle (see Figure 11–32). The following prompts illustrate the generation of a rotated dimension with the DIMLINEAR command and a known angle of 45°. If you do not know the angle, you could easily establish the angle by clicking the endpoints at "A" and "D" in Figure 11–32 when prompted to "Specify angle of dimension line <0>."

 Try It! - Open the drawing file 11_Dim Rotated. Verify that OSNAP is on and set to Endpoint. Follow the command sequence below and Figure 11–32 for performing this dimensioning task.

 Command: **DLI** *(For DIMLINEAR)*
Specify first extension line origin or <select object>: *(Select the endpoint of the line at "A")*
Specify second extension line origin: *(Select the endpoint of the line at "B")*
Specify dimension line location or [Mtext/Text/Angle/Horizontal/Vertical/Rotated]: **R** *(For Rotated)*
Specify angle of dimension line <0>: **45**
Specify dimension line location or [Mtext/Text/Angle/Horizontal/Vertical/Rotated]: *(Pick a point at "C" to locate the dimension)*
Dimension text = **2.8284**

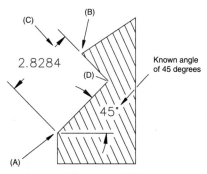

Figure 11–32

CONTINUE DIMENSIONS

The power of grouping dimensions for ease of reading has already been explained. Figure 11–33 shows yet another feature of dimensioning in AutoCAD: the practice of using Continue dimensions. With one dimension already placed, you issue the DIMCONTINUE command, which prompts you for the second extension line location. Picking the second extension line location strings the dimensions next to each other or continues the dimension.

Try It! - Open the drawing file 11_Dim Continue. Verify that Running OSNAP is on and set to Endpoint. Follow the command sequence below and Figure 11–33 for performing this dimensioning task.

 Command: **DLI** *(For DIMLINEAR)*
Specify first extension line origin or <select object>: *(Select the endpoint of the line at "A")*
Specify second extension line origin: *(Select the endpoint of the line at "B")*
Specify dimension line location or [Mtext/Text/Angle/Horizontal/Vertical/Rotated]: *(Locate the 1.75 horizontal dimension)*
Dimension text = 1.75

 Command: **DCO** *(For DIMCONTINUE)*
Specify a second extension line origin or [Undo/Select] <Select>: *(Select the endpoint of the line at "C")*
Dimension text = 1.25
Specify a second extension line origin or [Undo/Select] <Select>: *(Select the endpoint of the line at "D")*
Dimension text = 1.50
Specify a second extension line origin or [Undo/Select] <Select>: *(Select the endpoint of the line at "E")*
Dimension text = 1.00
Specify a second extension line origin or [Undo/Select] <Select>: *(Press ENTER when finished)*
Select continued dimension: *(Press ENTER to exit this command)*

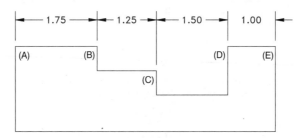

Figure 11–33

BASELINE DIMENSIONS

Yet another aid in grouping dimensions is the DIMBASELINE command. Continue dimensions place dimensions next to each other; Baseline dimensions establish a base or starting point for the first dimension, as shown in Figure 11–34. Any dimensions that follow in the DIMBASELINE command are calculated from the common base point already established. This is a very popular mode to use when one end of an object acts as a reference edge. When you place dimensions using the DIMBASELINE command, a baseline spacing setting of 0.38 units controls the spacing of the dimensions from each other.

 Try It! - Open the drawing file 11_Dim Baseline. Verify that Running OSNAP is on and set to Endpoint. Follow the command sequence below and Figure 11–34 for performing this dimensioning task.

 Command: **DLI** *(For DIMLINEAR)*
Specify first extension line origin or <select object>: *(Select the endpoint of the line at "A")*
Specify second extension line origin: *(Select the endpoint of the line at "B")*
Specify dimension line location or [Mtext/Text/Angle/Horizontal/Vertical/Rotated]:
    (Locate the 1.75 horizontal dimension)
Dimension text = 1.75

 Command: **DBA** *(For DIMBASELINE)*
Specify a second extension line origin or [Undo/Select] <Select>: *(Select the endpoint of the line at "C")*
Dimension text = 3.00
Specify a second extension line origin or [Undo/Select] <Select>: *(Select the endpoint of the line at "D")*
Dimension text = 4.50
Specify a second extension line origin or [Undo/Select] <Select>: *(Select the endpoint of the line at "E")*
Dimension text = 5.50
Specify a second extension line origin or [Undo/Select] <Select>: *(Press ENTER when finished)*
Select base dimension: *(Press ENTER to exit this command)*

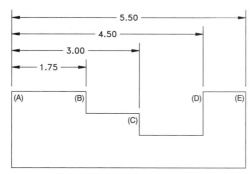

Figure 11–34

Note: To re-establish the base point for Continue or Baseline dimensions, press ENTER at the following prompt:

Specify first extension line origin or <select object>:

A new prompt asking you for the base extension will appear. Pick the extension line that will act as the new baseline or continue dimension.

 THE QDIM COMMAND

A more efficient means of placing a series of continued or baseline dimensions is the QDIM or Quick Dimension command. You identify a number of valid corners representing intersections or endpoints of an object and all dimensions are placed. In Figure 11–35, a crossing box is used to identify all corners to dimension to. When you have finished identifying the crossing box and pressing the ENTER key, a preview of the dimensioning mode appears in Figure 11–36. During this preview mode, you can right-click and have a shortcut menu appear. This allows you to select other dimension modes. When the dimension line is identified in Figure 11–37, all continued dimensions are placed.

Try It! - Open the drawing file 11_Qdim Continuous. Follow the command sequence below and Figures 11–35 through 11–37 for performing this dimensioning task.

 Command: **QDIM**
Select geometry to dimension: *(Pick a point at "A" in Figure 11–35)*
Specify opposite corner: *(Pick a point at "B")*
Select geometry to dimension: *(Press ENTER to continue)*
Specify dimension line position, or
[Continuous/Staggered/Baseline/Ordinate/Radius/Diameter/datumPoint/Edit]
<Continuous>: *(Change to a different mode or locate the dimension line at "C" in Figure 11–37)*

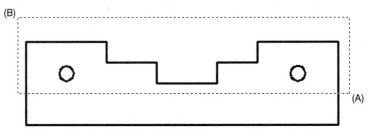

Figure 11–35

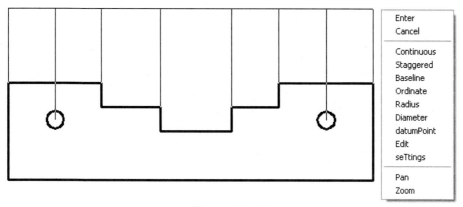

Figure 11–36

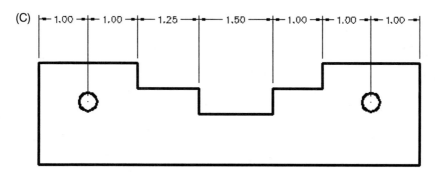

Figure 11–37

By default, the QDIM command places continued dimensions. Before locating the dimension line, you have the option of placing the Staggered dimensions in Figure 11–38. The process with this style begins with adding dimensions to inside details and continuing outward until all features are dimensioned.

 Try It! - Open the drawing file 11_Qdim Staggered. Activate the QDIM command and identify the same set of objects in Figure 11–35. Change to the Staggered mode. Your display should appear similar to the illustration in Figure 11–38.

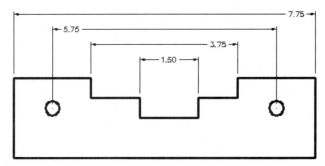

Figure 11–38

Another option of the QDIM command is the ability to place Baseline dimensions. As with all Baseline dimensions, an edge is used as the baseline or datum. All dimensions are calculated from the left edge as in Figure 11–39. The datumPoint option of the QDIM command would allow you to select the right edge of the object in Figure 11–39 as the new datum or baseline.

 Try It! - Open the drawing file 11_Qdim Baseline. Activate the QDIM command and identify the same set of objects in Figure 11–35. Change to the Baseline mode. Your display should appear similar to the illustration in Figure 11–39.

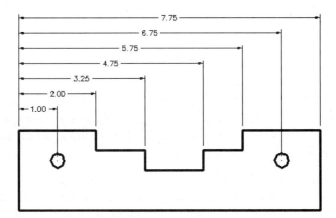

Figure 11–39

The Ordinate option of the QDIM command calculates dimensions from a known 0,0 corner. The main Ordinate dimensioning topic will be discussed in greater detail later in this chapter. A new User Coordinate System must first be created in the corner of the object. Then the QDIM command is activated along with the Ordinate options to display the results in Figure 11–40. Dimension lines are not used in Ordinate dimensions, in order to simplify the reading of the drawing.

Try It! - Open the drawing file 11_Qdim Ordinate. Activate the QDIM command and identify the same set of objects in Figure 11–35. Change to the Ordinate mode. Your display should appear similar to the illustration in Figure 11–40.

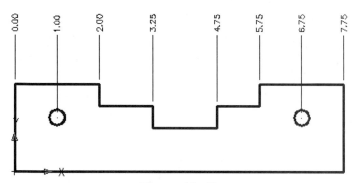

Figure 11–40

 Tip: The QDIM command can also be used to edit an existing group of dimensions. Activate the command, select all existing dimensions with a crossing box, and enter an option to change the style of all dimensions.

DIAMETER DIMENSIONING

Arcs and circles should be dimensioned in the view where their true shape is visible. The mark in the center of the circle or arc indicates its center point, as shown in Figure 11–41. You may place the dimension text either inside or outside the circle; you may also use grips to aid in the dimension text location of a diameter or radius dimension.

 Try It! - Open the drawing file 11_Dim Radial. Use Figure 11–41 and the following command sequence for placing diameter and radius dimensions.

 Command: **DDI** *(For DIMDIAMETER)*
Select arc or circle: *(Select the edge of the large circle)*
Dimension text = 2.50
Specify dimension line location or [Mtext/Text/Angle]: *(Pick a point to locate the diameter dimension)*

 Command: **DDI** *(For DIMDIAMETER)*
Select arc or circle: *(Select the edge of the small circle)*
Dimension text = 1.00
Specify dimension line location or [Mtext/Text/Angle]: *(Pick a point to locate the diameter dimension)*

Command: **DRA** *(For DIMRADIUS)*

Select arc or circle: *(Select the edge of the arc)*
Dimension text = 1.50
Specify dimension line location or [Mtext/Text/Angle]: *(Pick a point to locate the radius dimension)*

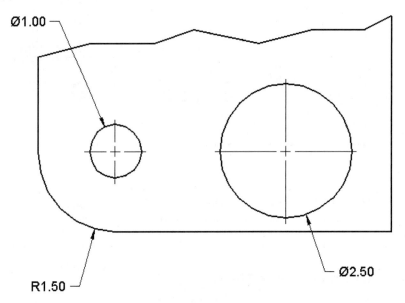

Ø1.00

R1.50

Ø2.50

Figure 11–41

USING QDIM FOR RADIUS AND DIAMETER DIMENSIONS

As with linear dimensions, the QDIM command has options that can be applied to radius and diameter dimensions. In Figures 11–42 and 11–43, all circles are selected. Activating either the Radius or Diameter option displays the results in the illustrations. When locating the radius or diameter dimensions, you can specify the angle of the leader. A predefined leader length is applied to all dimensions. Grips could be used to relocate the dimensions to better places.

Try It! - Open the drawing file 11_Qdim Radius. Use Figure 11–42 and the command sequence below for placing a series of radius dimensions using the QDIM command.

Command: **QDIM**

Select geometry to dimension: *(Select the polyline in Figure 11–42)*
Select geometry to dimension: *(Press ENTER to continue)*
Specify dimension line position, or
[Continuous/Staggered/Baseline/Ordinate/Radius/Diameter/datumPoint/Edit]
<Continuous>: **R** *(For Radius)*

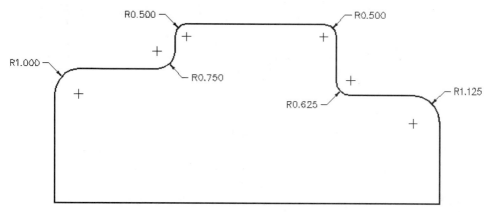

Figure 11–42

 Try It! - Open the drawing file 11_Qdim Diameter. Use Figure 11–43 and the command sequence below for placing a series of diameter dimensions using the QDIM command.

 Command: **QDIM**

Select geometry to dimension: *(Select the six circles in Figure 11–43)*
Select geometry to dimension: *(Press ENTER to continue)*
Specify dimension line position, or
[Continuous/Staggered/Baseline/Ordinate/Radius/Diameter/datumPoint/Edit]
<Continuous>: **D** *(For Diameter)*
Specify dimension line position, or
[Continuous/Staggered/Baseline/Ordinate/Radius/Diameter/datumPoint/Edit]
<Diameter>: *(Pick a point to locate the diameter dimension)*

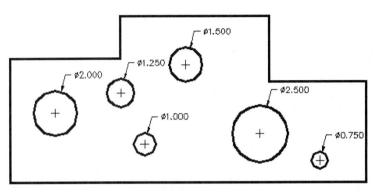

Figure 11–43

LEADER LINES

A leader line is a thin, solid line leading from a note or dimension ending with an arrowhead, illustrated at "A" in Figure 11–44. The arrowhead should always terminate at an object line such as the edge of a hole or arc. A leader to a circle or arc should be radial; this means it is drawn so that if extended, it would pass through the center of the circle illustrated at "B." Leaders should cross as few object lines as possible and should never cross each other. The short horizontal shoulder of a leader should meet the dimension illustrated at "A." It is poor practice to underline the dimension with the horizontal shoulder illustrated at "C." Example "C" also illustrates a leader not lined up with the center or radial. This may affect the appearance of the leader. Again, check your office's standard practices to ensure that this example is acceptable.

Yet another function of a leader is to attach notes to a drawing, illustrated at "D." Notice that the two notes attached to the view have different terminators: arrows and dots. It is considered good practice to adopt only one terminator for the duration of the drawing. The sequence for the LEADER command is as follows:

Command: **LEADER**
Specify leader start point: **Nea**
to *(Select the edge of the arc at "A")*
Specify next point: *(Pick a point to locate the end of the leader)*
Specify next point or [Annotation/Format/Undo] <Annotation>: *(Press* ENTER *to accept this default)*
Enter first line of annotation text or <options>: *(Press* ENTER *to accept this default)*
Enter an annotation option [Tolerance/Copy/Block/None/Mtext] <Mtext>: *(Press* ENTER *to display the Multiline Text Editor dialog box. Enter the desired text to make up the leader note and click the OK button to place the leader)*

 Try It! - The drawing file 11_Dim Leader is provided on the CD to allow you to practice using the LEADER command sequence and Figure 11–44.

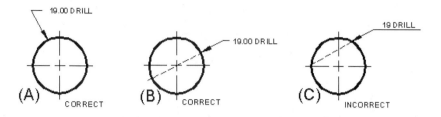

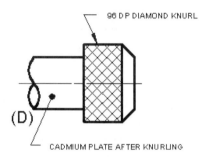

Figure 11–44

THE QLEADER COMMAND

The QLEADER command, or Quick Leader, provides numerous controls for placing leaders in your drawing. This command may appear to operate similarly to the LEADER command previously explained.

 Try It! - Open the drawing file 11_Qleader. Study the prompt sequence below and Figure 11–45 for this command.

Command: **LE** *(For QLEADER)*

Specify first leader point, or [Settings]<Settings>: **Nea**
to *(Pick the point nearest at "A")*
Specify next point: *(Pick a point at "B")*
Specify next point: *(Press ENTER to continue)*
Specify text width <0.00>: *(Press ENTER to accept this default)*
Enter first line of annotation text <Mtext>: *(Press ENTER to display the Multiline Text Editor dialog box. Enter "R0.20" and "2 PLACES" in two separate text lines. Click the OK button to place the leader)*

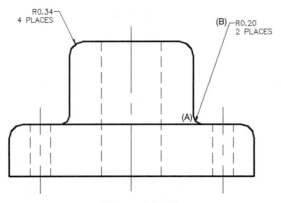

Figure 11–45

Pressing ENTER at the first Quick Leader prompt displays the Leader Settings dialog box in Figure 11–46. Various controls are available to add greater functionality to your leaders through three tabs: Annotation, Leader Line & Arrow, and Attachment.

The Annotation tab deals with the object placed at the end of the leader. By default, the MText radio button is selected, allowing you to add a note through the Multiline Text Editor dialog box. You could also copy an object at the end of the leader, have a geometric tolerancing symbol placed in the leader, have a pre-defined block placed in the leader, or leave the leader blank.

Command: **LE** *(For QLEADER)*

Specify first leader point, or [Settings]<Settings>: *(Press ENTER to accept the default value and display the Leader Settings dialog box in Figure 11–46)*

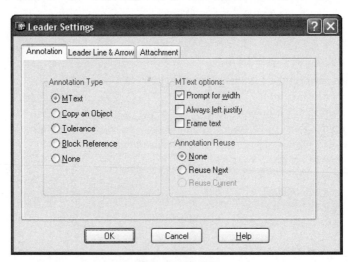

Figure 11–46

The Leader Line & Arrow tab in Figure 11–47 allows you to draw a leader line consisting of straight segments or in the form of a spline object. You can control the number of points used to define the leader; a maximum of three points will be more than enough to create your leader. You can even change the arrowhead through the list provided in Figure 11–48.

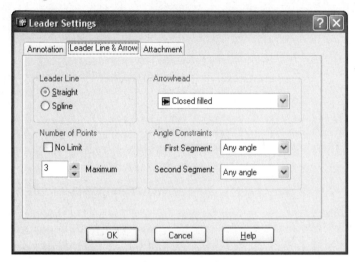

Figure 11–47

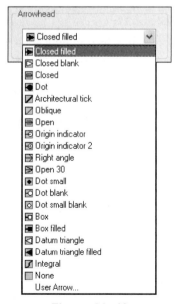

Figure 11–48

The Attachment tab in Figure 11–49 allows you to control how text is attached to the end of the leader through various justification modes. The settings in the figure are the default values. For example, back in Figure 11–45, a Text Attachment mode of Middle of top line is used because the text is placed on the left side of the leader.

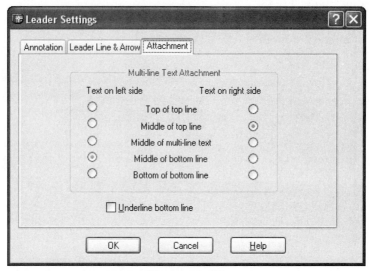

Figure 11–49

DIMENSIONING ANGLES

Dimensioning angles requires two lines forming the angle in addition to the location of the vertex of the angle along with the dimension arc location. Before going any further, understand how the curved arc for the angular dimension is derived. At "A" in Figure 11–50, the dimension arc is struck with its center at the vertex of the object.

 Try It! - Open the drawing file 11_Dim Angle. Use the following command sequence and Figure 11–50 for performing this operation on all three angles.

 Command: **DAN** *(For DIMANGULAR)*
Select arc, circle, line, or <specify vertex>: *(Select line "A")*
Select second line: *(Select line "B")*
Specify dimension arc line location or [Mtext/Text/Angle]: *(Pick a point at "C" to locate the dimension)*
Dimension text = 53

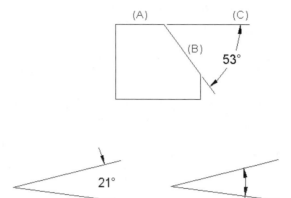

Figure 11–50

DIMENSIONING SLOTS

For slots, first select the view where the slot is visible. Two methods of dimensioning the slot are illustrated in Figure 11–51. With the first method, you call out a slot by locating the center-to-center distance of the two semicircles, followed by a radius dimension to one of the semicircles; which radius dimension is selected depends on the available room to dimension. A second method involves the same center-to-center distance followed by an overall distance designating the width of the slot. This dimension is the same as the diameter of the semicircles. It is considered good practice to place this dimension inside the slot.

 Try It! - The drawing file 11_Dim Slot1 is provided on the CD to allow you to practice placing dimensions on the slot in Figure 11–51.

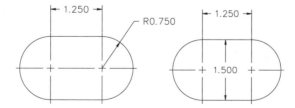

Figure 11–51

A more complex example in Figure 11–52 involves slots formed by curves and angles. Here, the radius of the circular center arc is called out. Angles reference each other for accuracy. As in Figure 11–51, the overall width of the slot is dimensioned, which happens to be the diameter of the semicircles at opposite ends of the slot. Use the

DIMALIGN command to place the 0.48 dimension. Use OSNAP-Nearest to identify the location at "A" and OSNAP-Perpendicular to identify the location at "B."

 Try It! - The drawing file 11_Dim Slot2 is provided on the CD to allow you to practice placing dimensions on the angular slot in Figure 11–52.

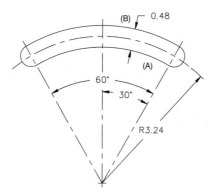

Figure 11–52

ORDINATE DIMENSIONING

The plate in Figure 11–53 consists of numerous drill holes with a few slots, in addition to numerous 90° angle cuts along the perimeter. This object is not considered difficult to draw or make because it mainly consists of drill holes. However, conventional dimensioning techniques make the plate appear complex because a dimension is required for the location of every hole and slot in both the X and Y directions. Add standard dimension components such as extension lines, dimension lines, and arrowheads, and it is easy to get lost in the complexity of the dimensions even on this simple object.

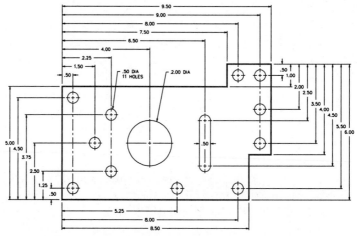

Figure 11–53

A better dimensioning method is illustrated in Figure 11–54, called Ordinate or Datum dimensioning. Here, dimension lines or arrowheads are not drawn; instead, one extension line is constructed from the selected feature to a location specified by you. A dimension is added to identify this feature in either the X or Y direction. It is important to understand that all dimension calculations occur in relation to the current User Coordinate System (UCS), or the current 0,0 origin. In Figure 11–54, with the 0,0 origin located in the lower-left corner of the plate, all dimensions in the horizontal and vertical directions are calculated in relation to this 0,0 location. Holes and slots are called out with the DIMDIAMETER command. The following illustrates a typical ordinate dimensioning command sequence:

Command: **DOR** *(For DIMORDINATE)*
Specify feature location: *(Select a feature using an Osnap option)*
Specify leader endpoint or [Xdatum/Ydatum/Mtext/Text/Angle]: *(Locate a point outside of the object)*
Dimension text = Calculated value

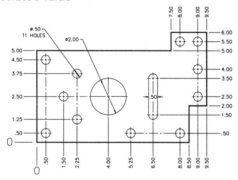

Figure 11–54

To understand how to place ordinate dimensions, see the example in Figure 11–55 and the prompt sequence below. Before you place any dimensions, a new User Coordinate System must be moved to a convenient location on the object with the UCS command and the Origin option. All ordinate dimensions will reference this new origin because it is located at coordinate 0,0. At the command prompt, enter DOR (for DIMORDINATE) to begin ordinate dimensioning. Select the quadrant of the arc at "A" as the feature. For the leader endpoint, pick a point at "B." Be sure Ortho mode is on. It is also helpful to snap to a convenient snap point for this and other dimensions along the direction. This helps in keeping all ordinate dimensions in line with each other.

Try It! - Open the drawing file 11_Dim Ordinate. Follow the next series of figures and the following command sequences to place ordinate dimensions.

Command: **DOR** *(For DIMORDINATE)*

Specify feature location: **Qua**
of *(Select the quadrant of the slot at "A")*
Specify leader endpoint or [Xdatum/Ydatum/Mtext/Text/Angle]: *(Locate a point at "B")*
Dimension text = 1.50

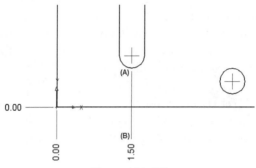

Figure 11–55

With the previous example in Figure 11–55 highlighting horizontal ordinate dimensions, placing vertical ordinate dimensions is identical (see Figure 11–56). With the UCS still located in the lower-left corner of the object, select the feature at "A" using either the Endpoint or Quadrant modes. Pick a point at "B" in a convenient location on the drawing. Again, it is helpful if Ortho is on and you snap to a grid dot.

Command: **DOR** *(For DIMORDINATE)*

Specify feature location: **Qua**
of *(Select the quadrant of the slot at "A")*
Specify leader endpoint or [Xdatum/Ydatum/Mtext/Text/Angle]: *(Locate a point at "B")*
Dimension text = 3.00

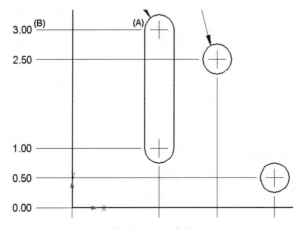

Figure 11–56

When spaces are tight to dimension to, two points not parallel to the X or Y axis will result in a "jog" being drawn (see Figure 11–57). It is still helpful to snap to a grid dot when performing this operation; however, be sure Ortho is turned off.

Command: **DOR** *(For DIMORDINATE)*

Specify feature location: **End**

of *(Select the endpoint of the line at "A")*

Specify leader endpoint or [Xdatum/Ydatum/Mtext/Text/Angle]: *(Locate a point at "B")*

Dimension text = 2.00

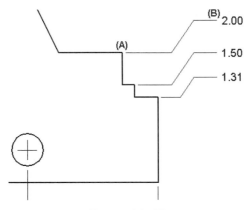

Figure 11–57

Ordinate dimensioning provides a neat and easy way of organizing dimensions for machine tool applications. Only two points are required to place the dimension that references the current location of the UCS. See Figure 11–58.

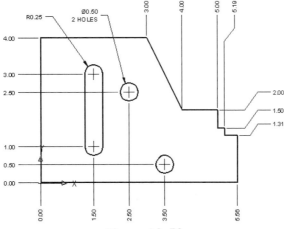

Figure 11–58

EDITING DIMENSIONS

Use the DIMEDIT command to add text to the dimension value, rotate the dimension text, rotate the extension lines for an oblique effect, or return the dimension text to its home position. Figure 11–59 shows the effects of adding text to a dimension and rotating the dimension to a user-specified angle.

 Try It! - Open the drawing file 11_Dim Dimedit. Follow the next series of figures and command prompt sequences for accomplishing this task.

 Command: **DED** *(For DIMEDIT)*

Enter type of dimension editing [Home/New/Rotate/Oblique] <Home>: **N** *(For New. This will display the Multiline Text Editor dialog box. Add the text "TYPICAL" on the other side of the <>. When finished, click the OK button)*
Select objects: *(Select the 5.00 dimension)*
Select objects: *(Press ENTER to perform the dimension edit operation)*

 Command: **DED** *(For DIMEDIT)*

Enter type of dimension editing [Home/New/Rotate/Oblique] <Home>: **R** *(For Rotate)*
Specify angle for dimension text: **10**
Select objects: *(Select the 5.00 dimension)*
Select objects: *(Press ENTER to perform the dimension edit operation)*

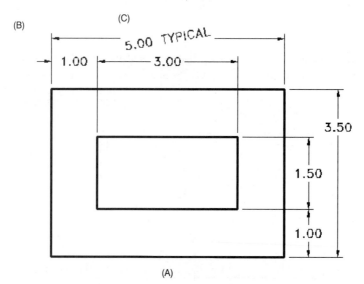

Figure 11–59

Regular AutoCAD modify commands can also affect dimensions. When you use the STRETCH command on the object in Figure 11–59, points "A" and "B" identify a crossing window. Point "C" is the base point of displacement. The results are displayed in Figure 11–60. Not only did the object lines stretch to the new position but also the dimensions all updated themselves to new values.

 Command: **S** *(For STRETCH)*

Select objects to stretch by crossing-window or crossing-polygon...
Select objects: *(Pick a point at "A" in Figure 11–59)*
Specify opposite corner: *(Pick a point at "B" to activate the crossing window)*
Select objects: *(Press* ENTER *to continue)*
Specify base point or displacement: *(Pick a point at "C" in Figure 11–59; it could also be anywhere on the screen)*
Specify second point of displacement: **@0.75<180**

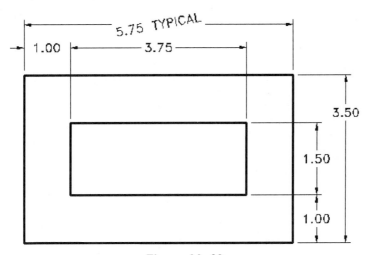

Figure 11–60

One of the other options of the DIMEDIT command is the ability for you to move and rotate the dimension text and still have the text return to its original or home location. This is the purpose of the Home option. Selecting the 5.75 dimension in Figure 11–61 will return it to its original position. However, you would have to use the New option of the DIMEDIT command to remove the text "TYPICAL."

 Command: **DED** *(For DIMEDIT)*

Enter type of dimension editing [Home/New/Rotate/Oblique] <Home>: **H** *(For Home)*
Select objects: *(Select the 5.00 dimension)*
Select objects: *(Press* ENTER *to perform the dimension edit operation)*

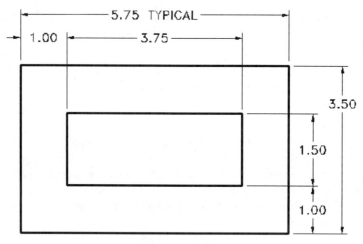

Figure 11–61

The DIMEDIT command also has an Oblique option that allows you to enter an obliquing angle, which will rotate the extension lines and reposition the dimension line. This option is useful if you are interested in placing dimensions on isometric drawings.

 Try It! - Open the drawing file 11_Dim Oblique. Follow the next series of figures and command prompt sequences for accomplishing this task.

Command: **DED** *(For DIMEDIT)*
Enter type of dimension editing [Home/New/Rotate/Oblique] <Home>: **O** *(For Oblique)*
Select objects: *(Select the 2.00 and 1.00 dimensions in Figure 11–62)*
Select objects: *(Press ENTER to continue)*
Enter obliquing angle (Press ENTER for none): **150**

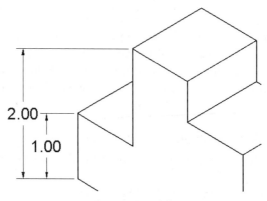

Figure 11–62

The results are illustrated in Figure 11–63, with both dimensions being repositioned with the Oblique option of the DIMEDIT command. Notice that the extension lines and dimension lines were affected; however, the dimension text stayed the same.

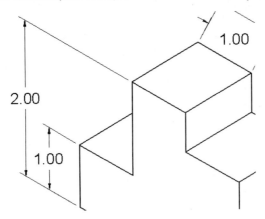

Figure 11–63

The Oblique option of the DIMEDIT command was used to rotate the dimension at "A" at an obliquing angle of 210° (see Figure 11–64). An obliquing angle of -30° was used to rotate the dimension at "B," and the dimensions at "C" required an obliquing angle of 90°. This represents proper isometric dimensions, except for the orientation of the text.

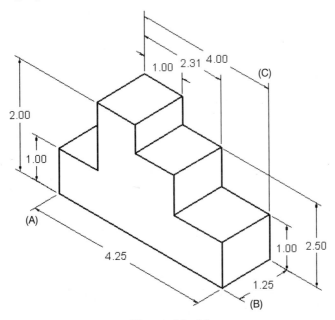

Figure 11–64

TOLERANCES

Interchangeability of parts requires replacement parts to fit in an assembly of an object no matter where the replacement part comes from. In Figure 11–65, the U-shaped channel of 2.000 units in width is to accept a mating part of 1.995 units. Under normal situations, there is no problem with this drawing or callout. However, what if the production person cannot make the channel piece exactly at 2.000? What if he or she is close and the final product measures 1.997? Again, the mating part will have no problems fitting in the 1.997 slot. What if the mating part is not made exactly 1.995 units but is instead 1.997? You see the problem. As easy as it is to attach a dimension to a drawing, some thought needs to go into the possibility that based on the numbers, maybe the part cannot be easily made. Instead of locking dimensions in using one number, a range of numbers would allow the production person the flexibility to vary in any direction and still have the parts fit together. This is the purpose of converting some basic dimensions to tolerances.

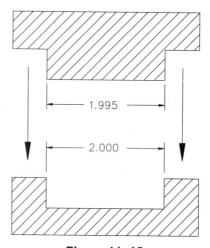

1.995

2.000

Figure 11–65

The example in Figure 11–66 shows the same two mating parts; this time, two sets of numbers for each part are assigned. For the lower part, the machinist must make the part anywhere between 2.002 and 1.998, which creates a range of 0.004 units of variance. The upper mating part must be made within 1.997 and 1.993 units in order for the two to fit correctly. The range for the upper part is also 0.004 units. As you will soon see, whether upper or lower numbers are used, the parts will always fit together. If the bottom part is made to 2.002 and the top part is made to 1.993, the parts will fit. If the bottom part is made to 1.998 and the upper part is made to 1.997, the parts will fit. In any case or combination, if the dimensions are followed exactly as stated by the tolerances, the pieces will always fit. If the bottom part is made to 1.998 and the upper part is made to 1.999, the upper piece is rejected. The method of assigning upper and lower values to dimensions is called limit dimensioning. Here, the larger value in all cases is placed above the smaller value. This is also called a clearance fit because any combinations of numbers may be used and the parts will still fit together.

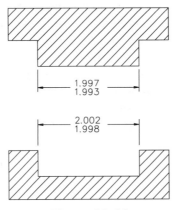

Figure 11–66

The object in Figure 11–67 has a different tolerance value assigned to it. The basic dimension is 2.000; in order for this part to be accepted, the width of this object may go as high as 2.002 units or as low as 1.999 units, giving a range of 0.003 units by which the part may vary. This type of tolerance is called a plus/minus dimension with an upper limit of 0.002 difference from the lower limit of -0.001.

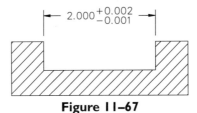

Figure 11–67

Illustrated in Figure 11–68 is yet another way to display tolerances. It is very similar to the plus/minus method except that both upper and lower limit values are the same; for this reason, this method is called plus and minus dimensions. The basic dimension is still 2.000 with upper and lower tolerance limits of 0.002 units.

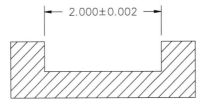

Figure 11–68

GEOMETRIC DIMENSIONING AND TOLERANCING (GDT)

In the object in Figure 11–69, approximately 1000 items need to be manufactured based on the dimensions of the drawing. Also, once constructed, all objects needed to be tested to be sure the 90° surface did not deviate over 0.005 units from its base. Unfortunately, the wrong base was selected as the reference for all dimensions. Which should have been the correct base feature? As all items were delivered, they were quickly returned because the long 8.80 surface drastically deviated from required 0.005 unit deviation or zone. This is one simple example that demonstrates the need for using geometric dimension and tolerancing techniques. First, this method deals with setting tolerances to critical characteristics of a part. Of course, the function of the part must be totally understood by the designer in order for tolerances to be assigned. The problem with the object in Figure 11–69 was the note, which did not really specify which base feature to choose as a reference for dimensioning. Figure 11–70 shows the same object complete with dimensioning and tolerancing symbols. The letter "A" inside the rectangle identifies the datum or reference surface. The tolerance symbol at the end of the long edge tells the individual making this part that the long edge cannot deviate more than 0.005 units using surface "A" as a reference. The datum triangles that touch the part are a form of dimension arrowhead.

Using geometric tolerancing symbols ensures more accurate parts with less error in interpreting the dimensions.

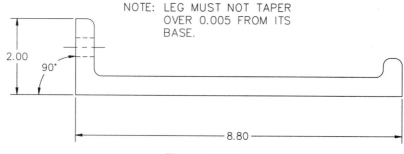

Figure 11–69

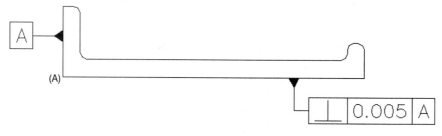

Figure 11–70

GDT SYMBOLS

Entering TOL at the command prompt brings up the main Geometric Tolerance dialog box in Figure 11–71. This box contains all tolerance zone boxes and datum identifier areas. Clicking in one of the dark boxes under the Sym area displays the Symbol dialog box illustrated in Figure 11–72. This dialog box contains all major geometric dimensioning and tolerancing symbols. Choose the desired symbol by clicking the specific symbol; this will return you to the main Geometric Tolerance dialog box.

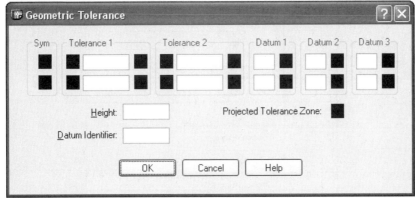

Figure 11–71

Figure 11–72

Figure 11–73 shows a chart outlining all geometric tolerancing symbols supported in the Symbol dialog box. Alongside each symbol is the characteristic controlled by the symbol in addition to a tolerance type. Tolerances of Form such as Flatness and Straightness can be applied to surfaces without a datum being referenced. On the other hand, tolerances of Orientation such as Angularity and Perpendicularity require datums as reference.

Symbol	Characteristic	Type of Tolerance
▱	Flatness	Form
—	Straightness	
◯	Roundness	
⌀	Cylindricity	
⌒	Profile of a Line	Profile
⌓	Profile of a Surface	
∠	Angularity	Orientation
⊥	Perpendicularity	
//	Parallelism	
⊕	Position	Location
◎	Concentricity	
⌯	Symmetry	
↗	Circular Runout	Runout
↗↗	Total Runout	

Figure 11–73

With the symbol placed inside this dialog box as in Figure 11–74, you now assign such items as tolerance values, maximum material condition modifiers, and datums. In Figure 11–74, the tolerance of Parallelism is to be applied at a tolerance value of 0.005 units to Datum "A."

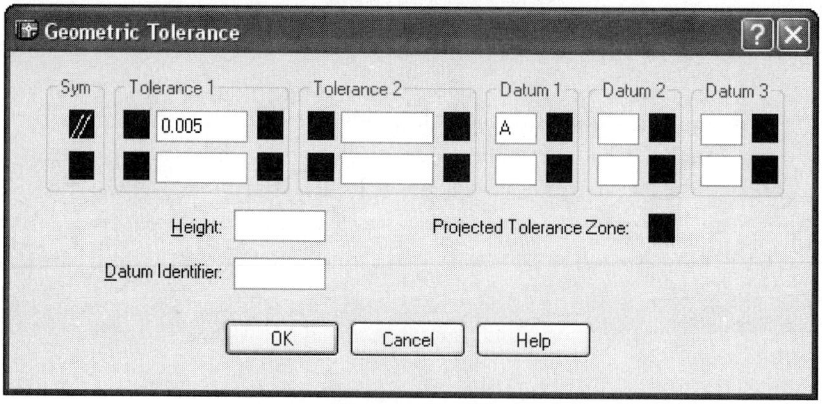

Figure 11–74

ADDITIONAL GD&T SYMBOLS

In the main Geometric Tolerance dialog box, shown in Figure 11–74, clicking the black box next to the 0.005 tolerance value activates the Material Condition dialog box in Figure 11–75. This dialog box is devoted to the concept of Material Condition and acts as a modifier to the main tolerance value.

Figure 11–75

Figure 11–76 shows the meaning of the symbols used in this dialog box. In brief, Maximum Material Condition refers to the condition of a characteristic such as a hole when the most material exists. Least Material Condition refers to the condition of a characteristic where the least material exists. Regardless of Feature Size indicates that the characteristic tolerance such as Flatness or Straightness must be maintained regardless of the actual produced size of the object.

Datums are represented by surfaces, points, lines, or planes and are used for referencing the features of an object. The symbols illustrated in Figure 11–76 are "flags" used to identify a datum. The next series of figures illustrates how datums are used in combination with tolerances of Orientation.

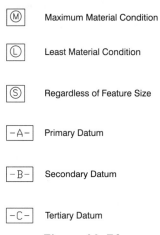

Figure 11–76

THE FEATURE CONTROL SYMBOL

Once you have completed the desired information in the Geometric Tolerance dialog box, the result is illustrated in Figure 11–77, in the form of a Feature Control Symbol.

This is the actual order in which all symbols, values, modifiers, and datums are placed. Follow the graphic and information in Figure 11–77 to view the contents of each area of the Feature Control Symbol box.

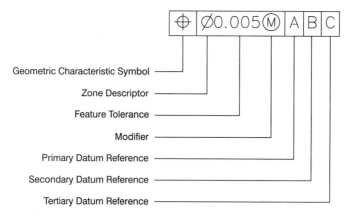

Figure 11–77

TOLERANCE OF ANGULARITY EXAMPLE

Illustrated in the Geometric Tolerance dialog box in Figure 11–78 is an example of applying the tolerance of Angularity to a feature. First the symbol is identified along with the tolerance value. Since this tolerance requires a datum, the letter "A" identifies primary datum "A." The results are illustrated in Figure 11–79. Datum "A" identifies the surface of reference as the base of the object. The feature control symbol is applied to the angle and reads: "The surface dimensioned at 60° cannot deviate more than 0.005 units from datum 'A'." The graphic in Figure 11–79 shows an exaggerated tolerance zone and is used for illustrative purposes only.

 Note: The file 11_GDT Angularity is provided on the CD to allow you to experiment with this tolerance tool.

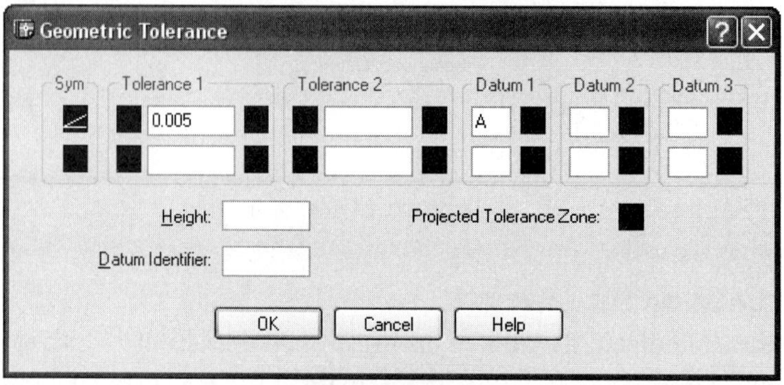

Figure 11–78

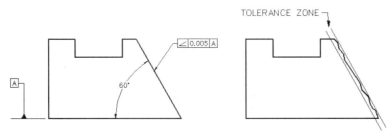

Figure 11–79

TOLERANCE OF FLATNESS EXAMPLE

The next example applies a tolerance of Flatness to a surface. First the Geometric Tolerance dialog box has the Flatness symbol assigned in addition to a tolerance value of 0.003 units in Figure 11–80. The results are illustrated in Figure 11–81, with the feature control box being applied to the top surface. The tolerance box reads: "The surface must be flat with a 0.003 unit tolerance zone." The tolerance range is displayed in the second graphic and is exaggerated. The tolerance of Flatness does not usually require a datum for reference.

 Note: The file 11_GDT Flatness is provided on the CD to allow you to experiment with this tolerance tool.

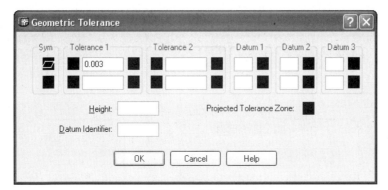

Figure 11–80

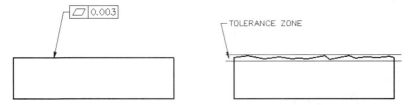

Figure 11–81

TOLERANCE OF PERPENDICULARITY EXAMPLE

This next example illustrates a tolerance of Perpendicularity. The Geometric Tolerance dialog box reflects the perpendicularity symbol along with a tolerance value of 0.003 units. See Figure 11–82. This tolerance characteristic requires a datum to be most effective. In Figure 11–83, the feature control box reads: "This surface must be perpendicular to datum 'A' within a tolerance zone of 0.003 units." The second graphic shows the tolerance zone and the amount of deviation that is acceptable. It is meant to be exaggerated and is used for illustrative purposes only.

 Note: The file 11_GDT Perpendicularity is provided on the CD to allow you to experiment with this tolerance tool.

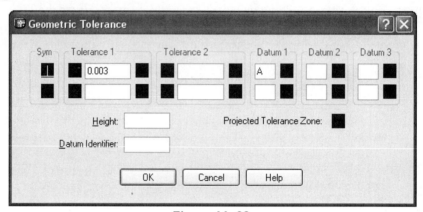

Figure 11–82

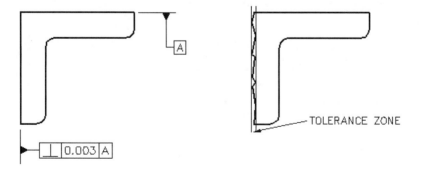

Figure 11–83

TOLERANCE OF POSITION EXAMPLE

This last example displays a tolerance of Position that is applied to a hole feature. Since a hole is centered from two edges, two datums are identified in the Geometric

Tolerance dialog box illustrated in Figure 11–84. This is sometimes called a circular tolerance. The feature control box illustrated in Figure 11–85 reads: "The center of the hole must lie within a circular tolerance zone of 0.050 units in relation to datums 'A' and 'B'."

 Note: The file 11_GDT Position is provided on the CD to allow you to experiment with this tolerance tool.

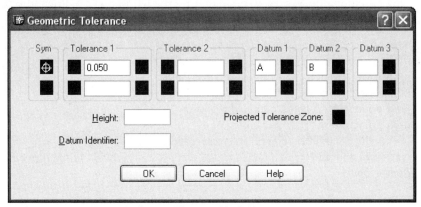

Figure 11–84

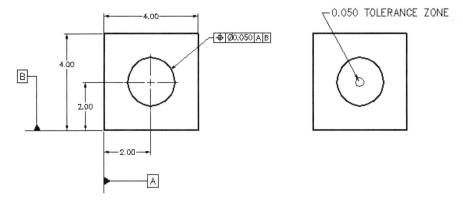

Figure 11–85

CHARACTER MAPPING FOR DIMENSION SYMBOLS

Illustrated in Figure 11–86 is a typical counterbore operation and the correct dimension layout. The LEADER command was used to begin the diameter dimension. The counterbore and depth symbols were generated with the MTEXT command along with the

Unicode Character Mapping dialog box. In any case, the first step in using the dimension symbols is to first create a new text style, for which you can use any name.

 Try It! - Open the drawing file 11_Character Map. Use the following command sequence and the next series of figures to perform this task.

First use the QLEADER command to place the leader. All parameters in the Leader Settings dialog box should be properly set, including the Middle of top line radio button for the right attachment location.

Command: **LE** *(For QLEADER)*
Specify first leader point, or [Settings]<Settings>: **Nea**
To *(Select the edge of the circle)*
Specify next point: *(Pick a point to extend the leader)*
Specify next point: *(Press ENTER to continue)*
Specify text width <0.0000>: *(Press ENTER to accept this default value)*
Enter first line of annotation text <Mtext>: *(Press ENTER to activate the Multiline Text Editor dialog box)*

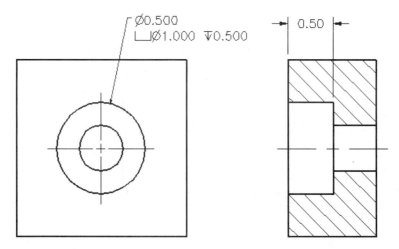

Figure 11–86

Enter the first listing for the hole in the Multiline Text Editor dialog box in Figure 11–87. Right-click over the Multiline Text Editor to find the Symbol menu. The characters "%%c" signify the diameter symbol, which is selected from the Symbol menu under Diameter. When finished, press ENTER to drop down to the second line.

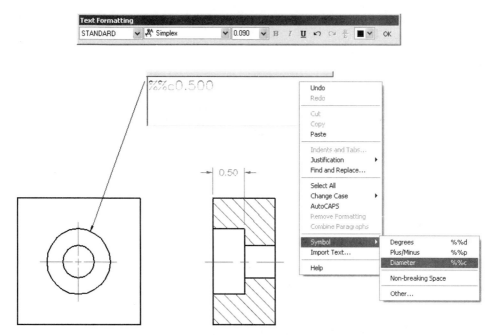

Figure 11–87

The remaining symbols are located in a special area called a Unicode Character Map. Clicking Other in Figure 11–88 displays this character mapping dialog box in Figure 11–89. Notice, in the upper-left corner, that the current font is GDT, which holds all geometric tolerancing symbols along with the special dimensioning symbols such as counterbore, deep, and countersink. Be sure your current font is set to GDT. Once you identify a symbol, double-click it. A box will appear around the symbol. Also, the symbol will appear in the Characters to Copy area in the upper-right corner of the dialog box. Click the Copy button to copy this symbol to the Windows Clipboard.

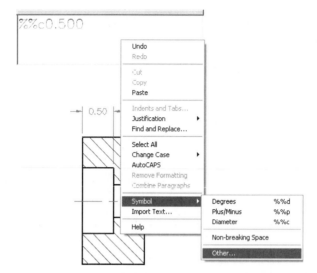

Figure 11–88

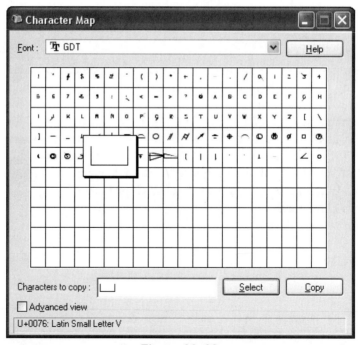

Figure 11–89

Return to the Multiline Text Editor dialog box and press CTRL+V, which performs a paste operation. Because the counterbore symbol was copied to the Windows Clipboard, it pastes into the dialog box in Figure 11–90.

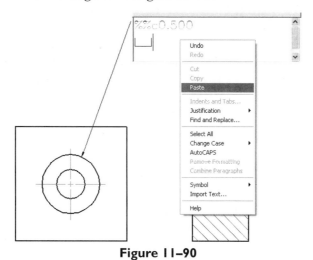

Figure 11–90

Enter text and repeat the procedure for using the diameter and depth symbols outlined in Figure 11–91.

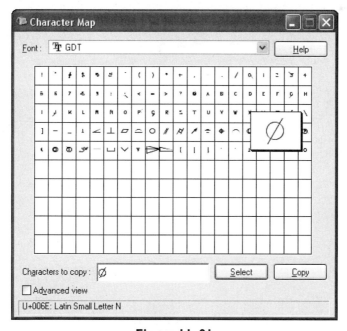

Figure 11–91

The finished results are displayed in Figure 11–92. You will notice the second row of dimension text appears smaller than the top row. Highlight the second row of text and symbols in Figure 11–93 and change the height to 0.09. This will keep all text the same size. Click the OK button to return to the drawing. The leader and notes are both displayed as in Figure 11–94.

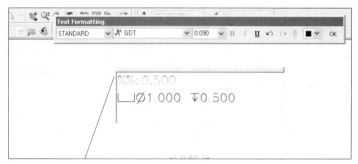

Figure 11–92

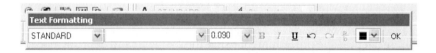

Figure 11–93

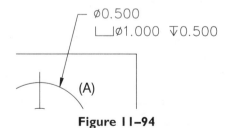

Figure 11–94

APPLICATIONS OF DIMENSIONING SYMBOLS

In today's global economy, the transfer of documents in the form of drawings is becoming a standard way of doing business. As drawings are shared with a subsidiary of an overseas company, two different forms of language may be needed to interpret the dimensions of the drawing. In the past, this has led to confusion in interpreting drawings, and as a result, a system of dimensioning symbols has been developed. It is hoped that a symbol will be easier to recognize and interpret than a note in a different language about the particular feature being dimensioned.

Figure 11–95 shows some of the more popular dimensioning symbols in use today on drawings. Notice how the symbols are designed to make as clear and consistent an interpretation of the dimension as possible. As an example, the Deep or Depth symbol displays an arrow pointing down. This symbol is used to identify how far into a part a drill hole goes. The Arc Length symbol identifies the length of an arc, and so on.

Symbol	Description
⌒	Arc Length
X.XX	Basic Dimension
▷	Conical Taper
⊔	Counterbore or Spotface
∨	Countersink
↧	Deep or Depth
⌀	Diameter
X.XX	Dimension Not to Scale
2X	Number of Times—Places
R	Radius
(X.XX)	Reference Dimension
S⌀	Spherical Diameter
SR	Spherical Radius
◁	Slope
□	Square

Figure 11–95

The size of all general dimensioning symbols is illustrated in Figure 11–96. All values are in relation to "h" or the relative height of the dimension numerals. As an example, with a dimension numeral height of 0.18, study the illustration in Figure 11–97 for finding the size of the counterbore dimension symbol. Because the height of the counterbore symbol is defined by "h," simply substitute the dimension numeral height of 0.18 for the height of the counterbore symbol. Because the width of the symbol is "2h," multiply 0.18 by 2 to obtain a value of 0.36. These symbols may be created and saved as blocks for insertion in drawings. They may also be part of an existing text font mapped to a particular key on the keyboard. Pressing that key produces the dimension symbol.

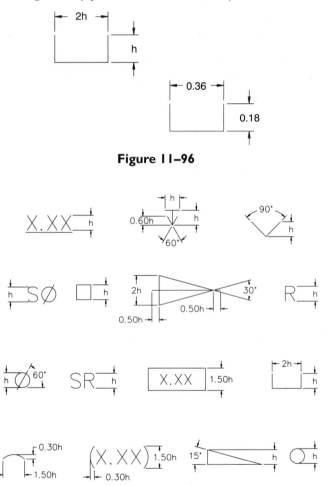

Figure 11–96

Figure 11–97

APPLYING THE BASIC DIMENSION SYMBOL

Drawing a rectangular box around the dimension numeral (see Figure 11–98) identifies a basic dimension. The basic dimension represents the theoretical distance of the feature, such as the distance from the center hole to the left hole as 4.00 units with the distance from the left hole to the right hole as 8.00 units. It is a known fact that it is impossible to maintain tolerances to obtain an exact value of 4.00 or 8.00 units. You can construct a rectangle; however, this may be a tedious task depending on the number of decimal places past the zero, which determines the size of the rectangle. To constrain the rectangle around dimension text, choose Dimension Style from the Format pull-down menu and then Modify. Click the Tolerances tab and in the Method area, change None to Basic to draw the rectangle around dimension text.

 Try It! - The file 11_Dim Basic is available on the CD. Open it and experiment with placing the basic dimension symbol around a number of dimension values.

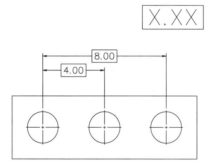

Figure 11–98

APPLYING THE REFERENCE DIMENSION SYMBOL

Illustrated in Figure 11–99 is a series of Continue dimensions strung together in one line. The dimension numeral 3.00 is enclosed in parentheses and is referred to as a reference dimension. Reference dimensions are not required drawing dimensions; they are placed for information purposes. The parentheses on the keyboard are used along with the dimension numeral to create a reference dimension.

 Try It! - The file 11_Dim Reference is available on the CD. Open it and experiment with surrounding a dimension value with the parenthesis using the DDEDIT command.

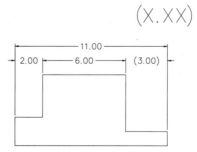

Figure 11–99

APPLYING THE NOT TO SCALE DIMENSION SYMBOL

At times it is necessary to place a dimension that is not to scale, as illustrated in Figure 11–100. To distinguish regular dimensions from a dimension that is not to scale, this dimension is identified in the drawing by a line drawn under the dimension. To do this in an AutoCAD drawing, enter the following text string for the dimension value: %%u5.00%%u. The "%%u" toggles on Underline Text mode followed by the dimension text of 5.00 units. To toggle underline off, complete the text string with another "%%u." When you use the Multiline Text Editor dialog box, an underline button is provided to simplify this task.

 Try It! - The file 11_Dim NTS is available on the CD. Open it and experiment with placing the underscore beneath a dimension value.

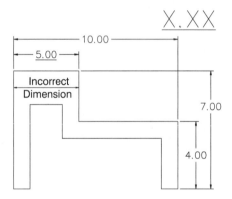

Figure 11–100

APPLYING THE SQUARE DIMENSION SYMBOL

Use this dimension symbol to identify the cross section of an object as a square. This will limit the need for two dimensions of the same value to be placed identifying a square. The square dimension symbol should be placed before the dimension numeral, as in the illustration in Figure 11–101.

 Try It! - The file 11_Dim Square is available on the CD. Open it and experiment with placing the square symbol found in the Unicode Character Map next to a dimension value.

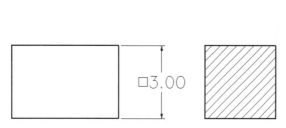

Figure 11–101

APPLYING THE SPHERICAL DIAMETER DIMENSION SYMBOL

The spherical illustrated in Figure 11–102 is in the form of a dome-shaped feature. Use the diameter symbol preceded by the letter "S" for spherical.

Try It! - The file 11_Dim Sph Dia is available on the CD. Open it and experiment by adding the letter "S" to the prefix of the dimension. Use the DDEDIT command to accomplish this.

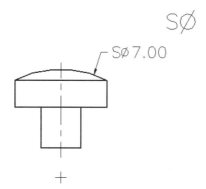

Figure 11–102

APPLYING THE SPHERICAL RADIUS DIMENSION SYMBOL

The illustration in Figure 11–103 is similar to the previous spherical example except that a radius identifies the spherical. Place the letter "S" in front of the "R" symbol.

Try It! - The file 11_Dim Sph Rad is available on the CD. Open it and experiment by adding the letter "R" to the prefix of the dimension. Use the DDEDIT command to accomplish this.

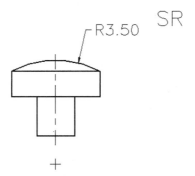

Figure 11–103

APPLYING THE DIAMETER DIMENSION SYMBOL

A circle with a diagonal line through it represents this symbol. This diameter symbol always precedes the dimension numeral with no space in between. This symbol is automatically placed before the dimension numeral when you use the DIMDIAMETER command. All holes represented by the diameter symbol are understood to pass completely through a part, as in Figure 11–104.

 Try It! - The file 11_Dim Dia Sym is available on the CD. Open it and experiment with placing the diameter symbol found in the Unicode Character Map next to a dimension value. Use the LEADER command (LE) to place the diameter value, then use DDEDIT command (ED) to add the diameter symbol.

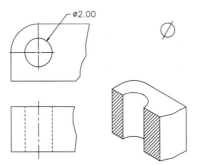

Figure 11–104

APPLYING THE DEPTH DIMENSION SYMBOL - METHOD #1

When a hole does not pass completely through a part, as in Figure 11–105, the depth dimension symbol is used. The diameter of the hole is placed first, followed by the depth symbol and the distance the hole is to be drilled. This figure also shows yet another method of identifying holes, although it is considered dated.

Try It! - The file 11_Dim Depth1 is available on the CD. Open it and experiment with placing the depth symbol found in the Unicode Character Map next to a dimension value. Be sure to make all text fonts and symbols the same height.

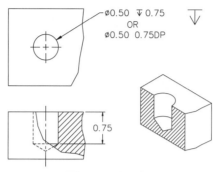

Figure 11–105

APPLYING THE DEPTH DIMENSION SYMBOL - METHOD #2

Illustrated in Figure 11–106 is yet another application of the depth dimension symbol. First the diameter of the hole is identified. Next, a counterdrill diameter with angle of the counterdrill is given. The depth of the counterdrill completes the dimension. This operation is used when a flat-head screw needs to be recessed below the surface of a part.

 Try It! - The file 11_Dim Depth2 is available on the CD. Open it and experiment with placing the depth symbol found in the Unicode Character Map next to a dimension value. You may have to change the text justification mode to left.

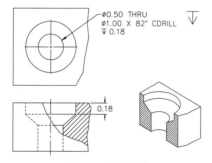

Figure 11–106

APPLYING THE COUNTERBORE DIMENSION SYMBOL

A counterbore is an enlarged portion of a previously drilled hole. The purpose of a counterbore is to receive the head of such screws as socket-head or fillister-head screws. The counterbore example in Figure 11–107 first identifies the diameter of the thru hole. On the next line come the counterbore symbol and its diameter. The final specification is the depth of the counterbore identified by the Depth dimension symbol. Figure 11–107 shows another method of identifying counterbores, although it is considered dated.

 Try It! - The file 11_Dim Cbore is available on the CD. Open it and experiment with placing the depth and counterbore symbols found in the Unicode Character Map next to a dimension value. Be sure to make all text fonts and symbols the same height.

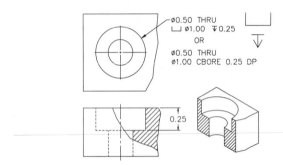

Figure 11–107

APPLYING THE SPOTFACE DIMENSION SYMBOL

A spotface is similar to the counterbore except that it is usually made quite a bit shallower than the counterbore. The purpose of the spotface is to seat a washer, preventing it from moving around along the surface of a part. The counterbore symbol is used to identify a spotface, as in Figure 11–108. The diameter of the thru hole is first given. On the next line, the counterbore symbol followed by diameter is given. Finally the depth of the counterbore is identified by the distance and the Depth dimension symbol. Figure 11–108 also shows another method of identifying spotfaces, although it is considered dated.

 Try It! - The file 11_Dim Spotface is available on the CD. Open it and experiment with placing the depth and counterbore symbol found in the Unicode Character Map next to a dimension value.

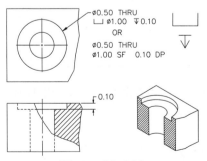

Figure 11–108

APPLYING THE SLOPE DIMENSION SYMBOL

The slope dimension symbol applies to an inclined surface and the amount of rise in the surface given by two dimensions, as in Figure 11–109. This rise is indicated by the change in height per unit distance along a baseline. The slope symbol is placed followed by the change in height and the change in unit distance. This makes the slope dimension a ratio of the height and unit distance.

 Try It! - The file 11_Dim Slope is available on the CD. Open it and experiment with placing the slope symbol found in the Unicode Character Map next to a dimension value.

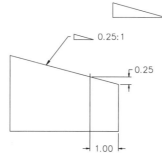

Figure 11–109

APPLYING THE COUNTERSINK DIMENSION SYMBOL

A countersink is a V-shaped conical taper at one end of a hole. The purpose is to accept a flat-head screw and make it flush with the top surface of the part. In Figure 11–110, three holes of 0.50 diameter are first identified. On the next line, the countersink symbol followed by the number of degrees in the countersink is specified. Below this example is yet another method of identifying countersinks although it is considered dated.

 Try It! - The file 11_Dim Csink is available on the CD. Open it and experiment with placing the countersink symbol found in the Unicode Character Map next to a dimension value.

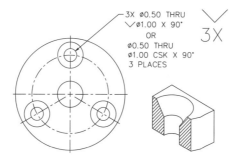

Figure 11–110

APPLYING THE ARC DIMENSION SYMBOL

Illustrated in Figure 11–111 is the difference between dimensioning the chord of an arc and the actual distance of the arc. The 7.00 unit dimension, in addition to defining the width of the part, specifies the distance of the chord of an arc. The 8.48 unit dimension is the distance of the arc and is specified by a small arc symbol above the distance. Accomplish this by drawing a small arc and moving it above the arc distance. Although AutoCAD does not dimension the length of an arc, it does allow you to use the LIST command to obtain the length of an arc. You then enter this value at the dimension distance prompt.

 Try It! - The file 11_Dim Arc is available on the CD. Open it and experiment with placing the arc symbol above the dimension value.

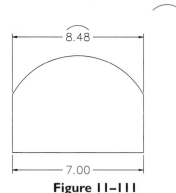

Figure 11–111

APPLYING THE ORIGIN DIMENSION SYMBOL

The origin dimension symbol is used to indicate the origin of the dimension. The small circle is substituted for an arrowhead as the termination of the dimension line, as in Figure 11–112. To construct the open circle and filled arrowhead, choose Dimension Style from the Format menu and then Modify. From there, go to the Arrowheads group in the Lines and Arrows dialog box, click the down arrow next to 1st and choose the Dot Blanked terminator. Next, click the down arrow next to 2nd and choose the Closed Filled terminator.

 Try It! - The file 11_Dim Origin is available on the CD. Open it and experiment with placing the origin terminator along the dimension line.

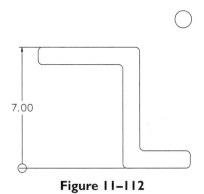

Figure 11–112

APPLYING THE CONICAL TAPER DIMENSION SYMBOL

Similar to the slope dimension symbol, the conical taper symbol is used when the amount of taper per unit of length is desired. The conical taper symbol is placed before the value of the taper. The value of the taper is based on a ratio of 1.00 units in length to a Delta diameter illustrated in Figure 11–113. The Delta diameter is based on the large shaft diameter minus the diameter taken at the area where the 1.00 unit length is located. This value becomes the ratio to 1.00 unit of length.

 Try It! - The file 11_Dim Taper is available on the CD. Open it and experiment with placing the taper symbol found in the Unicode Character Map next to a dimension value.

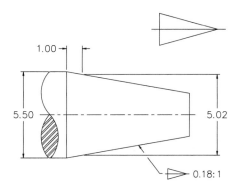

Figure 11–113

LOCATION OF UNIFORMLY SPACED FEATURES

Figure 11–114 shows an example of a plate consisting of eight holes equally spaced at 45° angles about a 7.30-diameter bolthole circle. Eliminating the need to dimension each individual angle, the dimension symbol "X" is used for the number of times a feature is repeated. Notice in Figure 11–114 that "8X" is used to call out the number of times the angle and diameter of the holes repeat. Use the DDEDIT command to add the "8X" prefix to the 45° and 1.28 diameter dimensions.

 Try It! - The file 11_Dim Spaced Holes1 is available on the CD. Open it and experiment with editing the dimension values with the DDEDIT command.

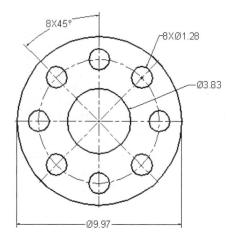

Figure 11–114

Figure 11–115 is similar to the previous illustration. Five 72° angles are used to locate five slots each with a width of 1.50 units. Use the DDEDIT command to add the "5X" prefix to the 72° and 1.50 dimensions.

 Try It! - The file 11_Dim Spaced Holes2 is available on the CD. Open it and experiment with editing the dimension values with the DDEDIT command.

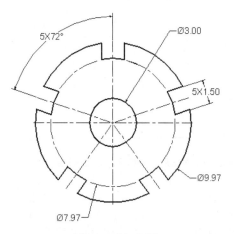

Figure 11–115

Figure 11–116 shows a different example of the location of features that are uniformly spaced. Six holes of 1.00 diameter are located along a bolt circle of 7.30 diameter. All six holes are spaced at 30° angles. Since the holes are not laid out in a full circle, only five angles are required to locate the six holes. The symbol "5X" is used to identify the number of angles, while the symbol "6X" is used to identify the number of holes. Use the DDEDIT command to add the "5X" prefix to the 30° dimension and the "6X" prefix to the 1.00 dimension.

 Try It! - The file 11_Dim Spaced Holes3 is available on the CD. Open it and experiment with editing the dimension values with the DDEDIT command.

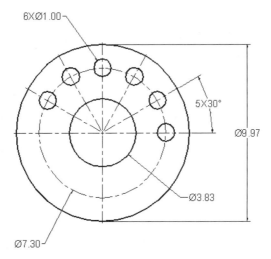

Figure 11–116

GRIPS AND DIMENSIONS

Grips have a tremendous amount of influence on dimensions. Grips allow dimensions to be moved to better locations; grips also allow the dimension text to be located at a better position along the dimension line.

 Try It! - Open the drawing file 11_Dimgrip. In the example in Figure 11–117, notice the various unacceptable dimension placements. The 2.88 horizontal dimension lies almost on top of another extension line. To relocate this dimension to a better position, click the dimension. Notice that the grips appear and the entire dimension highlights. Now click the grip near "A" in the illustration in Figure 11–117. (This grip is located at the left end of the dimension line.) When the grip turns red, the Stretch mode is currently active. Stretch the dimension above the extension line but below the 3.50 dimension. Press the ESC key to turn off the grips. The same results can be accomplished with the 4.50 horizontal dimension as it is stretched closer to the 3.50 dimension.

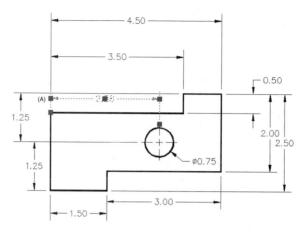

Figure 11–117

Notice that the two vertical dimensions in Figure 11–118 on the left side of the object do not line up with each other; this would be considered poor practice. Pick both dimensions and notice the appearance of the grips, in addition to both dimensions being highlighted. Click the upper grip at "A" of the 1.25 dimension. When this grip turns red and places you in Stretch mode, select the grip at "B" of the opposite dimension. The result will be that both dimensions now line up with each other. The same can be accomplished with the 1.50 and 3.00 horizontal dimensions.

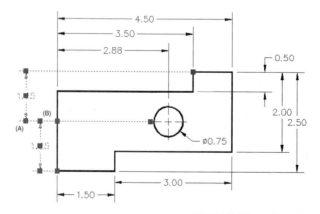

Figure 11–118

In the illustration in Figure 11–119, the 2.50 dimension text is too close to the 2.00 vertical dimension on the right side of the object. Click the 2.50 dimension; the grips appear and the dimension highlights. Click in the grip representing the text location at "A." When this grip turns red, stretch the dimension text to a better location.

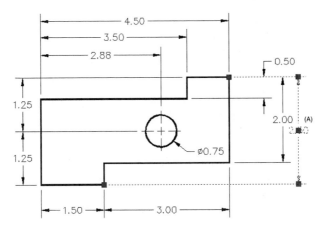

Figure 11–119

It is very easy to use grips to control the placement of diameter and radius text. As shown in Figure 11–120, click the diameter dimension; the grips appear, in addition to the diameter dimension being highlighted. Click the grip that locates the dimension text at "A." When this grip turns red, relocate the diameter dimension text to a better location using the Stretch mode.

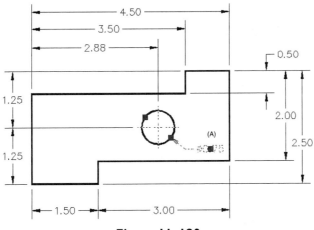

Figure 11–120

The completed object, with dimensions edited through grips, is displayed in Figure 11–121.

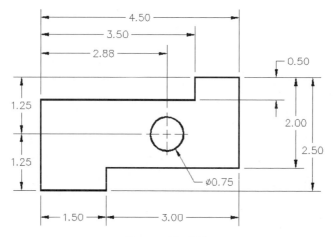

Figure 11–121

OBJECT SNAP AND DIMENSIONS

Proper selection of proper Object Snap modes is important in dimensioning. You can create a dimension from the endpoint of an object's corner to what you think is an opposite corner. However, incorrect dimensions result from accidentally selecting the endpoints of extension lines of other dimensions as shown in Figure 11–122. For this reason, it is highly recommended that you use the Object Snap Intersection mode for beginning dimensions. This will ensure that only valid intersections are selected.

Other problems involve dimensioning an object that has been crosshatched. Using the Object Snap Endpoint or Intersection mode could select an intersection created by an object line and the crosshatch pattern. It is therefore considered good practice to turn off the layer holding the crosshatch pattern before attempting dimensions.

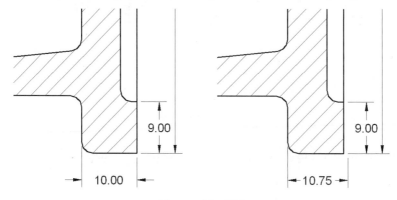

Figure 11–122

CONTROLLING THE ASSOCIATIVITY OF DIMENSIONS

THE DIMASSOC SYSTEM VARIABLE

The associativity of dimensions is controlled by the DIMASSOC variable. By default, this value is set to 2 for new drawings. This means the dimension is associated with the object being dimensioned. Associativity means that if the object changes size or if some element of an object changes location, the dimension associated with the object will change as well. You may find that for existing drawings, this variable is set to 1. This is called a non-associative dimension. This dimension is not associated with the object being dimensioned; however, it is possible to have the dimension value automatically updated using conventional AutoCAD editing commands. As in past versions of AutoCAD, the EXPLODE command can be used for breaking up associative and non-associative dimensions into individual objects (this is not recommended). This action is the same as setting the DIMASSOC variable to 0.

A setting of 0 (Zero) will create exploded dimensions. There is no association between the various elements of the dimension. All dimension lines, extension lines, arrowheads, and dimension text are drawn as separate objects.

Try It! - Open the drawing 11_Dimassoc0. This drawing in Figure 11–123 has dimensions placed with the DIMASSOC variable set to 0. The dimensions in this figure are all exploded. A polyline shape has various dimensions placed on its outside. Click the polyline to activate its grips. Click the grip at "A" and stretch the polyline vertex down. Notice the dimensions remain in their original positions. Also, the dimension values do not update. Pres ESC to turn off the grips. Now enter the STRETCH command, pick a crossing box from "B" to "C" and stretch the lower polyline vertex up. Notice how the extension lines connected to the vertex stretch along with the polyline. However, the remainder of the dimension components remains originally positioned.

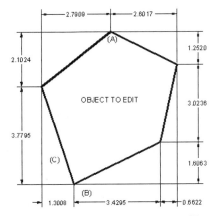
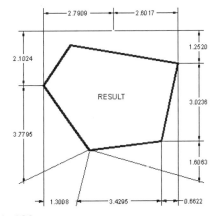

Figure 11–123

Setting the DIMASSOC variable to 1 creates a non-associative dimension. The elements of the dimension (dimension line, extension line, text, arrow, etc.) are formed into a single object. If the dimension's definition point on the object moves, the dimension value is updated.

Try It! - Open the drawing 11_Dimassoc1. This drawing in Figure 11–124 has dimensions placed with the DIMASSOC variable set to 1. Another polyline shape has various dimensions placed on its outside. Click the polyline to activate its grips. Click the grip at "A" and stretch the polyline vertex down and to the left. Notice the dimensions remain in their original positions. Press ESC to turn off the grips. Also, the dimension values are not updated. Now enter the STRETCH command, pick a crossing box from "B" to "C," and stretch the lower polyline vertex up and to the right. Notice that the dimensions affected by the crossing window are all updated. The values you get will be different from the ones shown in the figure. This is due to the definition points of the dimensions being included in the crossing window during the stretch operation. Definition points are the "links" that create the association between the object and the dimension.

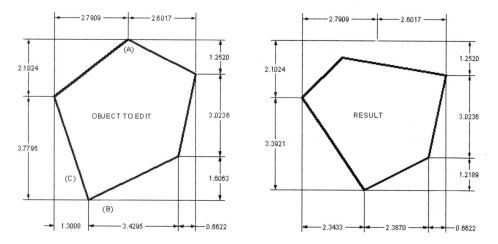

Figure 11–124

Setting the DIMASSOC variable to 2 creates an associative dimension. The dimension elements are formed into a single object. The definition points of the dimension are also linked with association points on the geometry. This means that if the association point on the geometry is moved, all elements of the dimension (dimension location, orientation, value, etc.) are updated.

Try It! - Open the drawing 11_Dimassoc2. This drawing in Figure 11–125 has dimensions placed with the DIMASSOC variable set to 2. Another polyline shape has various dimensions placed on its outside. Click the polyline to activate its grips. Click the grip at "A" and stretch the polyline vertex up and to the left. Notice all dimension locations, orientations, and values are automatically updated. Click the polyline vertex at "B" and stretch this vertex up and to the right. The same results occur with the dimension elements being updated based on the new location of the object. Use grips to realign any dimension that is not updated to a new location.

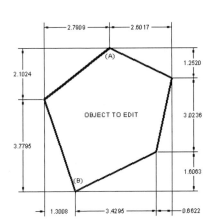 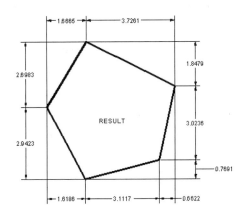

Figure 11–125

 Tip: When you make changes to the DIMASSOC variable, these changes are stored in the drawing file.

ASSOCIATIVE DIMENSIONS AND LAYOUTS

It is considered good practice to dimension all objects while inside model space. This will be demonstrated in the next chapter. The main reason is because there was no associativity with the model space objects and dimensions placed in a layout or paper space. This has changed. Once you scale the model space objects inside a layout, the scale of the layout has a direct bearing on the dimensions being placed.

 Try It! - Open the drawing file 11_Dim Layout. Two viewports are arranged in a single layout called Plan Views in Figure 11–126. The images inside of these viewports are scaled differently; the floor plan in the left viewport is scaled to 1/8"=1'-0". The image in the right viewport has been scaled 1/4"=1'-0". The DIMASSOC variable is currently set to 2. This will allow you to dimension the objects at two different scales and still have the correct dimensions appear. Use the DIMLINEAR command and place a few linear dimensions on the floor plan.

Then use the DIMRADIUS command and place a few radius dimensions on the image of the pool in the right viewport. Zoom into a few of the dimensions and see that they reflect the model space distances. Place more dimensions on the floor plan using the DIMCONTINUE command and observe the correct values being placed.

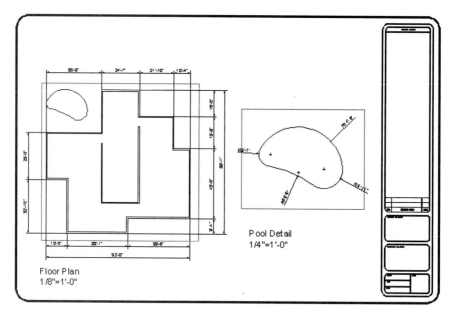

Figure 11–126

REASSOCIATING DIMENSIONS

As associative dimensions are tied to objects in a layout, cases may arise where you pan the object inside the viewport. Depending on how this is accomplished, the dimensions may or may not be updated to the new location of the object. If you use the PAN command, the dimensions should be updated to the new object position. If you press the wheel down on an Intellimouse and pan the object, the dimensions will not be updated. This is where you would regenerate all dimensions using the DIMREGEN command. The dimensions will be reestablished based on the new locations of the objects.

Try It! - Open the drawing 11_Dimregen. In Figure 11–127, a portion of the drawing and dimensions are obscured by the viewport. Notice you are in floating model space. Pan the image back to the center of the screen as in Figure 11–128. If the dimensions are not relocated to the new position of the model objects, use the DIMREGEN command.

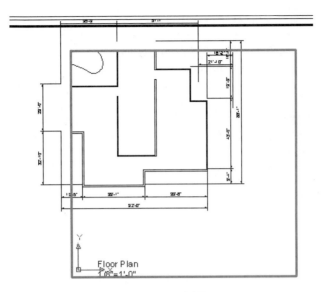

Figure 11–127

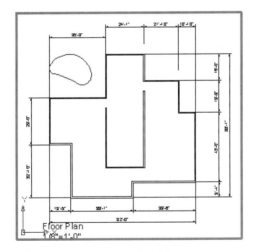

Figure 11–128

A few dimensions may not be updated. This is identified where the extension line origins are not tied to the object as in Figure 11–129. In this case, the DIMREASSOCIATE command can be used. It is found in the Dimension pull-down menu in the figure. Use the next series of figures and the following command sequence for performing this task. Make sure you are in Paper Space before completing the following command sequence.

Command: **DRE** (For DIMREASSOCIATE)

Select dimensions to reassociate ...

Select objects: *(Pick the dimension that has the value 21'-10")*

Select objects: *(Press ENTER to continue)*

Specify first extension line origin or [Select object] <next>: *(The endpoint of the associative dimension is identified. Use the OSNAP-Intersection mode and click the corner of the object in Figure 11–130)*

Specify second extension line origin <next>: *(Press ENTER to continue)*

Resuming DIMREASSOCIATE command.

Specify second extension line origin <next>: *(Press ENTER to continue)*

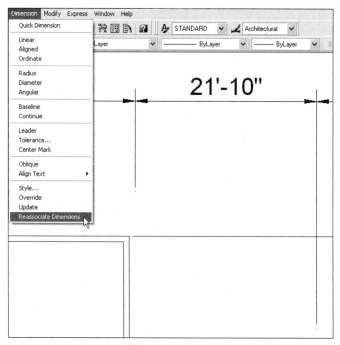

Figure 11–129

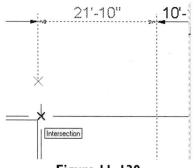

Figure 11–130

The results are displayed in Figure 11–131. With the dimension reassociated, notice the value has changed to 21'-7". Use the DIMREASSOCIATE command on the remaining two horizontal dimensions. The final results of this operation are illustrated in Figure 11–132.

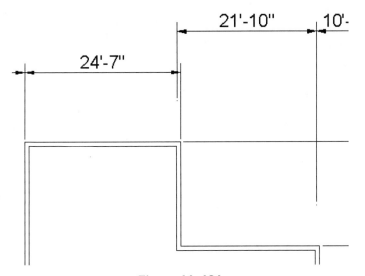

Figure 11–131

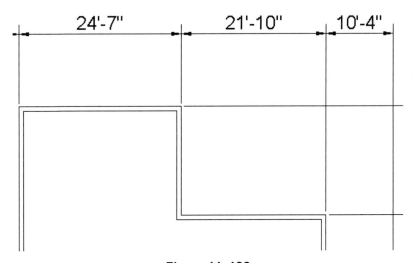

Figure 11–132

TUTORIAL EXERCISE: 11_FIXTURE.DWG

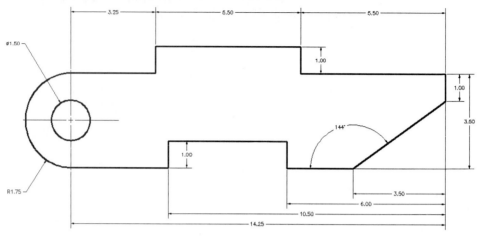

Figure 11–133

Purpose

The purpose of this tutorial is to add dimensions to the drawing of 11_Fixture.

System Settings

The drawing in Figure 11–133 is already constructed. Enter AutoCAD and follow the steps in this tutorial for adding dimensions. Be sure the following Object Snap modes are currently set: Endpoint and Intersection.

Layers

All layers have already been created:

Name	Color	Linetype
Cen	Yellow	Center
Defpoints	White	Continuous
Dim	Yellow	Continuous
Object	White	Continuous

Suggested Commands

Use the DIMCENTER, DIMLINEAR, DIMCONTINUE, DIMBASELINE, DIMDIAMETER, and DIMRADIUS commands for placing dimensions throughout this tutorial.

Whenever possible, substitute the appropriate command alias in place of the full AutoCAD command in each tutorial step; for example, use "CP" for the COPY command, "L" for the LINE command, and so on. The complete listing of all command aliases is located in Table 1–2.

STEP I

Open the drawing file 11_Fixture in Figure 11–134. A series of linear, baseline, continue, radius, and diameter dimensions will be added to this view. Before continu- ing, verify that running OSNAP is set to Endpoint and Intersection modes and that OSNAP is turned On.

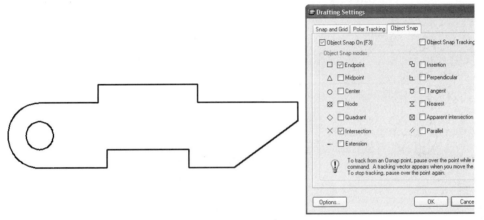

Figure 11–134

A number of dimension styles have been created to control various dimension properties. You will learn more about dimension styles in Chapter 12. For now, make the Center Mark dimension style current by clicking on its name in the Dimension Styles Control box shown in Figure 11–135.

Figure 11–135

STEP 2

Using the Dimension Center command (DCE), add a center mark to identify the center of the circular features. Touch the edge of the arc to place the center mark. The Center Mark dimension style already allows a small and long dash to be constructed. See Figure 11–136.

Command: **DCE** *(For DIMCENTER)*

Select arc or circle: *(Pick the edge of the arc in Figure 11–136)*

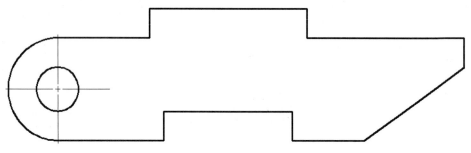

Figure 11–136

STEP 3

Now that the center mark is properly placed, it is time to begin adding the horizontal and vertical dimensions. Before performing these tasks, first make the Mechanical dimension style current by clicking on its name in the Dimension Styles Control box in Figure 11–137.

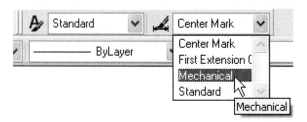

Figure 11–137

STEP 4

Verify in the Status bar that OSNAP is turned On. Then place a linear dimension of 5.50 units from the intersection at "A" to the intersection at "B" in Figure 11–138.

 Command: **DLI** *(For DIMLINEAR)*

Specify first extension line origin or <select object>: *(Pick the intersection at "A")*

Specify second extension line origin: *(Pick the intersection at "B")*

Specify dimension line location or [Mtext/Text/Angle/Horizontal/Vertical/Rotated]: *(Locate the dimension in Figure 11–138)*

Dimension text = 5.50

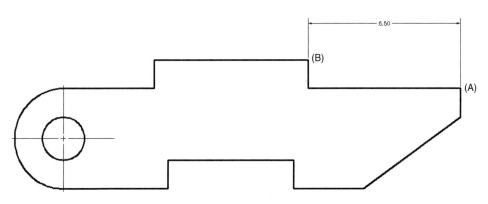

Figure 11–138

STEP 5

Place two continue dimensions at intersection "A" and endpoint "B" in Figure 11–139. These continue dimensions know to calculate the new dimension from the second extension line of the previous dimension.

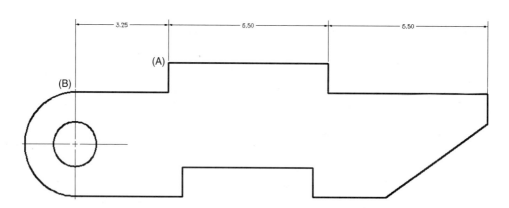 Command: **DCO** (For DIMCONTINUE)
Specify a second extension line origin or [Undo/Select] <Select>: (Pick the intersection at "A")

Dimension text = 5.50
Specify a second extension line origin or [Undo/Select] <Select>: (Pick the endpoint at "B")
Dimension text = 3.25
Specify a second extension line origin or [Undo/Select] <Select>: (Press ENTER)
Select continued dimension: (Press ENTER)

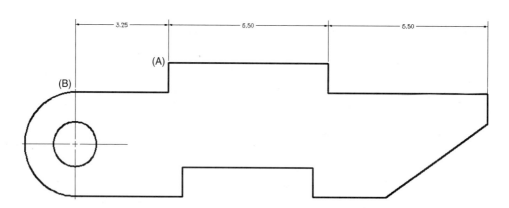

Figure 11–139

STEP 6

Place a linear dimension of 3.50 units from the intersection at "A" to the intersection at "B" in Figure 11–140.

 Command: **DLI** *(For DIMLINEAR)*
Specify first extension line origin or <select object>: *(Pick the intersection at "A")*

Specify second extension line origin: *(Pick the intersection at "B")*
Specify dimension line location or [Mtext/Text/Angle/Horizontal/Vertical/Rotated]: *(Locate the dimension in Figure 11–140)*
Dimension text = 3.50

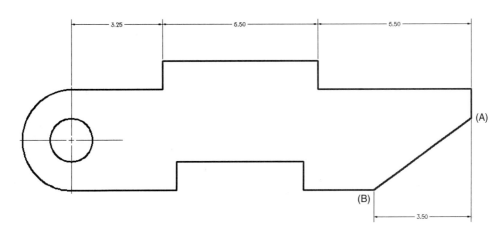

Figure 11–140

STEP 7

Add a series of baseline dimensions (see Figure 11–141). All baseline dimensions are calculated from the first extension line of the previous dimension (the linear dimension measuring 3.50 units).

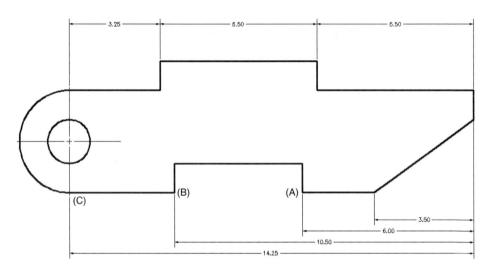

Command: **DBA** (For DIMBASELINE)
Specify a second extension line origin or [Undo/Select] <Select>: (Pick the intersection at "A")
Dimension text = 6.00

Specify a second extension line origin or [Undo/Select] <Select>: (Pick the intersection at "B")
Dimension text = 10.50
Specify a second extension line origin or [Undo/Select] <Select>: (Pick the endpoint at "C")
Dimension text = 14.25
Specify a second extension line origin or [Undo/Select] <Select>: (Press ENTER)
Select base dimension: (Press ENTER)

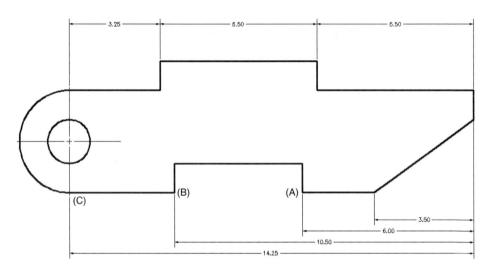

Figure 11–141

STEP 8

Add a diameter dimension to the circle using the DIMDIAMETER command (DDI). Then add a radius dimension to the arc using the DIMRADIUS command (DRA). See Figure 11–142.

 Command: **DDI** *(For DIMDIAMETER)*

Select arc or circle: *(Pick the edge of the circle at "A")*

Dimension text = 1.50

Specify dimension line location or [Mtext/Text/Angle]: *(Locate the diameter dimension in Figure 11–142)*

 Command: **DRA** *(For DIMRADIUS)*

Select arc or circle: *(Pick the edge of the arc at "B")*

Dimension text = 1.75

Specify dimension line location or [Mtext/Text/Angle]: *(Locate the diameter dimension in Figure 11–142)*

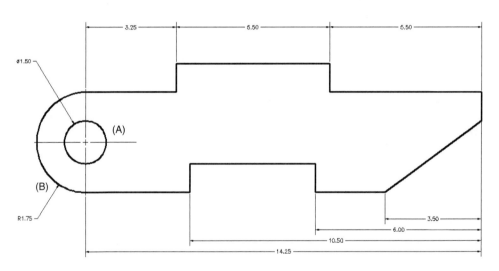

Figure 11–142

STEP 9

Place a vertical dimension of 1.00 units using the DIMLINEAR command (DLI). Pick the first extension line origin at the intersection at "A" and the second extension line origin at the intersection at "B". Then follow this by adding a baseline dimension using the DIMBASELINE command (DBA). See Figure 11–143.

Command: **DLI** *(For DIMLINEAR)*

Specify first extension line origin or <select object>: *(Pick the intersection at "A")*
Specify second extension line origin: *(Pick the intersection at "B")*

Specify dimension line location or [Mtext/Text/Angle/Horizontal/Vertical/Rotated]: *(Locate the dimension in Figure 11–143)*
Dimension text = 1.00

Command: **DBA** *(For DIMBASELINE)*

Specify a second extension line origin or [Undo/Select] <Select>: *(Pick the intersection at "C")*
Dimension text = 3.50
Specify a second extension line origin or [Undo/Select] <Select>: *(Press ENTER)*
Select base dimension: *(Press ENTER)*

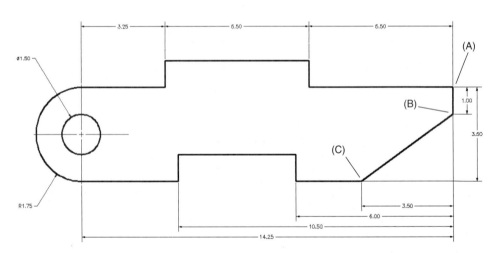

Figure 11–143

STEP 10

Add an angular dimension of 144 degrees using the DIMANGULAR command (DAN) as shown in Figure 11–144.

🔺 Command: **DAN** *(For DIMANGULAR)*
Select arc, circle, line, or <specify vertex>: *(Pick the line at "A")*

Select second line: *(Pick the line at "B")*
Specify dimension arc line location
    or [Mtext/Text/Angle]: *(Locate the dimension in Figure 11–144)*
Dimension text = 144

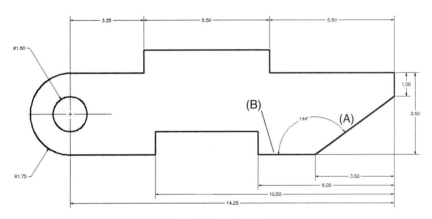

Figure 11–144

STEP 11

The next series of dimensions involve placing linear dimensions so that one extension line is constructed by the dimension and the other is actually an object line that already exists. You need to turn off one extension line while leaving the other extension turned on. This is accomplished by using an existing dimension style. Make the First Extension Off dimension style current by clicking on its name in the Dimension Styles Control box in Figure 11–145. This dimension style will suppress or turn off the first extension line while leaving the second extension line visible. This will prevent the extension line from being drawn on top of the object line.

Figure 11–145

STEP 12

Add the two linear dimensions to the slots by using the following command sequences and using Figure 11–146 as a guide.

Command: **DLI** *(For DIMLINEAR)*

Specify first extension line origin or <select object>: *(Pick the intersection at "A")*

Specify second extension line origin: *(Pick the intersection at "B")*

Specify dimension line location or [Mtext/Text/Angle/Horizontal/Vertical/Rotated]: *(Locate the dimension in Figure 11–146)*

Dimension text = 1.00

Command: **DLI** *(For DIMLINEAR)*

Specify first extension line origin or <select object>: *(Pick the intersection at "C")*

Specify second extension line origin: *(Pick the intersection at "D")*

Specify dimension line location or [Mtext/Text/Angle/Horizontal/Vertical/Rotated]: *(Locate the dimension in Figure 11–146)*

Dimension text = 1.00

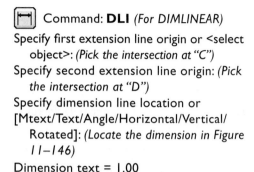

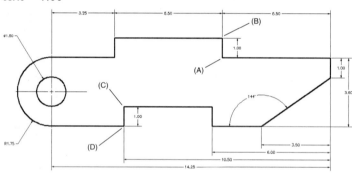

Figure 11–146

The completed fixture drawing with all dimensions is shown in Figure 11–147.

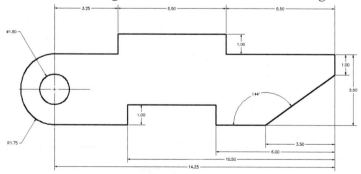

Figure 11–147

 Open the Exercise Manual PDF file for Chapter 11 on the accompanying CD for more tutorials and exercises.

 If you have the accompanying Exercise Manual, refer to Chapter 11 for more tutorials and exercises.

CHAPTER 12

The Dimension Style Manager

Dimensions have different settings that could affect the group of dimensions. These settings include the control of the dimension text height, the size and type of arrowhead used, and whether the dimension text is centered in the dimension line or placed above the dimension line. These are but a few of the numerous settings available to you. In fact, some settings are used mainly for architectural applications, while other settings are only for mechanical uses. As a means of managing these settings, dimension styles are used to group a series of dimension settings under a unique name to determine the appearance of dimensions. This chapter will cover in detail the Dimension Style Manager dialog and the following tabs associated with it: Lines and Arrows; Text; Fit; Primary Units; Alternate Units; and Tolerances. Additional topics include overriding a dimension style, comparing two dimension styles, and modifying the dimension style of an object.

THE DIMENSION STYLE MANAGER DIALOG BOX

Begin the process of creating a dimension style by choosing Style… from the Dimension pull-down menu as in Figure 12–1. This activates the Dimension Style Manager dialog box in Figure 12–2. Entering the keyboard command ddim can also activate this dialog box as can choosing Dimension Style from the Format pull-down menu. The current dimension style is listed as "Standard." This style is automatically available when you create any new drawing. In the middle of the dialog box is an image icon that displays how some dimensions will appear based on the current value of the dimension settings. Various buttons are also available to set a dimension style current, create a new dimension style, make modifications to an existing dimension style, display dimension overrides, and compare two dimension styles regarding their differences and similarities.

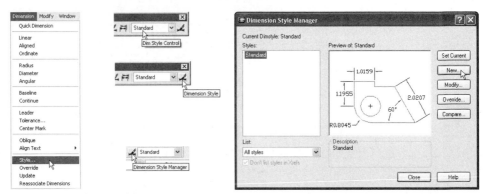

Figure 12–1   **Figure 12–2**

To create a new dimension style, click on the New... button. This activates the Create New Dimension Style dialog box in Figure 12–3. Enter a new name such as "Mechanical" in the New Style Name area. Then click the Continue button. This takes you to the New Dimension Style: Mechanical dialog box in Figure 12–4, where a number of tabs hold all of the settings needed in dimensioning. The Lines and Arrows tab deals with settings that control dimension lines, extension lines, arrowheads, and center marks used for circles. The Text tab contains the settings that control the appearance, placement, and alignment of dimension text. The Fit tab contains various fit options for placing dimension text and arrows, especially in narrow places. You control the units used in dimensioning through the Primary Units tab. If you need to display primary and secondary units in the same drawing, the Alternate Units tab is used. Finally for mechanical applications, various ways to show tolerances are controlled in the Tolerances tab. Making changes to any of the settings under the tabs will update the preview image, showing these changes. This provides a quick way of previewing how your dimensions will appear in your drawing. All of these areas will be explained in greater detail throughout this chapter.

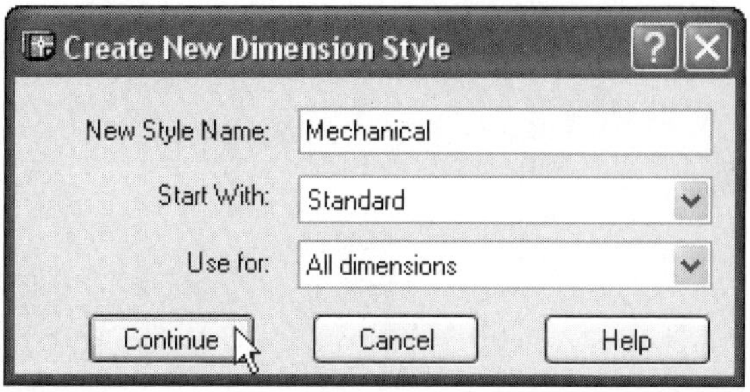

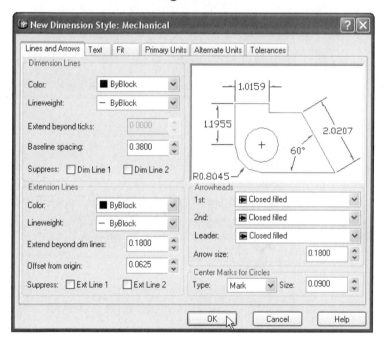

Figure 12–3

Figure 12–4

Click the OK button. This returns you to the Dimension Style Manager dialog box in Figure 12–5. Notice the Mechanical dimension style has been added to the list of styles. Also, any changes made in the tab areas are automatically saved to the dimension style you are creating or modifying. Click on Mechanical and then pick the Set Current button to make this style the current dimension style. Clicking the Close button returns you to your drawing. Let's get a closer look at all of the tabs and their settings. Type the letter "D" to reopen the Dimension Style Manager. Make sure Mechanical is highlighted and click on the Modify… button.

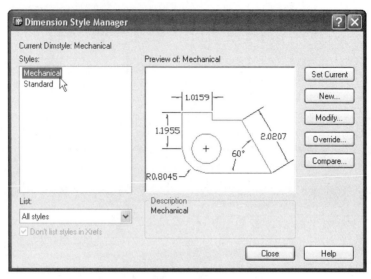

Figure 12–5

THE LINES AND ARROWS TAB

The Modify Dimension Style: Mechanical dialog box in Figure 12–6 displays the Lines and Arrows tab, which will now be discussed in greater detail. This tab consists of four main areas dealing with dimension lines, extension lines, arrowheads, and center markers.

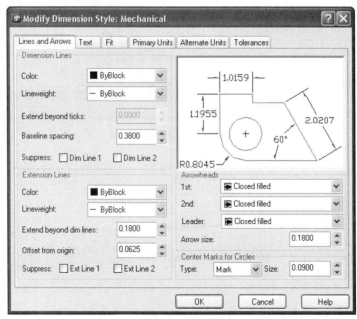

Figure 12–6

DIMENSION LINE SETTINGS

Use the Dimension Lines area, shown in Figure 12–7, to control the color, visibility, and spacing of the dimension line.

Clicking on the Color button of the Dimension Lines area allows you to assign a different color to the dimension line (see Figure 12–8). The color you assign here will override the color of the layer that is current when you place dimensions. The color of the current arrowhead terminator will also be affected by the dimension line color. A practical use of this operation is to assign color to the dimension line as a means of controlling line quality by assigning different pen weights for color-dependent plot styles.

The Lineweight button allows you to assign a different lineweight to the dimension line.

The Baseline spacing setting in the Dimension Lines area controls the spacing of baseline dimensions, because they are placed at a distance from each other similar to the illustration in Figure 12–9. This value affects the DIMBASELINE command.

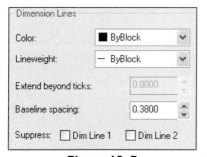

Figure 12–7

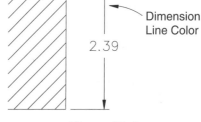

Figure 12–8

By default, dimension line suppression is turned off. To turn on suppression of dimension lines, place a check in the Dim Line 1 or Dim Line 2 box next to Suppress. This operation turns off the display of dimension lines, similar to the illustration in Figure 12–10. This may be beneficial where tight spaces require that only the dimension text to be placed.

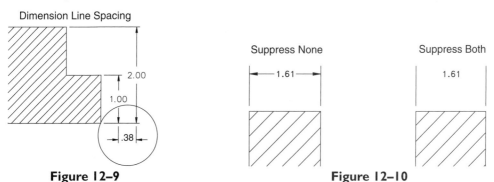

Figure 12–9

Figure 12–10

EXTENSION LINES

The Extension Lines area controls the color, the distance the extension extends past the arrowhead, the distance from the object to the end of the extension line, and the visibility of extension lines. See Figure 12–11.

As with dimension lines, you can set extension lines to a different color to control line quality by clicking on the Color button and assigning a new color that only affects extension lines. See Figure 12–12.

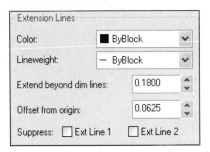

Figure 12–11

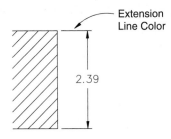

Figure 12–12

The Extend beyond dim lines setting controls how far the extension extends past the arrowhead or dimension line, as shown in Figure 12–13. By default, a value of 0.18 is assigned to this setting.

The Offset from Origin setting controls how far away from the object the extension line will start (see Figure 12–14). By default, a value of 0.06 is assigned to this setting.

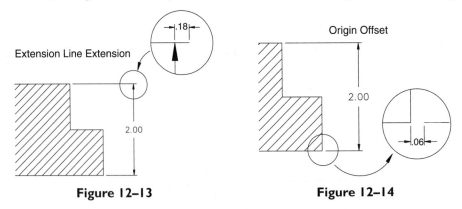

Figure 12–13　　　　　　　　　　**Figure 12–14**

The Suppress 1st and 2nd check boxes control the visibility of extension lines. They are useful when you dimension to an object line and for avoiding placing the extension line on top of the object line. Placing a check in the 1st box of Suppress turns off the first extension line. The same is true when you place a check in the 2nd box of Suppress—the second extension line is turned off. Study the examples in Figure 12–15 to get a better idea about how suppression of extension lines operates.

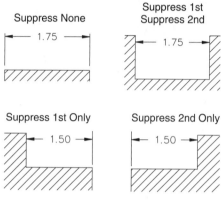

Figure 12–15

ARROWHEAD SETTINGS

Use the Arrowheads area to control the type of arrowhead terminator used for dimension lines and leaders (see Figure 12–16). This dialog box also controls the size of the arrowhead.

Figure 12–16

Clicking in the 1st: box displays a number of arrowhead terminators. Choose the desired terminator from the partial list in Figure 12–17. When you choose the desired arrowhead from the 1st: box, the 2nd: box automatically updates to the selection made in the 1st. If you choose an arrowhead from the 2nd: edit box, first and second arrowheads may be different at opposite ends of the dimension line; this is desired in some applications. Choosing a terminator in the Leader box displays the arrowhead that will be used whenever you place a leader.

Illustrated in Figure 12–18 is the complete set of arrowheads, along with their names. The last arrow type is a User Arrow, which allows you to define your own custom arrowhead.

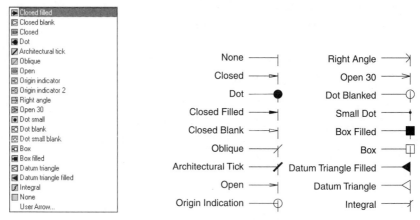

Figure 12–17 **Figure 12–18**

Use the Arrow size setting to control the size of the arrowhead terminator. Arrow types that appear filled in are controlled by the FILL command. With Fill turned on, all arrowheads are filled in as in Figure 12–19. With Fill turned off, only the outline of the arrow is displayed.

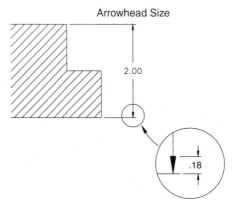

Figure 12–19

CENTER MARK SETTINGS

The Center Marks for Circles area allows you to control the type of center marker used when identifying the centers of circles and arcs. You can make changes to these settings by clicking on the three center mark modes: Mark, None, or Line, as in Figure 12–20. The Size box controls the size of the small plus mark (+). If Center Mark or Center Line is chosen, these lines will show up when placing radius and diameter dimensions.

The three types of center marks are illustrated in Figure 12–21. The Mark option places a plus mark in the center of the circle. The Line option places the plus mark and extends a line past the edge of the circle. The None option will display a circle with no center mark.

Tip: It is considered good practice to add center marks to all circles and arcs before adding radius and diameter dimensions. To do this, first pick the Center Line type and use the DCE (Dimension Center) command to place the center marks. When finished, pick the None type in the center mark type area. Now all diameter and radius dimensions can be added without duplicating the center marks.

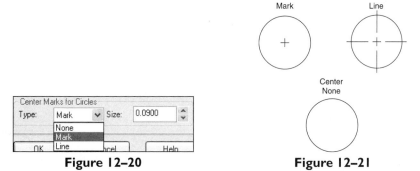

Figure 12–20

Figure 12–21

THE TEXT TAB

Use the Text tab in Figure 12–22 to change the text appearance, such as height, the text placement, such as centered vertically and horizontally, and the text alignment, such as always horizontal or parallel with the dimension line.

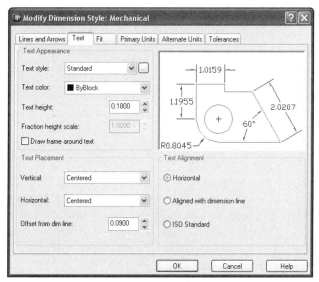

Figure 12–22

TEXT APPEARANCE

Use the Text Appearance area to make changes to settings such as the dimension Text style, dimension Text color, Text height, Fraction height scale, and the ability to draw a frame or box around the text. See Figure 12–23.

The Text color area is used to assign a different color exclusively to the dimension text (see Figure 12–24). This can prove very beneficial especially when you assign a medium lineweight to the dimension text and a thin lineweight to the dimension and extension lines.

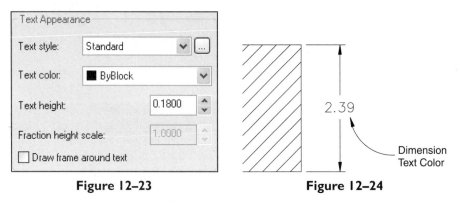

Figure 12–23   **Figure 12–24**

The Text height setting controls the size of the dimension text as in Figure 12–25. By default, a value of 0.18 is assigned to this setting.

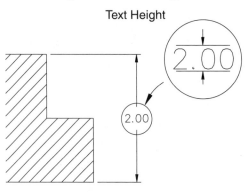

Figure 12–25

If the primary dimension units are set to architectural or fractional, the Fraction height scale activates. Changing this value affects the height of fractions that appear in the dimension. In Figure 12–26, a fractional height scale of 0.5000 will make the fractions as tall as the primary dimension number in the preview window.

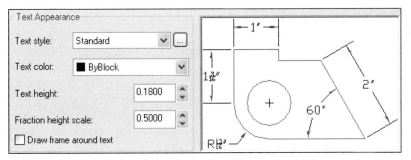

Figure 12–26

Placing a check in the Draw frame around text box will draw a rectangular box around all dimensions, as shown in the preview window in Figure 12–27. This can be done to emphasize the dimension text. It also is a way of calling out a basic dimension. This will be explained later in this chapter when the topic of tolerances is discussed.

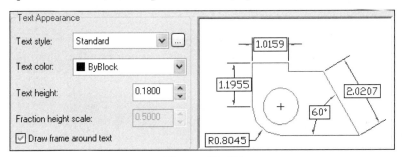

Figure 12–27

TEXT PLACEMENT

The Text Placement area in Figure 12–28 allows you to control the vertical and horizontal placement of dimension text. You can also set an offset distance from the dimension line for the dimension text.

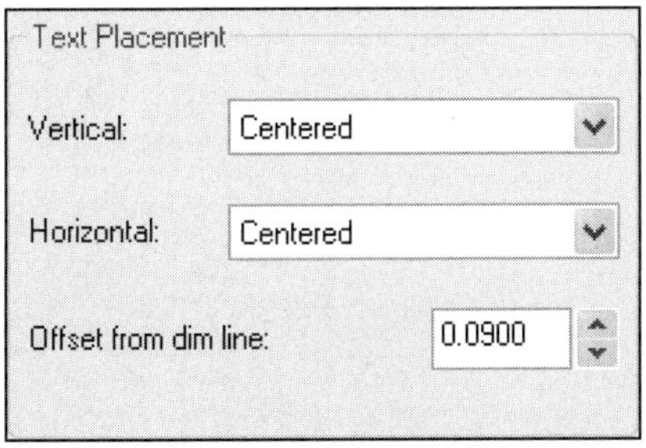

Text Placement	
Vertical:	Centered
Horizontal:	Centered
Offset from dim line:	0.0900

Figure 12–28

VERTICAL TEXT PLACEMENT

The Vertical area of Text Placement controls the vertical justification of dimension text. Clicking in the drop-down edit box allows you to set vertical justification modes. By default, dimension text is centered vertically in the dimension line. Other modes include justifying vertically above the dimension line, justifying vertically outside the dimension line, and using the JIS (Japan International Standard) for placing text vertically. See Figure 12–29.

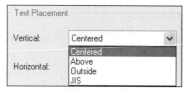

Figure 12–29

Illustrated in Figure 12–30 is the result of setting the Vertical Justification to Centered. The dimension line will automatically be broken to accept the dimension text.

Illustrated in Figure 12–31 is the result of setting the Vertical Justification to Above. Here the text is placed directly above a continuous dimension line. This mode is very popular for architectural applications.

Figure 12–30   **Figure 12–31**

Figure 12–32 illustrates the result of setting the Vertical Justification mode to outside. All text, including those contained in angular and radial dimensions, will be placed outside the dimension lines and leaders. Note the term JIS. This refers to a Japanese standard for vertically justifying dimension text.

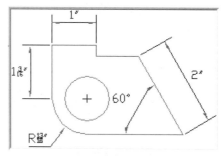

Figure 12–32

HORIZONTAL TEXT PLACEMENT

At times, dimension text needs to be better located in the horizontal direction; this is the purpose of the Horizontal justification area. Illustrated in Figure 12–33 are the five modes of justifying text horizontally. By default, the horizontal text justification is centered in the dimension.

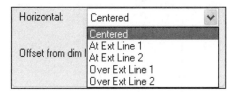

Figure 12–33

Clicking on the Centered option of the Horizontal Justification area displays the dimension text in Figure 12–34. This is the default setting because it is the most commonly used text justification mode in dimensioning.

Clicking on the At Ext Line 1 option displays the dimension text in Figure 12–35, where the dimension text slides close to the first extension line. Use this option to position the text out of the way of other dimensions. Notice the corresponding option to have the text positioned nearer to the second extension line.

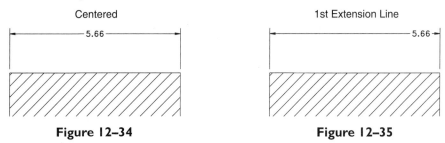

Figure 12–34   **Figure 12–35**

Clicking on the Over Ext Line 1 option displays the dimension text parallel to and over the first extension line in Figure 12–36. Notice the corresponding option to position dimension text over the second extension line.

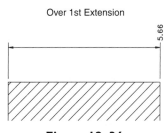

Figure 12–36

The last item of the Text Placement area deals with setting an offset distance from the dimension line. When you place dimensions, a gap is established between the inside ends of the dimension lines and the dimension text. Entering different values depending on the desired results can control this gap. Study Figures 12–37 and 12–38, which have different text offset settings. Entering a value of zero (0) will force the dimension to touch the edge of the dimension text. Negative values are not supported.

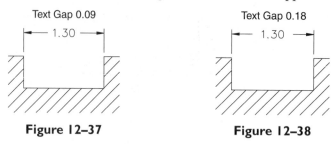

Figure 12–37   **Figure 12–38**

TEXT ALIGNMENT

Use the Text Alignment area in Figure 12–39 to control the alignment of text. Dimension text can either be placed horizontally or parallel (aligned) to the edge of the object being dimensioned. An ISO (International Standards Organization) Standard is also available for metric drawings. Click in the appropriate radio button to turn the desired text alignment mode on.

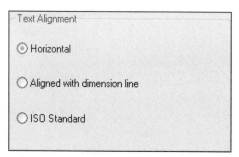

Figure 12–39

If the Horizontal radio button is clicked, all text will be read horizontally as shown in Figure 12–40. This includes text located inside and outside the extension lines.

Clicking on the Aligned with dimension line radio button displays the alignment results shown in Figure 12–41. Here all text is read parallel to the edge being dimensioned. Not only will vertical dimensions align the text vertically, but the 2.06 dimension used to dimension the incline is also parallel to the edge.

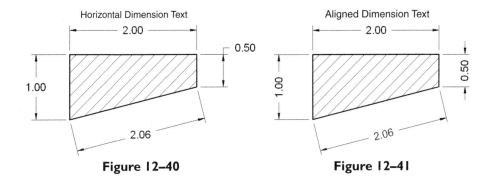

Figure 12–40   **Figure 12–41**

THE FIT TAB

Use the Fit tab in Figure 12–42 to control how text and/or arrows are displayed if there isn't enough room to place both inside the extension lines. You could place the text over the dimension with or without a leader line. The Scale for Dimension Features area is very important when you dimension in Model Space or when you scale dimensions to Paper Space units. You can even fine tune the placement of text manually and force the dimension line to be drawn between extension lines.

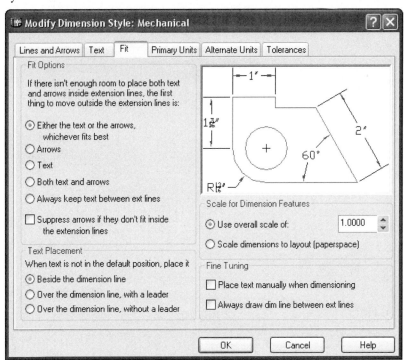

Figure 12–42

FIT OPTIONS

The Fit Options area in Figure 12–43 has the radio button set for Either the text or the arrows, whichever fits best. AutoCAD will decide to move either text or arrows outside extension lines. It will determine the item that fits the best. This setting is illustrated in the preview area in Figure 12–42. It so happens that the preview image is identical when you click on the radio button for Arrows. This tells AutoCAD to move the arrowheads outside the extension lines if there isn't enough room to fit both text and arrows.

Clicking on the Text radio button of the Fit Options area updates the preview image in Figure 12–44. Here you are moving the dimension text outside the extension lines if the text and arrows do not fit. Since the value 1.0159 is the only dimension that does not fit, it is placed outside the extension lines; but the arrows are drawn inside the extension lines.

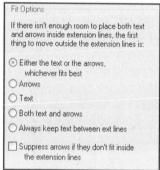

Figure 12–43

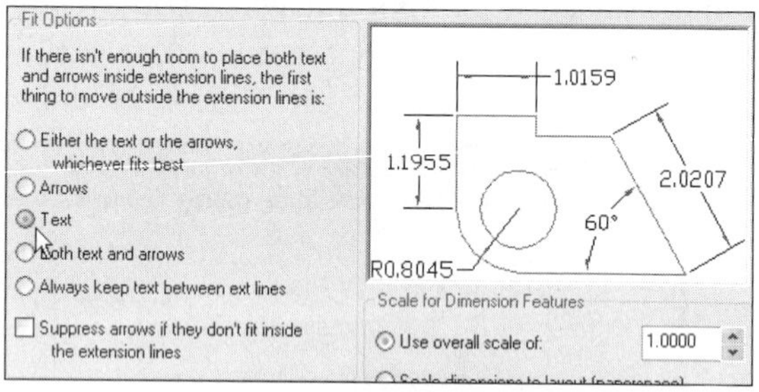

Figure 12–44

Clicking on the Both text and arrows radio button updates the preview image in Figure 12–45 where the 1.0159 dimension text and arrows are both placed outside of the extension lines.

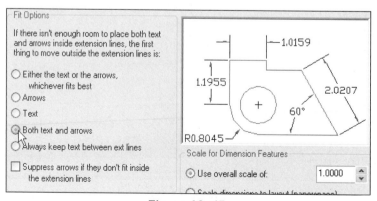

Figure 12–45

If you click on the radio button for Always keep text between ext lines, the result is illustrated in the preview image in Figure 12–46. Here all dimension text, including the radius dimension, is placed between the extension lines.

If you click on the radio button for "Either the text or the arrows, whichever fits best" and you also place a check in the box for "Suppress arrows if they don't fit inside the extension lines", the dimension line is turned off only for dimensions that cannot fit the dimension text and arrows. This is illustrated in Figure 12–47.

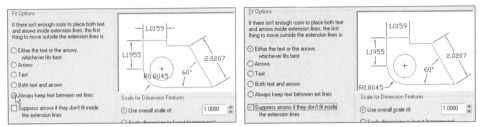

Figure 12–46   **Figure 12–47**

TEXT PLACEMENT

You control the placement of the text if it is not in the default position. Your choices, shown in Figure 12–48, are Beside the dimension line, which is the default, Over the dimension line, with a leader, or Over the dimension line, without a leader.

If you click on the radio button for Text in the Fit Options area and you click on the radio button for Over the dimension line, with a leader, the result is illustrated in Figure 12–49. For the 1.0159 dimension that does not fit, the text is placed outside the dimension line with the text connected to the dimension line with a leader.

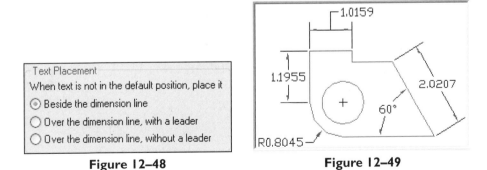

Figure 12–48   **Figure 12–49**

If you click on the radio button for Text in the Fit Options area and you click on the radio button for Over the dimension line, without a leader, the result is illustrated in

Figure 12–50. For the 1.0159 dimension that does not fit, the text is placed outside the dimension. No leader is used.

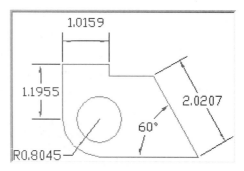

Figure 12–50

SCALE FOR DIMENSION FEATURES

This setting acts as a multiplier and globally affects all current dimension settings that are specified by sizes or distances. This means that if the current overall scale value is 1.00, other settings that require values will remain unchanged. There is also a radio button allowing you have dimensions automatically scaled to Paper Space units inside a layout. The scale of the viewport will control the scale of the dimensions. (See Figure 12–51.)

Figure 12–52 shows the effects of an overall scale factor of 1.00 and 2.00. The dimension text, arrows, origin offset, and extension beyond the arrow have all doubled in size.

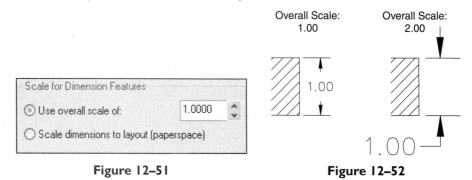

Figure 12–51   **Figure 12–52**

Try It! - Open the drawing file 12_Dimscale. A simple floor plan is displayed in Figure 12–53. This floor plan is designed to be plotted at a scale of 1/2"=1'-0". Also displayed in this floor plan are dimensions in the magenta color. However the dimensions are too small to be viewed. Set the Scale for Dimensions Features found under the Fit tab of the Dimension Style dialog box to 24 (1' (12") divided by 1/2 is 24). The dimension values and tick marks should now be visible.

DRAWING SCALE: 1/2"=1'-0"

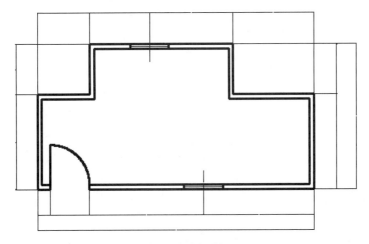

Figure 12–53

FINE TUNING

You have two options to add further control of the fitting of dimension text in the Fine Tuning area in Figure 12–54. You can have total control for horizontally justifying dimension text if you place a check in the box for Place text manually when dimensioning. You can also force the dimension line to be drawn between extension lines by placing a check in this box. The results are displayed in Figure 12–55.

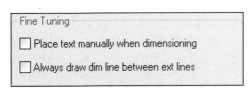

Figure 12–54

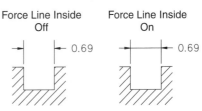

Figure 12–55

THE PRIMARY UNITS TAB

Use the Primary Units tab in Figure 12–56 to control settings affecting the primary units. This includes the type of units the dimensions will be constructed in (decimal, engineering, architectural, and so on), and whether the dimension text requires a prefix or suffix.

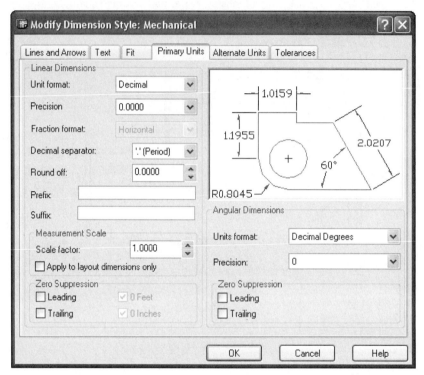

Figure 12–56

The Linear Dimensions area in Figure 12–57 has various settings that deal with primary dimension units. A few of the settings deal with the format when working with fractions. This area activates only if you are working in architectural or fractional units. Even though you may be drawing in architectural units, the dimension units are set by default to decimal. You can also designate the decimal separator as a Period, Comma, or Space.

Clicking in the box for Unit format in Figure 12–58 displays the types of units you can apply to dimensions. You also control the precision of the primary units by clicking in the Precision box.

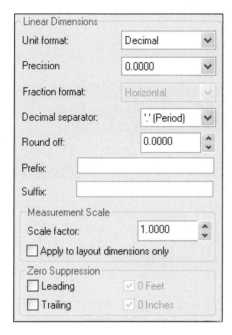

Figure 12–57

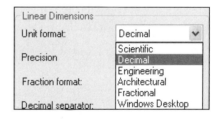

Figure 12–58

Use a Round Off value to round off all dimension distances to the nearest unit based on the round off value. With a round off value of 0.0000, the dimension text reflects the actual distance being dimensioned, as in Figure 12–59. With a round off value set to 0.2500, the dimension text reflects the next 0.2500 increment, namely 2.50. See Figure 12–60.

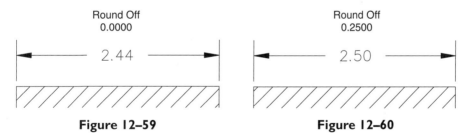

Figure 12–59   **Figure 12–60**

A prefix allows you to specify text that will be placed in front of the dimension text whenever you place a dimension. Use the Suffix box to control the placement of a character string immediately after the dimension value. In the illustration in Figure 12–61, the letters "IN." have been added to the dimension, signifying "inches."

682

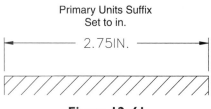

Figure 12–61

MEASUREMENT SCALE

The Measurement Scale area in Figure 12–62 acts as a multiplier for all linear dimension distances, including radius and diameter dimensions. When a dimension distance is calculated, the current value set in the Scale factor edit box is multiplied by the dimension to arrive at a new dimension value. In the illustration in Figure 12–63, and with a Linear Scale value of 1.00, the dimension distances are taken at their default values.

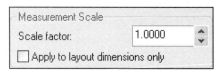

Figure 12–62

In Figure 12–64, the Linear Scale value has been changed to 2.00 units. This means 2.00 will multiply every dimension distance; the result is that the previous 3.00 and 2.00 dimensions are changed to 6.00 and 4.00. In a similar fashion, having a Linear Scale value set to 0.50 will cut all dimension values in half.

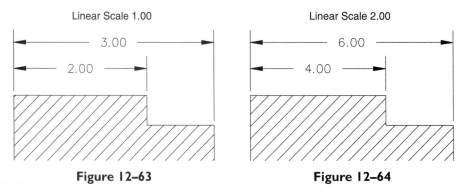

Figure 12–63   **Figure 12–64**

ANGULAR DIMENSIONS

Use the Angular Dimensions area in Figure 12–65 to control unit of measure for angles. Clicking in the Units format box displays four angular measurement modes.

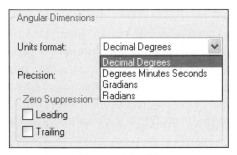

Figure 12–65

THE ALTERNATE UNITS TAB

Use the Alternate Units tab in Figure 12–66 to enable alternate units, set the units and precision of the alternate units, set a multiplier for all units, use a round off distance, and set a prefix and suffix for these units. You also have two placement modes for displaying these units. By default, alternate units are placed beside primary units. The alternate units are enclosed in square brackets. With two sets of units being displayed, your drawing could tend to become very busy. An application of using Alternate Units would be to display English and metric dimension values, since some design firms require both.

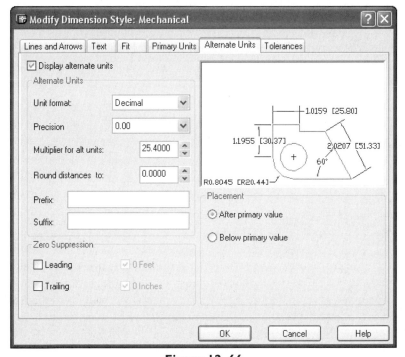

Figure 12–66

ALTERNATE UNITS

Once alternate units are enabled, all items in the Alternate Units area become active, as shown in Figure 12–67. Figure 12–68 shows the effect of turning off Alternate Units. In this example, a single dimension value is calculated and placed between dimension and extension lines. Figure 12–69 shows the effects of Alternate Units being enabled. Next to the calculated dimension value, the alternate value is placed in brackets. This value depends on the current setting in the Multiplier for all units edit box. With this factor set to 25.40, it is used as a multiplier for all calculated alternate dimension values. As with primary units, you can set a round off value, a prefix, and a suffix to be used exclusively for alternate units. For example, if the alternate unit of measure is millimeters, you could enter "mm" for a suffix.

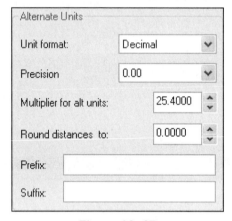

Figure 12–67

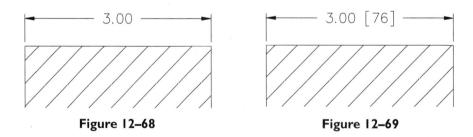

Figure 12–68   Figure 12–69

It has already been mentioned that alternate dimensions are placed inside square brackets and to the right of the primary dimension. You could click in the radio button for Below primary value in Figure 12–70. This places the primary dimension above the dimension line and the alternate dimension below it, which may help with the crowded look that alternate dimensions tend to add in a drawing.

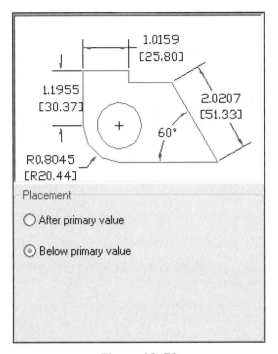

Figure 12–70

THE TOLERANCES TAB

The Tolerances tab in Figure 12–71 consists of various edit boxes used to control the five types of tolerance settings: none, symmetrical, deviation, limits, and basic. Depending on the type of tolerance being constructed, an Upper Value and Lower Value may be set to call out the current tolerance variance. The Vertical position setting allows you to determine where the tolerance will be drawn in relation to the location of the body text. The Scaling for height setting controls the text size of the tolerance. Usually this value is smaller than the body text size.

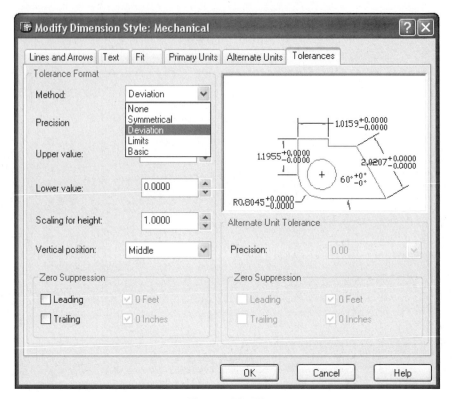

Figure 12–71

TOLERANCE FORMAT

The five tolerance types available in the drop-down list are illustrated in Figure 12–72. A tolerance setting of None uses the calculated dimension value without applying any tolerances. The Symmetrical tolerance uses the same value set in the Upper and Lower Value. The Deviation tolerance setting will have a value set in the Upper Value and an entirely different value set in the Lower Value. The Limits tolerance will use the Upper and Lower values and place the results with the larger limit dimension placed above the smaller limit dimension. The Basic tolerance setting does not add any tolerance value; instead, a box is drawn around the dimension value.

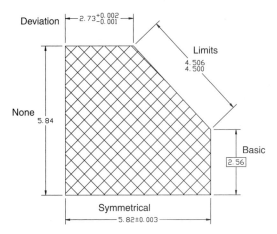

Figure 12–72

Try It! - A series of drawing files are available to dimension a shape using the appropriate tolerancing mode. The following drawing files can be opened: 12_Dim Basic, 12_Dim Deviation, 12_Dim Limits, and 12_Dim Symmetrical. When you open a drawing such as 12_Dim Basic, all settings have been saved to a dimension style. All you do is place a series of linear and aligned dimensions. Each tolerance mode is illustrated in Figure 12–73.

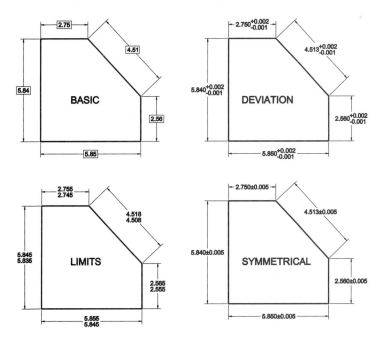

Figure 12–73

The Scaling for height setting affects the height of the tolerance text. In Figure 12–74, a tolerance height setting of 1.00 has no effect on the height of the tolerance text; in fact the tolerance text is the same height as the main dimension text. Figure 12–75 shows the effect of setting the tolerance height to a value of 0.70 units. Here, the tolerance height is noticeably smaller than the main dimension text height. It is good practice to have the main dimension text value set to a higher value than the tolerance heights, for greater emphasis on the main dimension.

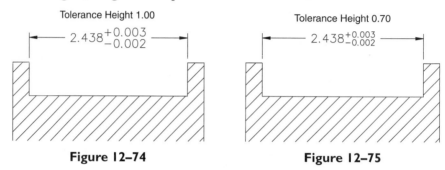

Figure 12–74   **Figure 12–75**

DIMASSOC—THE ASSOCIATIVE DIMENSIONING CONTROL SETTING

As mentioned in Chapter 11, use this dimension setting to set the associative dimensioning mode. By default, this setting is 2 where all objects that make up the dimension are considered one object and are associated with the object being dimensioned. If this setting is set to 1, all objects that make up the dimension are still considered one object; however the dimension is not associated with the object. If this setting is changed to 0, all dimension components such as arrowheads, extension lines, dimension lines, and dimension text will be considered single objects (see Figures 12–76 and 12–77). This has the same effect as using the EXPLODE command on an associative dimension. In fact, grips have no effect on a dimension considered exploded. It is considered poor practice to set this variable to 0 or to explode dimensions that are associative. The prompt for this dimension setting is as follows:

Command: **DIMASSOC**
Enter new value for DIMASSOC <2>:

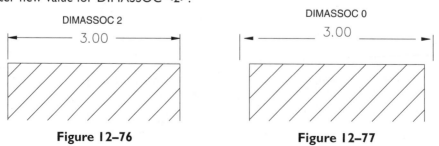

Figure 12–76   **Figure 12–77**

USING DIMENSION TYPES IN DIMENSION STYLES

In addition to creating dimension styles, you can assign dimension types to dimension styles. The purpose of using dimension types is to reduce the number of dimension styles defined in a drawing. For example, the object in Figure 12–78 consists of linear dimensions with three decimal place accuracy, a radius dimension with one decimal place accuracy and an angle dimension with a box surrounding the number. Normally you would have to create three separate dimension styles to create this effect. However, the linear, radius, and angular dimensions consist of what are called dimension types.

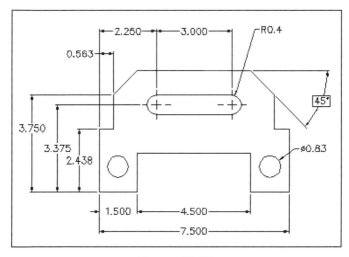

Figure 12–78

Dimension types appear similar to Figure 12–79 when assigned to the Mechanical dimension style. Before creating an angular dimension, you create the dimension type and make changes in various tabs located in the Modify Dimension Style dialog box. These changes will only apply to the angular dimension type. To expose the dimension types, click on the New button in the main Dimension Style Manager dialog box.

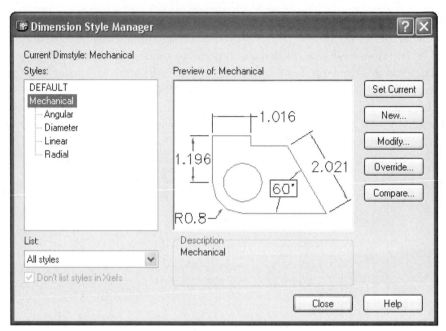

Figure 12–79

When the Create New Dimension Style dialog box appears in Figure 12–80, click in the Use for: box. This usually displays "All dimensions." Notice all dimension types appearing. You can make changes to dimension settings that will apply only for linear, angular, radius, diameter, and ordinate dimensions. Also, a dimension type for Leaders and Tolerances is available. Highlight this option and then click the Continue button.

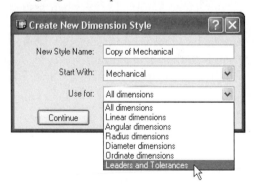

Figure 12–80

This takes you to the New Dimension Style: Mechanical: Leader dialog box in Figure 12–81. Any changes you make in the tabs will only apply whenever you place a leader dimension. The use of Dimension Types is an efficient means of keeping the number of dimension styles down to a manageable level.

Try It! - Open the drawing file 12_Dimension Types. A series of Dimension Types have already been created. Use the illustration in Figure 12–78 and add all dimensions to this object.

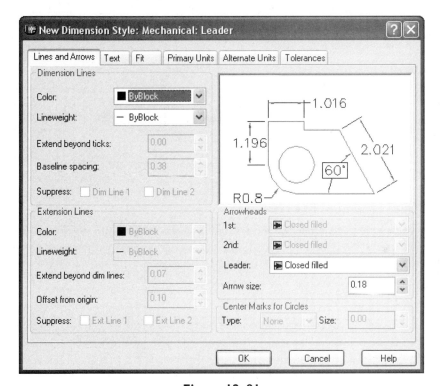

Figure 12–81

OVERRIDING A DIMENSION STYLE

Try It! - Open the drawing file 12_Dimension Override and activate the Dimension Style Manager dialog box. For special cases where you need to change the settings of one dimension, a dimension override would be used. First launch the Dimension Style Manager dialog box. Clicking on the Override button displays the Override Current Style: Mechanical dialog box in Figure 12–82. Check the Ext Line 1 and Ext Line 2 boxes in the Extension Lines area for suppression. One dimension needs to be constructed without displaying extension lines.

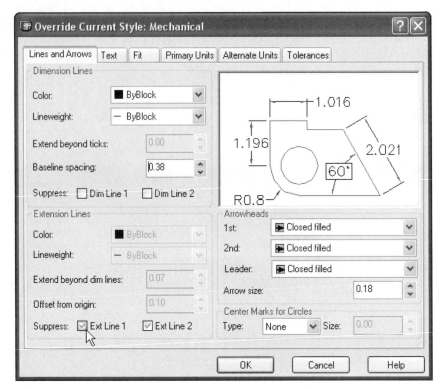

Figure 12–82

Clicking the OK button returns you to the main Dimension Style Manager dialog box in Figure 12–83. Notice that, under the Mechanical style, a new dimension type has been created called <style overrides> (Angular, Diameter, Linear, and Radial were already existing in this example). Also notice that the image in the Preview box shows sample dimensions displayed without extension lines. The <style overrides> dimension style is also the current style. Click the Close button to return to your drawing.

Note: If the Preview box does not show the correct sample image when you create the new <style overrides>, close and get right back into the Dimension Style Manager dialog box. The Preview box will show the correct sample image.

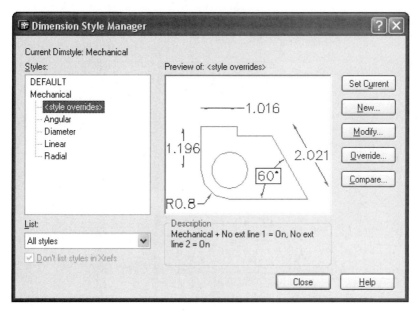

Figure 12–83

In Figure 12–84, a linear dimension is placed identifying the vertical distance of 1.250. To avoid the mistake of placing extension lines on top of existing object lines, the extension lines are not drawn due to the style override.

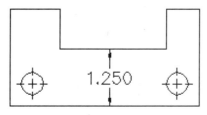

Figure 12–84

Unfortunately, if you place other linear dimensions, these will also lack extension lines. The Dimension Style Manager dialog box is once again activated. Clicking on an existing dimension style such as Mechanical and then clicking on the Set Current button displays the AutoCAD Alert box in Figure 12–85. If you click OK, the style override will disappear from the listing of dimension types. If you would like to save the overrides under a name, click the Cancel button, right-click on the <style overrides> listing, and rename this style to a new name. This will preserve the settings under this new name.

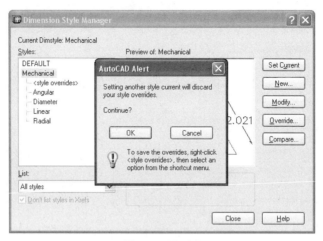

Figure 12–85

COMPARING TWO DIMENSION STYLES

Yet another feature of the Dimension Style Manager dialog box deals with the ability to compare the properties of two dimension styles or view all the properties of one style.

 Try It! - Open the drawing file 12_Dimension Compare. Activate the Dimension Style Manager dialog box and click on the Compare button. In Figure 12–86 when comparing the DEFAULT dimension style with the CONTINUE dimension style, which was previously created, AutoCAD lists the differences between the two styles by providing the description of the dimension setting, the dimension variable name, the setting value of the DEFAULT style, and the setting value of the CONTINUE style. This allows you to quickly preview the differences between two dimension styles.

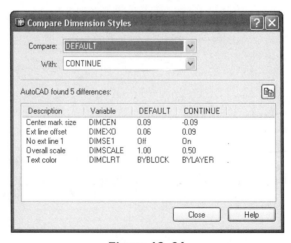

Figure 12–86

You can also obtain a complete listing of all settings and variables for single styles by comparing CONTINUE with CONTINUE in Figure 12–87. Use the scroll bar to view all variables, descriptions, and current values.

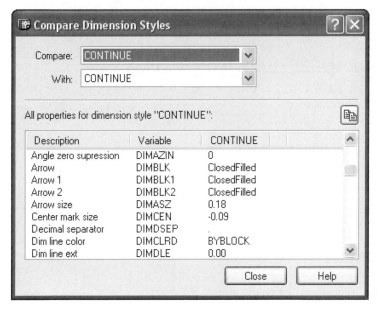

Figure 12–87

MODIFYING THE DIMENSION STYLE OF AN OBJECT

Existing dimensions in a drawing can easily be modified through a number of methods that will be outlined in this section. Figure 12–88 illustrates the first of these methods: the use of the Dim Style Control box, which is part of the Dimension toolbar. Be sure to have this toolbar displayed on your screen before continuing.

Try It! - Open the drawing file 12_Dimension Edit. First click on the dimension in Figure 12–88 and notice that the dimension highlights and the grips appear. The current dimension style is also displayed in the Dim Style Control box. Click in the control box to display all other styles currently defined in the drawing. Click on the style name TOLERANCE to change the highlighted dimension to that style. Press ESC to turn off the grips.

Tip: Be careful when docking this toolbar along the side of your screen. In lower screen resolutions, the Dim Style Control box will not display.

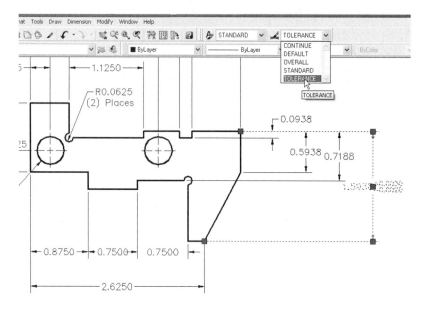

Figure 12–88

The second method of modifying dimensions is illustrated in Figure 12–89. In this example, select the dimension in Figure 12–89; it highlights and grips appear. Right-clicking the mouse button displays the cursor menu. Choosing Dim Style heading displays the cascading menu of all dimension styles defined in the drawing. Clicking on Other... displays the Apply Dimension Style dialog box in Figure 12–90. Click on the TOLERANCE style and then click the OK button to change the highlighted dimension to the new style.

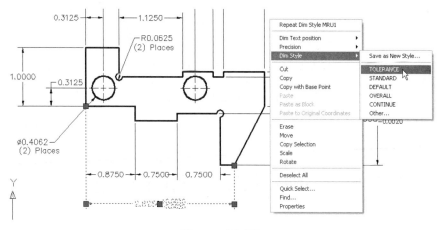

Figure 12–89

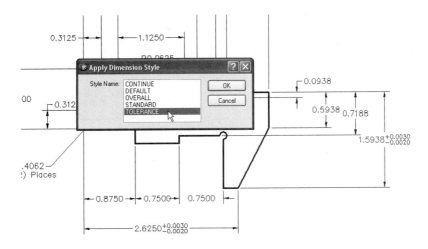

Figure 12–90

The third method of modifying dimensions begins with selecting the overall length dimension in Figure 12–91. Again, the dimension highlights and grips appear. Clicking on the Properties button displays the Properties Window in Figure 12–91. Click on the Alphabetic tab to display all dimension settings and current values. If you click in the name box next to Dim style, all dimension styles will appear in the edit box. Click on the OVERALL style to change the highlighted dimension to the new style. Close the Properties Palette and press the ESC key to turn off the grips.

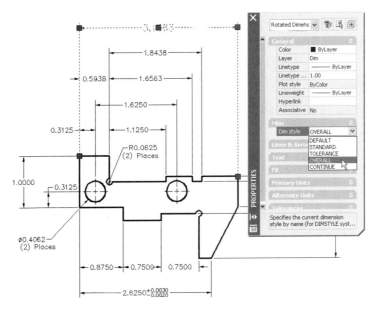

Figure 12–91

TUTORIAL EXERCISE: 12_DIMEX.DWG

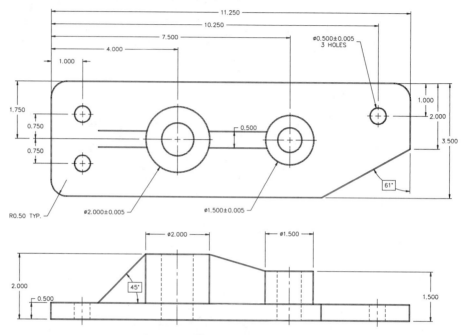

Figure 12–92

Purpose

The purpose of this tutorial is to place dimensions on the drawing of the two-view object illustrated In Figure 12–92.

System Settings

No special system settings need to be made for this drawing file.

Layers

The drawing file Dimex.Dwg has the following layers already created for this tutorial.

Name	Color	Linetype
Object	Magenta	Continuous
Hidden	Red	Hidden
Center	Yellow	Center
Dim	Yellow	Continuous

Suggested Commands

Open the drawing called 12_Dimex.Dwg. The following dimension commands will be used: DIMLINEAR, DIMCONTINUE, DIMBASELINE, DIMCENTER, DIMRADIUS, DIMDIAMETER, DIMANGULAR, and LEADER. All dimension commands may be chosen from the Dimension toolbar or the Dimension pull-down menu, or entered from the keyboard. Use the ZOOM command to get a closer look at details and features that are being dimensioned.

Whenever possible, substitute the appropriate command alias in place of the full AutoCAD command in each tutorial step; for example, use "CO" for the COPY command, "L" for the LINE command, and so on. The complete listing of all command aliases is located in Chapter 1, Table 1–2.

STEP I

To prepare for the dimensioning of the drawing, use the DDIM command to activate the Dimension Style Manager dialog box. Click on the New button, which activates the Create New Dimension Style dialog box in Figure 12–93. In the New Style Name area, enter "MECHANICAL." Click the Continue button to create the style. See Figure 12–93.

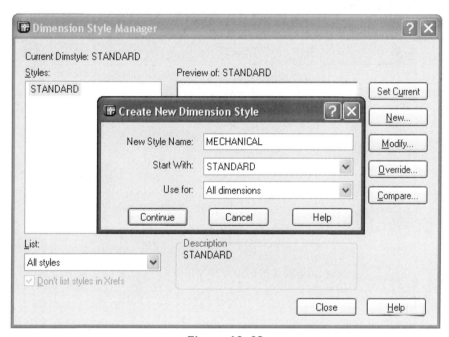

Figure 12–93

STEP 2

The New Dimension Style: MECHANI-CAL dialog box appears in Figure 12–94. Make the following changes in the Lines and Arrows tab: Change the size of the arrowheads from 0.18 to a new value of 0.12; Change the size of the Extension beyond dimension lines from a value of 0.18 to a new value of 0.07. Change the Offset from origin from a value of 0.06 to a new value of 0.12. Finally, be sure the Center Marks for Circles is set to Line. Your display should appear similar to Figure 12–94.

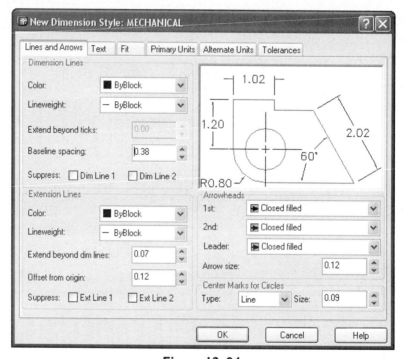

Figure 12–94

STEP 3

Click on the Text tab and change the Text color from Byblock to Green. Also change the Text height from a value of 0.18 to a new value of 0.12. Your display should appear similar to Figure 12–95. Click the OK button to return to the main Dimension Style Manager dialog box.

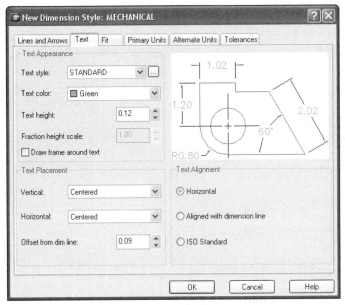

Figure 12–95

STEP 4

You will now create four dimension types and change various settings for each type. The four dimension types are Linear, Angular, Radial, and Diameter. Click the New button to display the Create New Dimen- sion Style dialog. In the Use for: box, click on Linear dimensions in Figure 12–96 and then click the Continue button. Any changes made to dimension settings will be applied only to linear dimensions.

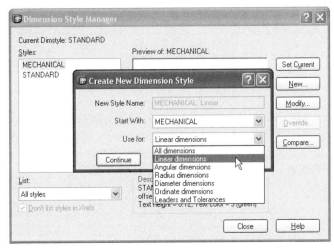

Figure 12–96

STEP 5

Click on the Primary Units tab and change the Precision for linear dimensions to 3 places. Your display should appear similar to Figure 12–97. Notice that the image in the preview box displays linear dimensions only to 3-decimal-place accuracy. Click the OK button to return to the main Dimension Style Manager dialog box.

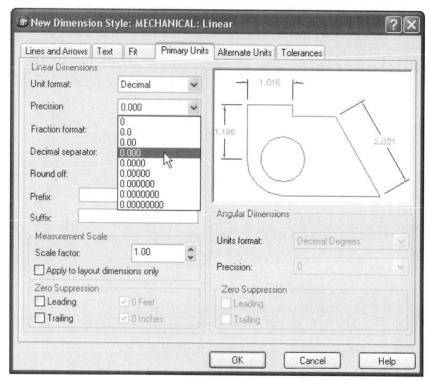

Figure 12–97

STEP 6

Notice in the Styles area of the Dimension Style Manager that a new style type called Linear is listed under MECHANICAL. Click the New button and create another dimension type by selecting Angular dimensions in Figure 12–98. Click the Continue button.

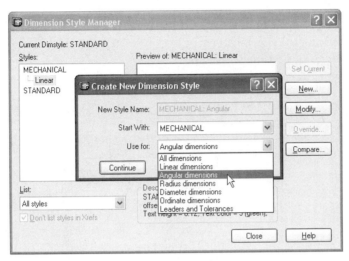

Figure 12–98

STEP 7

A rectangular box will be drawn around all angular dimensions to signify that this value is considered basic. Click on the Text tab and in the Text Appearance area, place a check for Draw frame around text. Notice the image in the Preview box shows only an angular dimension with a box placed around the text. Your display should appear similar to Figure 12–99. Click the OK button to return to the main Dimension Style Manager dialog box.

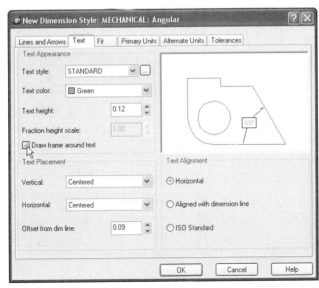

Figure 12–99

STEP 8

Create another new dimension type by clicking on Radius dimensions in Figure 12–100. Click the Continue button.

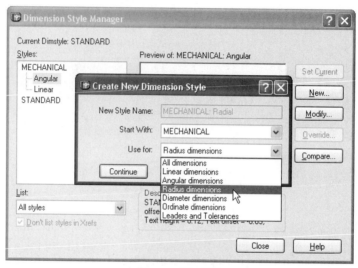

Figure 12–100

STEP 9

Click on the Lines and Arrows tab and change the Center Marks for Circles to None in Figure 12–101. Then click in the Fit tab and in the Fit options area, click on the radio button for Text in Figure 12–102 to have the text placed outside any radius dimension. Click the OK button and return to the main Dimension Style Manager dialog box.

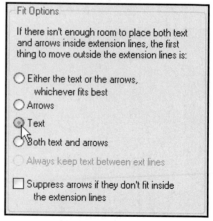

Figure 12–102

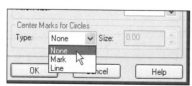

Figure 12–101

STEP 10

Create the last new dimension type by clicking on Diameter dimensions in Figure 12–103. Click the Continue button.

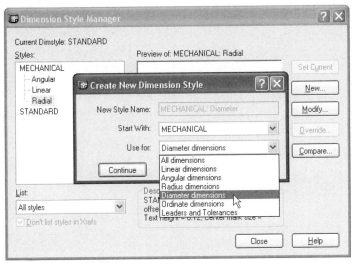

Figure 12–103

STEP 11

As with the Radius dimension type, click on the Lines and Arrows tab and change the Center Marks for Circles to None in Figure 12–104.

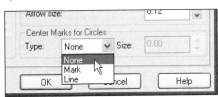

Figure 12–104

Then click in the Fit tab and in the Fit options area, click on the radio button for Both text and arrows in Figure 12–105, to have the text and arrows placed outside any diameter dimension.

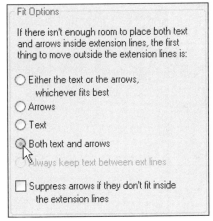

Figure 12–105

Click on the Primary Units tab and change the Precision to 3-decimal-place accuracy in Figure 12–106.

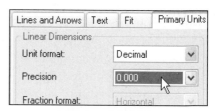

Figure 12–106

Finally, click on the Tolerances tab and change the Method to Symmetrical and the Upper value to 0.005 in Figure 12–107. Click the OK button and return to the main Dimension Style Manager dialog box.

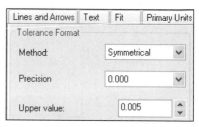

Figure 12–107

STEP 12

In the Dimension Style Manager dialog box, click on MECHANICAL in the Styles area. Then click on the Set Current button to make MECHANICAL the current dimension style. Your display should appear similar to Figure 12–108. Click the Close button to save all changes and return to the drawing.

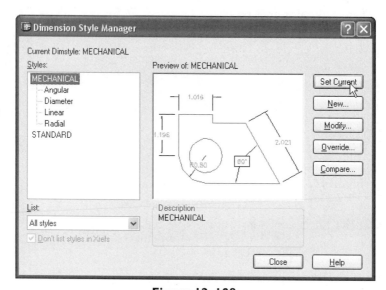

Figure 12–108

STEP 13

Begin placing center markers to identify the centers of all circular features in the Top view. Use the DIMCENTER command to perform this operation on circles "A" through "E" in Figure 12–109.

 Command: **DCE** *(For DIMCENTER)*

Select arc or circle: *(Select the edge of circle "A")*

Repeat the above procedure for circles "B," "C," "D," and "E" in Figure 12–109.

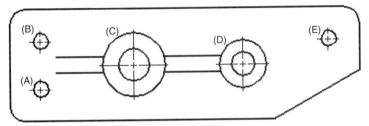

Figure 12–109

STEP 14

Use the QDIM command to place a string of baseline dimensions. First select the individual highlighted lines in Figure 12–110. When the group of dimensions previews as continued dimensions, change this grouping to Baseline and click a location to place the baseline group of dimensions, as shown in Figure 12–111.

 Command: **QDIM**

Select geometry to dimension: *(Select the lines "A" through "F" in Figure 12–110)*

Select geometry to dimension: *(Press ENTER to continue)*

Specify dimension line position, or [Continuous/Staggered/Baseline/ Ordinate/Radius/Diameter/ datumPoint/Edit]
<Continuous>: **B** *(For Baseline)*

Specify dimension line position, or [Continuous/Staggered/Baseline/ Ordinate/Radius/Diameter/ datumPoint/Edit]
<Baseline>: *(Pick a point on the screen to locate the baseline dimensions and exit the command)*

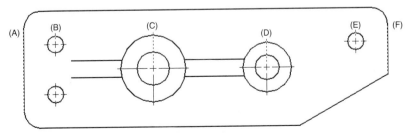

Figure 12–110

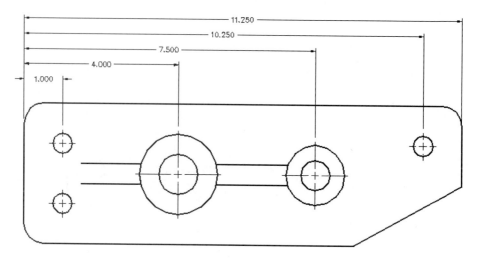

Figure 12–111

STEP 15

Magnify the left side of the Top view using the ZOOM command. Then use the DIMLINEAR command to place the 0.750 vertical dimension in Figure 12–112.

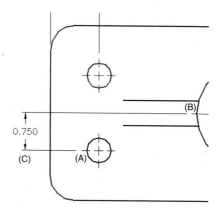

Figure 12–112

Command: **DLI** *(For DIMLINEAR)*

Specify first extension line origin or
    <select object>: *(Select the endpoint of the centerline at "A")*
Specify second extension line origin:
    (Select the endpoint of the centerline at "B")
Specify dimension line location or
[Mtext/Text/Angle/Horizontal/Vertical/
    Rotated]: *(Locate the dimension approximately at "C")*
Dimension text = 0.750

STEP 16

Use the DIMCONTINUE command to place the next dimension in line with the previous dimension in Figure 12–113.

 Command: **DCO** (For DIMCONTINUE)

Specify a second extension line origin
  or [Undo/Select] <Select>: (Select the
  endpoint of the centerline at "A")
Dimension text = 0.750
Specify a second extension line origin or
  [Undo/Select] <Select>: (Press ENTER to
  exit this mode)
Select continued dimension: (Press ENTER
  to exit this command)

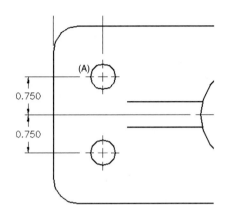

Figure 12–113

STEP 17

Use the DIMLINEAR command to place the 1.750 vertical dimension in Figure 12–114.

 Command: **DLI** (For DIMLINEAR)

Specify first extension line origin or
  <select object>: (Select the endpoint of
  the extension line at "A")
Specify second extension line origin:
  (Select the endpoint of the line at "B")
Specify dimension line location or
[Mtext/Text/Angle/Horizontal/Vertical/
  Rotated]: (Locate the dimension
  approximately at "C")
Dimension text = 1.750

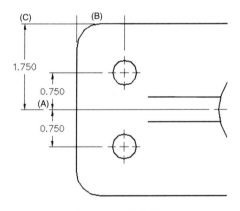

Figure 12–114

STEP 18

Use the PAN command to slide over to the right side of the Top view while keeping the same zoom percentage. Then use the QDIM command and select the highlighted lines in Figure 12–115. When the group of dimensions previews as continued dimensions, change this grouping to baseline and click a location to place the baseline group of dimensions in Figure 12–116.

Command: **QDIM**

Select geometry to dimension: *(Select lines "A" through "C" in Figure 12–115)*
Select geometry to dimension: *(Press ENTER to continue)*
Specify dimension line position, or [Continuous/Staggered/Baseline/ Ordinate/Radius/Diameter/datumPoint/ Edit] <Continuous>: **B** *(For Baseline)*
Specify dimension line position, or [Continuous/Staggered/Baseline/ Ordinate/Radius/Diameter/datumPoint/ Edit] <Baseline>: **P** *(For datumPoint)*

Select new datum point: (Select the endpoint at "A" again to identify a new datum point)
Specify dimension line position, or [Continuous/Staggered/Baseline/ Ordinate/Radius/Diameter/ datumPoint/Edit] <Baseline>: *(Pick a point on the screen to locate the baseline dimensions and exit the command)*

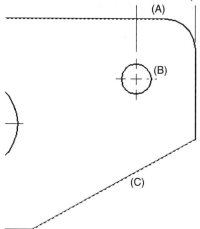

Figure 12–115

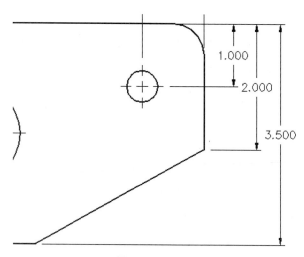

Figure 12–116

STEP 19

Use the DIMANGULAR command to place the 61° dimension in Figure 12–117.

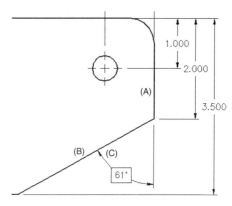

⬛ Command: **DAN** *(For DIMANGULAR)*

Select arc, circle, line, or <specify vertex>: *(Select the line at "A")*
Select second line: *(Select the line at "B")*
Specify dimension arc line location or [Mtext/Text/Angle]: *(Locate the angular dimension at "C")*
Dimension text = 61

Figure 12–117

STEP 20

Use the ZOOM command and the Extents option to display both the Front and Top views. Then use the DIMDIAMETER command to place two diameter dimensions in Figure 12–118.

⬛ Command: **DDI** *(For DIMDIAMETER)*

Select arc or circle: *(Select the circle at "A")*
Dimension text = 1.000

Specify dimension line location or [Mtext/Text/Angle]: *(Locate the diameter dimension at "B")*

⬛ Command: **DDI** *(For DIMDIAMETER)*

Select arc or circle: *(Select the circle at "C")*
Dimension text = 0.750
Specify dimension line location or [Mtext/Text/Angle]: *(Locate the diameter dimension at "D")*

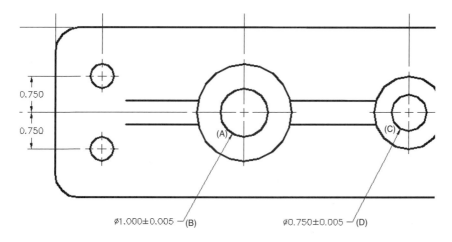

Figure 12–118

STEP 21

Place a radius dimension using the DIM-RADIUS command in Figure 12–119.

[icon] Command: **DRA** *(For DIMRADIUS)*

Select arc or circle: *(Select the arc at "A")*
Dimension text = 0.50
Specify dimension line location or
   [Mtext/Text/Angle]: *(Locate the radius
   dimension at "B")*

Because the two other arcs share the same radius value, use the DDEDIT command to edit this dimension value.

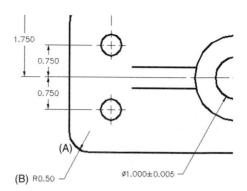

Figure 12–119

Clicking on the radius value activates the Multiline Text Editor dialog box. Add the note "TYP." for TYPICAL as the dimension suffix in Figure 12–120. Click the OK button to return to your drawing.

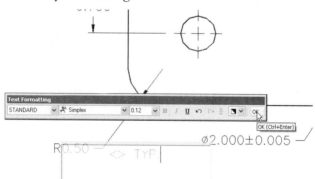

Figure 12–120

The results are displayed in Figure 12–121.

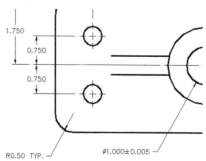

Figure 12–121

STEP 22

Place a diameter dimension using the DIM-DIAMETER command in Figure 12–122.

 Command: **DDI** *(For DIMDIAMETER)*

Select arc or circle: *(Select the circle at "A")*

Dimension text = 0.500

Specify dimension line location or [Mtext/ Text/Angle]: *(Locate the diameter dimension at "B")*

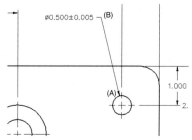

Figure 12–122

Since the two other smaller holes share the same diameter value, use the DDEDIT command to edit this dimension value. Clicking on the diameter value activates the Multiline Text Editor dialog box. Drop down to the next line and add the note "3 HOLES" in Figure 12–123. Click the OK button to return to your drawing.

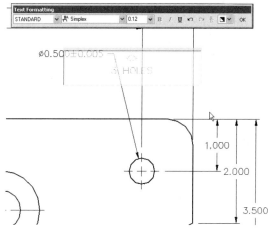

Figure 12–123

The results are illustrated in Figure 12–124.

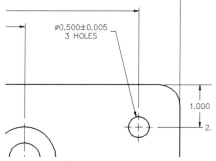

Figure 12–124

STEP 23

One more dimension needs to be placed in the Top view: the 0.50 width of the rib. Unfortunately, because of the placement of this dimension, extension lines will be drawn on top of the object's lines. This is considered poor practice. To remedy this, a dimension override will be created. To do this, activate the Dimension Style Manager dialog box and click on the Override button. This displays the Override Current Style: MECHANICAL dialog box. In the Lines and Arrows tab, place checks in the boxes to Suppress (turn off) Ext Line 1 and Ext Line 2 in the Extension Lines area (see Figure 12–125).

Click the OK button and notice the new <style overrides> listing under MECHANICAL in Figure 12–126. Notice also the preview image lacks extension lines. Click the Close button to return to the drawing.

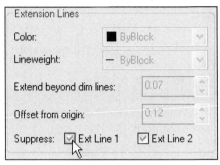

Figure 12–125

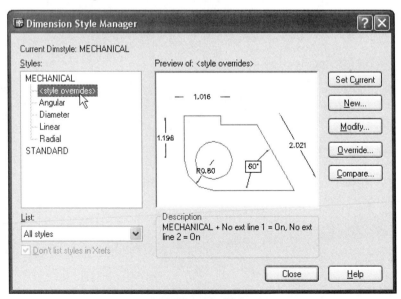

Figure 12–126

STEP 24

Pan to the middle area of the Top view and place the 0.500 rib width dimension in Figure 12–127 while in the dimension style override. This dimension should be placed without having any extension lines visible.

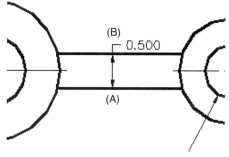

Figure 12–127

Command: **DLI** *(For DIMLINEAR)*
Specify first extension line origin or
   <select object>: **Nea**
to *(Select the line at "A')*
Specify second extension line origin: **Per**
to *(Select the line at "B")*
Specify dimension line location or
   [Mtext/Text/Angle/Horizontal/Vertical/
   Rotated]: *(Locate the dimension at "B")*
Dimension text = 0.500

STEP 25

The completed Top view, including dimensions, is illustrated in Figure 12–128. Use this figure to check that all dimensions have been placed and all features such as holes and fillets have been properly identified.

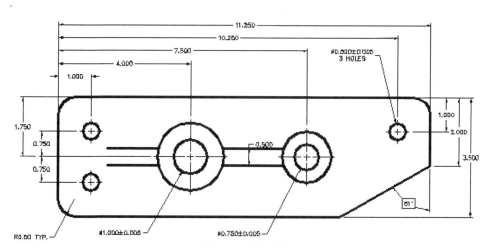

Figure 12–128

716

The Front view in Figure 12–129 will now be the focus for the next series of dimensioning steps. Again use the ZOOM and PAN commands whenever you need to magnify or slide to a better drawing view position.

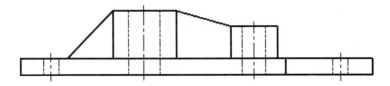

Figure 12–129

Before continuing, activate the Dimension Style Manager dialog box, click on the MECHANICAL style and then click on the Set Current button. An AutoCAD Alert box displays in Figure 12–130. Making MECHANICAL current will discard any style overrides. Click the OK button to discard the changes because you need to return to having extension lines visible in linear dimensions.

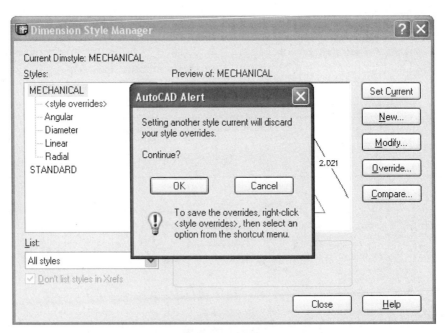

Figure 12–130

STEP 26

Place a horizontal linear dimension in Figure 12–131 using the DIMLINEAR command.

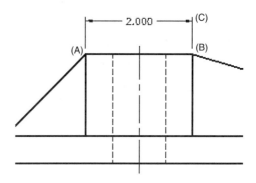

Figure 12–131

Command: **DLI** *(For DIMLINEAR)*
Specify first extension line origin or
  <select object>: *(Select the endpoint of the corner at "A")*
Specify second extension line origin:
  (Select the endpoint of the corner at "B")
Specify dimension line location or
  [Mtext/Text/Angle/Horizontal/Vertical/
  Rotated]: *(Locate the dimension at "C")*
Dimension text = 2.000

STEP 27

Place horizontal and vertical dimensions in Figure 12–132 using the DIMLINEAR command.

Command: **DLI** *(For DIMLINEAR)*
Specify first extension line origin or
  <select object>: *(Select the endpoint of the corner at "A")*
Specify second extension line origin:
  (Select the endpoint of the corner at "B")
Specify dimension line location or
  [Mtext/Text/Angle/Horizontal/Vertical/
  Rotated]: *(Locate the dimension at "C")*
Dimension text = 1.500

Command: **DLI** *(For DIMLINEAR)*
Specify first extension line origin or
  <select object>: *(Select the endpoint of the corner at "A")*
Specify second extension line origin:
  (Select the endpoint of the corner at "D")
Specify dimension line location or
  [Mtext/Text/Angle/Horizontal/Vertical/
  Rotated]: *(Locate the dimension at "E")*
Dimension text = 1.500

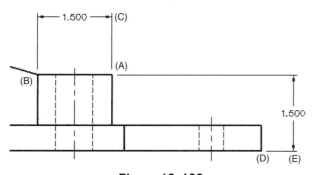

Figure 12–132

STEP 28

Use the QDIM command to place the vertical baseline dimensions in Figure 12–133. Place the 45° angular dimension.

 Command: **QDIM**

Select geometry to dimension: *(Select horizontal lines "A", "B", and "C")*
Select geometry to dimension: *(Press ENTER to continue)*
Specify dimension line position, or [Continuous/Staggered/Baseline/ Ordinate/Radius/Diameter/ datumPoint/Edit] <Baseline>: **P** *(For datumPoint)*
Select new datum point: *(Select the endpoint at "A" again to identify a new datum point)*
Specify dimension line position, or [Continuous/Staggered/Baseline/ Ordinate/Radius/Diameter/ datumPoint/Edit] <Baseline>: *(Locate the baseline dimensions at "D")*

 Command: **DAN** *(For DIMANGULAR)*

Select arc, circle, line, or <specify vertex>: *(Select the line at "E")*
Select second line: *(Select the line at "F")*
Specify dimension arc line location or [Mtext/Text/Angle]: *(Locate the angular dimension)*
Dimension text = 45

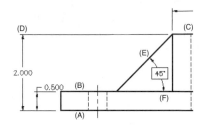

Figure 12–133

The remaining step is to add the diameter symbol to the 2.000 and 1.500 dimensions since the dimension is placed on a cylindrical object. Use the Multiline Text Editor in Figure 12–134 to accomplish this task.

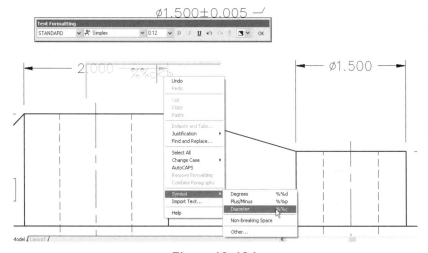

Figure 12–134

Once you have completed all dimensioning steps, your drawing should appear similar to Figure 12–135.

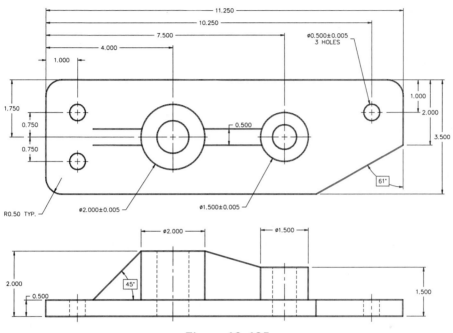

Figure 12–135

 Open the Exercise Manual PDF file for Chapter 12 on the accompanying CD for more tutorials and exercises.

 If you have the accompanying Exercise Manual, refer to Chapter 12 for more tutorials and exercises.

CHAPTER 13

Analyzing 2D Drawings

This chapter will show how a series of commands may be used to calculate distances and angles of selected objects. It will also show how surface areas may be calculated on complex geometric shapes. The following pages highlight all of AutoCAD's inquiry commands and show how you can use them to display useful information on an object or group of objects. You can invoke AutoCAD's inquiry commands by using the Inquiry toolbar, by entering the commands at the keyboard, or by choosing them from the Tools menu, as illustrated in Figures 13–1 and 13–2.

USING INQUIRY COMMANDS

The following is a list of the inquiry commands with a short description of each:

AREA   calculates the surface area after you specify a series of points or select a polyline or circle. You can add or subtract multiple objects to calculate the area with holes and cutouts.

DIST   calculates the distance between two points. Also provides the delta X,Y,Z coordinate values, the angle in the XY plane, and the angle from the XY plane.

HELP   provides online help for any command. You can type HELP at the keyboard, click on the F1 function key or choose the Help menu.

ID   displays the X,Y,Z absolute coordinate of a selected point.

LIST   displays key information on the object selected.

STATUS   displays important information on the current drawing.

TIME   displays the time spent in the drawing editor.

The following pages give a detailed description of how these commands are used in interaction with different AutoCAD objects.

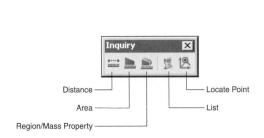

Figure 13–1

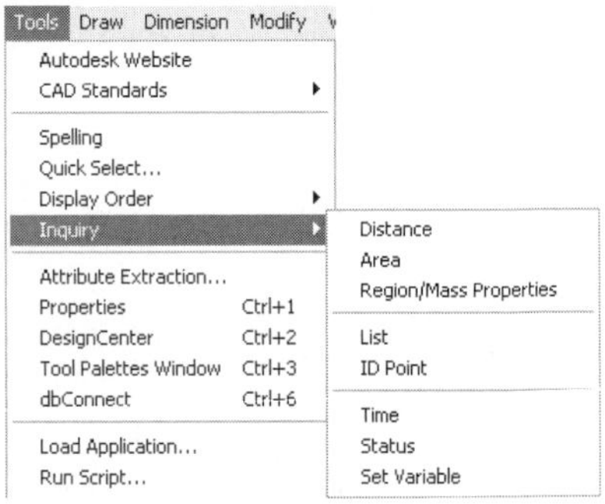

Figure 13–2

FINDING THE AREA OF AN ENCLOSED SHAPE

The AREA command is used to calculate the area through the selection of a series of points. Select the endpoints of all vertices of the image in Figure 13–3 with the OSNAP-Endpoint option. Once you have selected the first point along with the remaining points in either a clockwise or counterclockwise pattern, respond to the prompt "Next point:" by pressing ENTER to calculate the area of the shape. Along with the area is a calculation for the perimeter.

Try It! - Open the drawing file 13_Area1. Use Figure 13–3 and the prompt sequence below for finding the area by identifying a series of points.

Command: **AA** (For AREA)
Specify first corner point or [Object/Add/Subtract]: (Select the endpoint at "A")
Specify next corner point or press ENTER for total: (Select the endpoint at "B")
Specify next corner point or press ENTER for total: (Select the endpoint at "C")
Specify next corner point or press ENTER
    for total: (Select the endpoint at "D")
Specify next corner point or press
    ENTER for total: (Select the endpoint
    at "E")
Specify next corner point or press
    ENTER for total: (Select the endpoint
    at "A")
Specify next corner point or press
    ENTER for total: (Press ENTER to
    calculate the area)
Area = 25.25, Perimeter = 20.35

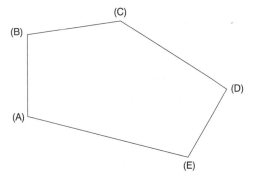

Figure 13–3

FINDING THE AREA OF AN ENCLOSED POLYLINE OR CIRCLE

The previous example showed how to find the area of an enclosed shape using the AREA command and identifying the corners and intersections of the enclosed area by a series of points. For a complex area, this could be a very tedious operation. As a result, the AREA command has a built-in Object option that will calculate the area and perimeter of a polyline and the area and circumference of a circle. Finding the area of a polyline can only be accomplished if one of the following conditions are satisfied:

- The shape must have already been constructed through the PLINE command.

- The shape must have already been converted to a polyline through the PEDIT command if originally constructed from individual objects.

 Try It! - Open the drawing file 13_Area2. Use Figure 13–4 and the following prompt sequence for finding the area of both shapes.

 Command: **AA** *(For AREA)*

Specify first corner point or [Object/Add/Subtract]: **O** *(For Object)*
Select objects: *(Select the polyline at "A")*
Area = 24.88, Perimeter = 19.51

 Command: **AA** *(For AREA)*

Specify first corner point or [Object/Add/Subtract]: **O** *(For Object)*
Select objects: *(Select the circle at "B")*
Area = 7.07, Circumference = 9.42

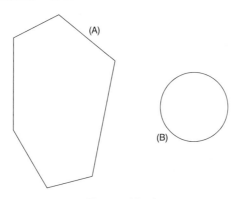

Figure 13–4

 Try It! - Open the drawing file 13_Extrude1. In Figure 13–5, either convert all line segments into a single polyline object or use the boundary command to trace a polyline on the top of all the line segments. In either case, answer Question 1 regarding 13_Extrude1.

724

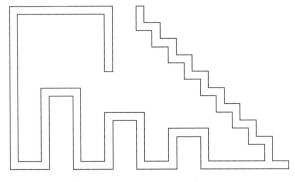

Figure 13–5

Question 1

What is the total surface area of 13_Extrude1?

(A) 9.1242

(B) 9.1246

(C) 9.1250

(D) 9.1254

(E) 9.1258

The total surface area of the 13_Extrude1 is "C," 9.1250.

 Try It! - Open the drawing file 13_Extrude2. In Figure 13–6, either convert all line segments into a single polyline object or use the BOUNDARY command to trace a polyline on the top of all the line segments. In either case, answer Question 1 regarding 13_Extrude2.

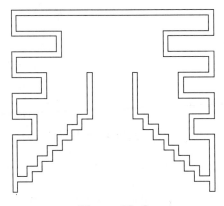

Figure 13–6

Question 1

What is the total surface area of 13_Extrude2?

(A) 28.9362

(B) 28.9366

(C) 28.9370

(D) 28.9374

(E) 28.9378

The total surface area of the 13_Extrude2 is "A," 28.9362.

 ## FINDING THE AREA OF A SURFACE BY SUBTRACTION

You may wish to calculate the area of a shape with holes cut through it. The steps you use to calculate the total surface area: (1) calculate the area of the outline and (2) subtract the objects inside the outline. All individual objects, except circles, must first be converted to polylines through the PEDIT command. Next, find the overall area and add it to the database using the Add mode of the AREA command. Exit the Add mode and remove the inner objects using the Subtract mode of the AREA command. Remember that all objects must be in the form of a circle or polyline. This means that the inner shape at "B" in Figure 13–7 must also be converted to a polyline through the PEDIT command before the area is calculated. Care must be taken when selecting the objects to subtract. If an object is selected twice, it is subtracted twice and may yield an inaccurate area in the final calculation.

For the image in Figure 13–7, the total area with the circle and rectangle removed is 30.4314.

 Try It! - Open the drawing file 13_Area3. Use Figure 13–7 and the following prompt sequence to verify this area calculation.

 Command: **AA** *(For AREA)*
Specify first corner point or [Object/Add/Subtract]: **A** *(For Add)*
Specify first corner point or [Object/Subtract]: **O** *(For Object)*
(ADD mode) Select objects: *(Select the polyline at "A")*
Area = 47.5000, Perimeter = 32.0000
Total area = 47.5000
(ADD mode) Select objects: *(Press ENTER to exit ADD mode)*
Specify first corner point or [Object/Subtract]: **S** *(For Subtract)*
Specify first corner point or [Object/Add]: **O** *(For Object)*
(SUBTRACT mode) Select objects: *(Select the polyline at "B")*
Area = 10.0000, Perimeter = 13.0000
Total area = 37.5000

(SUBTRACT mode) Select objects: *(Select the circle at "C")*
Area = 7.0686, Circumference = 9.4248
Total area = 30.4314
(SUBTRACT mode) Select objects: *(Press ENTER to exit SUBTRACT mode)*
Specify first corner point or [Object/Add]: *(Press ENTER to exit this command)*

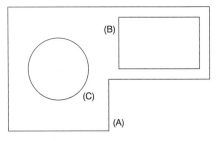

Figure 13–7

 Try It! - Open the drawing file 13_Shield. In Figure 13–8, use the BOUNDARY command to trace a polyline on the top of all the line segments. Subtract all four-sided shapes from the main shape using the AREA command. Answer Question 1 regarding 13_Shield.

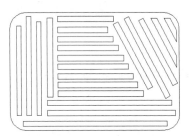

Figure 13–8

Question 1

What is the total surface area of 13_Shield with all inner four-sided shapes removed?

(A) 44.2242

(B) 44.2246

(C) 44.2250

(D) 44.2254

(E) 44.2258

The total surface area of the 13_Shield is "B," 44.2246.

THE DIST (DISTANCE) COMMAND

The DIST command calculates the linear distance between two points on an object, whether it be the distance of a line, the distance between two points, or the distance from the quadrant of one circle to the quadrant of another circle. The following information is also supplied when you use the DIST command: the angle in the XY plane, the angle from the XY plane, and the delta X,Y,Z coordinate values. The angle in the XY plane is given in the current angular mode set by the Drawing Units dialog box. The delta X,Y,Z coordinate is a relative coordinate value taken from the first point identified by the DIST command to the second point.

Try It! - Open the drawing file 13_Distance. Use Figure 13–9 and the prompt sequences below for the DIST command.

 Command: **DI** *(For DIST)*

Specify first point: *(Select the endpoint at "A")*
Specify second point: *(Select the endpoint at "B")*
Distance = 6.36, Angle in XY Plane = 45.00, Angle from XY Plane = 0.00
Delta X = 4.50, Delta Y = 4.50, Delta Z = 0.00

 Command: **DI** *(For DIST)*

Specify first point: *(Select the endpoint at "C")*
Specify second point: *(Select the endpoint at "D")*
Distance = 9.14, Angle in XY Plane = 192.75, Angle from XY Plane = 0.00
Delta X = -8.91, Delta Y = -2.02, Delta Z = 0.00

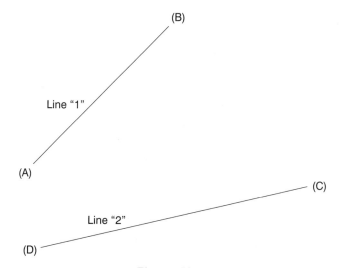

Figure 13–9

INTERPRETATION OF ANGLES USING THE DIST COMMAND

Previously, it was noted that the DIST command yields information regarding distance, delta X,Y coordinate values, and angle information. Of particular interest is the angle in the XY plane formed between two points. In Figure 13–10, picking the endpoint of the line segment at "A" as the first point followed by the endpoint of the line segment at "B" as the second point displays an angle of 42°. This angle is formed from an imaginary horizontal line drawn from the endpoint of the line segment at "A" in the zero direction.

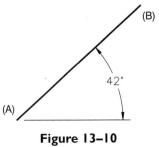

Figure 13–10

Take care when using the DIST command to find an angle on an identical line segment illustrated in Figure 13–11, as with the example in Figure 13–10. However, notice that the two points for identifying the angle are selected differently. With the DIST command, you select the endpoint of the line segment at "B" as the first point, followed by the endpoint of the segment at "A" for the second point. A new angle in the XY plane of 222° is formed. In Figure 13–11, the angle is calculated by the construction of a horizontal line from the endpoint at "B," the new first point of the DIST command. This horizontal line is also drawn in the zero direction. Notice the relationship of the line segment to the horizontal baseline. Be careful when identifying the order of line segment endpoints for extracting angular information.

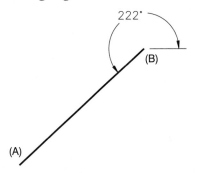

Figure 13–11

THE ID (IDENTIFY) COMMAND

The ID command is probably one of the more straightforward of the inquiry commands. ID can stand for "Identify" or "Locate Point" and allows you to obtain the current absolute coordinate listing of a point along or near an object.

In Figure 13–12, the coordinate value of the center of the circle at "A" was found through the use of ID and the OSNAP-Center mode. The coordinate value of the starting point of text string "B" was found with ID and the OSNAP-Insert mode. The coordinate value of the endpoint of line segment "C" was found with ID and the OSNAP-Endpoint mode. The coordinate value of the midpoint of line segment at "CD" was found with ID and the OSNAP-Midpoint mode. Finally, the coordinate value of the current position of point "E" was found with ID and the OSNAP-Node mode.

 Try It! - Open the drawing file 13_ID. Follow the prompt sequences below for calculating the X,Y,Z coordinate point of the objects in Figure 13–12.

 Command: **ID**
Specify point: **Cen**
of *(Select the edge of circle "A")*
X = 2.00 Y = 7.00 Z = 0.00

 Command: **ID**
Specify point: **Ins**
of *(Select the text at "B")*
X = 5.54 Y = 7.67 Z = 0.00

 Command: **ID**
Specify point: **End**
of *(Select the line at "C")*
X = 8.63 Y = 4.83 Z = 0.00

 Command: **ID**
Specify point: **Mid**
of *(Select line "CD")*
X = 5.13 Y = 3.08 Z = 0.00

 Command: **ID**
Specify point: **Nod**
of *(Select the point at "E")*
X = 9.98 Y = 1.98 Z = 0.00

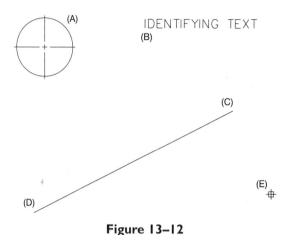

IDENTIFYING TEXT

Figure 13–12

THE LIST COMMAND

Use the LIST command to obtain information about an object or group of objects. In Figure 13–13, two rectangles are displayed along with a circle. However, are the rectangles made up of individual line segments or a polyline object? Using the LIST command on each object informs you that the first rectangle at "A" is a polyline, the circle lists as a circle, and the second rectangle is actually a block reference. In addition to the object type, you also can obtain key information such as the layer that the object resides on, area and perimeter information for polylines, and circumference information for circles.

Try It! - Open the drawing file 13_List. Study the prompt sequence below for using the LIST command.

Command: **LI** *(For LIST)*
Select objects: *(Select the objects at "A", "B", and "C" in order)*
Select objects: *(Press ENTER to list the information on each object)*
LWPOLYLINE Layer: "0"
Space: Model space
Handle = 2B
Closed
Constant width   0.0000
area   3.3835
perimeter   7.6235
at point X=   4.7002 Y=   4.1846 Z=   0.0000
at point X=   7.1049 Y=   4.1846 Z=   0.0000
at point X=   7.1049 Y=   5.5916 Z=   0.0000
at point X=   4.7002 Y=   5.5916 Z=   0.0000

CIRCLE   Layer:"0"
Space: Model space
Handle = 2B
center point, X= 7.6879 Y= 4.8881 Z= 0.0000
radius   1.4719
circumference   9.2479
area   6.8058
BLOCK REFERENCE Layer:"0"
Space: Model space
Handle = 31
Handle = 34
"rec"
at point, X= 10.6756 Y= 4.1846 Z= 0.0000
X scale factor   -1.0000
Y scale factor   1.0000
rotation angle   0
Z scale factor   1.0000

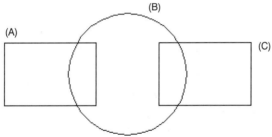

Figure 13–13

USING THE STATUS COMMAND

Once inside a large drawing, sometimes it becomes difficult to keep track of various settings that have been changed from their default values to different values required by the drawing. To obtain a listing of these important settings contained in the current drawing file, use the STATUS command. Once the STATUS command is invoked, the graphics screen changes to a text screen displaying the information shown in the following prompt sequence.

Command: **STATUS**
793 objects in D:\BUILDING.DWG

Model space limits are	X: 0'-0"	Y: 0'-0"	(Off)
X: 120'-0"	Y: 90'-0"		
Model space uses	X: 0'-0"	Y: 0'-0"	
X: 120'-0"	Y: 90'-0"		
Display shows	X: -0'-0 3/16"	Y: -0'-0 1/8"	
X: 173'-5 3/8"	Y: 90'-0 1/8"		

Insertion base is	X: 0'-0"	Y: 0'-0"	Z: 0'-0"
Snap resolution is	X: 3'-0"	Y: 3'-0"	
Grid spacing is	X: 0'-0"	Y: 0'-0"	
Current space:	Model space		
Current layout:	Model		
Current layer:	"0"		
Current color:	BYLAYER — 7 (white)		
Current linetype:	BYLAYER — "CONTINUOUS"		
Current lineweight:	BYLAYER		
Current plot style:	ByLayer		
Current elevation:	0'-0"	thickness: 0'-0"	

Fill on Grid off Ortho off Qtext off Snap off Tablet off

Object snap modes:	Center, Endpoint, Intersection, Quadrant, Extension
Free dwg disk (D:) space:	1119.2 MBytes
Free temp disk (C:) space:	422.0 MBytes
Free physical memory:	6.6 Mbytes (out of 63.4M).
Free swap file space:	76.1 Mbytes (out of 127.7M).

THE TIME COMMAND

The TIME command provides the following information:

- The current date and time.
- The date and time the drawing was created.
- The last time the drawing was updated.
- The total time spent editing the drawing so far.
- The total time spent in AutoCAD (not necessarily in a particular drawing).
- The current automatic save time interval.

CURRENT TIME

Displays the current date and time.

CREATED

This date and time value is set when you use the NEW command to create a new drawing file. This value is also set to the current date and time whenever a drawing is saved under a different name with the SAVE or SAVEAS commands.

LAST UPDATED

This data consists of the date and time the current drawing was last updated. This value updates itself when you use the SAVE command.

TOTAL EDITING TIME

This represents the total time spent editing the drawing. The timer is always updating itself and cannot be reset to a new or different value.

ELAPSED TIMER

This timer runs while AutoCAD is in operation, and you can turn it on or off or reset it.

NEXT AUTOMATIC SAVE IN

This timer displays when the next automatic save will occur. This value is controlled by the system variable SAVETIME. If this system variable is set to zero, the automatic save utility is disabled. If the timer is set to a nonzero value, the timer displays when the next automatic save will take place. The increment for automatic saving is in minutes.

Command: **TIME**
Current time: Wednesday, May 26, 1999 at 10:10:32:122 PM
Times for this drawing:
Created: Friday, November 12, 1993 at 3:03:11:980 PM
Last updated: Sunday, January 24, 1999 at 11:41:01:087 PM
Total editing time: 0 days 00:31:32.760
Elapsed timer (on):   0 days 00:31:32.760
Next automatic save in: <no modifications yet>

TUTORIAL EXERCISE: C-LEVER.DWG

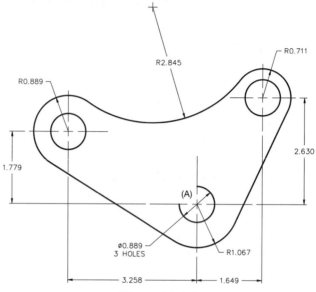

Figure 13–14

Purpose

This tutorial is designed to show you various methods of constructing the C-Lever object in Figure 13–14. Numerous questions will be asked about the object, requiring the use of the AREA, DIST, ID, and LIST commands.

System Settings

Use the Drawing Units dialog box and change the number of decimal places past the zero from four units to three units. Keep the default drawing limits at (0.000,0.000) for the lower-left corner and (12.000,9.000) for the upper-right corner. Check to see that the following Object Snap modes are already set: Endpoint, Extension, Intersection, Center.

Layers

Create the following layers with the format:

Name	Color	Linetype
Boundary	Magenta	Continuous
Object	Yellow	Continuous

Suggested Commands

Begin drawing the C-Lever with point "A" illustrated in Figure 13–14 at absolute coordinate 7.000,3.375. Begin laying out all circles. Then draw tangent lines and arcs. Use the TRIM command to clean up unnecessary objects. To prepare to answer the AREA command question, convert the profile of the C-Lever to a polyline using the BOUNDARY command. Other questions pertaining to distances, angles, and point identifications follow. Do not dimension this drawing.

Whenever possible, substitute the appropriate command alias in place of the full AutoCAD command in each tutorial step; for example, use "CP" for the COPY command, "L" for the LINE command, and so on. The complete listing of all command aliases is located in Chapter 1, Table 1–2.

STEP 1

Make the Object layer current. Then construct one circle of 0.889 diameter with the center of the circle at absolute coordinate 7.000,3.375 (see Figure 13–15). Construct the remaining circles of the same diameter by using the COPY command with the Multiple option. Use of the @ symbol for the base point in the COPY command identifies the last known point, which in this case is the center of the first circle drawn at coordinate 7.000,3.375.

Command: **C** *(For CIRCLE)*
Specify center point for circle or [3P/2P/ Ttr (tan tan radius)]: **7.000,3.375**
Specify radius of circle or [Diameter]: **D** *(For Diameter)*

Specify diameter of circle: **0.889**

Command: **CP** *(For COPY)*
Select objects: **L** *(For Last)*
Select objects: *(Press ENTER to continue)*
Specify base point or displacement, or [Multiple]: **M** *(For Multiple)*
Specify base point: **@**
Specify second point of displacement or <use first point as displacement>: **@1.649,2.630**
Specify second point of displacement or <use first point as displacement>: **@-3.258,1.779**
Specify second point of displacement or <use first point as displacement>: *(Press ENTER to exit this command)*

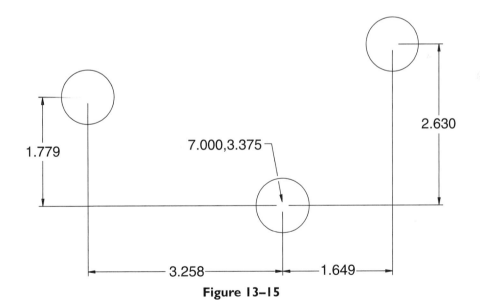

Figure 13–15

STEP 2

Construct three more circles (see Figure 13–16). Even though these objects actually represent arcs, circles will be drawn now and trimmed later to form the arcs.

⊙ Command: **C** *(For CIRCLE)*
Specify center point for circle or [3P/2P/ Ttr (tan tan radius)]: *(Select the edge of the circle at "A" to snap to its center)*
Specify radius of circle or [Diameter] <0.445>: **1.067**

⊙ Command: **C** *(For CIRCLE)*
Specify center point for circle or [3P/2P/ Ttr (tan tan radius)]: *(Select the edge of the circle at "B" to snap to its center)*
Specify radius of circle or [Diameter] <1.067>: **0.889**

⊙ Command: **C** *(For CIRCLE)*
Specify center point for circle or [3P/2P/ Ttr (tan tan radius)]: *(Select the edge of the circle at "C" to snap to its center)*
Specify radius of circle or [Diameter] <0.889>: **0.711**

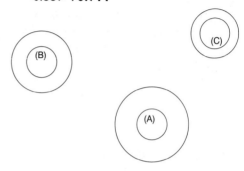

Figure 13–16

STEP 3

Construct lines tangent to the three outer circles as shown in Figure 13–17.

▱ Command: **L** *(For LINE)*
Specify first point: **Tan**
to *(Select the outer circle near "A")*
Specify next point or [Undo]: **Tan**
to *(Select the outer circle near "B")*
Specify next point or [Undo]: *(Press ENTER to exit this command)*

▱ Command: **L** *(For LINE)*
Specify first point: **Tan**
to *(Select the outer circle near "C")*
Specify next point or [Undo]: **Tan**

to *(Select the outer circle near "D")*
Specify next point or [Undo]: *(Press ENTER to exit this command)*

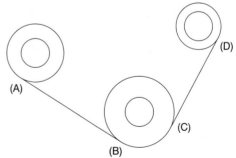

Figure 13–17

STEP 4

Construct a circle tangent to the two circles in Figure 13–18 using the CIRCLE command with the Tangent-Tangent-Radius option (TTR).

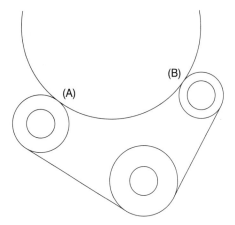

Figure 13–18

 Command: **C** *(For CIRCLE)*

Specify center point for circle or [3P/2P/ Ttr (tan tan radius)]: **TTR**

Specify point on object for first tangent of circle: *(Select the outer circle near "A")*

Specify point on object for second tangent of circle: *(Select the outer circle near "B")*

Specify radius of circle <0.711>: **2.845**

STEP 5

Use the TRIM command to clean up and form the finished drawing. Pressing ENTER at the "Select objects:" prompt will select all objects as cutting edges although they will not highlight (see Figure 13–19). Study the following prompts for selecting the objects to trim.

 Command: **TR** *(For TRIM)*

Current settings: Projection=UCS Edge=None

Select cutting edges ...

Select objects: *(Press ENTER which will select all objects cutting edges)*

Select object to trim or shift-select to extend or [Project/Edge/Undo]: *(Select the circle at "A")*

Select object to trim or shift-select to extend or [Project/Edge/Undo]: *(Select the circle at "B")*

Select object to trim or shift-select to extend or [Project/Edge/Undo]: *(Select the circle at "C")*

Select object to trim or shift-select to extend or [Project/Edge/Undo]: *(Select the circle at "D")*

Select object to trim or shift-select to extend or [Project/Edge/Undo]: *(Press ENTER to exit this command)*

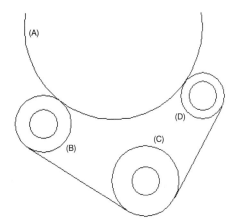

Figure 13–19

CHECKING THE ACCURACY OF C-LEVER.DWG

Once the C-Lever has been constructed, answer the following questions to determine the accuracy of this drawing. Use Figure 13–20 to assist in answering the questions.

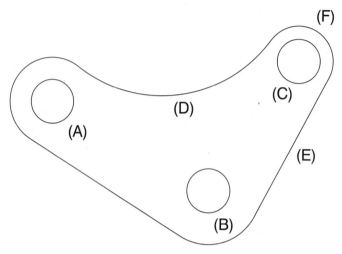

Figure 13–20

1. What is the total area of the C-Lever with all three holes removed?

   (A) 13.744

   (B) 13.749

   (C) 13.754

   (D) 13.759

   (E) 13.764

2. What is the total distance from the center of circle "A" to the center of circle "B"?

   (A) 3.692

   (B) 3.697

   (C) 3.702

   (D) 3.707

   (E) 3.712

3. What is the angle formed in the XY plane from the center of circle "C" to the center of circle "B"?

   (A) 223°

   (B) 228°

   (C) 233°

   (D) 238°

   (E) 243°

4. What is the delta X,Y distance from the center of circle "C" to the center of circle "A"?

   (A) -4.907,-0.851

   (B) -4.907,-0.856

   (C) -4.907,-0.861

   (D) -4.907,-0.866

   (E) -4.907,-0.871

5. What is the absolute coordinate value of the center of arc "D"?

(A) 5.869,8.218

(B) 5.869,8.223

(C) 5.869,8.228

(D) 5.869,8.233

(E) 5.869,8.238

6. What is the total length of line "E"?

(A) 3.074

(B) 3.079

(C) 3.084

(D) 3.089

(E) 3.094

7. What is the total length of arc "F"?

(A) 2.051

(B) 2.056

(C) 2.061

(D) 2.066

(E) 2.071

A solution for each question follows, complete with the method used to arrive at the answer. Apply these methods to any type of drawing that requires the use of inquiry commands.

 Tip: Before performing an area calculation that requires you to remove numerous objects, turn the BLIPMODE variable on. This will place a small mark on the object you select. In this way, you can keep better track of the objects you select. When finished, turn BLIPMODE off.

SOLUTIONS TO THE QUESTIONS ON C-LEVER

Question 1

What is the total area of the C-Lever with all three holes removed?

(A) 13.744   (D) 13.759

(B) 13.749   (E) 13.764

(C) 13.754

First make the Boundary layer current. Then use the BOUNDARY command and pick a point inside the object at "A" in Figure 13–21. This will trace a polyline around all closed objects on the Boundary layer.

Command: **BO** *(For BOUNDARY)*
(When the Boundary Creation dialog box appears, click on the Pick Points button.)
Select internal point: *(Pick a point inside of the object at "Y")*

Selecting everything...
Selecting everything visible...
Analyzing the selected data...
Analyzing internal islands...
Select internal point: *(Press* ENTER *to create the boundaries)*
BOUNDARY created 4 polylines

Next, turn off the Object layer. All objects on the Boundary layer should be visible. Then use the AREA command to add and subtract objects to arrive at the final area of the object.

 Command: **AA** *(For AREA)*
Specify first corner point or [Object/Add/Subtract]: **A** *(For Add)*
Specify first corner point or [Object/Subtract]: **O** *(For Object)*

(ADD mode) Select objects: *(Select the edge of the shape near "X")*
Area = 15.611, Perimeter = 17.771
Total area = 15.611
(ADD mode) Select objects: *(Press ENTER to continue)*
Specify first corner point or [Object/Subtract]: **S** *(For Subtract)*
Specify first corner point or [Object/Add]: **O** *(For Object)*
(SUBTRACT mode) Select objects: *(Select circle "A")*
Area = 0.621, Perimeter = 2.793
Total area = 14.991
(SUBTRACT mode) Select objects: *(Select circle "B")*
Area = 0.621, Perimeter = 2.793
Total area = 14.370
(SUBTRACT mode) Select objects: *(Select circle "C")*

Area = 0.621, Perimeter = 2.793
Total area = 13.749
(SUBTRACT mode) Select objects: *(Press ENTER to continue)*
Specify first corner point or [Object/Add]: *(Press ENTER to exit this command)*

The total area of the C-Lever with all three holes removed is (B), 13.749.

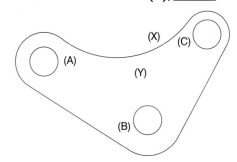

Figure 13–21

Question 2

What is the total distance from the center of circle "A" to the center of circle "B"?

(A) 3.692   (D) 3.707

(B) 3.697   (E) 3.712

(C) 3.702

Use the DIST (Distance) command to calculate the distance from the center of circle "A" to the center of circle "B" in Figure 13–22. Be sure to use the OSNAP-Center mode for locating the centers of all circles.

Command: **DI** *(For DIST)*
Specify first point: *(Select the edge of the circle at "A")*

Specify second point: *(Select the edge of the circle at "B")*
Distance = 3.712, Angle in XY Plane = 331, Angle from XY Plane = 0
Delta X = 3.258, Delta Y = -1.779, Delta Z = 0.000

The total distance from the center of circle "A" to the center of circle "B" is (E), 3.712.

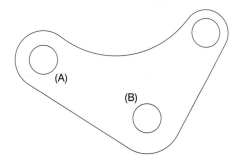

Figure 13–22

Question 3

What is the angle formed in the XY plane from the center of circle "C" to the center of circle "B"?

(A) 223° (D) 238°

(B) 228° (E) 243°

(C) 233°

Use the DIST (Distance) command to calculate the angle from the center of circle "C" to the center of circle "B" in Figure 13–23. Be sure to use the OSNAP-Center mode for locating the centers of all circles.

 Command: **DI** *(For DIST)*

Specify first point: *(Select the edge of the circle at "C")*

Specify second point: *(Select the edge of the circle at "B")*

Distance = 3.104, Angle in XY Plane = 238, Angle from XY Plane = 0

Delta X = -1.649, Delta Y = -2.630, Delta Z = 0.000

The angle formed in the XY plane from the center of circle "C" to the center of circle "B" is (D), 238°.

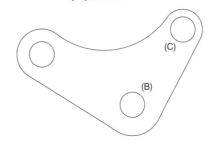

Figure 13–23

Question 4

What is the delta X,Y distance from the center of circle "C" to the center of circle "A"?

(A) -4.907,-0.851 (D) -4.907,-0.866

(B) -4.907,-0.856 (E) -4.907,-0.871

(C) -4.907,-0.861

Use the DIST (Distance) command to calculate the delta X,Y distance from the center of circle "C" to the center of circle "A" in Figure 13–24. Be sure to use the OSNAP-Center mode. Notice that additional information is given when you use the DIST command. For the purpose of this question, we will only be looking for the delta X,Y distance. The DIST command will display the relative X,Y,Z distances. Since this is a 2D problem, only the X and Y values will be used.

 Command: **DI** *(For DIST)*

Specify first point: *(Select the edge of the circle at "C")*

Specify second point: *(Select the edge of the circle at "A")*

Distance = 4.980, Angle in XY Plane = 190, Angle from XY Plane = 0

Delta X = -4.907, Delta Y = -0.851, Delta Z = 0.000

The delta X,Y distance from the center of circle "C" to the center of circle "A" is (A), -4.907,-0.851.

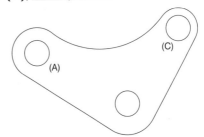

Figure 13–24

742

Question 5

What is the absolute coordinate value of the center of arc "D"?

    (A) 5.869,8.218   (D) 5.869,8.233

    (B) 5.869,8.223   (E) 5.869,8.238

    (C) 5.869,8.228

The ID command is used to get the current absolute coordinate information on a desired point (see Figure 13–25). This command will display the X,Y,Z coordinate values. Since this is a 2D problem, only the X and Y values will be used.

 Command: **ID**

Specify point: **Cen**

of *(Select the edge of the arc at "D")*

X = 5.869   Y = 8.223   Z = 0.000

The absolute coordinate value of the center of arc "D" is (B), 5.869,8.223.

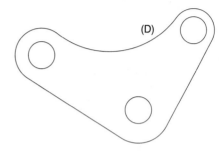

Figure 13–25

Question 6

What is the total length of line "E"?

    (A) 3.074   (D) 3.089

    (B) 3.079   (E) 3.094

    (C) 3.084

Use the DIST (Distance) command to find the total length of line "E" in Figure 13–26. Be sure to use the OSNAP-Endpoint mode. Notice that additional information is given when you use the DIST command. For the purpose of this question, we will only be looking for the distance.

 Command: **DI** *(For DIST)*

Specify first point: (Select the endpoint of the line at "X")

Specify second point: (Select the endpoint of the line at "Y")

Distance = 3.084, Angle in XY Plane = 64, Angle from XY Plane = 0

Delta X = 1.328, Delta Y = 2.783, Delta Z = 0.000

The total length of line "E" is (C), 3.084.

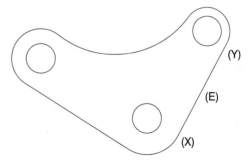

Figure 13–26

742

Question 7

What is the total length of arc "F"?

(A) 2.051   (D) 2.066

(B) 2.056   (E) 2.071

(C) 2.061

The LIST command is used to calculate the lengths of arcs. However, a little preparation is needed before you perform this operation. If arc "F" is selected as in Figure 13–27, notice that the entire outline is selected because it is a polyline. Use the EXPLODE command to break the outline into individual objects. Use the LIST command to get a listing of the arc length. See also Figure 13–28.

 Command: **X** *(For EXPLODE)*

Select objects: *(Select the edge of the dashed polyline in Figure 13–27)*
Select objects: *(Press ENTER to perform the explode operation)*

 Command: **LI** *(For LIST)*

Select objects: *(Select the edge of the arc at "F" in Figure 13–28)*
Select objects: *(Press ENTER to continue)*
ARC   Layer: "Boundary"
Space: Model space
Handle = 49
center point, X=   8.649 Y=   6.005 Z=
    0.000
radius   0.711
start angle   334
end angle   141
length   2.071

The total length of arc "F" is (E), 2.071.

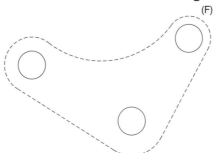

Figure 13–27

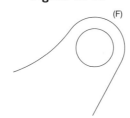

Figure 13–28

 Open the Exercise Manual PDF file for Chapter 13 on the accompanying CD for more tutorials and exercises.

 If you have the accompanying Exercise Manual, refer to Chapter 13 for more tutorials and exercises.

CHAPTER 14

Section Views

This chapter will cover Section Views, which are created by slicing an object along a cutting plane in order to view its interior details. The basics of section views will be covered first. Full, half, assembly, aligned, and offset sections will be shown. Other special section topics include sectioning ribs, creating broken sections, revolved sections, removed sections, and isometric sections. After covering the basics of sections, the chapter will shift to a discussion on how AutoCAD crosshatches objects through the Boundary Hatch dialog box (bhatch command). You will be able to select from a collection of many hatch patterns including gradient patterns for special effects when adding hatch patterns to your drawings. Hatch scaling and angle considerations will also be covered along with associative crosshatching. Once hatching exists in a drawing, it can be easily modified through the Hatch Edit dialog box. Automatic hatch scaling in a drawing layout (Paper Space) will also be covered.

SECTION VIEW BASICS

Principles of orthographic projections remain the key method for the production of engineering drawings. As these drawings get more complicated in nature, the job of the designer becomes more challenging in the interpretation of views, especially where hidden features are involved. The concept of slicing a view to expose these interior features is the purpose of performing a section. Figure 14–1 is a pictorial representation of a typical flange consisting of eight bolt holes and counterbore hole in the center.

The drawings in Figure 14–2 show a typical solution to a multiview problem complete with Front and Side views. The Front view displaying the eight bolt holes is obvious to interpret; however, the numerous hidden lines in the Side view make the drawing difficult to understand, and this is considered a relatively simple drawing. To relieve the confusion associated with a drawing too difficult to understand because of numerous hidden lines, a section drawing is made for the part.

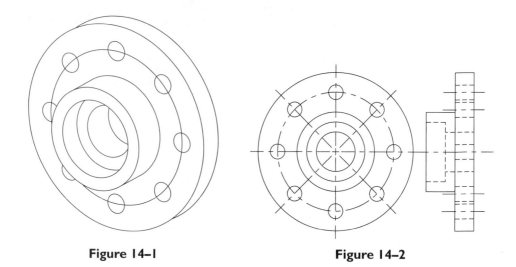

Figure 14–1   **Figure 14–2**

To understand section views better, see Figure 14–3. Creating a section view slices an object in such a way as to expose what used to be hidden features and convert them to visible features. This slicing or cutting operation can be compared to that of using a glass plate or cutting plane to perform the section. In the object in Figure 14–3, the glass plate cuts the object in half. It is the responsibility of the designer or CAD operator to convert one half of the object to a section and to discard the other half. Surfaces that come in contact with the glass plane are crosshatched to show where material was cut through.

A completed section view drawing is shown in Figure 14–4. Two new types of lines are also shown: namely a cutting plane line and section lines. The cutting plane line performs the cutting operation on the Front view. In the Side view, section lines show the surfaces that were cut. The counterbore edge is shown because this corner would be visible in the section view. Notice that holes are not section lined because the cutting plane passes across the center of the hole. Notice also that hidden lines are not displayed in the Side view. It is poor practice to merge hidden lines into a section view, although there are always exceptions. The arrows of the cutting plane line tell the designer to view the section in the direction of the arrows and discard the other half.

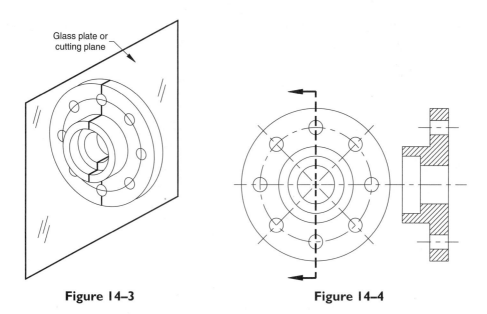

Figure 14–3   **Figure 14–4**

The cutting plane line consists of a very thick line with a series of dashes approximately 0.25" in length (see Figure 14–5). A polyline may be used to create this line of 0.05 thickness. The arrows point in the direction of sight used to create the section, with the other half generally discarded. Assign this line one of the dashed linetypes; the hidden linetype is reserved for detailing invisible features in views. The section line, in contrast with the cutting plane line, is a very thin line (see Figure 14–6). This line identifies the surfaces being cut by the cutting plane line. The section line is usually drawn at an angle and at a specified spacing.

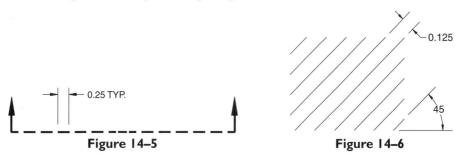

Figure 14–5   **Figure 14–6**

A wide variety of hatch patterns are already supplied with the software. One of these patterns, ANSI31, is displayed in Figure 14–7. This is one of the more popular patterns, with lines spaced 0.125 units apart and at a 45° angle.

The object in Figure 14–8 illustrates proper section lining techniques. Much of the pain of spacing the section lines apart from each other and at angles has been eased considerably with the use of the computer as a tool. However, the designer must still practice proper section lining techniques at all times for clarity of the section.

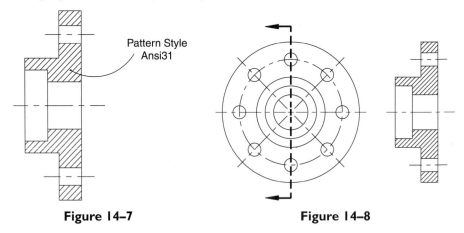

Figure 14–7   **Figure 14–8**

The next four examples illustrate common errors involved with section lines. In Figure 14–9, section lines run in the correct direction and at the same angle; however, the hidden lines have not been converted to object lines. This will confuse the more experienced designer because the presence of hidden lines in the section means more complicated invisible features.

Figure 14–10 shows yet another error encountered when sections are created. Again, the section lines are properly placed; however, all surfaces representing holes have been removed, which displays the object as a series of sectioned blocks unconnected, implying four separate parts.

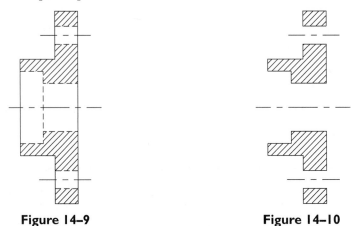

Figure 14–9   **Figure 14–10**

Figure 14–11 appears to be a properly sectioned object; however, upon closer inspection, we see that the angle of the crosshatch lines in the upper half differs from the angle of the same lines in the lower-left half. This suggests two different parts when, in actuality, it is the same part.

In Figure 14–12, all section lines run in the correct direction. The problem is that the lines run through areas that were not sliced by the cutting plane line. These areas in Figure 14–12 represent drill and counterbore holes and are left unsectioned.

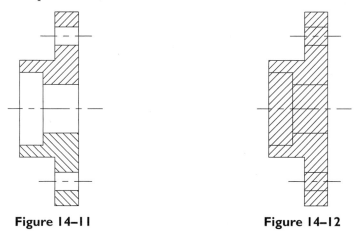

Figure 14–11   **Figure 14–12**

These have been identified as the most commonly made errors when an object is crosshatched. Remember just a few rules to follow:

- Section lines are present only on surfaces that are cut by the cutting plane line.
- Section lines are drawn in one direction when the same part is crosshatched.
- Hidden lines are usually omitted when a section view is created.
- Areas such as holes are not sectioned because the cutting line only passes across this feature.

FULL SECTIONS

When the cutting plane line passes through the entire object, a full section is formed. In Figure 14–13, a full section would be the same as taking an object and cutting it completely in half. Depending on the needs of the designer, one half is kept while the other half is discarded. The half that is kept is shown with section lines.

The multiview solution to the problem in Figure 14–13 is shown in Figure 14–14. The Front view is drawn with lines projected across to form the Side view. To show that the Side view is in section, a cutting plane line is added to the Front view. This line performs the physical cut. You have the option of keeping either half of the object.

This is the purpose of adding arrowheads to the cutting plane line. The arrowheads define the direction of sight in which you view the object to form the section. You must then interpret what surfaces are being cut by the cutting plane line in order to properly add crosshatching lines to the section that is located in the Right Side view. Hidden lines are not necessary once a section has been made.

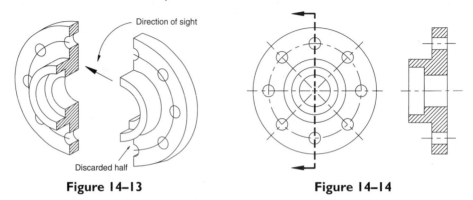

Figure 14–13   **Figure 14–14**

Numerous examples illustrate a cutting plane line with the direction of sight off to the left. This does not mean that a cutting plane line cannot have a direction of sight going to the right as in Figure 14–15. In this example, the section is formed from the Left Side view if the circular features are located in the Front view.

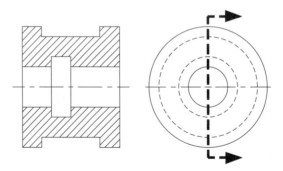

Figure 14–15

HALF SECTIONS

When symmetrical-shaped objects are involved, sometimes it is unnecessary to form a full section by cutting straight through the object. Instead, the cutting plane line passes only halfway through the object, which makes the drawing in Figure 14–16 a half section. The rules for half sections are the same as for full sections; namely, a direction of sight is established, part of the object is kept, and part is discarded.

The views are laid out in Figure 14–17 in the usual multiview format. To prepare the object as a half section, the cutting plane line passes halfway through the Front view before being drawn off to the right. Notice there is only one direction of sight arrow and a centerline is used to depict where the sectioned portion of the object ends in the Right Side view. The Right Side view is converted to a half section by the crosshatching of the upper half of the Side view while hidden lines remain in the lower half.

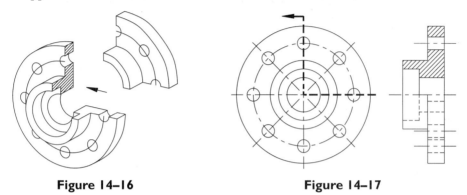

Figure 14–16   **Figure 14–17**

Depending on office practices, some designers prefer to omit hidden lines entirely from the Side view similar to the drawing in Figure 14–18. In this way, the lower half is drawn showing only what is visible.

Figure 14–19 shows another way of drawing the cutting plane line to conform to the Right Side view drawn in section. Hidden lines have been removed from the lower half; only those lines visible are displayed.

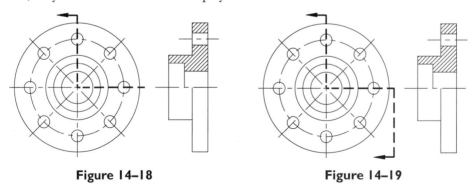

Figure 14–18   **Figure 14–19**

ASSEMBLY SECTIONS

It would be unfair to give designers the impression that section views are only used for displaying internal features of individual parts. Another advantage of using section views is that it permits the designer to create numerous objects, assemble them, and then slice the assembly to expose internal details and relationships of all parts. This type of section is an assembly section similar to Figure 14–20. For all individual parts, notice the section lines running in the same directions. This follows one of the basic rules of section views: keep section lines at the same angle for each individual part.

Figure 14–21 shows the difference between assembly sections and individual parts that have been sectioned. For parts in an assembly that contact each other, it is good practice to alternate the directions of the section lines to make the assembly much clearer and to distinguish the parts from each other. You can accomplish this by changing the angle of the hatch pattern or even the scale of the pattern.

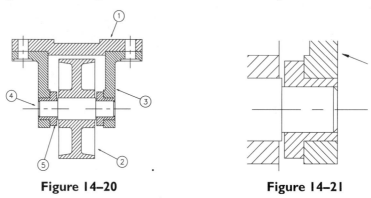

Figure 14–20   **Figure 14–21**

To identify parts in an assembly, an identifying part number along with a circle and arrowhead line are used, as shown in Figure 14–22. The line is very similar to a leader line used to call out notes for specific parts on a drawing. The addition of the circle highlights the part number. Sometimes this type of callout is referred to as a "bubble."

In the enlarged assembly shown in Figure 14–23, the large area in the middle is actually a shaft used to support a pulley system. With the cutting plane passing through the assembly including the shaft, refrain from crosshatching features such as shafts, screws, pins, or other types of fasteners. The overall appearance of the assembly is actually enhanced when these items are not crosshatched.

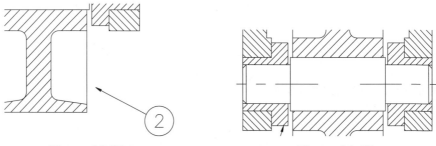

Figure 14–22   **Figure 14–23**

ALIGNED SECTIONS

Aligned sections take into consideration the angular position of details or features of a drawing. Instead of the cutting plane line being drawn vertically through the object as in Figure 14–24, the cutting plane is angled or aligned with the same angle as the elements. Aligned sections are also made to produce better clarity of a drawing. In Figure 14–24, with the cutting plane forming a full section of the object, it is difficult to obtain the true size of the angled elements. In the Side view, they appear foreshortened or not to scale. Hidden lines were added as an attempt to better clarify the view.

Instead of the cutting plane line being drawn all the way through the object, the line is bent at the center of the object before being drawn through one of the angled legs (see Figure 14–25). The direction of sight arrows on the cutting plane line not only determines the direction in which the view will be sectioned, but it also shows another direction for rotating the angled elements so they line up with the upper elements. This rotation is usually not more than 90°. As lines are projected across to form the Side view, the section appears as if it were a full section. This is only because the features were rotated and projected in section for greater clarity of the drawing.

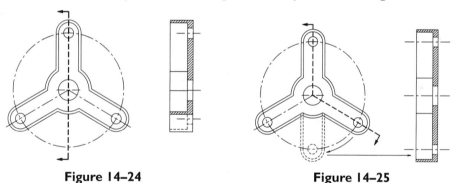

Figure 14–24   Figure 14–25

OFFSET SECTIONS

Offset sections take their name from the offsetting of the cutting plane line to pass through certain details in a view (see Figure 14–26). If the cutting plane line passes straight through any part of the object, some details would be exposed while others would remain hidden. By offsetting the cutting plane line, the designer controls its direction and which features of a part it passes through. The view to section follows the basic section rules. Notice the changes in direction of the cutting plane as it passes through the object are shown in the Top view but not in the Front view.

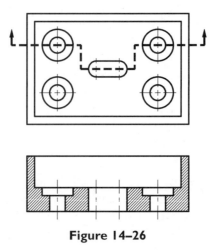

Figure 14–26

SECTIONING RIBS

Parts made out of cast iron with webs or ribs used for reinforcement do not follow basic rules of sections. In Figure 14–27, the Front view has the cutting plane line passing through the entire view; the Side view at "A" is crosshatched according to section view basics. However, it is difficult to read the thickness of the base because the crosshatching includes the base along with the web. A more efficient method is to leave off crosshatching webs as in "B." Therefore, not crosshatching the web exposes other important details such as thickness of bases and walls.

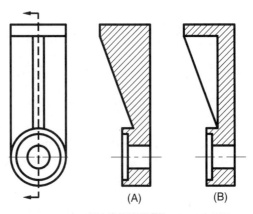

Figure 14–27

The object in Figure 14–28 is another example of performing a full section on an area consisting of webbed or ribbed features. If you do not crosshatch the webbed areas, more information is available, such as the thickness of the base and wall areas around the cylindrical hole. This may not be considered true projection; however, it is considered good practice.

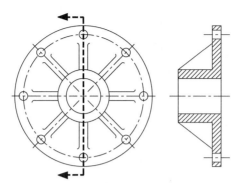

Figure 14–28

BROKEN SECTIONS

At times, only a partial section of an area needs to be created. For this reason, a broken section might be used. The object in Figure 14–29 shows a combination of sectioned areas and conventional areas outlined by the hidden lines. When converting an area to a broken section, you create a break line, crosshatch one area, and leave the other area in a conventional drawing. Break lines may take the form of short freehanded line segments as shown in Figure 14–30 or a series of long lines separated by break symbols as in Figure 14–31. You can use the LINE command with Ortho off to produce the desired effect as in Figures 14–29 and 14–30.

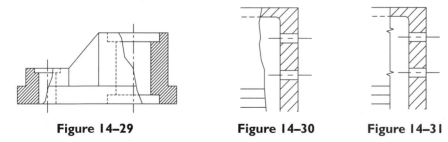

| **Figure 14–29** | **Figure 14–30** | **Figure 14–31** |

REVOLVED SECTIONS

Section views may be constructed as part of a view with revolved sections. In Figure 14–32, the elliptical shape is constructed while it is revolved into position and then crosshatched. The crosshatched shape represents the cross section shape of the arm

Figure 14–33 is another example of a revolved section where a cross section of the C-clamp was cut away and revolved to display its shape.

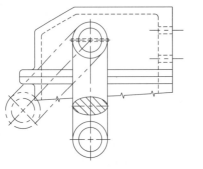

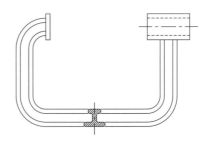

Figure 14–32   **Figure 14–33**

REMOVED SECTIONS

Removed sections are very similar to revolved sections except that instead of the section being drawn somewhere inside the object, as is the case with a revolved section, the section is placed elsewhere or removed to a new location in the drawing. The cutting plane line is present, with the arrows showing the direction of sight. Identifying letters are placed on the cutting plane and underneath the section to keep track of the removed sections, especially when there are a number of them on the same drawing sheet. See Figure 14–34.

Another way of displaying removed sections is to use centerlines as a substitute for the cutting plane line. In Figure 14–35, the centerlines determine the three shapes of the chisel and display the basic shapes. Identification numbers are not required in this particular example.

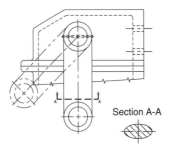

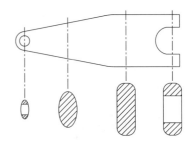

Figure 14–34   **Figure 14–35**

ISOMETRIC SECTIONS

Section views may be incorporated into pictorial drawings that appear in Figure 14–36 and 14–37. Figure 14–36 is an example of a full isometric section with the cutting plane passing through the entire object. In keeping with basic section rules, only those surfaces sliced by the cutting plane line are crosshatched. Isometric sections make it easy to view cut edges compared to holes or slots. Figure 14–37 is an example of an isometric drawing converted to a half section.

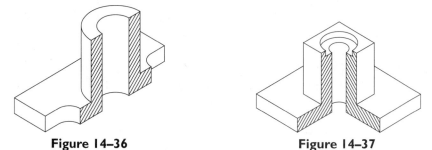

| **Figure 14–36** | **Figure 14–37** |

ARCHITECTURAL SECTIONS

Mechanical representations of machine parts are not the only type of drawings where section views are used. Architectural drawings rely on sections to show the type of building materials that go into the construction of foundation plans, roof details, or wall sections, as shown in Figure 14–38. Here numerous types of crosshatching symbols are used to call out the different types of building materials, such as brick veneer at "A," insulation at "B," finished flooring at "C," floor joists at "D," concrete block at "E," earth at "F," and poured concrete at "G." Section hatch patterns that are provided in AutoCAD may be used to crosshatch most building components.

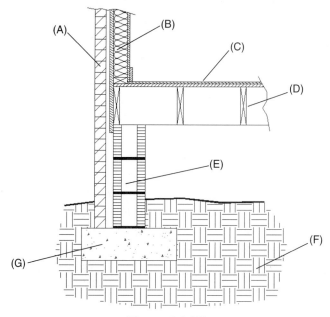

Figure 14–38

THE BHATCH COMMAND

Choosing Hatch… from the Draw pull-down menu or picking the Hatch button in Figure 14–39 activates the main Boundary Hatch and Fill dialog box in Figure 14–40. Use the Hatch tab to pick points identifying the area or areas to be crosshatched, select objects using one or more of the popular selection set modes, select hatch patterns supported in the software, and change the scale and angle of the hatch pattern. All areas grayed out in Figure 14–40, are currently inactive. Only when additional parameters have been satisfied will these buttons activate for use.

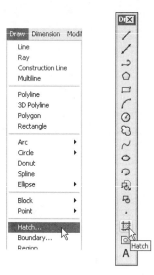

Figure 14–39

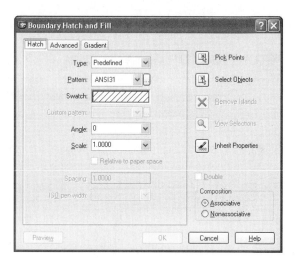

Figure 14–40

Try It! – Open the drawing file 14_Hatch Basics. The object shown in Figure 14–41 will be used to demonstrate the boundary crosshatching method. The object needs to have areas "A," "B," and "C" crosshatched.

The BREAK command is one technique used to isolate the three areas shown in Figure 14–42. Another technique is to trace a closed polyline on top of the three areas, perform the hatch operation, and then erase each polyline boundary. The BHATCH command is the more efficient method of automatically highlighting areas to be crosshatched.

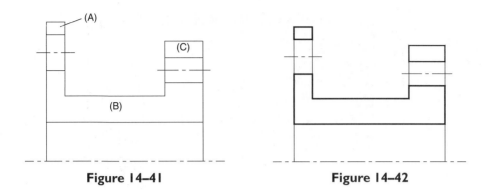

Figure 14–41   **Figure 14–42**

The BHATCH command opens the Hatch tab of the Boundary Hatch and Fill dialog box. Click on Pick Points. Figure 14–43 illustrates three areas with internal points identified by the "X's." The dashed areas identify the highlighted areas to be crosshatched.

In the final step of the BHATCH command, as the crosshatch pattern is applied to the areas, the highlighted outlines deselect, leaving the crosshatch patterns. See Figure 14–44.

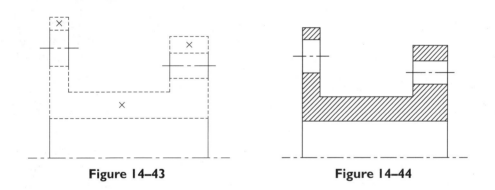

Figure 14–43   **Figure 14–44**

Choosing the Advanced tab in the main Boundary Hatch and Fill dialog box activates the dialog box shown in Figure 14–45. This dialog box holds a number of crosshatching options, such as Make a New Boundary Set, Boundary Style, Boundary Options, and an Island Detection style check box.

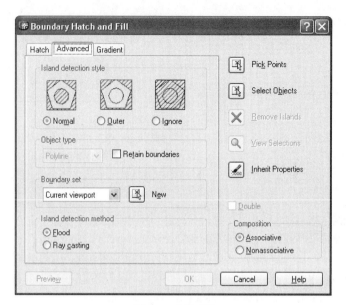

Figure 14–45

Figure 14–46 shows three boundary styles and how they affect levels of crosshatching. The Normal boundary style is the default hatching method in which, when you select all objects within a window, the hatching begins with the outermost boundary, skips the next inside boundary, hatches the next innermost boundary, and so on. The outermost boundary style hatches only the outermost boundary of the object. The Ignore style ignores the default hatching methods of alternating crosshatching and hatches the entire object.

You have the option of retaining the hatch boundary by clicking in the check box for Retain Boundaries. By default, Flood mode in the Island Detection area is turned on. In this way, selecting an internal point identifies all boundaries including islands (see Figure 14–47).

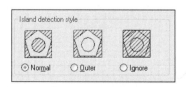

Figure 14–46

Figure 14–47

The Hatch tab holds numerous hatch patterns in a drop-down list box in Figure 14–48. Clicking on a pattern in this listing makes the pattern current.

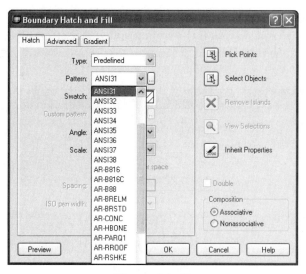

Figure 14–48

Selecting the pattern (…) button next to the pattern name in Figure 14–48 activates a series of tabs that display a number of crosshatching patterns already created and ready for use in Figure 14–49. Select a particular pattern by clicking on the pattern itself. If the wrong pattern was selected, simply choose the correct one or move to another tab to view other patterns for use.

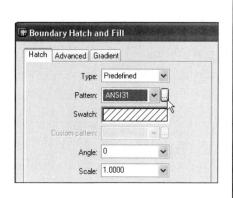

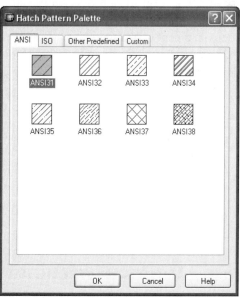

Figure 14–49

Once a hatch pattern is selected, the name is displayed in the Boundary Hatch and Fill dialog box in Figure 14–50. If the scaling and angle settings look favorable, click on the Pick Points button. The command prompt requires you to select an internal point or points.

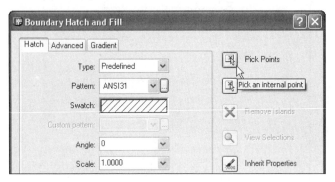

Figure 14–50

Continue working with the drawing file 14_Hatch Basics. In Figure 14–51, click in the three areas marked by the "X." Notice that these areas highlight. If these areas are correct, press ENTER to return to the Boundary Hatch and Fill dialog box.

When you're back in the Boundary Hatch and Fill dialog box, you now have the option of first previewing the hatch to see if all settings and the appearance of the pattern are desirable. Click on the Preview button in the lower-left corner of the dialog box to accomplish this. The results appear in Figure 14–52. At this point in the process, the pattern is still not permanently placed on the object. Pressing the ENTER key to exit preview mode returns you to the main Boundary Hatch and Fill dialog box. If the hatch pattern is correct in appearance, click the OK button to place the pattern with the drawing.

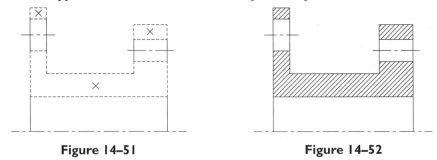

Figure 14–51   **Figure 14–52**

SOLID FILL HATCH PATTERNS

Activating the Boundary Hatch and Fill dialog box and clicking on the pattern (...) button followed by the Other Predefined tab displays the dialog box in Figure 14–53.

Additional patterns are available and include brick, roof, and other construction related materials. One powerful pattern is called SOLID.

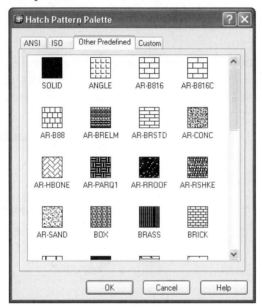

Figure 14–53

 Try It! – Open the drawing file 14_Hatch Solid1. The object in Figure 14–54 illustrates a thin wall that needs to be filled in. The BHATCH command used with the SOLID hatch pattern can perform this task very efficiently. When you pick an internal point, the entire closed area will be filled with a solid pattern and placed on the current layer. The color of the pattern will take on the same color as the one assigned to the current layer.

The solid hatch pattern in Figure 14–55 was placed with one internal point pick.

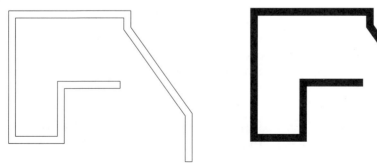

Figure 14–54   **Figure 14–55**

Try It! – Open the drawing file 14_Hatch Solid2. Applying the SOLID hatch pattern to the object in Figure 14–56 can easily be accomplished with the BHATCH command. This object also is a thin wall example with the addition of various curved edges and fillets.

The solid hatch pattern also works for curved outlines as well as outlines involving known vertex corners. Figure 14–57 displays the solid pattern, completely filling in the outline where all corners consist of some type of curve generated by the FILLET command.

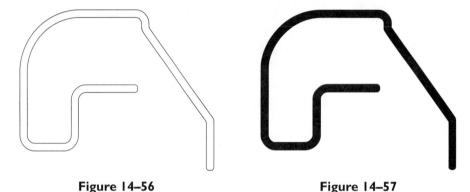

Figure 14–56   **Figure 14–57**

GRADIENT PATTERNS

A gradient hatch pattern is a solid hatch fill that makes a smooth transition from a lighter shade to a darker shade. Predefined patterns such as linear, spherical, and radial sweep are available to provide different effects. As with the vector hatch patterns that have always been supplied with AutoCAD, the angle of the gradient patterns can also be controlled. Gradient patterns can also be associative; the pattern will update if the boundary is changed. To edit a gradient pattern, simply double-click on it to launch the Hatch Edit dialog box. Gradient patterns can be selected and applied by picking the Gradient tab of the Boundary Hatch and Fill dialog box in Figure 14–58. A few of the controls for this tab of the dialog box are explained as follows:

- One Color—When this option is selected, a color swatch with a Browse button and a Shade and Tint slider appears. The One Color option designates a fill that uses a smooth transition between darker shades and lighter tints of a single color.

- Two Color—When this option is selected, a color swatch and Browse button display for colors 1 and 2. This option allows the fill to transition smoothly between two colors.

- Color Swatch—This is the default color displayed as set by the current color in the drawing. This option specifies the color to be used for the gradient fill. The presence of the three dots (...) next to the color swatch signifies the Browse button. Use this to display the Select Color dialog box, similar to the dialog used for assigning colors to layers. Use this dialog box to select color based on the AutoCAD Index Color, True Color, or Color Book.

- The Shade and Tint Slider—This option designates the amount of tint and shade applied to the gradient fill of one color. Tint is defined as the selected color mixed with white. Shade is defined as the selected color mixed with black.

- The Centered Option—Use this option for creating special effects with gradient patterns. When this option is checked, the gradient fill will appear symmetrical. If the option is not selected, AutoCAD shifts the gradient fill up and to the left. This position creates the illusion that a light source is located to the left of the object.

- Angle—Set an angle that affects the gradient pattern fill. This angle setting is independent of the angle used for regular hatch patterns.

- The Gradient Patterns—Nine gradient patterns are available for you to apply. These include linear sweep, spherical, and parabolic.

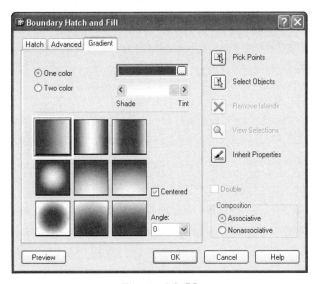

Figure 14–58

The example of the house elevation in Figure 14–59 illustrates parabolic linear sweep being applied above and below the house.

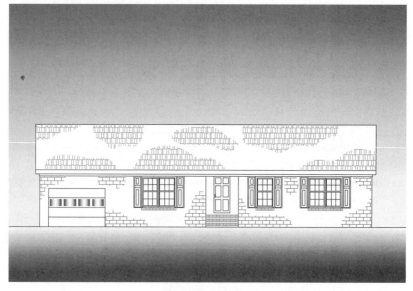

Figure 14–59

The garage detail in Figure 14–60 illustrates one of the parabolic gradient patterns applied to the windows.

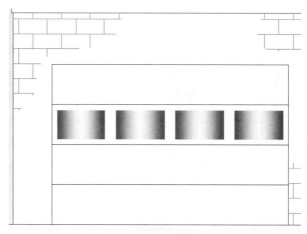

Figure 14–60

HATCH PATTERN SYMBOL MEANINGS

Below is a listing of all hatch patterns and their purpose. For example, if you are constructing a mechanical assembly where you want to distinguish plastic material from steel, use the patterns ANSI34 and ANSI32 respectively. Refer to the following list for the purpose of other materials and their associated hatch patterns.

SOLID	Solid fill
ANGLE	Angle steel
ANSI31	ANSI Iron, Brick, Stone masonry
ANSI32	ANSI Steel
ANSI33	ANSI Bronze, Brass, Copper
ANSI34	ANSI Plastic, Rubber
ANSI35	ANSI Fire brick, Refractory material
ANSI36	ANSI Marble, Slate, Glass
ANSI37	ANSI Lead, Zinc, Magnesium, Sound/Heat/Elec Insulation
ANSI38	ANSI Aluminum
AR-B816	8x16 Block elevation stretcher bond
AR-B816C	8x16 Block elevation stretcher bond with mortar joints
AR-B88	8x8 Block elevation stretcher bond
AR-BRELM	Standard brick elevation English bond with mortar joints
AR-BRSTD	Standard brick elevation stretcher bond
AR-CONC	Random dot and stone pattern
AR-HBONE	Standard brick herringbone pattern @ 45 degrees
AR-PARQ1	2x12 Parquet flooring: pattern of 12x12
AR-RROOF	Roof shingle texture
AR-RSHKE	Roof wood shake texture
AR-SAND	Random dot pattern
BOX	Box steel
BRASS	Brass material
BRICK	Brick or masonry-type surface
BRSTONE	Brick and stone
CLAY	Clay material
CORK	Cork material
CROSS	A series of crosses
DASH	Dashed lines
DOLMIT	Geological rock layering
DOTS	A series of dots
EARTH	Earth or ground (subterranean)
ESCHER	Escher pattern

FLEX	Flexible material
GRASS	Grass area
GRATE	Grated area
HEX	Hexagons
HONEY	Honeycomb pattern
HOUND	Houndstooth check
INSUL	Insulation material
ACAD_ISO02W100	dashed line
ACAD_ISO03W100	dashed space line
ACAD_ISO04W100	long dashed dotted line
ACAD_ISO05W100	long dashed double dotted line
ACAD_ISO06W100	long dashed triplicate dotted line
ACAD_ISO07W100	dotted line
ACAD_ISO08W100	long dashed short dashed line
ACAD_ISO09W100	long dashed double-short-dashed line
ACAD_ISO10W100	dashed dotted line
ACAD_ISO11W100	double-dashed dotted line
ACAD_ISO12W100	dashed double-dotted line
ACAD_ISO13W100	double-dashed double-dotted line
ACAD_ISO14W100	dashed triplicate-dotted line
ACAD_ISO15W100	double-dashed triplicate-dotted line
LINE	Parallel horizontal lines
MUDST	Mud and sand
NET	Horizontal / vertical grid
NET3	Network pattern 0-60-120
PLAST	Plastic material
PLASTI	Plastic material
SACNCR	Concrete
SQUARE	Small aligned squares
STARS	Star of David
STEEL	Steel material
SWAMP	Swampy area
TRANS	Heat transfer material
TRIANG	Equilateral triangles
ZIGZAG	Staircase effect

USING BHATCH TO HATCH ISLANDS

Try It! – Open the drawing file 14_Hatch Islands. When charged with the task of crosshatching islands, first activate the Boundary Hatch and Fill dialog box in Figure 14–61 and select a pattern. Select Pick Points in the upper-right corner of the dialog box. Next pick a point at "A" in Figure 14–62, which will define not only the outer perimeter of the object but the inner shapes as well. This result is due to Flood mode in the Island Detection area being activated in the Advanced tab of the Boundary Hatch and Fill dialog box.

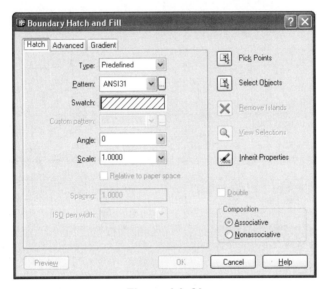

Figure 14–61

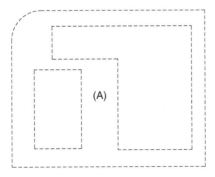

Figure 14–62

Selecting the Preview button will display the hatch pattern shown in Figure 14–63. If changes need to be made, such as a change in the hatch scale or angle, the preview allows these changes to be made. After making changes, be sure to preview the pattern once again to check if the results are desirable. Choosing OK places the pattern and exits to the command prompt.

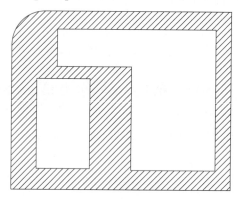

Figure 14–63

HATCH PATTERN SCALING

Hatching patterns are predefined in size and angle. When you use the BHATCH command, the pattern used is assigned a scale value of 1.00, which will draw the pattern exactly the way it was originally created.

 Try It! – Open the drawing file 14_Hatch Scale. Activate the Boundary Hatch and Fill dialog box, accept all default values, and pick internal points to hatch the object in Figure 14–64.

Entering a different scale value for the pattern in the Boundary Hatch and Fill dialog box will either increase or decrease the spacing between crosshatch lines. Figure 14–65 is an example of the ANSI31 pattern with a new scale value of 0.50.

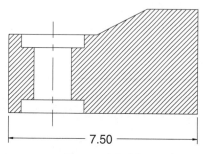

Figure 14–64

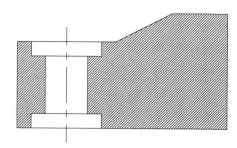

Figure 14–65

As you can decrease the scale of a pattern to hatch small areas, so also can you scale the pattern up for large areas. Figure 14–66 has a hatch scale of 2.00, which doubles all distances between hatch lines.

Use care when hatching large areas. In Figure 14–67, the total length of the object measures 190.50 millimeters. If the hatch scale of 1.00 were used, the pattern would take on a filled appearance. This results in numerous lines being generated, which will increase the size of the drawing file. A value of 25.4, the number of millimeters in an inch, is used to scale hatch lines for metric drawings.

 Try It! – Open the drawing file 14_Hatch Scale mm. Activate the Boundary Hatch and Fill dialog box and change the pattern scale to 25.4 units. Select the Pick Points button to pick internal points to crosshatch the object illustrated in Figure 14–67.

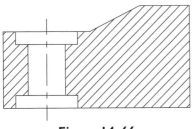

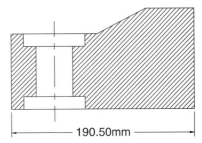

Figure 14–66   **Figure 14–67**

HATCH PATTERN ANGLE MANIPULATION

As with the scale of the hatch pattern, depending on the effect, you can control the angle for the hatch pattern within the area being hatched. By default, the BHATCH command displays a 0° angle for all patterns.

 Try It! – Open the drawing file 14_Hatch Angle. Activate the Boundary Hatch and Fill dialog box and hatch the object in Figure 14–68 keeping all default values. The angle for "ANSI31" is drawn at 45°—the angle in which the pattern was originally created.

Experiment with the angle setting of the Boundary Hatch and Fill dialog box by entering any angle different from the default value of "0" to rotate the hatch pattern by that value. This means that if a pattern were originally designed at a 45° angle like "ANSI31," entering a new angle for the pattern would begin rotating the pattern starting at the 45° position. In Figure 14–68, a new angle of 45° is entered in the Boundary Hatch and Fill dialog box. Since the original angle was already 45°, this new angle value is added to the original to obtain a vertical crosshatch pattern; positive angles rotate the hatch pattern in the counterclockwise direction.

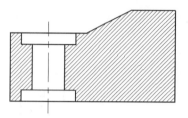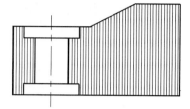

Figure 14–68

Again, entering an angle other than the default rotates the pattern from the original angle to a new angle. In Figure 14–69, an angle of 90° has been applied to the "ANSI31" pattern in the Boundary Hatch and Fill dialog box. Providing different angles for patterns is useful when you create section assemblies where different parts are in contact with each other and patterns are placed at different angles, because it makes the parts easy to see.

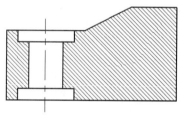

Figure 14–69

PRECISION HATCH PATTERN PLACEMENT

At times, you want to control where the hatch pattern begins inside a shape. Figure 14–70 shows two cases where a brick hatch pattern was applied. In the image on the left, the brick pattern was applied without any special insertion base points. In this image, the pattern seems to be laid out based on the center of the rectangle. As the brick pattern reaches the edges of the rectangle, the pattern just ends.

A system variable exists that allows you to start the hatch pattern at a designated point. This system variable is called SNAPBASE and must be entered from the keyboard. When the following prompt appears, you select a point on an object. You then use the BHATCH command to crosshatch the shape. In the image on the right in Figure 14–70, the snap base point was defined in the lower-left corner of the rectangle.

Command: **SNAPBASE**
Enter new value for SNAPBASE <0.00,0.00>: End
of (Select the lower-left corner of the rectangle)

When the brick pattern is applied to the rectangle on the right, the bricks align with the lower-left corner of the rectangle.

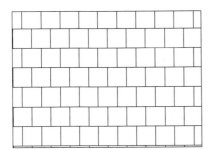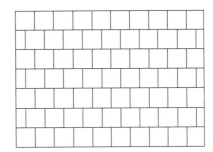

Figure 14–70

INHERIT HATCHING PROPERTIES

Try It! – Open the drawing file 14_Hatch Inherit. The image in Figure 14–71 consists of a simple assembly drawing. At least three different hatch patterns are displayed to define the different parts of the assembly. Unfortunately, a segment of one of the parts was not crosshatched and it is unclear what pattern, scale, and angle were used to place the pattern.

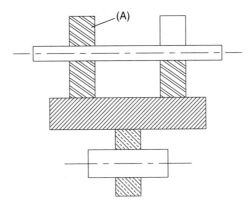

Figure 14–71

Whenever you are faced with this problem, click on the Inherit Properties button in the main Boundary Hatch and Fill dialog box in Figure 14–72. Clicking on the pattern at "A" in Figure 14–71 sets the pattern, scale, and angle from the selected pattern.

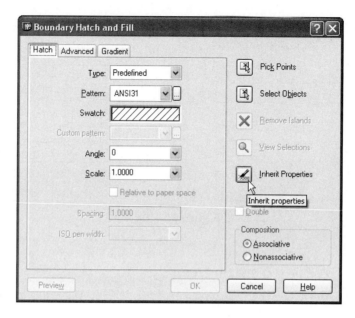

Figure 14–72

To complete the assembly, click an internal point in the empty area. Right-click and select Enter from the shortcut menu. The hatch pattern is placed in this area and it matches that of the other patterns to identify the common parts, as shown in Figure 14–73.

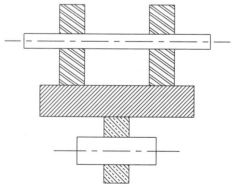

Figure 14–73

PROPERTIES OF ASSOCIATIVE HATCHING

Try It! – Open the drawing file 14_Hatch Assoc. The following is a review of associative crosshatching. In Figure 14–74, the plate needs to be hatched in the ANSI31 pattern at a scale of one unit and an angle of zero; the two slots and three holes are to be considered islands. Enter the BHATCH command, make the necessary changes in the Boundary Hatch and Fill dialog box, click on the Pick Points button, and mark an internal point somewhere inside the object (such as at "A").

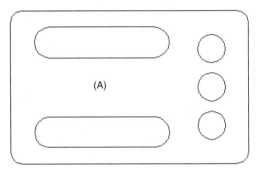

Figure 14–74

When all objects inside and including the outline highlight, press ENTER to return to the Boundary Hatch and Fill dialog box. It is very important to realize that all objects highlighted are tied to the associative crosshatch object. Each shape works directly with the hatch pattern to ensure that the outline of the object is being read by the hatch pattern and that the hatching is performed outside the outline. You have the option of first previewing the hatch pattern or applying the hatch pattern. In either case, the results appear similar to the object in Figure 14–75.

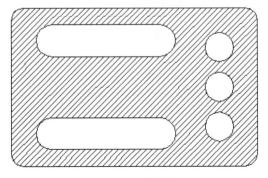

Figure 14–75

Associative hatch objects may be edited, and the results of this editing will have an immediate impact on the appearance of the hatch pattern. For example, the two outer holes need to be increased in size by a factor of 1.5; also, the middle hole needs to be repositioned to the other side of the object (see Figure 14–76). Using the MOVE command not only allows you to reposition the hole, but when the move is completed, the hatch pattern mends itself automatically to the moved circle. In the same manner, using the SCALE command to increase the size of the two outer circles by a value of 1.5 units makes the circles large and automatically mends the hatch pattern.

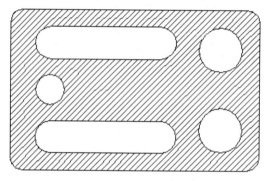

Figure 14–76

In Figure 14–77, the length of the slots needs to be decreased. Also, hole "A" needs to be deleted. Use the STRETCH command to shorten the slots. Use the crossing box at "B" to select the slots and stretch them one unit to the right. Use the ERASE command to delete hole "A."

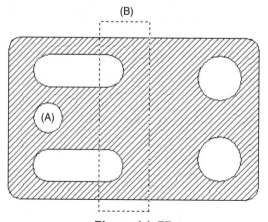

Figure 14–77

The result of the editing operations is shown in Figure 14–78. In the figure, associative hatching has been enhanced for members to be erased while the hatch pattern still maintains its associativity. The inner hole was deleted with the ERASE command. After the ERASE command was executed, the hatch pattern updated itself and hatched through the area once occupied by the hole.

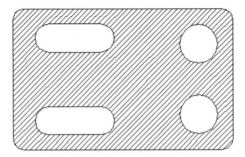

Figure 14–78

Once a hatch pattern is placed in the drawing, it will not update itself to any new additions in the form of closed shapes in the drawing. In Figure 14–79, a rectangle was added at the center of the object. Notice, however, that the hatch pattern cuts directly through the rectangle. Since the hatch pattern does not have the intelligence to recognize the new boundary, the entire hatch pattern must be deleted and hatched again. In this way, the rectangular boundary will be recognized.

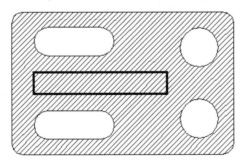

Figure 14–79

 THE HATCHEDIT COMMAND

 Try It! – Open the drawing file 14_Hatchedit. In Figure 14–80, the pattern needs to be increased to a new scale factor of 3 units and the angle of the pattern needs to be rotated 90°. The current scale value of the pattern is 1 unit and the angle is 0°. Issuing the HATCHEDIT command prompts you to select the hatch pattern to edit. Clicking on the hatch pattern anywhere inside the object in Figure 14–80 displays the Hatch Edit dialog box, shown in Figure 14–81.

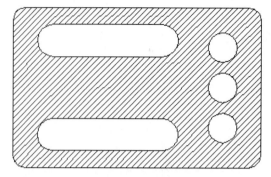

Figure 14–80

With the dialog box displayed as in Figure 14–81, click in the Scale edit box to change the scale from the current value of 1 unit to the new value of 3 units. Next, click in the Angle edit box and change the angle of the hatch pattern from the current value of 0° to the new value of 90°. This will increase the spacing between hatch lines and rotate the pattern 90° in the counterclockwise direction.

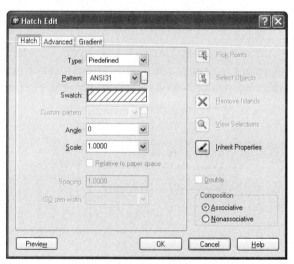

Figure 14–81

Clicking the OK button in the Hatch Edit dialog box returns you to the drawing editor and updates the hatch pattern to these changes (see Figure 14–82). Be aware of the following rules for the HATCHEDIT command to function: it only works on an associative hatch pattern. If the pattern loses associativity through the use of the MIRROR command or if the hatch pattern is exploded, the HATCHEDIT command will cease to function.

Note: Double-clicking on any associative hatch pattern will automatically launch the Hatch Edit dialog box. You could also pre-select the hatch pattern, right-click, and click on Hatch edit… to launch the Hatch Edit dialog box.

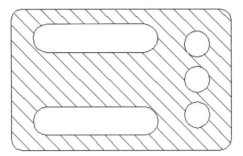

Figure 14–82

DRAGGING PATTERNS FROM DESIGNCENTER

Another efficient means of adding hatch patterns to your drawing is through the AutoCAD DesignCenter where you can display the pattern, drag it from the Design-Center, and drop it into a closed shape. (The DesignCenter will be discussed in greater detail in Chapter 16.)

 Try It! – Open the drawing file 14_Hatch Drag Drop. The inner walls of the object in Figure 14–83 need to be crosshatched. We have already seen how this is accomplished through the Boundary Hatch and Fill dialog box. We will now use the DesignCenter to accomplish this task.

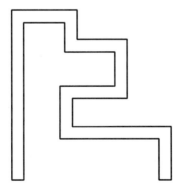

Figure 14–83

Activate the DesignCenter in Figure 14–84 by entering the ADCENTER command in from the keyboard or by pressing CTRL + 2. With the DesignCenter active, find the tool at the top of the DesignCenter labeled Favorites and click on this button. A number of predefined items are already present in this folder. Identify the file acad.pat and double-click on this item.

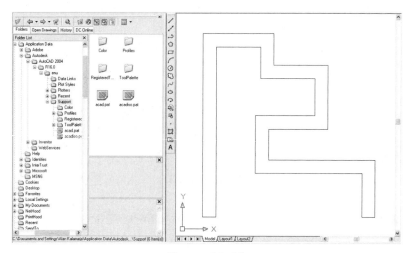

Figure 14–84

This will change the DesignCenter panel to reflect all hatch patterns illustrated in Figure 14–85 held inside this file. By default, these hatch patterns are identical to those found in the Boundary Hatch and Fill dialog box. Press and hold your mouse button down on a selected pattern, such as ANSI31, and drag the pattern into your drawing screen. Keep dragging the pattern until it is inside the closed shape. Then release your finger on the mouse button.

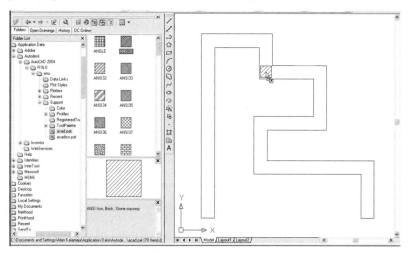

Figure 14–85

The results are illustrated in Figure 14–86 with the hatch pattern filling the entire closed shape of the object.

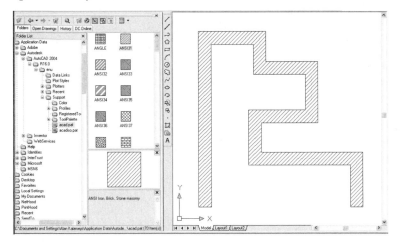

Figure 14–86

HATCH SCALING RELATIVE TO PAPER SPACE

While in a layout, you have the opportunity to scale a hatch pattern relative to the current paper space scale in a viewport. In this way, you can easily have AutoCAD calculate the hatch pattern scale, since this is determined by the scale of the image inside a viewport. While inside a floating model space viewport, activate the Boundary Hatch and Fill dialog box in Figure 14–87. Notice the Relative to paper space check box. Placing a check in this box activates this feature. This feature is grayed out if you try to hatch in model space.

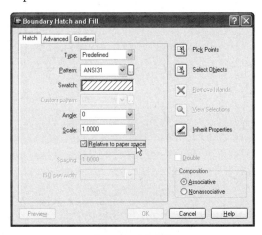

Figure 14–87

Figure 14–88 shows a valve gasket hatched in model space. Notice that the hatch pattern spacing is too large.

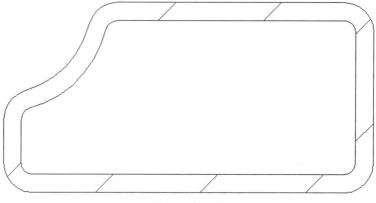

Figure 14–88

The same valve gasket is illustrated in paper space in Figure 14–89. Notice the scale of 10:1 inside the viewport.

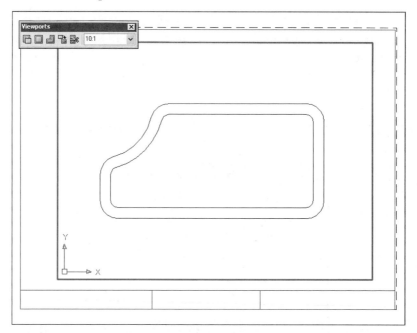

Figure 14–89

The finished gasket with hatching applied is illustrated in Figure 14–90. Notice how the hatching is at the correct scale because of the viewport scale.

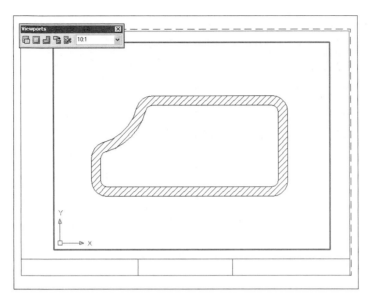

Figure 14–90

 Try It! – Open the drawing file 14_Hatch Partial Plan. The inner walls of the object in Figure 14–91 need to be crosshatched. We have already seen how this is accomplished through the Boundary Hatch and Fill dialog box. We will now use the DesignCenter to accomplish this task.

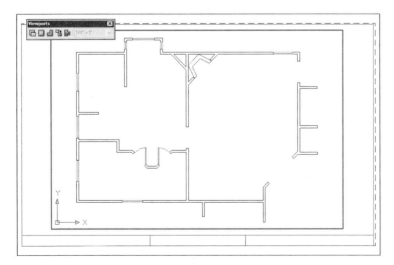

Figure 14–91

The finished partial plan is illustrated in Figure 14–92.

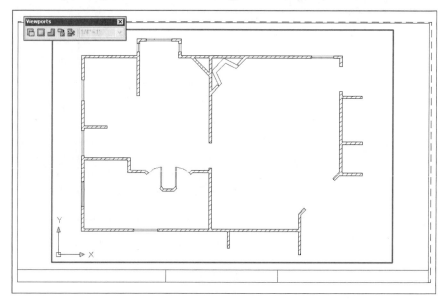

Figure 14–92

TUTORIAL EXERCISE: 14_COUPLER.DWG

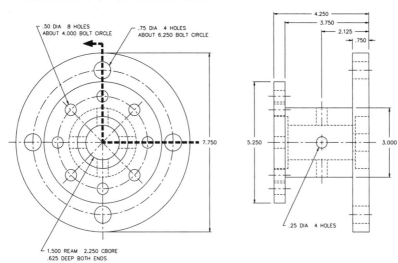

Figure 14–93

Purpose

This tutorial is designed to use the MIRROR and BHATCH commands to convert 14_Coupler to a half section as shown in figure 14–93.

System Settings

Since this drawing is provided on CD, edit an existing drawing called "14_Coupler." Follow the steps in this tutorial for converting the upper half of the object to a half section. All Units, Limits, Grid, and Snap values have been previously set.

Layers

The following layers are already created:

Name	Color	Linetype
Object	White	Continuous
Center	Yellow	Center
Hidden	Red	Hidden
Hatch2	Magenta	Continuous
Cutting Plane Line	Yellow	Dashed
Dimension	Yellow	Continuous

Suggested Commands

Convert one-half of the object to a section by erasing unnecessary hidden lines. Use the Layer Control box and change the remaining hidden lines to the Object layer. Issue the BHATCH command and use the ANSI31 hatch pattern to hatch the upper half of the 14_Coupler on the Hatch layer.

Whenever possible, substitute the appropriate command alias in place of the full AutoCAD command in each tutorial step. For example, use "CP" for the COPY command, "L" for the LINE command, and so on. The complete listing of all command aliases is located in Chapter 1, Table 1–2.

STEP I

Use the MIRROR command to copy and flip the upper half of the Side view and form the lower half. When in object selection mode, use the Remove option to deselect the main centerline and hole. If these objects are included in the mirror operation, a duplicate copy of these objects will be created. See Figure 14–94.

Command: **MI** *(For MIRROR)*
Select objects: *(Pick a point at "A")*
Specify opposite corner: (Pick a point at "B")
Select objects: *(Hold down the SHIFT key and pick centerlines "C" and "D" to remove them from the selection set)*

Select objects: *(With the SHIFT key held down, pick a point at "E")*
Specify opposite corner: *(With the SHIFT key held down, pick a point at "F")*
Select objects: *(Press ENTER to continue)*
Specify first point of mirror line: *(Select the endpoint of the centerline near "C")*
Specify second point of mirror line: *(Select the endpoint of the centerline near "D")*
Delete source objects? [Yes/No] <N>: *(Press ENTER to perform the mirror operation)*

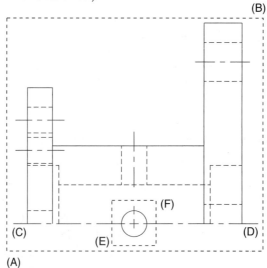

Figure 14–94

STEP 2

Begin converting the upper half of the
Side view to a half section by using the
ERASE command to remove any unnecessary hidden lines and centerlines from the
view. See Figure 14–95.

 Command: **E** *(For ERASE)*
Select objects: *(Carefully select the hidden
lines labeled "A," "B," "C," and "D")*
Select objects: *(Select the centerline labeled
"E")*
Select objects: *(Press ENTER to execute the
ERASE command)*

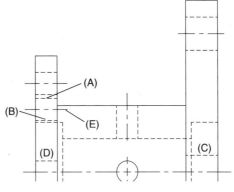

Figure 14–95

STEP 3

Since the remaining hidden lines actually
represent object lines when shown in a
full section, use the Layer Control box
in Figure 14–96 to convert all highlighted
hidden lines from the Hidden layer to the
Object layer.

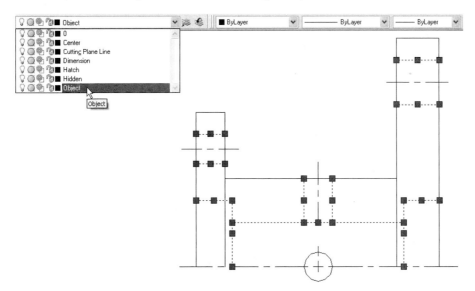

Figure 14–96

STEP 4

Remove unnecessary line segments from the upper half of the converted section using the TRIM command. Use the horizontal line at "A" as the cutting edge, and select the two vertical segments at "B" and "C" as the objects to trim. See Figure 14–97.

 Command: **TR** (For TRIM)
Current settings: Projection=UCS
 Edge=None
Select cutting edges ...
Select objects: (Select the horizontal line at "A")

Select objects: (Press ENTER to continue)
Select object to trim or shift-select to extend or [Project/Edge/Undo]: (Select the vertical line at "B")
Select object to trim or shift-select to extend or [Project/Edge/Undo]: (Select the vertical line at "C")
Select object to trim or shift-select to extend or [Project/Edge/Undo]: (Press ENTER to exit this command)

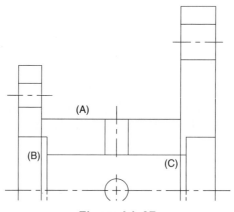

Figure 14–97

STEP 5

Make the Hatch layer the current layer. Then, use the BHATCH command to display the Boundary Hatch and Fill dialog box shown in Figure 14–98. Use the pattern "ANSI31" and keep all default settings. Click on the Pick Points button. When the drawing reappears, click inside areas "A," "B," "C," and "D" in Figure 14–99. When finished selecting these internal points, press ENTER to return to the Boundary Hatch and Fill dialog box.

Command: **BH** (For BHATCH)
(The Boundary Hatch and Fill dialog box appears. Make changes to match Figure 14–98. When finished, click on the Pick Points < button.)
Select internal point: (Pick a point at "A")
Selecting everything...
Selecting everything visible...
Analyzing the selected data...
Analyzing internal islands...

Select internal point: (Pick a point at "B")

Analyzing internal islands...
Select internal point: *(Pick a point at "C")*
Analyzing internal islands...
Select internal point: *(Pick a point at "D")*
Analyzing internal islands...
Select internal point: *(Press* ENTER *to exit this area and return to the Boundary Hatch and Fill dialog box)*

While in the Boundary Hatch and Fill dialog box, click on the Preview button to view the results. If the hatch results are acceptable, right-click to place the hatch pattern.

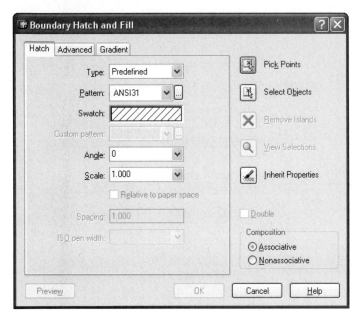

Figure 14–98

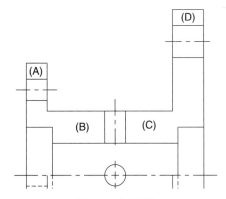

Figure 14–99

STEP 6

The complete hatched view is shown in Figure 14–100. The full drawing is also shown in Figure 14–101, along with dimensions. As with other half section examples, it is the option of the operator or designer to show all hidden lines in the lower half or delete all hidden and centerlines from the lower half and simply interpret the section in the upper half.

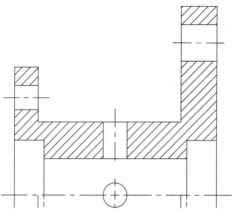

Figure 14–100

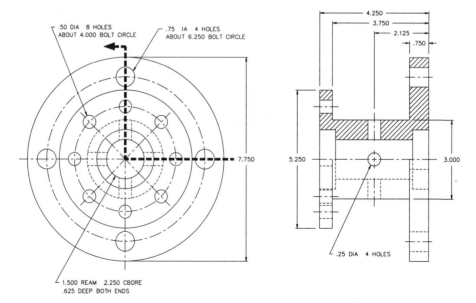

.50 DIA 8 HOLES
ABOUT 4.000 BOLT CIRCLE

.75 IA 4 HOLES
ABOUT 6.250 BOLT CIRCLE

7.750

1.500 REAM 2.250 CBORE
.625 DEEP BOTH ENDS

4.250
3.750
2.125
.750

5.250

3.000

.25 DIA 4 HOLES

Figure 14–101

TUTORIAL EXERCISE: 14_ELEVATION.DWG

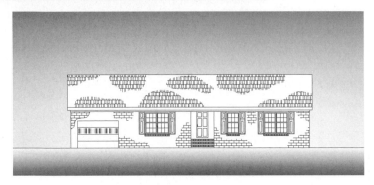

Figure 14–102

Purpose

This tutorial is designed to use the Inherit Properties button of the Boundary Hatch and Fill dialog box on 14_Elevation. Gradient hatch patterns will also be applied to this drawing for presentation purposes.

System Settings

Since this drawing is provided on CD, edit an existing drawing called "14_Elevation." Follow the steps in this tutorial for creating an elevation of the residence. All Units, Limits, Grid, and Snap values have been set previously.

Layers

The following layers are already created:

Name	Color	Linetype
Object	White	Continuous
Center	Yellow	Center
Hidden	Red	Hidden
Hatch2	Magenta	Continuous
Cutting Plane Line	Yellow	Dashed
Dimension	Yellow	Continuous

Suggested Commands

Add hatching to create an elevation with materials as shown.

Whenever possible, substitute the appropriate command alias in place of the full AutoCAD command in each tutorial step. For example, use "CP" for the COPY command, "L" for the LINE command, and so on. The complete listing of all command aliases is located in Chapter 1, Table 1–2.

STEP 1

Open the drawing 14_Elevation. Notice the appearance of a roof and brick hatch patterns in Figure 14–103. The roof and brick patterns need to be applied to the other irregular areas of this house elevation. First, make the Roof Hatch layer current through the Layer Control box in Figure 14–104.

Figure 14–103

Figure 14–104

STEP 2

Activate the Boundary Hatch and Fill dialog box and click on the Inherit Properties button in Figure 14–105. This button will allow you to click on an existing hatch pattern and have the pattern name, scale, and angle transfer to the proper fields in the dialog box. In other words, you will not have to figure out the name, scale, or angle of the pattern. The only requirement is that a hatch pattern already exists in the drawing.

When you return to the drawing in Figure 14–105, notice the appearance of a glyph that is similar to the Match Properties icon. Click on the existing roof pattern in the figure.

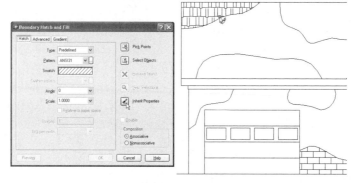

Figure 14–105

STEP 3

After you pick the existing hatch pattern in Figure 14–105, your cursor changes appearance again. Now pick internal points inside every irregular shape in Figure 14–106. Notice that each one highlights. When you're finished, press ENTER to return to the Boundary Hatch and Fill dialog box. Click the Preview button to examine the results. If the hatch pattern appearance is correct, right-click to place the roof hatch patterns.

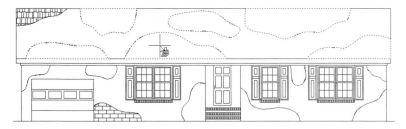

Figure 14–106

STEP 4

Next make the Exterior Brick layer current using the Layer Control box in Figure 14–107. Then activate the Boundary Hatch and Fill dialog box, click on the Inherit Properties button, pick the existing brick pattern, and click inside every irregular shape in Figure 14–108 until each one highlights. Preview the hatch pattern and right-click if the results are correct.

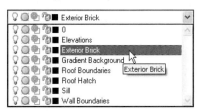

Figure 14–107

Figure 14–108

STEP 5

To give the appearance that the hatch patterns are floating on the roof and wall, use the Layer Control box in Figure 14–109 to turn off the two layers that control the boundaries of these patterns; namely Roof Boundaries and Wall Boundaries. Your display should appear similar to Figure 14–110.

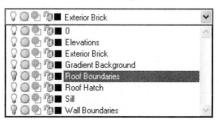

Figure 14–109

Figure 14–110

STEP 6

A number of gradient hatch patterns will now be applied to the outer portions of the elevation. First perform a ZOOM-EXTENTS. Then change the current layer to Gradient Background using the Layer Control box in Figure 14–111.

Command: **Z** *(For ZOOM)*
Specify corner of window, enter a scale factor (nX or nXP), or
[All/Center/Dynamic/Extents/Previous/Scale/Window] <real time>: **E** (For Extents)

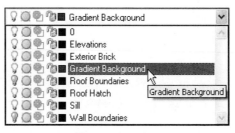

Figure 14–111

STEP 7

You will now apply a gradient hatch pattern to the upper portion of the elevation plan. Activate the Boundary Hatch and Fill dialog box, click on the Gradient tab, and pick the pattern as shown in Figure 14–112. Then click on the Pick Points button and click an internal point in the figure of the elevation. Once the boundary highlights, press ENTER, preview the hatch, and right-click to place the hatch pattern.

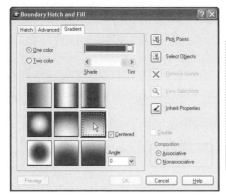

Figure 14–112

STEP 8

Apply another gradient hatch pattern to the lower portion of the elevation plan. Activate the Boundary Hatch and Fill dialog box, click on the Gradient tab, and pick the pattern as shown in Figure 14–113. Then click on the Pick Points button and click an internal point in the figure of the elevation. Once the boundary highlights, press ENTER, preview the hatch, and right-click to place the hatch pattern.

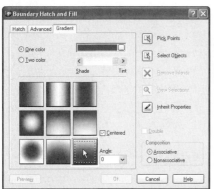

Figure 14–113

STEP 9

Apply gradient patterns to each of the garage windows in Figure 14–114 by using the Boundary Hatch command for each of the four windows.

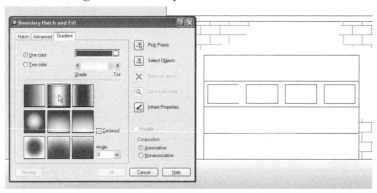

Figure 14–114

STEP 10

If the boundary lines that define the elevation disappear, send the gradient patterns back using the Draworder feature in Figure 14–115.

Figure 14–115

 Open the Exercise Manual PDF file for Chapter 14 on the accompanying CD for more tutorials and exercises.

 If you have the accompanying Exercise Manual, refer to Chapter 14 for more tutorials and exercises.

CHAPTER 15

Auxiliary Views

During the discussion of multiview drawings, we discovered that you need to draw enough views of an object to accurately describe it. In most cases, this requires a Front, Top, and Right Side view. Sometimes additional views are required, such as Left Side, Bottom, and Back views, to show features not visible in the three primary views. Other special views, like sections, are created to expose interior details for better clarity. Sometimes all of these views are still not enough to describe the object, especially when features are located on an inclined surface. To produce a view perpendicular to this inclined surface, an auxiliary view is drawn. This chapter will describe where auxiliary views are used and how they are projected from one view to another. A tutorial exercise is presented to show the steps in the construction of an auxiliary view.

AUXILIARY VIEW BASICS

Figure 15–1 presents interesting results if constructed as a multiview drawing or orthographic projection.

Figure 15–1

Figure 15–2 should be quite familiar; it represents the standard glass box with object located in the center. Again, the purpose of this box is to prove how orthographic views are organized and laid out. Figure 15–2 is no different. First the primary views, Front, Top, and Right Side views are projected from the object to intersect perpendicular with the glass plane. Under normal circumstances, this procedure would satisfy most multiview drawing cases. Remember, only those views necessary to describe the object are drawn. However, upon closer inspection, we notice that the object in the Front view consists of an angle forming an inclined surface.

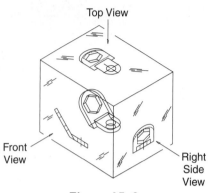

Figure 15–2

When you lay out the Front, Top, and Right Side views, a problem occurs. The Front view shows the basic shape of the object, the angle of the inclined surface (see Figure 15–3). The Top view shows the true size and shape of the surface formed by the circle and arc. The Right Side view shows the true thickness of the hole from information found in the Top view. However, there does not exist a true size and shape of the features found in the inclined surface at "A." We see the hexagonal hole going through the object in the Top and Right Side views. These views, however, show the detail not actual size; instead the views are foreshortened. For that matter, the entire inclined surface is foreshortened in all views. Dimensioning the size and location of the hexagonal hole would be challenging with these three views. This is one case where the Front, Top, and Right Side views are not enough to describe the object. An additional view, or auxiliary view, is used to display the true shape of surfaces along an incline.

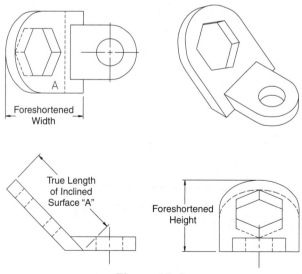

Figure 15–3

To demonstrate the creation of an auxiliary view, let's create another glass box; this time with an inclined plane on one side. This plane is always parallel to the inclined surface of the object. Instead of just the Front, Top, and Right Side views being projected, the geometry describing the features along the inclined surface is projected to the auxiliary plane similar to Figure 15–4.

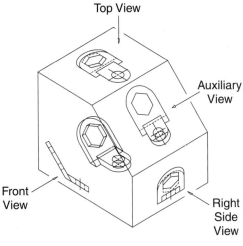

Figure 15–4

As in multiview projection, the edges of the glass box are unfolded with the edges of the Front view plane as the pivot. See Figure 15–5.

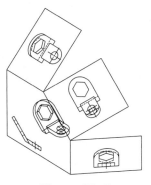

Figure 15–5

All planes are extended perpendicular to the Front view where the rotation stops. The result is the organization of the multiview drawing complete with an auxiliary viewing plane, as shown in Figure 15–6.

800

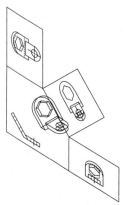

Figure 15–6

Figure 15–7 represents the final layout complete with auxiliary view. This figure shows the auxiliary being formed as a result of the inclined surface present in the Front view. An auxiliary view may be created in relation to any inclined surface located in any view. Also, the figure displays circles and arcs in the Top view that appear as ellipses in the auxiliary view.

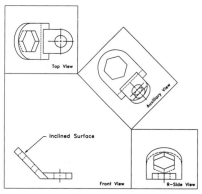

Figure 15–7

It is usually not required that elliptical shapes be drawn in one view where the feature is shown true size and shape in another. The resulting view minus these elliptical features is called a partial view, used extensively in auxiliary views. An example of the Top view converted to a partial view is shown in Figure 15–8.

Figure 15–8

A few rules to follow when constructing auxiliary views are displayed pictorially in Figure 15–9. First, the auxiliary plane is always constructed parallel to the inclined surface. Once this is established, visible as well as hidden features are projected from the incline to the auxiliary view. These projection lines are always drawn perpendicular to the inclined surface and the auxiliary view.

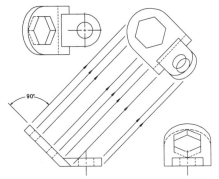

Figure 15–9

CONSTRUCTING AN AUXILIARY VIEW

Try It! – Open the drawing file 15_Aux Basics. Figure 15–10 is a multiview drawing consisting of Front, Top, and Right Side views. The inclined surface in the Front view is displayed in the Top and Right Side views; however, the surface appears foreshortened in both adjacent views. An auxiliary view of the incline needs to be made to show its true size and shape. Currently the display screen has the grid on in addition to the position of the typical AutoCAD cursor. Follow Figures 15–11 through 15–20 for one suggested method for projecting to find auxiliary views.

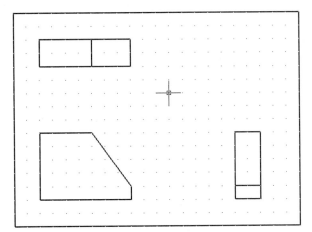

Figure 15–10

To assist with the projection process, it would help if the current grid display could be rotated parallel and perpendicular to the inclined surface (see Figure 15–11). This is accomplished through the SNAP command and the Rotate option.

Command: **SN** *(For SNAP)*
Specify snap spacing or [ON/OFF/Aspect/Rotate/Style/Type] <0.5000>: **R** *(For Rotate)*
Specify base point <0.0000,0.0000>: *(Select the endpoint of the line at "A")*
Specify rotation angle <0>: *(Select the endpoint of the line at "B")*

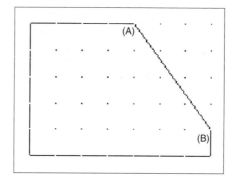

Figure 15–11

Figure 15–12 shows the result of rotating the grid through the use of the SNAP command. This operation has no effect on the already existing views; however, the grid is now placed rotated in relation to the incline located in the Front view. Notice that the appearance of the standard AutoCAD cursor has also changed to conform to the new grid orientation. In addition to snapping to these new grid dots, lines are easily drawn perpendicular to the incline with Ortho on, which will draw lines in relation to the current cursor.

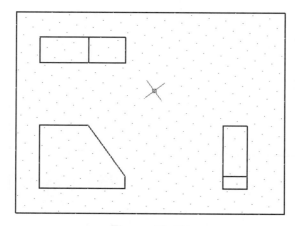

Figure 15–12

Use the OFFSET command to construct a reference line at a specified distance from the incline in the Front view (see Figure 15–13). This reference line becomes the start for the auxiliary view.

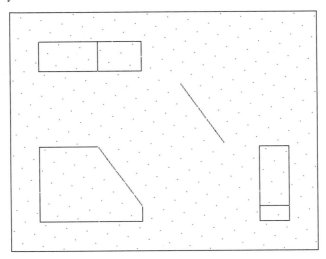

Figure 15–13

Use the LENGTHEN command to extend the two endpoints of the previous line (see Figure 15–14). The exact distances are not critical; however, the line should be long enough to accept projector lines from the Front view.

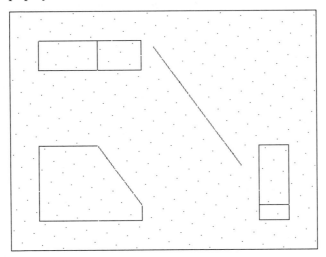

Figure 15–14

Use the OFFSET command to copy the auxiliary reference line the thickness of the object (see Figure 15–15). This distance may be retrieved from the depth of the Top or Right Side views because they both contain the depth measurement of the object.

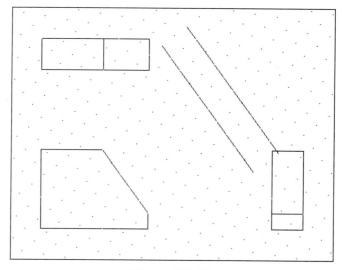

Figure 15–15

Use the LINE command and connect each intersection on the Front view perpendicular to the outer line on the auxiliary view (see Figure 15–16). Draw the lines starting with the OSNAP-Intersect option and ending with the OSNAP-Perpend option.

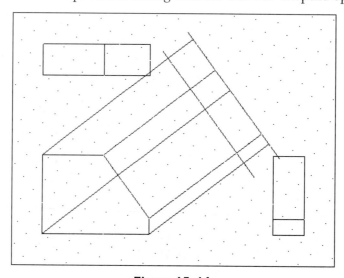

Figure 15–16

Before editing the auxiliary view, analyze the drawing to see if any corners in the Front view are to be represented as hidden lines in the auxiliary view (see Figure 15–17). It turns out that the lower-left corner of the Front view is hidden in the auxiliary view. Use the Layer Control box to convert this projection line from a linetype of continuous to hidden.

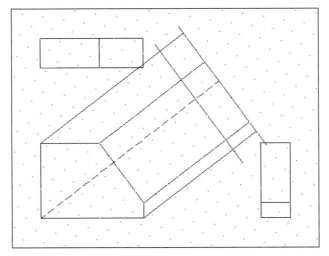

Figure 15–17

Use the TRIM command to partially delete all projection lines and corners of the auxiliary view. Another method would be to use the FILLET command set to a radius of 0. Selecting two lines in a corner automatically trims the excess lines away (see Figure 15–18).

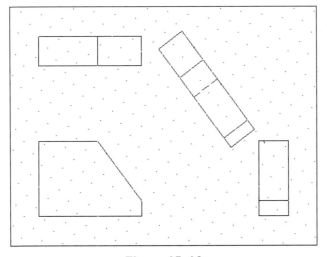

Figure 15–18

The result is a multiview drawing complete with auxiliary view displaying the true size and shape of the inclined surface. See Figure 15–19.

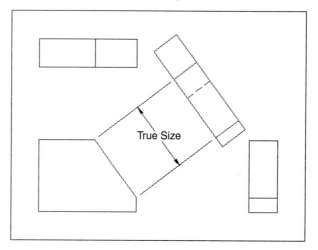

Figure 15–19

For dimensioning purposes, use the SNAP command and set the grid and cursor appearance back to normal. See Figure 15–20.

Command: **SN** *(For SNAP)*
Specify snap spacing or [ON/OFF/Aspect/Rotate/Style/Type] <0.5000>: **R** *(For Rotate)*
Specify base point <Current default>: **0,0**
Specify rotation angle <Current default>: **0**

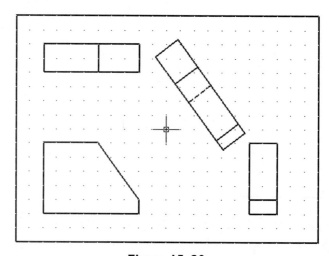

Figure 15–20

TUTORIAL EXERCISE: 15_BRACKET.DWG

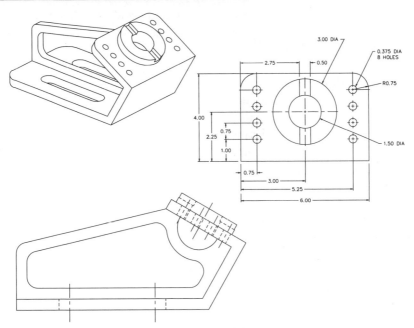

Figure 15–21

Purpose

This tutorial is designed to allow you to construct an auxiliary view of the inclined surface in the bracket shown in Figure 15–21.

System Settings

Since this drawing is provided on CD, edit an existing drawing called "15_Bracket." Follow the steps in this tutorial for the creation of an auxiliary view. All Units, Limits, Grid, and Snap values have been previously set.

Layers

The following layers have already been created with the following format:

Name	Color	Linetype
CEN	Yellow	Center
DIM	Yellow	Continuous
HID	Red	Hidden
OBJ	Cyan	Continuous

Suggested Commands

Begin this tutorial by using the OFFSET command to construct a series of lines parallel to the inclined surface containing the auxiliary view. Next construct lines perpendicular to the inclined surface. Use the CIRCLE command to begin laying out features that lie in the auxiliary view. Use ARRAY to copy the circle in a rectangular pattern. Add centerlines using the DIMCENTER command. Insert a predefined view called "Top." A three-view drawing consisting of Front, Top, and auxiliary views is completed.

Whenever possible, substitute the appropriate command alias in place of the full AutoCAD command in each tutorial step. For example, use "CP" for the COPY command, "L" for the LINE command, and so on. The complete listing of all command aliases is located in Table 1–2.

STEP 1

Before you begin, understand that an auxiliary view will be taken from a point of view illustrated in Figure 15–22. This direction of sight is always perpendicular to the inclined surface. This perpendicular direction ensures that the auxiliary view of the inclined surface will be of true size and shape. Begin this tutorial by checking to see that Ortho mode is turned off for the next few steps using the command from the keyboard or by clicking on the ortho button located in the status bar of the display screen. Also, make the OBJ layer current. Activate the View dialog box in Figure 15–23 and restore a previously saved view called "FRONT".

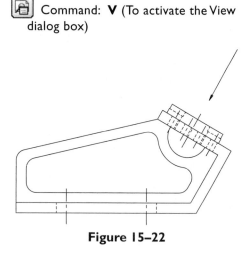

Command: **V** (To activate the View dialog box)

Figure 15–22

ORTHO Command: **ORTHO**
Enter mode [ON/OFF] <OFF>: *(Press* ENTER *to accept this default)*

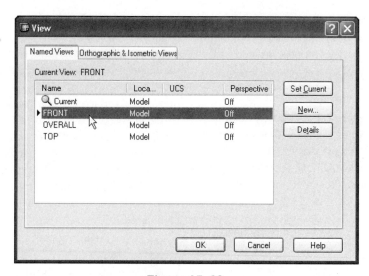

Figure 15–23

STEP 2

Use the snap command to rotate the grid perpendicular to the inclined surface. For the base point, identify the endpoint of the line at "A" in Figure 15–24. For the rotation angle, use the rubber-band cursor and mark a point at the endpoint of the line at "B." The grid should change along with the appearance of the standard cursor. Activate the View dialog box in Figure 15–25 and restore the view called "OVERALL".

Command: **SN** *(For SNAP)*
Specify snap spacing or [ON/OFF/Aspect/ Rotate/Style/Type] <0.50>: **R** *(For Rotate)*
Specify base point <0.00,0.00>: *(Select the endpoint at "A")*
Specify rotation angle <0>: *(Select the endpoint at "B")*

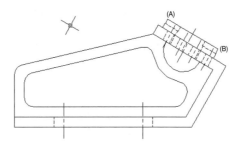 Command: **V** (To activate the View dialog box)

Figure 15–24

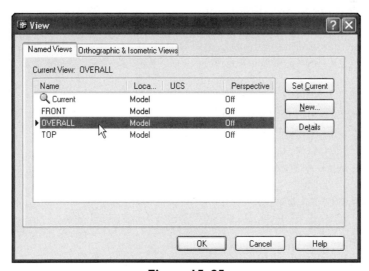

Figure 15–25

STEP 3

Turn snap off by pressing F9. Begin
the construction of the auxiliary view
by using the OFFSET command to copy
a line parallel to the inclined line (see
Figure 15–26). Use an offset distance of
8.50 as the distance between the Front
and auxiliary views.

 Command: **O** *(For OFFSET)*

Specify offset distance or [Through]
    <Through>: **8.50**
Select object to offset or <exit>: *(Select
    the inclined line at "A")*
Specify point on side to offset: *(Pick a
    point anywhere near "B")*
Select object to offset or <exit>: *(Press*
    ENTER *to exit this command)*

(B)

(A)

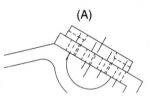

Figure 15–26

STEP 4

Refer to the working drawing in Figure
15–21 for the necessary dimensions
required to construct the auxiliary view.
Use the OFFSET command again to begin
constructing the depth of the auxiliary
view in Figure 15–27. Remember that
the depth of the auxiliary view is the same
dimension as the depth found in the Top
and Right Side views. Set the offset distance to 6.00. Then set the current layer
to OBJ using the Layer Control box.

(B)

(A)

 Command: **O** *(For OFFSET)*

Specify offset distance or [Through]
    <8.50>: *6.00*
Select object to offset or <exit>: *(Select
    the inclined line at "A")*
Specify point on side to offset: *(Pick a
    point anywhere near "B")*
Select object to offset or <exit>: *(Press*
    ENTER *to exit this command)*

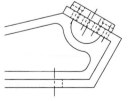

Figure 15–27

STEP 5

Project two lines from the endpoints of the Front view at "A" and "C" in Figure 15–28. These lines should extend past the outer line of the auxiliary view. Turn the Snap off and Ortho on. This should aid in this operation.

ORTHO Command: **ORTHO**
Enter mode [ON/OFF] <OFF>: **ON**

Command: **L** *(For LINE)*
Specify first point: *(Select the endpoint of the line at "A")*
Specify next point or [Undo]: *(Pick a point anywhere near "B")*
Specify next point or [Undo]: *(Press ENTER to exit this command)*

Command: **L** *(For LINE)*
Specify first point: *(Select the endpoint of the line at "C")*
Specify next point or [Undo]: *(Pick a point anywhere near "D")*
Specify next point or [Undo]: *(Press ENTER to exit this command)*

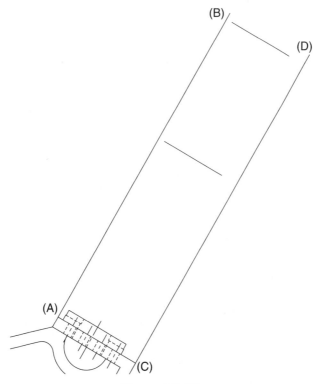

Figure 15–28

STEP 6

Use the ZOOM-WINDOW option to magnify the display of the auxiliary view similar to Figure 15–29. Then use the Multiple option of the FILLET command to create four corners of the view in Figure 15–29. When you're finished with this operation, activate the View dialog box in Figure 15–30 and click on the New button. When the New View dialog box appears in Figure 15–31, save the display to a new name of "AUX".

 Command: **F** *(For FILLET)*

Current settings: Mode = TRIM, Radius = 0.00

Select first object or [Polyline/Radius/Trim/mUltiple]: **U** *(For Multiple)*

Select first object or [Polyline/Radius/Trim/mUltiple]: *(Select line "A")*
Select second object: *(Select line "B")*
Select first object or [Polyline/Radius/Trim/mUltiple]: *(Select line "B")*
Select second object: *(Select line "C")*
Select first object or [Polyline/Radius/Trim/mUltiple]: *(Select line "C")*
Select second object: *(Select line "D")*
Select first object or [Polyline/Radius/Trim/mUltiple]: *(Select line "D")*
Select second object: *(Select line "A")*
Select first object or [Polyline/Radius/Trim/mUltiple]: *(Press ENTER to exit this command)*

 Command: **V** (To activate the View dialog box)

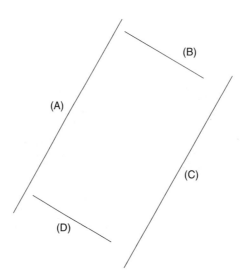

Figure 15–29

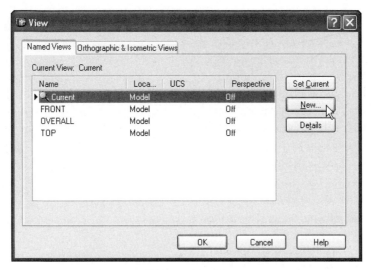

Figure 15–30

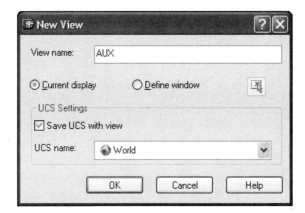

Figure 15–31

STEP 7

Use the ZOOM-Previous option to demagnify the screen back to the original display. Then draw a line from the endpoint of the centerline in the Front view at "A" to a point past the auxiliary view at "B" in Figure 15–32. Check to see that Ortho mode is on. This line will assist in constructing this line in the auxiliary view.

 Command: **Z** *(For ZOOM)*

Specify corner of window, enter a scale factor (nX or nXP), or

[All/Center/Dynamic/Extents/Previous/Scale/Window] <real time>: **P** *(For Previous)*

 Command: **L** (For LINE)

Specify first point: *(Select the endpoint of the centerline at "A")*

Specify next point or [Undo]: *(Pick a point anywhere near "B")*

Specify next point or [Undo]: *(Press ENTER to exit this command)*

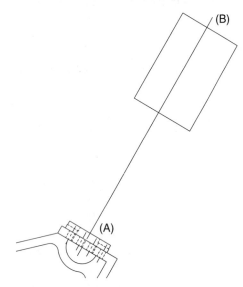

Figure 15–32

STEP 8

Use the OFFSET command and offset the line at "A" a distance of 3.00 units in Figure 15–33. The intersection of this line and the previous line form the center for placing two circles.

 Command: **O** *(For OFFSET)*

Specify offset distance or [Through] <6.00>: **3.00**

Select object to offset or <exit>: *(Select the line at "A")*

Specify point on side to offset: *(Pick a point anywhere near "B")*

Select object to offset or <exit>: *(Press ENTER to exit this command)*

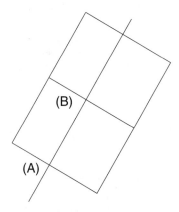

Figure 15–33

STEP 9

Activate the View dialog box in Figure 15–34 and restore the view "AUX." Then, draw two circles of diameters 3.00 and 1.50 from the center at "A" in Figure 15–35 using the CIRCLE command. For the center of the second circle, you can use the @ option to pick up the previous point that was the center of the 3.00-diameter circle.

Command: **V** (To activate the View dialog box)

Command: **C** *(For CIRCLE)*

Specify center point for circle or [3P/ 2P/Ttr (tan tan radius)]: *(Select the intersection at "A")*

Specify radius of circle or [Diameter]: **D** *(For Diameter)*

Specify diameter of circle: **3.00**

Command: **C** *(For CIRCLE)*

Specify center point for circle or [3P/2P/ Ttr (tan tan radius)]: **@** *(For the last point)*

Specify radius of circle or [Diameter] <1.50>: **D** *(For Diameter)*

Specify diameter of circle <3.00>: **1.50**

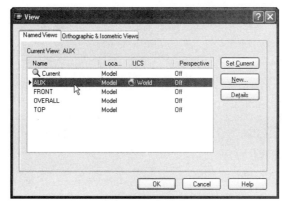

Figure 15–34

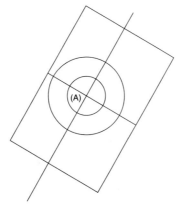

Figure 15–35

STEP 10

Use the OFFSET command to offset the centerline the distance of 0.25 units in Figure 15–36. Perform this operation on both sides of the centerline. Both offset lines form the width of the 0.50 slot.

 Command: **O** *(For OFFSET)*

Specify offset distance or [Through] <3.00>: **0.25**

Select object to offset or <exit>: *(Select the middle line at "A")*

Specify point on side to offset: *(Pick a point anywhere near "B")*

Select object to offset or <exit>: *(Select the middle line at "A" again)*

Specify point on side to offset: *(Pick a point anywhere near "C")*

Select object to offset or <exit>: *(Press ENTER to exit this command)*

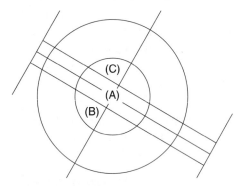

Figure 15–36

STEP 11

Use the TRIM command to trim away portions of the lines using the circles as cutting edges in Figure 15–37.

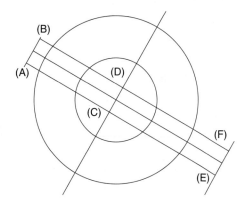 Command: **TR** *(For TRIM)*

Current settings: Projection=UCS Edge=None

Select cutting edges ...

Select objects: *(Select both circles as cutting edges)*

Select objects: *(Press ENTER to continue)*

Select object to trim or shift-select to extend or [Project/Edge/Undo]: *(Select the line at "A")*

Select object to trim or shift-select to extend or [Project/Edge/Undo]: *(Select the line at "B")*

Select object to trim or shift-select to extend or [Project/Edge/Undo]: *(Select the line at "C")*

Select object to trim or shift-select to extend or [Project/Edge/Undo]: *(Select the line at "D")*

Select object to trim or shift-select to extend or [Project/Edge/Undo]: *(Select the line at "E")*

Select object to trim or shift-select to extend or [Project/Edge/Undo]: *(Select the line at "F")*

Select object to trim or shift-select to extend or [Project/Edge/Undo]: *(Press ENTER to exit this command)*

Figure 15–37

STEP 12

Use the ERASE command to delete the two lines at "A" and "B" in Figure 15–38. Standard centerlines will be placed here later, marking the center of both circles.

 Command: **E** *(For ERASE)*

Select objects: *(Select the lines at "A" and "B")*

Select objects: *(Press ENTER to execute this command)*

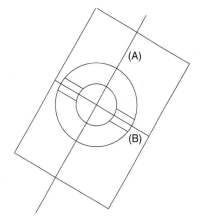

Figure 15–38

STEP 13

To identify the center of the small 0.375-diameter circle, use the OFFSET command to copy parallel the line at "A" a distance of 0.75 and the line at "C" the distance of 1.00 in Figure 15–39.

 Command: **O** *(For OFFSET)*

Specify offset distance or [Through] <0.25>: **0.75**

Select object to offset or <exit>: *(Select the line at "A")*

Specify point on side to offset: *(Pick a point anywhere near "B")*

Select object to offset or <exit>: *(Press ENTER to exit this command)*

 Command: **O** *(For OFFSET)*

Specify offset distance or [Through] <0.75>: **1.00**

Select object to offset or <exit>: *(Select the line at "C")*

Specify point on side to offset: *(Pick a point anywhere near "D")*

Select object to offset or <exit>: *(Press ENTER to exit this command)*

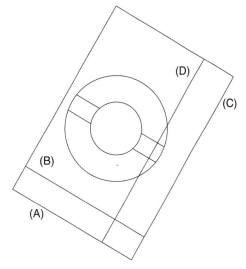

Figure 15–39

STEP 14

Draw a circle of 0.375 diameter from the intersection of the two lines created in the last OFFSET command in Figure 15–40. Use the ERASE command to delete the two lines labeled "B" and "C." A standard center marker will be placed at the center of this circle.

 Command: **C** *(For CIRCLE)*

Specify center point for circle or [3P/ 2P/Ttr (tan tan radius)]: *(Select the intersection at "A")*

Specify radius of circle or [Diameter] <0.75>: **D** *(For Diameter)*

Specify diameter of circle <1.50>: **0.375**

 Command: **E** *(For ERASE)*

Select objects: *(Select the two lines at "B" and "C")*

Select objects: *(Press ENTER to execute this command)*

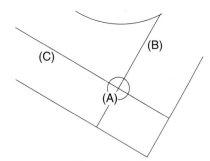

Figure 15–40

STEP 15

Set the current layer to CEN using the Layer Control box. Prepare the following parameters before placing a center marker at the center of the 0.375 diameter circle in Figure 15–41. Set the dimension variable DIMCEN to a value of -0.07 units. The negative value will construct lines that are drawn outside the circle. Use the DIMCENTER command to place the center marker.

Command: **DIMCEN**

Enter new value for DIMCEN <0.09>: **-0.07**

 Command: **DCE** *(For DIMCENTER)*

Select arc or circle: *(Select the small circle at "A")*

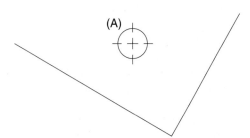

Figure 15–41

STEP 16

Use the ROTATE command to rotate the center marker parallel to the edges of the auxiliary view in Figure 15–42. Select the center marker and circle as the objects to rotate. Check to see that Ortho is on.

 Command: **RO** *(For ROTATE)*

Current positive angle in UCS:
ANGDIR=counterclockwise
ANGBASE=0
Select objects: *(Select the intersection at the small circle and all objects that make up the center marker; the Window option is recommended here)*
Select objects: *(Press ENTER to continue)*

Specify base point: *(Select the center of the small circle)*
Specify rotation angle or [Reference]: *(Pick a point anywhere near "B")*

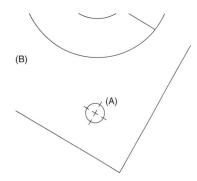

Figure 15–42

STEP 17

Since the remaining seven holes form a rectangular pattern, use the Array dialog box in Figure 15–43 and perform a rectangular array based on the objects in Figure 15–44. The number of rows is two and number of columns four. Distance between rows is -4.50 units and between columns is 0.75 units. You must also specify an angle for the array. Clicking on the angle measure button in Figure

15–43 will return you to the display screen where you can pick an endpoint at "A" and an endpoint at "B" in Figure 15–44 to specify the angle if it is not known. Since the angle of 150 degrees is given in this dialog box example, you could easily enter it into this field.

 Command: **AR** (To activate the Array dialog box)

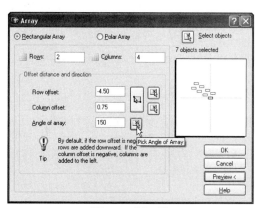

Figure 15–43

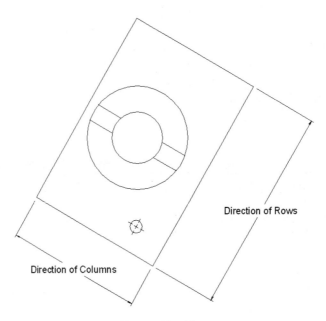

Figure 15–44

STEP 18

Use the FILLET command set to a radius of 0.75 to place a radius along the two corners of the auxiliary, following the prompts and Figure 15–45.

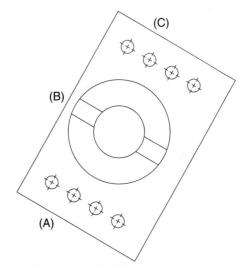

Figure 15–45

⌐ Command: **F** *(For FILLET)*
Current settings: Mode = TRIM, Radius = 0.00
Select first object or [Polyline/Radius/Trim/mUltiple]: **R** (For Radius)
Specify fillet radius <0.00>: **0.75**
Select first object or [Polyline/Radius/Trim/mUltiple]: *(Select line "A")*
Select second object: *(Select line "B")*

⌐ Command: **F** *(For FILLET)*
Current settings: Mode = TRIM, Radius = 0.75
Select first object or [Polyline/Radius/Trim/mUltiple]: *(Select line "B")*
Select second object: *(Select line "C")*

STEP 19

Place a center marker in the center of the two large circles using the existing value of the dimension variable DIMCEN (see Figure 15–46). Since the center marker is placed in relation to the World coordinate system, use the ROTATE command to rotate it parallel to the auxiliary view. Ortho should be on.

⊕ Command: **DCE** *(For DIMCENTER)*

Select arc or circle: *(Select the large circle at "A")*

↺ Command: **RO** *(For ROTATE)*

Current positive angle in UCS:
    ANGDIR=counterclockwise
    ANGBASE=0
Select objects: *(Select all lines that make up the large center marker)*

Select objects: *(Press* ENTER *to continue)*
Specify base point: **Cen**
of *(Select the edge of the large circle at "A")*
Specify rotation angle or [Reference]:
    (Pick a point anywhere near "B")

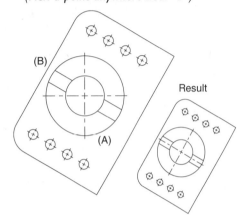

Figure 15–46

STEP 20

Activate the View dialog box in Figure 15–47 and restore the view named "OVERALL". Complete the drawing of the bracket by activating the Insert dialog box in Figure 15–48 and inserting an existing block called "TOP" into the drawing. This block represents the complete Top view of the drawing (see Figure 15–49). Use an insertion point of 0,0 for placing this view in the drawing.

⊟ Command: **V** (To activate the View dialog box)

⊡ Command: **I** (To activate the Insert dialog box)

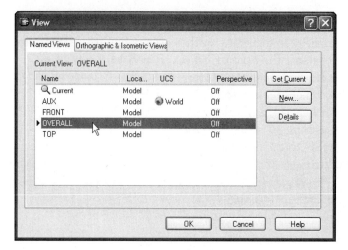

Figure 15–47

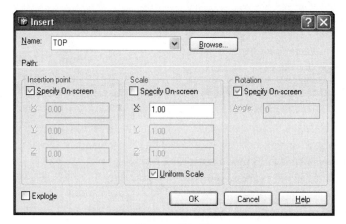

Figure 15–48

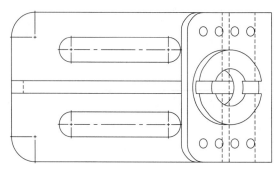

Figure 15–49

STEP 21

Return the grid to its original ortho-graphic form using the Rotate option of the SNAP command. Use a base point of 0,0 and a rotation angle of 0 degrees. You could also rotate the snap back to an angle of 0 through the Drafting Settings dialog box in Figure 15–50. The com-pleted drawing should appear similar to Figure 15–51.

Command: **SN** *(For SNAP)*
Specify snap spacing or [ON/OFF/Aspect/
   Rotate/Style/Type] <0.50>: **R** *(For
   Rotate)*
Specify base point <17.33,5.85>: **0,0**
Specify rotation angle <330>: **0**

 Note: You cannot use the Drafting Settings dialog box to rotate the snap unless you know the exact angle. A measuring tool for angles is not built into the dialog box. This is why the command version of the SNAP command must be used.

Figure 15–50

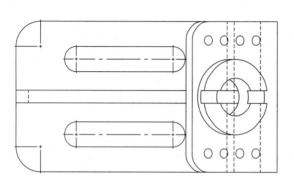
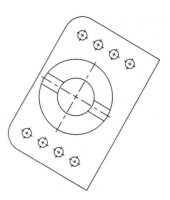

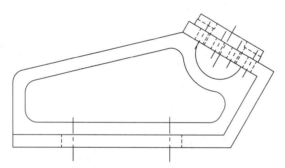

Figure 15–51

 Open the Exercise Manual PDF file for Chapter 15 on the accompanying CD for more tutorials and exercises.

 If you have the accompanying Exercise Manual, refer to Chapter 15 for more tutorials and exercises.

CHAPTER 16

Block Creation, AutoCAD DesignCenter, and MDE

This chapter begins the study of how blocks are created and merged into drawing files. This is a major productivity enhancement and is considered an electronic template similar to those methods used with block templates in manual drafting practices. The first segment of this chapter will discuss what blocks are and how they are created. Blocks are typically inserted in the current drawing but can be inserted in any drawing by utilizing the proper commands and techniques. Next, this chapter continues with a discussion about using the Insert dialog box and the AutoCAD DesignCenter to bring blocks into drawings. The DesignCenter is a special feature that allows blocks to be inserted in drawings with drag and drop techniques. In addition, blocks found in other drawings can easily be inserted in the current drawing from the DesignCenter. Yet another feature, the Tool palette, will allow you to organize blocks and hatch patterns in one convenient area. These object types can then be shared with the current drawing through drag and drop techniques. Finally, the chapter will discuss the use of MDE (Multiple Design Environment). This feature allows the opening of multiple drawings within a single AutoCAD session and provides a convenient method of exchanging data, such as blocks, between one drawing and another. As an added bonus to AutoCAD users, a series of block libraries is supplied with the package. These block libraries include such application areas as mechanical, architectural, electrical, piping, and welding just to name a few.

WHAT ARE BLOCKS?

Blocks usually consist of smaller components of a larger drawing. Typical examples include doors and windows for floor plans, nuts and bolts for mechanical assemblies, and resistors and transistors for electrical schematics. In Figure 16–1 showing an electrical schematic, all resistors, capacitors, tetrodes, and diodes are considered blocks that make up the total drawing of the electrical schematic. The capacitor is highlighted as one of these components.

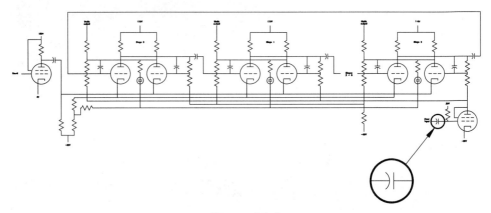

Figure 16–1

Blocks are created and then inserted in a drawing. When creating the block, you must first provide a name for the block. The capacitor in Figure 16–2 was assigned the name "CAPACITOR." Also, when you create a block, an insertion point is required. This acts as a reference point from which the block will be inserted. In Figure 16–2, the insertion point of the block is the left end of the line.

BLOCK NAME:
CAPACITOR

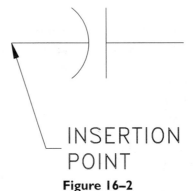

INSERTION
POINT

Figure 16–2

At times, blocks have to be rotated into position. In Figure 16–3, notice that one capacitor is rotated at a 45° angle while the other capacitor is rotated 270°. In this way the same block can be used numerous times even though it is positioned differently in the drawing.

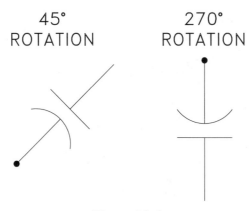

Figure 16–3

METHODS OF SELECTING BLOCK COMMANDS

Commands that deal with blocks can be selected from one of two pull-down menu areas. The first one is illustrated in Figure 16–4; selecting Block from the Draw pull-down menu exposes Make (the Block Definition dialog box) and Base, which defines a new base or insertion point for a block.

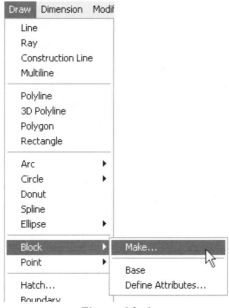

Figure 16–4

The second area that contains block-related commands is found on the Insert pull-down menu in Figure 16–5. This displays Block…, which activates the Insert dialog box.

Figure 16–5

Another convenient way of selecting Block-related commands is through the Draw toolbar or the Insert toolbar shown in Figure 16–6. You can also enter block-related commands from the keyboard, either using their entire name or through command aliases, as in the following examples:

Enter B to activate the Block Definition dialog box (BLOCK).

Enter I to activate the Insert dialog box (INSERT).

Enter W to activate the Write Block dialog box (WBLOCK).

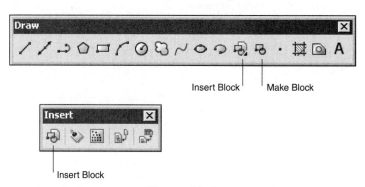

Figure 16–6

For a complete listing of other command aliases, refer to Chapter 1, Table 1–2.

The following commands will be discussed in greater detail in the following pages:

BLOCK	(Block Definition dialog box)
WBLOCK	(Write Block dialog box)
INSERT	(Insert dialog box)
ADCENTER	(AutoCAD DesignCenter)

CREATING A LOCAL BLOCK USING THE BLOCK COMMAND

Figure 16–7 is a drawing of a hex head bolt. This drawing consists of one polygon representing the hexagon, a circle indicating that the hexagon is circumscribed about the circle, and two centerlines. The centerlines were constructed with the dimension variable DIMCEN set to a -0.09 value and the DIMCENTER command. Rather than copy these individual objects numerous times throughout the drawing, you can create a block using the BLOCK command. Selecting Block from the Draw pull-down menu and then selecting Make, as in Figure 16–8, activates the Block Definition dialog box illustrated in Figure 16–9. Entering the letter B from the keyboard also activates the Block Definition dialog box.

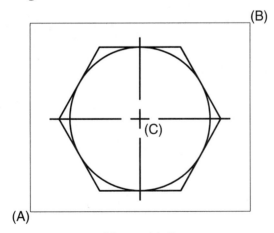

Figure 16–7

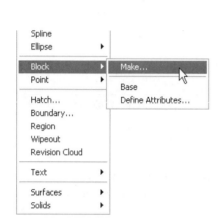

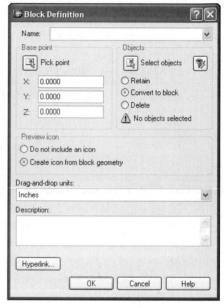

Figure 16–8   **Figure 16–9**

When you create a block, numerous objects that make up the block are all considered a single object. This dialog box allows for the newly created block to be merged into the current drawing file. This means that a block is only available in the drawing it was created in; it cannot be shared directly with other drawings. (The WBLOCK command is used to create global blocks that can be inserted in any drawing file. This command will be discussed later on in this chapter.)

To create a block through the Block Definition dialog box, enter the name of the block in the Name edit box, such as "Hexbolt" (see Figure 16–10). You can use up to 255 alphanumeric characters (only 31 if the system variable EXTNAMES is set to 0) when naming a block.

Figure 16–10

Next, click on the Select objects button; this will return you to the drawing editor (the Quick Select button provides a means of using filters to select objects). Create a window from "A" to "B" around the entire hex bolt to select all objects (refer to Figure 16–7).

When finished, press ENTER. This will return you to the Block Definition dialog box, where a previewed image of the block you are about to create is present (see Figure 16–11). The image will appear only if the Create icon from block geometry radio button is selected. Whenever you create a block, you can elect to allow the original objects that made up the hex bolt to remain on the screen (select Retain button), to be replaced with an instance of the block (select Convert to Block button) or to be removed after the block is created (select Delete button). Select the Delete radio button to erase the original objects after the block is created. If the original objects that made up the hex bolt are unintentionally removed from the screen during this creation process, the OOPS command can be used to retrieve all original objects to the screen.

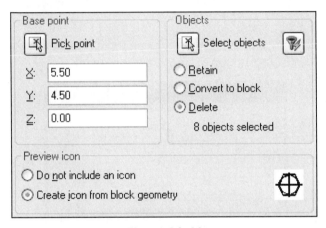

Figure 16–11

The next step is to create a base point or insertion point for the block. This is considered a point of reference and should be identified at a key location along the block. By default, the Base point in the Block Definition dialog box is located at 0,0,0. To enter a more appropriate base point location, click on the Pick point button; this will return to the drawing editor and allow you to pick the center of the hex bolt at "C" in Figure 16–7 using the OSNAP-Intersect mode. Once selected, this returns you to the Block Definition dialog box; notice how the Base Point information now reflects the key location along the block (see Figure 16–12).

Yet another feature of the Block Definition dialog box allows you to add a description for the block. Many times, the name of the block hides the real meaning of the block, especially if the block name consists of only a few letters. Click in the Description edit box and add the following statement: "This is a hexagonal head bolt." This will allow you to refer back to the description in case the block name does not indicate the true intended purpose for the block (see Figure 16–13). Specifying the Insert units determines the type of units utilized for scaling the block when it is inserted from the AutoCAD DesignCenter.

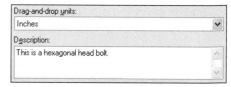

Figure 16–12   **Figure 16–13**

The complete dialog box is illustrated in Figure 16–14. If you are satisfied with the block name, base point, and description, click on the OK button at the bottom of the dialog box to add the block to the database of the drawing.

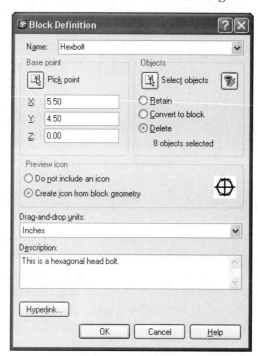

Figure 16–14

As the block is written to the database of the current drawing, the individual objects used to create the block automatically disappear from the screen (due to our earlier selection of the Delete radio button). When the block is later inserted in a drawing, it will be placed in relation to the insertion point.

CREATING A DRAWING FILE USING THE WBLOCK COMMAND

In the previous example of using the BLOCK command, several objects were grouped into a single object for insertion in a drawing. As this type of block is created, its intended purpose is for insertion in the drawing it was created in. Using the WBLOCK (Write block) command creates a block by a method that is similar to the process used in the BLOCK command. However, this will write the objects to a separate file, which allows the block to be inserted in any drawing. Entering WBLOCK or just W at the command prompt activates the Write Block dialog box illustrated in Figure 16–15.

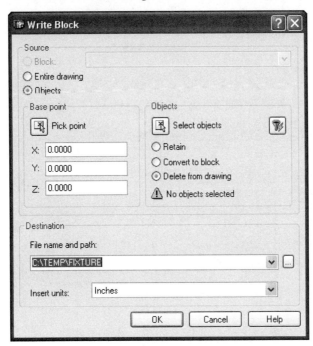

Figure 16–15

In Figure 16–16 the source for the write (global) block is Objects (the Objects radio button is selected). Follow the same steps demonstrated earlier for creating blocks by selecting the objects that will make up the global block. The Block radio button (grayed out because no blocks exist in the current drawing) would allow you to select an existing block to create the global block. The Entire drawing radio button would allow you to select the complete current drawing for creation of the block file. Next, click on the Base point area's Pick point button to select an appropriate insertion point for the global block.

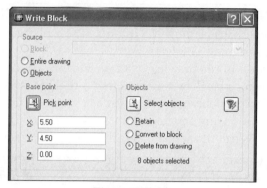

Figure 16–16

It may be unclear where the block file is actually being written. Be sure to carefully provide the destination information (see Figure 16–17) for the Write Block dialog box. Provide an appropriate file name in the edit box. To give a location for the block, click on the Browse button and choose a suitable drive and folder in which to store the file (see Figure 16–18). Clicking on the OK button in the Write Block dialog box writes the objects to the drive location with the file name specified by the user.

Figure 16–17

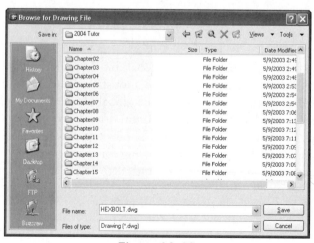

Figure 16–18

THE INSERT DIALOG BOX

Once blocks are created through the BLOCK and WBLOCK commands, they are merged or inserted in the drawing through the INSERT command. Enter INSERT or I from the keyboard or choose the command from the Insert pull-down, the Insert toolbar, or the Draw toolbar, as shown in Figure 16–19, to activate the Insert dialog box in Figure 16–20. This dialog box is used to dynamically insert blocks.

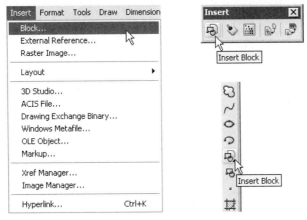

Figure 16–19

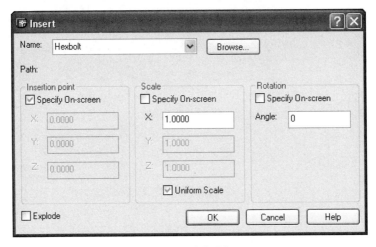

Figure 16–20

First, by clicking on the Name drop-down list box, select a block from the current drawing (clicking on the Browse button locates global blocks or drawing files). After you identify the name of the block to insert, the point where the block will be inserted must be specified, along with its scale and rotation angle. By default, the Insertion point area's Specify on Screen box is checked. This means you will be prompted for the insertion point at the command prompt area of the drawing editor. The default values for the scale and rotation insert the block at their original size and orientation. It was already mentioned that a block consists of several objects combined into a single object. The Explode box determines if the block is inserted as one object or if the block is inserted and then exploded back to its individual objects. Once the name of the block is selected, such as Hexbolt, click on the OK button; the following prompts complete the block insertion operation:

Specify insertion point or [Scale/X/Y/Z/Rotate/PScale/PX/PY/PZ/PRotate]: *(Mark a point at "A" in Figure 16–21 to insert the block)*

If the Specify On-Screen boxes are also checked for scale and rotation, the following prompts will complete the block insertion operation.

Specify insertion point or [Scale/X/Y/Z/Rotate/PScale/PX/PY/PZ/PRotate]: *(Mark a point at "A" in Figure 16–21 to insert the block)*
Enter X scale factor, specify opposite corner, or [Corner/XYZ] <1>: *(Press ENTER to accept default X scale factor)*
Enter Y scale factor <use X scale factor>:): *(Press ENTER to accept default)*
Specify rotation angle <0>: *(Press ENTER to accept the default rotation angle and insert the block in Figure 16–21)*

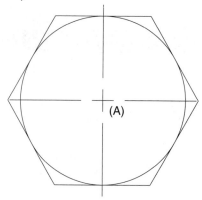

(A)

Figure 16–21

If blocks are defined as part of the database of the current drawing, they may be selected from the Name drop-down list box (see Figure 16–22).

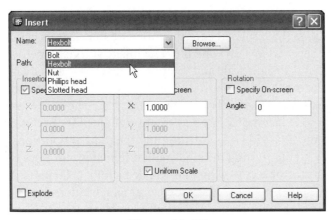

Figure 16–22

For inserting global blocks in a drawing, select the Insert dialog box Browse… button also shown in Figure 16–22, which displays the Select Drawing File dialog box illustrated in Figure 16–23. This is the same dialog box associated with opening drawing files. In fact, you will be unable to distinguish between global blocks and any other AutoCAD drawing file. Global blocks are simply drawing files created in a unique way and either can be inserted in the current drawing. Select the desired folder where the global block or drawing file is located followed by the name of the drawing. This will return you to the main Insert dialog box with the file now available in the Name drop-down list box and ready for insertion.

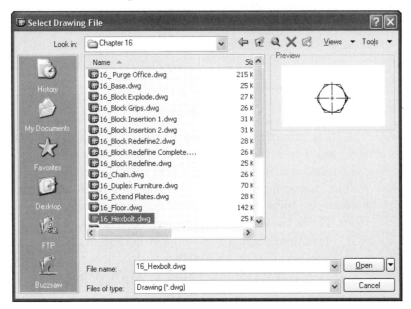

Figure 16–23

APPLICATIONS OF BLOCK INSERTIONS

Figure 16–24 shows the results of entering different scale factors and rotation angles when blocks are inserted in a drawing file. The image at "A" shows the block inserted in a drawing with its default scale and rotation angle values. The image at "B" shows the result of inserting the block with a scale factor of 0.50 and a rotation angle of 0°. The image appears half its normal size. The image at "C" shows the result of inserting the block with a scale factor of 1.75 and a rotation angle of 0°. In this image, the scale factor increases the block in size while keeping the same proportions. The image at "D" shows the result of inserting the block with different X and Y scale factors. In this image, the X scale factor is 0.50 while the Y scale factor is 2.00 units. Notice how out of proportion the block appears. There are certain applications where different scale factors are required to produce the desired effect. The image at "E" shows the result of inserting the block with the default scale factor and a rotation angle of 30°. As with all rotations, a positive angle rotates the block in the counterclockwise direction; negative angles rotate in the clockwise direction.

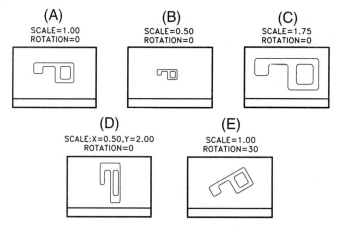

Figure 16–24

Figure 16–25 shows the results of entering different negative scale factors when inserting blocks in a drawing file. The image at "A" shows the block inserted in a drawing with its default scale and rotation angle values. The image at "B" shows the result of inserting the block with an X scale factor of −1.00 units; notice how the image flips in the X direction. This result is similar to using the MIRROR command with the Y-axis as the axis to mirror about. The image at "C" shows the result of inserting the block with a Y scale factor of −1.00 units; this time, notice how the image flips in the Y direction. The image at "D" shows the result of inserting the block with both X and Y scale factors of −1.00 units. Here, the object is flipped in both the X and Y directions.

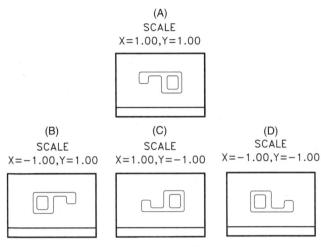

Figure 16–25

TRIMMING TO BLOCK OBJECTS

The ability to trim to a block object is possible through the TRIM command. As you select cutting edges on the block to trim to, the command isolates the cutting edges from the remainder of objects that make up the block. In Figure 16–26, the outside edges of the bolt act as cutting edges.

 Try It! - To test this feature, open the drawing 16_Trim Plates in Figure 16–26 and follow the command sequence and illustrations below.

 Command: **TR** *(For TRIM)*
Current settings: Projection=UCS, Edge=None
Select cutting edges ...
Select objects: *(Pick the edge of the bolt at "A")*
Select objects: *(Pick the edge of the bolt at "B")*
Select objects: *(Press ENTER to continue)*
Select object to trim or shift-select to extend or [Project/Edge/Undo]: *(Pick the line at "C")*
Select object to trim or shift-select to extend or [Project/Edge/Undo]: *(Pick the line at "D")*
Select object to trim or shift-select to extend or [Project/Edge/Undo]: *(Pick the line at "E")*
Select object to trim or shift-select to extend or [Project/Edge/Undo]: *(Press ENTER to exit this command)*

The result of using the TRIM command is illustrated in Figure 16–27.

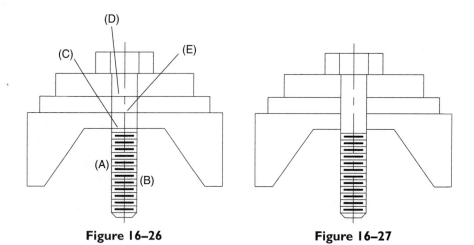

| Figure 16–26 | Figure 16–27 |

EXTENDING TO BLOCK OBJECTS

As with the TRIM command, objects can be extended to boundary edges that are part of a block with the EXTEND command. Selecting boundary edges that consist of block components isolates these objects. This enables you to extend objects to these boundary edges.

 Try It! - To test this feature, open the drawing 16_Extend Plates in Figure 16–28 and follow the command sequence and illustrations below.

Command: **EX** *(For EXTEND)*

Current settings: Projection=UCS, Edge=None

Select boundary edges ...

Select objects: *(Pick the edge of the bolt at "A")*

Select objects: *(Pick the edge of the bolt at "B")*

Select objects: *(Press ENTER to continue with this command)*

Select object to extend or shift-select to trim or [Project/Edge/Undo]: *(Pick the line at "C")*

Select object to extend or shift-select to trim or [Project/Edge/Undo]: *(Pick the line at "D")*

Select object to extend or shift-select to trim or [Project/Edge/Undo]: *(Pick the line at "E")*

Select object to extend or shift-select to trim or [Project/Edge/Undo]: *(Pick the line at "F")*

Select object to extend or shift-select to trim or [Project/Edge/Undo]: *(Pick the line at "G")*

Select object to extend or shift-select to trim or [Project/Edge/Undo]: *(Pick the line at "H")*

Select object to extend or shift-select to trim or [Project/Edge/Undo]: *(Press* ENTER *to exit this command)*

The result of using the EXTEND command on block objects as cutting edges is illustrated in Figure 16–29.

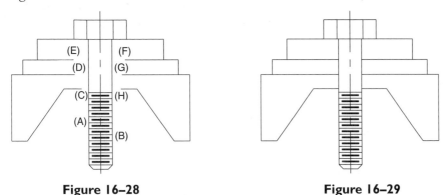

Figure 16–28   **Figure 16–29**

EXPLODING BLOCKS

It has already been mentioned that as blocks are inserted in a drawing file, they are considered one object even though they consist of numerous individual objects. At times it is necessary to break a block up into its individual parts. The EXPLODE command is used for this. Figure 16–30 shows three blocks that have been inserted with different scale factors. The block at "A" was inserted with the default scale and rotation angle values. The block at "B" was inserted with an X scale factor of 0.50 and a Y scale factor of 2.00 units. Finally, the block at "C" was inserted with an X scale factor of –1.00 and a rotation angle of 30°. In all three cases, the figures will appear the same even after the blocks are broken up into individual objects through the EXPLODE command.

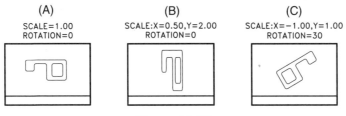

Figure 16–30

MANAGING UNUSED BLOCK DATA WITH THE PURGE DIALOG BOX

AutoCAD stores named objects (blocks, dimension styles, layers, layouts, linetypes, multiline styles, shapes, and text styles) in with the drawing. When the drawing is opened, AutoCAD determines whether other objects in the drawing reference each named object. If a named object is unused and not referenced, you can remove the definition of the named object from the drawing by using the PURGE command. This is a very important productivity technique used for compressing or cleaning up the database of the drawing. Picking File from the pull-down menu followed by Drawing Utilities displays Purge... in Figure 16–31. Clicking on Purge... displays the Purge dialog box in Figure 16–32.

Figure 16–31

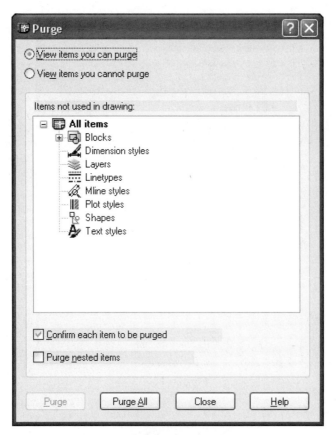

Figure 16–32

The Blocks category is expanded by clicking on the "+". This produces the list all the items that are currently unused in the drawing that can be removed (see Figure 16–33.) Clicking on the item "asesmp" and then picking the Purge button at the bottom of the dialog box will display the Confirm Purge dialog box in Figure 16–34. Click the Yes button and the block is removed from the listed items in Figure 16–35.

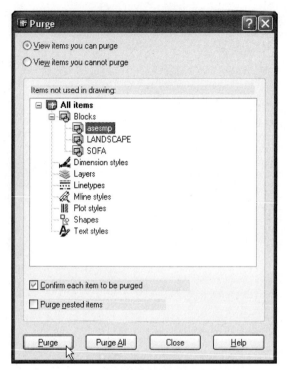

Figure 16–33

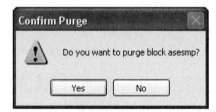

Figure 16–34

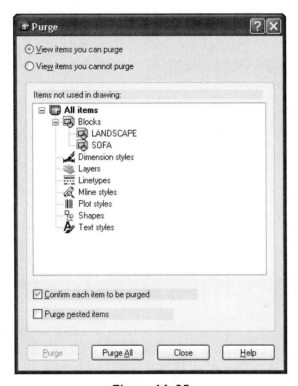

Figure 16–35

Other controls are available in the Purge dialog box such as the ability to view items that cannot be purged and the capability of purging nested items. An example of purging a nested item would be purging a block definition that lies inside another block definition. Entering -PURGE or PU at the Command prompt also activates the Purge dialog box.

Tip: The layer "0", Standard text and dimension styles, and the Continuous linetype cannot be purged from a drawing.

Try It! - Open the drawing file 16_Purge in Figure 16–36. Activate the Purge dialog box and click the Purge All button. You will be prompted to remove all items individually through the Confirm Purge dialog box in Figure 16–37. Answer yes to each item you want purged until all items including blocks, layers, linetypes, text styles, dimension styles, and multiline styles have been removed from the database of the drawing.

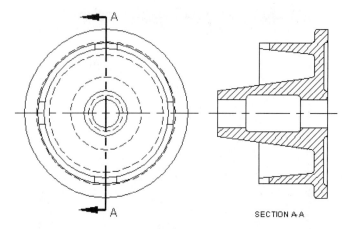

Figure 16–36

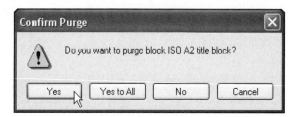

Figure 16–37

When finished, the following items were removed or purged from this drawing:

"ISO A2 title block"
"ISO-25"
"Front View"
"Isometric View"
"Phantom"
"Section View"
"BORDER2"
"CENTER2"
"DASHED"
"DIVIDE2"
"HIDDEN2"
"PARALLEL_5"
"ITALICC"
"ROMANS"

GRIPS AND BLOCKS

Grips have interesting results when blocks are selected. When a block is selected as shown in Figure 16–38, only one grip appears. The location of this grip happens to be at the insertion point of the block. The system variable GRIPBLOCK controls this and can be set in the Options dialog box, Selection tab (see Figure 16–39). Notice that, in the Grips area, the check box for Enable grips within blocks is not checked. A grip will only appear at the insertion point of the block as a result of this setting in the dialog box.

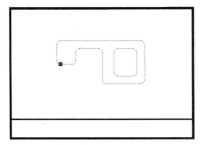

Figure 16–38

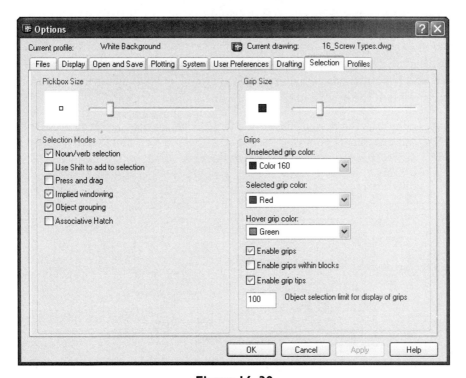

Figure 16–39

An entirely different result is achieved when you place a check in the Enable Grips within blocks check box, as in Figure 16–40. Now, whenever a block is selected, the grips will appear not only at the insertion point but on all objects that make up the block (see Figure 16–41).

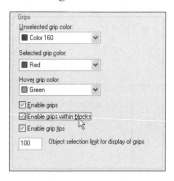

Figure 16–40

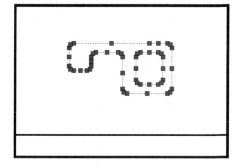

Figure 16–41

REDEFINING BLOCKS

At times, a block needs to be edited. Rather than erase all occurrences of the block in a drawing, you can redefine it. This is considered a major productivity technique because all blocks that share the same name as the block being redefined will automatically update to the latest changes. Figure 16–42 shows various blocks inserted in a drawing; one is inserted at normal size, another at half size, another with different X and Y scale factors, and two others rotated at different angles. The block name is "Brick" and the insertion point of the brick is at the center of the middle circle.

 Try It! - The drawing file 16_Block Redefine has been provided on CD for you to experiment with redefining blocks. Open it now and follow the next series of illustrations used to accomplish this task.

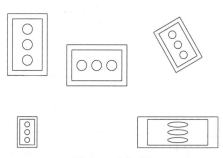

Figure 16–42

In order for a block in the current drawing to be redefined, another block is inserted and exploded to break it down into individual objects. In the example of the Brick, the inside rectangle needs to be deleted. The name of the block will still be "Brick" and its insertion point will remain at the center of the middle circle. Perform the exploding and erasing operations. The new brick is illustrated in Figure 16–43.

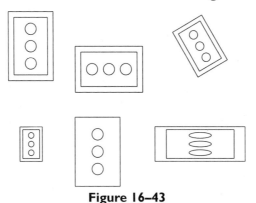

Figure 16–43

Activate the Block Definition dialog box through the BLOCK command; name the block "Brick," select the proper objects needed to define the new brick, identify the insertion point at the center of the middle circle, and give a description of the brick as illustrated in Figure 16–44.

After you click on the OK button, another dialog box appears similar to Figure 16–45. A block called "Brick" is already defined in the drawing. This dialog box needs confirmation to redefine all other blocks in the drawing.

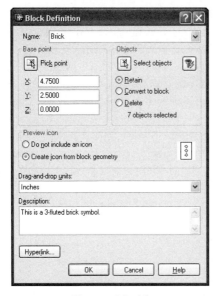

Figure 16–44

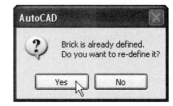

Figure 16–45

Clicking the Yes button in the AutoCAD Alert box redefines all blocks in the drawing to reflect the changes to the new brick design. These changes are illustrated in Figure 16–46. However, notice one of the bricks did not update to the latest changes. This brick was exploded sometime during the drawing process. Since it already consists of individual objects and is no longer a block definition, it will not update to any new changes made to the brick.

 Note: While it is popular to merge complete drawings with other drawings and then explode them, care should be taken when exploding small component blocks, especially when global changes are desired.

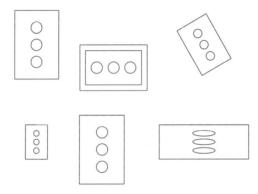

Figure 16–46

THE BASE COMMAND

Entire drawings can easily be inserted in other drawings through the Insert dialog box. This is a popular technique even when you want to use only a portion of the inserted drawing. Since the drawing file is considered one object when inserted, it can be exploded and any unwanted segments can then be removed. Also, by default, a drawing file is inserted in relation to the 0,0 screen location (usually the lower-left corner of the display screen). In Figure 16–47, the title block was constructed from point 2,2; however, when it is inserted in another drawing, the title will be inserted in relation to point 0,0. To provide better control over the insertion of drawing files, the BASE command can be used. This command identifies a new insertion point. In Figure 16–48, a new base point is identified at the lower-left corner of the title block. Now when the title block is inserted, it will be brought into the drawing in relation to its lower left corner rather than point 0,0. Study the figures and the following command sequence for using the BASE command:

Command: **BASE**
Enter base point <0.0000,0.0000,0.0000>: **End**
of (Pick the endpoint of the title block at "A" in Figure 16–48)

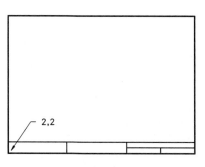

Figure 16–47

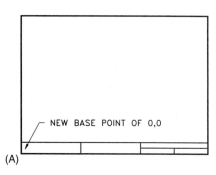

(A)

Figure 16–48

BLOCKS AND THE DIVIDE COMMAND

The DIVIDE command was already covered in Chapter 5. It allows you to select an object, give the number of segments, and place point objects at equally spaced distances depending on the number of segments. The DIVIDE command has a Block option that allows you to place blocks at equally spaced distances.

 Try It! - Open the drawing file 16_Speaker. In Figure 16–49, a counter-bore hole and centerline needs to be copied 12 times around the elliptical centerline so that each hole is equally spaced from others. (The illustration of the block "CBORE" is displayed at twice its normal size.) Because the ARRAY command is used to copy objects in a rectangular or circular pattern, that command cannot be used for an ellipse. The DIVIDE command's Block option allows you to specify the name of the block and the number of segments. In Figure 16–49, the elliptical centerline is identified as the object to divide. Follow the command sequence to place the block "CBORE" in the elliptical pattern:

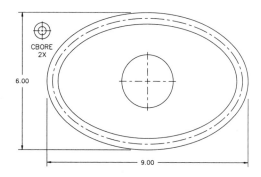

Figure 16–49

Command: **DIV** *(For DIVIDE)*
Select object to divide: *(Select the elliptical centerline)*
Enter the number of segments or [Block]: **B** *(For Block)*
Enter name or block to insert: **CBORE**

Align block with object? [Yes/No] <Y>:*(Press* ENTER *to accept)*
Enter the number of segments: **12**

The results are illustrated in Figure 16–50. Notice how the elliptical centerline is divided into 12 equal segments by 12 blocks called "CBORE." In this way, any object may be divided through the use of blocks.

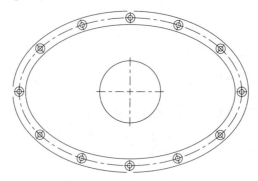

Figure 16–50

BLOCKS AND THE MEASURE COMMAND

As with the DIVIDE command, the MEASURE command offers increased productivity when you measure an object and insert blocks at the same time.

 Try It! - Open the drawing file 16_Chain. In Figure 16–51, a polyline path will be divided into 0.50 length segments using the block "CHAIN2." The perimeter of the polyline was calculated by the LIST command to be 22.00 units, which is evenly divisible by 0.50 and will allow the insertion of 44 blocks. Follow the command prompt sequence below for placing a series of blocks called "CHAIN2" around the polyline path to create a linked chain:

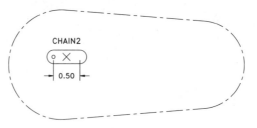

Figure 16–51

Command: **ME** *(For MEASURE)*
Select object to measure: *(Pick the polyline path in Figure 16-51)*
Specify length of segment or [Block]: **B** *(For Block)*
Enter name of block to insert: **CHAIN2**

Align block with object? [Yes/No] <Y> *(Press* ENTER *to accept)*
Specify length of segment: **0.50**

The result is illustrated in Figure 16–52, with all chain links being measured along the polyline path at increments of 0.50 units.

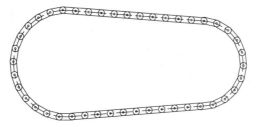

Figure 16–52

Answering "No" to the prompt "Align block with object? <Y>" displays the results in Figure 16–53. Here all blocks are inserted horizontally and travel in 0.50 increments. While the polyline path has been successfully measured, the results are not acceptable for creating the chain.

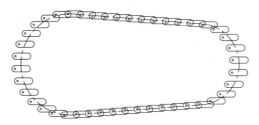

Figure 16–53

RENAMING BLOCKS

Blocks can be renamed to make their meanings more clear through the rename command. This command can be found in the Format pull-down menu as shown in Figure 16–54, and when selected will display a dialog box similar to the one illustrated in Figure 16–55. Clicking on Blocks in the dialog box lists all blocks defined in the current drawing. One block with the name "REFRIG" was abbreviated and we wish to give it a full name. Clicking on the name "REFRIG" will paste it in the Old Name edit box. Type the desired full name REFRIGERATOR in the Rename To edit box and click on the Rename To button to rename the block.

Figure 16–54

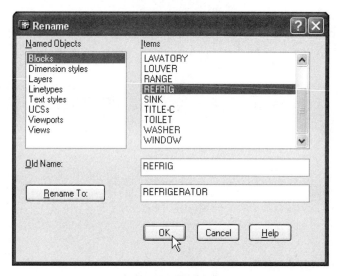

Figure 16–55

ADDITIONAL TIPS FOR WORKING WITH BLOCKS

CREATE BLOCKS ON LAYER 0

Blocks are best controlled, when dealing with layer colors, linetypes and lineweights, by being drawn on Layer 0, because it is considered a neutral layer. By default, Layer 0 is assigned the color White and the Continuous linetype. Objects drawn on Layer 0 and then converted to blocks will take on the properties of the current layer when inserted in the drawing. The current layer will control color, linetype and lineweight.

CREATE BLOCKS FULL SIZE IF APPLICABLE

Figure 16–56 illustrates a drawing of a refrigerator complete with dimensions. In keeping with the concept of drawing in real world units in CAD or at full size, individual blocks must also be drawn at full size in order for them to be inserted in the drawing at the correct proportions. For this block, construct a rectangle 28 units in the X direction and 24 units in the Y direction. Create a block called Refrigerator by picking the rectangle. When testing out this block, be sure to be in a drawing that is set to the proper units based on a scale such as ½"=1'-0" or ¼"=1'-0". These are typical architectural scales and the block of the refrigerator will be in the correct proportions on these types of sheet sizes.

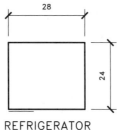

REFRIGERATOR

Figure 16–56

USE GRID WHEN PROPORTIONALITY, BUT NOT SCALE, IS IMPORTANT

Sometimes blocks represent drawings where the scale of each block is not important. In the previous example of the Refrigerator, scale was very important in order for the refrigerator to be drawn according to its full size dimensions. This is not the case in Figure 16–57 of the drawing of the resistor block. Electrical schematic blocks are generally not drawn to any specific scale: however, it is important that all blocks are proportional to each other. Setting up a grid is considered good practice in keeping all blocks of the same proportions. Whatever the size of the grid, all blocks are designed around the same grid size. The result is in Figure 16–58; with four blocks being drawn with the same grid, their proportions look acceptable.

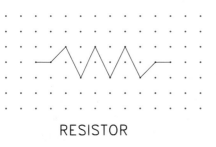

RESISTOR

Figure 16–57

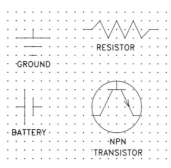

Figure 16–58

CREATE BLOCK OUT OF 1-UNIT PROPORTIONS

Figure 16–59 shows a door block with its insertion point at the bottom of the line that represents the door. Rather than create each door block separately to account for different door sizes, create the door in Figure 16–59 to fit into a 1-unit by 1-unit square. The purpose of drawing the door block inside a 1-unit square is to create only one block of the door and insert it at a scale factor matching the required door size. For example, for a 2'-8" door, enter 2'8 or 32 when prompted for the X and Y scale factors. For a 3'-0" door, enter 3' or 36 when prompted for the scale factors. Numerous doors of different types can be inserted in a drawing using only one block of the door. Also try using a negative scale factor to mirror the door as it is inserted.

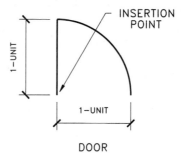

Figure 16–59

LISTING BLOCK INFORMATION

Using the LIST command on a block displays the following information:

BLOCK REFERENCE Layer: "0"
Space: Model space
Handle = 760
"RESISTOR"
at point, X=6.86 Y=2.33 Z=0.00
X scale factor 1.00
Y scale factor 1.00
rotation angle   0
Z scale factor 1.00

Listing a block identifies the object as a Block Reference. Notice that the name of the block is listed, along with insertion point, scale factors, and rotation angle.

Another way of listing block information is through the Properties palette as shown in Figure 16–60. This gives a more graphical approach to the block information. Also, various items can be changed in the Properties palette, which will automatically affect the inserted block in the drawing.

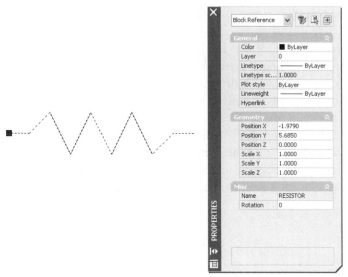

Figure 16–60

THE AUTOCAD DESIGNCENTER

A revolutionary tool called the AutoCAD DesignCenter has been developed as a simple, efficient device for sharing drawing data. It provides a means of inserting blocks and drawings even more efficiently than through the Insert dialog box. This feature has the distinct advantage of inserting specific blocks internal to one drawing into another drawing. If the DesignCenter is not present, you can load it by choosing AutoCAD DesignCenter from the Tools pull-down menu, as shown in Figure 16–61, selecting the AutoCAD DesignCenter button from the Standard toolbar also in Figure 16–61 or typing ADCENTER at the keyboard.

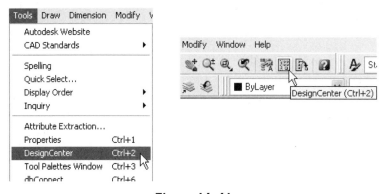

Figure 16–61

When used for the first time, the DesignCenter loads in the middle of the screen. A new Auto-hide feature for the DesignCenter, allows it to hide (collapse) once you move the cursor outside of the DesignCenter window. This helps clear the drawing area when it's not being used. To expand it again, simply move the cursor over the DesignCenter title bar. A shortcut menu allows you to turn the Auto-hide and docking features on or off as desired as shown in Figure 16–62. If you prefer that the DesignCenter remain on the screen, you can still dock it to the left or right side of the AutoCAD drawing screen. The DesignCenter can be resized as the user requires. You can unload it by clicking the X in the title bar, selecting DesignCenter from the Tools pull-down menu, clicking on the DesignCenter button on the Standard toolbar, or entering the ADCCLOSE command at the keyboard.

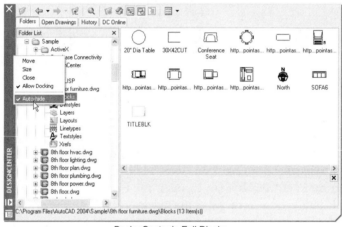

DesignCenter in Full Display   DesignCenter in Auto-hide Mode

Figure 16–62

Tip: The DesignCenter can also be launched by pressing and holding down the CTRL key and typing 2 (CTRL+2).

Block libraries may be prepared in different formats in order for them to be used through the AutoCAD DesignCenter. One method is to place all global blocks in one folder. The AutoCAD DesignCenter identifies this folder and graphically lists all drawing files to be inserted. Another method of organizing blocks is to create one drawing containing all local blocks. When this drawing is identified through the AutoCAD DesignCenter, all blocks internal to this drawing display in the AutoCAD DesignCenter palette area.

AUTOCAD DESIGNCENTER COMPONENTS

The AutoCAD DesignCenter is isolated in Figure 16–63. The following components of the dialog box are identified below and in the figure: Control buttons, Tree View or Navigation Pane, Palette or Content Pane, and Current Path.

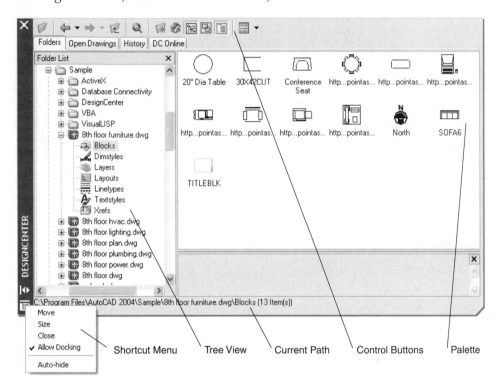

Figure 16–63

A more detailed illustration of the DesignCenter Control buttons is found in Figure 16–64. The following buttons are identified: Load, Back, Forward, Up, Search, Favorites, Home, Tree View Toggle, Preview, Descriptions, and Views. It may be necessary to resize the DesignCenter to see all the buttons.

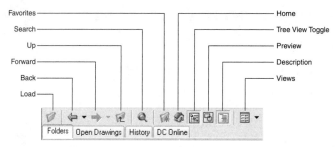

Figure 16–64

In Figure 16–63, the Folders tab has been selected to list the local and network drives in the tree view. If you select the Open Drawings tab (see Figure 16–65), only the drawings that are currently open in AutoCAD will be displayed. The History tab lists the last 20 locations accessed through the DesignCenter. DC Online (DesignCenter Online) provides access to symbols and product information by way of online service.

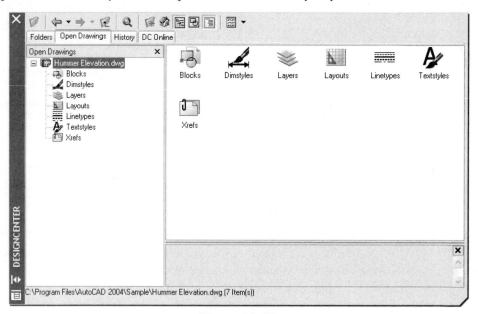

Figure 16–65

In Figure 16–63, the Tree View Toggle button is selected to show all the folders in a typical Windows Explorer style. If the toggle is off, the Tree View is no longer displayed (see Figure 16–66), providing a larger Palette area. The Load, Back, Forward, Up, Search and Home buttons provide additional methods to assist with locating and selecting drawings.

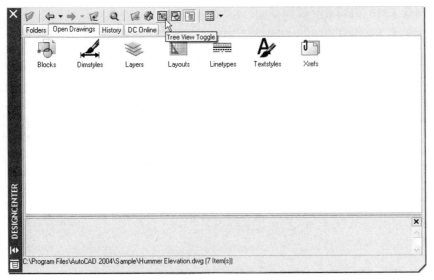

Figure 16–66

Clicking on the Preview button displays an enlarged version of the selected block in the Palette area (see Figure 16–67). This gives you a better idea of what the block will look like once it is inserted in the drawing. The Description button displays a description for the selected item in the Palette area. The View button changes the display in the palette area (large icons, small icons, list, or details). Large icons are displayed in Figure 16–67.

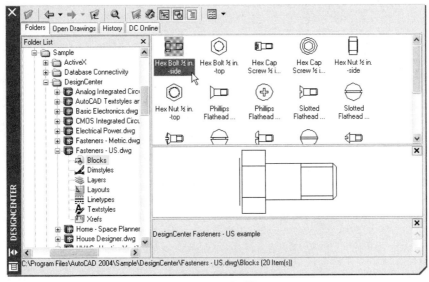

Figure 16–67

USING THE TREE VIEW

As mentioned earlier, clicking the Tree View button expands the AutoCAD DesignCenter to look similar to the illustration in Figure 16–68. The AutoCAD DesignCenter divides into two major areas; on the right is the familiar Palette area; on the left is the tree view (a layout of network and local drive folders). Clicking on a folder displays its contents in the Palette area as shown in Figure 16–68. Clicking on a drawing in the Tree View displays the drawing objects, as shown in Figure 16–69, that can be shared through the DesignCenter (Blocks, Dimstyles, Layers, Layouts, Linetypes, Textstyles, and Xrefs). Double-click the Blocks icon (or click the "+" symbol and choose Blocks in the Tree View) to display the blocks available in the selected drawing (see Figure 16–70).

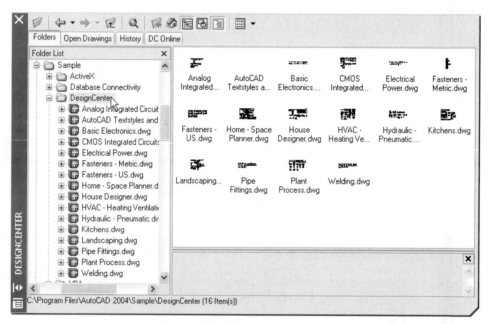

Figure 16–68

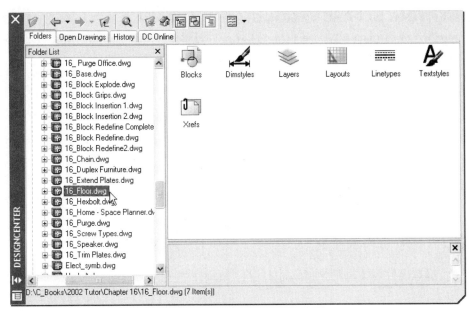

Figure 16–69

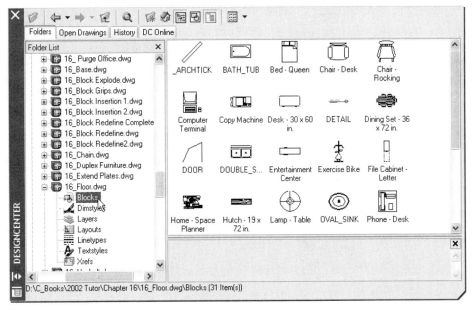

Figure 16–70

INSERTING BLOCKS THROUGH THE AUTOCAD DESIGNCENTER

Figure 16–71 displays a typical floor plan drawing along with the AutoCAD Design-Center (docked to the left side of the screen) showing the blocks identified in the current drawing. If no blocks are found internal to the drawing, the Palette area will be empty.

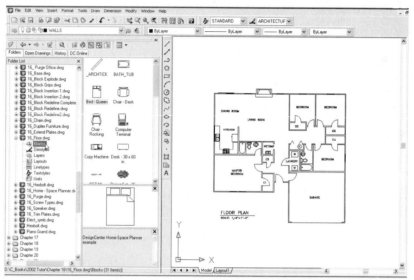

Figure 16–71

The AutoCAD DesignCenter operates on the "drag and drop" principle. Select the desired block located in the palette area of the DesignCenter, drag it out, and drop it into the desired location of the drawing. Figure 16–72 shows a queen-size bed dragged and dropped into the bedroom area of the floor plan.

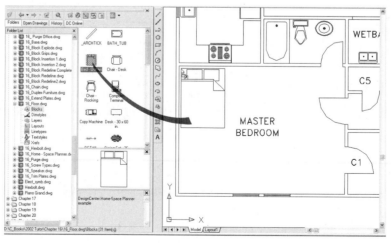

Figure 16–72

When performing a basic drag and drop operation with the left mouse button, all you have to do is identify where the block is located and drop it into that location (the use of running object snaps can ensure that the blocks are dropped in a specific location). What if the block needs to be scaled or rotated? If the drag and drop method is performed with the right mouse button, a shortcut menu is provided (see Figure 16–73). Selecting Insert Block from the shortcut menu provides the Insert dialog box in Figure 16–74, which allows you to specify different scale and rotation values. Figure 16–75 shows a rocking chair inserted and rotated into position in the corner of the wall. Instead of dragging the block with the right mouse button, double-click on the block in the palette area and the Insert dialog box will be provided immediately.

Figure 16–73

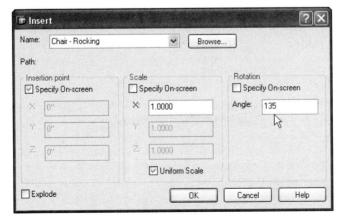

Figure 16–74

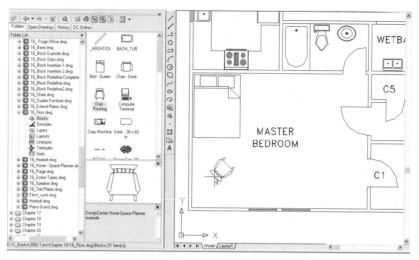

Figure 16–75

INSERTING GLOBAL BLOCKS THROUGH THE AUTOCAD DESIGNCENTER

Once a block or drawing file is available in the palette area of the AutoCAD Design-Center, it can easily be inserted in the current AutoCAD drawing. In the Tree View area of the DesignCenter, select the appropriate drive and folder where your file is located. The drawing files in this folder will display in the palette area. Now drag the desired file from the palette area into the current AutoCAD drawing and drop it (see Figure 16–76). Drag and drop with the right mouse button, just as with blocks, to make the Insert dialog box available for changing the scale or rotation values.

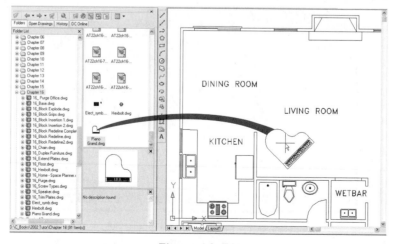

Figure 16–76

INSERTING BLOCKS INTERNAL TO OTHER DRAWINGS

Numerous times throughout the design process, you need to insert a block that belongs to another drawing file. Prior to the AutoCAD Design Center you had to insert the whole drawing in order to get to its internal blocks. Then, you had to remember to use the PURGE command or risk the threat of increasing the size of the drawing database tremendously.

Inserting blocks internal to other drawings with the AutoCAD DesignCenter is again a simple drag and drop operation. The drawing selected in the DesignCenter Tree View does not have to be the same drawing opened in AutoCAD. That means that any drawing blocks can be made available in the palette area and can be dragged and dropped into another drawing. In Figure 16–77, FloorPlan is the current drawing open in AutoCAD. Blocks from a Landscape drawing are inserted in the FloorPlan drawing, with the blocks dragged from the palette area and dropped into the FloorPlan drawing.

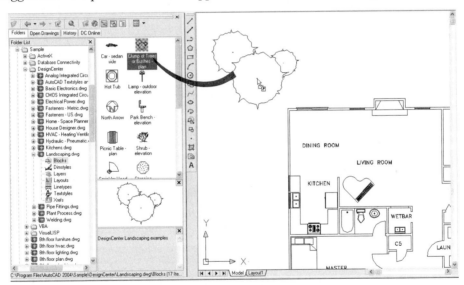

Figure 16–77

PERFORMING SEARCHES FOR INTERNAL BLOCKS

Sometimes, you know a block exists in a particular drawing file somewhere on the hard drive. You do not know the name of the drawing file; however, you think the name of the block has something to do with a north arrow. AutoCAD has a remarkable search and find mechanism that can search out blocks that are internal to a drawing.

 Try It! - Start a new drawing and activate the AutoCAD DesignCenter. Click on the Search button to display the Search dialog box illustrated in Figure 16–78. Select the type of object you want to search for in the Look for drop-down list box. Select "Blocks" to display the Blocks tab as shown in Figure 16–79.

868

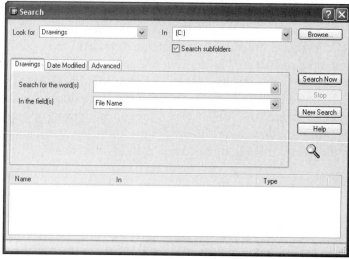

Figure 16–78

Next select the drive you want to search, from the In drop-down list box (use the Browse button to limit the search to a particular folder if that information is known). For this example you should select the drive where AutoCAD is installed. Type the name of the block you are searching for in the Search for the name edit box. After entering "North" in the edit box, click the Search Now button and AutoCAD will search all the drawings in the selected folders for all instances of the "North" block.

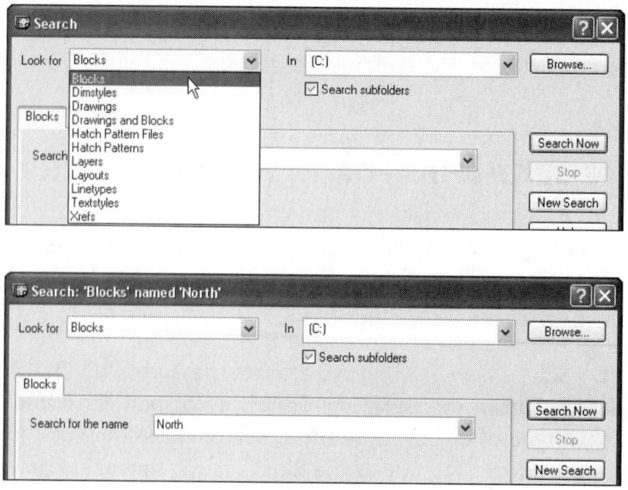

Figure 16–79

Figure 16—80 displays the results of the search on the block name "North." A number of drawings have been identified in the edit box at the bottom of the AutoCAD Search dialog box.

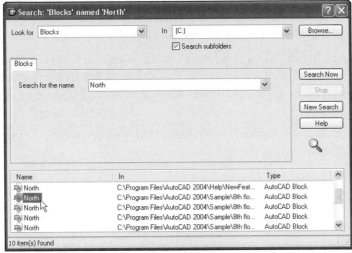

Figure 16–80

Double-clicking on the drawing name (8th floor HVAC.Dwg) at the bottom of the AutoCAD Find: dialog box in Figure 16–80 expands the AutoCAD DesignCenter. Select the Blocks object type (see Figure 16–81) to display all internal blocks in the drawing. The block "North" can now be dragged and dropped into the current drawing file.

 Tip: If you are not sure of the exact block name when performing a search, use an asterisk (wildcard) in the block name. For example, a search for "No*" would locate block names such as North, Node, None, and North Arrow.

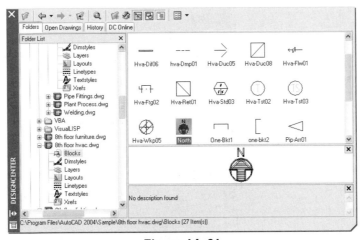

Figure 16–81

USING THE TOOL PALETTE

Tool palettes allow you to organize blocks and hatch patterns for insertion into your drawing. This feature is somewhat similar to the DesignCenter with the ability to drag and drop blocks, layers, dimension styles, text styles, and hatch patterns into a drawing. The Tool palette, however, is specific to blocks and hatch patterns. You can also customize the Tool palette to meet your individual drawing needs.

By default when AutoCAD is first loaded, the Tool palette is positioned in the upper right corner of your display screen. If the Tool palette is not visible, it can be activated by selecting it from the Tools pull-down menu or from the Standard toolbar as shown in Figure 16–82.

 Tip: The Tool palette can also be activated by pressing down the CTRL key and typing 3 (CTRL+3).

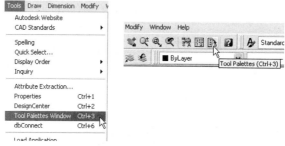

Figure 16–82

Once displayed, the Tool palette provides a number of sample tabs for you to experiment with as shown in Figure 16–83. Notice in this figure how blocks and hatch patterns have been organized under the same Sample Office tab.

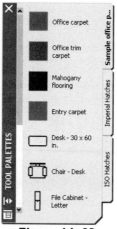

Figure 16–83

As with the DesignCenter, placing hatch patterns is just a matter of dragging the pattern from the Tool palette in Figure 16–84 and dropping it into a closed area of your drawing as in the fireplace hearth illustrated in the figure. The results are also shown in Figure 16–84 with the hatching pattern applied to the closed area.

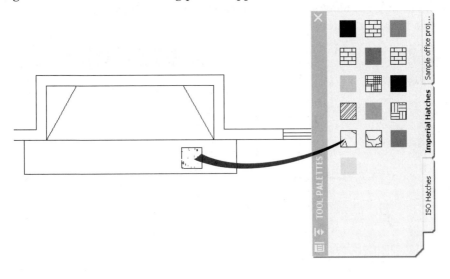

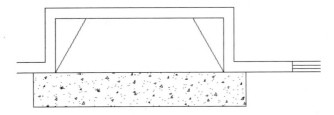

Figure 16–84

The Tool palette has a number of very powerful features to automate its operation. For example, in Figure 16–85, a chair symbol is selected. Right-clicking on this block activates a shortcut menu. The following options are available:

Cut   Cutting the block from the Tool palette to the Windows clipboard

Copy   Copying the block to the Windows clipboard

Delete Tool   Deleting this block from the Tool palette

Rename   Changing the name of this block in the Tool palette

You could also choose the Properties option in Figure 16–85. This activates the Tool Properties dialog box. All of the information in the fields can be changed and applied to this symbol. For example, suppose you need to change the insertion scale of this block for a number of drawings. Changing the scale in the Tool Properties dialog box will change the block's scale as it is inserted into the drawing. This feature is also available for hatch patterns when using the Tool palette.

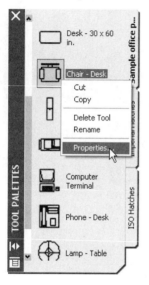

Figure 16–85

The process for creating new Tool palettes is very easy and straightforward. First, right click anywhere inside the Tool palette to display the menu illustrated in Figure 16–86. Then click on the New Tool Palette option. This automatically creates a blank tool palette. In Figure 16–86, a new tool palette name, "Electrical," has been entered in the edit box.

Other options of this Tool Palette menu include:

Allow Docking—Allows the Tool palette to be docked to the sides of your display screen. Removing the check disables this feature.

Auto-Hide—When checked, this feature collapses the Tool palette so only the thin blue strip is displayed. When you move your cursor over the blue strip, the Tool palette re-displays.

Transparency—Activates a Transparency dialog box, which controls the opaqueness of the Tool palette.

View Options—Activates the View Options dialog box, which controls the size of the hatch and block icons and whether the icon is labeled or not.

Delete Tool Palette—Deletes the Tool palette. A warning dialog box appears asking if you really want to perform this operation.

Rename Tool Palette—Renames the Tool palette.

Customize—Activates the Customize dialog box, which allows you to create a new tool palette.

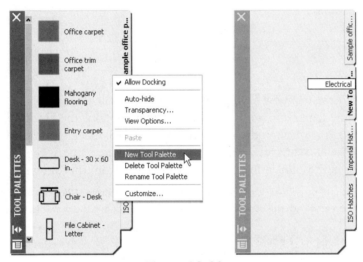

Figure 16–86

Once the Tool palette name is entered, a tab is created for this palette as shown in Figure 16–86. To add blocks and hatch patterns to this new palette, activate the DesignCenter, search for the folder that contains the symbols you wish to place in the Tool palette, and drag and drop these blocks or hatch patterns from DesignCenter into the Tool palette as shown in Figure 16–87.

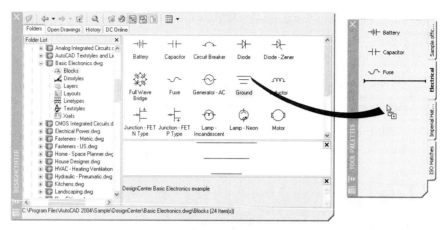

Figure 16–87

If you want to create a new Tool palette from one whole drawing in DesignCenter, activate DesignCenter and go to the drawing that contains the blocks. Right-clicking on this drawing displays the menu in Figure 16–88. Clicking on the Create Tool Palette option will create the Tool palette using the same name as the DesignCenter drawing. The new tool palette will contain all blocks from this drawing.

 Tip: If you right-click a folder and select Create Tool Palette of Blocks, a new Tool palette will be created with the name of the folder and it will contain the drawings from that folder.

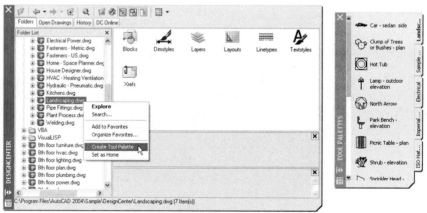

Figure 16–88

As stated earlier, the Transparency feature of the menu illustrated in Figure 16–89 allows you to control the opaqueness of the Tool palette. Picking Transparency activates the Transparency dialog box illustrated in the figure. By default, the slider bar is set to the Less position. This means if the Tool palette is positioned on top of your drawing, you will not be able to view what is underneath it. Changing the slider bar to the More position in the figure will make the Tool palette transparent to the point that you will be able to see objects under the tool palette.

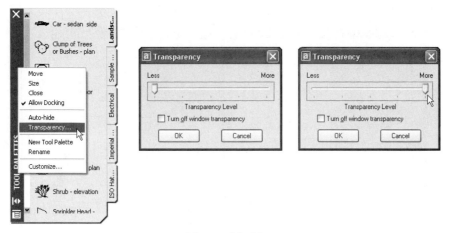

Figure 16–89

The results of setting the Transparency dialog box to More are illustrated in Figure 16–90. You can see both the Tool palette and the drawing at the same time. This is a very powerful option for displaying more information in your drawing.

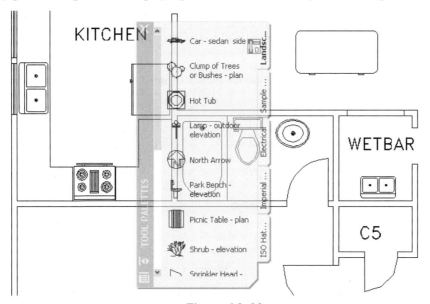

Figure 16–90

MDE — MULTIPLE DESIGN ENVIRONMENT

The Multiple Design Environment allows users to open multiple drawings within a single session of AutoCAD (see Figure 16–91). This feature, like AutoCAD Design-Center, allows sharing of data between drawings. You can easily copy and move objects, such as blocks, from one drawing to another.

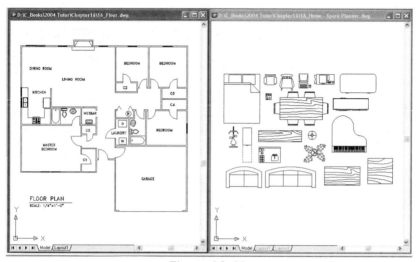

Figure 16–91

OPENING MULTIPLE DRAWINGS

Repeat the OPEN command as many times as necessary to open all drawings you will need. In fact, you can select multiple drawings in the Select File dialog box by holding down CTRL or SHIFT as you select the files (see Figure 16–92).

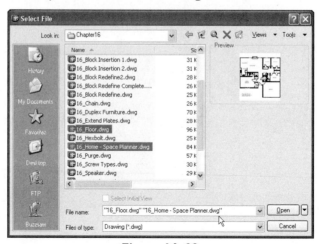

Figure 16–92

Once the drawings are open, use CTRL+F6 or CTRL+TAB to switch back and forth between the drawings. To efficiently work between drawings, you may wish to tile (see Figure 16–91) or cascade (see Figure 16–93) the drawing windows utilizing the Window pull-down menu (see Figure 16–94). A list of all the open drawings are displayed at the bottom of the Window pull-down. Selecting one of the file names is another convenient way to switch between drawings. Remember to use the CLOSE command (Window or File pull-down) to individually close any drawings that are not being used. To close all drawings in a single operation, the CLOSEALL command (Window pull-down) can be used. If changes were made to any of the drawings, you will be prompted to save those changes before the drawing closes.

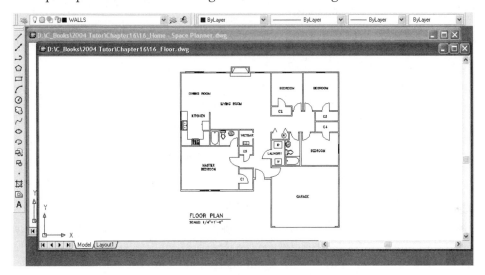

Figure 16–93

Figure 16–94

WORKING BETWEEN DRAWINGS

Once your drawings are opened and arranged on the screen, you are now ready to cut and paste, copy and paste, or drag and drop objects between drawings. The first step is to cut or copy objects from a drawing. The object information is stored on the Windows clipboard until you are ready for the second step, which is to paste the objects into that same drawing or any other open drawing. These operations are not limited to AutoCAD. In fact, you can cut, copy and paste between different Windows applications.

Use one of the following commands to cut and copy your objects:

CUTCLIP   (to remove selected objects from a drawing and store them on the clipboard)

COPYCLIP   (to copy selected objects from a drawing and store them on the clipboard)

COPYBASE   (similar to the COPYCLIP command but allows the selection of a base point for locating your objects when they are pasted)

Use one of the following commands to paste your objects:

PASTECLIP   (pastes the objects at the location selected)

PASTEBLOCK   (similar to PASTECLIP command but objects are inserted as a block and an arbitrary block name is assigned)

The commands listed can be typed at the keyboard, selected from the Edit pull-down menu (see Figure 16–95), selected from the Standard toolbar (see Figure 16–96), or selected from a shortcut menu (activated by a right-click in the drawing screen area; see Figure 16–97).

Figure 16–95

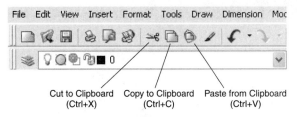

Figure 16–96

Figure 16–97

Objects may also be copied between drawings with drag and drop operations. After selecting the objects, place the cursor over the objects (without selecting a grip) and then drag and drop the objects in the new location. Dragging with the right mouse button depressed will provide a shortcut menu allowing additional control over pasting operations (see Figure 16–98).

Figure 16–98

TUTORIAL EXERCISE: ELECTRICAL SCHEMATIC

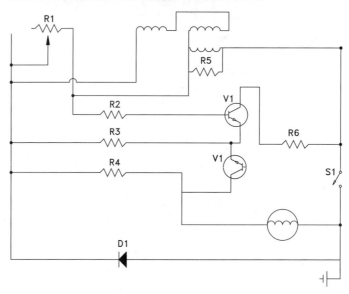

Figure 16–99

Purpose

This tutorial is designed to lay out electrical blocks such as resistors, transistors, diodes, and switches to form an electrical schematic (see Figure 16–99).

System Settings

Create a new drawing called ELECT_SYMB to hold the electrical blocks. Keep all default units and limits settings. Set the GRID and SNAP commands to 0.0750 units. This will be used to assist in the layout of the blocks. Once this drawing is finished, create another new drawing called ELECTRICAL_SCHEM1. This drawing will show the layout of an electrical schematic. Keep the default units but use the LIMITS command to set the upper-right corner of the display screen to (17.0000,11.0000). Grid and snap do not have to be set for this drawing.

Layers

Create the following layer for ELECT_SYMB with the format:

Name	Color	Linetype
0	White	Continuous

Create the following layers for ELECTRICAL_SCHEM1 with the format:

Name	Color	Linetype
Border	White	Continuous
Blocks	Red	Continuous
Wires	Blue	Continuous

Suggested Commands

Begin this tutorial by creating a new drawing called ELECT_SYMB, which will hold all electrical blocks. Use Figure 16–100 as a guide in drawing all electrical blocks. A grid with spacing 0.0750 would provide further assistance with the drawing of the blocks. Once all blocks are drawn, the BLOCK command is used to create blocks out of the individual blocks. Save this drawing and create a new drawing file called ELECTRICAL_SCHEM1. Use the AutoCAD DesignCenter to drag and drop the internal blocks from the drawing ELECT_SYMB into the new drawing ELECTRICAL_SCHEM1. Connect all blocks with lines that represent wires and electrical connections. Add block identifiers, and save the drawing.

Whenever possible, substitute the appropriate command alias in place of the full AutoCAD command in each tutorial step. For example, use "CP" for the COPY command, "L" for the LINE command, and so on. The complete listing of all command aliases is located in Chapter 1, Table 1–2.

STEP 1

Begin a new drawing and call it ELECT_SYMB. Be sure to save this drawing to a convenient location. It will be used along with the AutoCAD DesignCenter to bring the internal blocks into another drawing file.

STEP 2

Using a grid/snap of 0.075 units, construct each block using Figure 16–100 as a guide. The grid/snap will help keep all blocks proportional to each other. Create all the blocks on the neutral layer 0. Also, the "X" located on each block signifies its insertion point (do not draw the "X").

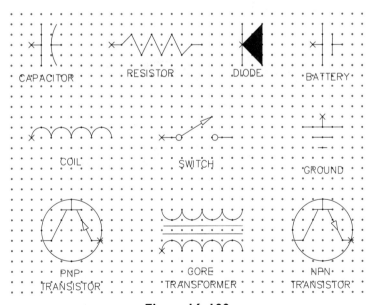

Figure 16–100

STEP 3

For a more detailed approach to creating the blocks, the resistor block will be created. Using Figure 16–101 as a guide, first construct the resistor with the LINE command while using the grid/snap to connect points until the resistor is created. Be sure Snap mode is active for this procedure. Then issue the BLOCK command, which will bring up the dialog box illustrated in Figure 16–102. This will be an internal block to the drawing ELECT_SYMB; name the block RESISTOR. In the Select objects area of the dialog box, select all lines that represent the resistor (if you labeled the symbols, do not select the text as part of the block). Select the Retain radio button so that the symbol will remain visible in your drawing. Returning to the dialog box will show the objects selected in a small image icon on the right of the dialog box. Next, identify the insertion point of the resistor at the "X" located in Figure 16–101. It is very important to use an Object Snap mode when picking the insertion point. Finally, add a description for the block. When finished with this dialog box, click on the OK button to create the internal block. Repeat this procedure for the remainder of the blocks in Figure 16–100. When complete, save the drawing as ELECT_SYMB.

RESISTOR

Figure 16–101

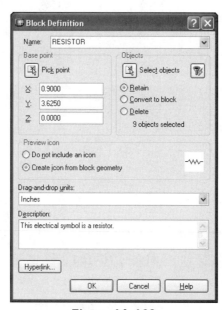

Figure 16–102

STEP 4

Begin another new drawing called ELEC-TRICAL_SCHEM1. With the blocks identified inside the drawing named ELECT_SYMB, activate the Auto-CAD DesignCenter. Click on the Tree View icon and expand the AutoCAD DesignCenter to display the folders on your hard drive. Locate the correct folder in which the drawing was saved and click on it. Double-click on ELECT_SYMB and select Blocks to display the internal blocks illustrated in Figure 16–103. These internal blocks can now be inserted in any AutoCAD drawing file.

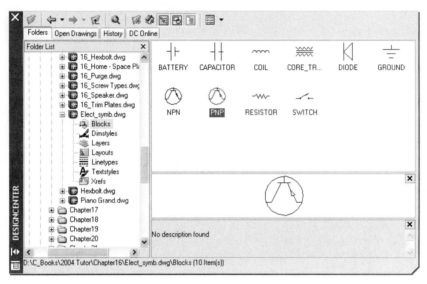

Figure 16–103

STEP 5

If you cannot find the drawing file, an alternate step would be to click on the Search button in the AutoCAD Design-Center. This will activate the AutoCAD Search dialog box. Ensure that "drawings" is selected in the Look for edit box and that the correct drive is selected in the In edit box. Enter ELECT_SYMB for the Drawing name and click on the Search Now button. The result, illustrated in Figure 16–104, shows the subdirectory location of the drawing file. Double-clicking on this drawing file expands the AutoCAD DesignCenter to display the ELECT_SYMB.DWG in the Tree View. Double-clicking on this drawing file and selecting Blocks will display all blocks internal to it (see Figure 16–105).

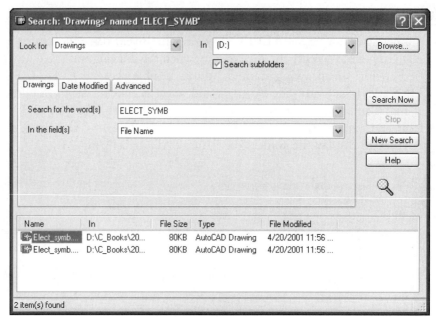

Figure 16–104

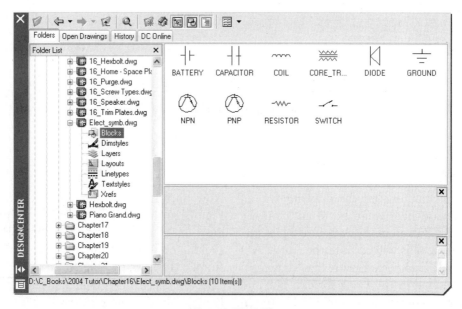

Figure 16–105

STEP 6

Make the Blocks layer current and begin dragging and dropping the blocks into the drawing shown in Figure 16–106. Double-click an image of a block located in the DesignCenter will launch the Insert dialog box should you need to change the scale or rotation angle of the block. In the case of the electrical schematic, it is not critical to place the block exactly, because they are moved to better locations when connected to lines that represent wires.

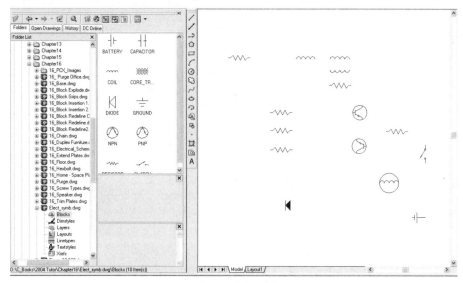

Figure 16–106

STEP 7

Make the Wires layer current and begin connecting all the blocks to form the schematic (use Figure 16–107 as a guide). Some type of Object Snap mode such as Endpoint or Insert must be used for the wire lines to connect exactly with the endpoints of the blocks. If spaces get tight or if the blocks look too crowded, move the block to a better location and continue drawing lines to form the schematic. Hint: The STRETCH command is a fast and easy way to reposition the blocks without having to reconnect the wires.

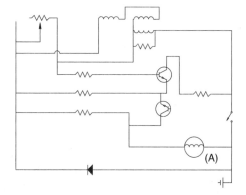

Figure 16–107

STEP 8

Add text (place it on the Blocks layer) identifying all resistors, transistors, diodes, and switches by number. The text height used was 0.12; however, this could vary depending on the overall size of your schematic. Also, certain areas of the schematic show connections with the presence of a dot at the intersection of the connection. Use the DONUT command with an inside diameter of 0.00 and an outside diameter of 0.075. Place donuts on the Wires layer in all locations shown in Figure 16–108.

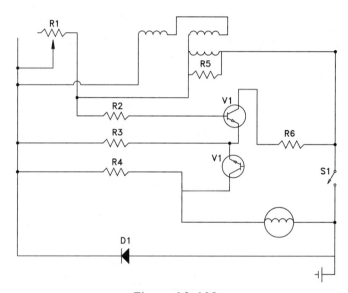

Figure 16–108

 Open the Exercise Manual PDF file for Chapter 16 on the accompanying CD for more tutorials and exercises.

 If you have the accompanying Exercise Manual, refer to Chapter 16 for more tutorials and exercises.

Using Attributes

This chapter begins the study of how to create, display, edit, and extract attributes. Attributes consist of intelligent text data that is attached to a block. The data could consist of part description, part number, catalog number, and price. Whenever the block is inserted in a drawing, you are prompted for information that once entered becomes attribute data. Attributes could also be associated with a title block for entering such items as drawing name, who created, checked, revised, and approved the drawing, drawing date, and drawing scale. Once included in the drawing, the attributes can be extracted and shared with other programs.

ABOUT ATTRIBUTES

An attribute may be considered a label that is attached to a block. This label is called a tag and can contain any type of information that you desire. Examples of attribute tags are illustrated in Figure 17–1. The tags are RESISTANCE, PART_NAME, WATTAGE, and TOLERANCE and relate to the particular symbol they are attached to, in this case an electrical symbol of a resistor. Attribute tags are placed in the drawing with the symbol. When the block, which includes the tags and symbol, is inserted, AutoCAD requests the values for the attributes. The same resistor with attribute values is illustrated in Figure 17–2. When you create the attribute tags, you determine what information is requested, the actual prompts, and the default values for the information requested. Once the values are provided and inserted in the drawing, the information contained in the attributes can be displayed, extracted, and even shared with spreadsheet or database programs.

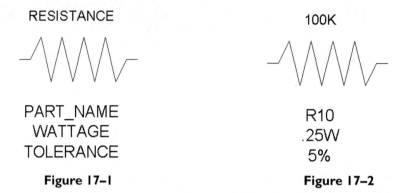

RESISTANCE

100K

PART_NAME
WATTAGE
TOLERANCE

R10
.25W
5%

Figure 17–1

Figure 17–2

Once attributes are inserted in the drawing, the following additional tasks can be performed on attributes:

1. Attributes can be turned on, turned off, or displayed normally through the use of the ATTDISP command.

2. The individual attributes can be changed through the Attribute Edit dialog box, which is activated through the ATTEDIT command.

3. Attributes can be globally edited using the -ATTEDIT command.

4. Once edited, attributes can be extracted into various text file modes through the Extract Attributes dialog box, which is activated by the ATTEXT command.

5. The text file created by the attribute extraction process can be exported to a spreadsheet or database program.

The following commands, which will be used throughout this chapter, assist in the creation and manipulation of attributes:

ATTDEF—Activates the Attribute Definition dialog box used for the creation of attributes.

ATTDISP—Used to control the visibility of attributes in a drawing.

ATTEDIT—Activates a dialog box used to edit attribute values.

-ATTEDIT—Used to edit attributes singly or globally. The hyphen (-) in front of the command activates prompts from the command line. There is no dialog box available for globally editing attributes.

ATTEXT—Activates the Attribute Extraction dialog box used for extracting attributes in CDF, SDF, or text file formats.

The following additional commands are available in AutoCAD to better assist with managing and manipulating attributes:

BATTMAN—Activates the Block Attribute Manager dialog box used to edit attribute properties of a block definition.

EATTEDIT—Activates the Enhanced Attribute Editor dialog box used to edit the attributes of a block.

EATTEXT—Activates the Attribute Extraction Wizard used to extract attributes.

CREATING ATTRIBUTES THROUGH THE ATTRIBUTE DEFINITION DIALOG BOX

Selecting Define Attributes from the Block cascading menu on the Draw pull-down menu shown in Figure 17–3 activates the Attribute Definition dialog box in Figure 17–4. The following components of this dialog box will be explained in this section: Attribute Tag, Attribute Prompt, Attribute Value, and Attribute Mode.

Figure 17–3

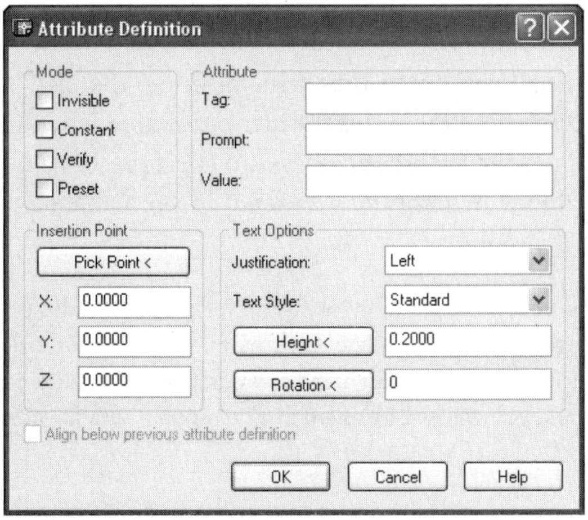

Figure 17–4

ATTRIBUTE TAG

A tag is the name given to an attribute. Typical attribute tags could include PART_NAME, CATALOG_NUMBER, PRICE, DRAWING_NAME, and SCALE. The underscore is used to separate words because spaces are not allowed in the tag names.

ATTRIBUTE PROMPT

The attribute prompt is the text that appears on the text line when the block containing the attribute is inserted in the drawing. If you want the prompt to be the same as the tag name, enter a null response by leaving the Prompt edit box blank. If the Constant mode is specified for the attribute, the prompt area is not available.

ATTRIBUTE VALUE

The attribute value is the default value displayed when the attribute is inserted in the drawing. This value can be accepted or a new value entered as desired. The attribute value is handled differently if the attribute mode selected is Constant or Preset.

ATTRIBUTE MODES

Invisible—This mode is used to determine whether the label is invisible when the block containing the attribute is inserted in the drawing. If you later want to make the attribute visible, you can use the ATTDISP command.

Constant—Use this mode to give every attribute the same value. This might be very useful when the attribute value is not subject to change. However, if you designate an attribute to contain a constant value, you cannot change it later in the design process.

Verify—Use this mode to verify that every value is correct. This is accomplished by prompting the user twice for the attribute value.

Preset—This allows for the presetting of values that can be changed. However, you are not prompted to enter the attribute value when inserting a block. The attribute values are automatically set to their preset values. The preset option is not available if an attribute is entered in the Value edit box of the Attribute Definition dialog box.

 Note: The effects caused by invoking the Verify and Preset modes will only be apparent when the attdia system variable is Off (the Enter Attributes dialog box is not displayed). When entering data at the command prompt, you will be asked twice for data that is to be verified and you will not be asked at all to supply data for attributes that are preset.

TUTORIAL EXERCISE: B TITLE BLOCK.DWG

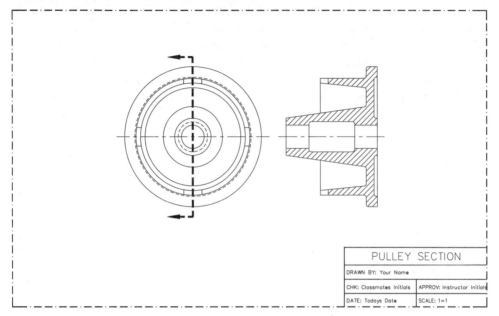

DRAWN BY: Your Name

CHK: Classmates Initials — APPROV: Instructor Initials

DATE: Todays Date — SCALE: 1=1

PULLEY SECTION

Figure 17–5

Purpose

This tutorial is designed to assign attributes to a title block. The completed drawing is illustrated in Figure 17–5.

System Settings

Because the drawings B Title Block and 17_Pulley Section are provided on the CD, all units and drawing limits have already been set.

Layers

All layers have already been created for this drawing.

Suggested Commands

Open the drawing B Title Block. Using the Attribute Definition dialog box (activated through the ATTDEF command), assign attributes consisting of the following tag names: DRAWING_NAME, DRAWN_BY, CHECKED_BY, APPROVED_BY, DATE, and SCALE. Save this drawing file with its default name. Open the drawing 17_Pulley Section and insert the title block in this drawing and answer the questions designed to complete the title block information.

Whenever possible, substitute the appropriate command alias in place of the full AutoCAD command in each tutorial step. For example, use "CP" for the COPY command, "L" for the LINE command, and so on. The complete listing of all command aliases is located in Chapter 1, Table 1–2.

STEP I

Open the drawing B Title Block and observe the title block area in Figure 17–6. Various point objects are present to guide you in the placement of the attribute information. All points are located on the layer "Points." To further assist you when creating attributes, the illustration in Figure 17–7 will be used throughout this tutorial exercise. All points are identified by a letter. You will need to refer back to this figure numerous times for placing all attributes and completing this tutorial.

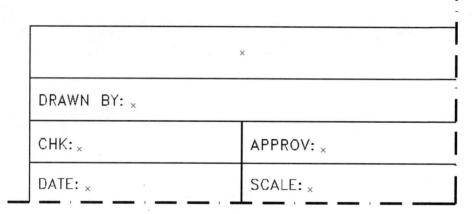

Figure 17–6

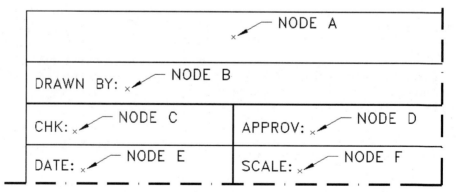

Figure 17–7

STEP 2

Activate the Attribute Definition dialog box in Figure 17–8. Leave all items unchecked in the Mode area of this dialog box. In the Attribute area, make the following changes: Enter "DRAWING_NAME" in the Tag edit box. In the Prompt edit box, enter "What is the name of this drawing?" For the Value, enter "UNNAMED" in this edit box. In the Text Options area, change the Justification to Middle and the Height to 0.25 units. In the Insertion Point area, click the Pick Point< button. This will return you to your drawing. Using OSNAP-Node, pick the point at "A" in Figure 17–7. This will return you to the Attribute Definition dialog box. Notice the X and Y boxes reflect the point you just picked. When finished with all areas of this dialog box, click the OK button. The DRAWING_NAME tag is added to the title block area in Figure 17–9.

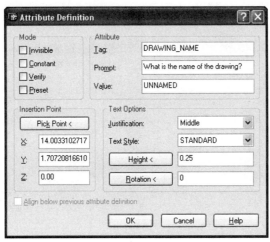

Figure 17–8

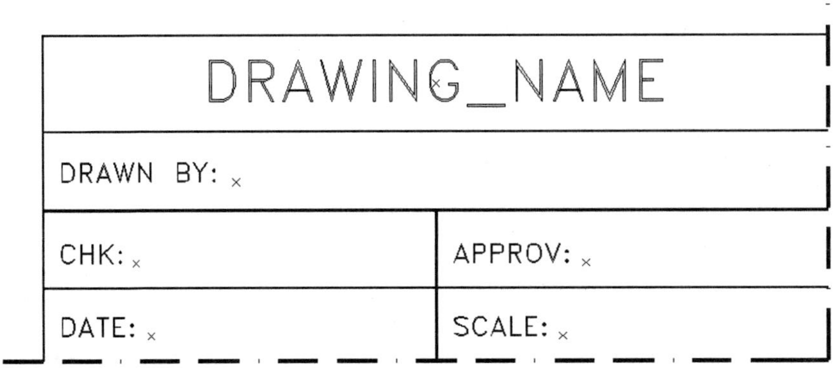

DRAWING_NAME

DRAWN BY: x

CHK: x   APPROV: x

DATE: x   SCALE: x

Figure 17–9

STEP 3

To define the next attribute, activate the Attribute Definition dialog box in Figure 17–10. Leave all items unchecked in the Mode area of this dialog box. In the Attribute area, make the following changes: Enter "DRAWN_BY" in the Tag edit box. In the Prompt edit box, enter "Who created this drawing?" For the Value, enter "UNNAMED" in this edit box. In the Text Options area, verify the Justification is set to Left and change the Height to 0.12 units. In the Insertion Point area, click the Pick Point< button. This will return you to your drawing. Using OSNAP-Node, pick the point at "B" in Figure 17–7. This will return you to the Attribute Definition dialog box. Notice the X and Y boxes reflect the point you just picked. When finished with all areas of this dialog box, click the OK button. The DRAWN_BY tag is added to the title block area in Figure 17–11.

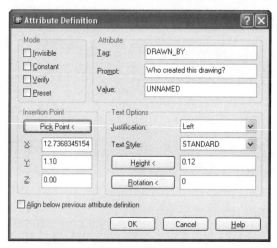

Figure 17–10

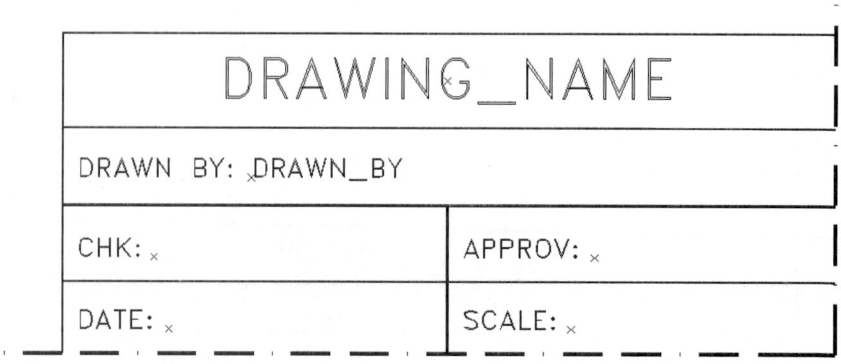

DRAWING_NAME

DRAWN BY: DRAWN_BY

CHK: x

APPROV: x

DATE: x

SCALE: x

Figure 17–11

STEP 4

Activate the Attribute Definition dialog box in Figure 17–12. Leave all items unchecked in the Mode area of this dialog box. In the Attribute area, make the following changes: Enter "CHECKED_BY" in the Tag edit box. In the Prompt edit box, enter "Who will be checking this drawing?" For the Value, enter "CHIEF DESIGNER" in this edit box. In the Text Options area, verify the Justification is set to Left and the Height is 0.12 units. In the

Insertion Point area, click the Pick Point< button. This will return you to your drawing. Using OSNAP-Node, pick the point at "C" in Figure 17–7. This will return you to the Attribute Definition dialog box. Notice the X and Y boxes reflect the point you just picked. When finished with all areas of this dialog box, click the OK button. The CHECKED_BY tag is added to the title block area in Figure 17–13.

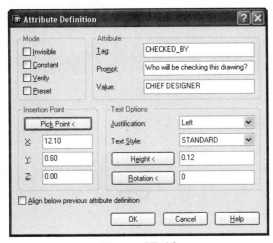

Figure 17–12

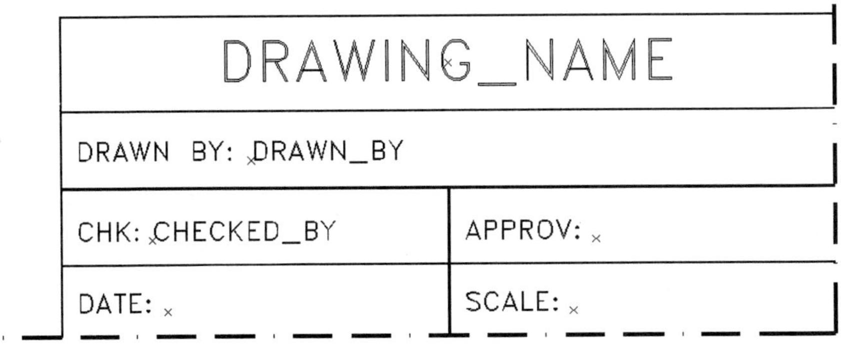

DRAWING_NAME

DRAWN BY: DRAWN_BY

CHK: CHECKED_BY   APPROV: x

DATE: x   SCALE: x

Figure 17–13

STEP 5

Activate the Attribute Definition dialog box in Figure 17–14. Leave all items unchecked in the Mode area of this dialog box. In the Attribute area, make the following changes: Enter "APPROVED_BY" in the Tag edit box. In the Prompt edit box, enter "Who will be approving this drawing?" For the Value, enter "CHIEF ENGINEER" in this edit box. In the Text Options area, verify the Justification is set to Left and the Height is 0.12 units. In the Insertion Point area, click the Pick Point< button. This will return you to your drawing. Using OSNAP-Node, pick the point at "D" in Figure 17–7. This will return you to the Attribute Definition dialog box. Notice the X and Y boxes reflect the point you just picked. When finished with all areas of this dialog box, click the OK button. The APPROVED_BY tag is added to the title block area in Figure 17–15.

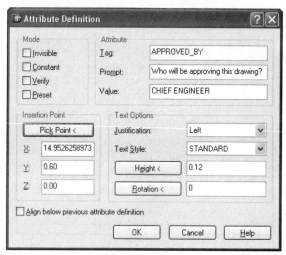

Figure 17–14

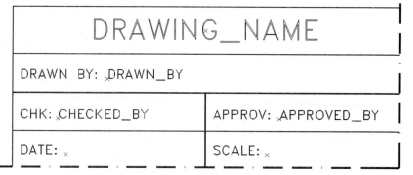

Figure 17–15

STEP 6

Activate the Attribute Definition dialog box in Figure 17–16. Leave all items unchecked in the Mode area of this dialog box. In the Attribute area, make the following changes: Enter "DATE" in the Tag edit box. In the Prompt edit box, enter "When was this drawing completed?" For the Value, enter "UNDATED" in this edit box. In the Text Options area, verify the Justification is set to Left and the Height is 0.12 units. In the Insertion Point area,

click the Pick Point< button. This will return you to your drawing. Using OSNAP-Node, pick the point at "E" in Figure 17–7. This will return you to the Attribute Definition dialog box. Notice the X and Y boxes reflect the point you just picked. When finished with all areas of this dialog box, click the OK button. The DATE tag is added to the title block area in Figure 17–17.

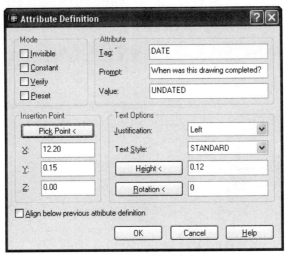

Figure 17–16

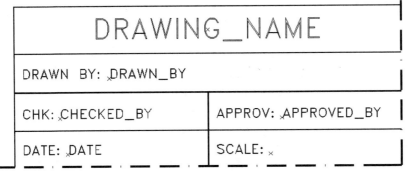

Figure 17–17

STEP 7

Activate the Attribute Definition dialog box in Figure 17–18. Leave all items unchecked in the Mode area of this dialog box. In the Attribute area, make the following changes: Enter "SCALE" in the Tag edit box. In the Prompt edit box, enter "What is the scale of this drawing?" For the Value, enter 1=1 in this edit box. In the Text Options area, verify the Justification is set to Left and the Height is 0.12 units. In the Insertion Point area, click the Pick Point< button. This will return you to your drawing. Using

OSNAP-Node, pick the point at "F" in Figure 17–7. This will return you to the Attribute Definition dialog box. Notice the X and Y boxes reflect the point you just picked. When finished with all areas of this dialog box, click the OK button. The SCALE tag is added to the title block area in Figure 17–19.

You have now completed creating the attributes for the title block; turn off the "Points" layer. Close and save the changes to this drawing with its default name.

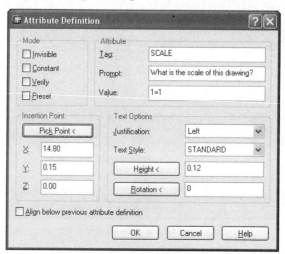

Figure 17–18

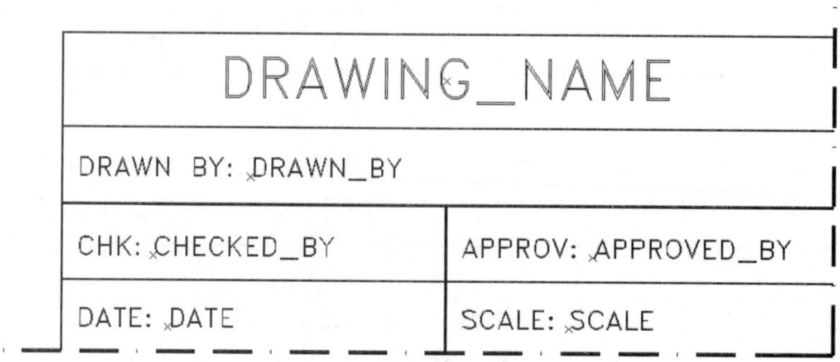

```
DRAWING_NAME

DRAWN BY: DRAWN_BY

CHK: CHECKED_BY          APPROV: APPROVED_BY

DATE: DATE               SCALE: SCALE
```

Figure 17–19

STEP 8

To test out the attributes and see how they function in a drawing, first open the drawing file 17_Pulley Section in Figure 17–20. Notice this image is viewed from inside a layout (or paper space). The current page setup is based on the DWF6-ePlot.pc3 file, with the current sheet size as ANSI Expand B (17.00 x 11.00 inches). The viewport that holds the two views of the Pulley has the No

Plot state assigned in the Layer Properties Manager dialog box. You could turn the Viewports layer off to hide the viewport if you prefer. You will now insert the title block with attributes in this drawing.

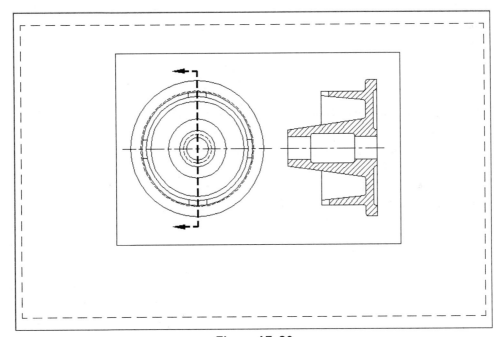

Figure 17–20

STEP 9

Verify that the system variable ATTDIA is set to 1. This will allow you to enter your attribute values through the use of a dialog box. Make the Border layer current and activate the Insert dialog box in Figure 17–21. Clear the Specify On-screen check box in the Insertion point area. This will automatically place the title block at the 0,0,0 location of the layout, in the lower-left corner of the printable area indicators. Before the title block can be placed, you must first fill in the boxes in the Edit

Attributes dialog box illustrated in Figure 17–22 (if this dialog box does not appear and you are prompted for values at the command line, it is because ATTDIA is set to 0). Complete all boxes—enter appropriate names and initials as directed. (HINT: Pressing the TAB key is a quick way of moving from one box to another while inside any dialog box.)

Command: I *(For INSERT)*

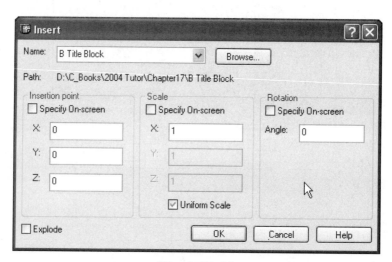

Figure 17–21

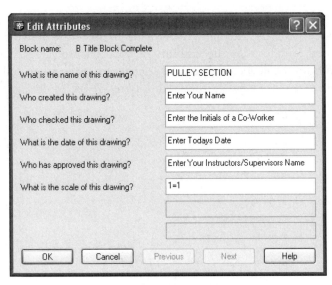

Figure 17–22

The completed drawing with title block and attributes inserted is illustrated in Figure 17–23.

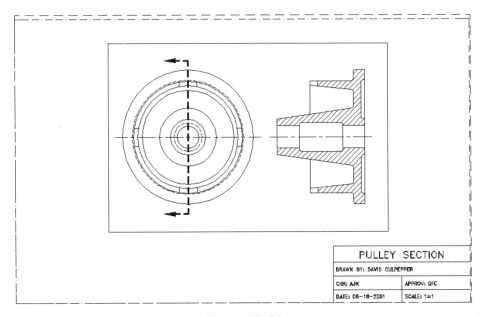

Figure 17–23

CONTROLLING THE DISPLAY OF ATTRIBUTES

You don't always want attribute values to be visible in the entire drawing. The ATTDISP command is used to determine the visibility of the attribute values. This command can be entered from the keyboard or can be selected from the View pull-down menu under Display > Attribute Display, shown in Figure 17–24.

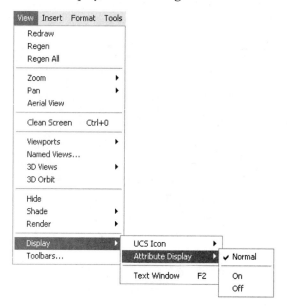

Figure 17–24

Command: **ATTDISP**
Enter attribute visibility setting [Normal/ON/OFF] <Normal>: *(Enter the desired option)*

The following three modes are used to control the display of attributes:

VISIBILITY NORMAL

This setting displays attributes based on the mode set through the Attribute Definition dialog box. If some attributes were created with the Invisible mode turned on, these attributes will not be displayed through this setting, which makes this setting popular for displaying certain attributes and hiding others.

VISIBILITY ON

Use this setting to force all attribute values to be displayed on the screen. This affects even attribute values with the Invisible mode turned on.

VISIBILITY OFF

Use this setting to force all attribute values to be turned off on the screen. This is especially helpful in busy drawings that contain lots of detail and text.

All three settings are illustrated in Figure 17–25. With ATTDISP set to Normal, attribute values defined as invisible will not be displayed. This is the case for the WATTAGE and TOLERANCE tags. The Invisible mode was turned on inside the Attribute Definition dialog box when they were created. As a result, these values will not be visible. With ATTDISP set to On, all resistor values are forced to be visible. When set to Off, this command makes all resistor attribute values invisible.

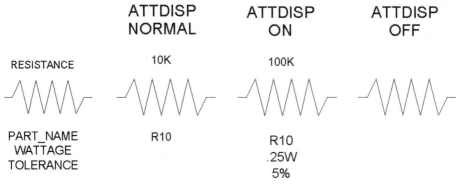

Figure 17–25

After the ATTDISP command is used, the display is regenerated to show the new state (unless REGENAUTO is turned off).

TUTORIAL EXERCISE: 17_CIRCUIT COMPONENTS.DWG

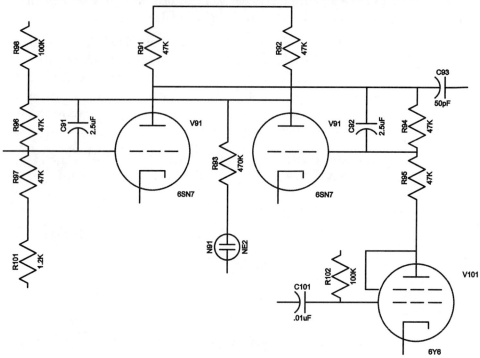

Figure 17–26

Purpose

This tutorial is designed to control the display of attributes in a drawing. The completed drawing is illustrated in Figure 17–26.

System Settings

Open the drawing 17_Circuit Components. All units and drawing limits have already been set. Attributes have also been assigned to various electrical components.

Layers

All layers have already been created for this drawing.

Suggested Commands

Observe the results when you execute the ATTDISP command along with the Normal, On, and Off options.

Whenever possible, substitute the appropriate command alias in place of the full AutoCAD command in each tutorial step. For example, use "CP" for the COPY command, "L" for the LINE command, and so on. The complete listing of all command aliases is located in Chapter 1, Table 1–2.

STEP 1

Open the drawing 17_Circuit Components. Each electrical symbol component has been defined with the following attribute tags:

PNAME
PTYPE
PSPEC1
PSPEC2
PVALUE

During the creation of the attribute tags, the Invisible modifier was applied to the

PTYPE, PSPEC1, and PSPEC2 tags. The ATTDISP command will be used to observe the result of displaying attributes On, Off, or in Normal setting.

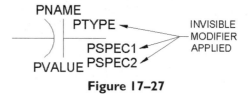

Figure 17–27

STEP 2

Activate the ATTDISP command and set the attribute display to On. Observe in Figure 17–28 that all attributes are now visible, even those attributes that were originally defined with the Invisible modifier.

Command: **ATTDISP**
Enter attribute visibility setting [Normal/
ON/OFF] <Normal>: **ON**

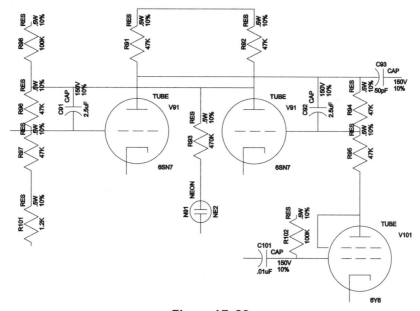

Figure 17–28

STEP 3

Activate the ATTDISP command and set the attribute display to Off. This action forces all attributes to be made invisible, as shown in Figure 17–29.

Command: **ATTDISP**
Enter attribute visibility setting [Normal/ ON/OFF] <ON>: **OFF**

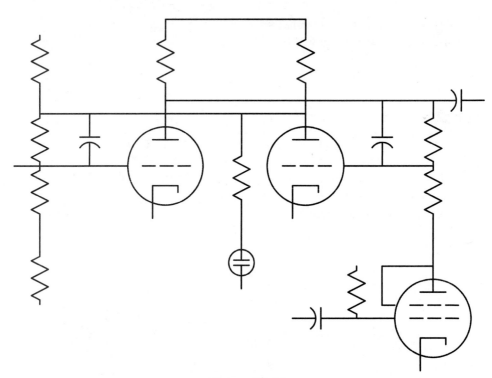

Figure 17–29

STEP 4

Activate the ATTDISP command and set the attribute display back to Normal. Observe the final results in Figure 17–30. This completes the tutorial exercise.

Command: **ATTDISP**
Enter attribute visibility setting [Normal/ ON/OFF] <OFF>: **N** *(For Normal)*

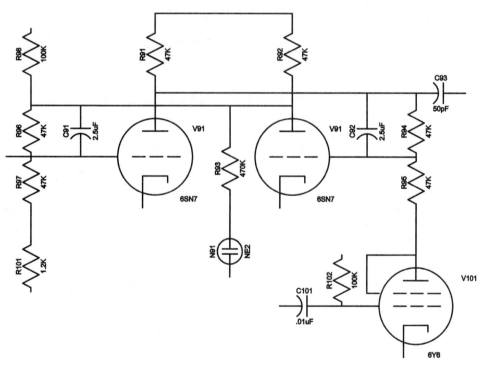

Figure 17–30

CHANGING AN ATTRIBUTE VALUE WITH THE EDIT ATTRIBUTES DIALOG BOX

Once part of a drawing, attributes can be modified by using the Edit Attributes dialog box. To display the dialog box, enter the ATTEDIT command from the keyboard.

Command: **ATTEDIT**
Select block reference: *(Select a block with attributes, and the Edit Attributes dialog box appears)*

Selecting the block with attributes to be edited displays the dialog box in Figure 17–31. Make any changes to the attribute values in the boxes provided. Clicking the OK button dismisses the dialog box and updates the attribute values.

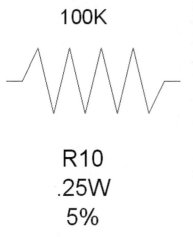

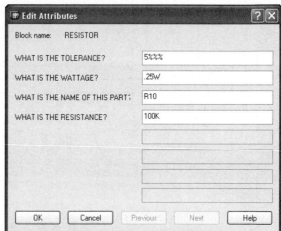

Figure 17–31

EDITING ATTRIBUTE PROPERTIES

After attributes have been placed in a drawing, changes to properties such as attribute position, text height, and angle can be made through the -ATTEDIT command.

Command: **-ATTEDIT** *(The hyphen (-) in front of the command activates prompts from the command line)*
Edit attributes one at a time? [Yes/No] <Y>

The response determines the string of options, which will follow:

Entering "Yes" selects attributes individually (one by one) for editing. The attributes that are currently visible on the screen can be edited. The attributes to be edited can be further restricted by object selection, block names, tags, or values of the attributes to be edited.

Entering "No" prepares the attributes for global editing (all attributes). You can also restrict the editing to block names, tags, values, and on-screen visibility. Global editing will be discussed in greater detail later in this chapter.

After choosing to edit attributes one at a time, you are next asked to select the block name to be edited. You can choose the blocks to be edited by entering a specific block name (specify more than one by separating the names with a comma), using the global wildcard (*), or by using the (?) symbol to replace common characters.

Enter block name specification <*>: *(Type a specific block name or press ENTER to affect all blocks in the drawing)*

AutoCAD prompts for the parts of the attributes to be edited:

Enter attribute tag specification <*>: *(Type a specific tag or press ENTER to affect all attribute tags)*
Enter attribute value specification <*>: *(Type a specific value or press ENTER to affect all attribute values)*

Your reply to each prompt targets the specific attributes to be edited.

You then select the attributes. Only those attributes that satisfy the criteria specified can be selected.

Select Attributes: *(Select individually or by window)*

After you have selected each attribute to be edited with the object selection process, an "X" marks the first attribute value that can be edited, illustrated in Figure 17–32. Notice also that the attribute value is currently highlighted as an additional means of identifying the attribute value being modified.

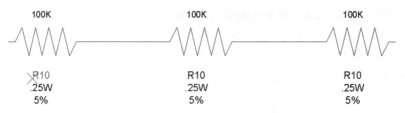

Figure 17–32

The "X" mark stays with the current attribute to be edited until you enter the Next option (or press ENTER) and a new attribute is marked and highlighted in Figure 17–33. You are then prompted for the following options:

Enter an option [Value/Position/Height/Angle/Style/Layer/Color/Next] <N>: *(Enter the desired option)*

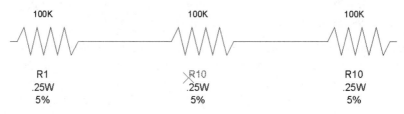

Figure 17–33

These are the various properties of an attribute that can be modified. The first letter can be used to select the appropriate option, or press ENTER to move to the next attribute to edit. A brief description of each property is listed below:

Value	Enter a new attribute value.
Position	Locate a new position for the attribute text.
Height	Change the attribute text height.
Angle	Change the attribute text angle.
Style	Change the text style of the attribute text.
Layer	Change the layer of the attribute text.
Color	Change the attribute text color.
Next	Move to the next attribute to be edited.

Figure 17–34 illustrates a number of the properties that can be changed when attribute values are edited individually.

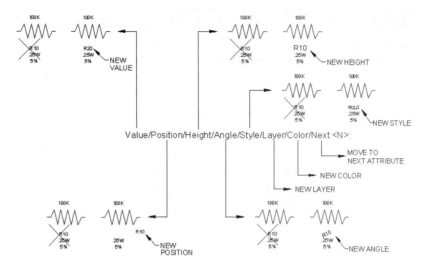

Figure 17–34

CHANGING OR REPLACING VALUES

When you enter the Value option, the following prompt appears:

Enter type of value modification [Change/Replace] <R>:

The Change option is used to modify a few characters in the attribute text, as in a misspelling. If you choose "C," the following prompt appears:

Enter string to change: (Enter those letters that make up the string to change)
Enter new string: (Enter the new letters to replace the old string)

You should respond to the first prompt with the string of characters to be changed and to the second prompt with the string you want it replaced with.

The Replace option is used to change the entire attribute value. You are prompted:

Enter new Attribute value: (Enter the new attribute value)

TUTORIAL EXERCISE: 17_LAUNDRY ROOM.DWG

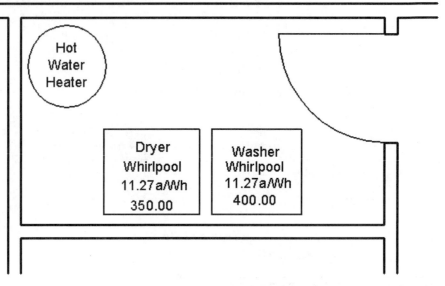

Figure 17–35

Purpose

This tutorial is designed to edit various properties of attributes individually. The completed drawing is illustrated in Figure 17–35.

System Settings

Open the drawing 17_Laundry Room. All units and drawing limits have already been set. Attributes have also been assigned to various electrical components.

Layers

All layers have already been created for this drawing.

Suggested Commands

You will be using the -ATTEDIT command to move and reposition a number of attributes for better reading in this drawing.

Whenever possible, substitute the appropriate command alias in place of the full AutoCAD command in each tutorial step. For example, use "CP" for the COPY command, "L" for the LINE command, and so on. The complete listing of all command aliases is located in Chapter 1, Table 1–2.

STEP 1

Open the drawing 17_Laundry Room. In Figure 17–36, a washer and dryer complete with attributes have been inserted in this drawing. The attributes are properly organized in the dryer block. However, the attribute values assigned to the washer have been moved, rotated, had the color changed, and even had one of its values entered incorrectly. The purpose of this exercise is to move all attributes back to their original positions in their original colors and to edit one of the attribute values.

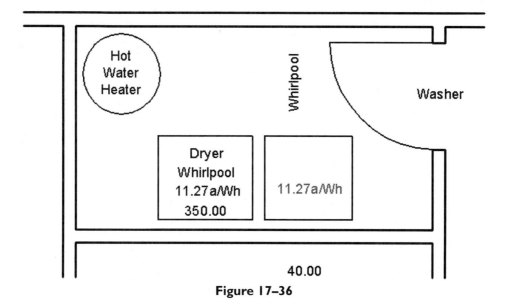

Figure 17–36

STEP 2

Begin editing the attributes individually by first rotating the text "Whirlpool" back to its 0° angle in Figure 17–37 and then repositioning the text above the red catalog number in Figure 17–38. When using the Position option of the -ATTEDIT command, use your cursor to move the text into the desired location.

Command: **-ATTEDIT**

Edit attributes one at a time? [Yes/No]
   <Y>: *(Press* ENTER *to accept the default)*

Enter block name specification <*>:
   WASHER
Enter attribute tag specification <*>:
   MANUFACTURER
Enter attribute value specification <*>:
   (Press ENTER *to accept the default)*
Select Attributes: *(Pick "Whirlpool" located above the washer)*
Select Attributes: *(Press* ENTER *to continue)*
I attributes selected.

Enter an option [Value/Position/Height/
Angle/Style/Layer/Color/Next] <N>: **A**
(For Angle)

Specify new rotation angle <90>: **0** *(This
rotates the text back to its original angle
in Figure 17–37)*

Enter an option [Value/Position/Height/
Angle/Style/Layer/Color/Next] <N>: **P**
(For Position)

Specify new text insertion point <no
change>: *(Move your cursor above and
on center of the red catalog number; the
result is displayed in Figure 17–38)*

Enter an option [Value/Position/Height/
Angle/Style/Layer/Color/Next] <N>:
(Press ENTER *to exit this command)*

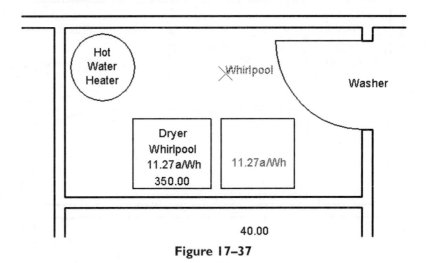

Figure 17–37

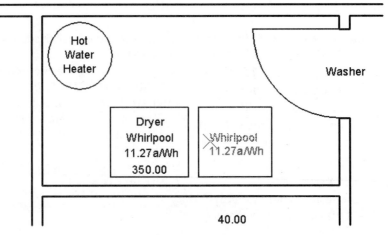

Figure 17–38

STEP 3

Next, move the "Washer" text above the "Whirlpool" text in Figure 17–39 using the Position option of the -ATTEDIT command.

Command: **-ATTEDIT**
Edit attributes one at a time? [Yes/No]
    <Y>: *(Press* ENTER *to accept the default)*
Enter block name specification <*>:
    WASHER
Enter attribute tag specification <*>:
    PRODUCT_NAME
Enter attribute value specification <*>:
    (Press ENTER *to accept the default)*
Select Attributes: *(Pick the "Washer" text item in Figure 17–39)*

Select Attributes: *(Press* ENTER *to continue)*
1 attributes selected.
Enter an option [Value/Position/Height/
    Angle/Style/Layer/Color/Next] <N>: **P**
    (For Position)
Specify new text insertion point <no
    change>: *(Move your cursor above and
    on center of the word "Whirlpool". The
    result is displayed in Figure 17-40)*
Enter an option [Value/Position/Height/
    Angle/Style/Layer/Color/Next] <N>:
    (Press ENTER *to exit this command)*

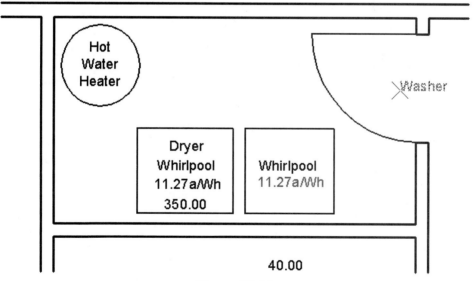

Figure 17–39

STEP 4

Use the -ATTEDIT command to change the color of the catalog number in Figure 17–40 from Red to Bylayer.

Command: **-ATTEDIT**
Edit attributes one at a time? [Yes/No]
 <Y>: *(Press* ENTER *to accept the default)*
Enter block name specification <*>:
 WASHER
Enter attribute tag specification <*>:
 CATALOG_NO.
Enter attribute value specification <*>:
 (Press ENTER *to accept the default)*

Select Attributes: *(Pick the catalog number in Figure 17–40)*
Select Attributes: *(Press* ENTER *to continue)*
1 attributes selected.
Enter an option [Value/Position/Height/
 Angle/Style/Layer/Color/Next] <N>: **C**
 (For Color)
Enter new color [Truecolor/COlorbook]
 <1 (red)>: BYLAYER
Enter an option [Value/Position/Height/
 Angle/Style/Layer/Color/Next] <N>:
 (Press ENTER *to exit this command)*

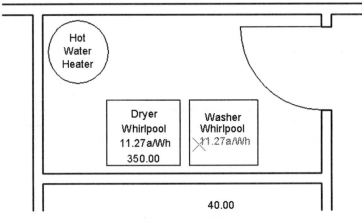

Figure 17–40

STEP 5

Finally, move the price attribute value, "40.00" to a new location under the catalog number (see Figure 17–41) and change the value from "40.00" to "400.00" with the -ATTEDIT command. The final results are displayed in Figure 17–42.

Command: **-ATTEDIT**
Edit attributes one at a time? [Yes/No]
 <Y>: *(Press* ENTER *to accept the default)*
Enter block name specification <*>:
 WASHER

Enter attribute tag specification <*>:
 PRICE
Enter attribute value specification <*>:
 (Press ENTER *to accept the default)*
Select Attributes: *(Pick the "40.00" text)*
Select Attributes: *(Press* ENTER *to continue)*
1 attributes selected.
Enter an option [Value/Position/Height/
 Angle/Style/Layer/Color/Next] <N>: **P**
 (For Position)
Specify new text insertion point <no
 change>: *(Pick a point below the center
 of the catalog number)*

Enter an option [Value/Position/Height/ Angle/Style/Layer/Color/Next] <N>: **V** *(For Value)*

Enter type of value modification [Change/ Replace] <R>: **R** *(For Replace)*

Enter new attribute value: **400.00**

Enter an option [Value/Position/Height/ Angle/Style/Layer/Color/Next] <N>: *(Press* ENTER *to exit this command)*

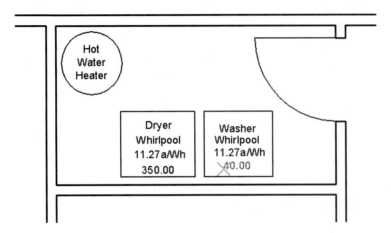

Figure 17–41

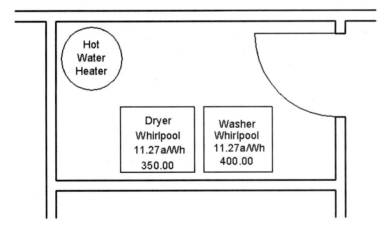

Figure 17–42

EDITING ATTRIBUTES GLOBALLY

Global editing is used to edit multiple attributes at one time. As usual, the criteria you specify will limit the set of attributes selected for editing.

You can choose global editing by responding to the initial prompt:

Command: **-ATTEDIT**
Edit Attributes one at a time? [Yes/No] <Y>: **N** *(For No)*

Entering "No" to this prompt will result in the following comment:

Performing global editing of attribute values.

The next prompt asks if you will be editing only attributes visible on the screen.

Edit only attributes visible on screen? [Yes/No] <Y>:

You are then prompted:

Enter block name specification <*>:
Enter attribute tag specification <*>:
Enter attribute value specification <*>:

The next prompt asks you to select the attributes to globally edit:

Select Attributes:

Use the standard object selection process to choose the group of attributes to edit. The objects selected will be highlighted.

Respond to the next prompts with the string you wish to change and the changes you wish to make.

Enter string to change:
Enter new string:

Let's examine how this process of globally editing attributes works. In Figure 17–43, suppose the tolerance value of "5%" needs to be changed to "10%" on all blocks containing this attribute value. You could accomplish this through the Edit Attributes dialog box. However, this would only allow you to edit each attribute value one at a time. A more productive way would be to use the -ATTEDIT command and edit the group of attributes globally. Follow the examples and prompt sequence below for accomplishing this task:

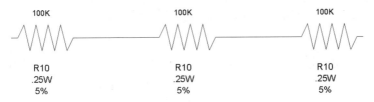

Figure 17–43

Command: **-ATTEDIT**
Edit attributes one at a time? [Yes/No] <Y>: **N**
Performing global editing of attribute values.
Edit only attributes visible on screen? [Yes/No] <Y>: *(Press* ENTER *to accept the default)*
Enter block name specification <*>: **RESISTOR**
Enter attribute tag specification <*>: **TOLERANCE**
Enter attribute value specification <*>: *(Press* ENTER *to affect all value specifications)*
Select Attributes: *(Pick a point at "A" in Figure 17–44)*
Specify opposite corner: *(Pick a point at "B" in Figure 17–44)*
Select Attributes: *(Press* ENTER *to continue)*
3 attributes selected.
Enter string to change: **5%**
Enter new string: **10%**

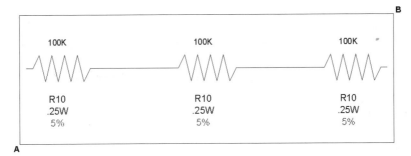

Figure 17–44

The result is shown in Figure 17–45, where all attributes were changed in a single command.

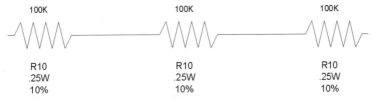

Figure 17–45

TUTORIAL EXERCISE: 17_COMPUTERS.DWG

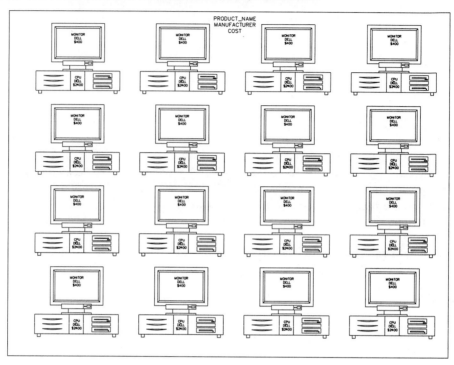

Figure 17–46

Purpose

This tutorial is designed to globally edit attribute values using the -ATTEDIT command on the group of computers in Figure 17–46.

System Settings

Open the drawing 17_Computers. All units and drawing limits have already been set. Attributes have also been assigned to various computer components.

Layers

All layers have already been created for this drawing.

Suggested Commands

Activate the -ATTEDIT command. Switch to global attribute editing mode and remove all dollar signs from the COST attribute values. This editing operation would be necessary if you needed to change the data in the COST box from character to numeric values for an extraction operation (extracting attribute data will be discussed in more detail later in this chapter).

Whenever possible, substitute the appropriate command alias in place of the full AutoCAD command in each tutorial step. For example, use "CP" for the COPY command, "L" for the LINE command, and so on. The complete listing of all command aliases is located in Chapter 1, Table 1–2.

STEP 1

Open the drawing 17_Computers. Figure 17–47 shows an enlarged view of four computer workstations. For each workstation, "$400.00" needs to be changed to "400.00", and each "$2400.00" needs to be changed to "2400.00". In other words, the dollar sign needs to be removed from all values for each workstation.

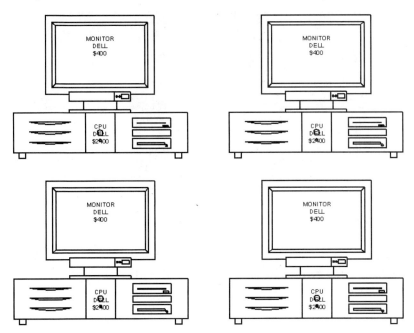

Figure 17–47

STEP 2

Use the -ATTEDIT command and the illustration below to globally edit all attribute values and remove all dollar signs from all values.

Command: **-ATTEDIT**
Edit attributes one at a time? [Yes/No]
    <Y>: **N** *(For No)*
Performing global editing of attribute
    values.
Edit only attributes visible on screen?
    [Yes/No] <Y>: *(Press ENTER to accept*
    this default)
Enter block name specification <*>: *(Press*
    ENTER *to accept this default)*

Enter attribute tag specification <*>:
    COST
Enter attribute value specification <*>:
    (Press ENTER *to accept this default)*
Select Attributes: *(Pick a point at "A" in*
    Figure 17–48)
Specify opposite corner: *(Pick a point at*
    "B" in Figure 17–48)
Select Attributes: *(Press* ENTER*)*
32 attributes selected.
Enter string to change: **$**
Enter new string: *(Press* ENTER*; this*
    will remove the dollar sign from all
    attribute values)

(B)

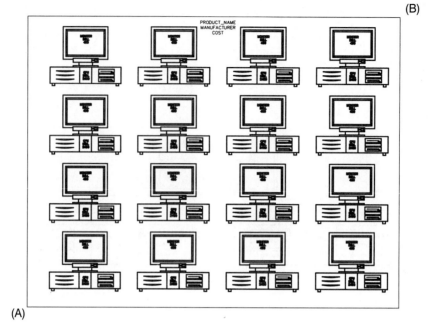

(A)

Figure 17–48

The results are illustrated in Figure 17–49 with all dollar signs removed from all attribute values. The attributes can now be prepared for extraction.

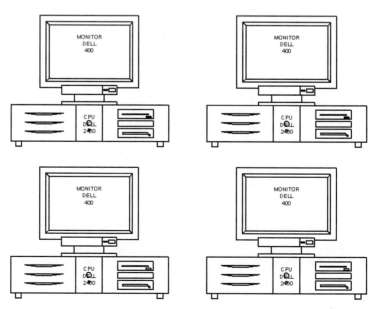

Figure 17–49

EXTRACTING ATTRIBUTES

Extracting attributes is great for keeping records of your inserted blocks. You could maintain a database on furniture; model and cost figures on parts used in a design; or the number, type and cost of windows in a plan. You could then print out all these items in a report or transfer this data to a database or spreadsheet program.

The ATTEXT command is used to extract attribute data from the drawing in a specified form. Entering the command activates the Attribute Extraction dialog box in Figure 17–50.

Command: **ATTEXT**

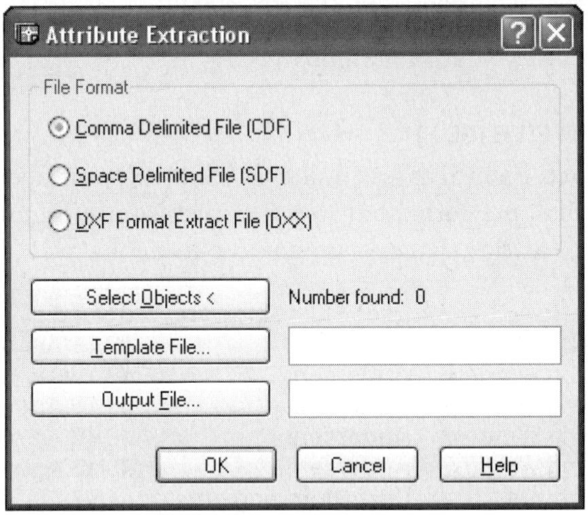

Figure 17–50

Before using this dialog box to extract the data, you must first select the format used for displaying the attributes. Next you must identify a template file that will have to be created prior to using this command. This template file is key in the extraction process. Finally, you supply a name for the extraction file.

ATTRIBUTE EXTRACTION FORMATS

Attributes can be extracted in three formats. All three formats are in the form of text files that can be read by applications such as Notepad.

COMMA DELIMITED FILE (CDF)

The Comma Delimited File (CDF) produces a file that contains delimiters (commas) that separate the data fields illustrated in Figure 17–51. The character fields are enclosed in quotes. This format can be read directly by some database programs.

```
'Bathtub','Eljer Plumbingware','#15.7/Ej', 225.00
'32" Double Hung Window','Andersen','#8.16/An', 125.00
'32" Double Hung Window','Andersen','#8.16/An', 125.00
'32" Double Hung Window','Andersen','#8.16/An', 125.00
'32" Double Hung Window','Andersen','#8.16/An', 125.00
'32" Double Hung Window','Andersen','#8.16/An', 125.00
'Bathroom Lavatory','Eljer Plumbingware','#15.7/Ej', 125.00
'Kitchen Sink','Universal Rundle','#15.7/Uni', 175.00
'Refrigerator','Frigidaire','#11.27a/Fr', 450.00
'Kitchen Range','White-Westinghouse','#11.27a/Whi', 600.00
'Toilet','Eljer Plumbingware','#15.7/Ej', 140.00
'Dryer','Whirlpool','#11.27a/Wh', 350.00
'Washer','Whirlpool','#11.27a/Wh', 400.00
```

Figure 17–51

SPACE DELIMITED FILE (SDF)

The Space Delimited File (SDF) is similar to CDF except it does not use commas; rather the data field is pre-formatted to a standard length. Figure 17–52 shows an example of extracted attribute information in the form of a SDF text file:

```
Bathtub                 Eljer Plumbingware    #15.7/Ej      225.00
32" Double Hung Window   Andersen              #8.16/An      125.00
32" Double Hung Window   Andersen              #8.16/An      125.00
32" Double Hung Window   Andersen              #8.16/An      125.00
32" Double Hung Window   Andersen              #8.16/An      125.00
32" Double Hung Window   Andersen              #8.16/An      125.00
Bathroom Lavatory        Eljer Plumbingware    #15.7/Ej      125.00
Kitchen Sink             Universal Rundle      #15.7/Uni     175.00
Refrigerator             Frigidaire            #11.27a/Fr    450.00
Kitchen Range            White-Westinghouse    #11.27a/Whi   600.00
Toilet                   Eljer Plumbingware    #15.7/Ej      140.00
Dryer                    Whirlpool             #11.27a/Wh    350.00
Washer                   Whirlpool             #11.27a/Wh    400.00
```

Figure 17–52

DXF FORMAT EXTRACTION FILE (DXX)

The DXF Format Extraction File (DXX) produces a subset of the AutoCAD Drawing Interchange File format containing only block reference, attribute, and end-of-sequence objects. DXF format extraction requires no template. The file extension .DXX distinguishes the output file from normal DXF files. Figure 17–53 shows a partial example of extracted attribute information in the form of a DXX text file. As shown in this figure, this format contains the block reference, attribute, and end-of-sequence objects.

```
Bathtub
  2
PRODUCT_NAME
 70
    0
  0
ATTRIB
  5
443
100
AcDbEntity
  8
0
100
AcDbAttribute
  1
Eljer Plumbingware
  2
MANUFACTURER
```

Figure 17–53

CREATING TEMPLATE FILES

To extract attribute information, you must first create a template file. This file controls the content and the display format for the information that will be extracted from the drawing. A text editor can be used to prepare the template file. Each line of the template represents one field to be listed in the attribute output file. You can also specify the width of the field (in characters), and the number of decimal places to be displayed in numerical fields. Each field will be listed in the order shown in the template file. Figure 17–54 shows an example of an attribute template file created in Notepad.

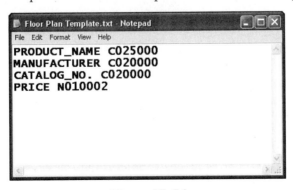

Figure 17–54

Figure 17–55 shows a breakdown of what is contained in a template file. The entries of NAME, DEPARTMENT, TITLE, and PHONE represent the attribute tags. The template file will search for these exact names. As a result, spelling is important. If you misspell an attribute tag in the template file, AutoCAD will be unable to locate the attribute information and complete the extraction process.

The second column in the template file contains formatting information used for extracting and displaying the data. In Figure 17–55, next to the NAME tag is the entry C010000. The leading **C** stands for Character. This tells AutoCAD to search for character data during the extraction process (character data can consist of numbers, letters and symbols). In a similar way, if the leading entry in the template file is **N**, this instructs AutoCAD to search for numeric data. Letters and symbols (other than a decimal point) are not allowed. If you define the leading character as N but your first attribute value begins with a symbol such as the $ sign, AutoCAD ignores this entry and no extraction takes place.

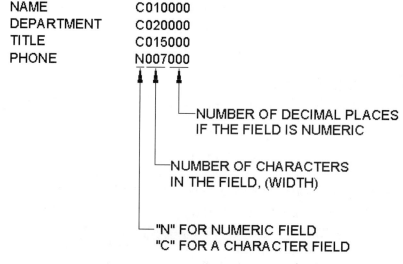

NAME   C010000
DEPARTMENT   C020000
TITLE   C015000
PHONE   N007000

—NUMBER OF DECIMAL PLACES
IF THE FIELD IS NUMERIC

—NUMBER OF CHARACTERS
IN THE FIELD, (WIDTH)

—"N" FOR NUMERIC FIELD
"C" FOR A CHARACTER FIELD

Figure 17–55

The next series of numbers next to the C in the NAME tag is **010**. This tells the extraction process to allocate 10 spaces in the output file to hold the attribute value, which is used mainly by the SDF format. This is illustrated in Figure 17–56 where four columns identified by the attribute tag names show the number of characters allotted to create a column effect. The last series, consisting of three zeros, **000**, deals with the number of decimal places. While this is not appropriate for a character attribute, it is important for a numeric attribute value. You could have an attribute tag called PRICE along with the following information: N020002. The **002** tells AutoCAD to add two decimal places with every numeric value it extracts under the tag of PRICE.

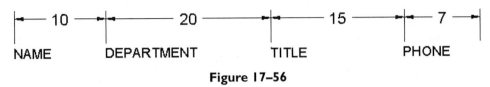

Figure 17–56

Each line of the template file specifies one field to be written to the attribute output file. This includes name of the field (the attribute tag), the character width, and the numeric precision. Each record of the attribute output file contains all information given in the order of the template file. While Figure 17–55 displays a typical template file designed to operate on the attribute tags, Figure 17–57 displays other types of information that can be extracted and placed in the output file, such as block name, block insertion point, and the layer the block was inserted on.

BL:LEVEL	Nwww000	Block nesting level
BL:NAME	Cwww000	Block name
BL:X	Nwwwddd	X coordinate of block insertion
BL:Y	Nwwwddd	Y coordinate
BL:Z	Nwwwddd	Z coordinate
BL:NUMBER	Nwww000	Block counter
BL:HANDLE	Cwww000	Block's handle
BL:LAYER	Cwww000	Block insertion layer
BL:ORIENT	Nwwwddd	Block rotation angle
BL:XSCALE	Nwwwddd	X scale factor of block
BL:YSCALE	Nwwwddd	Y scale factor
BL:ZSCALE	Nwwwddd	Z scale factor
BL:XEXTRUDE	Nwwwddd	X component of extrusion direction
BL:YEXTRUDE	Nwwwddd	Y component
BL:ZEXTRUDE	Nwwwddd	Z component
character	Cwww000	Character attribute tag
numeric	Nwww000	Numeric attribute tag

Figure 17–57

REDEFINING ATTRIBUTES

If more sweeping changes need to be made to attributes, a mechanism exists that allows you to redefine the attribute tag information globally. You follow a process similar to redefining a block, and the attribute values are also affected. A few examples of why you would want to redefine attributes might be to change their mode status (Invisible, Constant, etc.), change the name of a tag, reword a prompt, change a value to something completely different, add a new attribute tag or delete an existing tag entirely.

Any new attributes assigned to existing block references will use their default values. Old attributes in the new block definition retain their old values. If you delete an attribute tag, AutoCAD deletes any old attributes that are not included in the new block definition.

EXPLODING A BLOCK WITH ATTRIBUTES

Before redefining an attribute, you might first have to copy an existing block with attributes and explode it. This will return the block to its individual objects and return the attribute values to their original tag information. Figure 17–58 shows a kitchen sink with attribute values on the left and, on the right, the same sink but this time with attribute tags. The EXPLODE command was used on the right block to return the attribute values to their tags.

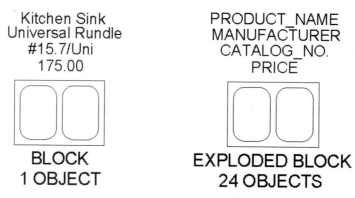

Figure 17–58

USING THE PROPERTIES PALETTE TO EDIT TAGS

A useful tool in making changes to attribute information is the Properties palette. For example, clicking the CATALOG_NO. tag and then activating the Properties palette in Figure 17–59 allows you to make changes to the attribute prompt and values. You could even replace the attribute tag name with something completely different without having to use the Attribute Definition dialog box. Scrolling down the Properties palette in Figure 17–60 exposes the four attribute modes. If an attribute was originally created to be visible, you could make it invisible by changing the Invisible modifier here from No to Yes.

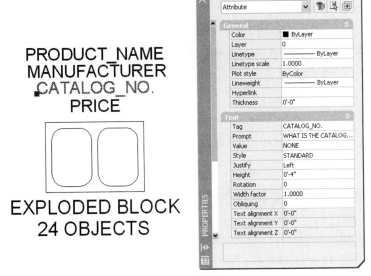

Figure 17–59

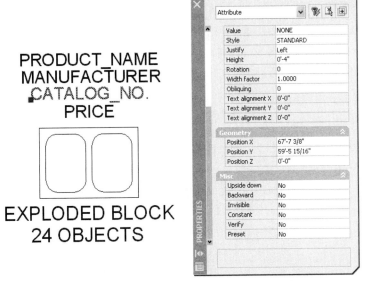

Figure 17–60

You could also double-click on the attribute tag in order to make any changes. Double-clicking on the "CATALOG_NO." attribute tag illustrated in Figure 17–61 displays the Edit Attribute Definition dialog box (DDEDIT command). Use this dialog box to make changes to the attribute tag, prompt, and default.

PRODUCT_NAME
MANUFACTURER
CATALOG_NO.
PRICE

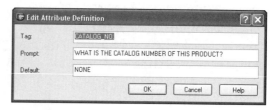

Edit Attribute Definition

Tag:	CATALOG_NO
Prompt:	WHAT IS THE CATALOG NUMBER OF THIS PRODUCT?
Default:	NONE

OK   Cancel   Help

Figure 17–61

REDEFINING THE BLOCK WITH ATTRIBUTES

Once you have made changes to the attribute tags, you are now ready to redefine the block along with the attributes with the ATTREDEF command. Follow the prompt sequence and illustration below to accomplish this task.

Command: **ATTREDEF**
Enter name of the block you wish to redefine: **SINK**
Select objects for new Block...
Select objects: *(Pick a point at "A" in Figure 17–62)*
Specify opposite corner: *(Pick a point at "B" in Figure 17–62)*
Select objects: *(Press ENTER to continue)*
Specify insertion base point of new Block: **MID**
of *(Pick the midpoint of the sink at "C" in Figure 17–62)*

B

PRODUCT_NAME
MANUFACTURER
CATALOG_NO.
PRICE

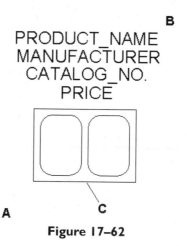

A   **C**

Figure 17–62

TUTORIAL EXERCISE: 17_RESISTORS.DWG

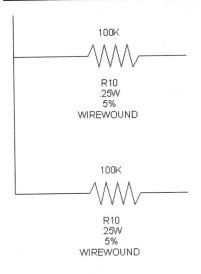

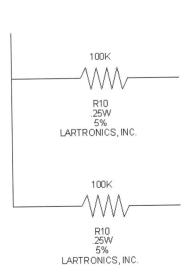

Figure 17–63

Purpose

This tutorial is designed to redefine attribute tags and update existing attribute values using the drawing illustrated in Figure 17–63.

System Settings

All units and drawing limits have already been set. Attributes have also been assigned to computer components.

Layers

All layers have already been created for this drawing.

Suggested Commands

The PROPERTIES and DDEDIT commands will be used to edit various characteristics of attributes.

Whenever possible, substitute the appropriate command alias in place of the full AutoCAD command in each tutorial step. For example, use "CP" for the COPY command, "L" for the LINE command, and so on. The complete listing of all command aliases is located in Chapter 1, Table 1–2.

STEP I

Open the drawing file 17_Resistors in Figure 17–64. A series of resistors is arranged in a partial circuit design. A number of changes need to be made to the original attribute definitions. Once the changes are made, the block and attributes will be redefined and all the blocks s will automatically be updated.

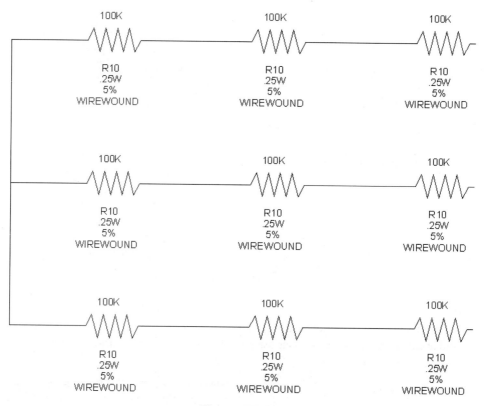

Figure 17–64

STEP 2

Copy one of the resistor symbols to a blank part of your screen. Then explode this block. This should break the block down into individual objects and return the attribute values to their tags (see Figure 17–65).

 Command: **CP** (For COPY)

Select objects: *(Pick resistor "A" in Figure 17–65)*

Select objects: *(Press* ENTER *to continue)*

Specify base point or displacement, or [Multiple]: *(Pick a point anywhere on Resistor "A")*

Specify second point of displacement or <use first point as displacement>: *(Pick a point anywhere to the left of Resistor "A")*

 Command: **X** *(For EXPLODE)*

Select objects: *(Pick resistor "B" in Figure 17–65)*

Select objects: *(Press* ENTER *to perform the explode operation)*

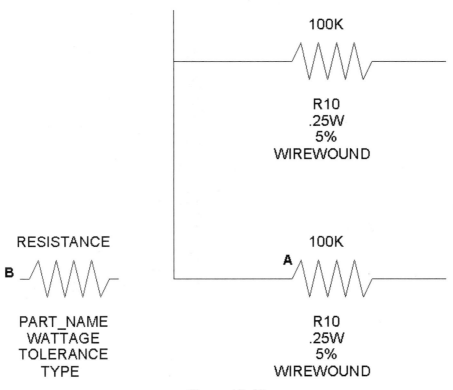

Figure 17–65

STEP 3

Let's examine what needs to be accomplished. The ATTDISP command is set to Normal; however, all the attribute values are displayed. Some of this information will be made invisible to simplify the drawing. Once invisible, this data could still be extracted with the ATTEXT command or viewed by setting the ATTDISP command to On. In this step, the tags WATTAGE, TOLERANCE, and TYPE will have their attribute mode changed to Invisible.

First select the WATTAGE, TOLER-ANCE, and TYPE tags. These items will be highlighted, along with grips being displayed. Activate the Properties palette and observe at the top that three Attributes are currently selected. Scroll down to the bottom of this window and under the Misc section, take note of the Invisible attribute mode. Click "No," pick the down arrow in this edit box, and set the Invisible mode to "Yes" in Figure 17–66. Move the cursor into the drawing area and press ESC to remove the object highlight and the grips. These three tags have now been changed to Invisible.

RESISTANCE

PART_NAME

WATTAGE

TOLERANCE

TYPE

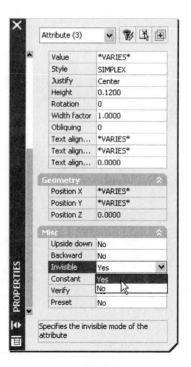

Figure 17–66

STEP 4

In this step, the tag TYPE needs to be replaced with a new tag. This also means creating a new prompt and default value. Double-click on the TYPE tag to display the Edit Attribute Definition dialog box as shown in Figure 17–67 and change the Tag from "TYPE" to "SUPPLIER". In the Prompt edit box, change the existing prompt to "Supplier Name?" In the Default edit box, change "WIRE-WOUND" to "LARTRONICS, INC," as shown in Figure 17–67. When finished, click the OK button to accept the changes and dismiss the dialog box.

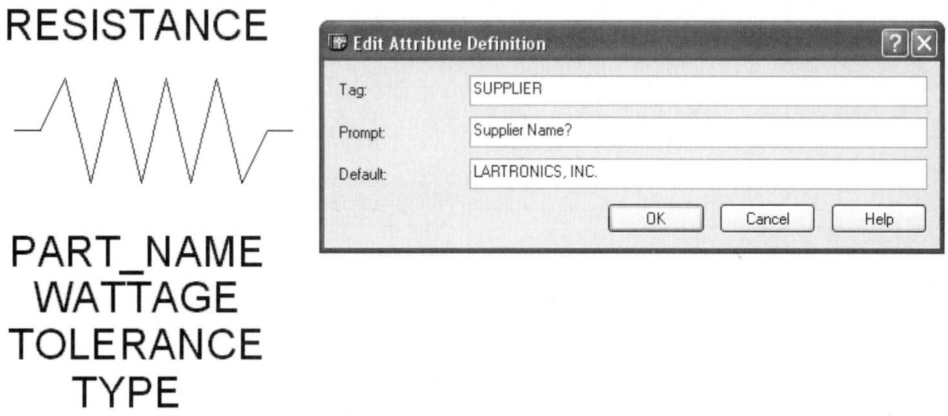

RESISTANCE

PART_NAME
WATTAGE
TOLERANCE
TYPE

Figure 17–67

STEP 5

After these changes are made to the exploded block, the ATTREDEF command will be used to redefine the block and automatically update all resistor blocks to their new values and states. Activate the ATTREDEF command. Enter "RESISTOR" as the name of the block to redefine. Select the block and attributes in Figure 17–68. Pick the new insertion point at "C" in Figure 17–68. As the drawing is regenerated, all blocks are updated to the new attribute values, as shown in Figure 17–69.

Command: **ATTREDEF**
Enter name of the block you wish to redefine: **RESISTOR**
Select objects for new Block...
Select objects: *(Pick a point at "A" in Figure 17–68)*
Specify opposite corner: *(Pick a point at "B" in Figure 17–68)*
16 found
Select objects: *(Press ENTER to continue)*
Specify insertion base point of new Block: **END**
of *(Pick the endpoint of the resistor at "C" in Figure 17–68)*

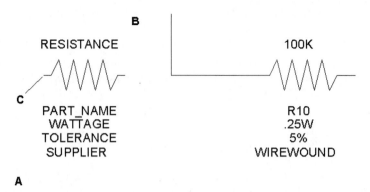

Figure 17–68

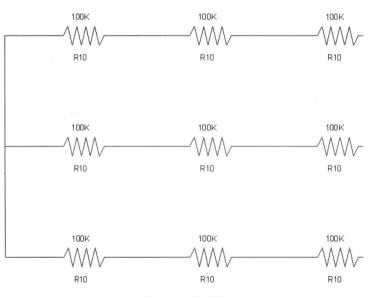

Figure 17–69

STEP 6

To see if the SUPPLIER tag replaced the TYPE tag, activate the ATTDISP command and turn all attribute values on. The results are displayed in Figure 17–70.

Command: **ATTDISP**
Enter attribute visibility setting [Normal/
ON/OFF] <Normal>: **ON**

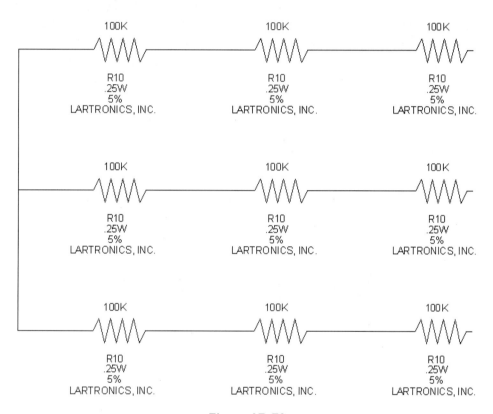

Figure 17–70

ENHANCED ATTRIBUTE EDIT AND EXTRACTION TOOLS

A number of extra attribute tools in the form of dialog boxes are available to better assist in the modification and extraction of attributes. These are the Enhanced Attribute Editor dialog box, the Block Attribute Manager dialog box, and the Attribute Extraction Wizard. These tools can all be selected from the Modify II toolbar illustrated in Figure 17–71.

The Enhanced Attribute Editor dialog box can also be displayed by selecting Object from the Modify pull-down menu, followed by Attribute, and then Single in Figure 17–72. The Block Attribute Manager dialog box can also be selected from this area of the Object > Attribute cascading menu. Activate the Attribute Extraction Wizard by selecting Attribute Extraction from the Tools pull-down menu, in Figure 17–73. All three of these tools will be explained in greater detail.

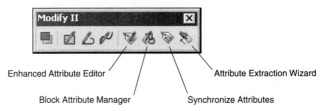

Enhanced Attribute Editor

Attribute Extraction Wizard

Block Attribute Manager

Synchronize Attributes

Figure 17–71

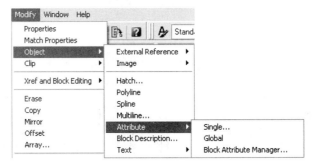

Figure 17–72

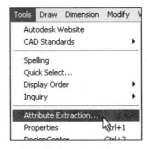

Figure 17–73

THE ENHANCED ATTRIBUTE EDITOR DIALOG BOX

Click the Edit Attribute toolbar button to display the Enhanced Attribute Editor dialog box in Figure 17–74. This dialog box can also be activated by entering EATTEDIT at the Command prompt. In the figure, the Dryer block was selected. Notice the attribute value is based on the selected tag. The value can then be modified in the Value edit box. To change to a different tag, select it with your mouse cursor. Notice that this dialog box has three tabs. The Attribute tab allows you to select the attribute tag that will be edited.

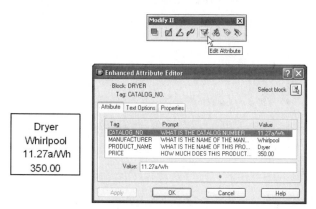

Figure 17–74

 Note: Double-clicking on a block containing attributes values will also launch the Enhanced Attribute Editor dialog box illustrated in Figure 17–74.

Clicking the Text Options tab displays the image illustrated in Figure 17–75. Use this area to change the properties of the text associated with the attribute tag selected, such as text style, justification, height, rotation and so on. Once changes are made, click the Apply button.

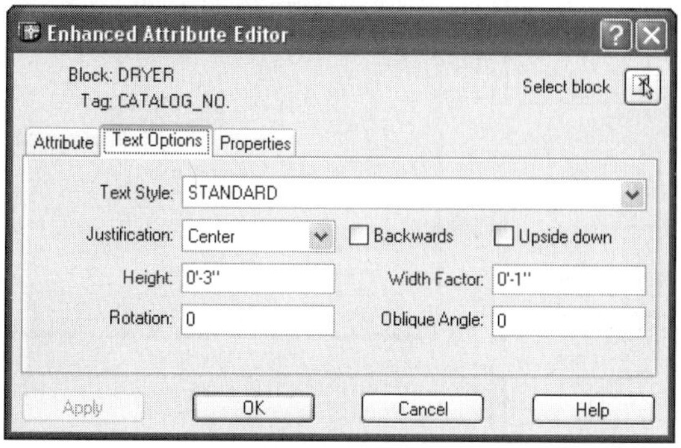

Figure 17–75

Clicking the Properties tab displays the image illustrated in Figure 17–76. Use this area to change such properties as layer, color, lineweight, and so on. When you have completed the changes to the attribute, click the Apply button.

All three tabs of the Enhanced Attribute Editor dialog box have a Select block button visible in the upper-right corner of each dialog box. Once you have edited a block, use this button to select a different block for editing, if desired.

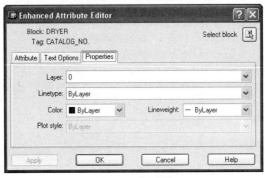

Figure 17–76

THE BLOCK ATTRIBUTE MANAGER

Clicking the Block Attribute Manager toolbar button or entering BATTMAN at the Command prompt displays the Block Attribute Manager dialog box in Figure 17–77. This dialog box is displayed as long as attributes are defined in your drawing. It does not require you to pick any blocks because it searches the database of the drawing and automatically lists all blocks with attributes. These are illustrated in the drop-down list in Figure 17–77. As will be shown, the Block Attribute Manager dialog box allows you to edit attributes globally.

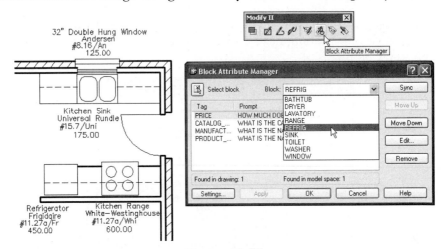

Figure 17–77

Clicking the Edit button displays the Edit Attribute dialog box in Figure 17–78. Three tabs similar to the Enhanced Attribute Editor dialog box are displayed. Use these tabs to edit the attribute characteristics, the text options (style, justification, etc.), or the properties (layer, etc.) of the attribute. Notice at the bottom of the dialog box in Figure 17–78 that as changes are made, you will automatically see these changes because the Auto preview changes check box is selected.

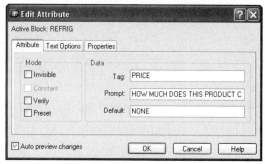

Figure 17–78

Clicking the Settings button in Figure 17–77 displays the Settings dialog box in Figure 17–79. If you click one of the properties such as Height and click the OK button in the figure, this property will be displayed in the main Block Attribute Manager dialog box in Figure 17–80.

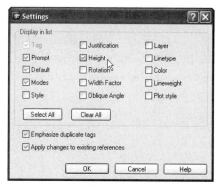

Figure 17–79

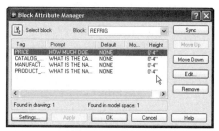

Figure 17–80

TUTORIAL EXERCISE: 17_BLOCK ATTRIB MGR.DWG

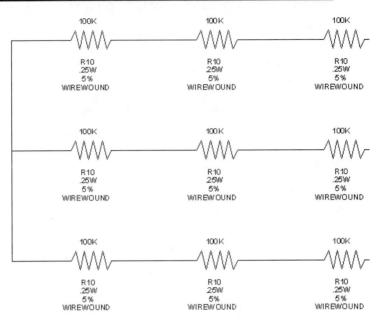

Figure 17–81

Purpose

This tutorial is designed to modify the properties of all attributes with the Block Attribute Manager dialog box using the drawing illustrated in Figure 17–81.

System Settings

All units and drawing limits have already been set. Attributes have also been assigned to the resistor block symbols.

Layers

All layers have already been created for this drawing.

Suggested Commands

The Block Attribute Manager dialog box will be used to turn the Invisible mode On for a number of attribute values. These changes will be automatically seen as they are made.

Whenever possible, substitute the appropriate command alias in place of the full AutoCAD command in each tutorial step. For example, use "CP" for the COPY command, "L" for the LINE command, and so on. The complete listing of all command aliases is located in Chapter 1, Table 1–2.

STEP 1

Open the drawing file 17_Block Attrib Mgr in Figure 17–81. The following attribute tags need to have their Invisible modifier turned On: TYPE, TOLERANCE, and WATTAGE. Activate the Block Attribute Manager dialog box (BATTMAN command) in Figure 17–82 and select the TYPE tag. Then select the Edit button.

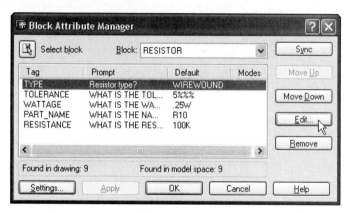

Figure 17–82

STEP 2

This takes you to the Edit Attribute dialog box in Figure 17–83. In the Attribute tab, place a check next to the Invisible mode and click the OK button. Notice that the changes are global and take place automatically. The attribute value WIREWOUND is no longer visible for any of the resistor blocks in Figure 19-84.

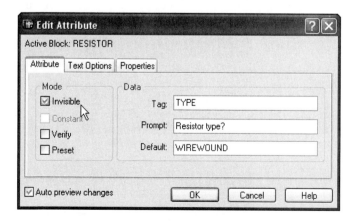

Figure 17–83

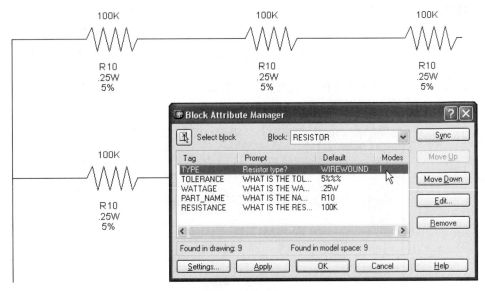

Figure 17–84

STEP 3

Turn the Invisible mode On for the TOLERANCE and WATTAGE tags. Your display should appear similar to Figure 17–85, with all attributed tags invisible, except for PART_NAME and RESISTANCE.

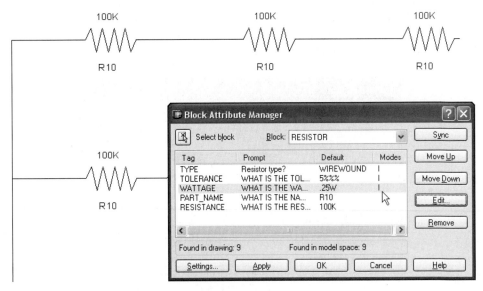

Figure 17–85

THE ATTRIBUTE EXTRACTION WIZARD

Clicking the Attribute Extraction Wizard toolbar button or entering EATTEXT at the Command prompt activates the Attribute Extraction Wizard in Figure 17–86. Use this wizard to step you through the attribute extraction process. You will be prompted to pick a new type of template with the BLK extension. If one is not already available, you can create it here in this dialog box. You will be able to individually select the attributes as well as specific block information to extract. You can preview the extraction file, save the template, and export the results in either TXT (tab-delimited file), CSV (comma delimited), or XLS (Microsoft Excel) format.

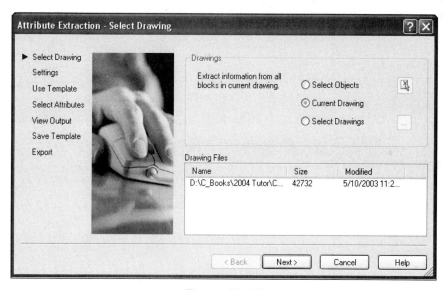

Figure 17–86

SYSTEM VARIABLES THAT CONTROL ATTRIBUTES

ATTREQ

Determines whether the INSERT command uses default attribute settings during insertion of blocks. The following settings can be used:

    0 No attribute values are requested; all attributes are set to their default values.

Turns on prompts or a dialog box for attribute values, as specified by ATTDIA.

Command: **ATTREQ**
Enter new value for ATTREQ <1>:

ATTMODE

Controls the display of attributes. The following settings can be used:

    0 Off: Makes all attributes invisible.

    1 Normal: Retains current visibility of each attribute; visible attributes are displayed, invisible attributes are not.

On: Makes all attributes visible.

Command: **ATTMODE**
Enter new value for ATTMODE <1>:

ATTDIA

Controls whether the INSERT command uses a dialog box for attribute value entry. The following settings can be used:

    0 Issues prompts on the command line.

Uses a dialog box.

Command: **ATTDIA**
Enter new value for ATTDIA <1>:

Open the Exercise Manual PDF file for Chapter 17 on the accompanying CD for more tutorials and exercises.

If you have the accompanying Exercise Manual, refer to Chapter 17 for more tutorials and exercises.

CHAPTER 18

Working with External References and Raster Image Files

This chapter begins the study of External References and how they differ from blocks. When changes are made to a drawing that has been externally referenced into another drawing, these changes are updated automatically. There is no need to redefine the external reference in the same way blocks are redefined. Additionally, a section on using the AutoCAD DesignCenter to attach external references is included in this chapter. When sharing of data is important, the ETRANSMIT command is used to collect all support files that make up a drawing. This will guarantees to the individual reviewing the drawing that all support files come with the drawing. Working with raster images is also discussed in this chapter. Through the Image Manager dialog box, raster images can be attached to a drawing file for presentation purposes. The brightness, contrast, and fade factor of the image can also be adjusted. To control the order in which images display in your drawing, the DRAWORDER command will be explained.

COMPARING EXTERNAL REFERENCES AND BLOCKS

It has been seen so far in Chapter 16 that it is very popular to insert blocks and drawing files in other drawing files. One advantage in performing these insertions is a smaller file size, which results because all objects connected with each block are grouped into a single object. It is only when blocks are exploded that the drawing size once again increases because the objects that made up the block have been separated into their individual objects. Another advantage of blocks is their ability to be redefined in a drawing.

External references are similar to blocks in that they appear to be considered one object. However, external references are attached to the drawing file, whereas blocks are inserted in the drawing. This attachment actually sets up a relationship between the current drawing file and the external referenced drawing. For example, take a floor plan and externally reference it into a current drawing file. All objects associated with the floor plan are brought in; including their current layer and linetype qualities. With the floor plan acting as a guide, such items as electrical symbols, furniture, and dimensions can be added to the floor plan and saved to the current drawing file. Now a design change needs to be made to the floor plan. You open the floor plan, stretch a few doors and walls to new locations, and save the file. The next time the drawing holding the electrical symbols and furniture arrangement is opened, the changes to the floor plan are made automatically to the current drawing file. This is one of the primary advantages of external references over blocks. Also, even though blocks are known to reduce file size; using external references reduces the file size even more than using blocks. One disadvantage of an external reference is that items such as layers and other blocks that belong to the external reference can be viewed but have limited capabilities for manipulation.

CHOOSING EXTERNAL REFERENCE COMMANDS

One way to select external reference commands is from the toolbars. Select External Reference from the Insert or Reference toolbar, illustrated in Figures 18–1 and 18–2. Yet another way is through the keyboard by entering XREF or XR. A third way is to choose Xref Manager from the Insert pull-down menu, as shown in Figure 18–3. Whichever method you use, the Xref Manager dialog box displays, as illustrated in Figure 18–4.

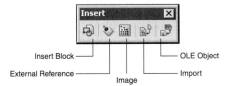

Figure 18–1

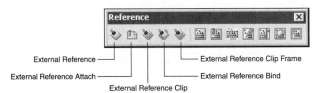

Figure 18–2

Figure 18–3

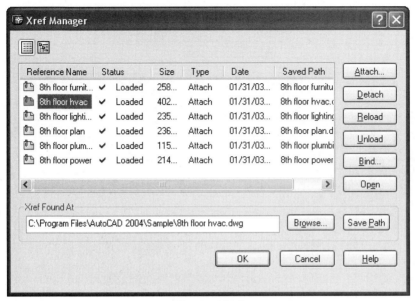

Figure 18–4

ATTACHING AN EXTERNAL REFERENCE

The primary mode in the Xref Manager dialog box in Figure 18–4 is the Attach option. This can be compared with the INSERT command for merging blocks into drawing files. The Attach option sets up a path, which looks for the external reference every time the drawing containing it is loaded.

Try It! - Start a new drawing file from scratch. Then, open the Xref Manager dialog box and click on the Attach button, shown in Figure 18–4, to activate the Select Reference File dialog box illustrated in Figure 18–5. This dialog box is very similar to the one for selecting drawing files to initially load into Auto-CAD. Click on the file 18_Asesmp1 to attach and click the Open button.

Note: To accomplish the Try Its in this chapter, copy the Chapter 18 files to a convenient location/folder on your hard drive. This way you will be able to modify and save the drawings as required.

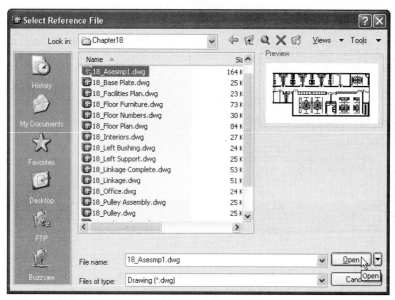

Figure 18–5

Once the file 18_Asesmp1 is selected, the External Reference dialog box appears in Figure 18–6. Some of the information contained in this dialog box is similar to the information in the Insert dialog box. The insertion point, scale and rotation angle of the external reference can be specified in this dialog box or on the screen. Of importance, in the upper part of the dialog box, is the path information associated with the external reference. If, during file management, an externally referenced file is moved to a new location, the new path of the external reference must be re-established. Otherwise it does not load into the drawing it was attached to. Take the check out of the box under "Insertion point". This will insert this external reference at absolute coordinate 0,0,0. Clicking the OK button returns you to the drawing editor and attaches the file to the current drawing. Since you will not see the floor plan on the small drawing sheet, perform a ZOOM-ALL. Your results should appear similar to the illustration in Figure 18–7.

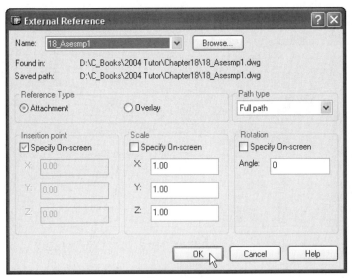

Figure 18–6

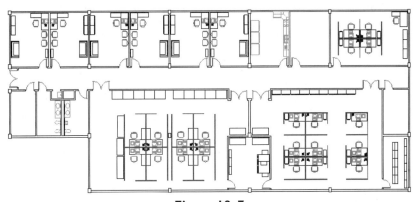

Figure 18–7

With the external reference attached to the drawing file, layers and blocks that belong to the external reference take on a new meaning. Illustrated in Figure 18–8 are the current layers that are part of the drawing. However, notice how a number of layers begin with the same name (18_Asesmp1); also, what appears to be a vertical bar separates 18_Asesmp1 from the names of the layers. Actually, the vertical bar represents the "Pipe" symbol on the keyboard, and it designates that the layers belong to the external reference, namely 18_Asesmp1. These layers can be turned off, frozen, or even have the color changed. However, you cannot make this layer current for drawing on because it belongs to the external reference.

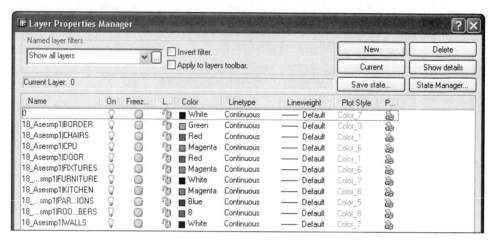

Figure 18–8

Just as layers are affected by the presence of an external reference, blocks display the same behavior. When blocks are listed, as in Figure 18–9, the name of the external reference is first given, with the "Pipe" symbol following, and finally the actual name of the block, as in 18_Asesmp1|DESK2. As with layers, these blocks cannot be used in the drawing because of the presence of the "Pipe" symbol, which is not supported in the name of the block. This does, however, provide a quick way of identifying the valid blocks and those belonging to an external reference. Study the following prompt sequence and Figure 18–9 for identifying blocks that belong to external references. To view the blocks defined in this drawing, follow the command sequence below to perform this operation.

Command: **-B** *(for -BLOCK. The "-" preceding the command provides a command line version of the block command)*
Enter block name or [?]: **?**
Enter block(s) to list <*>: *(Press ENTER to accept the default and display the list in Figure 18–9)*

```
Defined blocks.
    "18_Asesmp1"                          Xref: resolved
    "18_Asesmp1|ADCADD_ZZ"                Xdep: "18_Asesmp1"
    "18_Asesmp1|CB30"                     Xdep: "18_Asesmp1"
    "18_Asesmp1|CB36"                     Xdep: "18_Asesmp1"
    "18_Asesmp1|CC30"                     Xdep: "18_Asesmp1"
    "18_Asesmp1|CCBASE"                   Xdep: "18_Asesmp1"
    "18_Asesmp1|CHAIR7"                   Xdep: "18_Asesmp1"
    "18_Asesmp1|COPIER2"                  Xdep: "18_Asesmp1"
    "18_Asesmp1|CW1236"                   Xdep: "18_Asesmp1"
    "18_Asesmp1|CW3012"                   Xdep: "18_Asesmp1"
    "18_Asesmp1|CW3030"                   Xdep: "18_Asesmp1"
    "18_Asesmp1|CW3036"                   Xdep: "18_Asesmp1"
    "18_Asesmp1|DESK2"                    Xdep: "18_Asesmp1"
    "18_Asesmp1|DESK3"                    Xdep: "18_Asesmp1"
    "18_Asesmp1|DESK4"                    Xdep: "18_Asesmp1"
    "18_Asesmp1…

                                          …18_Asesmp1"

    "18_Asesmp1|SOFA2"                    Xdep: "18_Asesmp1"
    "18_Asesmp1|TABLE1"                   Xdep: "18_Asesmp1"
    "18_Asesmp1|TABLE2"                   Xdep: "18_Asesmp1"
    "18_Asesmp1|VENDING"                  Xdep: "18_Asesmp1"


    User        External      Dependent     Unnamed
    Blocks      References    Blocks        Blocks
      0            1            42             0
```

Figure 18–9

One of the real advantages of using external references is the way they affect drawing changes. To demonstrate this, first save your drawing file under the name 18_Facilities. Close this drawing and open the original floor plan called 18_Asesmp1. Use the Insert dialog box to insert a block called "ROOM NUMBERS" into the 18_Asesmp1 file. Use an insertion point of 0,0,0 for placing this block and click OK in the Insert dialog box in Figure 18–10. Save this drawing file and open up the drawing 18_Facilities. In Figure 18–11, the image of the drawing file 18_Asesmp1 is brought up again. Notice how room tags have been added to the floor plan. Once the drawing holding the external reference is loaded, all room tags, which belong to the external reference, are automatically displayed.

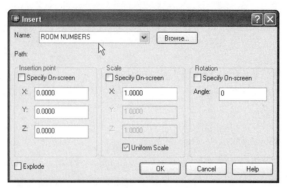

Figure 18–10

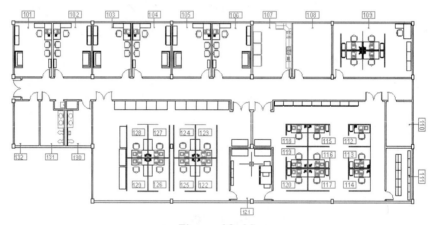

Figure 18–11

OVERLAYING AN EXTERNAL REFERENCE

Suppose in the last example of the facilities floor plan that some design groups need to see the room number labels while other design groups do not. This is the purpose of overlaying an external reference instead of attaching it. All design groups will see the information if the external reference is attached. If information is overlaid and the entire drawing is externally referenced, the overlaid information does not display; it's as if it is invisible. This option is illustrated in the next Try It! exercise.

Try It! - Open the drawing 18_Floor Plan. Then use the External Reference dialog box in Figure 18–12 to attach the file 18_Floor Furniture (the steps for accomplishing this were demonstrated in the pervious "Try It." Use an insertion point of 0,0,0 and be sure the reference type is Attachment. Your display should be similar to the image in Figure 18–13. You want all design groups to view this information.

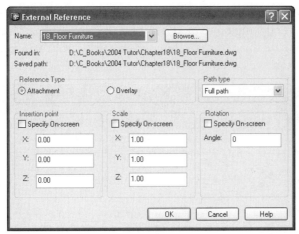

Figure 18–12

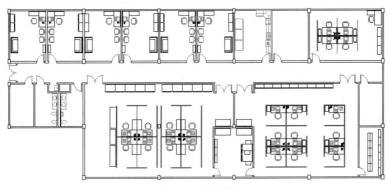

Figure 18–13

Now we will overlay an external reference. With Layer 0 current, click on the Attach button in the Xref Manager dialog box in Figure 18–4. When the Select Reference File dialog box appears in Figure 18–14, pick the file 18_Floor Numbers and click the Open button.

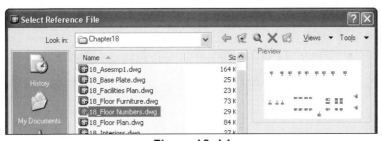

Figure 18–14

When the External Reference dialog box appears in Figure 18–15, change the Reference Type by clicking on the radio button next to Overlay. Under the Insertion Point heading, be sure this external reference will be inserted at 0,0,0. When finished, click the OK button.

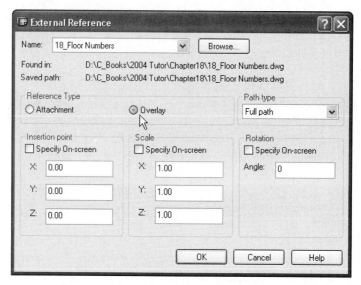

Figure 18–15

Your display should appear similar to Figure 18–16. Save this drawing file under its original name of 18_Floor Plan and then close the drawing file.

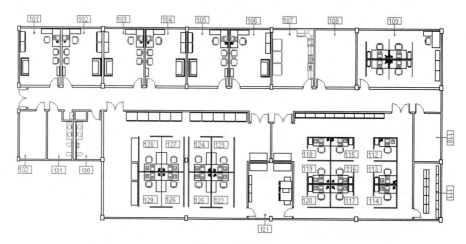

Figure 18–16

Now start a new drawing file from scratch. In the Xref Manager dialog box, click on the Attach button and attach the drawing file 18_Floor Plan. When the External Reference dialog box appears in Figure 18–17, be sure the Reference Type is set to Attachment and that the drawing will be inserted at 0,0,0. When finished, click the OK button.

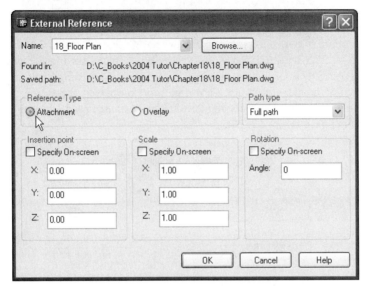

Figure 18–17

Your drawing will not be displayed because the default limits are too small, so perform a ZOOM-ALL operation and observe the results in Figure 18–18. Notice that the room numbers do not display because they were originally overlaid in the file 18_Floor Plan. This completes this Try It! exercise.

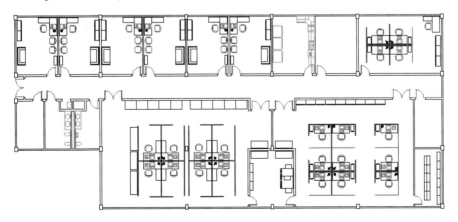

Figure 18–18

THE XBIND COMMAND

Earlier, it was mentioned that blocks and layers belonging to external references cannot be used in the drawing they were externally referenced into. There is a way, however, to convert a block or layer into a useful object through the XBIND command.

Try It! - Open the drawing file 18_Facilities Plan. Then select the xbind command from the Reference toolbar or the Modify pull-down as shown in Figure 18–19. This activates the Xbind dialog box, which lists the external reference 18_ASESMP1. Clicking on the "+" sign expands the listing of all named objects such as blocks and layers associated with the external reference. Clicking on the "+" sign in Block lists all individual blocks associated with the external reference (see Figure 18–20).

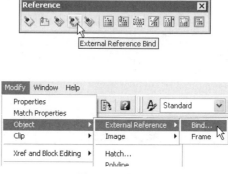

Figure 18–19

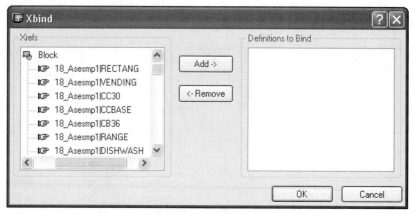

Figure 18–20

Click on the block 18_Asesmp1|DESK2 from the listing on the left; then click on the Add-> button. This moves the block name over to the right under the listing of Definitions to Bind. Do the same for 18_Asesmp1|DESK3 and 18_Asesmp1|DESK4. When finished adding these items, click on the OK button to bind the blocks to the current drawing file (see Figure 18–21).

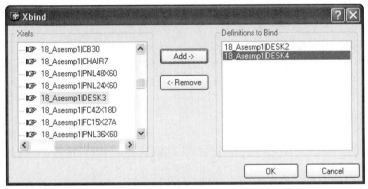

Figure 18–21

Test to see that new blocks have in fact been bound to the current drawing file. Activate the Insert dialog box, click on the Name drop-down list box, and notice the display of the blocks, as in Figure 18–22. The three symbols just bound from the external reference still have the name of the external reference; namely 18_Asesmp1. However, instead of the "Pipe" symbol separating the name of the external reference and block names, the characters 0 are now used. This is what makes the blocks valid in the drawing. Now these three blocks can be inserted in the drawing file even though they used to belong only to the external reference 18_Asesmp1.

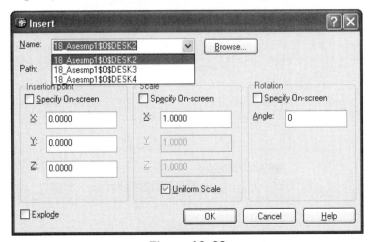

Figure 18–22

IN-PLACE REFERENCE EDITING

In-Place Reference Editing allows you to edit a reference drawing from the current drawing, which is externally referencing it. You then save the changes back to the original drawing file. This becomes an efficient way of making a change to a drawing file from an externally referenced file.

Try It! - Open the drawing file 18_Pulley Assembly in Figure 18–23. Both holes located in the Base Plate, Left Support, and Right Support need to be stretched 0.20 units to the left and right. This will center the holes along the Left and Right Supports. Since all images that make up the Pulley Assembly are external references, the In-Place Reference Editing feature will be illustrated.

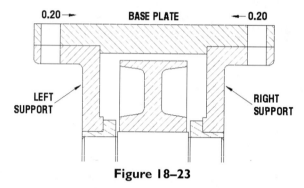

Figure 18–23

Begin by accessing the REFEDIT command from the Refedit toolbar (see Figure 18–24) or by choosing Edit Reference In-place from the Modify pull-down menu, followed by Xref and Block Editing (see Figure 18–25).

Figure 18–24

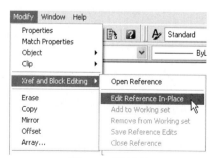

Figure 18–25

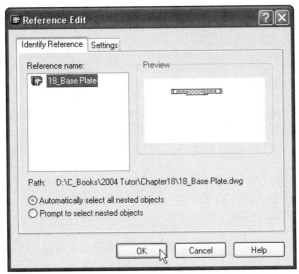

Figure 18–26

Once in the REFEDIT command, you are prompted to select the reference you wish to modify. Pick the top line of the Base Plate in your drawing. This displays the Reference Edit dialog in Figure 18–26. The reference to be edited, which is 18_Base Plate in our case, should be selected. Nested references may also be displayed; one of these could be selected for editing instead, if desired. Clicking on the OK button returns you to the screen. The Refedit toolbar is automatically displayed (see Figure 18–27) and you are returned to the command prompt. You can now make modifications to the Base Plate by using the STRETCH command to stretch both holes a distance of 0.20 units to the inside.

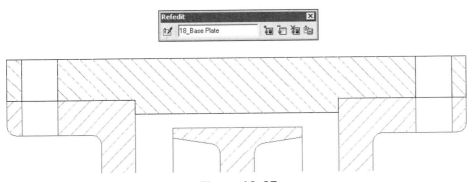

Figure 18–27

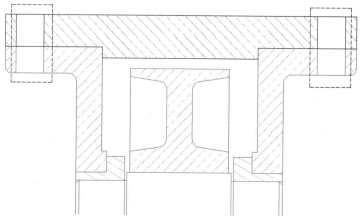

Figure 18–28

When performing the stretch operation, notice that even if the crossing window were to extend across both parts (see Figure 18–28), only the holes in the base plate will be modified (see Figure 18–29). When satisfied with the changes, select the Save Back Changes to Reference button on the toolbar. An AutoCAD alert box (see Figure 18–30) will ask you to confirm the saving of reference changes.

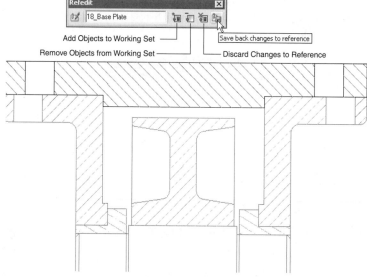

Figure 18–29

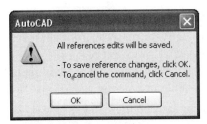

Figure 18–30

After you click OK, the results can be seen in Figure 18–31. Notice that only the objects that belong to the external reference (in this case, the holes that are part of the Base Plate) are affected. Perform the same series of steps using In-Place Reference Editing separately for the Left and Right supports. Stretch the hole located on the Left Support a distance of 0.20 units to the right. Stretch the hole located on the Right Support a distance of 0.20 units to the left. The final results of this In-Place Reference Editing Try It! exercise are displayed in Figure 18–32.

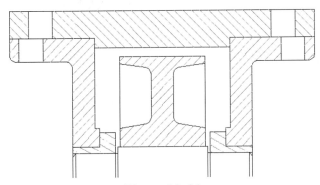

Figure 18–31

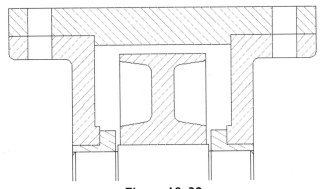

Figure 18–32

BINDING AN EXTERNAL REFERENCE

Use the XREF command to activate the Xref Manager dialog box in Figure 18–33. Select the external reference that you want to bind to the current drawing and click the Bind button. This option activates the Bind Xrefs dialog box illustrated in Figure 18–34. Two options are available inside this dialog box: Bind and Insert.

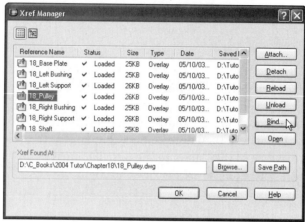

Figure 18–33

Figure 18–34

The Bind option binds to the current drawing file all blocks, layers, dimension styles, etc. that belonged to an external reference. After you perform this operation, layers can be made current and blocks inserted in the drawing. For example, a typical block definition belonging to an external reference is listed in the symbol table as XREFname|BLOCKname. Once the external reference is bound, all block definitions are converted to XREFname\$0\$BLOCKname. In the example in Figure 18–35, the following referenced layers were converted to usable layers with the Bind option:

Referenced Layers	Converted Layers
18_Pulley|Hatch	18_Pulley\$0\$Hatch
18_Pulley|Object	18_Pulley\$0\$Object
18_Pulley|Text	18_Pulley\$0\$Text

Note: The same naming convention is true for blocks, dimension styles, and other named items. The result of binding an external reference is similar to a drawing that was inserted into another drawing.

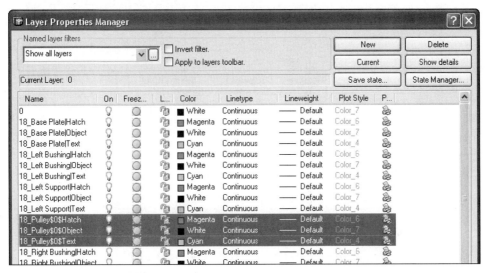

Figure 18–35

The Insert option of the Bind Xrefs dialog box in Figure 18–34 is similar to the Bind option. However, instead of named items such as blocks and layers being converted to the format XREFname0BLOCKname, the name of the external reference is stripped, leaving just the name of the block, layer, or other named item (BLOCKname, LAYERname, etc.) In the example in Figure 18–36, the following referenced layers were converted to usable layers with the Insert option:

Referenced Layers	Converted Layers
18_Pulley|Hatch	Hatch
18_Pulley|Object	Object
18_Pulley|Text	Text

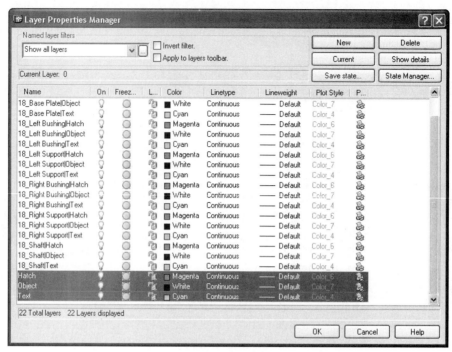

Figure 18–36

 Note: It is considered an advantage to use the Bind option over the Insert option. This way, you will be able to identify the layer that was tied to the previously used external reference.

USING LIST VIEW AND TREE VIEW

The Xref Manager dialog box makes it easy to keep track of external references in a drawing. By default, external references are listed as shown in Figure 18–37. Clicking on the Tree View button lists external references in hierarchical form similar to the illustration in Figure 18–38.

Figure 18–37

Figure 18–38

CLIPPING AN EXTERNAL REFERENCE

Since attaching an external reference displays the entire reference file, you have the option of displaying a portion of the file. This is accomplished by clipping the external reference with the XCLIP command. Choose this command from either the Reference toolbar or the Modify > Clip > Xref pull-down menu as shown in Figure 18–39.

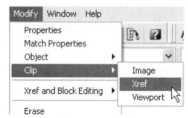

Figure 18–39

This operation is very useful when you want to emphasize a particular portion of your external reference file. Clipping boundaries include polylines in the form of rectangles, regular polygonal shapes, or even irregular polyline shapes. All polylines must form closed shapes.

 Try It! - Open the drawing file 18_Facilities Plan. Follow the illustration in Figure 18–40 and Command prompt sequence below for performing this operation.

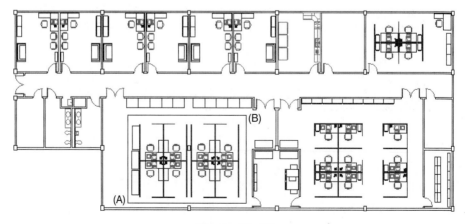

Figure 18–40

Command: **XCLIP**

Select objects: *(Pick the external reference)*
Select objects: *(Press ENTER to continue)*
Enter clipping option
[ON/OFF/Clipdepth/Delete/generate Polyline/New boundary] <New>: **N** *(For New)*
Specify clipping boundary:
[Select polyline/Polygonal/Rectangular] <Rectangular>: *(Press ENTER)*
Specify first corner: *(Pick a point at "A" in Figure 18–40)*
Specify opposite corner: *(Pick a point at "B" in Figure 18–40)*

The results are displayed in Figure 18–41. If you want to return the clipped image back to the full external reference, use the XCLIP command and use the OFF clipping option. This temporarily turns off the clipping frame. To permanently remove the clipping frame, use the Delete clipping option.

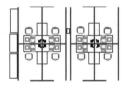

Figure 18–41

 Command: **XCLIP**

Select objects: *(Pick the external reference)*
Select objects: *(Press ENTER)*
Enter clipping option
[ON/OFF/Clipdepth/Delete/generate Polyline/New boundary] <New>: **Off** *(To temporarily turn the clipping frame off and display the entire external reference)*

 Command: **XCLIP**

Select objects: *(Pick the external reference)*
Select objects: *(Press ENTER)*
Enter clipping option
[ON/OFF/Clipdepth/Delete/generate Polyline/New boundary] <New>: **On** *(To turn the clipping frame on and display the clipped image of the external reference)*

 Command: **XCLIP**

Select objects: *(Pick the external reference)*
Select objects: *(Press ENTER)*
Enter clipping option
[ON/OFF/Clipdepth/Delete/generate Polyline/New boundary] <New>: **Delete** (This will permanently delete the clipping frame and display the entire external reference)

OTHER OPTIONS OF THE XREF MANAGER DIALOG BOX

The following additional options of the Xref Manager dialog box will now be discussed, using Figure 18–42 as a guide.

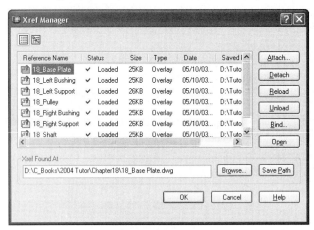

Figure 18–42

DETACH

Use this option to permanently detach or remove an external reference from the database of a drawing.

RELOAD

This option loads the most current version of an external reference. It is used when changes to the external reference are made while the external reference is currently being used in another drawing file. This option works well in a networked environment where all files reside on a file server.

UNLOAD

Unload is similar to the Detach option with the exception that the external reference is not permanently removed from the database of the drawing file. Since this option suppresses the external reference from any drawing regenerations, it is used as a productivity technique. Reload the external reference when you want it returned to the screen.

EXTERNAL REFERENCE NOTIFICATION TOOLS

A series of tools and icons are available to assist with managing external references. When a drawing consisting of external references opens, an icon appears in the extreme lower-right corner of your screen as shown in Figure 18–43. Clicking on this icon will launch the Xref Manager dialog box. In the event the source file was changed or modified, the next time you return to the external reference drawing, a notification message appears in Figure 18–44 informing you that the external reference was changed. This example refers to the Pulley Assembly in Figure 18–23. The Base Plate was modified and saved. When you return to the Pulley Assembly, the notification message appears announcing the change to the file 18_Base Plate.

Figure 18–43

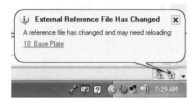

Figure 18–44

Clicking on the name of the modified external reference (18_Base Plate) in the notification message box will launch the Xref Manager dialog box. You will notice the appearance of a red exclamation point in Figure 18–45 alerting you that 18_Base Plate needs to be reloaded in order for the change to be reflected in the Pulley Assembly drawing.

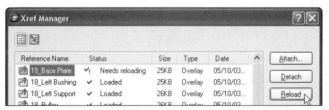

Figure 18–45

Clicking on the Reload button while the external reference is highlighted will change the icon to a series of arrows in Figure 18–46. Notice in the Xref Manager dialog box that the status has changed to Reload. Clicking the OK button will reload the external reference and update the drawing to the changes.

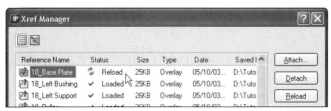

Figure 18–46

If you re-enter the Xref Manager dialog box after updating the external reference, notice that an icon consisting of the letter "i" in a circle appears as shown in Figure 18–47. This icon informs you that this external reference (18_Base Plate) was recently changed. When you save this drawing file, the icon with the letter "i" will disappear and be replaced by the familiar checkmark as shown in Figure 18–47.

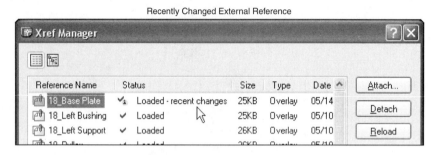

Figure 18–47

Illustrated in Figure 18–48 is another feature of external references. The left support of the pulley assembly was selected. When you right-click, the menu in Figure 18–48 appears. Use this menu to perform the following tasks:

Edit Xref in-place—This option activates the Reference Edit dialog box for the purpose of editing the external reference in-place.

Open Xref—This option will open the selected external reference in a separate window.

Clip Xref—This option launches the XCLIP command for the purpose of clipping a portion of the external reference

Xref Manager—This option launches the Xref Manager dialog box.

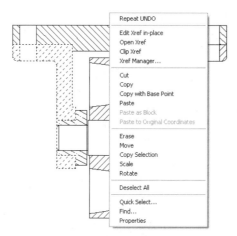

Figure 18–48

ATTACHING EXTERNAL REFERENCES WITH AUTOCAD DESIGNCENTER

In Chapter 16, the AutoCAD DesignCenter was used to insert blocks and drawing files into the current drawing file. Just as blocks can be inserted, so also can External References be attached to a drawing file. The same rules apply for external references as for blocks. In the AutoCAD DesignCenter, select the desired folder in the tree view (the left column) to load the palette (the right column) with the file you want to externally reference (see Figure 18–49). Using the right mouse button, drag and drop the file symbol into the drawing. The shortcut menu shown in Figure 18–50 lets you insert the file as a block or attach it as an external reference. Clicking on Attach as Xref will display the External Reference dialog box in Figure 18–51. Click OK and select an insertion point to complete attaching the external reference to the drawing.

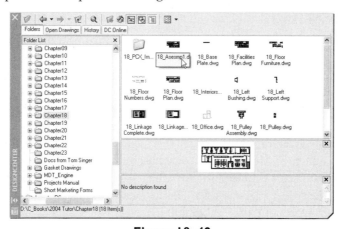

Figure 18–49

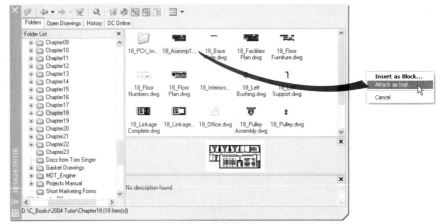

Figure 18–50

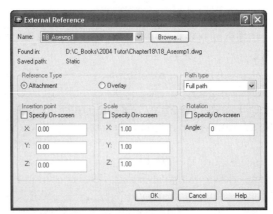

Figure 18–51

USING ETRANSMIT

This AutoCAD utility is helpful in reading the database of your drawing and listing all support files needed. Once these files are identified, you can have this utility gather all files into one EXE or ZIP file. You can then copy these files to a disk, CD, or transmit the files over the Internet. Clicking on eTransmit... located in the File pull-down menu in Figure 18–52 displays the Create Transmittal dialog box in Figure 18–53. Notice in the General tab that you can supply a series of notes to be used to document the contents of your eTransmittal file. Under the Type: category, you specify the EXE or ZIP file types. The EXE file type is especially useful since it automatically self-extracts the information without the need for any third-party software. Under the Location: category, pick the folder you want the EXE or ZIP files to be located. If you know these files are going to an AutoCAD 2000 user, you place a check in the box next to "Convert drawings to:"

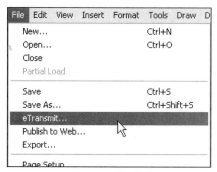

Figure 18–52

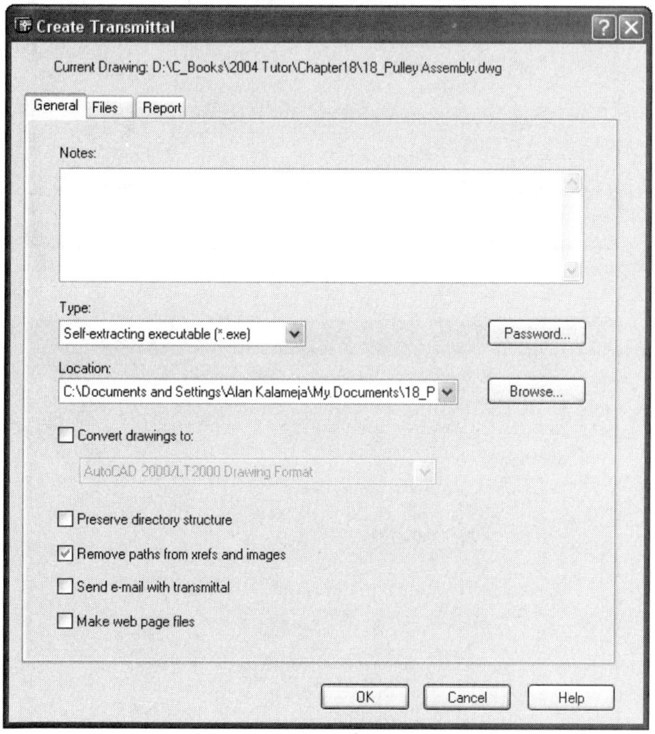

Figure 18–53

Clicking on the Files tab in Figure 18–54 displays a list of all files that will be included in the transmittal set. The Tree View mode is displayed in this figure. Buttons at the top of the dialog box are used to switch between the List View and Tree View modes. Clicking on the Report tab displays a list of all files in Figure 18–55. Click the OK button to create the EXE or ZIP transmittal file.

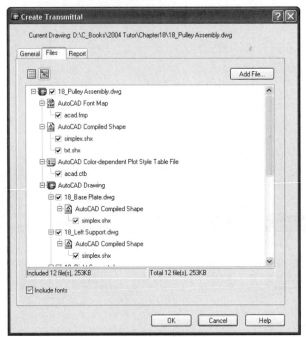

Figure 18–54

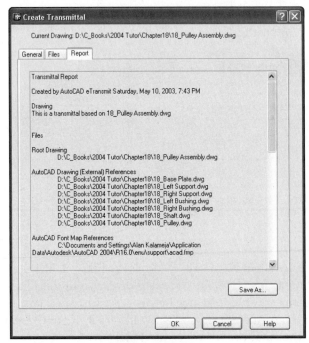

Figure 18–55

WORKING WITH RASTER IMAGES

Raster images in the form of JPG, GIF, PCX and so on can easily be merged with your vector-based AutoCAD drawings. The addition of raster images gives a new dimension to your drawings. Typical examples of raster images are digital photographs of an elevation of a house or an isometric view of a machine part. Whatever the application, working with raster images is very similar to what was just covered with external references. You attach the raster image to your drawing file. Once part of your drawing, additional tools are available to manipulate and fine-tune the image for better results. As with external references, an Image Manager exists to control various aspects of your image. Clicking Image Manager… from the Insert pull-down menu in Figure 18–56 displays the Image Manager dialog box in Figure18–57. Image commands may also be selected from the Reference toolbar in Figure 18–58.

Figure 18–56

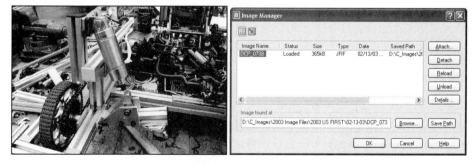

Figure 18–57

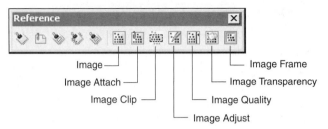

Figure 18–58

Try It! - Begin this exercise on working with raster images by opening the drawing file 18_Linkage illustrated in Figure 18–59. In this exercise a raster image will be attached and placed in the blank area to the right of the two-view drawing.

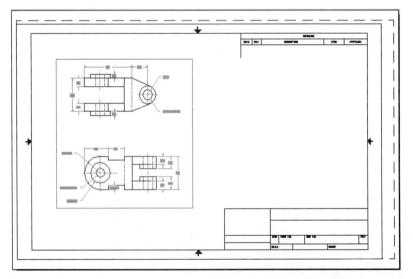

Figure 18–59

Click on Image Manager… located under the Insert pull-down menu in Figure 18–56. When the Image Manager dialog box displays in Figure 18–57, click on the Attach… button. When the Select Image File dialog box appears in Figure 18–60, click on the file Linkage1.jpg. Then click on the Open button.

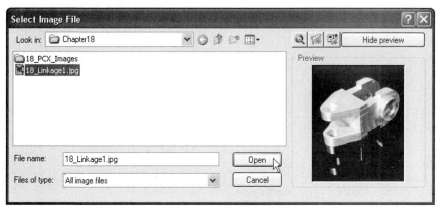

Figure 18–60

When the Image dialog box appears in Figure 18–61, leave all default settings. You will be specifying the insertion point on the drawing screen. Click the OK button to continue.

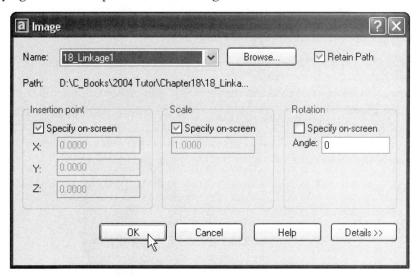

Figure 18–61

Once back in the drawing file, pick a point anchoring the lower-left corner of the image in Figure 18–62. Now move your cursor in an upward right direction to scale the image as necessary. Pick a second point at a convenient location to display the image.

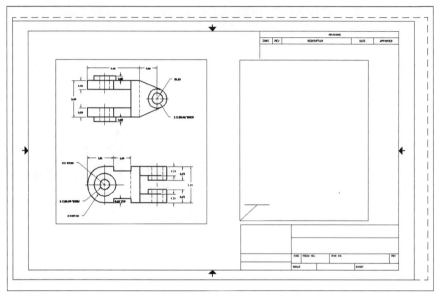

Figure 18–62

As the image displays on your screen, you can adjust its size very easily. Click on the edge of the image (the image frame) and notice the grips appearing at the four corners of the image. Clicking on a grip and then moving your cursor will increase or decrease its size (see Figure 18–63.)

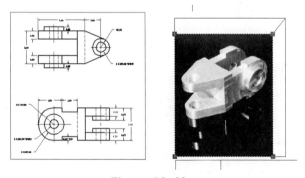

Figure 18–63

From the Reference toolbar in Figure 18–64, identify the Image Adjust button and pick it now. Picking the edge of the raster image frame will display the Image Adjust dialog box in Figure 18–64. Adjust the Brightness, Contrast, and Fade settings and notice that each of these affects the image previewed to the right of this dialog box. You can easily return to the original image by clicking the Reset button in the lower-left corner of this dialog box.

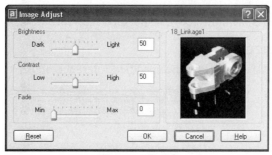

Figure 18–64

Clicking the OK button in the Image Adjust dialog box will return you back to the drawing. The final results are illustrated in Figure 18–65 with the vector drawing and raster image sharing the same layout.

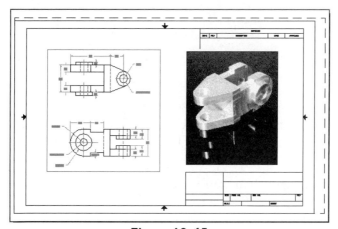

Figure 18–65

USING THE DRAWORDER COMMAND

The DRAWORDER command can be selected from the Modify II toolbar or from the Tools pull-down menu as shown in Figure 18–66. With the enhancements made to raster images and the ability to merge raster images with vector graphics, it is important to control the order in which these images are displayed. The four options are described below.

BRING TO FRONT

The selected object is brought to the top of the drawing order.

SEND TO BACK

The selected object is sent to the bottom of the drawing order.

BRING ABOVE OBJECT

The selected object is brought above a specified reference object.

SEND UNDER OBJECT

The selected object is sent below the specified reference object.

Figure 18–66

In the illustration in Figure 18–67, the image of North America has been merged with the Multiline text object "NORTH AMERICA". Unfortunately, because of the drawing order of operations, the image is masking part of the mtext object.

Display Order is selected from the Tools area of the pull-down menu area followed by the Bring to Front option. The mtext object is selected as the object to bring to the top of the drawing order, and the results are illustrated in Figure 18–68. The mtext object is now at the top of the drawing order and can be read in full.

Figure 18–67

Figure 18–68

TUTORIAL EXERCISE: EXTERNAL REFERENCES

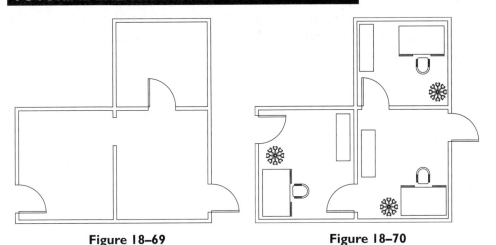

Figure 18–69   **Figure 18–70**

Purpose

This tutorial is designed to use the office floor plan in Figure 18–69 to create an interior plan in Figure 18–70 consisting of various interior symbols, such as desks, chairs, shelves, and plants. The office floor plan will be attached to another drawing file through the XREF command.

System Settings

Since these drawings are provided on the CD, all system settings have been made.

Layers

The creation of layers is not necessary, because layers already exist for both drawing files you will be working on.

Suggested Commands

Begin this tutorial by opening the drawing 18_Office.Dwg, which should be located on the CD, and view its layers and internal block definitions. Then open the drawing 18_Interiors.Dwg, also located on the CD, and view its layers and internal blocks. The file 18_Office.Dwg will now be attached to 18_Interiors.Dwg. Once this is accomplished, chairs, desks, shelves, and plants will be inserted in the office floor plan for laying out the office furniture. Once 18_Interiors.Dwg is saved, a design change needs to be made to the original office plan; open 18_Office.Dwg and stretch a few doors to new locations. Save this file and open 18_Interiors.Dwg; notice how the changes to the doors are automatically made. The Xbind dialog box will also be shown as a means for making a symbol that had previously belonged to an external reference usable in the file 18_Interiors.Dwg.

Whenever possible, substitute the appropriate command alias in place of the full AutoCAD command in each tutorial step. For example, use "CP" for the COPY command, "L" for the LINE command, and so on. The complete listing of all command aliases is located in Chapter 1, Table 1–2.

STEP 1

Open 18_Office.Dwg, which can be found on the CD, and observe a simple floor plan consisting of three rooms.

Furniture will be laid out using the floor plan as a template (see Figure 18–71).

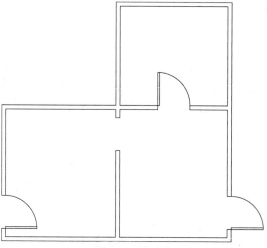

Figure 18–71

STEP 2

While in 18_Office.Dwg, use the Layer Properties Manager dialog box and observe the layers that exist in the drawing for such items as doors, walls,

and floor (see Figure 18–72). These layers will change once the office plan is attached to another drawing through the XREF command.

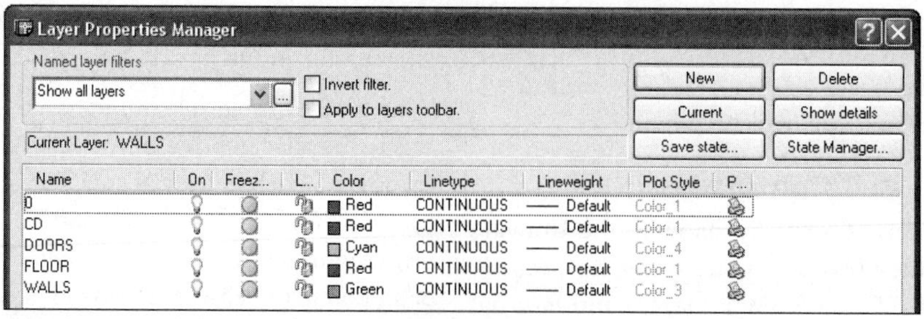

Name	On	Freez..	L..	Color	Linetype	Lineweight	Plot Style	P...
0				■ Red	CONTINUOUS	—— Default	Color_1	
CD				■ Red	CONTINUOUS	—— Default	Color_1	
DOORS				□ Cyan	CONTINUOUS	—— Default	Color_4	
FLOOR				■ Red	CONTINUOUS	—— Default	Color_1	
WALLS				■ Green	CONTINUOUS	—— Default	Color_3	

Layer Properties Manager — Named layer filters — Show all layers — Invert filter. — Apply to layers toolbar. — New — Delete — Current — Show details — Current Layer: WALLS — Save state... — State Manager...

Figure 18–72

STEP 3

While in the office plan, activate the Insert dialog box through the INSERT command. At times, this dialog box is useful for displaying all valid blocks in a drawing. Clicking on the Name drop-down list box displays the results in Figure 18–73. Two blocks are currently defined in this drawing; as with the layers, once the office plan is merged into another drawing through the XREF command, these block names will change. When finished viewing the defined blocks, close 18_Office.Dwg.

Figure 18–73

STEP 4

This next step involves opening 18_Interiors.Dwg, also found on the CD, and looking at the current layers found in this drawing. Once this drawing is open, use the Layer Properties Manager dialog box to observe that layers exist in this drawing for such items as floor and furniture (see Figure 18–74).

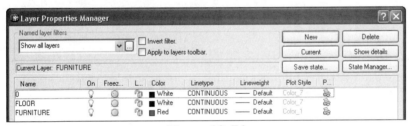

Figure 18–74

STEP 5

As with the office plan, activate the Insert dialog box through the INSERT command to view the blocks internal to the drawing (see Figure 18–75). The four blocks listed consist of various furniture items and will be used to lay out the interior plan.

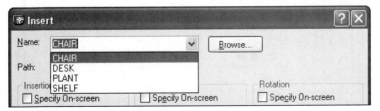

Figure 18–75

STEP 6

Close 18_Office.Dwg and verify that 18_ Interiors.Dwg is still open and active. The office floor plan will now be attached to the interior plan in order for the furniture to be properly laid out. Make the Floor layer current. Rather than insert the office plan as a block, use the Xref Manager dialog box to attach the drawing. This dialog box will activate when you enter the XREF command from the keyboard or choose Xref Manager from the Insert pull-down menu, as in Figure 18–76. After the dialog box displays, click on the Attach button, shown in Figure 18–77.

Figure 18–76

Figure 18–77

STEP 7

Clicking on the Attach button shown in Figure 18–77 displays the Select Reference File dialog box in Figure 18–78. Find the appropriate folder and click on the drawing file 18_Office.Dwg.

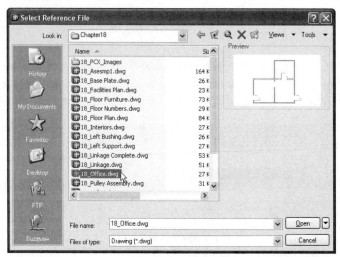

Figure 18–78

STEP 8

Selecting 18_Office.Dwg back in Figure 18–78 displays the External Reference dialog box in Figure 18–79. Notice that 18_OFFICE is the name of the external reference file chosen for attachment in the current drawing. Verify that the Reference Type is an attachment. If selected, remove the check marks from the Specify On-screen boxes for the Insertion point, Scale, and Rotation. In the External Reference dialog box, click on the OK button shown in Figure 18–79 to attach 18_Office.Dwg to 18_Interiors.Dwg.

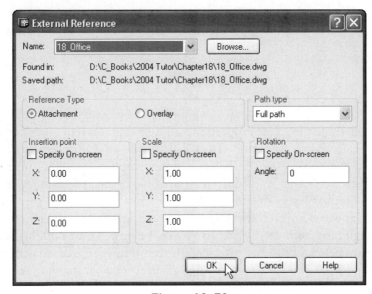

Figure 18–79

STEP 9

The floor plan is now attached to the Interiors drawing at the insertion point 0,0 as shown in Figure 18–80. This file looks identical to the original 18_Office.Dwg file; one way to determine the status of the external reference is to use the LIST command and select the edge of one of the walls that make up the office plan. The result is illustrated in Figure 18–81. Upon closer inspection, the image of 18_ OFFICE is actually an External Reference, as verified by the LIST command.

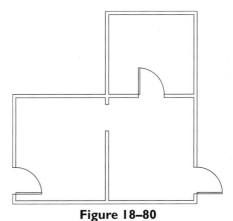

Figure 18–80

```
BLOCK REFERENCE   Layer: "FLOOR"
    Space: Model space
    Handle = 221
    "18_Office"
    External reference
    at point, X=   0.0000  Y=   0.0000  Z=   0.0000
    X scale factor    1.0000
    Y scale factor    1.0000
    rotation angle       0
    Z scale factor    1.0000
```

Figure 18–81

STEP 10

Once again, activate the Layer Properties Manager dialog box, paying close attention to the display of the layers. Using Figure 18–82 as a guide, you can see the familiar layers of Floor and Furniture. However, notice the group of layers beginning with 18_Office; the layers actually belonging to the external reference file have the designation of XREF|LAYER. For example, the layer 18_Office|doors represents a layer located in the file 18_Office.Dwg that holds all door symbols. The "|" or pipe symbol is used to separate the name of the external reference from the layer. The layers belonging to the external reference file may be turned on or off, locked, or even frozen. However, these layers cannot be made current for drawing on.

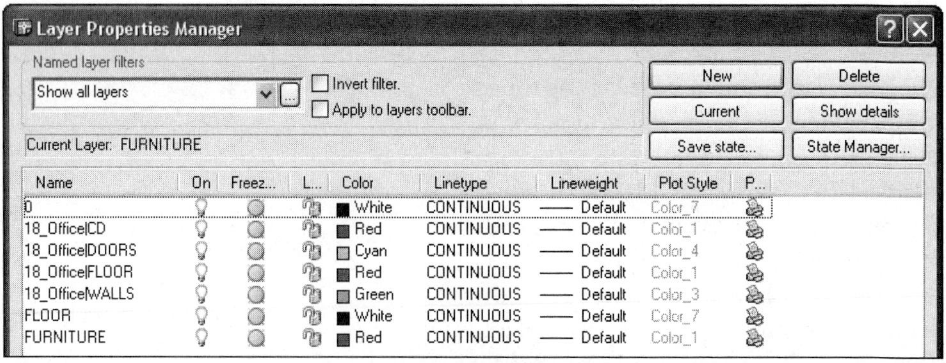

Figure 18–82

STEP 11

Look at the blocks that are currently defined in the drawing and notice how they appear. The Insert dialog box will show only those blocks that can be inserted in the drawing; instead, use the -BLOCK command, which will identify all blocks in the drawing; those you can and cannot insert (see Figure 18–83.) Notice how the DOOR and REVCLOUD appear in the list of blocks. As with the layers in the last step, the presence of the "|" or "pipe" symbol means the block is not valid and therefore cannot be inserted in the drawing file.

Command: **-BLOCK** *(The "-" preceding the command provides a command line version of the block command)*
Enter block name or [?]: **?**
Enter block(s) to list <*>: *(Press* ENTER *to list all blocks)*

```
Command: -BLOCK
Enter block name or [?]: ?
Enter block(s) to list <*>: (Press Enter for all)
Defined blocks.
 "18_Office"                      Xref: resolved
   "18_Office|DOOR"               Xdep: "18_Office"
   "18_Office|REVCLOUD"           Xdep: "18_Office"
   "CHAIR"
   "DESK"
   "PLANT"
   "SHELF"
User       External      Dependent    Unnamed
Blocks     References    Blocks       Blocks
  4            1            2            0
```

Figure 18–83

STEP 12

Make the Furniture layer current and begin inserting the desk, chair, shelf, and plant symbols in the drawing using the INSERT command or through the AutoCAD DesignCenter. The external reference file 18_OFFICE is to be used as a guide throughout this layout. It is not important that your drawing match exactly the image in Figure 18–84. After positioning all symbols in the floor plan, close and save your drawing under its original name of 18_Interiors.Dwg.

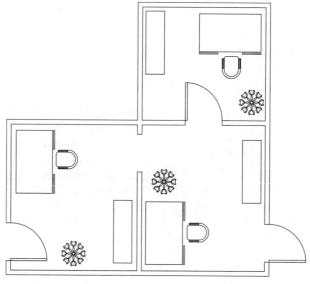

Figure 18–84

STEP 13

Now open the original drawing file 18_Office.Dwg and make the following modifications: stretch all three doors as indicated to new locations and mirror one of the doors so it is positioned closer to the wall; stretch the wall opening over to the other end of the room (see Figure 18–85). Finally, close and save these changes under the original name of 18_Office.Dwg.

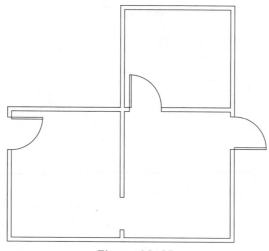

Figure 18–85

STEP 14

Now open the drawing file 18_ Interiors.Dwg and notice the results in Figure 18–86. Because 18_Office.Dwg was attached through the External Reference dialog box, any changes to the original drawing file are automatically reflected. In this case, observe how some of the furniture is now in the way of the doorways. If the office plan had been inserted in the interiors drawing as a block, these changes would not have occurred automatically.

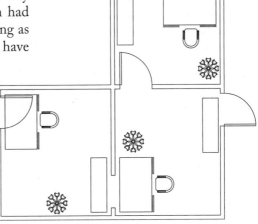

Figure 18–86

STEP 15

Because of the changes in the door openings, edit the drawing by moving the office furniture to better locations. Figure 18–87 can be used as a guide, although your drawing may appear different.

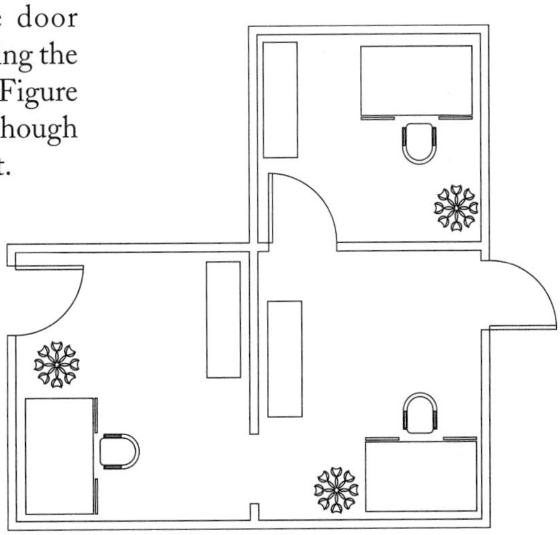

Figure 18–87

STEP 16

A door needs to be added to an opening in one of the walls of the office plan. However, the door symbol belongs to the externally referenced file 18_Office.Dwg. The door block is defined as 18_Office|DOOR; the "|" character is not valid in the naming of the block and therefore cannot be used in the current drawing. The block must first be bound to the current drawing before it can be used. Use the XBIND command to make the external reference's Door block available in the 18_Interiors.Dwg. This command can be selected from the Reference toolbar or the Modify pull-down as shown in Figure 18–88. This displays the Xbind dialog box in Figure 18–89. While in this dialog box, click on the + symbol next to the file 18_Office and then click on the + symbol next to Block. This will display all blocks that belong to the external reference (see Figure 18–89).

Figure 18–88

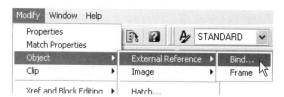

Figure 18–89

STEP 17

Clicking on 18_OFFICE|DOOR followed by the Add-> button in Figure 18–89 moves the block of the door to the Definitions to Bind area in Figure 18–90.

Click OK to dismiss the dialog box and the door symbol is now a valid block that can be inserted in the drawing.

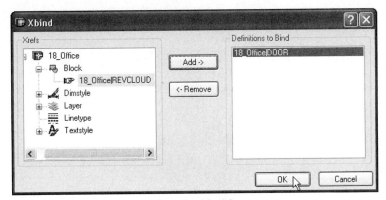

Figure 18–90

STEP 18

Activate the Insert dialog box through the INSERT command, click on the Name drop-down list box, and notice the block of the door in Figure 18–91. It is now listed as 18_Office0DOOR; the "|" character was replaced by the "0," making the block valid in the current drawing. This is AutoCAD's standard way of converting blocks that belong to external references to blocks that can be used in the current drawing file. This same procedure works on layers belonging to external references as well.

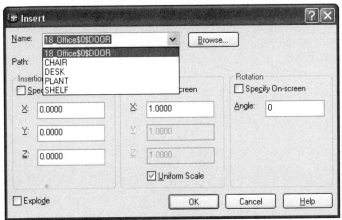

Figure 18–91

STEP 19

Make the Floor layer current and insert the door symbol into the open gap as illustrated in Figure 18–92 to complete the drawing.

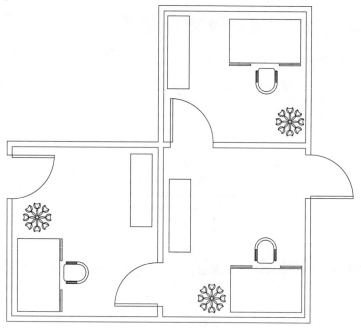

Figure 18–92

 Open the Exercise Manual PDF file for Chapter 18 on the accompanying CD for more tutorials and exercises.

 If you have the accompanying Exercise Manual, refer to Chapter 18 for more tutorials and exercises.

CHAPTER 19

Multiple Viewport Drawing Layouts

This chapter is a continuation of Chapter 9, "An Introduction to Drawing Layouts." First, multiple images of the same drawing file at different scales will be demonstrated. This will include a number of layering and dimensioning techniques to achieve success. This chapter continues by demonstrating how to arrange multiple details of different drawing files in the same layout. External references will be used to perform this task. Also, a new layer technique of freezing layers in certain viewports will be utilized. The command-line-driven vplayer command will also be explained as an alternate means of freezing multiple layers in selected viewports.

ARRANGING DIFFERENT VIEWS OF THE SAME DRAWING

Try It! – Begin this chapter on multiple viewport layouts by opening the drawing file 19_Roof Plan. This drawing in Figure 19–1 represents a roof plan of a structure and is drawn in Model Space at full size. The roof plan needs to be plotted out at a scale of 1/8" = 1'-0". Figure 19–2 illustrates the Fit tab of the Modify Dimension Style: dialog box that appears when you click on the Modify button of the Dimension Style Manager dialog box. The important point to focus on in this illustration is the Overall Scale value in the dialog box. Notice that it is set to a value of 96, which is arrived at when 1' or 12 is divided by 1/8" or 0.125 (1/8"=1' is the same scale as 1=96). This value will also be used shortly to scale the roof plan once it is brought into Paper Space. A second viewport will be created to provide a detail view of Area "A" (see Figure 19–1). A scale of 1/2" = 1'-0" (scale value of 24) will be utilized in this new viewport.

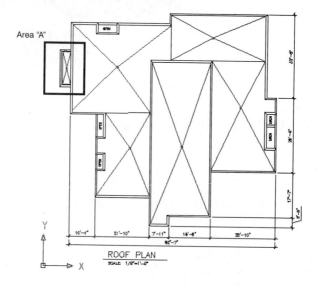

Figure 19–1

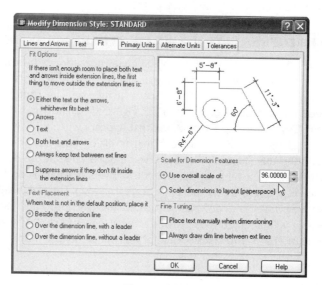

Figure 19–2

To begin the layout, select the Layout1 tab at the bottom of the AutoCAD screen. The Page Setup dialog box in Figure 19–3 will be provided. Change the layout name to Roof Plan. With the Layout Setting tab selected, choose a "C" size sheet of paper from the drop-down list box. If "C" size is not available in the list box, select the Plot Device tab and choose a printer that supports the plotting of "C" size drawings (select

DWF6 ePlot.pc3 if an appropriate plotter is not available). Verify that the plotting scale is 1:1 and click OK to accept the settings. You will be returned to the screen and provided a view of your new drawing sheet layout (see Figure 19–4). The "C" size sheet is shown with a dashed line representing the printable area. You have also been provided with a floating viewport, which allows you to see the roof plan.

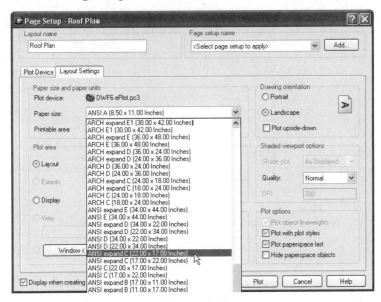

Figure 19–3

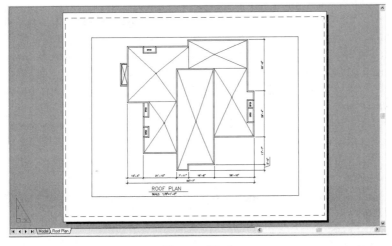

Figure 19–4

Verify from the status bar or the UCS icon that you are now in Paper Space. Make the Title Block layer current and then use the INSERT command to insert a "C" size title block (one is provided for you as a block in this drawing). The results are shown in Figure 19–5. This title block will be used as a guide to lay out both views of the roof plan at different scales.

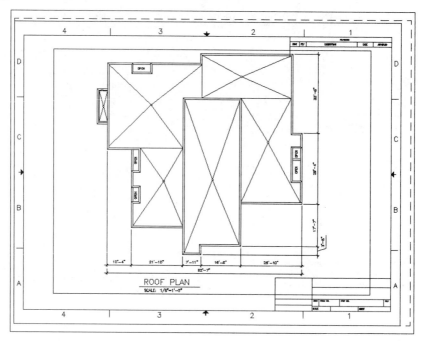

Figure 19–5

The total roof plan will fit inside the original large viewport; the detail area will fit in a new smaller viewport. Use grips to change the shape of the original large viewport. Locate the lower left corner near point "A" and the upper right corner near point "B" as shown in Figure 19–6. Next use the vports command to create the small viewport. Move both viewports to the Vports layer so that you will be able to turn them off later. Do not worry about the exact size of the viewports. The images may not fit inside the viewports when they are scaled later and grips can stretch the viewports to the desired size at any time.

Command: **-VPORTS**
Specify corner of viewport or
[ON/OFF/Fit/Shadeplot/Lock/Object/Polygonal/Restore/2/3/4] <Fit>: *(Pick at "C" in Figure 19–6)*
Specify opposite corner: *(Pick at "D" in Figure 19–6)*
Regenerating model.

Notice that the image of the roof plan is visible in both viewports. This is typical of the Layout environment. Demonstrated later on in this chapter is a way to freeze certain layers in one viewport while keeping layers in other viewports visible. This control is not needed for this segment.

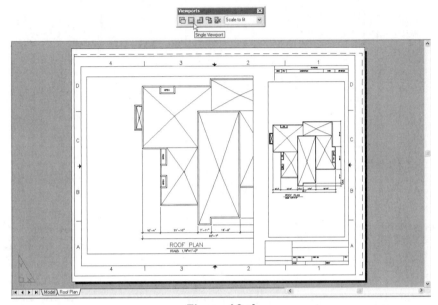

Figure 19–6

Switch to Model Space by clicking on the PAPER button on the status bar (to change it to MODEL) or use the MSPACE command at the keyboard.

Command: **MS** *(For MSPACE)*

Note: While in Paper Space, you can double-click in a viewport to switch to floating Model Space and at the same time make that viewport active. You can also switch back to Paper Space by double-clicking outside of any viewports.

In floating Model Space, click inside of the large viewport to make it active. Because the scale of this image is 1/8" = 1'-0", use the ZOOM command and a factor of 1/96XP to scale the image to Paper Space units (see Figure 19–7). You could also use the Viewports toolbar and click on the 1/8" = 1' scale.

Command: **Z** *(For ZOOM)*
Specify corner of window, enter a scale factor (nX or nXP), or
[All/Center/Dynamic/Extents/Previous/Scale/Window] <real time>: **1/96XP**

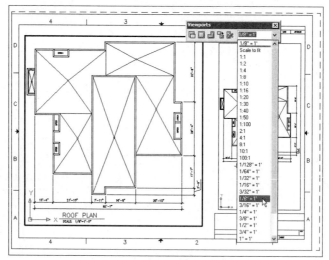

Figure 19–7

The image may get larger or smaller depending on the scale factor and the size of the viewport. Use the PAN command to center the drawing in the viewport. Be prepared to use grips to size the viewport relative to the image of the roof plan if required (this must be accomplished in paper space). Next, while still in floating Model Space, activate the smaller viewport. Pan the area of interest to the center of the viewport before scaling it (see Figure 19–8). Centering your image before scaling is a good habit; this will prevent you from losing the image you wanted to an area outside the viewport as it is scaled.

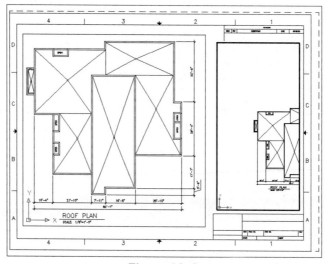

Figure 19–8

This area will be enlarged; to do this, use a larger scale, such as 1/2" = 1'-0". Instead of using the ZOOM command and a scale factor of 1/24XP (24 is the value found when 1' or 12 is divided by 1/2" or 0.50), use the Viewports toolbar to scale the detail. Once the Viewports toolbar is displayed (see Figure 19–9), note that the scale of the active viewport is displayed in the drop-down list box. With the small viewport active, click on the drop-down list box and select 1/2" = 1'-0".

 Remember: To size or modify a viewport, you must be in Paper Space. To scale or pan your image in a viewport, you must be in floating Model Space. A common beginner's mistake is to click on the viewport toolbar scales while in Paper Space.

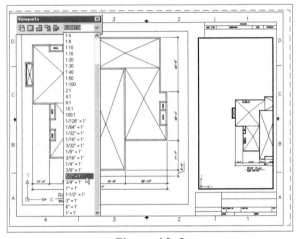

Figure 19–9

The image is panned over further to the right, as in Figure 19–10, to accommodate dimensions. Multiple viewports provide a quick method of arranging multiple details of the same drawing. Save the drawing 19_Roof Plan at this point.

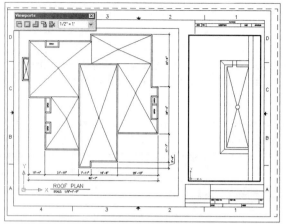

Figure 19–10

SCALING DIMENSIONS TO PAPER SPACE VIEWPORTS

In this segment we will continue using the drawing file 19_Roof Plan (reopen the drawing). Back in Figure 19–10, two images of the same drawing were laid out in Paper Space at two different scale factors. Now, the task is to have the dimension size, such as its text height match throughout the entire drawing. The roof plan in the larger viewport already has dimensions. The dimension scale of 96 is used for the drawing scale of 1/8" = 1'-0". However, the image in the smaller viewport is scaled differently. The dimension scale in this viewpoint must be adjusted to reflect the new drawing scale.

First a new layer is created and made current in the Layer Properties Manager dialog box; this new layer will be called Detail Dim, and it will hold all dimensions that pertain to the detail of the roof plan (see Figure 19–11).

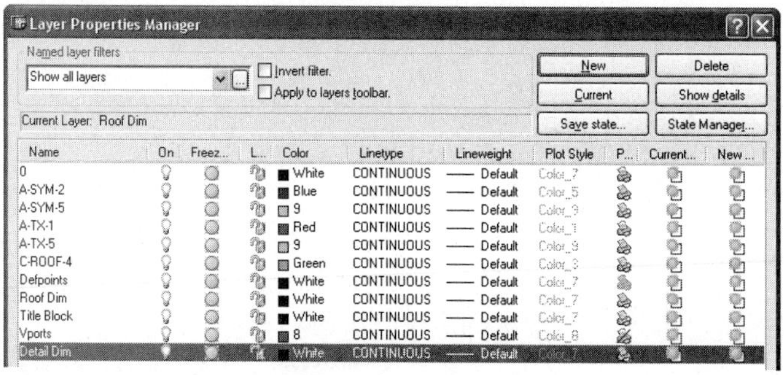

Figure 19–11

Next, a new dimension style called "Detail" is created. Clicking on the New... button in the Dimension Style Manager dialog box in Figure 19–12 displays the Create New

Dimension Style dialog box shown in Figure 19–13. Enter " Detail" in the New Style Name edit box and click the Continue button.

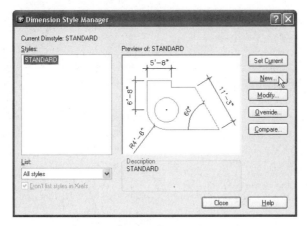

Figure 19–12

Figure 19–13

Click on the Fit tab in the New Dimension Style: dialog box to display the Scale for Dimension Features area shown in Figure 19–14. Notice in this illustration, that the original Overall scale factor is still set to 96 to match the scale of 1/8" = 1'-0". The scale of the roof plan detail is 1/2" = 1'-0". One technique would be to change the Overall scale value from 96 to a new value of 24. A better method (see Figure 19–15) is to select the radio button for Scale dimensions to layout (paperspace). This will automatically set the Overall dimension scale based on the scale inside the current floating Model Space viewport. After selecting the radio button, click on the OK button to return to the main Dimension Style Manager dialog box, shown in Figure 19–16. With the Detail style selected, click the Set Current button and then click on the Close button. The new dimension style has been saved and is current.

 Note: It would not be necessary to create a separate dimension style when utilizing the Scale dimensions to layout (paperspace) button. When dimensions are created in floating model space, they are automatically scaled per that viewport's scale factor.

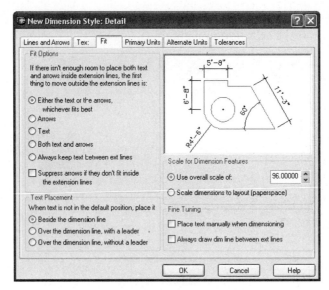

Figure 19–14

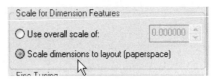

Figure 19–15

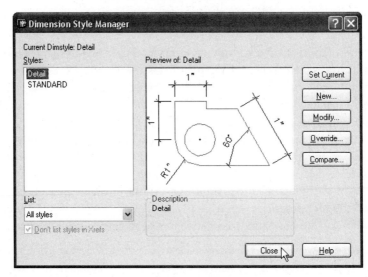

Figure 19–16

With the small viewport active, set the Detail Dim layer as the current layer and add vertical and horizontal dimensions using the dimlinear command in Figure 19–17.

Command: **DLI** *(For DIMLINEAR)*
Specify first extension line origin or <select object>: **Int**
of *(Pick the intersection at "A")*
Specify second extension line origin: **Int**
of (Pick the intersection at "B")
Specify dimension line location or
[Mtext/Text/Angle/Horizontal/Vertical/Rotated]: *(Locate the dimension)*
Dimension text = 14'-6"

Command: **DLI** *(For DIMLINEAR)*
Specify first extension line origin or <select object>: **Int**
of *(Pick the intersection at "A")*
Specify second extension line origin: **Int**
of *(Pick the intersection at "C")*
Specify dimension line location or
[Mtext/Text/Angle/Horizontal/Vertical/Rotated]: *(Locate the dimension)*
Dimension text = 4'-5"

Notice in Figure 19–17 that all dimensions in both viewports have the same size even though the scales of both images are different. The ability to scale dimensions to Paper Space viewports through the Dimension Styles dialog box is a very important and powerful feature of AutoCAD. Save your drawing at this point.

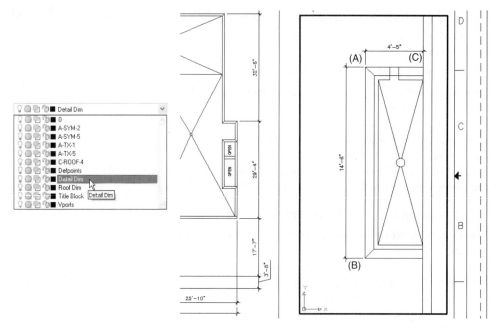

Figure 19–17

CONTROLLING THE VISIBILITY OF LAYERS IN PAPER SPACE VIEWPORTS

In this segment we will continue using the drawing file 19_Roof Plan (reopen the drawing). It appears that the drawing in Figure 19–18 is ready to be plotted. However, notice the upper left corner of the large viewport; as dimensions were added to the detail of the roof plan in the smaller viewport, these same dimensions also appear in the overall roof plan in the large viewport. Again, this is a typical occurrence in the Layout environment.

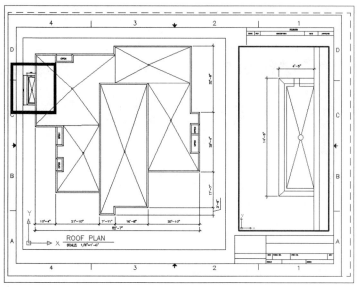

Figure 19–18

Another powerful tool while you work inside Paper Space is the ability to freeze layers in certain viewports. In Figure 19–18, the two dimensions placed in the detail need to be frozen in the viewport of the overall roof plan. Before continuing, be sure the large viewport holding the overall roof plan is the current viewport. Activate the Layer Properties Manager dialog box in Figure 19–19 and notice the various layer states such as Name, On, and Lock. Also notice that there are three states that deal with freezing viewports (Freeze in all VP, Current VP Freeze, and New VP Freeze). You may have to resize the dialog box to see all the states available.

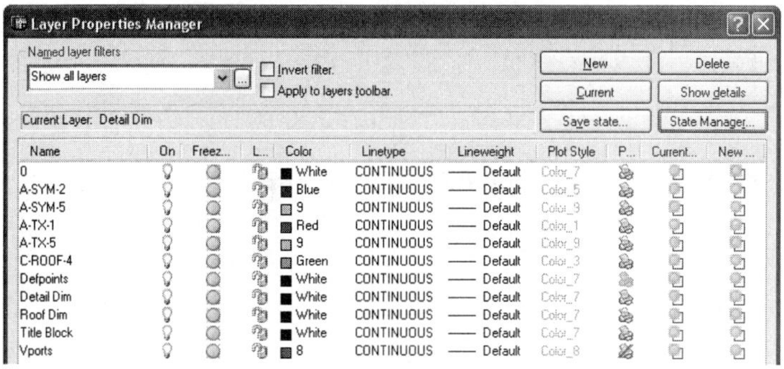

Figure 19–19

Expanding the heading in the third column in Figure 19–20 displays the title and purpose of the function of this state, namely to freeze layers in all viewports. This is the Layer state most commonly used in Model Space for freezing layers and then thawing them. It would not be a good idea to use this for the layouts of the roof plan and detail since freezing the dimension layer would freeze the dimensions in both viewports. We want the dimensions to be visible only in the detail viewport.

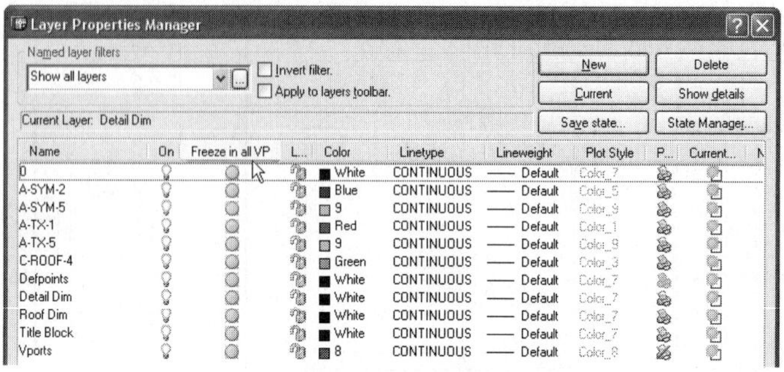

Figure 19–20

Expanding the heading in the tenth column in Figure 19–21 displays the title and purpose of the next function, namely to freeze layers in the current viewport. This is the Layer state that will be used to freeze the Detail Dim layer only in the current viewport, which should be the large viewport. As shown in Figure 19–22, clicking on the icon to freeze the Detail Dim layer in the current viewport displays a snowflake, signifying that the layer is frozen only in the current viewport.

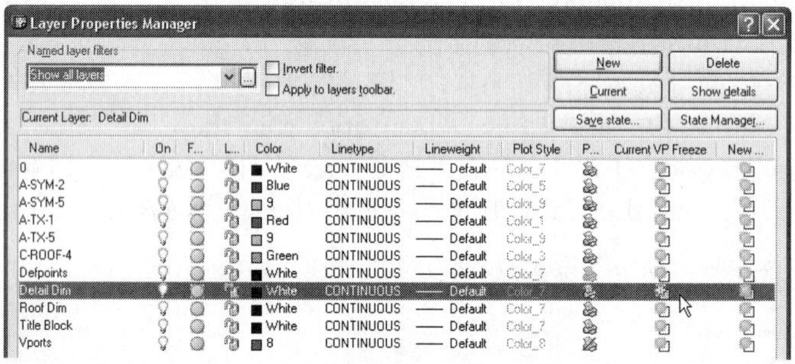

Figure 19–21

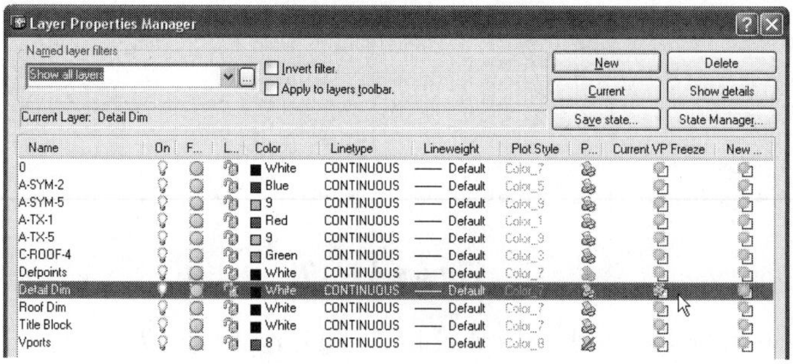

Figure 19–22

Clicking the OK button at the bottom of the Layer Properties Manager dialog box exits the dialog box and returns you to the drawing. Notice, in Figure 19–23, that the layer holding the detail dimensions has disappeared from the large viewport; however, the Detail Dim layer is still visible in the detail viewport.

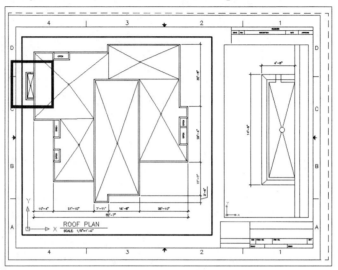

Figure 19–23

Expanding the heading in the eleventh column in Figure 19–24 displays the title and purpose of the next function, namely to freeze layers in any new viewports. This Layer function will not be used for this drawing. However it too is very useful if more viewports need to be created. It has already been demonstrated that when you set up multiple viewports, the same image appears in all viewports. What this function does is freeze selected layers in any newly created viewports. For example, if you don't want any dimensions showing up when a new viewport is created, you could click on the New Viewport Freeze icon associated with the Roof Dim and Detail Dim layers. Then, when any new viewports are created, the image of the roof plan will appear without the dimensions.

Name	On	F...	L...	Color	Linetype	Lineweight	Plot Style	P...	Current...	New VP Freeze
0				White	CONTINUOUS	Default	Color_7			
A-SYM-2				Blue	CONTINUOUS	Default	Color_5			
A-SYM-5				9	CONTINUOUS	Default	Color_9			
A-TX-1				Red	CONTINUOUS	Default	Color_1			
A-TX-5				9	CONTINUOUS	Default	Color_9			
C-ROOF-4				Green	CONTINUOUS	Default	Color_3			
Defpoints				White	CONTINUOUS	Default	Color_7			
Detail Dim				White	CONTINUOUS	Default	Color_7			
Roof Dim				White	CONTINUOUS	Default	Color_7			
Title Block				White	CONTINUOUS	Default	Color_7			
Vports				8	CONTINUOUS	Default	Color_8			

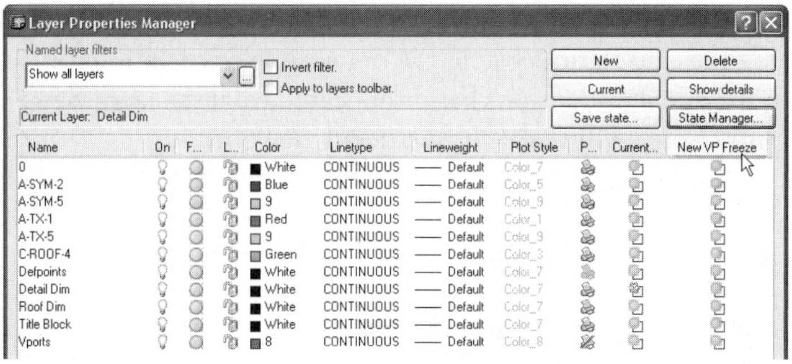

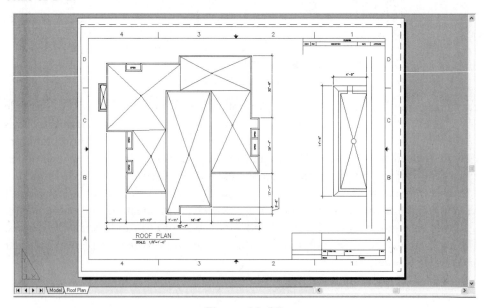

Figure 19–24

With the Detail Dim layer frozen in the overall roof plan viewport, we have completed our multiple view layout drawing. You could now switch back to Paper Space, turn off the layer holding all viewports (see Figure 19–25), and plot the drawing out at a scale of 1=1.

Figure 19–25

Layers have always had the capability of being frozen or thawed. However, this was accomplished globally while in Model Space. Freezing layers in current and new viewports allows freeze and thaw operations to be viewport-specific.

CREATING A DETAIL PAGE IN LAYOUT MODE WITH THE AID OF EXTERNAL REFERENCES

This next discussion focuses on laying out on the same sheet a series of details composed of different drawings, which can be at different scales. As the viewports are laid out in Paper Space and images of the drawings are inserted in floating Model Space, all images will appear in all viewports. This is not a major problem, because Paper Space allows for layers to be frozen in certain viewports and not in others. What if you are not sure of the exact layer names to freeze? This becomes a big problem. If images of drawings are not inserted, but instead attached as External References, this affords greater control over layers, as shown in the next Paper Space example.

Try It! – Open the drawing file 19_Bearing Details. Figure 19–26 shows three objects: a body, a bushing, and a bearing. The body and bearing will be laid out at a scale of 1=1. The bushing will be laid out at a scale of 2=1 (enlarged to twice its normal size). The dimension scales have been set for all objects in order for all dimension text to be displayed at the same size. Also, layers have been created for each object. Each object exists as a single file on the hard drive. The three drawing files you will use to complete this segment are 19_Body, 19_Bearing, and 19_Bushing.

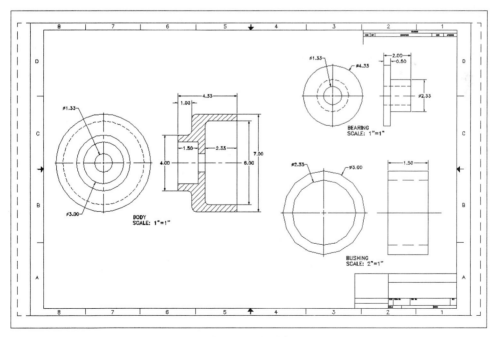

Figure 19–26

The following has already been done for you:

- A Vports layer was created. This layer will hold all viewport information and when the drawing is completed can be frozen or turned off.

- A Xref layer was created. This is where all external reference information will be attached.

- In Layout Mode, an ANSI D size sheet was selected for Page Setup.

- A "D" size title block was inserted onto the Title Block layer.

- The VPORTS command was used to create three viewports on the Vports layer.

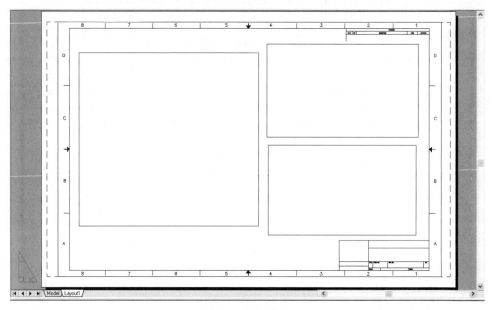

Figure 19–27

In Figure 19–27, notice how all viewports are blank or empty, because the drawing database contains no objects. This will change shortly—the drawing files will now be attached. Before beginning this process, first switch to floating Model Space by double-clicking in the large viewport at the left to make it current, and make the Xref layer the new current layer. The display should be similar to Figure 19–28.

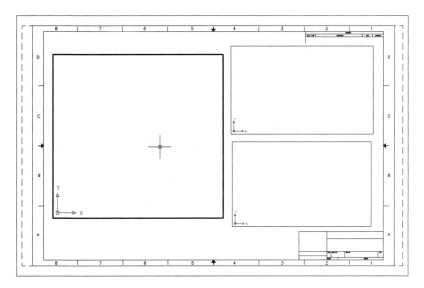

Figure 19–28

Activate the Xref Manager dialog box through the XREF command. Click on the Attach button shown in Figure 19–29 and search for the desired file to attach inside the current viewport; the drawing file 19_Body will be used. After you locate this drawing file, it appears in the External Reference dialog box in Figure 19–30. Click the OK button and click inside the viewport at a convenient location to attach the drawing file.

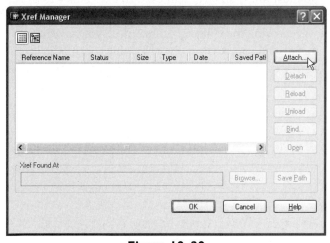

Figure 19–29

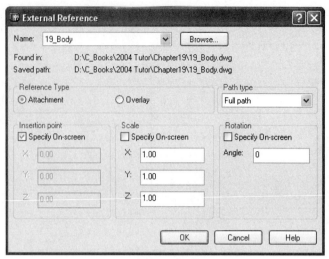

Figure 19–30

Notice how the attached drawing of 19_Body.Dwg appears in all three viewports in Figure 19–31. The Layer Properties Manager dialog box could be used to freeze the layers; however, each viewport must be made active before the layers can be frozen, which would take two steps. Another method, utilizing the vplayer command, will now be introduced. This command stands for Viewport Layer and is another way to freeze layers in one viewport and not in others. However, for this command to function properly, you must know all layer names to freeze. This was the purpose of using External References to attach to the viewports; the External Reference file controls all layers. To freeze layers using the vplayer command, all you have to know is the name of the External Reference file. Since the name of this file controls all layers, you can freeze all layers that begin with the name of the External Reference at the same time.

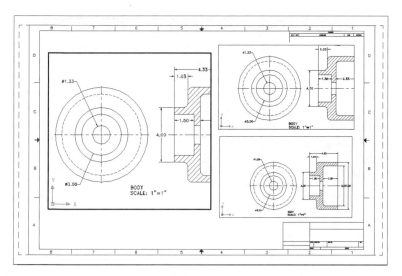

Figure 19–31

Before using the VPLAYER command, first activate Layer Properties Manager dialog box to view the layers created through the XREF command and the Attach option (see Figure 19–32). Notice the grouping of layers that begins with 19_Body and the pipe (|) symbol. The Layer Properties Manager dialog box can be closed at this point. These layers need to be frozen in the other two viewports through the VPLAYER command and the following sequence:

Command: **VPLAYER**
Enter an option [?/Freeze/Thaw/Reset/Newfrz/Vpvisdflt]: **F** *(For Freeze)*
Enter layer name(s) to freeze or <select objects>: **19_Body*** *(For all layers that begin with 19_Body)*
Enter an option [All/Select/Current] <Current>: **S** *(For Select)*
Switching to Paper space.
Select objects: *(Pick the upper right viewport in Figure 19–31)*
Select objects: *(Pick the lower right viewport in Figure 19–31)*
Select objects: *(Press ENTER to continue)*
Switching to Model space.
Enter an option [?/Freeze/Thaw/Reset/Newfrz/Vpvisdflt]: *(Press ENTER to exit this command)*

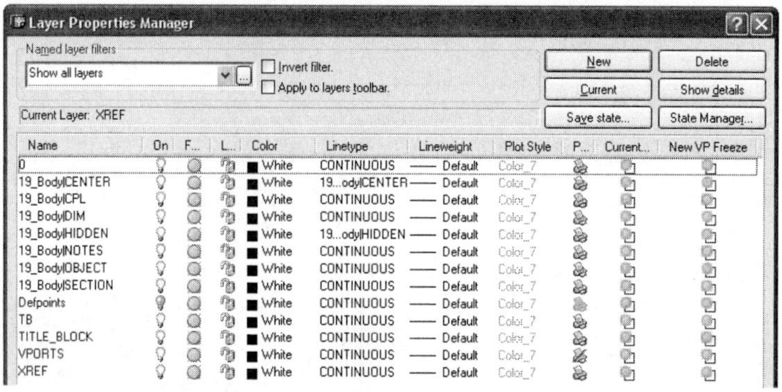

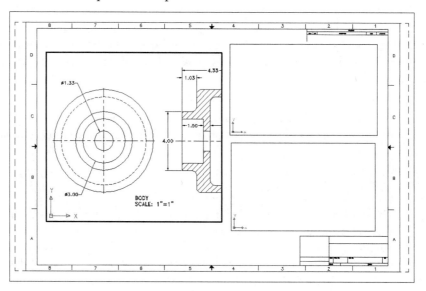

Figure 19–32

The results are illustrated in Figure 19–33, with all layers associated with 19_Body frozen in the upper right and lower right viewports. This is the reason "19_Body*" was entered as the name of the layer to freeze. Notice, in the previous prompt sequence, that the command automatically switched to Paper Space to select the viewports and then back to Model Space to complete the command.

Figure 19–33

With the large viewport at the left still active, as in Figure 19–33, issue the ZOOM command along with the XP option to scale the image at a factor of 1XP, which is another way of saying full size. The Viewports toolbar, shown in Figure 19–34, provides an alternate method for zooming to Paper Space scale factors—simply choose the desired scale from the drop-down list box.

Command: **Z** *(For ZOOM)*
Specify corner of window, enter a scale factor (nX or nXP), or [All/Center/Dynamic/
Extents/Previous/Scale/Window] <real time>: **1XP**

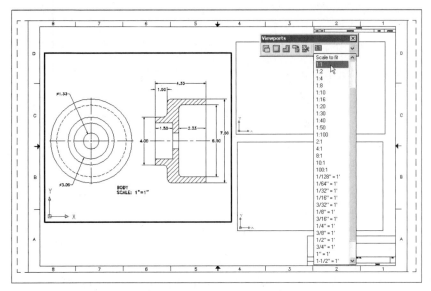

Figure 19–34

If the image appears too large for the viewport, switch to Paper Space and adjust the size of the viewport using grips.

Next, make the upper right viewport current and use the External Reference dialog box to attach the drawing file 19_Bearing to this viewport. Again, the image appears in all viewports, illustrated in Figure 19–35. Use the VPLAYER command to freeze all layers beginning with 19_Bearing in the large viewport on the left and in the lower right viewport.

Command: **VPLAYER**
Enter an option [?/Freeze/Thaw/Reset/Newfrz/Vpvisdflt]: **F** *(For Freeze)*
Enter layer name(s) to freeze or <select objects>: **19_Bearing*** *(For all layers that begin with 19_Bearing)*
Enter an option [All/Select/Current] <Current>: **S** *(For Select)*
Switching to Paper space.
Select objects: *(Pick the left viewport in Figure 19–35)*
Select objects: *(Pick the lower right viewport in Figure 19–35)*
Select objects: *(Press ENTER to continue)*
Switching to Model space.
Enter an option [?/Freeze/Thaw/Reset/Newfrz/Vpvisdflt]: *(Press ENTER to exit this command)*

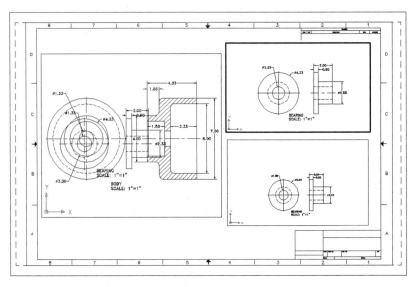

Figure 19–35

With the upper right viewport still active, as in Figure 19–35, use the Viewports toolbar or issue the ZOOM command along with the XP option to scale the image at a factor of 1XP or full size. The results are illustrated in Figure 19–36.

Command: **Z** *(For ZOOM)*
Specify corner of window, enter a scale factor (nX or nXP), or [All/Center/Dynamic/ Extents/Previous/Scale/Window] <real time>: **1XP**

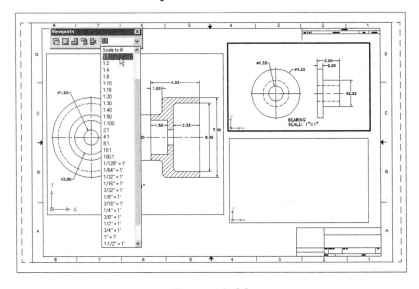

Figure 19–36

The lower right viewport is now made current and the External Reference dialog box is used to attach the drawing file 19_Bushing to this viewport. The image again appears in all viewports, illustrated in Figure 19–37. The VPLAYER command is used to freeze all layers beginning with 19_Bushing in the large viewport on the left and in the upper right viewport.

Command: **VPLAYER**
Enter an option [?/Freeze/Thaw/Reset/Newfrz/Vpvisdflt]: **F** *(For Freeze)*
Enter layer name(s) to freeze or <select objects>: **19_Bushing*** *(For all layers that begin with 19_Bushing)*
Enter an option [All/Select/Current] <Current>: **S** *(For Select)*
Switching to Paper space.
Select objects: *(Pick the left viewport in Figure 19–37)*
Select objects: *(Pick the upper right viewport in Figure 19–37)*
Select objects: *(Press ENTER to continue)*
Switching to Model space.
Enter an option [?/Freeze/Thaw/Reset/Newfrz/Vpvisdflt]: *(Press ENTER to exit this command)*

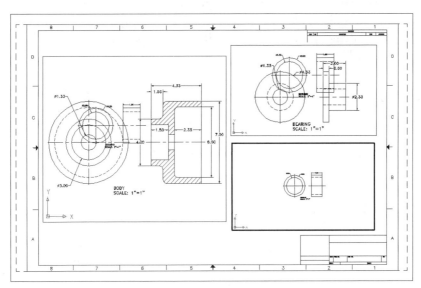

Figure 19–37

With the lower right viewport still active, as in Figure 19–37, use the Viewports toolbar or issue the ZOOM command along with the XP option to scale the image at a factor of 2XP or twice its normal size. The results are illustrated in Figure 19–38.

Command: **Z** *(For ZOOM)*
Specify corner of window, enter a scale factor (nX or nXP), or [All/Center/Dynamic/Extents/Previous/Scale/Window] <real time>: **2XP**

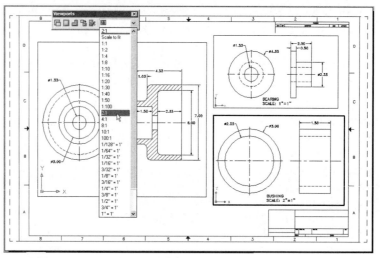

Figure 19–38

If necessary, switch back to Paper Space and use grips to adjust all viewports so that the entire image is displayed inside each viewport. You may also find it necessary to use the PAN command in floating Model Space to reposition the images. The results are illustrated in Figure 19–39.

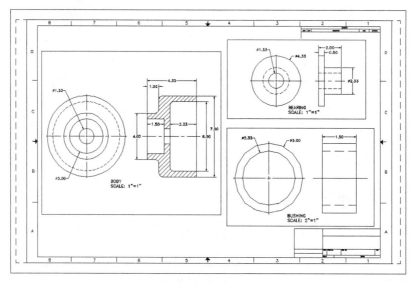

Figure 19–39

The detail sheet is almost ready to be plotted. Switch to floating Model Space and make the left viewport active. Activate the Layer Properties Manager dialog box to observe the layer states, shown in Figure 19–40. Notice, in the left viewport, how layers that begin with 19_Body are thawed and visible while all layers that begin with 19_Bearing and 19_Bushing are frozen in the viewport. This is the result of using the VPLAYER command.

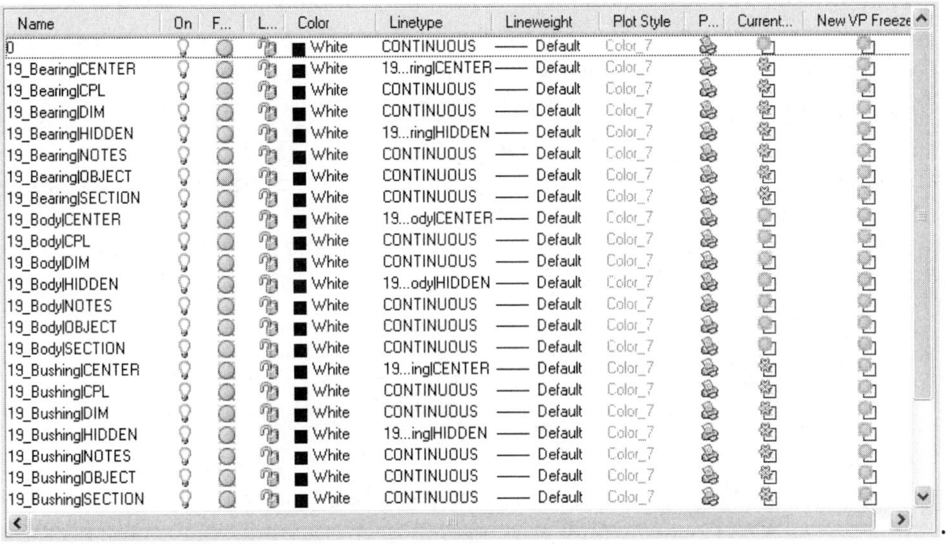

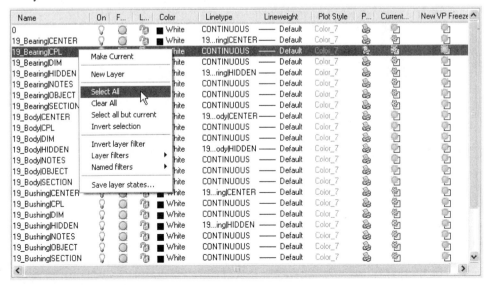

Figure 19–40

While inside the Layer Properties Manager dialog box, right-click one of the layer names to bring up the shortcut menu in Figure 19–41. Click on Select All to select all layers.

Figure 19–41

With all layers highlighted as in Figure 19–42, click on the icon to Freeze/Thaw in new viewports. Now if any new viewports are created for extra details, the images of the 19_Body, 19_Bearing, and 19_Bushing will not appear in the new viewports. This is yet another productivity tool available for Paper Space layout through the Layer Properties Manager dialog box. This same effect can be accomplished through the VPLAYER command and the following sequence:

Command: **VPLAYER**
Enter an option [?/Freeze/Thaw/Reset/Newfrz/Vpvisdflt]: **V** *(For Vpvisdflt)*
Enter layer name(s) to change viewport visibility or <select objects>: **19_Body*,19_ Bearing*,19_Bushing***
Enter a viewport visibility option [Frozen/Thawed] <Thawed>: **F** *(For Frozen)*
Enter an option [?/Freeze/Thaw/Reset/Newfrz/Vpvisdflt]: *(Press ENTER to exit this command)*

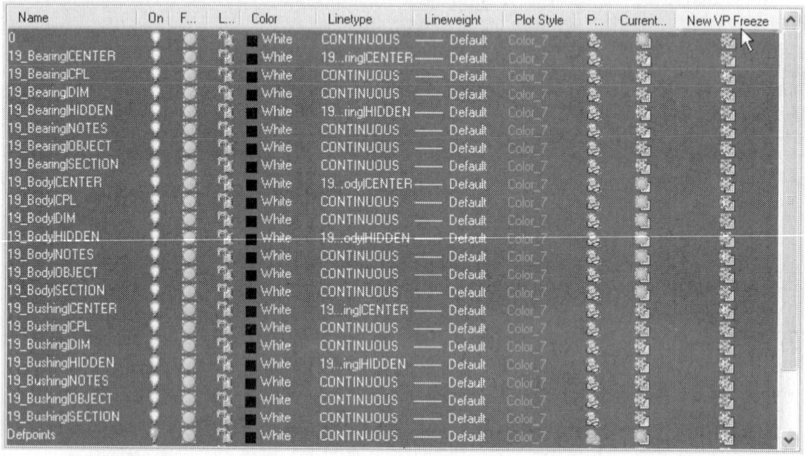

Figure 19–42

The Vpvisdflt option used above stands for Viewport Visibility Default and allows for layers to be frozen or thawed in any new viewport.

Finally, turn off the Vports layer to view the details. This detail sheet is complete.

ADDITIONAL VIEWPORT CREATION METHODS

When constructing viewports, you are not limited to rectangular or square shapes. Although you have been using the Scale drop-down list box in the Viewports toolbar in Figure 19–43 to scale the image inside of the viewport, other buttons are available and act on the shape of the viewport. You can clip an existing viewport to reflect a different shape, convert an existing closed object into a viewport, construct a multisided closed or polygonal viewport, or display the Viewports dialog box (you have already constructed a single viewport in a previous exercise). See Figure 19–43 for the location of these tools.

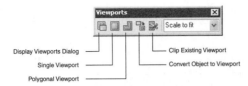

Figure 19–43

 Try It! – Open the drawing file 19_Floor Viewports. Notice that Vports is the current layer. You will convert the large rectangular viewport in Figure 19–44 into a polygonal viewport by a clipping operation. First pick on the Clip Existing Viewport button. Pick the rectangular viewport and begin picking points to construct a polygonal viewport around the perimeter of the floor plan dimensions in Figure 19–45. You can turn ORTHO on to assist with this operation. It is not critical that all lines are orthogonal (horizontal or vertical). The following command sequence will also aid with this operation.

Command: **VPCLIP**

Select viewport to clip: *(Select the rectangular viewport in Figure 19–45)*
Select clipping object or [Polygonal] <Polygonal>: *(Press ENTER)*
Specify start point: *(Pick at "A")*
Specify next point or [Arc/Length/Undo]: *(Pick at "B")*
Specify next point or [Arc/Close/Length/Undo]: *(Pick at "C")*
Specify next point or [Arc/Close/Length/Undo]: *(Pick at "D")*
Specify next point or [Arc/Close/Length/Undo]: *(Pick at "E")*
Specify next point or [Arc/Close/Length/Undo]: *(Pick at "F")*
Specify next point or [Arc/Close/Length/Undo]: **C** *(To close the shape)*

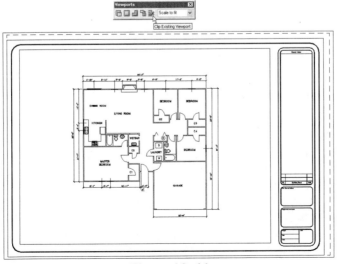

Figure 19–44

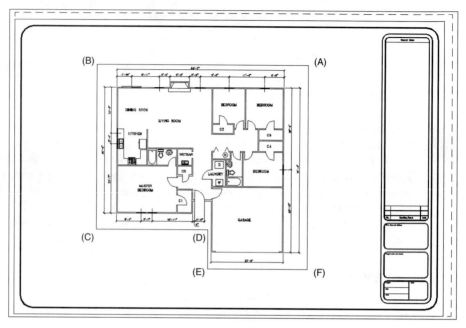

Figure 19–45

When finished, move the viewport with the image of the floor plan to the right of the screen. Then construct a circle in the upper left corner of the title block (see Figure 19–46). Pick the Convert Object to Viewport button and select the circle you just constructed. Notice the circle converts to a viewport with the entire floor plan displayed inside of its border. Double-click inside of this new viewport to make it current and change the scale of the image to the 1/2"=1'-0" scale using either the Viewports toolbar or the ZOOM command.

Command: **-VPORTS**
Specify corner of viewport or
[ON/OFF/Fit/Shadeplot/Lock/Object/Polygonal/Restore/2/3/4] <Fit>: **O** *(For Object)*
Specify object to clip viewport: *(Pick the circle in Figure 19–46)*

Command: **Z** *(For ZOOM)*
Specify corner of window, enter a scale factor (nX or nXP), or
[All/Center/Dynamic/Extents/Previous/Scale/Window] <real time>: **1/24XP**

Pan inside of the circular viewport until the laundry and bathroom appear. Adjust the size of the circular viewport with grips if the image is too large or small. When finished, double-click outside of the edge of the viewport to switch to Paper Space. Your drawing should appear similar to Figure 19–46.

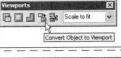

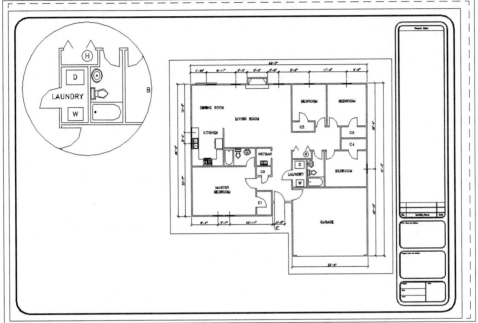

Figure 19–46

Click on the Polygonal Viewport button and construct a multisided viewport similar to the one located in Figure 19–47. Close the shape and do not be concerned that a few lines may be inclined.

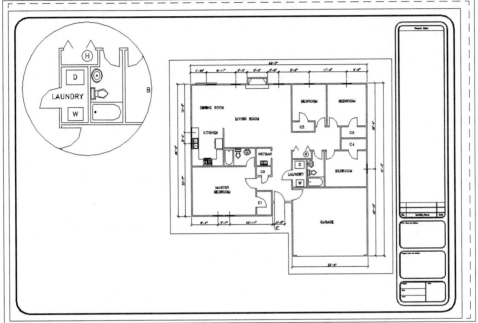 Command: **-VPORTS**

Specify corner of viewport or
[ON/OFF/Fit/Shadeplot/Lock/Object/Polygonal/Restore/2/3/4] <Fit>: **P** *(For Polygonal)*
Specify start point: *(Pick at "A" in Figure 19–47)*
Specify next point or [Arc/Length/Undo]: *(Pick at "B")*
Specify next point or [Arc/Close/Length/Undo]: *(Pick at "C")*
Specify next point or [Arc/Close/Length/Undo]: *(Pick at "D")*
Specify next point or [Arc/Close/Length/Undo]: *(Pick at "E")*
Specify next point or [Arc/Close/Length/Undo]: *(Pick at "F")*
Specify next point or [Arc/Close/Length/Undo]: **C** *(To close the shape)*
Regenerating model.

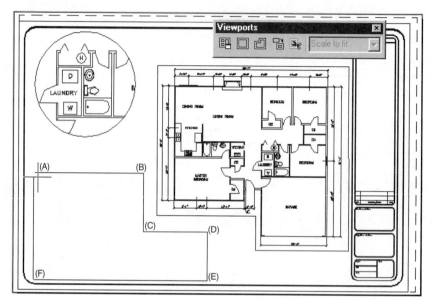

Figure 19–47

As the image of the floor plan appears in this new viewport, double click inside of the new viewport to make it current. Scale the image inside of the viewport to the scale 3/8"=1'-0". Pan until the kitchen area and master bathroom are visible. Your display should appear similar to Figure 19–48.

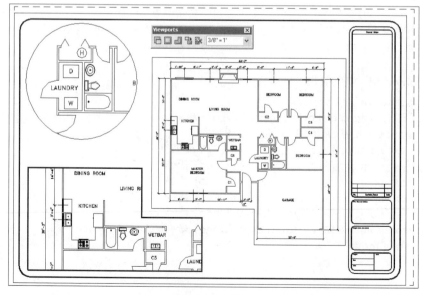

Figure 19–48

Double-click outside of this viewport to return to Paper Space. Make the layer Notes current. Add text identifying the name and scale of each viewport. Make any final adjustments to viewports using grips. When finished, turn off the Vports layer. Your display should appear similar to Figure 19–49.

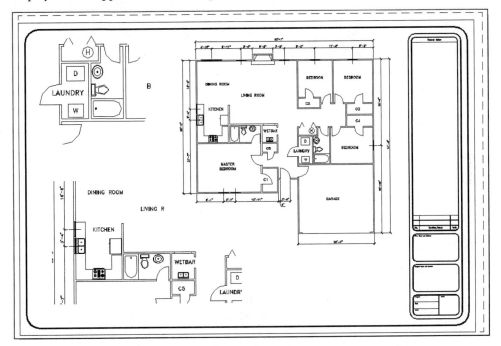

Figure 19–49

MATCHING THE PROPERTIES OF VIEWPORTS

In Chapter 7, the MATCHPROP (Match Properties) command was introduced as a means of transferring all or selected properties from a source object to a series of destination objects. In addition to transferring layer, color, and linetype information, dimension styles, hatch properties, and text styles can also be transferred from one object to another. You can also transfer viewport information from a source viewport to other viewports. Information such as viewport layer, whether the viewport is locked or unlocked, and the viewport scale are a few of the properties to transfer to other viewports. When you enter the MATCHPROP command and select the Settings option, the dialog box in Figure 19–50 will appear. When transferring viewport properties, be sure the Viewport option of this dialog box is checked.

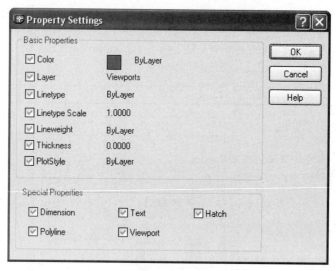

Figure 19–50

 Try It! – Open the drawing file 19_Matchprop Viewports as shown in Figure 19–51. This drawing consists of four viewports holding different object types. The object in the first viewport (labeled "A" in Figure 19–51) is already scaled to 1=1. This viewport belongs to the Viewport layers; this viewport is also locked to prevent accidental zooming while in floating model space. All other viewports do not belong to the Viewports layer. They are all scaled differently. Also, none of these viewports are locked. These three viewports need to have the same properties as the first; namely all viewports need to belong to the Viewports layer; all viewports need to be locked; and all images inside all viewports need to be scaled to 1=1. Rather than perform these operations on each individual viewport, the MATCHPROP command will be used to accomplish this task.

 Command: **MA** *(For MATCHPROP)*
Select source object: *(Pick the edge of Viewport "A")*
Current active settings: Color Layer Ltype Ltscale Lineweight Thickness
PlotStyle Text Dim Hatch Polyline Viewport
Select destination object(s) or [Settings]: *(Pick the edge of Viewport "B")*
Select destination object(s) or [Settings]: *(Pick the edge of Viewport "C")*
Select destination object(s) or [Settings]: *(Pick the edge of Viewport "D")*
Select destination object(s) or [Settings]: *(Press ENTER to exit this command)*

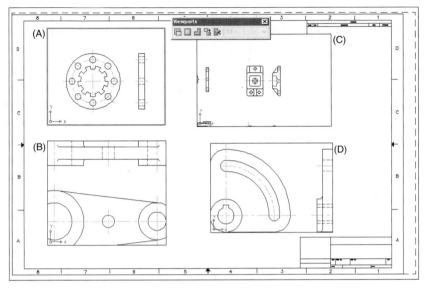

Figure 19–51

The results are illustrated in Figure 19–52. All viewports share the same layer, are scaled to 1=1, and are locked after using the MATCHPROP command.

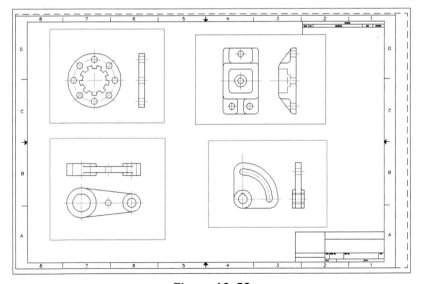

Figure 19–52

LAYER 0 IN MODEL SPACE

When you construct objects in Model Space, Layer 0 is considered neutral. It is recommended that you not draw any geometry or objects such as lines, circles, or arcs on Layer 0. Still, you might accidentally construct on Layer 0 and get away with it in Model Space, but not in Paper Space, especially when you create a detail sheet consisting of numerous drawing that have been attached through the XREF command.

Figure 19–53 consists of a typical layout in Paper Space; two views of different drawings have been laid out. Activating the Xref Manager dialog box shows that both drawings were attached to the viewports in Figure 19–54; namely Flange1 and Guide_A5. Also, both drawings were drawn on Layer 0. Notice in Figure 19–53 that both Xrefs appear in each viewport. To correct this, certain layers need to be frozen in their designated viewports.

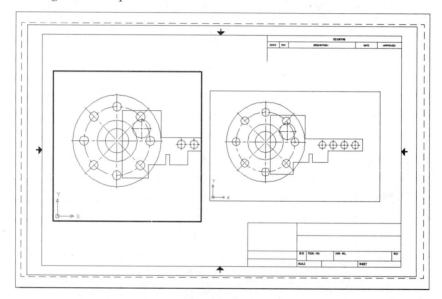

Figure 19–53

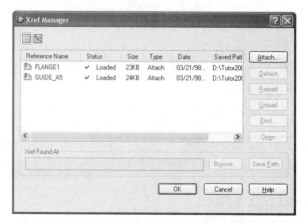

Figure 19–54

When the Layer Properties Manager dialog box opens, observe the layers that appear in the list in Figure 19–55. Instead of the familiar layers of Flange1|0 and Guide_A5|0, all that displays is Layer 0. There is no way to freeze the Layer 0 in the left viewport without affecting the layers in the right viewport. For this reason it is highly recommended that you keep all geometry off Layer 0. At this point, each drawing must be opened, layers assigned, and geometry changed to the new layers. Finally, each External Reference needs to be reattached.

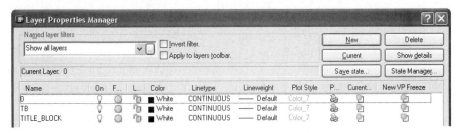

Figure 19–55

NOTES TO THE STUDENT AND INSTRUCTOR

As with the previous "Try It" exercises, one tutorial exercise has been designed around the topic of creating multiple viewports in Paper Space. As with all tutorials, you will tend to follow the steps very closely, taking care not to make a mistake. This is to be expected. However, most individuals rush through the tutorials to get the correct solution only to forget the steps used to complete the tutorial. This is also to be expected.

Of all tutorial exercises in this book, this Paper Space tutorial will probably pose the greatest challenge. Certain operations have to be performed in Paper Space while other operations require floating Model Space. For example, it was already illustrated that viewports constructed with the VPORTS command must be accomplished in Paper Space; scaling an image to Paper Space units using the XP option of the ZOOM command must be accomplished in floating Model Space, and so on.

It is recommended to both student and instructor that all tutorial exercises, and especially those tutorials that deal with Paper Space, be performed two or even three times. Completing the tutorial the first time will give you the confidence that it can be done; however, you may not understand all of the steps involved. Completing the tutorial a second time will allow you to focus on where certain operations are performed and why things behave the way they do. This still may not be enough. Completing the tutorial exercise a third time will allow you to anticipate each step and have a better idea of which operation to perform in each step.

This recommendation should be exercised with all tutorials; however, it should especially be practiced when you work on the following Paper Space tutorials.

TUTORIAL EXERCISE: 19_ARCH DETAILS.DWG

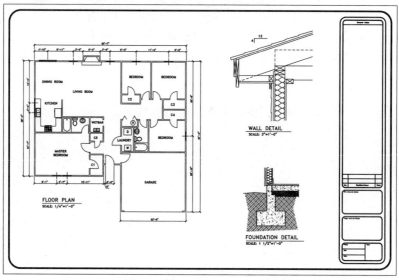

Figure 19–56

Purpose:

This tutorial exercise is designed to lay out three architectural details displayed in Figure 19–56 in the layout mode (Paper Space environment). The three details consist of a floor plan, cornice detail, and foundation detail. External references will also be used to assist in the control of layers.

System Settings:

Begin a new drawing called "19_Arch Details." Use the Drawing Units dialog box (activated through the UNITS command) to change the system of units to Architectural. Keep all default values for the units of the drawing.

Layers:

Create the following layers with the format:

Name	Color	Linetype
Xref	White	Continuous
Title_block	White	Continuous
Viewport	Gray (8)	Continuous

Suggested Commands:

A new drawing will be started using layout mode (Paper Space environment) and three viewports will be created through the VPORTS command. Three architectural details provided on the CD will then be attached to each viewport through the External Reference dialog box. The VPLAYER command will be used to freeze layers in certain viewports while keeping them visible in others. Each detail will be scaled to the Paper Space viewport through the XP option of the ZOOM command. This will enable the drawing to be plotted out at a scale factor of 1=1 because the scaling was already performed in each viewport.

Whenever possible, substitute the appropriate command alias in place of the full AutoCAD command in each tutorial step; for example, use "CP" for the COPY command, "L" for the LINE command, and so on. The complete listing of all command aliases is located in Table 1–2.

Phase I—Individual Drawing Preparation

While three drawings have been provided to step you through this tutorial, it is important to be aware of certain settings inside each drawing based on the scale of the drawing. These usually affect the linetype scale, dimension scale, and individual text height for notes. All three drawings will be explained in the steps that follow.

STEP I

Open the drawing file 19_Floor shown in Figure 19–57. This drawing will be plotted out at a scale of 1/4"=1'-0".

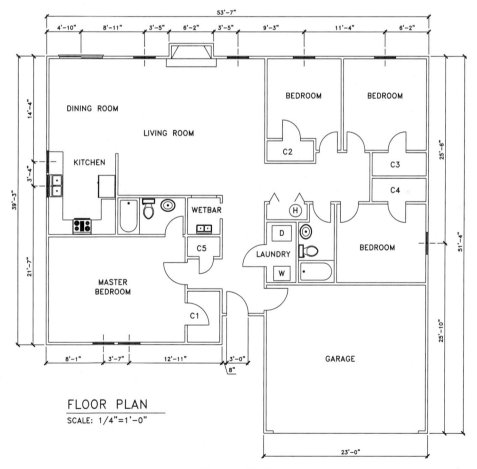

Figure 19–57

Issue the DDIM command and click on the Modify button to display the settings for the Architectural dimension style. Select the Fit tab and notice the overall scale value is set to 48, which is the multiplier found when 1' or 12 is divided by 1/4 or 0.25 units. All dimension parameters such as text and arrow height will be multiplied by a value of 48. Also take note of the layers assigned to this drawing (see Figure 19–58). Close this drawing when you are finished viewing the layers.

Layer name	State	Color	Linetype
0	On	7(white)	CONTINUOUS
CENTER	On	2(yellow)	CENTER2
COUNTERTOP	On	7(white)	CONTINUOUS
DEFPOINTS	On	7(white)	CONTINUOUS
DIM	On	2(yellow)	CONTINUOUS
DOORS	On	1(red)	CONTINUOUS
FIREPLACE	On	1(red)	CONTINUOUS
FRAMING	On	3(green)	CONTINUOUS
NOTES	On	4(cyan)	CONTINUOUS
SYMBOLS	On	7(white)	CONTINUOUS
TEXT	On	4(cyan)	CONTINUOUS
WALLS	On	7(white)	CONTINUOUS
WINDOWS	On	6 (magenta)	CONTINUOUS

Name	On	F...	L...	Color	Linetype	Lineweight	Plot Style	P...
0				White	CONTINUOUS	Default	Color_7	
CENTER				Yellow	CENTER2	Default	Color_2	
COUNTERTOP				White	CONTINUOUS	Default	Color_7	
Defpoints				White	CONTINUOUS	Default	Color_7	
DIM				Yellow	CONTINUOUS	Default	Color_2	
DOORS				Red	CONTINUOUS	Default	Color_1	
FIREPLACE				Red	CONTINUOUS	Default	Color_1	
FRAMING				Green	CONTINUOUS	Default	Color_3	
NOTES				Cyan	CONTINUOUS	Default	Color_4	
SYMBOLS				White	CONTINUOUS	Default	Color_7	
TEXT				Cyan	CONTINUOUS	Default	Color_4	
WALLS				White	CONTINUOUS	Default	Color_7	
WINDOWS				Magenta	CONTINUOUS	Default	Color_6	

Figure 19–58

STEP 2

Open the drawing file 19_Cornice (see Figure 19–59). This drawing will be plotted out at a scale of 3"=1'-0" or 1:4.

Issue the DDIM command and click on the Modify button to display the settings for the Standard dimension style. Select the Fit tab and notice the overall scale value is set to 4, which is the multiplier found when 1' or 12 is divided by 3. All dimension parameters such as text and arrow height will be multiplied by a value of 4. Also take note of the layers assigned to this drawing (see Figure 19–60). Close this drawing when you are finished viewing the layers.

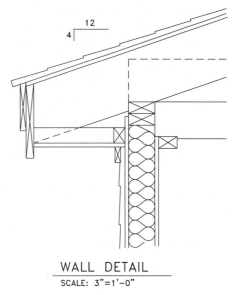

WALL DETAIL
SCALE: 3"=1'-0"

Figure 19–59

Layer name	State	Color	Linetype
0	On	7(white)	CONTINUOUS
FRAMING	On	3(green)	CONTINUOUS
HIDDEN	On	1(red)	HIDDEN
INSULATION	On	7(white)	CONTINUOUS
NOTES	On	4(cyan)	CONTINUOUS
SECTION	On	2 (yellow)	CONTINUOUS

Name	On	F..	L..	Color	Linetype	Lineweight	Plot Style	P...
0	♀	○	⬛	■ White	CONTINUOUS	—— Default	Color_7	🖨
FRAMING	♀	○	⬛	☐ Green	CONTINUOUS	—— Default	Color_3	🖨
HIDDEN	♀	○	⬛	■ Red	HIDDEN	—— Default	Color_1	🖨
INSULATION	♀	○	⬛	■ White	CONTINUOUS	—— Default	Color_7	🖨
NOTES	♀	○	⬛	☐ Cyan	CONTINUOUS	—— Default	Color_4	🖨
SECTION	♀	○	⬛	☐ Yellow	CONTINUOUS	—— Default	Color_2	🖨

Figure 19–60

STEP 3

Open the drawing file 19_Fndation. (see Figure 19–61). This drawing will be plotted out at a scale of 1 1/2"=1'-0" or 1:8.

Issue the DDIM command and click on the Modify button to display the settings for the Standard dimension style. Select the Fit tab and notice the overall scale value is set to 8, which is the multiplier found when 1' or 12 is divided by 1 1/2 or 1.50 units. All dimension parameters such as text and arrow height will be multiplied by 8. Also take note of the layers assigned to this drawing (see Figure 19–62). Close this drawing when you are finished viewing the layers.

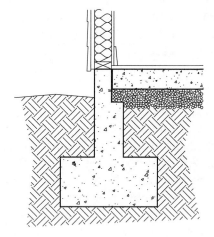

FOUNDATION DETAIL
SCALE: 1 1/2"=1'-0"

Figure 19–61

Layer name	State	Color	Linetype
0	On	7(white)	CONTINUOUS
CONCRETE	On	2(yellow)	CONTINUOUS
CONSTRUCTION	On	1(red)	CONTINUOUS
EARTH	On	3(green)	CONTINUOUS
FRAMING	On	3(green)	CONTINUOUS
GRAVEL	On	4(cyan)	CONTINUOUS
INSULATION	On	7(white)	CONTINUOUS
NOTES	On	4 (cyan)	CONTINUOUS

Name	On	F...	L...	Color	Linetype	Lineweight	Plot Style	P...
0				White	CONTINUOUS	—— Default	Color_7	
CONCRETE				Yellow	CONTINUOUS	—— Default	Color_2	
CONSTRUCTION				Red	CONTINUOUS	—— Default	Color_1	
EARTH				Green	CONTINUOUS	—— Default	Color_3	
FRAMING				Green	CONTINUOUS	—— Default	Color_3	
GRAVEL				Cyan	CONTINUOUS	—— Default	Color_4	
INSULATION				White	CONTINUOUS	—— Default	Color_7	
NOTES				Cyan	CONTINUOUS	—— Default	Color_4	

Figure 19–62

Phase II—Drawing Layout Setup

STEP 4

Begin a new drawing file starting from scratch. Save this drawing as 19_Arch Details. All three details that will make up the drawing 19_Arch Details are designed to fit on a "D" size sheet of paper. Use the Layout Wizard to setup the drawing from scratch. To begin creating a layout, enter LAYOUTWIZARD at the command prompt, choose it from the Tools pull-down menu under Wizards > Create Layout in Figure 19–63 to begin creating a layout (this command can also be selected from the Insert pull-down Layout > Layout Wizard). Type "Detail Sheet" as in Figure 19–63 for the name of the layout, and select the Next button.

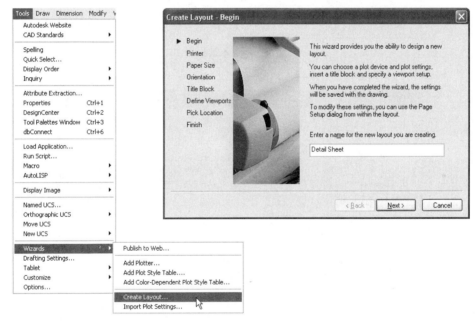

Figure 19–63

In the Create Layout - Printer dialog box, select the DWF6 ePlot.pc3 device, which will support a "D" size paper, and click the Next button. In Figure 19–64, the Create Layout - Paper Size dialog box, select the ARCH expand D sheet size. Click the Next button. Choose Landscape in the Create Layout - Orientation dialog box and click the Next button. In Figure 19–65, the Create Layout - Title Block dialog box, select Architectural Title Block.dwg as the title block to insert. To insert the drawing as a block rather than attach it as an external reference, verify that the Block radio button is selected in the Type area. Click the Next button.

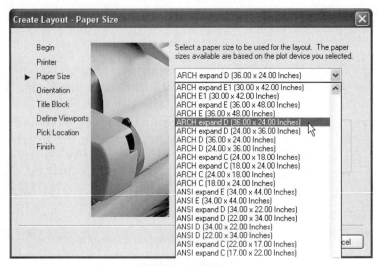

Figure 19–64

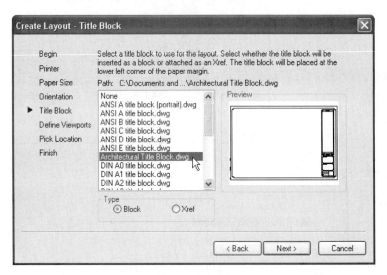

Figure 19–65

In Figure 19–66, the Create Layout - Define Layouts dialog box, you have several options for creating viewports. Select Single as the viewport setup, with a scale of 1/4" = 1'0". This viewport will be used for attaching the drawing file 19_Floor. The VPORTS command will be used later to create the other two viewports we will need. Click the Next button.

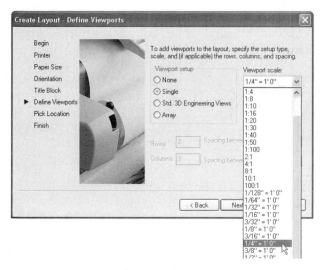

Figure 19–66

In the Create Layout - Pick Location dialog box, click the Select Location button and select the viewport corners using the command prompt sequence shown below and Figure 19–67 as a guide.

Click the Finish button in the last dialog box to complete the layout wizard.

Specify first corner: **1,3.75** *(At "A")*
Specify opposite corner: **18,20.50** *(At "B")*

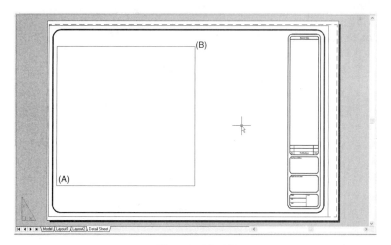

Figure 19–67

STEP 5

Notice three layers that were created earlier in the Layer Properties Manager dialog box in Figure 19–68: a layer called Title_block for all title block information, a layer called Viewport for the viewports used to aid in the layout of the drawing, and the Xref layer for external references. If any of these layers are missing, create them now in the Layer Properties Manager dialog box. Use the Layer Control box to change the title block and viewport created by the wizard to their respective layers.

Name	On	F...	L...	Color	Linetype	Lineweight	Plot Style	P...	Current...	New VP Freeze
0				■ White	Continuous	— Default	Color_7			
Title Block				■ White	Continuous	— Default	Color_7			
Viewport				■ 8	Continuous	— Default	Color_8			
Xref				■ White	Continuous	— Default	Color_7			

Figure 19–68

STEP 6

In this step, first make Viewport the current layer. All viewports will be drawn on this layer. Then, use the -VPORTS command to make two additional viewports on the Viewport layer. These are illustrated in Figure 19–69. One of these viewports will hold the drawing file 19_Cornice, the other 19_Fndation.

 Command: **-VPORTS**

Specify corner of viewport or [ON/OFF/Fit/Shadeplot/Lock/Object/ Polygonal/Restore/2/3/4] <Fit>:
19.00,11.50 *(At "A")*
Specify opposite corner: **28.50,21.50** *(At "B")*

Regenerating model.

 Command: **-VPORTS**

Specify corner of viewport or [ON/OFF/Fit/Hideplot/Lock/Object/ Polygonal/Restore/2/3/4] <Fit>:
19.00,1.25 *(At "C")*
Specify opposite corner: **28.50,10.50** *(At "D")*

Regenerating model.

Although the two new viewports were created with coordinates, in practice, the viewports are drawn to any size and scaled once the images are placed inside each.

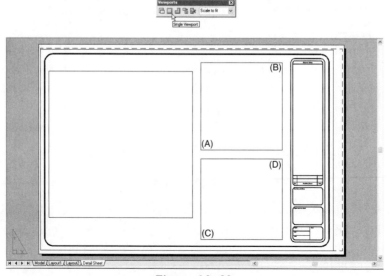

Figure 19–69

Phase III—Drawing Attachment

STEP 7

The Xref layer should have already been created; make this layer the new current layer, as shown in Figure 19–70. Double-click inside the large viewport to make it current and switch to floating Model Space. Your display should appear similar to Figure 19–71.

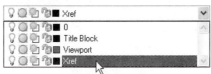

Figure 19–70

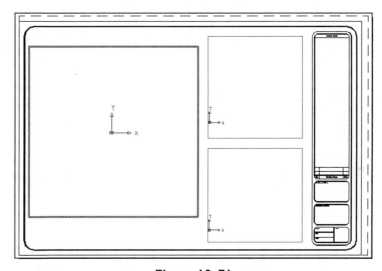

Figure 19–71

STEP 8

Use the XREF command and click on the Attach button to attach the drawing file 19_Floor. In the External Reference dialog box verify that the Insertion point Specify On-screen check box is selected; keep all remaining default values in Figure 19–72. Click OK and pick the insertion point. Use the PAN command to center the floor plan in the viewport as shown in Figure 19–73. Examine all layers of the drawing through the Layer Properties Manager dialog box shown in Figure 19–74. Notice how layers associated with the floor plan begin with the name "19_FLOOR." As a drawing is attached through the XREF command, all layers in the drawing are dependent on the original file name. This is AutoCAD's way of keeping track of all layers that belong to an external reference file. Notice how the file name and layer name are separated with the vertical bar (|) called the pipe symbol.

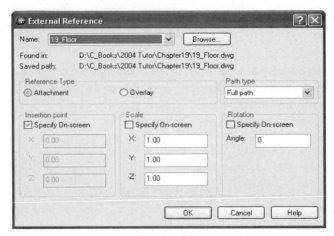

Figure 19–72

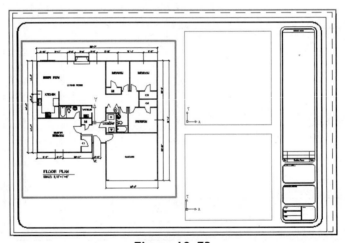

Figure 19–73

Name	On	F...	L...	Color	Linetype	Lineweight	Plot Style	P...	Current...	New VP Freeze
0				White	Continuous	—— Default	Color_7			
19_Floor|CENTER				Yellow	19...loor|CE...ER2—— Default	Color_2			
19_Floor|COUNTERTOP				White	Continuous	—— Default	Color_7			
19_Floor|DIM				Yellow	Continuous	—— Default	Color_2			
19_Floor|DOORS				Red	Continuous	—— Default	Color_1			
19_Floor|FIREPLACE				Red	Continuous	—— Default	Color_1			
19_Floor|FRAMING				Green	Continuous	—— Default	Color_3			
19_Floor|NOTES				Cyan	Continuous	—— Default	Color_4			
19_Floor|SYMBOLS				White	Continuous	—— Default	Color_7			
19_Floor|TEXT				Cyan	Continuous	—— Default	Color_4			
19_Floor|WALLS				White	Continuous	—— Default	Color_7			
19_Floor|WINDOWS				Magenta	Continuous	—— Default	Color_6			
Defpoints				White	Continuous	—— Default	Color_7			
Title Block				White	Continuous	—— Default	Color_7			
Viewport				8	Continuous	—— Default	Color_8			
Xref				White	Continuous	—— Default	Color_7			

Figure 19–74

STEP 9

From the previous step, the floor plan is in all three viewports, even though it doesn't appear in the two small viewports. Activate each of these other viewports and perform a ZOOM-EXTENTS in each to verify that they are there. Your display should appear similar to Figure 19–75.

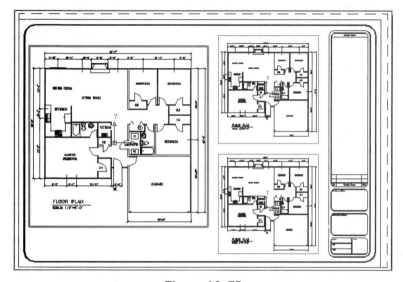

Figure 19–75

STEP 10

One of the advantages of using external references to arrange details in numerous viewports is that it is easy to freeze layers in certain viewports while keeping layers visible in others. Use the VPLAYER (Viewport Layer) command to accomplish this. Freeze all layers that begin with "19_Floor." As you proceed, this command automatically switches to Paper Space to select the viewports for freezing layers; the command switches back to Model Space when completed. The results of this operation are illustrated in Figure 19–76.

Command: **VPLAYER**
Enter an option [?/Freeze/Thaw/Reset/ Newfrz/Vpvisdflt]: **F** *(For Freeze)*

Enter layer name(s) to freeze or <select objects>: **19_Floor*** *(For all layers that begin with 19_Floor)*
Enter an option [All/Select/Current] <Current>: **S** *(For Select)*
Switching to Paper space.
Select objects: *(Pick the upper right viewport in Figure 19–75)*
Select objects: *(Pick the lower right viewport in Figure 19–75)*
Select objects: *(Press ENTER to continue)*
Switching to Model space.
Enter an option [?/Freeze/Thaw/Reset/ Newfrz/Vpvisdflt]: *(Press ENTER to exit this command and freeze all layers beginning with 19_Floor)*

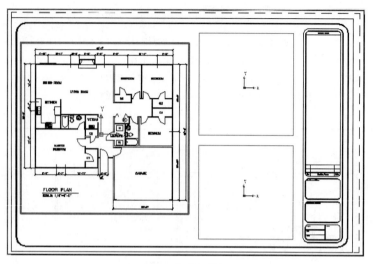

Figure 19–76

STEP 11

Be sure the large viewport, which contains the floor plan, is active. The floor plan is to be plotted out at a scale of 1/4"=1'-0". Normally, you would use the ZOOM command and enter a value of 1/48XP to properly scale the image of the floor plan to reflect Paper Space units. This is not required, due to the scaling of this viewport during utilization of the layout wizard earlier (see Figure 19–66). To verify that the image of the floor plan is properly scaled, open the Viewports toolbar. The toolbar displays the scale of the active viewport in Figure 19–77.

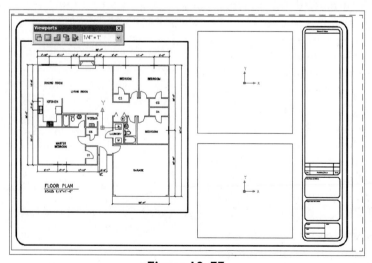

Figure 19–77

STEP 12

Verify that you are in floating Model Space (the MODEL button should be displayed in the status area), and activate the upper right viewport. Then use the XREF command to attach the drawing file 19_Cornice inside this viewport in Figure 19–78. It is important to attach this drawing file in the middle of the upper right viewport. The reason is that when you scale the image to size using the ZOOM command, the image or a portion of the image will be visible in the viewport, enabling you to pan the image into position. If this image is inserted at 0,0 and the image scaled, it may not be visible in the viewport. If an image is lost, performing a ZOOM-EXTENTS will quickly locate it. As in all cases involving floating Model Space, notice the cornice, although small, is also visible in the other two viewports.

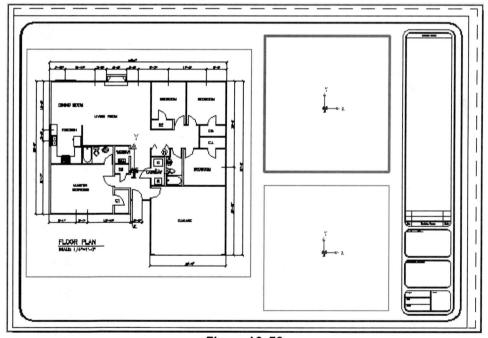

Figure 19–78

1046

STEP 13

As with the floor plan, use the VPLAYER (Viewport Layer) command to freeze all layers that begin with "19_Cornice" in the left and lower right viewports. The results are illustrated in Figure 19–79.

Command: **VPLAYER**
Enter an option [?/Freeze/Thaw/Reset/
   Newfrz/Vpvisdflt]: **F** (For Freeze)
Enter layer name(s) to freeze or <select
   objects>: **19_Cornice*** (For all layers
   that begin with 19_Cornice)
Enter an option [All/Select/Current]
   <Current>: **S** (For Select)

Switching to Paper space.
Select objects: (Pick the large left viewport
   in Figure 19–79)
Select objects: (Pick the lower right
   viewport in Figure 19–79)
Select objects: (Press ENTER to continue)
Switching to Model space.
Enter an option [?/Freeze/Thaw/Reset/
   Newfrz/Vpvisdflt]: (Press ENTER to
   exit this command and freeze all layers
   beginning with 19_Cornice)

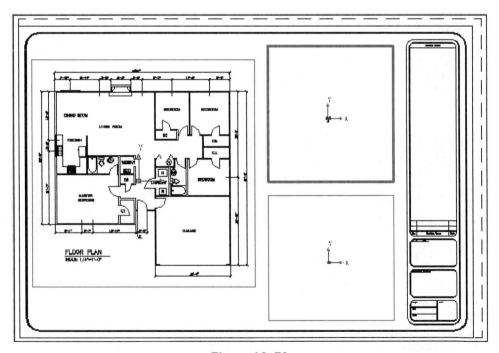

Figure 19–79

STEP 14

Be sure the upper right viewport is still active. Although the image of the cornice is visible, it is certainly not at the proper scale. The cornice is to be plotted out at a scale of 3"=1'-0". Use the Viewports toolbar and select the scale from the drop-down list box or use the ZOOM command and enter a value of 1/4XP to properly scale the image of the cornice detail (see Figure 19–80).

Command: **Z** *(For ZOOM)*

Specify corner of window, enter a scale factor (nX or nXP), or [All/Center/Dynamic/Extents/Previous/Scale/Window] <real time>: **1/4XP**

If the image does not display in its entirety in the viewport, you can use the PAN command to pan the image across the viewport until it is positioned properly. You can easily do this, especially since this is still the active viewport in floating Model Space.

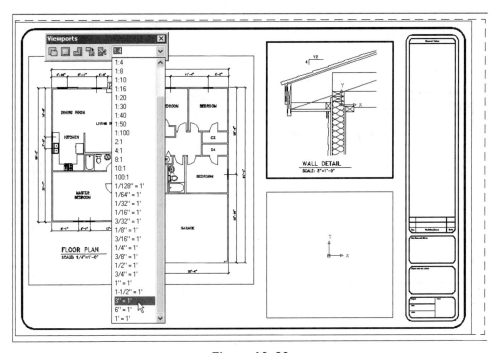

Figure 19–80

STEP 15

Again, verify that you are in floating Model Space (the MODEL button should be displayed in the status area), and activate the lower right viewport. Then use the XREF command to attach the file 19_Fndation.Dwg inside this viewport as in Figure 19–81. As with the cornice, attach the drawing file 19_Fndation in the middle of the lower right viewport. As in all cases involving floating Model Space, notice that the file 19_Fndation is also visible in the other two viewports.

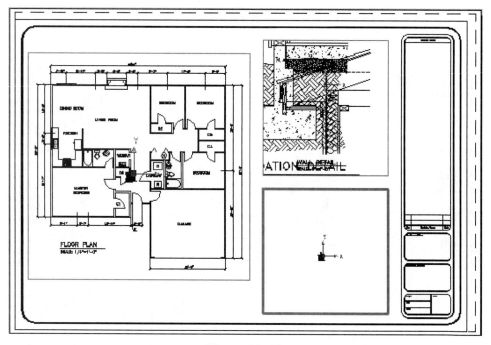

Figure 19–81

STEP 16

Use the VPLAYER (Viewport Layer) command to freeze all layers that begin with "19_Fndation" in the left and upper right viewports. The results are illustrated in Figure 19–82.

Command: **VPLAYER**
Enter an option [?/Freeze/Thaw/Reset/
    Newfrz/Vpvisdflt]: **F** *(For Freeze)*
Enter layer name(s) to freeze or <select
    objects>: **19_Fndation*** *(For all layers
    that begin with 19_Fndation)*
Enter an option [All/Select/Current]
    <Current>: **S** *(For Select)*

Switching to Paper space.
Select objects: *(Pick the large left viewport
    in Figure 19–82)*
Select objects: *(Pick the upper right
    viewport in Figure 19–82)*
Select objects: *(Press* ENTER *to continue)*
Switching to Model space.
Enter an option [?/Freeze/Thaw/Reset/
    Newfrz/Vpvisdflt]: *(Press* ENTER *to
    exit this command and freeze all layers
    beginning with 19_Fndation)*

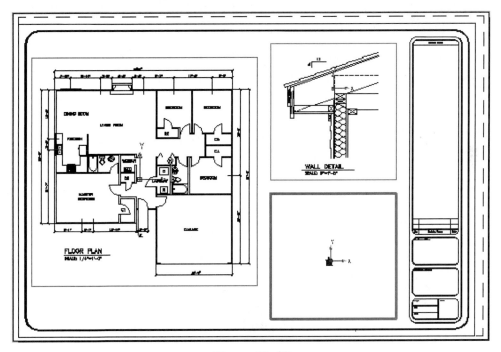

Figure 19–82

STEP 17

Check to see that the lower right viewport is active. Although the image of the foundation is visible, it is not at the proper scale. The foundation is to be plotted out at a scale of 1 1/2"=1'-0". Use the Viewports toolbar to select the scale from the drop-down list box or use the ZOOM command and enter a value of 1/8XP to properly scale the image of the foundation detail to reflect Paper Space units (see Figure 19–83).

Command: **Z** *(For ZOOM)*
Specify corner of window, enter a scale factor (nX or nXP), or [All/Center/Dynamic/Extents/Previous/Scale/Window] <real time>: **1/8XP**

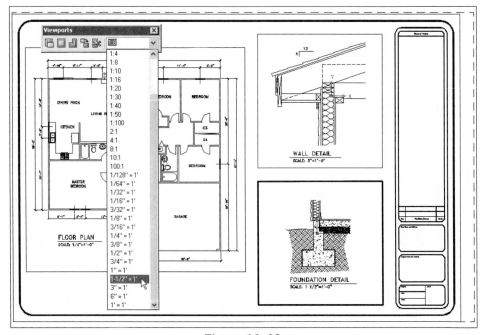

Figure 19–83

If the image does not display in its entirety in the viewport, you can use the PAN command to pan the image across the viewport until it is positioned properly. You can do this easily, especially since this is still the active viewport in floating Model Space.

STEP 18

All views are now properly positioned with images frozen in certain viewports and visible in others. If, however, another viewport were to be constructed through the -VPORTS command, all three images would appear in the new viewport. To prevent existing images from displaying in any new viewports, use the VPLAYER command along with the Vpvisdflt option. This options stands for Viewport Visibility Default. If certain layers are identified through this option and the layers are frozen, they will not appear in any new viewport created.

Command: **VPLAYER**
Enter an option [?/Freeze/Thaw/Reset/
    Newfrz/Vpvisdflt]: **V** *(For Vpvisdflt)*
Enter layer name(s) to change viewport
    visibility or <select objects>: **19_**
    Floor*,19_Cornice*,19_Fndation*

Enter a viewport visibility option [Frozen/
    Thawed] <Thawed>: **F** *(For Frozen)*
Enter an option [?/Freeze/Thaw/Reset/
    Newfrz/Vpvisdflt]: *(Press* ENTER *to exit
    this command)*

Yet another way to freeze all layers in any new viewports is to activate the Layer Properties Manager dialog box shown in Figure 19–84. First, select all layers that are associated with 19_Floor, 19_Cornice, and 19_Fndation. With these layers selected, click on one of the icons in the "New VP Freeze" column. This will change all sun symbols to the snowflake (see Figure 19–84).

Name	On	F...	L...	Color	Linetype	Lineweight	Plot Style	P...	Current...	New VP Freeze
19_Cornice|FRAMING	♀	○		■ Green	Continuous	—— Default	Color_3			
19_Cornice|HIDDEN	♀	○		■ Red	19...ice|HIDDEN	—— Default	Color_1			
19_Cornice|INS...ATION	♀	○		■ White	Continuous	—— Default	Color_7			
19_Cornice|NOTES	♀	○		□ Cyan	Continuous	—— Default	Color_4			
19_Cornice|SECTION	♀	○		□ Yellow	Continuous	—— Default	Color_2			
19_Floor|CENTER	♀	○		□ Yellow	19...loor|CE...ER2	—— Default	Color_2			
19_Floor|COUNTERTOP	♀	○		■ White	Continuous	—— Default	Color_7			
19_Floor|DIM	♀	○		□ Yellow	Continuous	—— Default	Color_2			
19_Floor|DOORS	♀	○		■ Red	Continuous	—— Default	Color_1			
19_Floor|FIREPLACE	♀	○		■ Red	Continuous	—— Default	Color_1			
19_Floor|FRAMING	♀	○		■ Green	Continuous	—— Default	Color_3			
19_Floor|NOTES	♀	○		□ Cyan	Continuous	—— Default	Color_4			
19_Floor|SYMBOLS	♀	○		■ White	Continuous	—— Default	Color_7			
19_Floor|TEXT	♀	○		□ Cyan	Continuous	—— Default	Color_4			
19_Floor|WALLS	♀	○		■ White	Continuous	—— Default	Color_7			
19_Floor|WINDOWS	♀	○		■ Magenta	Continuous	—— Default	Color_6			
19_...dation|CONCRETE	♀	○		□ Yellow	Continuous	—— Default	Color_2			
19_...dation|CON...TION	♀	○		■ Red	Continuous	—— Default	Color_1			
19_Fndation|EARTH	♀	○		■ Green	Continuous	—— Default	Color_3			
19_Fndation|FRAMING	♀	○		■ Green	Continuous	—— Default	Color_3			
19_Fndation|GRAVEL	♀	○		□ Cyan	Continuous	—— Default	Color_4			
19_...dation|INS...TION	♀	○		■ White	Continuous	—— Default	Color_7			
19_Fndation|NOTES	♀	○		□ Cyan	Continuous	—— Default	Color_4			
Defpoints	♀	○		■ White	Continuous	—— Default	Color_7			

Figure 19–84

STEP 19

Turn off the Vports layer to display only the three details. Be sure the PAPER button is displayed in the Status bar shown in Figure 19–85. The drawing of the three details can now be plotted out at to a scale of 1=1 because the ZOOM XP was used to scale each detail to the specific viewport it occupies. One final operation would be to delete the Layout1 and Layout2 tabs at the bottom of the display screen. Right-click on the Layout1 tab and pick the Delete option from the menu. Do the same for the Layout2 tab. Only the Model and Detail Sheet tabs should appear in the completed layout illustrated in Figure 19–86.

| SNAP | GRID | ORTHO | POLAR | OSNAP | OTRACK | LWT | PAPER |

Figure 19–85

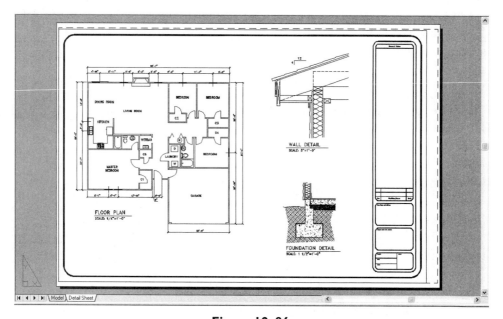

Figure 19–86

 Open the Exercise Manual PDF file for Chapter 10 on the accompanying CD for more tutorials and exercises.

 If you have the accompanying Exercise Manual, refer to Chapter 10 for more tutorials and exercises.

CHAPTER 20

Solid Modeling Fundamentals

It is said that humans see, hear, and exist in a 3D world. Why not design in three dimensions, and visualize the object before placing objects on the computer screen? Part of learning the art of visualization is the study and construction of models in 3D in order to obtain as accurate as possible an image of an object undergoing design. Solid models are mathematical models of actual objects that can be analyzed through the calculation of such items as mass properties, center of gravity, surface area, moments of inertia, and much more. The solid model starts the true design process by defining objects as a series of primitives. Boxes, cones, cylinders, spheres, and wedges are all examples of primitives. These building blocks are then joined together or subtracted from each other through certain modifying commands. Fillets and chamfers can be created to give the solid model a more realistic appearance and functionality. Two-dimensional views can be extracted from the solid model along with a cross section of the model. First, the methods of representing drawings will be reviewed and discussed briefly.

ORTHOGRAPHIC PROJECTION

The heart of any engineering drawing is the ability to break up the design into multiview projections, normally the three standard orthographic views, Front, Top, and Right sides (see Figure 20–1). The machinist or builder is then required to interpret the views and their dimensions, to paint a mental picture of a graphic version of the object as if it were already made or constructed.

Although most engineers and designers have the skill to easily convert the multiview drawing to a picture in their minds, a vast majority of us can be confused by the numerous hidden lines and centerlines of a drawing and their meaning to the overall design. We need some type of picture to help us interpret the multiview drawing and get a feel for what the part looks like, including the functionality of the part. This may be the major advantage of constructing an object in 3D.

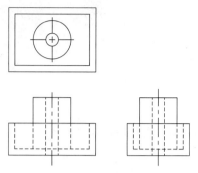

Figure 20–1

ISOMETRIC DRAWINGS

The isometric drawing is the easiest 3D representation to produce. This type of drawing is actually a 2D pictorial view, which gives the appearance of being 3D by laying out lines for depth on a 30° axis, height on a 90° (vertical) axis and width on a 150° axis. As easy as the isometric is to produce, it is also one of the most inaccurate methods of producing a 3D image.

Figure 20–2 shows an isometric view aligned to show the top, front, and right side views of the object. If we wished to see a different view of the object in isometric, a new drawing would have to be produced from scratch. This is the major disadvantage of using a 2D pictorial view such as an isometric drawing. Still, because of ease in drawing, isometric views remain popular in many school and technical drafting rooms.

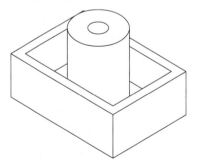

Figure 20–2

3D WIREFRAME MODELS

A 3D wireframe model is shown in Figure 20–3. This is the simplest type of 3D model and unlike the 2D isometric drawing, it can be viewed from any direction by simply changing our viewpoint. A wireframe model, however, as lines intersect from the front and back of the object, is sometimes difficult to visualize and interpret.

In 2D drawings the "Z" value of a XYZ coordinate remains set to zero. In 3D modeling height is shown by providing values for "Z". Values for height can be entered at the command prompt, selected using object snaps or selected through the use of aids, such as XYZ point filters. In this method, values are temporarily saved for later use inside a command. The values can be retrieved and a coordinate value entered to complete the command. Manipulation of a user-defined coordinate system (UCS) is another especially useful method for providing height when creating a three-dimensional object. A new coordinate system can be defined along any plane on which to place objects.

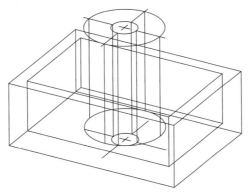

Figure 20–3

3D SURFACED MODELS

Surfacing picks up where the wireframe model leaves off. Because of the complexity of the wireframe and the number of intersecting lines, surfaces are applied to the wireframe. A major advantage of a surfaced model over a wireframe model is the capability of performing a hidden line removal operation. A HIDE is performed to view the model without interference of other faces. The surfaces created in Figure 20–4 are in the form of opaque objects, called 3D faces. 3D faces are created by selecting either three or four points in space, as required, to create the surface needed. Surfacing an existing wireframe can be easily accomplished by utilizing OSNAP options to assist in point selection. The VPOINT command is used to view the model in different 3D positions to make sure all sides of the model have been surfaced. The format of the 3DFACE command is outlined with the following prompts and Figure 20–4.

Command: **3DFACE**
Specify first point or [Invisible]: *(Select the endpoint at "A")*
Specify second point or [Invisible]: *(Select the endpoint at "B")*
Specify third point or [Invisible] <exit>: *(Select the endpoint at "C")*
Specify fourth point or [Invisible] <create three-sided face>: *(Select the endpoint at "D")*
Specify third point or [Invisible] <exit>: *(Press ENTER to exit this command)*

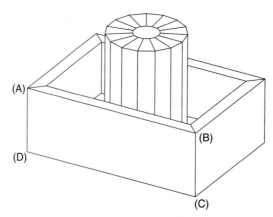

Figure 20–4

SOLID MODELS

The creation of a solid model remains the most complete way to represent a model in 3D (see Figure 20–5). It is also the most versatile 3D representation of an object. Wireframe models and surfaced models can be analyzed by taking distance readings and the identification of key points of the model. Key orthographic views such as Front, Top, and Right Side views can be extracted from wireframe models. Surfaced models can be imported into shading packages for increased visualization. Solid models do all of these operations and more (see Figure 20–6). This is because the solid model, as it is called, is a solid representation of the actual object. From cylinders to spheres, wedges to boxes, all objects that go into the creation of a solid model have volume. This allows a model to be constructed of what are referred to as primitives. These primitives are then merged into one using addition, subtraction, and intersection operations. What remains is the most versatile of 3D drawings. This method of creating models will be the main focus of this chapter.

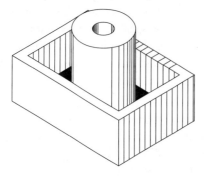

Figure 20–5

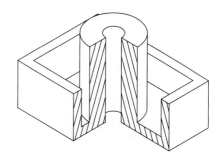

Figure 20–6

SOLID MODELING BASICS

All objects, no matter how simple or complex, are composed of simple geometric shapes or primitives. The shapes include boxes, wedges, cylinders, cones, spheres, and tori. Solid modeling allows for the creation of these primitives. Once created, the shapes are either merged or subtracted to form the final object. Follow the next series of steps to form the object in Figure 20–7.

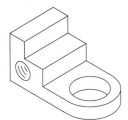

Figure 20–7

Begin the process of solid modeling by constructing a solid slab that will represent the base of the object. You can do this by using the BOX command. Supply the length, width, and height of the box, and the result is a solid slab, as shown in Figure 20–8.

Next, construct a cylinder using the CYLINDER command. This cylinder will eventually form the curved end of the object. One of the advantages of constructing a solid model is the ability to merge primitives together to form composite solids. Using the UNION command, combine the cylinder and box to form the complete base of the object in Figure 20–9.

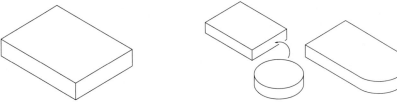

Figure 20–8   **Figure 20–9**

As the object progresses, create another solid box and move it into position on top of the base. There, use the UNION command to join this new block with the base. As you add new shapes during this process, they all become part of the same solid (see Figure 20–10).

Yet another box is created and moved into position. However, instead of being combined with the block, this new box is removed from the solid. Use the

SUBTRACT command to perform this subtraction operation and create the step shown in Figure 20–11.

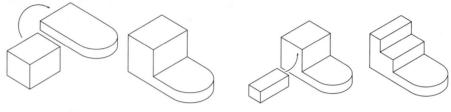

Figure 20–10   **Figure 20–11**

You can form holes in a similar fashion. First, use the CYLINDER command to create a cylinder with the diameter and depth of the desired hole. Next, move the cylinder into the solid and then subtract it using the SUBTRACT command. Again, the object in Figure 20–12 represents a composite solid object.

Using existing AutoCAD tools such as User Coordinate Systems (discussed in detail later in this chapter), create another cylinder with the CYLINDER command. It too is moved into position where it is subtracted through the SUBTRACT command. The complete solid model of the object is shown in Figure 20–13.

Figure 20–12   **Figure 20–13**

Advantages of constructing a solid model out of an object come in many forms. 2D profiles of different surfaces of the solid model can be taken to create orthographic views. Section views of solids can be automatically formed and crosshatched, as in Figure 20–14.

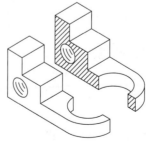

Figure 20–14

A very important analysis tool is the MASSPROP command, which is short for mass property. Information such as the mass and volume of the solid object is automatically calculated, along with centroids and moments of inertia, components that are used in computer aided engineering (CAE) and in finite element analysis (FEA) of the model. See Figure 20–15.

Mass: 223.2 gm
Volume: 28.39 cu cm (Err: 3.397)
Bounding box: X:-1.501—6.001 cm
                  Y:-0.0009944—2.75 cm
                  Z:-3.251—3.251 cm

Centroid: X: 2.36 cm
                  Y: 0.7067 cm
                  Z: 0.07384 cm

Moments
                  of inertia: X: 612.2 gm sq cm (Err: 76.16)
                  Y: 2667 gm sq cm (Err: 442.6)
                  Z: 2444 gm sq cm (Err: 383.2)

Figure 20–15

THE UCS COMMAND

Two-dimensional computer-aided design still remains the most popular form of representing drawings for most applications. However, in applications such as manufacturing, 3D models are becoming increasingly popular for creating rapid prototype models or for creating tool paths from the 3D model. To assist in this creation process, a series of User Coordinate Systems is used to create construction planes where features such as holes and slots are located. In Figure 20–16, a model of a box is displayed along with the User Coordinate System icon. Figure 20–17 identifies the directions of the three User Coordinate System axes. Notice that as X and Y are identified in the User Coordinate System icon, the positive Z direction is straight up; a negative Z direction is down. The UCS command is used to create different user-defined coordinate systems. The command line sequence follows along with all options; the command and options can also be selected from the UCS toolbar illustrated in Figure 20–18.

Command: **UCS**
Current ucs name: *WORLD*
Enter an option [New/Move/orthoGraphic/Prev/Restore/Save/Del/Apply/?/World]
   <World>: **N** *(For New)*
Specify origin of new UCS or [ZAxis/3point/OBject/Face/View/X/Y/Z] <0,0,0>:

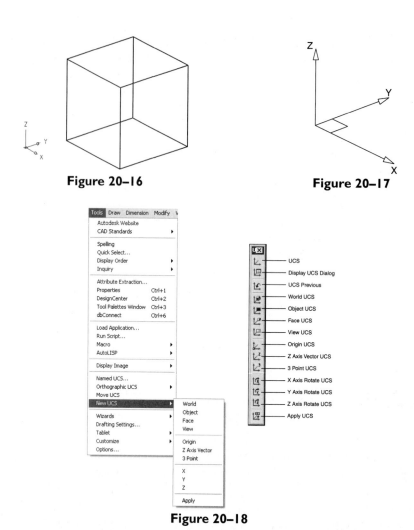

Figure 20–16

Figure 20–17

Figure 20–18

The following options are available when you use the UCS command:

> **New**—Identifies a new User Coordinate System using one of the following:
>> **ZAxis**—From two points defining the Z axis
>>
>> **3point**—By three points
>>
>> **Object**—In relation to an object selected
>>
>> **Face**—From the selected face of a solid object
>>
>> **View**—By the current display
>>
>> **X/Y/Z**—By rotation along the X, Y, or Z axis
>>
>> **<0,0,0>**—By identifying origin (0,0,0) at any identified point
>
> **Move**—Identifies a new User Coordinate System by shifting origin in the XY plane or changing the Z coordinate.

orthoGraphic—Identifies new User Coordinate System from 6 preset systems (Top/Bottom/Front/Back/Left/Right).

Prev—Sets the User Coordinate System icon to the previously defined User Coordinate System.

Restore—Restores a previously saved User Coordinate System.

Save—Saves the position of a User Coordinate System under a unique name given by the CAD operator.

Del—Deletes a User Coordinate System from the database of the current drawing.

Apply—Sets the current User Coordinate System setting to a specific viewport(s).

?—Lists all previously saved User Coordinate Systems.

World—Switches to the World Coordinate System from any previously defined User Coordinate System.

 ## THE UCS-NEW—<0,0,0> DEFAULT OPTION

The New option of the ucs command defines a new User Coordinate System by one of several offered methods. The default method is moving the current UCS to a new 0,0,0 position while leaving the direction of the X, Y, and Z axes unchanged. Figure 20–19 is a sample model with the current coordinate system being the World Coordinate System. Follow Figures 20–20 and 20–21 to define a new User Coordinate System using the <0,0,0> option.

 Try It! – Open the drawing file 20_Ucs Origin. Activate the ucs command. Using the New option of the ucs command, identify a new origin point for 0,0,0 at "A" as shown in Figure 20–20. This should move the User Coordinate System icon to the point that you specify. If the icon remains in its previous location and does not move, use the UCSICON command with the Origin option to display the icon at its new origin point. To prove that the corner of the box is now 0,0,0, construct a circle at the bottom of the 5" cube in Figure 20–21.

 Command: **UCS**
Current ucs name: *WORLD*
Enter an option [New/Move/orthoGraphic/Prev/Restore/Save/Del/Apply/?/World]
    <World>: **N** *(for New)*
Specify origin of new UCS or [ZAxis/3point/OBject/Face/View/X/Y/Z] <0,0,0>: **End**
of (Select the endpoint of the line at "A")

 Command: **C** *(For CIRCLE)*
Specify center point for circle or [3P/2P/Ttr (tan tan radius)]: **2.5,2.5,0**
Specify radius of circle or [Diameter]: **2**

Notice a small "plus" at the corner of the User Coordinate System icon (see Figure 20–21). This indicates that the UCS icon is displayed at the origin (0,0,0). A small box, as in Figure 20–19, indicates that the icon is displaced at the World Coordinate System (WCS).

It is also important to note that the circle in Figure 20–21 is a 2D object and can only be drawn in or parallel to the XY plane (the bottom or top of the box). To draw a circle in the side of the box, we will need to not only move (translate) our coordinate system but be able to rotate it as well.

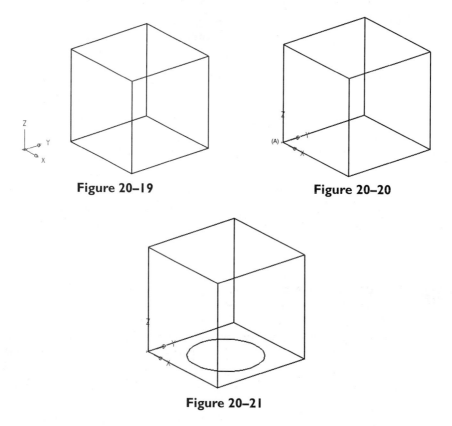

Figure 20–19   **Figure 20–20**

Figure 20–21

THE UCS-NEW—3POINT OPTION

Use the 3point option of the UCS command to specify a new User Coordinate System by identifying an origin and new directions of its positive X and Y axes (translate and rotate).

Try It! – Open the drawing file 20_Ucs 3p. Figure 20–22 shows a 3D cube in the World Coordinate System. To construct objects on the front panel, first define a new User Coordinate System parallel to the front. Use the following command sequence to accomplish this task.

Command: **UCS**
Current UCS name: *WORLD*
Enter an option [New/Move/orthoGraphic/Prev/Restore/Save/Del/Apply/?/World]
   <World>: **N** *(for New)*
Specify origin of new UCS or [ZAxis/3point/OBject/Face/View/X/Y/Z] <0,0,0>: **3** *(For 3point)*
Specify new origin point <0,0,0>: **End**
of *(Select the endpoint of the model at "A" as shown in Figure 20–23)*
Specify point on positive portion of X-axis <>: **End**
of *(Select the endpoint of the model at "B")*
Specify point on positive-Y portion of the UCS XY plane <>: **End**
of *(Select the endpoint of the model at "C")*

With the Y axis in the vertical position and the X axis in the horizontal position, any type of object can be constructed along this plane, such as the lines in Figure 20–24.

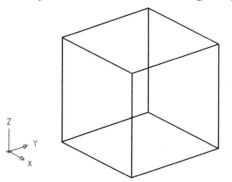

Figure 20–22

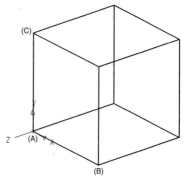

Figure 20–23

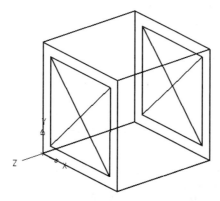

Figure 20–24

Try It! – Open the drawing file 20_Incline. The 3point method of defining a new User Coordinate System is quite useful in the example shown in Figure 20–25, where a UCS needs to be aligned with the inclined plane. Use the intersection at "A" as the origin of the new UCS, the intersection at "B" as the direction of the positive X axis, and the intersection at "C" as the direction of the positive Y axis.

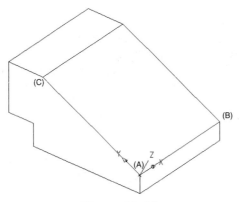

Figure 20–25

﹇ THE UCS–NEW—X/Y/Z ROTATION OPTIONS

Using the X/Y/Z rotation options will rotate the current user coordinate around the specific axis. Once a letter is selected as the pivot, a prompt appears asking for the rotation angle about the pivot axis. The right-hand rule is used to determine the positive direction of rotation around an axis. Think of the right hand gripping the pivot axis with the thumb pointing in the positive X, Y, or Z direction. The curling of the fingers on the right hand determines the positive direction of rotation. All positive rotations occur in the counter-clockwise directions. Figure 20–26 illustrates positive rotation about each axis. By viewing down the selected axis, the positive rotation is seen to be counterclockwise.

 Try It! – Open the drawing file 20_Ucs Rotate. Given the cube shown in Figure 20–27 in the World Coordinate System, the X option of the UCS command will be used to stand the icon straight up by entering a 90° rotation value, as in the following prompt sequence.

 Command: **UCS**
Current ucs name: *NO NAME*
Enter an option [New/Move/orthoGraphic/Prev/Restore/Save/Del/Apply/?/World] <World>: **N** *(for New)*
Specify origin of new UCS or [ZAxis/3point/OBject/Face/View/X/Y/Z] <0,0,0>: **X**
Specify rotation angle about X axis <90>: *(Press ENTER to accept 90° of rotation)*

The X-axis is used as the pivot of rotation; entering a value of 90° rotates the icon the desired degrees in the counterclockwise direction as in Figure 20–28.

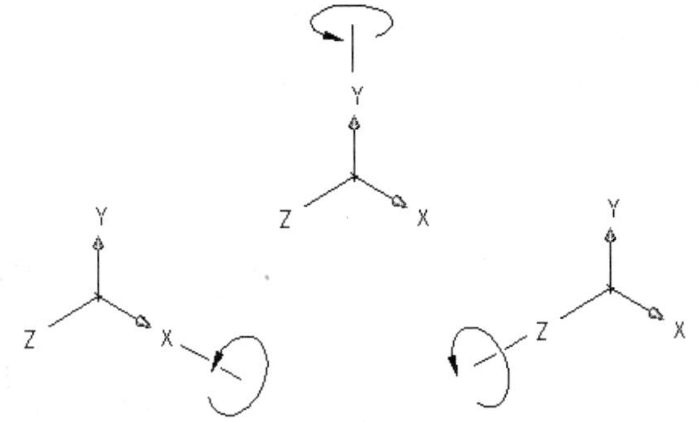

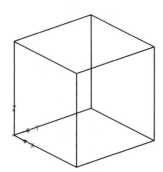

Figure 20–26

Figure 20–27

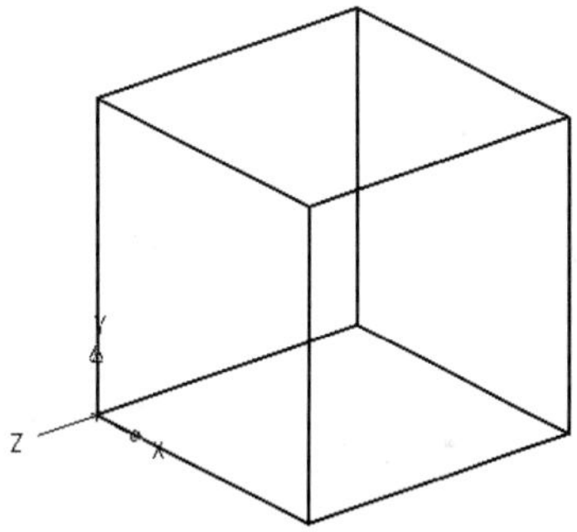

Figure 20–28

 THE UCS–NEW—OBJECT OPTION

Another option for defining a new User Coordinate System is to select an object and have the User Coordinate System align to that object (translate and rotate).

 Try It! – Open the drawing file 20_Ucs Object. Given the 3D cube in Figure 20–29, use the following command sequence and Figure 20–30 to accomplish this.

Command: UCS
Current ucs name: *WORLD*
Enter an option [New/Move/orthoGraphic/Prev/Restore/Save/Del/Apply/?/World]
    <World>: **N** *(for New)*
Specify origin of new UCS or [ZAxis/3point/OBject/Face/View/X/Y/Z] <0,0,0>: **OB**
    (for the Object option)
Select object to align UCS: *(Select the circle in Figure 20–30)*

The type of object selected determines the alignment (translation and rotation) of the User Coordinate System. In the case of the circle, the center of the circle becomes the origin of the User Coordinate System. Where the circle was selected becomes the point through which the positive X axis aligns. Other types of objects that can be selected include arcs, dimensions, lines, points, plines, solids, traces, 3dfaces, text, and blocks.

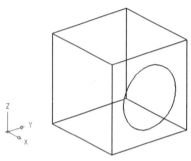
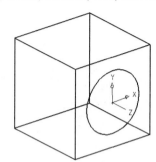

| Figure 20–29 | Figure 20–30 |

THE UCS–NEW–FACE OPTION

The Face option of the UCS command allows you to establish a new coordinate system aligned to the selected face of a solid object.

 Try It! – Open the drawing file 20_Ucs Face. Given the current User Coordinate System in Figure 20–31 and a solid box, follow the command sequence below to align the User Coordinate System with the Face option.

 Command: **UCS**
Current ucs name: *WORLD*
Enter an option [New/Move/orthoGraphic/Prev/Restore/Save/Del/Apply/?/World] <World>: **N** *(for New)*
Specify origin of new UCS or [ZAxis/3point/OBject/Face/View/X/Y/Z] <0,0,0>: **F** *(for Face)*
Select face of solid object: *(Select the edge of the model at point "A" in Figure 20–32)*
Enter an option [Next/Xflip/Yflip] <accept>: **N** *(the Next option relocates the UCS to the other face shared by the edge; see Figure 20–33)*
Enter an option [Next/Xflip/Yflip] <accept>: **N** *(the Next option returns the UCS to the original face in Figure 20–32)*
Enter an option [Next/Xflip/Yflip] <accept>: **X** *(the Xflip option rotates the UCS about the X axis 180° in Figure 20–34)*
Enter an option [Next/Xflip/Yflip] <accept>: **X** *(the Xflip option rotates the UCS back to its location in Figure 20–32)*
Enter an option [Next/Xflip/Yflip] <accept>: **Y** *(the Yflip option rotates the UCS about the Y axis 180° in Figure 20–35)*
Enter an option [Next/Xflip/Yflip] <accept>: **Y** *(the Yflip option rotates the UCS back to its location in Figure 20–32)*
Enter an option [Next/Xflip/Yflip] <accept>: *(Press ENTER to accept the UCS position)*

The results are displayed in Figure 20–36 with the User Coordinate System aligned to the right face. To select a face in this command you can click within the edges of a face or on one of its edges. Where you click directly affects the location of the origin. Use of the Next, Xflip, and Yflip options, which were demonstrated in the command sequence, were not required to obtain the results in Figure 20–36.

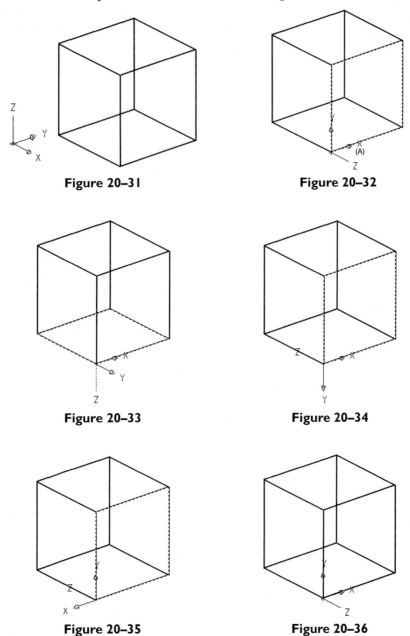

Figure 20–31   Figure 20–32

Figure 20–33   Figure 20–34

Figure 20–35   Figure 20–36

1068

THE UCS-NEW-VIEW OPTION

The View option of the UCS command allows you to establish a new coordinate system where the XY plane is perpendicular to the current screen viewing direction; in other words, it is parallel to the display screen.

 Try It! – Open the drawing file 20_Ucs View. Given the current User Coordinate System in Figure 20–37, follow the prompts below along with Figure 20–38 to align the User Coordinate System with the View option.

Command: **UCS**
Current ucs name: *WORLD*
Enter an option [New/Move/orthoGraphic/Prev/Restore/Save/Del/Apply/?/World]
    <World>: **N** *(for New)*
Specify origin of new UCS or [ZAxis/3point/OBject/Face/View/X/Y/Z] <0,0,0>: **V** *(for View)*

The results are displayed in Figure 20–38 with the User Coordinate System aligned parallel to the display screen.

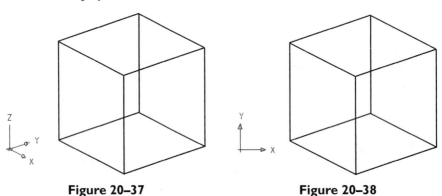

Figure 20–37   Figure 20–38

APPLICATIONS OF ROTATING THE UCS

 Try It! – Open the drawing file 20_Ucs Apps. Follow the next series of steps to perform numerous rotations of the User Coordinate System in the example of the cube shown in Figure 20–39.

Rotate the User Coordinate System along the X axis at a rotation angle of 90° using the following prompt sequence. This will align the User Coordinate System with the front of the cube shown in Figure 20–40.

 Command: **UCS**
Current ucs name: *NO NAME*
Enter an option [New/Move/orthoGraphic/Prev/Restore/Save/Del/Apply/?/World]
    <World>: **N** *(for New)*

Specify origin of new UCS or [ZAxis/3point/OBject/Face/View/X/Y/Z] <0,0,0>: **X**
Specify rotation angle about X axis <90>: *(Press ENTER to accept 90° of rotation)*

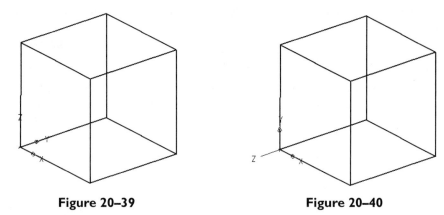

Figure 20–39   **Figure 20–40**

Next, align the User Coordinate System with one of the sides of the cube by performing the rotation using the Y axis as the pivot axis (see Figure 20–41). The positive
angle rotates the icon in the counterclockwise direction.

[icon] Command: **UCS**
Current ucs name: *NO NAME*
Enter an option [New/Move/orthoGraphic/Prev/Restore/Save/Del/Apply/?/World]
    <World>: **N** *(for New)*
Specify origin of new UCS or [ZAxis/3point/OBject/Face/View/X/Y/Z] <0,0,0>: **Y**
Specify rotation angle about Y axis <90>: *(Press ENTER to accept 90° of rotation)*

Next, rotate the User Coordinate System using the Z axis as the pivot axis. The degree
of rotation entered at 45° tilts the User Coordinate System shown in Figure 20–42.

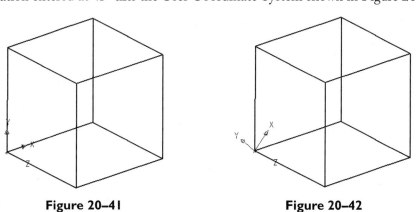

Figure 20–41   **Figure 20–42**

Command: **UCS**
Current ucs name: *NO NAME*
Enter an option [New/Move/orthoGraphic/Prev/Restore/Save/Del/Apply/?/World]
   <World>: **N** *(for New)*
Specify origin of new UCS or [ZAxis/3point/OBject/Face/View/X/Y/Z] <0,0,0>: **Z**
Specify rotation angle about Z axis <90>: **45**

To tilt the UCS icon about the Z axis so that the X axis is pointing down at a 45°
angle, enter a rotation angle of -90°. The results are shown in Figure 20–43.

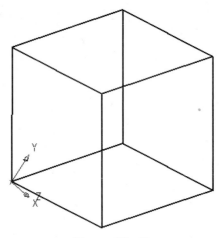

Figure 20–43

 Command: **UCS**
Current ucs name: *NO NAME*
Enter an option [New/Move/orthoGraphic/Prev/Restore/Save/Del/Apply/?/World]
   <World>: **N** *(for New)*
Specify origin of new UCS or [ZAxis/3point/OBject/Face/View/X/Y/Z] <0,0,0>: **Z**
Specify rotation angle about Z axis <90>: **-90**

USING THE UCS DIALOG BOX

It is considered good practice to assign a name to a User Coordinate System once it has
been created. Once numerous User Coordinate Systems have been defined in a drawing,
using their name instead of re-creating each coordinate system easily restores them.
You can accomplish this by using the Save and Restore options of the UCS command.
Another method is to choose Named UCS from the Tools pull-down menu illustrated
in Figure 20–44 (the UCSMAN command). This displays the UCS dialog box in Figure
20–45 with the Named UCSs tab selected. All User Coordinate Systems previously

defined in the drawing are listed here. To make one of these coordinate systems current, highlight the desired UCS name and select the Set Current button shown in Figure 20–45. A named UCS can also be made current by simply double-clicking it. To define (save) a coordinate system you must have first translated and rotated the UCS into a desired new position; then use the dialog box to select the "Unnamed" UCS and rename it. The UCS dialog box provides a quick method of restoring previously defined coordinate systems without entering them at the keyboard.

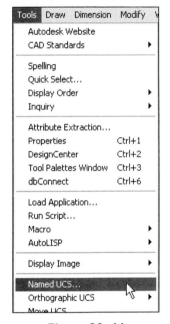

Figure 20–44

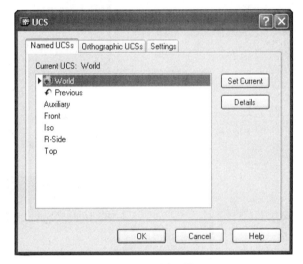

Figure 20–45

Choosing Preset from the Orthographic UCS cascading menu on the Tools pull-down menu (see Figure 20–46) also displays the UCS dialog box, but with the Orthographic UCSs tab selected, as shown in Figure 20–47. Use this tab to automatically rotate the User Coordinate System so that the XY plane is parallel and oriented to one of the six orthographic views: Front, Top, Back, Right Side, Left Side, and Bottom.

The Settings tab of the UCS dialog box in Figure 20–48 controls the setting of UCS system variables. In this tab, you can control such things as whether the UCS icon is displayed at the origin point, in the lower left corner of the screen, or if it is displayed at all.

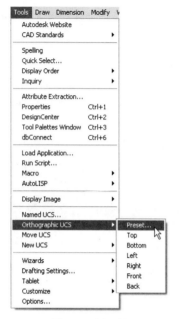

Figure 20–46

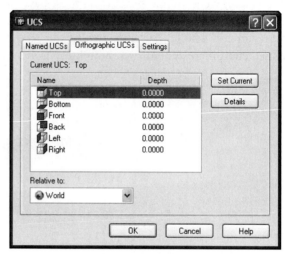

Figure 20–47

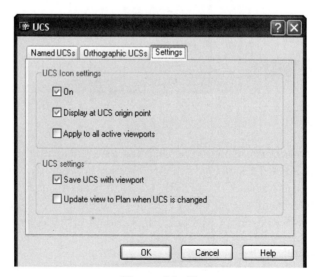

Figure 20–48

THE UCS ICON DIALOG BOX

A dialog box is provided to control the appearance of the User Coordinate System icon. Choosing UCS Icon from the Display cascading menu on the View menu followed by Properties (see Figure 20–49) displays the UCS Icon dialog box in Figure 20–50 (UCSICON command, Properties option). You can control the icon style, size, and color in this dialog box. Two different controls are provided to change the color of the icon in model space and in layout mode. The 3D radio button is selected in Figure 20–50 to choose the 3D UCS icon style shown. You can select the old style of icon by clicking the 2D radio button as shown in Figure 20–51. This older 2D icon does not have a visual representation of the Z axis. It is considered good practice to leave the UCS icon style setting in the 3D mode.

Figure 20–49

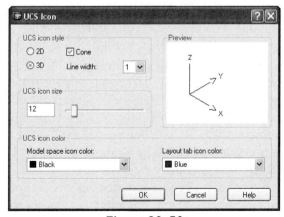

Figure 20–50

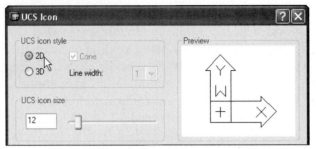

Figure 20–51

VIEWING 3D MODELS WITH THE 3DORBIT COMMAND

Various methods can be used to view a model in 3D. One of the more efficient ways is through the 3DORBIT command, which can be selected from the View menu shown in Figure 20–52. Command options as well as other viewing tools are located in the 3D Orbit toolbar in Figure 20–53.

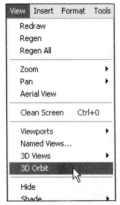

Figure 20–52

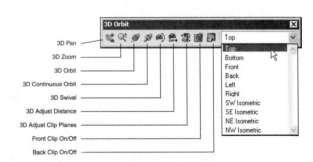

Figure 20–53

Choosing this command displays your model similar to Figure 20–54. Use the large circle to guide your model through a series of dynamic rotation maneuvers. For instance, moving your cursor outside the large circle at "A" allows you to dynamically rotate (drag) your model only in a circular direction. Moving your cursor to either of the circle quadrant identifiers at "B" and "C" allows you to rotate your model horizontally. Moving your cursor to either of the circle quadrant identifiers at "D" or "E" allows you to rotate your model vertically. Moving your cursor inside the large circle at "F" allows you to dynamically rotate your model to any viewing position. Right-click while in the 3DORBIT command to gain access to the numerous options such as Pan, Zoom, and Shading Modes provided by this command (see Figure 20–55).

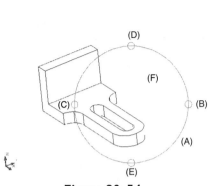

Figure 20–54

Figure 20–55

Other components of the 3D Orbit toolbar:

3D Continuous Orbit—Allows you to view an object in a continuous orbit motion.

3D Swivel—Allows you to view an object with a motion that is similar to looking through a camera viewfinder.

3D Adjust Distance—Allows you to view an object closer or farther away.

3D Adjust Clip Planes—Allows you to view an object after removing a portion of that object from the view with a front or back clipping plane.

Front Clip On/Off—Allows you to turn the front clipping plane on or off.

Back Clip On/Off—Allows you to turn the back clipping plane on or off.

THE VPOINT COMMAND

The object in Figure 20–56 represents the plan view of a 3D model. Unfortunately, it becomes very difficult to understand what the model looks like from the viewpoint shown. Use the VPOINT command to view a wireframe, surfaced, or solid model from any position in three dimensions. The following is a typical prompt sequence for the VPOINT command:

Command: **VPOINT**
Switching to the WCS
Current view direction: VIEWDIR=0.0000,0.0000,1.0000
Specify a view point or [Rotate] <display compass and tripod>:

The VPOINT command stands for "View point." A point is identified in 3D space. This point identifies where the model is viewed from. The object shown in Figure 20–57 is a typical result of the use of the VPOINT command.

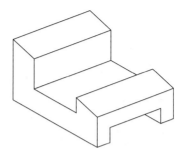

Figure 20–56

Figure 20–57

The VIEW command as well as the VPOINT command can be used to select a new viewpoint from which the model can be seen. Choosing 3D Views from the View menu exposes the 3D Views cascading menu in Figure 20–58. Choose from numerous viewpoint options. Items listed as Top, Bottom, SW Isometric, and NE Isometric are standard, stored viewpoints used to view a 3D model from these standard locations. They provide a fast way to view a model in a primary or isometric view.

Figure 20–58

A View toolbar is also available as shown in Figure 20–59. It contains options similar to the 3D Views cascading menu on the View menu. The View toolbar has the extra advantage of displaying icons that guide you in picking the desired viewpoint.

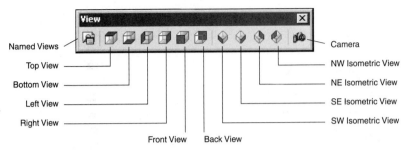

Figure 20–59

When you enter the VPOINT command at the command prompt, there are three methods of choosing a viewpoint. The first is through the use of an X,Y,Z coordinate. In the following command sequence, a new viewing point is established 1 unit in the positive X direction, 1 unit in the negative Y direction, and 1 unit in the positive Z direction. From this point in 3D space looking back at 0,0,0 determines the viewing direction (southeast isometric). Note that the previous viewing direction 0,0,1 is a top (plan) view.

Command: **VPOINT**

Switching to the WCS
Current view direction: VIEWDIR=0.0000,0.0000,1.0000
Specify a view point or [Rotate] <display compass and tripod>: **1,-1,1**

The second method of defining a viewpoint is through the Rotate option. The viewing direction is determined by two angles. The first angle defines the viewpoint in the XY plane as measured from the X axis. However, this is only 2D. The second angle defines the viewpoint from the XY plane. This tilts the viewing point up for a positive angle or down for a negative angle.

Command: **VPOINT**

Switching to the WCS
Current view direction: VIEWDIR=1.0000,-1.0000,1.0000
Specify a view point or [Rotate] <display compass and tripod>: **R** *(For Rotate)*
Enter angle in XY plane from X axis <315>: **45**
Enter angle from XY plane <35>: **30**

The third method of choosing a viewpoint is by use of the compass and dynamic viewing axis shown in Figure 20–60. Although the compass appears two-dimensional, it enables you to pick a 3D viewpoint depending on how you read the compass. The intersection of the horizontal and vertical lines forms the North Pole of the compass. The inner circle forms the equator and the outer circle forms the South Pole. The following examples illustrate numerous viewing points.

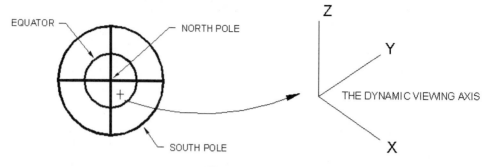

Figure 20–60

Command: **VPOINT**

Switching to the WCS

Current view direction: VIEWDIR=1.0000,-1.0000,1.0000

Specify a view point or [Rotate] <display compass and tripod>: *(Press ENTER to display the viewpoint compass in Figure 20–60)*

VIEWING ALONG THE EQUATOR

 Try It! – Open the drawing file 20_Vpoint. This file will allow you to experiment with using the VPOINT command. To obtain the results in Figure 20–61, use the VPOINT command and pick a point at "A" to select the Front view, pick a point at "B" to select the Top view, and pick a point at "C" to select the Right Side view. Coordinates could also have been entered to achieve the same results:

Command: **VPOINT**

Switching to the WCS

Current view direction: VIEWDIR=1.0000,-1.0000,1.0000

Specify a view point or [Rotate] <display compass and tripod>: **0,-1,0** *(or press ENTER and select point "A" on the compass)*

Command: **VPOINT**

Switching to the WCS

Current view direction: VIEWDIR=1.0000,-1.0000,1.0000

Specify a view point or [Rotate] <display compass and tripod>: **0,0,1** *(or press ENTER and select point "A" on the compass)*

Command: **VPOINT**

Switching to the WCS

Current view direction: VIEWDIR=1.0000,-1.0000,1.0000

Specify a view point or [Rotate] <display compass and tripod>: **1,0,0** *(or press ENTER and select point "A" on the compass)*

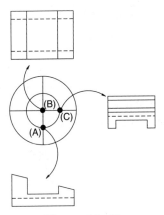

Figure 20–61

VIEWING NEAR THE NORTH POLE

Picking the four points shown in Figure 20–62 results in the different viewing points for the object. Since all points are inside the inner circle, the results are aerial views, or views from above. Remember that the inner circle symbolizes the Equator. Depending on which quadrant you select, you will look up at the object from the right corner, left corner, or either of the rear corners.

VIEWING NEAR THE SOUTH POLE

Picking the four points shown in Figure 20–63 results in underground views, or viewing the object from underneath. This is true because all points selected lie between the small and large circles. Again, remember that the small circle symbolizes the Equator, while the large circle symbolizes the South Pole.

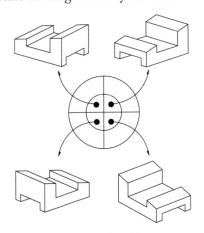

Figure 20–62

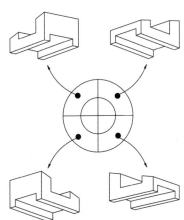

Figure 20–63

USING THE VIEWPOINT PRESETS DIALOG BOX

Choosing Viewpoint Presets from the 3D Views cascading menu on the View menu, shown in Figure 20–64, displays the Viewpoint Presets dialog box shown in Figure 20–65 (the DDVPOINT command). Use this dialog box to help define a new point from which to view a 3D model. The square image with a circle in the center allows you to define a viewing point in the XY plane. The semicircular-shaped image allows you to define a viewing point from the XY plane. The combination of both directions forms the 3D viewpoint.

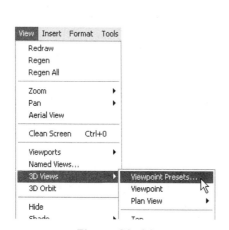

Figure 20–64

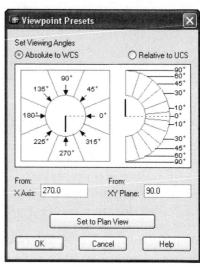

Figure 20–65

When selecting a viewing point from the X axis (in the XY plane), you have two options for selecting the desired point of view. When you pick outside the circle near 0, 45, 90, 135, 180, etc., as in "A" in Figure 20–66, the viewpoint snaps from one angle to another in increments of 45°. The resulting angle is displayed at the bottom of the square. If you want a more precise viewpoint in the XY plane, pick a point inside the circle, such as at "B." The resulting angle will display all values in between the standard 45° increments. If you already know the angle from the X axis, you can also enter it directly in the X-axis edit box.

Figure 20–66 illustrates setting a viewing angle in the XY plane. Figure 20–67 shows the second half of the Viewpoint Presets dialog box, which defines an angle from the XY plane. Selecting a point in between both semicircles at "A" snaps the viewpoint in 10°, 30°, 45°, 60°, and 90° increments. If you want a more precise selection, pick the viewing point inside the smaller semicircle, such as "B." This will allow you to select an angle different from the five default values listed above. An alternate method of selecting an angle from the XY plane would be to place the value in the XY Plane edit box.

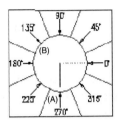

Figure 20–66

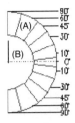

Figure 20–67

THE 3D COORDINATE SYSTEM ICON

The 3D Coordinate System icon can take on different forms during the drawing process. Viewing the icon in plan view or in an aerial view displays the image in Figure 20–68. The box indicates the World Coordinate System (WCS). In Figure 20–69 a "+" inside the box indicates that the WCS icon is displayed at 0,0,0. The icon must be totally displayed on the screen; if the icon is unable to be totally shown at its origin, the icon locates itself in the lower left corner of the display screen and no "+" is shown.

 Note: The UCSICON command can be used to turn the coordinate system icon on or off, shown in 3D or 2D style, and determine if it will be displayed at the Origin or always in the lower left corner (the Noorigin option).

The icon displayed in Figure 20–70 represents the paper space environment or layout mode. This environment is strictly two-dimensional and is normally used as a layout tool for arranging views and displaying title block information.

A User Coordinate System (UCS) icon (notice the box is missing) is shown in Figure 20–71. This coordinate system has been translated and/or rotated from the World Coordinate System. In Figure 20–72 the "+" indicates that the UCS icon is displayed at 0,0,0.

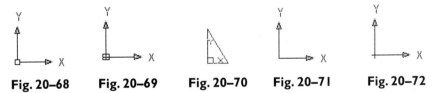

Fig. 20–68   **Fig. 20–69**   **Fig. 20–70**   **Fig. 20–71**   **Fig. 20–72**

SELECTING SOLID MODELING COMMANDS

Figures 20–73 and 20–74 illustrate two of the more popular methods for accessing solid modeling commands. In Figure 20–73, solid modeling commands can be chosen from the Draw menu. Choosing Solids displays four groupings of commands. The first grouping displays Box, Sphere, Cylinder, Cone, Wedge, and Torus, which are considered the building blocks of the solid model and are used to construct basic "primitives." The second grouping displays the "sweep" commands—EXTRUDE and REVOLVE, which provide an additional way to construct solid models. Polyline outlines or circles can be extruded (swept) to a thickness that you designate. You can also revolve (sweep) other polyline outlines about an object representing a centerline of rotation. The third grouping of solid modeling commands (SLICE, SECTION, and INTERFERE) enables you to slice the solid model into separate pieces, cut the solid model to provide a 2D section (in the form of a Region or 2D Solid), and perform an interference check where two solids are constructed near each other but must not touch or intersect. The last grouping, Setup, displays three commands designed to extract orthographic views from a solid model. The three commands are SOLDRAW, SOLVIEW, and SOLPROF. Notice also in Figure 20–74 the complete grouping of solid modeling commands on a toolbar. This is another convenient way to select solid modeling commands.

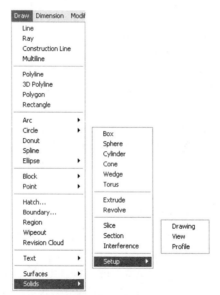

Figure 20–73

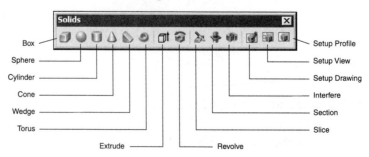

Figure 20–74

USING THE BOX COMMAND

Use the BOX command to create a 3D solid box. Choose this command in one of the following ways:

- From the Solids toolbar
- From the pull-down menu (Draw > Solids > Box)
- From the keyboard (BOX)

Methods for creating this type of solid include selecting 3D diagonal corners, selecting diagonal corners in the XY plane and then specifying the height, or by simply entering values for its length, width, and height. A Cube option simplifies construction if all three dimensions are the same.

 Try It! – Open the drawing file 20_Box. Use Figure 20–75 and the following command sequences to show examples of creating solid boxes.

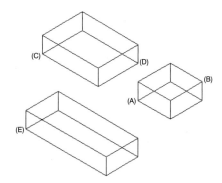

Figure 20–75

 Command: **BOX**
Specify corner of box or [CEnter] <0,0,0>: **4.00,7.00** *(At "A")*
Specify corner or [Cube/Length]: **6.00,9.00,1.00** *(At "B")*

 Command: **BOX**
Specify corner of box or [CEnter] <0,0,0>: **-1.50,6.50** *(At "C")*
Specify corner or [Cube/Length]: **2.00,9.00** *(At "D")*
Specify height: **1**

 Command: **BOX**
Specify corner of box or [CEnter] <0,0,0>: **2.00,2.00** *(At "E")*
Specify corner or [Cube/Length]: **L** *(To define the length of the box)*
Specify length: **5** *(In X direction)*
Specify width: **2** *(In Y direction)*
Specify height: **1** *(In Z direction)*

USING THE CONE COMMAND

A solid cone can be created by first defining a circular or elliptical base. This base tapers to a point perpendicular to the base. Choose this command in one of the following ways:

- From the Solids toolbar

- From the pull-down menu (Draw > Solids > Cone)

- From the keyboard (CONE)

 Try It! – Open the drawing file 20_Cone. Use the CONE command and Figure 20–76 to construct a cone of specified height with the base of the cone either circular or elliptical.

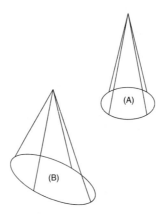

Figure 20–76

 Command: **CONE**
Current wire frame density: ISOLINES=4
Specify center point for base of cone or [Elliptical] <0,0,0>: **5.00,5.00** *(At "A")*
Specify radius for base of cone or [Diameter]: **1**
Specify height of cone or [Apex]: **4**

 Command: **CONE**
Current wire frame density: ISOLINES=4
Specify center point for base of cone or [Elliptical] <0,0,0>: **E** *(For an elliptical base)*
Specify axis endpoint of ellipse for base of cone or [Center]: **C** *(To define the center of the ellipse)*
Specify center point of ellipse for base of cone <0,0,0>: **6.50,-0.50** *(At "B")*
Specify axis endpoint of ellipse for base of cone: **@2<0**
Specify length of other axis for base of cone: **@1<90**
Specify height of cone or [Apex]: **4**

USING THE WEDGE COMMAND

The WEDGE command is very similar to the BOX command. Choose this command in one of the following ways:

- From the Solids toolbar

- From the pull-down menu (Draw > Solids > Wedge)

- From the keyboard (WEDGE)

This primitive is simply a box that has been diagonally cut. Use the WEDGE command to create a wedge with the base parallel to the current User Coordinate System and the slope of the wedge constructed along the X axis.

 Try It! – Open the drawing file 20_Wedge. Use the following prompts and Figure 20–77 as examples of how to use this command.

 Command: **WEDGE**
Specify first corner of wedge or [CEnter] <0,0,0>: **5.00,5.00** *(At "A")*
Specify corner or [Cube/Length]: **11.50,7.75** *(At "B")*
Specify height: **3**

 Command: **WEDGE**
Specify first corner of wedge or [CEnter] <0,0,0>: **8.50,-1.50** *(At "C")*
Specify corner or [Cube/Length]: **L** *(To define the length of the wedge)*
Specify length: **5** *(In X direction)*
Specify width: **1** *(In Y direction)*
Specify height: **3** *(In Z direction)*

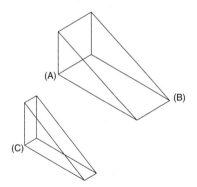

Figure 20–77

USING THE CYLINDER COMMAND

The CYLINDER command is similar to the CONE command except that the cylinder is drawn without a taper. The central axis of the cylinder is along the Z axis of the current User Coordinate System. Choose this command in one of the following ways:

- From the Solids toolbar

- From the pull-down menu (Draw > Solids > Cylinder)

- From the keyboard (CYLINDER)

 Try It! – Open the drawing file 20_Cylinder. Use this command and Figure 20–78 to construct a cylinder with either a circular or elliptical base.

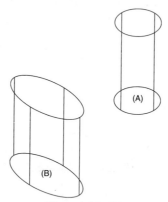

Figure 20–78

 Command: **CYLINDER**
Current wire frame density: ISOLINES=4
Specify center point for base of cylinder or [Elliptical] <0,0,0>: **7.00,7.00** *(At "A")*
Specify radius for base of cylinder or [Diameter]: **D** *(For Diameter)*
Specify diameter for base of cylinder: **2**
Specify height of cylinder or [Center of other end]: **4**

 Command: **CYLINDER**
Current wire frame density: ISOLINES=4
Specify center point for base of cylinder or [Elliptical] <0,0,0>: **E** *(For an elliptical base)*
Specify axis endpoint of ellipse for base of cylinder or [Center]: **C** *(To define the center of the ellipse)*
Specify center point of ellipse for base of cylinder <0,0,0>: **8.00,0.50** *(At "B")*
Specify axis endpoint of ellipse for base of cylinder: **@2<0**
Specify length of other axis for base of cylinder: **@1<90**
Specify height of cylinder or [Center of other end]: **4**

USING THE SPHERE COMMAND

Use the SPHERE command to construct a sphere by defining the center of the sphere along with a radius or diameter. As with the cylinder, the central axis of a sphere is along the Z axis of the current User Coordinate System. Choose this command in one of the following ways:

- From the Solids toolbar
- From the pull-down menu (Draw > Solids > Sphere)
- From the keyboard (SPHERE)

In Figure 20–79, the sphere is constructed in wireframe mode and does not look like much of a sphere. Perform a hidden line removal using the HIDE command to get a better view of the sphere.

An application of using a sphere is illustrated in Figure 20–80 where the sphere is used to construct a 45° elbow pipe fitting.

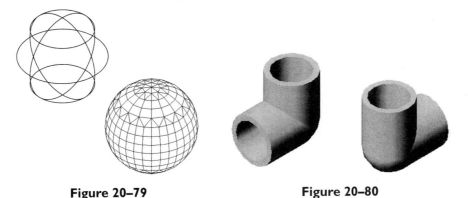

Figure 20–79   **Figure 20–80**

 Try It! – Open the drawing file 20_Sphere. Use the following command sequence for constructing a sphere.

 Command: **SPHERE**
Current wire frame density: ISOLINES=4
Specify center of sphere <0,0,0>: **7.00,7.00**
Specify radius of sphere or [Diameter]: **D** *(For Diameter)*
Specify diameter: **4**
Command: **HIDE**

1088

 THE TORUS COMMAND

A torus is formed when a circle is revolved about a line in the same plane as the circle. In other words, a torus is similar to a 3D donut. Choose this command in one of the following ways:

- From the Solids toolbar
- From the pull-down menu (Draw > Solids > Torus)
- From the keyboard (TORUS)

The torus can be constructed by either the radius or diameter method. When you use the radius method, two radius values must be used to define the torus; one for the radius of the tube and the other for the distance from the center of the torus to the center of the tube. Use two diameter values when specifying a torus by the diameter method. Once you construct the torus, it lies parallel to the current User Coordinate System.

 Try It! – Open the drawing file 20_Torus. Use the following prompts to construct a torus similar to Figure 20–81.

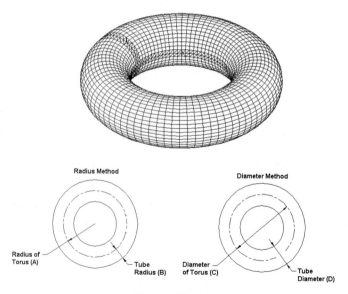

Figure 20–81

 Command: **TOR** *(For TORUS)*
Current wire frame density: ISOLINES=4
Specify center of torus <0,0,0>: **10.00,10.00**
Specify radius of torus or [Diameter]: **5** *(For the radius of the torus at "A")*
Specify radius or tube or [Diameter]: **1** *(For the radius of the tube at "B")*

Command: **TOR** *(For TORUS)*

Current wire frame density: ISOLINES=4

Specify center of torus <0,0,0>: **20.00,12.00** *(Identify the center of the torus through coordinate entry or by picking)*

Specify radius of torus or [Diameter]: **D** *(For diameter of the torus at "C")*

Specify diameter: **4**

Specify radius or tube or [Diameter]: **D** *(For diameter of the tube at "D")*

Specify diameter: **2**

Command: **HI** *(For HIDE)*

 Try It! – Open the drawing file 20_O Ring. A half section is illustrated in Figure 20–82. Using this figure and the following command prompt sequence, create two O-rings using the TORUS command.

 Command: **TOR** *(For TORUS)*

Current wire frame density: ISOLINES=4

Specify center of torus <0,0,0>: *(Press ENTER to accept this default)*

Specify radius of torus or [Diameter]: **D** *(For Diameter)*

Specify diameter: **3.00**

Specify radius of tube or [Diameter]: **D** *(For Diameter)*

Specify diameter: **0.125**

 Command: **CP** *(For COPY)*

Select objects: *(Pick the O Ring you just constructed)*

Select objects: *(Press ENTER to continue)*

Specify base point or displacement, or [Multiple]: **End**

of *(Pick the endpoint at "A")*

Specify second point of displacement or <use first point as displacement>: **End**

of *(Pick the endpoint at "B")*

Command: **HIDE** *(For hidden line removal)*

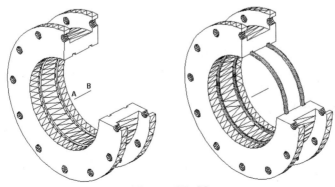

Figure 20–82

USING BOOLEAN OPERATIONS

To combine one or more primitives to form a composite solid, a Boolean operation is performed. Boolean operations must act on at least a pair of primitives, regions, or solids. These operations in the form of commands are located in the Modify menu under Solids Editing. They can also be obtained through the Solids Editing toolbar. Boolean operations allow you to add two or more objects together, subtract a single object or group of objects from another, or find the overlapping volume—in other words, to form the solid common to both primitives. Displayed in Figure 20–83 are the UNION, SUBTRACT, and INTERSECT commands that you use to perform these Boolean operations.

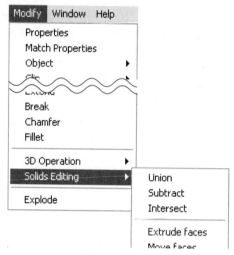

Figure 20–83

In Figure 20–84, a cylinder has been constructed along with a box. Depending on which Boolean operation you use, the results could be quite different. In Figure 20–85, both the box and cylinder are considered one solid object. This is the purpose of the UNION command: to join or unite two solid primitives into one. Figure 20–86 illustrates the intersection of the two solid primitives or the area that both solids have in common. This solid is obtained through the INTERSECT command. Figure 20–87 shows the result of removing or subtracting the cylinder from the box—a hole is formed inside the box as a result of using the SUBTRACT command. All Boolean operation commands can work on numerous solid primitives, that is, if you want to subtract numerous cylinders from a box, you can subtract all of the cylinders at one time.

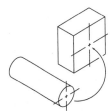

Figure 20–84

Figure 20–85

Figure 20–86

Figure 20–87

 ## THE UNION COMMAND

This Boolean operation joins two or more selected solid objects together into a single solid object. Choose this command in one of the following ways:

- From the Solids Editing toolbar
- From the pull-down menu (Modify > Solids Editing > Union)
- From the keyboard (UNI or UNION)

 Try It! – Open the drawing file 20_Union. Use the following command sequence and the image in Figure 20–88 for performing this task.

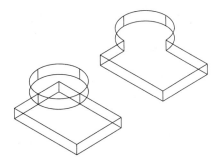

Figure 20–88

 Command: **UNI** *(For UNION)*

Select objects: *(Pick the box and cylinder)*

Select objects: *(Press ENTER to perform the union operation)*

THE SUBTRACT COMMAND

Use this command to subtract one or more solid objects from a source object. See Figure 20–89. Choose this command in one of the following ways:

- From the Solids Editing toolbar
- From the pull-down menu (Modify > Solids Editing > Subtract)
- From the keyboard (SU or SUBTRACT)

Try It! – Open the drawing file 20_Subtract. Use the following command sequence and the image in Figure 20–89 for performing this task.

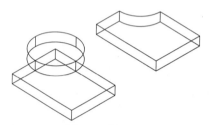

Figure 20–89

 Command: **SU** *(For SUBTRACT)*
Select solids and regions to subtract from...
Select objects: *(Pick the box)*
Select objects: *(Press ENTER to continue with this command)*
Select solids and regions to subtract...
Select objects: *(Pick the cylinder)*
Select objects: *(Press ENTER to perform the subtraction operation)*

THE INTERSECT COMMAND

Use this command to find the solid common to a group of selected solid objects. See Figure 20–90. Choose this command in one of the following ways:

- From the Solids Editing toolbar
- From the pull-down menu (Modify > Solids Editing > Intersect)
- From the keyboard (IN or INTERSECT)

Try It! – Open the drawing file 20_Intersection. Use the following command sequence and the image in Figure 20–90 for performing this task.

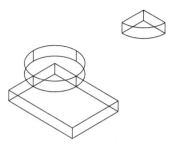

Figure 20–90

 Command: **IN** *(For INTERSECT)*

Select objects: *(Pick the box and cylinder)*
Select objects: *(Press ENTER to perform the intersection operation)*

3D APPLICATIONS OF UNIONING SOLIDS

Figure 20–91 shows an object consisting of one horizontal solid box, two vertical solid boxes, and two extruded semi-circular shapes. All primitives have been positioned with the MOVE command. To join all solid primitives into one solid object, use the UNION command. The order of selection of these solids for this command is not important.

 Try It! – Open the drawing file 20_3D App Union. Use the following prompts and Figure 20–91 for performing these tasks.

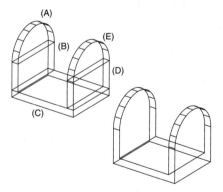

Figure 20–91

 Command: **UNI** *(For UNION)*

Select objects: *(Select the solid extrusion at "A")*
Select objects: *(Select the vertical solid box at "B")*
Select objects: *(Select the horizontal solid box at "C")*
Select objects: *(Select the vertical solid box at "D")*
Select objects: *(Select the solid extrusion at "E")*
Select objects: *(Press ENTER to perform the union operation)*

3D APPLICATIONS OF MOVING SOLIDS

Using the same problem from the previous example, let us now add a hole in the center of the base. The cylinder will be created through the CYLINDER command. It will then be moved to the exact center of the base. You can use the MOVE command along with the OSNAP-Tracking mode to accomplish this. Tracking mode will automatically activate the ORTHO mode when it is in use. See Figure 20–92.

Command: **CYLINDER**
Current wire frame density: ISOLINES=4
Specify center point for base of cylinder or [Elliptical] <0,0,0>: **3.00,3.00**
Specify radius for base of cylinder or [Diameter]: **.75**
Specify height of cylinder or [Center of other end]: **.25**

Command: **M** *(For MOVE)*
Select objects: *(Select the cylinder at "A")*
Select objects: *(Press ENTER to continue with this command)*
Specify base point or displacement: **Cen**
of *(Select the bottom of the cylinder at "A")*
Specify second point of displacement or <use first point as displacement>: **TK** *(To activate tracking mode)*
First tracking point: **Mid**
of *(Select the midpoint of the bottom of the base at "B")*
Next point (Press ENTER to end tracking): **Mid**
of *(Select the midpoint of the bottom of the base at "C")*
Next point (Press ENTER to end tracking): *(Press ENTER to end tracking and perform the move operation)*

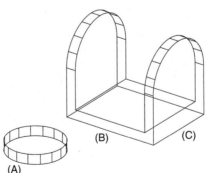

Figure 20–92

3D APPLICATIONS OF SUBTRACTING SOLIDS

Now that the solid cylinder is in position, use the SUBTRACT command to remove the cylinder from the base of the main solid and create a hole in the base. See Figure 20–93.

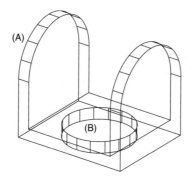

Figure 20–93

 Command: **SU** *(For SUBTRACT)*
Select solids and regions to subtract from...
Select objects: *(Select the main solid as source at "A")*
Select objects: *(Press* ENTER *to continue with this command)*
Select solids and regions to subtract...
Select objects: *(Select the cylinder at "B")*
Select objects: *(Press* ENTER *to perform the subtraction operation)*

Command: **HIDE** *(For hidden line removal view)*

Command: **RE** *(For REGEN to return to wireframe view)*

3D APPLICATIONS OF ALIGNING SOLIDS

Two more holes need to be added to the vertical sides of the solid object. A cylinder is constructed; however, it is in the vertical position. This object needs to be rotated and moved into position. The ALIGN command would be a good command to use in this situation. A series of source and destination points guides the placement of one object on another. The object is rotated and moved into position. Use the following command prompt sequence and Figure 20–94 for the ALIGN command.

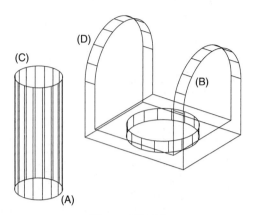

Figure 20–94

 Command: **CYLINDER**

Current wire frame density: ISOLINES=4
Specify center point for base of cylinder or [Elliptical] <0,0,0>: **3.00,3.00**
Specify radius for base of cylinder or [Diameter]: **.50**
Specify height of cylinder or [Center of other end]: **2.5**

Command: **AL** *(For ALIGN)*

Select objects: *(Select the cylinder)*
Select objects: *(Press ENTER to continue with this command)*
Specify first source point: **Cen**
of *(Select the bottom of the cylinder at "A")*
Specify first destination point: **Cen**
of *(Select the outside the main solid at "B")*
Specify second source point: **Cen**
of *(Select the top of the cylinder at "C")*
Specify second destination point: **Cen**
of *(Select the outside the main solid at "D")*
Specify third source point or <continue>: *(Press ENTER to continue)*
Scale objects based on alignment points? [Yes/No] <N>: *(Press ENTER to accept the
default value and perform the align operation)*

The cylinder is removed to create holes in each of the vertical sides of the solid object
through the SUBTRACT command. The completed object is shown in Figure 20–95.

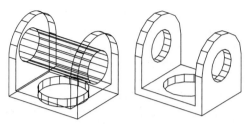

Figure 20–95

 Command: **SU** (For SUBTRACT)

Select solids and regions to subtract from...
Select objects: (Select the main solid as source)
Select objects: (Press ENTER to continue with this command)
Select solids and regions to subtract...
Select objects: (Select the cylinder)
Select objects: (Press ENTER to perform the subtraction operation)

Command: **HI** (For HIDE)

A BETTER UNDERSTANDING OF INTERSECTIONS

Try It! – Open the drawing file 20_Int1. One of the most misunderstood yet powerful Boolean operations is the Intersection. In Figure 20–96, the object at "C" represents the finished model. Look at the sequence beginning at "A" to see how to prepare the solid primitives for an Intersection operation. Two separate 3D Solid objects are created at "A." One object represents a block that has been filleted along with the placement of a hole drilled through. The other object represents the U-shaped extrusion. With both objects modeled, they are moved on top of each other at "B." The OSNAP-Midpoint was used to accomplish this. Finally the INTERSECT command is used to find the common volume shared by both objects; namely the illustration at "C."

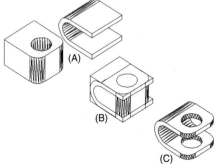

Figure 20–96

Try It! – Open the drawing file 20_Int2. The object in Figure 20–97 is another example of how the INTERSECT command may be applied to a solid model. For the results at "B" to be obtained of a cylinder that has numerous cuts, the cylinder is first created as a separate model. Then the cuts are made in another model at "A". Again, both models are moved together (use the Quadrant and Midpoint OSNAP modes) and then the INTERSECT command is used to achieve the results at "B." Before undertaking any solid model, first analyze how the model is to be constructed. Using intersections can create dramatic results, which would normally require numerous union and subtraction operations.

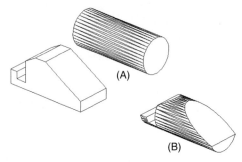

Figure 20–97

 Try It! – Open the drawing file 20_Int3. In Figure 20–98, use the MOVE command to move object "A" from the midpoint at "A" to the quadrant at "B." Run the INTERSECT command on both objects and observe the results in Figure 20–98.

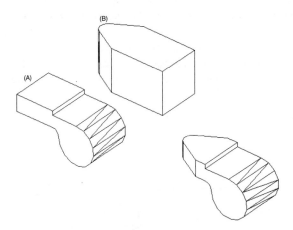

Figure 20–98

Try It! – Open the drawing file 20_Int4. In Figure 20–99, use the MOVE command to move object "A" from the midpoint at "A" to the midpoint at "B." Run the INTERSECT command on both objects and observe the results in Figure 20–99.

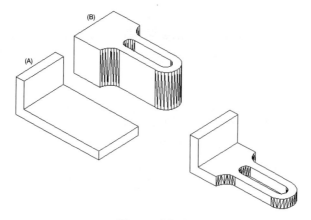

Figure 20–99

 Try It! – Open the drawing file 20_Int5. In Figure 20–100, use the MOVE command to move object "A" from the midpoint at "A" to the midpoint at "B." Run the INTERSECT command on both objects and observe the results in Figure 20–100.

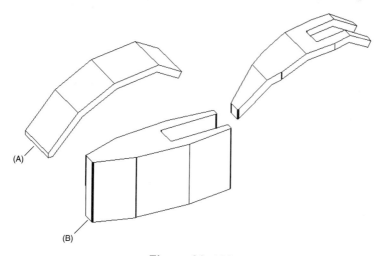

Figure 20–100

CREATING SOLID EXTRUSIONS

The EXTRUDE command creates a solid by extrusion. Choose this command in one of the following ways:

- From the Solids toolbar
- From the pull-down menu (Draw > Solids > Extrude)
- From the keyboard (EXT or EXTRUDE)

Only regions or closed, single entity objects such as circles and closed polylines can be extruded. Use the following prompts to construct a solid extrusion of the closed polyline object in Figure 20–101. For the height of the extrusion, you can enter a numeric value or you can specify the distance by picking two points on the display screen.

 Try It! – Open the drawing file 20_Extrude1. Use the following command sequence and Figure 20–101 for performing this task.

 Command: **EXT** *(For EXTRUDE)*
Current wire frame density: ISOLINES=4
Select objects: *(Select the polyline object at "A" in Figure 20–101)*
Select objects: *(Press ENTER to continue with this command)*
Specify height of extrusion or [Path]: **1.00**
Specify angle of taper for extrusion <0>: *(Press ENTER to accept the default and perform the extrusion operation)*

You can create an optional taper along with the extrusion by entering an angle value at the prompt "Specify angle of taper for extrusion."

 Try It! – Open the drawing file 20_Extrude2. Use the following command sequence and Figure 20–102 for performing this task.

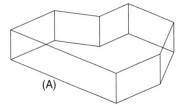

(A)

Figure 20–101

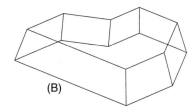

(B)

Figure 20–102

Command: **EXT** *(For EXTRUDE)*
Current wire frame density: ISOLINES=4
Select objects: *(Select the polyline object at "B" in Figure 20–102)*
Select objects: *(Press ENTER to continue with this command)*
Specify height of extrusion or [Path]: **1.00**
Specify angle of taper for extrusion <0>: **15**

You can also create a solid extrusion by selecting a path to be followed by the object being extruded. Typical paths include regular and elliptical arcs, 2D and 3D polylines, or splines. The extruded pipe in Figure 20–103 was created using the following steps: First the polyline path was created. Then, a new User Coordinate System was established through the UCS command along with the Z axis option; the new UCS was positioned at the end of the polyline with the Z axis extending along the polyline. A circle was constructed with its center point at the end of the polyline. Finally, the circle was extruded along the polyline path.

Try It! – Open the drawing file 20_Extrude Pipe. Use the following command sequence and Figure 20–103 for performing this task.

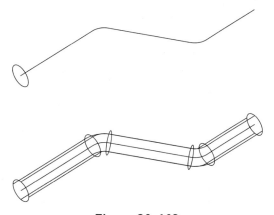

Figure 20–103

 Command: **EXT** *(For EXTRUDE)*
Current wire frame density: ISOLINES=4
Select objects: *(Select the small circle as the object to extrude)*
Select objects: *(Press ENTER to continue)*
Specify height of extrusion or [Path]: **P** *(For Path)*
Select extrusion path: *(Select the polyline object representing the path)*

CREATING REVOLVED SOLIDS

The REVOLVE command creates a solid by revolving an object about an axis of revolution. Choose this command in one of the following ways:

- From the Solids toolbar
- From the pull-down menu (Draw > Solids > Revolve)
- From the keyboard (REV or REVOLVE)

Only regions or closed, single entity objects such as polylines, polygons, circles, ellipses, and 3D polylines can be revolved. If a group of objects are not in the form of a single entity, group them together using the PEDIT command or create a closed polyline with the BOUNDARY command. The resulting image in Figure 20–104 represents a revolved 3D Solid object.

 Try It! – Open the drawing file 20_Revolve1. Use the following command sequence and Figure 20–104 for performing this task.

 Command: **REV** *(For REVOLVE)*
Current wire frame density: ISOLINES=4
Select objects: *(Select profile "A" as the object to revolve)*
Select objects: *(Press ENTER to continue with this command)*
Specify start point for axis of revolution or
Define axis by [Object/X (axis)/Y (axis)]: **O** *(For Object)*
Select an object: *(Select line "B")*
Specify angle of revolution <360>: *(Press ENTER to accept the default and perform the revolving operation)*

 Try It! – Open the drawing file 20_Revolve2. A practical application of this type of solid would be to first construct an additional solid consisting of a cylinder using the CYLINDER command. Be sure this solid is larger in diameter than the revolved solid. Existing OSNAP options are fully supported in solid modeling. Use the Center option of OSNAP along with the MOVE command to position the revolved solid inside the cylinder. See Figure 20–105.

 Command: **M** *(For MOVE)*
Select objects: *(Select the revolved solid in Figure 20–105)*
Select objects: *(Press ENTER to continue with this command)*
Specify base point or displacement: **Cen**
of *(Select the center of the revolved solid at "A")*
Specify second point of displacement or <use first point as displacement>: **Cen**
of *(Select the center of the cylinder at "B")*

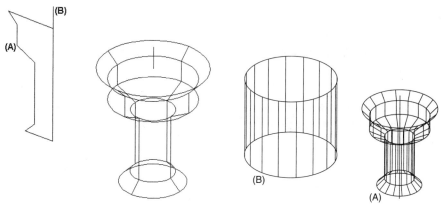

Figure 20–104   **Figure 20–105**

Once the revolved solid is positioned inside the cylinder, use the SUBTRACT command to subtract the revolved solid from the cylinder (see Figure 20–106). Use the HIDE command to perform a hidden line removal at "B" to check that the solid is correct (this would be difficult to interpret in wireframe mode).

⌾ Command: **SU** *(For SUBTRACT)*
Select solids and regions to subtract from...
Select objects: *(Select the cylinder as source)*
Select objects: *(Press* ENTER *to continue with this command)*
Select solids and regions to subtract...
Select objects: *(Select the revolved solid)*
Select objects: *(Press* ENTER *to perform the subtraction operation)*
Command: **HI** *(For HIDE)*

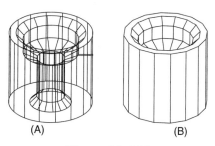

Figure 20–106

SYSTEM VARIABLES FOR SOLID MODELS

Three system variables are available that control the appearance of your solid model whenever performing shading or hidden line removal operations. They are ISOLINES, FACETRES, and DISPSILH. The following text describes each system variable in detail.

THE ISOLINES SYSTEM VARIABLE

Tessellation refers to the lines that are displayed on any curved surface to help you in visualizing the surface shown in Figure 20–107. Tessellation lines are automatically formed when you construct solid primitives such as cylinders and cones. These lines are also calculated when you perform solid modeling operations such as SUBTRACT and UNION.

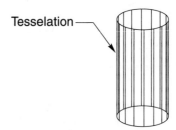

Figure 20–107

The number of tessellation lines per curved object is controlled by the system variable called ISOLINES. By default, this variable is set to a value of 4. Figure 20–108 shows the results of setting this variable to other values such as 9 and 20. After the isolines have been changed, regenerate the screen to view the results. The more lines used to describe a curved surface, the more accurate the surface will look in wireframe mode; however, it will take longer to process screen regenerations.

 Try It! – Open the drawing file 20_Tee Isolines. Experiment by changing the ISOLINES system variable to numerous values. Perform a drawing regeneration using the REGEN command each time that you change the number of isolines.

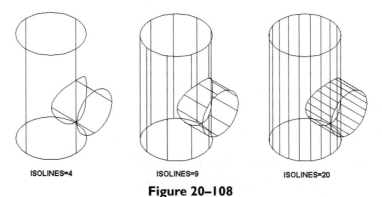

ISOLINES=4   ISOLINES=9   ISOLINES=20

Figure 20–108

THE FACETRES SYSTEM VARIABLE

As shown in Figure 20–108, tessellation lines on cylinders affect the display of solid models in wireframe mode. When you perform hidden line removals on these objects, the results are displayed in Figure 20–109. The curved surfaces are now displayed as flat triangular surfaces (faces) and the display of these faces is controlled by the FACETRES system variable. The cylinder with FACETRES set to 0.50 processes much more quickly than the cylinder with FACETRES set to 2, because there are fewer surfaces to process in such operations as hidden line removals. However, the image with FACETRES set to 10 shows a more defined circle. The default value for FACETRES is 0.50, which seems adequate for most applications.

 Try It! – Open the drawing file 20_Tee Facetres. Experiment by changing the FACETRES system variable to numerous values. Perform a HIDE each time you change the number of facet lines to view the results.

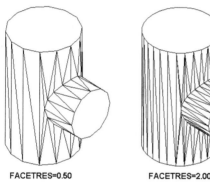 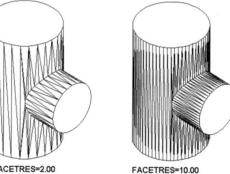

FACETRES=0.50   FACETRES=2.00   FACETRES=10.00

Figure 20–109

THE DISPSILH SYSTEM VARIABLE

To have the edges of your solid model take on the appearance of an isometric drawing when displayed as a wireframe, use the DISPSILH system variable. This system variable means "display silhouette" and is used to control the display of silhouette curves of solid objects while in either the wireframe or hide modes. Silhouette edges are either turned on or off with the results displayed in Figure 20–110; by default they are turned off. When a HIDE is performed, this system variable controls whether faces are drawn or suppressed on a solid model.

 Try It! – Open the drawing file 20_Tee Dispsilh. Experiment by turning the display of silhouette edges on and off. Regenerate your display each time you change this mode to view the results. Also use the HIDE command to see how the hidden line removal image is changed.

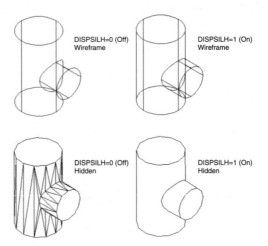

Figure 20–110

FILLETING SOLID MODELS

Filleting of simple or complex objects is easily handled with the FILLET command. This is the same command as the one used to create a 2D fillet.

 Try It! – Open the drawing file 20_Tee Fillet. Use the following prompt sequence and Figure 20–111 for performing this task.

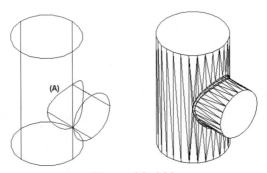

Figure 20–111

 Command: **F** *(For FILLET)*
Current settings: Mode = TRIM, Radius = 0.5000
Select first object or [Polyline/Radius/Trim/mUltiple]: *(Select the edge at "A," which represents the intersection of both cylinders)*
Enter fillet radius <0.5000>: **0.25**
Select an edge or [Chain/Radius]: *(Press ENTER to perform the fillet operation)*
I edge(s) selected for fillet.

Try It! – Open the drawing file 20_Slab Fillet. A group of objects with a series of edges can be filleted with the Chain option of the FILLET command. Use the following prompt sequence and Figure 20–112 for performing this task.

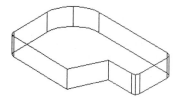 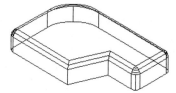

Figure 20–112

 Command: **F** *(For FILLET)*
Current settings: Mode = TRIM, Radius = 0.5000
Select first object or [Polyline/Radius/Trim/mUltiple]: *(Select the edge at "A")*
Enter fillet radius <0.5000>: *(Press ENTER to accept the default value)*
Select an edge or [Chain/Radius]: **C** *(For chain mode)*
Select an edge chain or [Edge/Radius]: *(Select edge "A" again; notice how the selection is chained until it reaches an abrupt corner)*
Select an edge chain or [Edge/Radius]: *(Select the edge at "B")*
Select an edge chain or [Edge/Radius]: *(Press ENTER when finished selecting all edges to perform the fillet operation)*
10 edge(s) selected for fillet.

CHAMFERING SOLID MODELS

Just as the FILLET command uses the Chain mode to group a series of edges together, the CHAMFER command uses the Loop option to perform the same type of operation.

 Try It! – Open the drawing file 20_Slab Chamfer. Use the following prompt sequence and Figure 20–113 for performing this task.

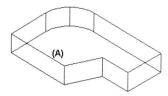 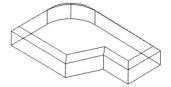

Figure 20–113

 Command: **CHA** *(For CHAMFER)*
(TRIM mode) Current chamfer Dist1 = 0.5000, Dist2 = 0.5000
Select first line or [Polyline/Distance/Angle/Trim/Method/mUltiple]: *(Pick the edge at "A")*
Base surface selection...

Enter surface selection option [Next/OK (current)] <OK>: **N** *(This option selects the other surface shared by edge "A"; the top surface should highlight. If not, use this option again)*
Enter surface selection option [Next/OK (current)] <OK>: *(Press ENTER to accept the top surface)*
Specify base surface chamfer distance <0.5000>: *(Press ENTER to accept the default)*
Specify other surface chamfer distance <0.5000>: *(Press ENTER to accept the default)*
Select an edge or [Loop]: **L** *(To loop all edges together into one)*
Select an edge loop or [Edge]: *(Pick any top edge, notice the loop option does not stop at abrupt corners)*
Select an edge loop or [Edge]: *(Press ENTER to perform the chamfering operation)*

MODELING TECHNIQUES USING VIEWPORTS

Using viewports can be very important and beneficial when laying out a solid model. Use the VPORTS command to lay out the display screen in a number of individually different display areas. Select this command by choosing Viewports from the View menu shown in Figure 20–114. A cascading menu appears with the different types of viewports available. Choosing New Viewports displays the Viewports dialog box shown in Figure 20–115. A number of viewport configurations can be selected that will convert the display screen to the viewport configuration desired. Selecting 3D in the Setup area automatically sets up 3D view points for the viewports.

Figure 20–114

Figure 20–116 shows the results of the dialog box settings and the way viewports can be used in the construction of a 3D solid model. The image in the right viewport represents the solid model viewed from a southeast isometric viewpoint displaying all three dimensions. The image on the left shows the model as seen from the top. The images are viewed based on the World Coordinate System. This gives you two viewports to help visualize the construction of the solid model.

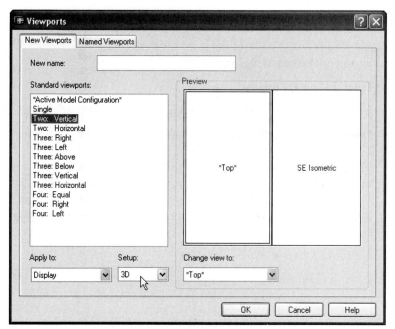

Figure 20–115

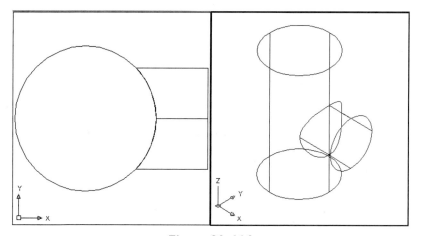

Figure 20–116

Try It! – Open the drawing file 20_Tee Vports. Use the previous images and text to set up a series of 3D viewports of the intersecting tee pipe object.

Tip: The VPORTS command can be used to create either tiled viewports in model space or floating viewports while in layout mode.

OBTAINING MASS PROPERTIES

The MASSPROP command calculates the mass properties of a solid model. Choose this command in one of the following ways:

- From the Inquiry toolbar
- From the pull-down menu (Tools > Inquiry > Region/Mass Properties)
- From the keyboard (MASSPROP)

Try It! – Open the drawing file 20_Tee Massprop. Use the MASSPROP command to calculate the mass properties of a selected solid (see Figure 20–117). All calculations in Figure 20–118 are based on the current position of the User Coordinate System. You will be given the option of writing this information to a file if desired.

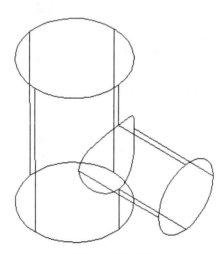

Figure 20–117

 Command: **MASSPROP**

Select objects: *(Select the model)*

```
---------------  SOLIDS  ---------------
Mass:                50.1407
Volume:              50.1407
Bounding box:        X: 5.0974 -- 10.6984
                     Y: 1.3497 --  4.5516
                     Z: 0.0000 --  5.0000
Centroid:            X: 7.2394
                     Y: 2.9507
                     Z: 2.5000
Moments of inertia:  X: 865.8985
                     Y: 3118.9425
                     Z: 3184.2145
Products of inertia: XY: 1071.0602
                     YZ: 369.8728
                     ZX: 907.4699
Radii of gyration:   X: 4.1556
                     Y: 7.8869
                     Z: 7.9690
Principal moments and X-Y-Z directions about centroid:
                     I: 115.9688 along [1.0000 0.0000 0.0000]
                     J: 177.7542 along [0.0000 1.0000 0.0000]
                     K: 119.8558 along [0.0000 0.0000 1.0000]
```

Figure 20–118

THE INTERFERE COMMAND

The INTERFERE command will identify any interference and highlight the solid models that overlap. If requested, it will create a composite solid based on the common volume between the interfering solids. Choose this command in one of the following ways:

- From the Solids toolbar

- From the pull-down menu (Draw > Solids > Interference)

- From the keyboard (INTERFERE)

Try It! – Open the drawing file 20_Pipe Interference. Use this command to find any interference shared by a series of solids. When an interference is identified, a solid can be created out of the common volume, as shown in Figure 20–119.

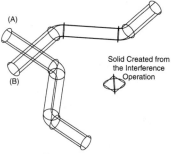

Figure 20–119

 Command: **INTERFERE**
Select first set of solids:
Select objects: *(Select solid "A")*
Select objects: *(Press* ENTER *to continue)*
Select second set of solids:
Select objects: *(Select solid "B")*
Select objects: *(Press* ENTER *to continue)*
Comparing 1 solid against 1 solid.
Interfering solids (first set): 1
                         (second set): 1
Interfering pairs: 1
Create interference solids? [Yes/No] <N>: **Y**

 ## CUTTING SECTIONS FROM SOLID MODELS

THE SECTION COMMAND

The SECTION command creates a region based on the intersection of a plane and the solid model. Choose this command in one of the following ways:

- From the Solids toolbar
- From the pull-down menu (Draw > Solids > Section)
- From the keyboard (SEC or SECTION)

 Try It! – Open the drawing file 20_Tee Section. Use this command to create a cross section of a selected 3D solid object similar to Figure 20–120. The section that is created is actually a Region (2D solid). Note that section lines (cross hatch lines) are not created. The command provides numerous options to allow control over where a section will be cut; in Figure 20–120, the location of the User Coordinate System defines the cutting plane. Once the section is created on the object, you can move it to a more convenient location and modify it (provide missing object and section lines) as necessary.

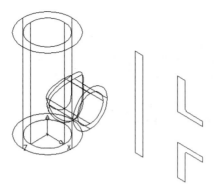

Figure 20–120

 Command: **SEC** *(For SECTION)*
Select objects: *(Select the object in Figure 20–120)*
Select objects: *(Press* ENTER *to continue with this command)*
Specify first point on Section plane by [Object/Zaxis/View/XY/YZ/ZX/3points]
    <3points>: **XY**
Specify a point on the XY-plane <0,0,0>: *(Press* ENTER *to accept the default and perform
    the sectioning operation)*

 Command: **M** *(For MOVE)*
Select objects: **L** *(Selects the last object created—the region)*
Select objects: *(Press* ENTER *to continue with this command)*
Specify base point or displacement: *(Pick a point on the screen)*
Specify second point of displacement or <use first point as displacement>: **@12<0**

 SLICING SOLID MODELS

THE SLICE COMMAND

Choose the SLICE command in one of the following ways:

- From the Solids toolbar
- From the pull-down menu (Draw > Solids > Slice)
- From the keyboard (SL or SLICE)

 Try It! – Open the drawing file 20_Tee Slice. This command is similar to the
SECTION command except that the solid model is actually cut or sliced at a
location that you define. In the example in Figure 20–121, this location is
defined by the User Coordinate System. Before the slice is made, you also have
the option of keeping either one or both halves of the object. The MOVE command is used to separate both halves in Figure 20–122.

Command: **SL** *(For SLICE)*

Select objects: *(Select the solid object in Figure 20–121)*

Select objects: *(Press ENTER to continue with this command)*

Specify first point on slicing plane by [Object/Zaxis/View/XY/YZ/ZX/3points] <3points>: **XY**

Specify a point on XY-plane <0,0,0>: *(Press ENTER to accept this default value)*

Specify a point on desired side of the plane or [keep Both sides]: **B** *(To keep both sides)*

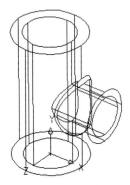

Figure 20–121

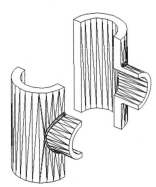

Figure 20–122

SHADING SOLID MODELS

Various shading modes are available to help you better visualize the solid model you are constructing. Access the seven shading modes by choosing Shade from the View pull-down menu in Figure 20–123. There is also a Shade toolbar dedicated to all shading modes, shown in Figure 20–124. Each one of these modes is explained in the next series of paragraphs.

Figure 20–123

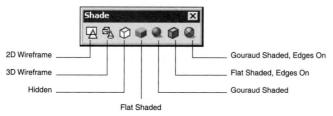

Figure 20–124

2D WIREFRAME

This shading mode, illustrated in Figure 20–125, displays a wireframe image of a solid model. Note that the User Coordinate System icon is displayed in the normal 3D or 2D style as determined by the property settings of the UCSICON command

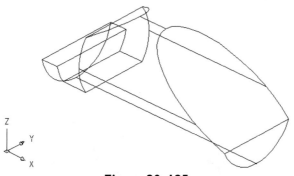

Figure 20–125

3D WIREFRAME

The 3D Wireframe mode is identical to the 2D mode regarding the wireframe mode. Notice however, that the User Coordinate System icon in Figure 20–126 has switched to a better-defined, color-coded icon that illustrates the directions of the X, Y, and Z axes.

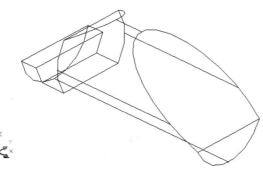

Figure 20–126

🔲 HIDDEN

The Hidden mode performs a hidden line removal, which is illustrated by the model in Figure 20–127. Only those surfaces in front of your viewing plane are visible.

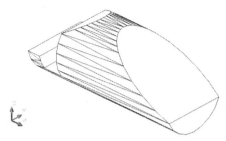

Figure 20–127

🔲 FLAT SHADE

You can achieve more dramatic viewing results by performing a Flat Shade of a solid model in Figure 20–128. Here the model is shaded based on the current color and screen background.

Figure 20–128

🔲 GOURAUD SHADED

This shade mode provides the highest quality because it not only applies color to the model but also smoothes all edges as well (see Figure 20–129).

Figure 20–129

FLAT SHADED, EDGES ON

At times, you need to stay in flat shaded mode yet still select edges of the model for performing other operations. To better select edges of a shaded model, turn the edges on using this mode. The results are illustrated in Figure 20–130, with the wireframe lines showing through the shaded model.

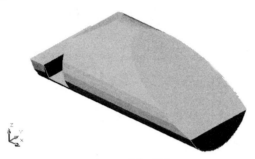

Figure 20–130

GOURAUD SHADED, EDGES ON

As with the flat shaded mode, you can have the wireframe edges show through a Gouraud Shaded model, as in Figure 20–131.

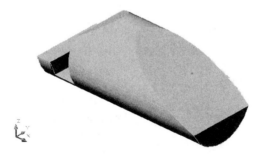

Figure 20–131

 Try It! – Open the drawing file 20_Shading Modes. Experiment with the series of shading modes. Your results should be similar to the images provided in the illustrations.

In addition, when you perform such commands as 3DORBIT, the current shade mode remains persistent. This means that if you are in Gouraud Shaded mode and you rotate your model using 3DORBIT, the model remains shaded through the rotation operation.

TUTORIAL EXERCISE: GUIDE.DWG

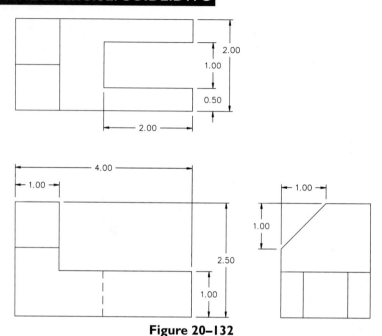

Figure 20–132

Purpose

This tutorial exercise is designed to produce a 3D solid model of the Guide from the information supplied in the orthographic drawing in Figure 20–132.

System Settings

Use the Drawing Units dialog box, keep the system of measurement set to decimal but change the number of decimal places past the zero from a value of 4 to 2. Leave the current limits of (0,0) by (12,9) as the default setting.

Layers

Special layers do not have to be created for this tutorial exercise.

Suggested Commands

Begin this drawing by constructing solid primitives of all components of the Guide using the BOX and WEDGE commands. Move the components into position and begin merging solids using the UNION command. To form the rectangular slot, move a solid box into position and use the SUBTRACT command to subtract the rectangle from the solid, thus forming the slot. Do the same procedure for the wedge. Perform a hidden line removal and view the solid.

Whenever possible, substitute the appropriate command alias in place of the full AutoCAD command in each tutorial step. For example, use "CP" for the COPY command, "L" for the LINE command, and so on. The complete listing of all command aliases is located in Table 1–2.

STEP 1

Begin this tutorial by constructing a solid box 4 units long by 2 units wide and 1 unit in height using the BOX command. See Figure 20–133. Begin this box at absolute coordinate 4.00,5.50. This slab will represent the base of the guide.

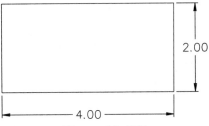

Figure 20–133

 Command: **BOX**

Specify corner of box or [CEenter]
    <0,0,0>: **4.00,5.50**
Specify corner or [Cube/Length]: **L** *(For Length)*
Specify length: **4.00**
Specify width: **2.00**
Specify height: **1.00**

STEP 2

Construct a solid box 1 unit long by 2 units wide and 1.5 units in height using the BOX command (see Figure 20–134). Begin this box at absolute coordinate 2.00,1.50. This slab will represent the vertical column of the guide.

 Command: **BOX**

Specify corner of box or [CEenter]
    <0,0,0>: **2.00,1.50**
Specify corner or [Cube/Length]: **L** *(For Length)*
Specify length: **1.00**
Specify width: **2.00**
Specify height: **1.50**

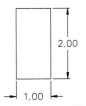

Figure 20–134

STEP 3

Construct a solid box 2 units long by 1 unit wide and 1 unit in height using the BOX command (see Figure 20–135). Begin this box at absolute coordinate 5.50,1.50. This slab will represent the rectangular slot made into the slab that will be subtracted at a later time.

 Command: **BOX**
Specify corner of box or [CEenter]
    <0,0,0>: **5.50,1.50**
Specify corner or [Cube/Length]: **L** *(For Length)*
Specify length: **2.00**
Specify width: **1.00**
Specify height: **1.00**

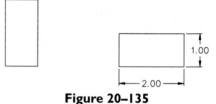

Figure 20–135

STEP 4

Use the WEDGE command to draw a wedge 1 unit in length, 1 unit wide, and 1 unit in height (see Figure 20–136). Begin this primitive at absolute coordinate 9.50,1.50. This wedge will be subtracted from the vertical column to form the inclined surface.

 Command: **WEDGE**
Specify first corner of wedge or
    [CEenter] <0,0,0>: **9.50,1.50**
Specify corner or [Cube/Length]: **L** *(For Length)*
Specify length: **1.00**
Specify width: **1.00**
Specify height: **1.00**

Figure 20–136

STEP 5

Use the VPOINT command to view the four solid primitives in three dimensions (see Figure 20–137). Use a new view-point of 1,-1,1. You can use a preset viewpoint by choosing 3D Views from the View menu and then SE Isometric. Next use the MOVE command to move the vertical column at "A" to the top of the base at "B."

Command: **VPOINT**

Current view direction:
    VIEWDIR=0.00,0.00,1.00
Specify a view point or [Rotate] <display
    compass and tripod>: **1,-1,1**

Command: **M** *(For MOVE)*

Select objects: *(Select the solid box at "A")*
Select objects: *(Press ENTER to continue
    with this command)*

Specify base point or displacement: **End**
of *(Select the endpoint of the solid at "A")*
Specify second point of displacement or
    <use first point as displacement>: **End**
of *(Select the endpoint of the base at "B")*

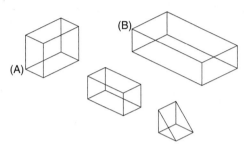

Figure 20–137

STEP 6

Use the UNION command to join the base and vertical column into one object. See Figure 20–138.

 Command: **UNI** *(For UNION)*

Select objects: *(Select the base at "A" and
    column at "B")*
Select objects: *(Press ENTER to perform the
    union)*

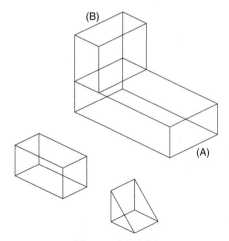

Figure 20–138

STEP 7

Use the move command to position the rectangle from its midpoint at "A" to the midpoint of the base at "B" in Figure 20–139. In the next step, the small rectangle will be subtracted to form the rectangular slot in the base.

 Command: **M** *(For MOVE)*
Select objects: *(Select box "A")*
Select objects: *(Press* ENTER *to continue with this command)*
Specify base point or displacement: **Mid**
of *(Select the midpoint of the rectangle at "A")*
Specify second point of displacement or <use first point as displacement>: **Mid**
of *(Select the midpoint of the base at "B")*

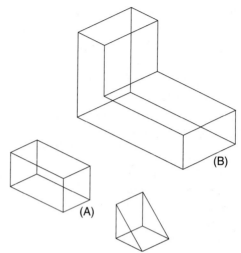

Figure 20–139

STEP 8

Use the SUBTRACT command to subtract the small box from the base of the solid object. See Figure 20–140.

⊚ Command: **SU** *(For SUBTRACT)*
Select solids and regions to subtract from...
Select objects: *(Select solid object "A")*
Select objects: *(Press* ENTER *to continue with this command)*
Select solids and regions to subtract...
Select objects: *(Select box "B" to subtract)*
Select objects: *(Press* ENTER *to perform the subtraction operation)*

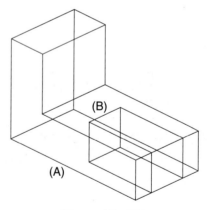

Figure 20–140

STEP 9

Use the ALIGN command to match points along a source object (wedge) with points along a destination object (guide). The selection of three sets of points will guide the placement and rotation of the wedge into the guide. See Figure 20–141.

Command: **AL** *(For ALIGN)*
Select objects: *(Select the wedge)*
Select objects: *(Press ENTER to continue with this command)*
Specify first source point: **End**
of *(Pick the endpoint of the wedge at "A")*
Specify first destination point: **End**
of *(Pick the endpoint of the guide at "B")*
Specify second source point: **End**
of *(Pick the endpoint of the wedge at "C")*
Specify second destination point: **End**
of *(Pick the endpoint of the guide at "D")*
Specify third source point or <continue>: **End**
of *(Pick the endpoint of the wedge at "E")*
Specify third destination point: **End**
of *(Pick the endpoint of the guide at "F")*

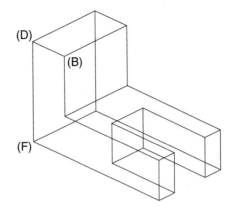

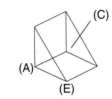

Figure 20–141

STEP 10

Use the SUBTRACT command to subtract the wedge from the guide, as shown in Figure 20–142.

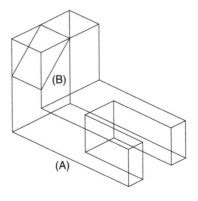

Command: **SU** *(For SUBTRACT)*
Select solids and regions to subtract from...
Select objects: *(Select guide "A")*
Select objects: *(Press ENTER to continue with this command)*
Select solids and regions to subtract...
Select objects: *(Select the wedge at "B")*
Select objects: *(Press ENTER to perform the subtraction operation)*

Figure 20–142

STEP 11

An alternate method of creating the inclined surface in the guide is to use the CHAMFER command on the vertical column of the guide. See Figure 20–143.

Command: **CHA** *(For CHAMFER)*
(TRIM mode) Current chamfer Dist1 = 0.00, Dist2 = 0.00
Select first line or [Polyline/Distance/ Angle/Trim/Method/mUltiple]: *(Select the line at "A")*
Base surface selection...
Enter surface selection option [Next/ OK (current)] <OK>: *(Press ENTER to accept the base surface)*
Specify base surface chamfer distance: I
Specify other surface chamfer distance <1.00>: I
Select an edge or [Loop]: *(Select the edge at "A")*
Select an edge or [Loop]: *(Press ENTER to perform the chamfer operation)*

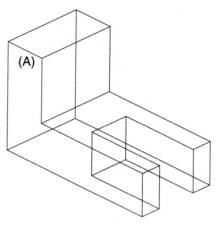

Figure 20–143

STEP 12

Use the HIDE command to perform a hidden line removal on all surfaces of the object. The results are shown in Figure 20–144.

Command: **HI** *(For HIDE)*
Regenerating model.

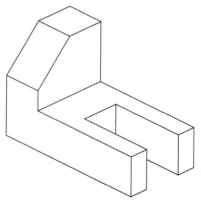

Figure 20–144

TUTORIAL EXERCISE: COLLAR.DWG

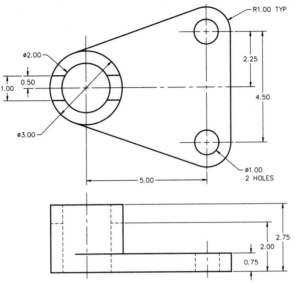

Figure 20–145

Purpose

This tutorial is designed to construct a solid model of the Collar using the dimensions in Figure 20–145.

System Settings

Use the current limits set to 0,0 for the lower left corner and (12,9) for the upper right corner. Change the number of decimal places from four to two using the Drawing Units dialog box. Snap and Grid values may remain as set by the default.

Layers

Create the following layer:

Name	Color	Linetype
Model	Cyan	Continuous

Suggested Commands

Begin this tutorial by laying out the Collar in plan view and drawing the basic shape out-lined in the Top view. Convert the objects to a polyline and extrude the objects to form a solid. Draw a cylinder and combine this object with the base. Add another cylinder and then subtract it to form the large hole through the model. Add two small cylinders and subtract them from the base to form the smaller holes. Construct a solid box, use the MOVE command to move the box into position, and subtract it to form the slot across the large cylinder.

Whenever possible, substitute the appropriate command alias in place of the full AutoCAD command in each tutorial step. For example, use "CP" for the COPY command, "L" for the LINE command, and so on. The complete listing of all command aliases is located in Table 1–2.

STEP 1

Begin the Collar by setting the Model layer current. Then draw the three circles shown in Figure 20–146 using the CIRCLE command. Place the center of the circle at "A" at 0,0. Perform a ZOOM-All after all three circles have been constructed. The center marks identifying the centers of the circles are used for illustrative purposes and do not need to be placed in the drawing.

Command: **C** *(For CIRCLE)*
Specify center point for circle or [3P/2P/Ttr (tan tan radius)]: **0,0**
Specify radius of circle or [Diameter]: **D** *(For Diameter)*
Specify diameter of circle: **3.00**

Command: **C** *(For CIRCLE)*
Specify center point for circle or [3P/2P/Ttr (tan tan radius)]: **5.00,2.25**
Specify radius of circle or [Diameter] <1.50>: **1.00**

Command: **C** *(For CIRCLE)*
Specify center point for circle or [3P/2P/Ttr (tan tan radius)]: **5.00,-2.25**
Specify radius of circle or [Diameter] <1.00>: *(Press ENTER to accept default)*

Command: **Z** *(For ZOOM)*
Specify corner of window, enter a scale factor (nX or nXP), or [All/Center/Dynamic/Extents/Previous/Scale/Window] <real time>: **All**

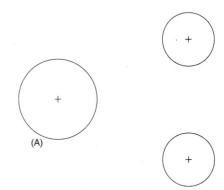

Figure 20–146

STEP 2

Draw lines tangent to the three arcs using the LINE command and the OSNAP-Tangent mode. Also see Figure 20–147.

Command: **L** *(For LINE)*
Specify first point: **Tan**
to *(Select the circle at "A")*
Specify next point or [Undo]: **Tan**
to *(Select the circle at "B")*
Specify next point or [Undo]: *(Press the ENTER key to exit the command)*

Repeat the above procedure to draw lines from "C" to "D" and "E" to "F."

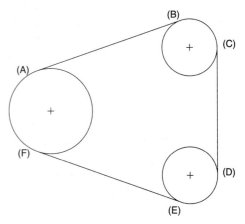

Figure 20–147

STEP 3

Use the TRIM command to trim the circles. When prompted to select the cutting edge object, press ENTER; this will make cutting edges out of all objects in the drawing. See Figure 20–148.

 Command: **TR** *(For TRIM)*

Current settings: Projection=UCS
    Edge=None
Select cutting edges ...
Select objects: *(Press ENTER to create cutting edges out of all objects)*
Select object to trim or shift-select to extend or [Project/Edge/Undo]: *(Select the circle at "A")*
Select object to trim or shift-select to extend or [Project/Edge/Undo]: *(Select the circle at "B")*
Select object to trim or shift-select to extend or [Project/Edge/Undo]: *(Select the circle at "C")*
Select object to trim or shift-select to extend or [Project/Edge/Undo]: *(Press ENTER to exit this command)*

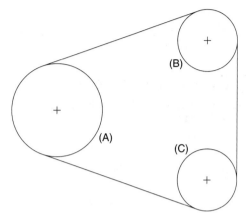

Figure 20–148

STEP 4

Prepare to construct the bracket by viewing the object in 3D using the VPOINT command and the coordinates 1,-1,1. You may also select a viewpoint by choosing 3D Views from the View menu and then SE Isometric. See Figure 20–149.

Command: **VPOINT**
Current view direction:
    VIEWDIR=0.00,0.00,1.00
Specify a view point or [Rotate] <display compass and tripod>: **1,-1,1**

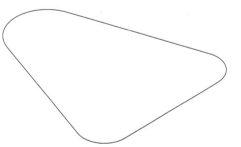

Figure 20–149

STEP 5

Convert all objects to a polyline using the Join option of the PEDIT command. See Figure 20–150.

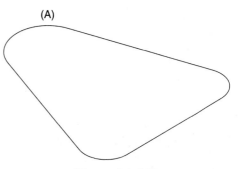

(A)

 Command: **PE** *(For PEDIT)*

Select polyline or [Multiple]: *(Select the arc at "A")*

Object selected is not a polyline

Do you want to turn it into one? <Y>
(Press ENTER to accept the default value)

Enter an option [Close/Join/Width/Edit vertex/Fit/Spline/Decurve/Ltype gen/Undo]: **J** *(For Join)*

Select objects: *(Select the three lines and remaining two arcs shown in Figure 20–150)*

Select objects: *(Press ENTER to perform the joining operation)*

5 segments added to polyline

Enter an option [Open/Join/Width/Edit vertex/Fit/Spline/Decurve/Ltype gen/Undo]: *(Press ENTER to exit this command)*

Figure 20–150

STEP 6

Use the EXTRUDE command to extrude the base to a thickness of 0.75 units. See Figure 20–151.

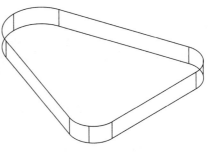

 Command: **EXT** *(For EXTRUDE)*

Current wire frame density: ISOLINES=4

Select objects: *(Select the polyline)*

Select objects: *(Press ENTER to continue with this command)*

Specify height of extrusion or [Path]: **0.75**

Specify angle of taper for extrusion <0>: *(Press ENTER to perform the extrusion operation)*

Figure 20–151

STEP 7

Create a cylinder using the CYLINDER command. Begin the center point of the cylinder at 0,0,0 with a diameter of 3.00 units and a height of 2.75 units. You may have to perform a ZOOM-All to display the entire model. See Figure 20–152.

 Command: **CYLINDER**

Current wire frame density: ISOLINES=4
Specify center point for base of cylinder
   or [Elliptical] <0,0,0>: *(Press* ENTER *to*
   accept the default of 0,0,0)
Specify radius for base of cylinder or
   [Diameter]: **D** *(For Diameter)*
Specify diameter for base of cylinder: **3**
Specify height of cylinder or [Center of
   other end]: **2.75**

 Command: **Z** *(For ZOOM)*

Specify corner of window, enter a scale
   factor (nX or nXP), or [All/Center/
   Dynamic/Extents/Previous/Scale/
   Window] <real time>: **All**

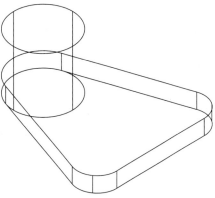

Figure 20–152

STEP 8

Merge the cylinder just created with the extruded base using the UNION command. See Figure 20–153.

 Command: **UNI** *(For UNION)*

Select objects: *(Select the extruded base
   and cylinder)*
Select objects: *(Press* ENTER *to perform the
   union operation)*

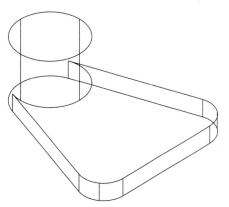

Figure 20–153

STEP 9

Use the CYLINDER command to create a 2.00-unit diameter cylinder representing a through hole, as shown in Figure 20–154. The height of the cylinder is 2.75 units with the center point at 0,0,0.

 Command: **CYLINDER**

Current wire frame density: ISOLINES=4
Specify center point for base of cylinder
   or [Elliptical] <0,0,0>: *(Press* ENTER *to accept the default of 0,0,0)*
Specify radius for base of cylinder or
   [Diameter]: **D** *(For Diameter)*
Specify diameter for base of cylinder:
   2.00
Specify height of cylinder or [Center of
   other end]: **2.75**

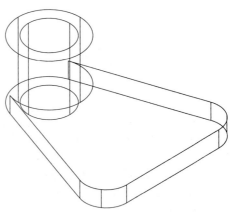

Figure 20–154

STEP 10

To cut the hole through the outer cylinder, use the SUBTRACT command. Select the base as the source object; select the inner cylinder as the object to subtract. Use the HIDE command to view the results. Your display should appear similar to Figure 20–155.

 Command: **SU** *(For SUBTRACT)*

Select solids and regions to subtract
   from...
Select objects: *(Select the base of the Collar)*
Select objects: *(Press* ENTER *to continue
   with this command)*
Select solids and regions to subtract...
Select objects: *(Select the 2.00 diameter
   cylinder just created)*
Select objects: *(Press* ENTER *to perform the
   subtraction operation)*

Command: **HI** *(For HIDE)*
Command: **RE** *(For REGEN)*
   (To return to wireframe mode)

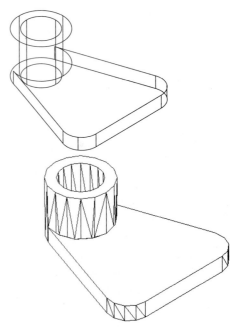

Figure 20–155

STEP 11

Begin placing the two small drill holes in the base using the CYLINDER command. Use the OSNAP-Center mode to place each cylinder at the center of arcs "A" and "B" in Figure 20–156.

 Command: **CYLINDER**

Current wire frame density: ISOLINES=4
Specify center point for base of cylinder or [Elliptical] <0,0,0>: **Cen**
of (Pick the bottom edge of the arc at "A")
Specify radius for base of cylinder or [Diameter]: **D** (For Diameter)
Specify diameter for base of cylinder: **1.00**
Specify height of cylinder or [Center of other end]: **0.75**

 Command: **CYLINDER**

Current wire frame density: ISOLINES=4
Specify center point for base of cylinder or [Elliptical] <0,0,0>: **Cen**
of (Pick the bottom edge of the arc at "B")

Specify radius for base of cylinder or [Diameter]: **D** (For Diameter)
Specify diameter for base of cylinder: **1.00**
Specify height of cylinder or [Center of other end]: **0.75**

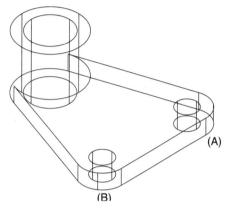

(A)

(B)

Figure 20–156

STEP 12

Subtract both 1.00-diameter cylinders from the base of the model using the SUBTRACT command. Use the HIDE command to view the results. Your display should appear similar to Figure 20–157.

 Command: **SU** (For SUBTRACT)

Select solids and regions to subtract from...
Select objects: (Pick the base of the Collar)
Select objects: (Press ENTER to continue with this command)
Select solids and regions to subtract...
Select objects: (Select both 1.00 diameter cylinders at the right)
Select objects: (Press ENTER to perform the subtraction operation)

Command: **HI** (For HIDE)

Command: **RE** (For REGEN)
(To return to wireframe mode)

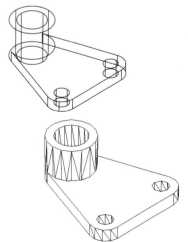

Figure 20–157

STEP 13

Begin constructing the rectangular slot that will pass through the two cylinders (see Figure 20–158). Use the BOX command to accomplish this. Locate the center of the box at 0,0,0 and make the box 4 units long, 1 unit wide, and 0.75 units high. Then move the box to the top of the cylinder. Use the geometry calculator to select the base point of displacement at the top of the box with the MEE function. After you locate two endpoints, the function will calculate the midpoint.

 Command: **BOX**

Specify corner of box or [CEnter]
   <0,0,0>: **CE** *(For the center of the box)*
Specify center of box <0,0,0>: *(Press*
   ENTER *to accept the default center value*
   of 0,0,0).

Specify corner or [Cube/Length]: **L**
Specify length: **4**
Specify width: **I**
Specify height: **0.75**

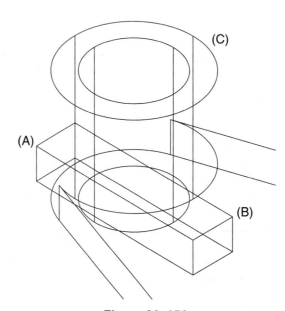 Command: **M** *(For MOVE)*

Select objects: *(Select the box just*
   constructed)
Select objects: *(Press* ENTER *to continue*
   with this command)
Specify base point or displacement: **'CAL**
   (To activate the geometry calculator)
Initializing…
>> Expression: **MEE**
>> Select one endpoint for MEE: *(Pick a*
   point on one corner of the box at "A")
>> Select another endpoint for MEE: *(Pick*
   a point on the opposite corner of the box
   at "B")
Specify second point of displacement or
   <use first point as displacement>: **Cen**
of *(Pick the edge of the cylinder at "C")*

Figure 20–158

STEP 14

Use the SUBTRACT command to subtract the rectangular box from the solid model, as in Figure 20–159.

 Command: **SU** *(For SUBTRACT)*

Select solids and regions to subtract from...
Select objects: *(Select the model at "A")*
Select objects: *(Press* ENTER *to continue with this command)*
Select solids and regions to subtract...
Select objects: *(Select the rectangular box at "B")*
Select objects: *(Press* ENTER *to perform the subtraction operation)*

The completed solid mode should appear similar to Figure 20–160.

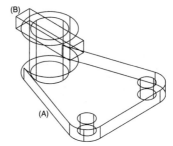

Figure 20–159

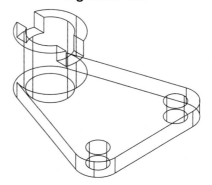

Figure 20–160

STEP 15

Change the facet resolution to a higher value. Then perform a hidden line removal to see the appearance of the model as shown in Figure 20–161.

Command: **FACETRES**

Enter new value for FACETRES
  <0.5000>: **1**

Command: **HI** *(For HIDE)*

Command: **RE** *(For REGEN)*
  (To return to wireframe mode)

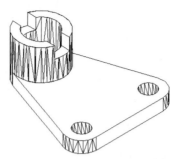

Figure 20–161

STEP 16

Important information can be extracted from the solid model to be used for design and analysis. The following properties are calculated by using the MASSPROP command: Mass, Volume, Bounding Box, Centroid, Moments of Inertia, Products of Inertia, Radii of Gyration, and Principal Moments about Centroid. Variances may occur in a few of the data fields when using this command.

 Command: **MASSPROP**

Select objects: *(Select the model of the Collar)*

Select objects: *(Press* ENTER *to perform the mass property calculation)*

————————SOLIDS————————

Mass:	29.12
Volume:	29.12
Bounding box:	X: -1.50–6.00
	Y: -3.25–3.25
	Z: 0.00–2.75
Centroid:	X: 2.36
	Y: 0.00
	Z: 0.69
Moments of inertia:	X: 82.90
	Y: 319.73
	Z: 349.82
Products of inertia:	XY: 0.00
	YZ: 0.00
	ZX: 25.72
Radii of gyration:	X: 1.69
	Y: 3.31
	Z: 3.47

Principal moments and X-Y-Z directions about centroid:
I: 65.07 along [0.98 0.00 -0.17]
J: 144.19 along [0.00 1.00 0.00]
K: 192.11 along [0.17 0.00 0.98]

 Open the Exercise Manual PDF file for Chapter 20 on the accompanying CD for more tutorials and exercises.

 If you have the accompanying Exercise Manual, refer to Chapter 20 for more tutorials and exercises.

CHAPTER 21

Editing Solid Models

Once a solid model is created, various methods are available to edit the model. The methods discussed in this chapter include the 3D Operation commands (ALIGN, ROTATE3D, MIRROR3D, and 3DARRAY) and the solidedit command. Various 3D drawings will be opened and used to illustrate these commands.

THE ALIGN COMMAND

We will begin by examining the 3D operation commands, which can be accessed by the Modify pull down menu as shown in Figure 21–1. The first command we will use to edit a solid model is the ALIGN command.

Figure 21–1

Try It! – Open the drawing file 21_Align3D. The objects in Figure 21–2 need to be positioned or aligned to form the object in "A." At this point, it is unclear at what angle the objects are currently rotated. When you use the ALIGN command, it is not necessary to know this information. Rather, you line up source points with destination points. When the three sets of points are identified, the object moves and rotates into position. The first source point acts as a base point for a move operation. The first destination point acts as a base point for rotation operations. The second and third sets of source and destination points establish the direction and amount of rotation required to align the objects.

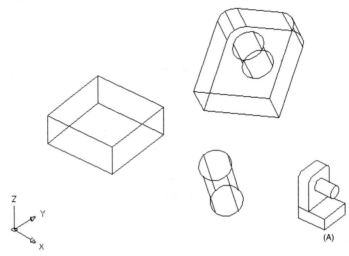

(A)

Figure 21–2

Follow the prompt sequence below and the illustration in Figure 21–3 for aligning the hole plate with the bottom base. The first destination point acts as a base point to which the cylinder locates.

Command: **HI** *(For HIDE)*
Command: **AL** *(For ALIGN)*
Select objects: *(Select the object with the hole)*
Select objects: *(Press* ENTER *to continue)*
Specify first source point: *(Select the endpoint at "A")*
Specify first destination point: *(Select the endpoint at "B")*
Specify second source point: *(Select the endpoint at "C")*
Specify second destination point: *(Select the endpoint at "D")*
Specify third source point or <continue>: *(Select the endpoint at "E")*
Specify third destination point: *(Select the endpoint at "F")*

The results of the previous step are illustrated in Figure 21–4. Next, align the cylinder with the hole. Circular shapes need only two sets of source and destination points for the shapes to be properly aligned.

Command: **AL** *(For ALIGN)*
Select objects: *(Select the cylinder)*
Select objects: *(Press* ENTER *to continue)*
Specify first source point: *(Select the top center of the cylinder at "A")*
Specify first destination point: *(Select the outer center of the hole at "B")*

Specify second source point: *(Select the bottom center of the cylinder at "C")*
Specify second destination point: *(Select the inner center of the hole at "D")*
Specify third source point or <continue>: *(Press* ENTER *to continue)*
Scale objects based on alignment points? [Yes/No] <N>: *(Press* ENTER *to accept the default and align the cylinder with the hole)*

The completed 3D model is illustrated in Figure 21–5. The ALIGN command provides an easy means of putting solid objects together to form assembly models.

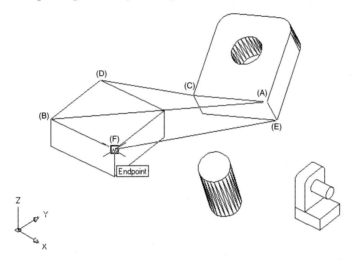

Figure 21–3

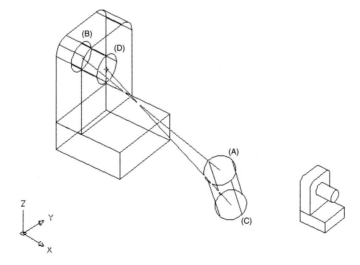

Figure 21–4

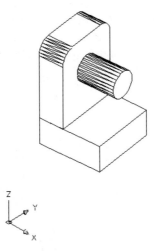

Figure 21–5

You are not limited to using the ALIGN command on 3D objects. There are 2D applications of the command, as shown in Figure 21–6.

 Try It! – Open the drawing file 21_Align2D. Here the edge of the triangle needs to the aligned with the edge of the rectangle. Follow the prompt sequence for performing this task. Again, pay attention to the importance of the first set of points in aligning the object.

Command: **AL** *(For ALIGN)*
Select objects: *(Select the triangle)*
Select objects: *(Press ENTER to continue)*
Specify first source point: *(Select the endpoint at "A")*
Specify first destination point: *(Select the endpoint at "B")*
Specify second source point: *(Select the endpoint at "C")*
Specify second destination point: *(Select the endpoint at "D")*
Specify third source point or <continue>: *(Press ENTER to continue)*
Scale objects based on alignment points? [Yes/No] <N>: *(Press ENTER to accept the default and align the triangle with the rectangle)*

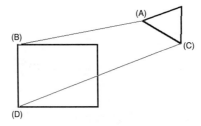

Figure 21–6

The results are illustrated in 21–7. Notice that the first set of points relocated the triangle and the second set of points rotated the triangle so that it is aligned with the edge of the rectangle. A third set of points is unnecessary for 2D operations. We could also have scaled the object during the align operation.

 Try It! – Open the drawing file 21_Align2D again. Answering Yes to the prompt "Scale objects based on alignment points?" displays the results in Figure 21–8. Here the triangle is scaled in size based on the two destination points of the rectangle.

Command: **AL** *(For ALIGN)*
Select objects: *(Select the triangle)*
Select objects: *(Press* ENTER *to continue)*
Specify first source point: *(Select the endpoint at "A" in Figure 21–6)*
Specify first destination point: *(Select the endpoint at "B" in Figure 21–6)*
Specify second source point: *(Select the endpoint at "C" in Figure 21–6)*
Specify second destination point: *(Select the endpoint at "D" in Figure 21–6)*
Specify third source point or <continue>: *(Press* ENTER *to continue)*
Scale objects based on alignment points? [Yes/No] <N>: **Y** *(To scale the edge of the triangle with the edge of the rectangle)*

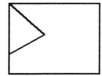

Figure 21–7

Figure 21–8

THE ROTATE3D COMMAND

The ROTATE, MIRROR, and ARRAY commands can all be used to modify 3D objects, however, the User Coordinate System must be carefully manipulated each time to get the desired results. ROTATE3D, MIRROR3D, and 3DARRAY provide a better 3D tool by providing 3D specific options that do not require the UCS to be changed constantly.

Besides the ALIGN command, another convenient way to position objects in 3D is by rotating them into position using the ROTATE3D command. In this command you will select an option that establishes an axis, about which the rotation will take place. A thorough understanding of the User Coordinate System and the right hand rule is a must in properly operating this command.

 Try It! – Open the drawing 21_Rot3D. In Figure 21–9, a base containing a slot needs to be joined with the two rectangular boxes to form a back and side. First select box "A" as the object to rotate in 3D. You will be prompted to define the axis of rotation. This axis will serve as a pivot point where the rotation occurs. In Figure 21–9, the axis of rotation is the Y axis, from the current

position of the User Coordinate System icon. Entering a positive angle of 90°
will rotate the box in the counterclockwise direction (as you would look down
the Y axis). Negative angles rotate in the clockwise direction (using the right
hand rule, point your thumb in the axis direction and your fingers will curl in
the positive rotation direction).

Command: **ROTATE3D**
Current positive angle: ANGDIR=counterclockwise ANGBASE=0
Select objects: *(Select box "A")*
Select objects: *(Press* ENTER *to continue)*
Specify first point on axis or define axis by
[Object/Last/View/Xaxis/Yaxis/Zaxis/2points]: **Y** *(For Y axis)*
Specify a point on the Y axis <0,0,0>: *(Select the endpoint at "A")*
Specify rotation angle or [Reference]: **90**

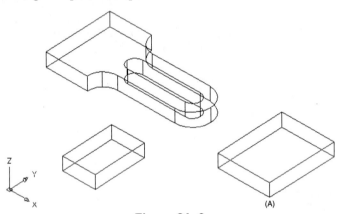

Figure 21–9

Next, the second box in Figure 21–10 is rotated 90° with the X axis selected as
the axis of rotation.

Command: **ROTATE3D**
Current positive angle: ANGDIR=counterclockwise ANGBASE=0
Select objects: *(Select box "A")*
Select objects: *(Press* ENTER *to continue)*
Specify first point on axis or define axis by
[Object/Last/View/Xaxis/Yaxis/Zaxis/2points]: **X** *(For X axis)*
Specify a point on the X axis <0,0,0>: *(Select the endpoint at "A")*
Specify rotation angle or [Reference]: **90**

The results of these operations are illustrated in Figure 21–11. Once the boxes are
rotated to the correct angles, they are moved into position using the MOVE command
and the appropriate Object Snap modes. Box "A" is moved from the endpoint of the
corner at "A" to the endpoint of the corner at "C." Box "B" is moved from the endpoint
of the corner at "B" to the endpoint of the corner at "C." Once moved, they are then

joined to the model through the UNION command. Finally use the HIDE command to obtain the results illustrated in Figure 21–12.

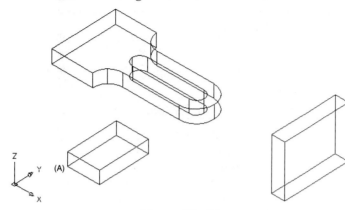

Figure 21–10

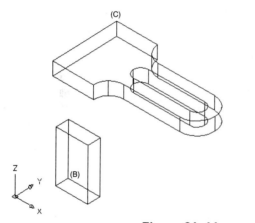

Figure 21–11

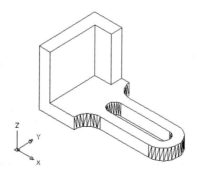

Figure 21–12

THE MIRROR3D COMMAND

The MIRROR3D command is a 3D version of the MIRROR command and behaves very much like the ROTATE3D command. In this command, however, instead of rotating about an axis, you mirror over a plane. Again, a thorough understanding of the User Coordinate System is a must in properly operating this command.

 Try It! – Open the drawing 21_Mirror3D. In Figure 21–13, only half of the object is created. The symmetrical object in Figure 21–14 is needed and can easily be created by using the MIRROR3D and UNION commands.

Command: **MIRROR3D**
Select objects: *(Select the part)*
Select objects: *(Press ENTER to continue)*
Specify first point of mirror plane (3 points) or
[Object/Last/Zaxis/View/XY/YZ/ZX/3points] <3points>: **YZ** *(For YZ plane)*
Specify point on YZ plane <0,0,0>: *(Select the endpoint at "A" in Figure 21–13)*
Delete source objects? [Yes/No] <N>: **N** *(Keep both objects)*

Command: **UNI** *(For UNION)*
Select objects: *(Select both solid objects)*
Select objects: *(Press ENTER to perform the union operation)*

Command: **HI** *(For HIDE, the solid should appear as in Figure 21–14)*

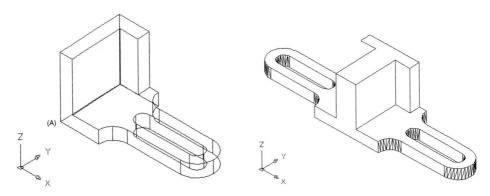

Figure 21–13	Figure 21–14

THE 3DARRAY COMMAND

The 3DARRAY command, like the ARRAY command, allows you to create both polar and rectangular arrays. For the 3D polar array, however, you select any 3D axis to rotate about and in the 3D version of the rectangular array you not only have rows and columns but levels as well.

Try It! – Open the drawing 21_3DArray. In Figure 21–15, six arms need to be arrayed around the hub in the center. To accomplish this we will perform a polar array about an axis running through the hub. We will rotate throughout a full 360°.

 Note: If you perform a polar array at an angle other than 360°, take note of the direction in which you select the axis because this will affect the direction of rotation (right hand rule).

Command: **3A** *(For 3DARRAY)*
Select objects: *(Select the Arm "A" in Figure 21–15)*
Select objects: *(Press ENTER to continue)*
Enter the type of array [Rectangular/Polar] <R>: **P** *(For Polar)*
Enter the number of items in the array: **6**
Specify the angle to fill (+=ccw, -=cw) <360>: *(Press ENTER to accept the default)*
Rotate arrayed objects? [Yes/No] <Y>: **Y**
Specify center point of array: **Cen**
of *(Specify the center of the back hub circle at "B" in Figure 21–15)*
Specify second point on axis of rotation: **Cen**
of *(Specify the center of the front hub circle in Figure 21–15)*

 Command: **UNI** *(For UNION)*
Select objects: *(Select the six arms and the hub)*
Select objects: *(Press ENTER to perform the union operation)*

Command: **HI** *(For HIDE, the solid should appear as in Figure 21–16)*

Command: **RE** *(For REGEN; this will convert the image back to wireframe mode)*

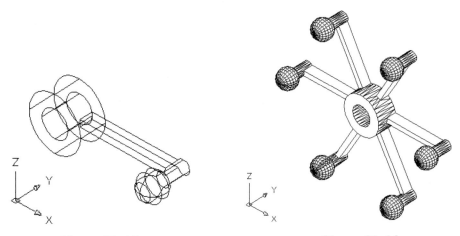

Figure 21–15   **Figure 21–16**

Finally, to demonstrate a rectangular array, use the solid just created in Figure 21–16 to create the array of solids in Figure 21–17. This array consists of two rows (in the y direction), three columns (in the x direction), and four levels (in the z direction). As with the regular ARRAY command you may specify negative or positive distances to control the direction of the array.

Command: **3A** *(For 3DARRAY)*
Select objects: *(Select the solid in Figure 21–16)*
Select objects: *(Press ENTER to continue)*
Enter the type of array [Rectangular/Polar] <R>: **R** *(For Rectangular)*
Enter the number of rows (—) <1>: **2**
Enter the number of columns (||||) <1>: **3**
Enter the number of levels (...) <1>: **4**
Specify the distance between rows (—): **10**
Specify the distance between columns (||||): **20**
Specify the distance between levels (...): **30**

 Command: **Z** *(For ZOOM)*
Specify corner of window, enter a scale factor (nX or nXP), or
[All/Center/Dynamic/Extents/Previous/Scale/Window] <real time>: **E** *(For Extents)*

Command: **HI** *(For HIDE, the solid should appear as in Figure 21–17)*

Figure 21–17

THE SOLIDEDIT COMMAND

Once features such as holes, slots, and extrusions are constructed in a solid model, the time may come to make changes to these features. This is the function of the SOLIDEDIT command. Select this command from the Solids Editing toolbar shown in Figure 21–18 or choose Solids Editing from the Modify pull-down menu shown in Figure 21–19. Both menus are arranged in three groupings; namely Face, Edge, and Body editing. These groupings will be discussed in the pages that follow. Also, it is recommended that instead of entering the SOLIDEDIT command at the command prompt, that you use the Solids Editing toolbar or pull-down menu to perform editing operations. This will eliminate a number of steps and make it easier to locate the appropriate option under the correct grouping.

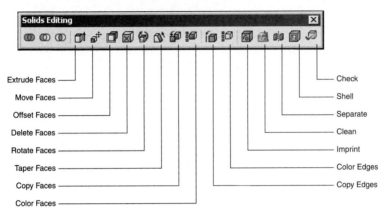

Figure 21–18

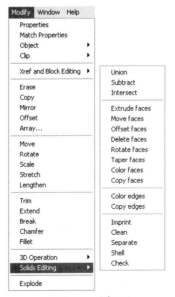

Figure 21–19

EXTRUDING (FACE EDITING)

Faces may be lengthened or shortened through the Extrude option of the SOLIDEDIT command. A positive distance extrudes the face in the direction of its normal. A negative distance extrudes the face in the opposite direction.

Try It! – Open the drawing file 21_Extrude. In Figure 21–20, the highlighted face at "A" needs to be decreased in height.

 Tip: You will achieve better results when selecting a face for the SOLIDEDIT command if you pick on the inside of the face rather than on the edge of the face.

 Command: **SOLIDEDIT**

Solids editing automatic checking: SOLIDCHECK=1
Enter a solids editing option [Face/Edge/Body/Undo/eXit] <eXit>: **F** *(For Face)*
Enter a face editing option
[Extrude/Move/Rotate/Offset/Taper/Delete/Copy/coLor/Undo/eXit] <eXit>: **E** *(For Extrude)*
Select faces or [Undo/Remove]: *(Select the face inside the area represented by "A")*
Select faces or [Undo/Remove/ALL]: *(Press ENTER to continue)*
Specify height of extrusion or [Path]: **-10.00**
Specify angle of taper for extrusion <0>: *(Press ENTER)*
Solid validation started.
Solid validation completed.
Enter a face editing option
[Extrude/Move/Rotate/Offset/Taper/Delete/Copy/coLor/Undo/eXit] <eXit>: *(Press ENTER)*
Solids editing automatic checking: SOLIDCHECK=1
Enter a solids editing option [Face/Edge/Body/Undo/eXit] <eXit>: *(Press ENTER)*

The result is illustrated at "B."

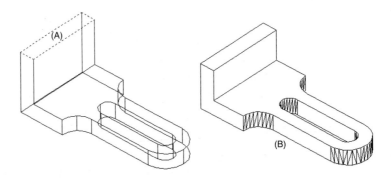

Figure 21–20

MOVING (FACE EDITING)

Try It! – Open the drawing 21_Move. The object in Figure 21–21 illustrates two intersecting cylinders. The two horizontal cylinders need to be moved 1 unit up from their current location. The cylinders are first selected at "A" and "B" through the SOLIDEDIT command along with the Move option.

Command: **SOLIDEDIT**

Solids editing automatic checking: SOLIDCHECK=1
Enter a solids editing option [Face/Edge/Body/Undo/eXit] <eXit>: **F** *(For Face)*
Enter a face editing option
[Extrude/Move/Rotate/Offset/Taper/Delete/Copy/coLor/Undo/eXit] <eXit>: **M** *(For Move)*
Select faces or [Undo/Remove]: *(Select both highlighted faces at "A" and "B")*
Select faces or [Undo/Remove/ALL]: *(Press ENTER to continue)*
Specify a base point or displacement: *(Pick the center of the bottom vertical cylinder)*
Specify a second point of displacement: **@0,0,1**
Solid validation started.
Solid validation completed.
Enter a face editing option
[Extrude/Move/Rotate/Offset/Taper/Delete/Copy/coLor/Undo/eXit] <eXit>: *(Press ENTER)*
Solids editing automatic checking: SOLIDCHECK=1
Enter a solids editing option [Face/Edge/Body/Undo/eXit] <eXit>: *(Press ENTER)*

The results are illustrated at "C."

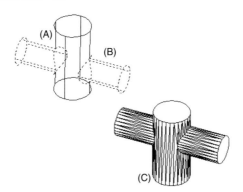

Figure 21–21

ROTATING (FACE EDITING)

Try It! – Open the drawing file 21_Rotate. In Figure 21–22, the triangular cutout needs to be rotated 45° in the clockwise direction. Use the Rotate Face option of the SOLIDEDIT command to accomplish this. You must select all faces of the triangular cutout at "A," "B," and "C." You will also have to remove the face that makes up the top of the rectangular base at "D" before proceeding.

Command: **SOLIDEDIT**

Solids editing automatic checking: SOLIDCHECK=1

Enter a solids editing option [Face/Edge/Body/Undo/eXit] <eXit>: **F** *(For Face)*

Enter a face editing option

[Extrude/Move/Rotate/Offset/Taper/Delete/Copy/coLor/Undo/eXit] <eXit>: **R** *(For Rotate)*

Select faces or [Undo/Remove]: *(Select all faces that make up the triangular extrusion at "A", "B", and "C", select "C" twice to highlight it)*

Select faces or [Undo/Remove/ALL]: **R** *(For Remove)*

Remove faces or [Undo/Add/ALL]: *(Select the face at "D" to remove)*

Remove faces or [Undo/Add/ALL]: *(Press ENTER to continue)*

Specify an axis point or [Axis by object/View/Xaxis/Yaxis/Zaxis] <2points>: **Z** *(For Zaxis)*

Specify the origin of the rotation <0,0,0>: *(Select the endpoint at "E")*

Specify a rotation angle or [Reference]: **-45** *(To rotate the triangular extrusion 45° in the clockwise direction)*

Solid validation started.

Solid validation completed.

Enter a face editing option

[Extrude/Move/Rotate/Offset/Taper/Delete/Copy/coLor/Undo/eXit] <eXit>: *(Press ENTER)*

Solids editing automatic checking: SOLIDCHECK=1

Enter a solids editing option [Face/Edge/Body/Undo/eXit] <eXit>: *(Press ENTER)*

The results are illustrated at "F."

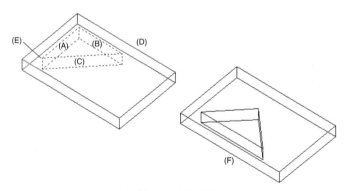

Figure 21–22

 OFFSETTING (FACE EDITING)

 Try It! – Open the drawing file 21_Offset. In the image in Figure 21–23 the holes need to be resized. Use the Offset Face option of the SOLIDEDIT command to increase or decrease the size of selected faces. Using positive values actually increases the volume of the solid. Therefore, the feature being offset gets smaller, similar to the illustration at "B." Entering negative values reduces the volume of the solid; this means the feature being offset gets larger, as in the figures at "C." Study the prompt sequence and Figure 21–23 for the mechanics of this command option.

Command: **SOLIDEDIT**

Solids editing automatic checking: SOLIDCHECK=1

Enter a solids editing option [Face/Edge/Body/Undo/eXit] <eXit>: **F** *(For Face)*

Enter a face editing option

[Extrude/Move/Rotate/Offset/Taper/Delete/Copy/coLor/Undo/eXit] <eXit>: **O** *(For Offset)*

Select faces or [Undo/Remove]: *(Select inside the edges of the two holes at "D" and "E"; you can select the edges of the holes twice, but if you do, you will need to use the Remove option to de-select the bottom and top edges at "F" and "G")*

Select faces or [Undo/Remove/ALL]: *(Press* ENTER *to continue)*

Specify the offset distance: **0.50**

Solid validation started.

Solid validation completed.

Enter a face editing option

[Extrude/Move/Rotate/Offset/Taper/Delete/Copy/coLor/Undo/eXit] <eXit>: *(Press* ENTER*)*

Solids editing automatic checking: SOLIDCHECK=1

Enter a solids editing option [Face/Edge/Body/Undo/eXit] <eXit>: (Press enter)

Figure 21–23

TAPERING (FACE EDITING)

 Try It! – Open the drawing 21_Taper. The object at "A" in Figure 21–24 represents a solid box that needs to have tapers applied to its sides. Using the Taper Face option of the SOLIDEDIT command allows you to accomplish this task. Entering a positive angle moves the location of the second point into the part as shown at "F." Entering a negative angle moves the location of the second point away from the part as shown at "G."

Command: **SOLIDEDIT**

Solids editing automatic checking: SOLIDCHECK=1

Enter a solids editing option [Face/Edge/Body/Undo/eXit] <eXit>: **F** *(For Face)*

Enter a face editing option

[Extrude/Move/Rotate/Offset/Taper/Delete/Copy/coLor/Undo/eXit] <eXit>: **T** *(For Taper)*

Select faces or [Undo/Remove]: *(Select faces "A" through "D")*

Select faces or [Undo/Remove/ALL]: *(Press* ENTER *to continue)*

Specify the base point: *(Select the endpoint at "E")*

Specify another point along the axis of tapering: *(Select the endpoint at "B")*

Specify the taper angle: **10** *(For the angle of the taper)*

Solid validation started.

Solid validation completed.

Enter a face editing option
[Extrude/Move/Rotate/Offset/Taper/Delete/Copy/coLor/Undo/eXit] <eXit>: *(Press ENTER)*
Solids editing automatic checking: SOLIDCHECK=1
Enter a solids editing option [Face/Edge/Body/Undo/eXit] <eXit>: *(Press ENTER)*

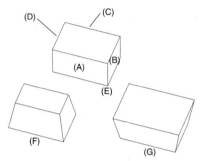

Figure 21–24

DELETING (FACE EDITING)

Faces can be erased through the Delete Face option of the solidedit command.

 Try It! – Open the drawing file 21_Delete. In Figure 21–25, select the hole at "A" as the face to erase.

 Command: **SOLIDEDIT**

Solids editing automatic checking: SOLIDCHECK=1
Enter a solids editing option [Face/Edge/Body/Undo/eXit] <eXit>: **F** *(For Face)*
Enter a face editing option
[Extrude/Move/Rotate/Offset/Taper/Delete/Copy/coLor/Undo/eXit] <eXit>: **D** *(For Delete)*
Select faces or [Undo/Remove]: *(Select inside the hole at "A")*
Select faces or [Undo/Remove/ALL]: *(Press ENTER to continue))*
Solid validation started.
Solid validation completed.
Enter a face editing option
[Extrude/Move/Rotate/Offset/Taper/Delete/Copy/coLor/Undo/eXit] <eXit>: *(Press ENTER)*
Solids editing automatic checking: SOLIDCHECK=1
Enter a solids editing option [Face/Edge/Body/Undo/eXit] <eXit>: *(Press ENTER)*

The results are illustrated in "C."

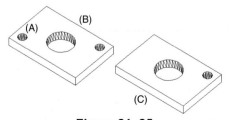

Figure 21–25

COPYING (FACE EDITING)

You can copy a face for use in the creation of another solid model using the Copy Face option of the SOLIDEDIT command.

 Try It! – Open the drawing file 21_Copy. In Figure 21–26 the face at "A" is selected to copy. Notice that all objects making up the face, such as the rectangle and circles, are highlighted. You will have to remove the face at "B." Picking a base point and second point copies the face at "C." The rectangle and circles are considered a single object called a region. The region could be exploded back into individual lines and circles, which could then be used to create a new object. A region can also be extruded, using the EXTRUDE command, to create another solid model such as the one illustrated at "D."

 Command: **SOLIDEDIT**

Solids editing automatic checking: SOLIDCHECK=1
Enter a solids editing option [Face/Edge/Body/Undo/eXit] <eXit>: **F** *(For Face)*
Enter a face editing option
[Extrude/Move/Rotate/Offset/Taper/Delete/Copy/coLor/Undo/eXit] <eXit>: **C** *(For Copy)*
Select faces or [Undo/Remove]: *(Select the bottom face at edge "A")*
Select faces or [Undo/Remove/ALL]: **R** *(For Remove)*
Remove faces or [Undo/Add/ALL]: *(Select the edge of the face at "B" to deselect it)*
Remove faces or [Undo/Add/ALL]: *(Press ENTER to continue)*
Specify a base point or displacement: *(Pick a point to copy from)*
Specify a second point of displacement: *(Pick a point to copy to)*
Enter a face editing option
[Extrude/Move/Rotate/Offset/Taper/Delete/Copy/coLor/Undo/eXit] <eXit>: *(Press ENTER)*
Solids editing automatic checking: SOLIDCHECK=1
Enter a solids editing option [Face/Edge/Body/Undo/eXit] <eXit>: *(Press ENTER)*

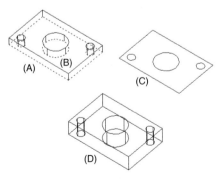

Figure 21–26

 IMPRINTING (BODY EDITING)

An interesting and powerful method of adding construction geometry to a solid model is though the process of imprinting.

Try It! – Open the drawing file 21_Imprint. In the imprinting illustration in Figure 21–27 at "A," a rectangle along with slot is already modeled in 3D. A line was constructed from the midpoints of the top surface of the solid model." The SOLIDEDIT command will be used to imprint this line to the model, which results in dividing the top surface into two faces.

Command: **SOLIDEDIT**
Solids editing automatic checking: SOLIDCHECK=1
Enter a solids editing option [Face/Edge/Body/Undo/eXit] <eXit>: **B** *(For Body)*
Enter a body editing option
[Imprint/seParate solids/Shell/cLean/Check/Undo/eXit] <eXit>: **I** *(For Imprint)*
Select a 3D solid: *(Select the solid model)*
Select an object to imprint: *(Select line "B")*
Delete the source object <N>: **Y** *(For Yes, this erases the line)*
Select an object to imprint: *(Press ENTER to perform the imprint operation)*
Enter a body editing option
[Imprint/seParate solids/Shell/cLean/Check/Undo/eXit] <eXit>: *(Press ENTER)*
Solids editing automatic checking: SOLIDCHECK=1
Enter a solids editing option [Face/Edge/Body/Undo/eXit] <eXit>: *(Press ENTER)*

The segments of the lines that come in contact with the 3D solid remain on the part's surface. However, these are no longer line segments; rather these lines now belong to the part. The lines actually separate the top surface into two faces.

Use the Extrude Face option of the SOLIDEDIT command on one of the newly created faces at "C" and increase the face in height by an extra 1.50 units. The results are illustrated in "D".

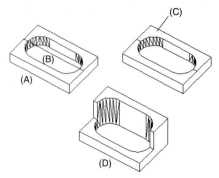

Figure 21–27

 SEPARATING SOLIDS (BODY EDITING)
Sometimes when performing union and subtraction operations on solid models, you can actually end up with models that don't actually touch (intersect) but act as a single object. The Separating Solids option of the SOLIDEDIT command is used to correct this condition.

Try It! – Open the drawing file 21_Separate. In Figure 21–28, the model at "A" is about to be sliced in half with a thin box created at the center of the circle and spanning the depth of the rectangular shelf.

Use the SUBTRACT command and subtract the rectangular box from the solid object. After you subtract the box, pick the solid at "B" and notice that both halves of the object highlight even though they appear separate. To convert the single solid model into two separate models, use the SOLIDEDIT command followed by the Body and Separate options.

Command: **SOLIDEDIT**
Solids editing automatic checking: SOLIDCHECK=1
Enter a solids editing option [Face/Edge/Body/Undo/eXit] <eXit>: **B** *(For Body)*
Enter a body editing option
[Imprint/seParate solids/Shell/cLean/Check/Undo/eXit] <eXit>: **P** *(For Separate)*
Select a 3D solid: *(Pick the solid model at "C")*
Enter a body editing option
[Imprint/seParate solids/Shell/cLean/Check/Undo/eXit] <eXit>: *(Press ENTER)*
Solids editing automatic checking: SOLIDCHECK=1
Enter a solids editing option [Face/Edge/Body/Undo/eXit] <eXit>: *(Press ENTER)*

This action separates the single model into two. When you select the model in "C," only one half highlights.

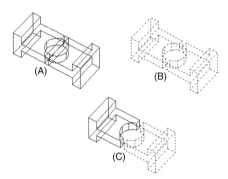

(A)   (B)

(C)

Figure 21–28

SHELLING (BODY EDITING)

Shelling is the process of constructing a thin wall inside or outside of a solid model. Positive thickness will produce the thin wall inside; negative values for thickness will produce the thin wall outside. This wall thickness remains constant throughout the entire model. Faces may be removed during the shelling operation to create an opening.

Try It! – Open the drawing file 21_Shell. For clarity, it is important to first rotate your model or your viewpoint such that any faces to be removed are visible. Use the 3DORBIT command and Figure 21–29 to change the model view from the one shown at "A" to the one shown at "B". Use the Hidden Shading Mode option in the 3DORBIT command to ensure that the bottom surface at "C" is visible. Now the Shell option of the SOLIDEDIT command can be used to "hollow out" the part and remove the bottom face. An additional note: only one shell is permitted in a model.

Command: **SOLIDEDIT**
Solids editing automatic checking: SOLIDCHECK=1
Enter a solids editing option [Face/Edge/Body/Undo/eXit] <eXit>: **B** *(For Body)*
Enter a body editing option
[Imprint/seParate solids/Shell/cLean/Check/Undo/eXit] <eXit>: **S** *(For Shell)*
Select a 3D solid: *(Select the solid model at "B")*
Remove faces or [Undo/Add/ALL]: *(Pick a point at "C")*
Remove faces or [Undo/Add/ALL]: *(Press* ENTER *to continue)*
Enter the shell offset distance: **0.20**
Solid validation started.
Solid validation completed.
Enter a body editing option
[Imprint/seParate solids/Shell/cLean/Check/Undo/eXit] <eXit>: *(Press* ENTER*)*
Solids editing automatic checking: SOLIDCHECK=1
Enter a solids editing option [Face/Edge/Body/Undo/eXit] <eXit>: *(Press* ENTER*)*

The results are illustrated at "D."

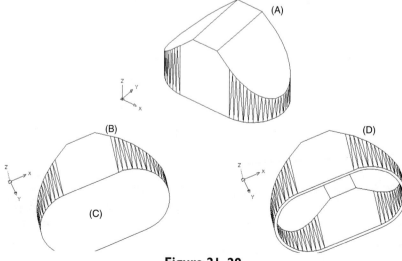

Figure 21–29

CLEANING (BODY EDITING)

When imprinted lines that form faces are not used, they can be deleted from a model by the Clean (Body Editing) option of the solidedit command.

 Try It! – Open the drawing file 21_Clean. The object at "A" in Figure 21–30 illustrates lines originally constructed on the top of the solid model. These lines were then imprinted at "B." Since these lines now belong to the model, the Clean option is used to remove them. The results are illustrated at "C."

 Command: **SOLIDEDIT**
Solids editing automatic checking: SOLIDCHECK=1
Enter a solids editing option [Face/Edge/Body/Undo/eXit] <eXit>: **B** *(For Body)*
Enter a body editing option
[Imprint/seParate solids/Shell/cLean/Check/Undo/eXit] <eXit>: **L** *(For Clean)*
Select a 3D solid: *(Select the solid model at "B")*
Enter a body editing option
[Imprint/seParate solids/Shell/cLean/Check/Undo/eXit] <eXit>: *(Press ENTER)*
Solids editing automatic checking: SOLIDCHECK=1
Enter a solids editing option [Face/Edge/Body/Undo/eXit] <eXit>: *(Press ENTER)*

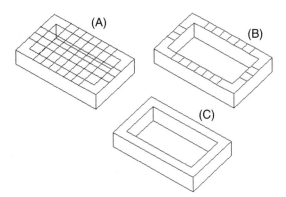

Figure 21–30

Additional options of the SOLIDEDIT command include the ability to apply a color to a selected face or edge of a solid model. You can even copy the edges from one model so that they can be used for construction purposes on other models.

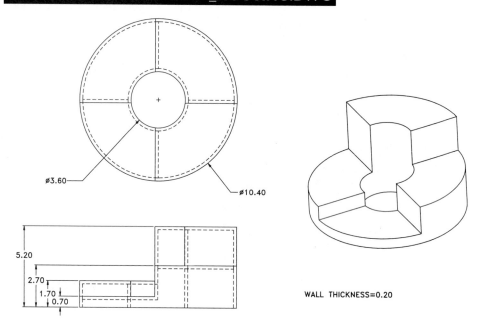

TUTORIAL EXERCISE: STEP_HOUSING.DWG

Ø3.60

Ø10.40

5.20

2.70

1.70

0.70

WALL THICKNESS=0.20

Figure 21–31

Purpose

This tutorial exercise is designed to create the solid model shown in Figure 21–31 using the SOLIDEDIT command.

System Settings

Use the Drawing Units dialog box to change the number of decimal places from four to two. Keep the default settings for limits.

Layers

Create a layer called Model for this tutorial exercise.

Suggested Commands

Begin a new drawing called Step Housing. Two circles will be drawn and extruded to form two cylinders. The smaller cylinder will then be subtracted from the larger cylinder to create a hole. Construction lines will be drawn from each quadrant of the large circle. These lines will then be imprinted on the model. This will enable the four faces to be extruded separately through the SOLIDEDIT command. After this is performed, the model will be rotated to display the base surface and a thin wall will be created through the Shell option of the SOLIDEDIT command.

Whenever possible, substitute the appropriate command alias in place of the full AutoCAD command in each tutorial step. For example, use "CP" for the COPY command, "L" for the LINE command, and so on. The complete listing of all command aliases is located in chapter 1, table 1–2.

STEP 1

Begin the Step Housing by making the Model layer current. Then draw one circle at a 5.00 diameter and another at a 2.00 diameter as in Figure 21–32.

⊘ Command: **C** *(For CIRCLE)*

Specify center point for circle or [3P/2P/ Ttr (tan tan radius)]: **7.00,4.00**
Specify radius of circle or [Diameter]: **D** *(For Diameter)*
Specify diameter of circle: **5.00**

⊘ Command: **C** *(For CIRCLE)*

Specify center point for circle or [3P/2P/ Ttr (tan tan radius)]: **@** *(To locate the last point)*
Specify radius of circle or [Diameter] <2.50>: **D** *(For Diameter)*
Specify diameter of circle <5.00>: **2.00**

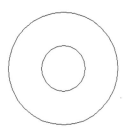

Figure 21–32

STEP 2

View the model from the South West position by picking the SW Isometric View button located in the View toolbar illustrated in Figure 21–33. You could also select this viewpoint by choosing 3D Views from the View pull-down menu and then SW Isometric. The results are illustrated in Figure 21–34.

Command: **-VIEW**
Enter an option [?/Orthographic/Delete/ Restore/Save/Ucs/Window]: **Swiso** *(For South West Isometric)*
Regenerating model.

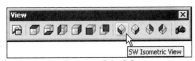

Fgure 21–33

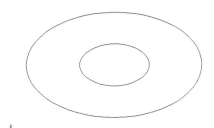

Figure 21–34

STEP 3

Extrude both circles to a height of 0.50 units using the EXTRUDE command (see Figure 21–35).

⊡ Command: **EXT** *(For EXTRUDE)*
Current wire frame density: ISOLINES=4
Select objects: *(Select both circles)*
Select objects: *(Press ENTER to continue)*
Specify height of extrusion or [Path]: **0.50**
Specify angle of taper for extrusion <0>: *(Press ENTER to perform the extrusion operation)*

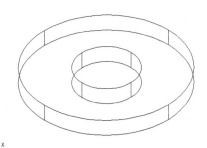

Figure 21–35

STEP 4

Subtract the small cylinder from the large cylinder to form a single solid model. Perform a hidden line removal using the HIDE command. Your model should appear similar to Figure 21–36.

 Command: **SU** *(For SUBTRACT)*

Select solids and regions to subtract from
Select objects: *(Select the large cylinder)*
Select objects: *(Press* ENTER *to continue)*
Select solids and regions to subtract ..
Select objects: *(Select the small cylinder)*
Select objects: *(Press* ENTER *to perform the subtraction operation)*
Command: **HI** *(For HIDE)*
Regenerating model.

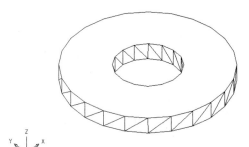

Figure 21–36

STEP 5

Draw lines at the top of the large cylinder in Figure 21–37. Use the OSNAP-Quadrant mode to accomplish this task.

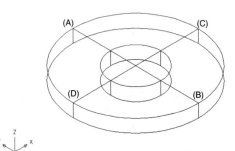

Figure 21–37

Command: **L** *(For LINE)*
Specify first point: **Qua**
of *(Pick the quadrant of the model at "A")*
Specify next point or [Undo]: **Qua**
of *(Pick the quadrant of the model at "B")*
Specify next point or [Undo]: *(Press* ENTER *to exit this command)*

Command: **L** *(For LINE)*
Specify first point: **Qua**
of *(Pick the quadrant of the model at "C")*
Specify next point or [Undo]: **Qua**
of *(Pick the quadrant of the model at "D")*
Specify next point or [Undo]: *(Press* ENTER *to exit this command)*

STEP 6

Use the Imprint option of the SOLIDEDIT command to create four faces at the top of the model in Figure 21–38.

 Command: **SOLIDEDIT**

Solids editing automatic checking:
  SOLIDCHECK=1
Enter a solids editing option [Face/Edge/
  Body/Undo/eXit] <eXit>: **B** *(For Body)*
Enter a body editing option
[Imprint/seParate solids/Shell/cLean/
  Check/Undo/eXit] <eXit>: **I** *(For
  Imprint)*
Select a 3D solid: *(Select the model)*
Select an object to imprint: *(Select line
  "A")*
Delete the source object [Yes/No] <N>:
  (Press ENTER*)*
Select an object to imprint: *(Select line
  "B")*
Delete the source object [Yes/No] <N>:
  (Press ENTER*)*
Select an object to imprint: *(Press* ENTER
  to end imprint selection)

Enter a body editing option
[Imprint/seParate solids/Shell/cLean/
  Check/Undo/eXit] <eXit>: *(Press*
  ENTER*)*
Solids editing automatic checking:
  SOLIDCHECK=1
Enter a solids editing option [Face/Edge/
  Body/Undo/eXit] <eXit>: *(Press* ENTER*)*

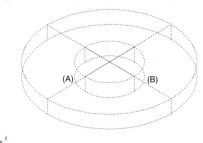

Figure 21–38

STEP 7

Use the face at "A" formed by the imprinted lines in Figure 21–39 and extrude it 2.50 units.

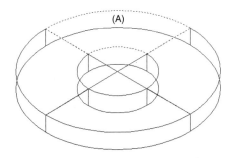

Figure 21–39

 Command: **SOLIDEDIT**

Solids editing automatic checking:
  SOLIDCHECK=1
Enter a solids editing option [Face/Edge/
  Body/Undo/eXit] <eXit>: **F** *(For Face)*
Enter a face editing option
[Extrude/Move/Rotate/Offset/Taper/
  Delete/Copy/coLor/Undo/eXit]
  <eXit>: **E** *(For Extrude)*
Select faces or [Undo/Remove]: *(Pick the
  face at A" in Figure 21–39)*
Select faces or [Undo/Remove/ALL]:
  (Press ENTER *to continue)*
Specify height of extrusion or [Path]:
  2.50

1160

Specify angle of taper for extrusion <0>:
   (Press ENTER *to continue*)
Solid validation started.
Solid validation completed.
Enter a face editing option
[Extrude/Move/Rotate/Offset/Taper/
   Delete/Copy/coLor/Undo/eXit]
   <eXit>: *(Press* ENTER*)*
Solids editing automatic checking:
   SOLIDCHECK=1

Enter a solids editing option [Face/Edge/
   Body/Undo/eXit] <eXit>: *(Press* ENTER*)*

Your model should appear similar to
Figure 21–40.

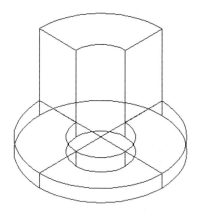

Figure 21–40

STEP 8

After performing the extrusion in Step
7, the top surface is no longer divided
into separate faces. We can use the lines
(source object), that we elected not to
delete earlier, to divide this surface into
separate faces again. Use the SOLIDEDIT
command and imprint the highlighted
lines in Figure 21–41. Use the same
prompt sequence as back in Step 6 to
perform this operation.

 Command: **SOLIDEDIT**
Solids editing automatic checking:
   SOLIDCHECK=1
Enter a solids editing option [Face/Edge/
   Body/Undo/eXit] <eXit>: **B** *(For Body)*
Enter a body editing option
[Imprint/seParate solids/Shell/cLean/
   Check/Undo/eXit] <eXit>: **I** *(For
   Imprint)*
Select a 3D solid: *(Select the model)*
Select an object to imprint: *(Select line "A")*
Delete the source object [Yes/No] <N>:
   (Press ENTER*)*
Select an object to imprint: *(Select line "B")*

Delete the source object [Yes/No] <N>:
   (Press ENTER*)*
Select an object to imprint: *(Press* ENTER
   to end imprint selection)
Enter a body editing option
[Imprint/seParate solids/Shell/cLean/
   Check/Undo/eXit] <eXit>: *(Press* ENTER*)*
Solids editing automatic checking:
   SOLIDCHECK=1
Enter a solids editing option [Face/Edge/
   Body/Undo/eXit] <eXit>: *(Press* ENTER*)*

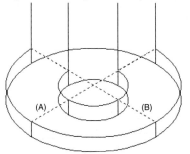

Figure 21–41

STEP 9

Extrude the highlighted face in Figure 21–42 using a negative distance of -0.25 units. This will drop the face down that distance as shown in Figure 21–43.

 Command: **SOLIDEDIT**

Solids editing automatic checking: SOLIDCHECK=1

Enter a solids editing option [Face/Edge/ Body/Undo/eXit] <eXit>: **F** *(For Face)*

Enter a face editing option [Extrude/Move/Rotate/Offset/Taper/ Delete/Copy/coLor/Undo/eXit] <eXit>: **E** *(For Extrude)*

Select faces or [Undo/Remove]: *(Pick the face at A" in Figure 21–42)*

Select faces or [Undo/Remove/ALL]: *(Press* ENTER *to continue)*

Specify height of extrusion or [Path]: **-0.25**

Specify angle of taper for extrusion <0>: *(Press* ENTER *to continue)*

Solid validation started.

Solid validation completed.

Enter a face editing option [Extrude/Move/Rotate/Offset/Taper/ Delete/Copy/coLor/Undo/eXit] <eXit>: *(Press* ENTER*)*

Solids editing automatic checking: SOLIDCHECK=1

Enter a solids editing option [Face/Edge/ Body/Undo/eXit] <eXit>: *(Press* ENTER*)*

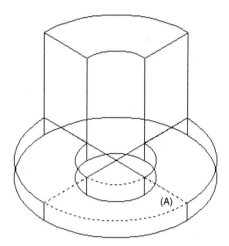

Figure 21–42

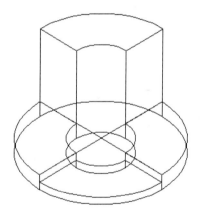

Figure 21–43

STEP 10

Extrude the face in Figure 21–44 to a height of 0.50 using the SOLIDEDIT command. Your display should appear similar to Figure 21–45.

Command: **SOLIDEDIT**

Solids editing automatic checking:
  SOLIDCHECK=1
Enter a solids editing option [Face/Edge/
  Body/Undo/eXit] <eXit>: **F** *(For Face)*
Enter a face editing option
[Extrude/Move/Rotate/Offset/Taper/
  Delete/Copy/coLor/Undo/eXit]
  <eXit>: **E** *(For Extrude)*
Select faces or [Undo/Remove]: *(Select
  the face at "A" in Figure 21–44)*
Select faces or [Undo/Remove/ALL]:
  (Press ENTER *to continue)*
Specify height of extrusion or [Path]:
  0.50
Specify angle of taper for extrusion <0>:
  (Press ENTER*)*
Solid validation started.
Solid validation completed.
Enter a face editing option
[Extrude/Move/Rotate/Offset/Taper/
  Delete/Copy/coLor/Undo/eXit]
  <eXit>: *(Press* ENTER*)*
Solids editing automatic checking:
  SOLIDCHECK=1
Enter a solids editing option [Face/Edge/
  Body/Undo/eXit] <eXit>: *(Press* ENTER*)*

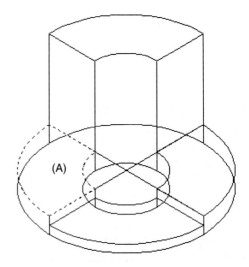

(A)

Figure 21–44

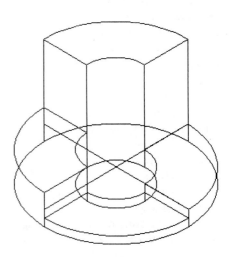

Figure 21–45

STEP 11

Extrude the face in Figure 21–46 to a height of 1.00 using the SOLIDEDIT command. Your display should appear similar to Figure 21–47.

Command: **SOLIDEDIT**

Solids editing automatic checking:
  SOLIDCHECK=1
Enter a solids editing option [Face/Edge/
  Body/Undo/eXit] <eXit>: **F** *(For Face)*
Enter a face editing option
[Extrude/Move/Rotate/Offset/Taper/
  Delete/Copy/coLor/Undo/eXit]
  <eXit>: **E** *(For Extrude)*
Select faces or [Undo/Remove]: *(Select
  the face at "A" in Figure 21–46)*
Select faces or [Undo/Remove/ALL]:
  (Press ENTER *to continue)*
Specify height of extrusion or [Path]:
  1.00
Specify angle of taper for extrusion <0>:
  (Press ENTER*)*
Solid validation started.
Solid validation completed.
Enter a face editing option
[Extrude/Move/Rotate/Offset/Taper/
  Delete/Copy/coLor/Undo/eXit]
  <eXit>: *(Press* ENTER*)*
Solids editing automatic checking:
  SOLIDCHECK=1
Enter a solids editing option [Face/Edge/
  Body/Undo/eXit] <eXit>: *(Press* ENTER*)*

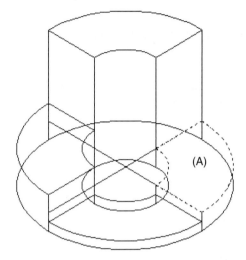

Figure 21–46

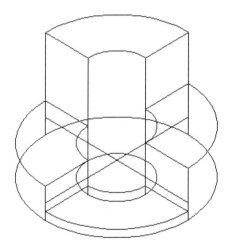

Figure 21–47

STEP 12

Erase the two lines originally used to construct the imprints. Change the Facet resolution to 5 using the FACETRES command. Perform a hidden line removal using the HIDE command. Your display should appear similar to Figure 21–48.

 Command: **E** *(For ERASE)*
Select objects: *(Select the 2 lines)*
Select objects: *(Press* ENTER)
Command: **FACETRES**
Enter new value for FACETRES <0.5000>:
  5.00
Command: **HI** (For HIDE)
Regenerating model.

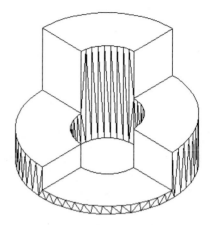

Figure 21–48

STEP 13

Before continuing with changes, save the current viewpoint of your model as a named view. Entering the view command displays the View dialog box in Figure 21–49. With the Named Views tab selected, click on the New button to display the New View dialog box. Enter "SW Iso" as the name of the view. Click the OK button in the New View dialog box to create the view. Click the OK button in the main View dialog box to return to your drawing. Now this view can be called up no matter how you change the display of the model.

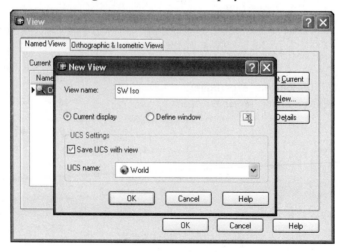

Figure 21–49

STEP 14

We will be creating a hollow shell of your model. To do this, you need to view the model from below. Selecting Viewpoint Presets from the View pull-down or entering VP (DDVPOINT command) at the command prompt displays the Viewpoint Presets dialog box shown in Figure 21–50.

Click in the right image until the XY Plane is set to -60°. You could also enter -60 in the XY Plane: edit box. This will allow you to create a viewing position underneath your model. Clicking the OK button displays your model as in Figure 21–51.

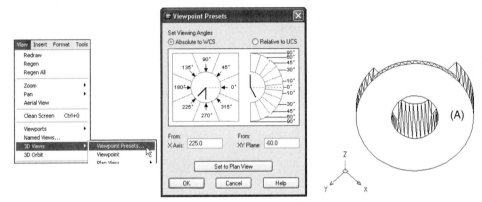

Figure 21–50   Figure 21–51

STEP 15

Now use the SOLIDEDIT command and Shell option to create a thin wall on the inside of your model. During the command sequence we will elect to remove the bottom face so that the bottom of our solid will be open. It may be helpful to use the HIDE command prior to performing a shell operation. This makes it easier to identify and select any faces that you want to remove.

 Command: **SOLIDEDIT**

Solids editing automatic checking:
   SOLIDCHECK=1
Enter a solids editing option [Face/Edge/
   Body/Undo/eXit] <eXit>: **B** *(For Body)*
Enter a body editing option
[Imprint/sePrate solids/Shell/cLean/
   Check/Undo/eXit] <eXit>: **S** *(For Shell)*
Select a 3D solid: *(Select the edge of the
   model)*
Remove faces or [Undo/Add/ALL]: *(Select
   the face at "A" in Figure 21–51)*

Remove faces or [Undo/Add/ALL]: *(Press* ENTER *to continue)*
Enter the shell offset distance: **0.20**
Solid validation started.
Solid validation completed.
Enter a body editing option
[Imprint/sePerate solids/Shell/cLean/Check/ Undo/eXit] <eXit>: *(Press* ENTER*)*
Solids editing automatic checking: SOLIDCHECK=1
Enter a solids editing option [Face/Edge/ Body/Undo/eXit] <eXit>: *(Press* ENTER*)*

Perform another hidden line removal to view the shell results using the HIDE command. Your display should appear similar to Figure 21–52.

Command: **HI** *(For HIDE)*
Regenerating model.

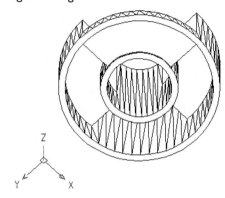

Figure 21–52

STEP 16

Use the VIEW command to display the View dialog box. Select the "SW Iso" view and make it current as shown in Figure 21–53. After clicking OK, your display should appear similar to Figure 21–54.

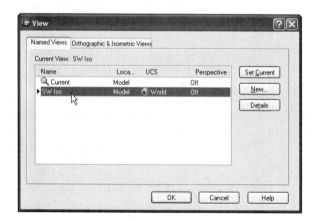

Figure 21–53

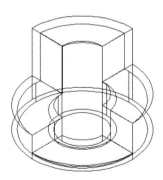

Figure 21–54

 Open the Exercise Manual PDF file for Chapter 21 on the accompanying CD for more tutorials and exercises.

 If you have the accompanying Exercise Manual, refer to Chapter 21 for more tutorials and exercises.

CHAPTER 22

Creating 2D Multiview Drawings From a Solid Model

One of the advantages of creating a 3D image in the form of a solid model is the ability to use the data of the solid model numerous times for other purposes. The purpose of this chapter is to generate 2D multiview drawings from the solid model. Two commands are used together to perform this operation: SOLVIEW and SOLDRAW. The SOLVIEW command is used to create a layout for the 2D views. This command automatically creates layers used to organize visible lines, hidden lines, and dimensions. The SOLDRAW command draws the requested views on the specific layers which were created by the SOLVIEW command (this includes the drawing of hidden lines to show hidden features, and even hatching if a section is requested). Included in the layout of 2D multiview projections will be the creation of orthographic, auxiliary, section, and isometric views for a drawing. A tutorial is available at the end of this chapter to demonstrate how to dimension orthographic views laid out with the SOLVIEW command.

THE SOLVIEW AND SOLDRAW COMMANDS

Once you create the solid model, you can lay out and draw the necessary 2D views using the Solids Setup commands. Choosing Solids from the Draw pull-down menu and then Setup, as shown in Figure 22–1, exposes the Drawing (SOLDRAW), View (SOLVIEW), and Profile (SOLPROF) commands. The Solids toolbar in Figure 22–2 also exposes these three commands. Only SOLDRAW and SOLVIEW will be discussed in this chapter. SOLPROF, although a powerful 2D multiview layout tool, does not include layout features that are available with SOLVIEW and SOLDRAW

Once the SOLVIEW command is entered, the display screen automatically switches to the first layout or Paper Space environment. Using SOLVIEW will lay out a view based on responses to a series of prompts, depending on the type of view you want to create. Usually, the initial view that serves as the starting point for other orthogonal views is based on the current user coordinate system. This needs to be determined before you begin this command. Once an initial view is created, it is very easy to create Orthographic, Section, Isometric, and even Auxiliary views.

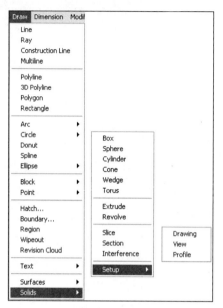

Figure 22–1

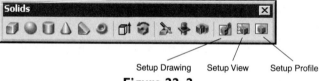

Setup Drawing   Setup View   Setup Profile
Figure 22–2

As SOLVIEW is used as a layout tool, the images of the views created are still simply plan views of the original solid model. In other words, after you lay out a view, it does not contain any 2D features, such as hidden lines. As shown in Figure 22–1, clicking on Drawing activates the SOLDRAW command, which is used to actually create the 2D profiles once it has been laid out through the SOLVIEW command.

 Try It! - Open the drawing file 22_Solview. Before using the SOLVIEW command, study the illustration of this solid model in Figure 22–3. In particular, pay close attention to the position of the User Coordinate System icon. The current position of the User Coordinate System will start the creation process of the base view of the 2D drawing.

 Tip: Before you start using the SOLVIEW command, remember to load the Hidden linetype. This will automatically assign this linetype to any new layer that requires hidden lines for the drawing mode. If the linetype is not loaded at this point, it must be manually assigned later to each layer that contains hidden lines through the Layer Properties Manager dialog box.

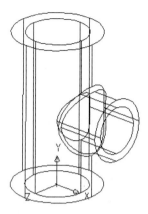

Figure 22–3

Activating SOLVIEW automatically switches the display to the layout or Paper Space environment. Since this is the first view to be laid out, the UCS option will be used to create the view based on the current User Coordinate System. The view produced is similar to looking down the Z axis of the UCS icon. A scale value may be entered for the view. For the View Center, click anywhere on the screen and notice the view being constructed. You can pick numerous times on the screen until the view is in a desired location. The placement of this first view is very important because other views will most likely be positioned from this one. When completed, press ENTER to place the view. Next, you are prompted to construct a viewport around the view. Remember to make this viewport large enough for dimensions to fit inside. Once the view is given a name, it is laid out similar to Figure 22–4.

 Command: **SOLVIEW**

Enter an option [Ucs/Ortho/Auxiliary/Section]: **U** *(For Ucs)*
Enter an option [Named/World/?/Current] <Current>: *(Press* ENTER*)*
Enter view scale <1.0000>: *(Press* ENTER*)*
Specify view center: *(Pick a point near the center of the screen to display the view; keep picking until the view is in the desired location)*
Specify view center <specify viewport>: *(Press* ENTER *to place the view)*
Specify first corner of viewport: *(Pick a point at "A" in Figure 22–4)*
Specify opposite corner of viewport: *(Pick a point at "B" in Figure 22–4)*
Enter view name: **FRONT**
UCSVIEW = 1 UCS will be saved with view
Enter an option [Ucs/Ortho/Auxiliary/Section]: *(Press* ENTER *to exit this command)*

Once the view has been laid out through the SOLVIEW command, use SOLDRAW to actually draw the view in two dimensions. If the Hidden linetype was loaded prior to using the SOLVIEW command, hidden lines will automatically be assigned to layers that contain hidden line information. The result of using the SOLDRAW command is shown in Figure 22–5. You are no longer looking at a 3D solid model but at a 2D drawing created by this command.

 Command: **SOLDRAW**

(If in Model Space, you are switched to Paper Space)
Select viewports to draw..
Select objects: *(Pick anywhere on the viewport in Figure 22–5)*
Select objects: *(Press ENTER to perform the SOLDRAW operation)*
One solid selected.

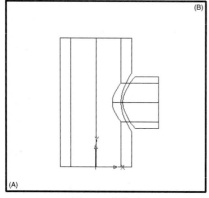

Figure 22–4

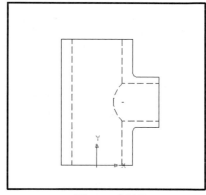

Figure 22–5

The use of layers in 2D-view layout is so important that when you run the SOLVIEW command, the layers shown in Figure 22–6 are created automatically. With the exception of Model and 0, the layers that begin with "FRONT" and the VPORTS layer were all created by the SOLVIEW command. The FRONT-DIM layer is designed to hold dimension information for the Front view. FRONT-HID holds all hidden lines information for the Front view; FRONT-VIS holds all visible line information for the Front view. All Paper Space viewports are placed on the VPORTS layer. The Model layer has automatically been frozen in the current viewport to hide the 3D model and show the 2D visible and hidden lines.

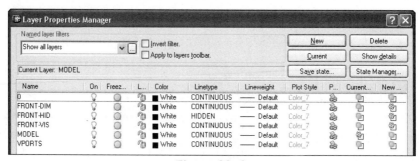

Figure 22–6

In order for the view shown in Figure 22–7 to be dimensioned, three operations must be performed. First, double-click inside the Front view to be sure it is the current floating Model Space viewport. Next, make FRONT-DIM the current layer. Finally, if it is not already positioned correctly, set the User Coordinate System to the current view using the View option of the UCS command. The UCS icon should be similar to the illustration in Figure 22–7 (you should be looking straight down the Z axis). Now add all dimensions to the view using conventional dimensioning commands with the aid of Object snap modes. When you work on adding dimensions to another view, the same three operations must be made in the new view: make the viewport active by double-clicking inside it, make the dimension layer current, and update the UCS, if necessary, to the current view with the View option.

When you draw the views using the SOLDRAW command and then add the dimensions, switching back to the solid model by clicking on the Model tab displays the image shown in Figure 22–8. In addition to the solid model of the object, the constructed 2D view and dimensions are also displayed. All drawn views from Paper Space will display with the model. To view just the solid model you would have to use the Layer Properties Manager dialog box along with the Freeze option and freeze all drawing-related layers.

 Tip: Any changes made to the solid model will not update the drawing views. If changes are made, you must erase the previous views and run SOLVIEW and SOLDRAW again to generate a new set of views.

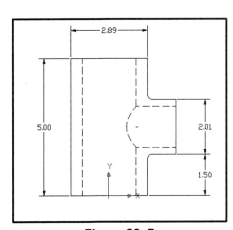

Figure 22–7

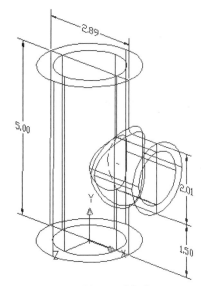

Figure 22–8

CREATING ORTHOGRAPHIC VIEWS

Once the first view is created, orthographic views can easily be created with the Ortho option of the SOLVIEW command.

Tip: To ensure predictable results, be sure to display the solid model in 2D wireframe mode (through the SHADEMODE command) prior to performing SOLVIEW/SOLDRAW operations. This can easily be verified—the solid model is not shaded and the UCS icon is not displayed in tri-color.

 Try It! - Open the drawing file 22_Ortho. Notice that you are in a layout and a Front view is already created. Follow the command prompt sequence below to create two orthographic views. When finished, your drawing should appear similar to Figure 22–9.

 Command: **SOLVIEW**
Enter an option [Ucs/Ortho/Auxiliary/Section]: **O** *(For Ortho)*
Specify side of viewport to project: *(Select the midpoint at "A")*
Specify view center: *(Pick a point above the front view to locate the top view)*
Specify view center <specify viewport>: *(Press ENTER to place the view)*
Specify first corner of viewport: *(Pick a point at "B")*
Specify opposite corner of viewport: *(Pick a point at "C")*
Enter view name: **TOP**
UCSVIEW = 1 UCS will be saved with view
Enter an option [Ucs/Ortho/Auxiliary/Section]: **O** *(For Ortho)*
Specify side of viewport to project: *(Select the midpoint at "D")*
Specify view center: *(Pick a point to the right of the front view to locate the right side view)*
Specify view center <specify viewport>: *(Press ENTER to place the view)*
Specify first corner of viewport: *(Pick a point at "E")*
Specify opposite corner of viewport: *(Pick a point at "F")*
Enter view name: **R_SIDE**
UCSVIEW = 1 UCS will be saved with view
Enter an option [Ucs/Ortho/Auxiliary/Section]: *(Press ENTER to exit this command)*

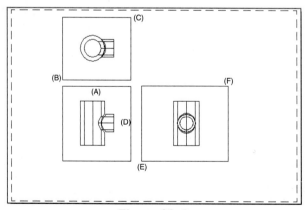

Figure 22–9

Running the SOLDRAW command on the three views displays the image as in Figure 22–10. Notice the appearance of the hidden lines in all views. The VPORTS layer is turned off to display only the three views.

Command: **SOLDRAW**

Select viewports to draw..

Select objects: *(Select the three viewports that contain the front, top, and right side view information)*

Select objects: *(Press* ENTER *to perform the* SOLDRAW *operation)*

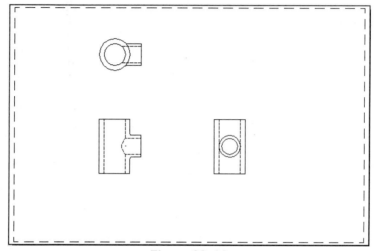

Figure 22–10

CREATING AN AUXILIARY VIEW

In Figure 22–11 the true size and shape of the inclined surface containing the large counterbore hole cannot be shown with the standard orthographic view. An auxiliary view must be used to properly show these features.

 Try It! - Open the drawing file 22_Auxiliary. From the 3D model in Figure 22–11 use the SOLVIEW command to create a Front view based on the current User Coordinate System. The results are shown in Figure 22–12.

 Command: **SOLVIEW**

Enter an option [Ucs/Ortho/Auxiliary/Section]: **U** *(For Ucs)*

Enter an option [Named/World/?/Current] <Current>: *(Press* ENTER*)*

Enter view scale <1.0000>: *(Press* ENTER*)*

Specify view center: *(Pick a point to locate the view–see Figure22–12)*

Specify view center <specify viewport>: *(Press* ENTER *to place the view)*

Specify first corner of viewport: *(Pick a point at "A" in Figure 22–12)*

Specify opposite corner of viewport: *(Pick a point at "B" in Figure 22–12)*

Enter view name: **FRONT**

UCSVIEW = 1 UCS will be saved with view

Enter an option [Ucs/Ortho/Auxiliary/Section]: *(Press* ENTER *to exit this command)*

Tip: Normally you do not end the SOLVIEW command after each view is laid out. Once you finish creating a view you simply enter the appropriate option (Ucs, Ortho, Auxiliary, or Section) and create the next one. This process can continue until all necessary views are provided.

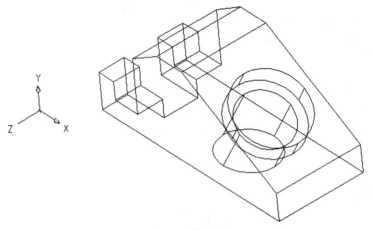

Figure 22–11

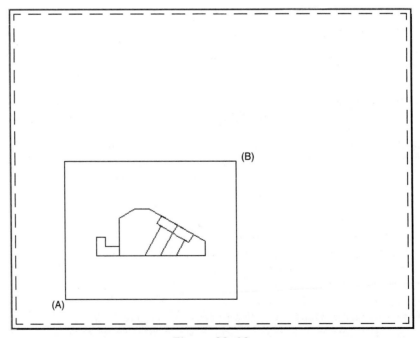

Figure 22–12

Now begin the process of constructing an auxiliary view as shown in Figure 22–13. After selecting the Auxiliary option of the SOLVIEW command, click on the endpoints at

"A" and "B" to establish the edge of the surface to view. Pick a point at "C" to indicate the side from which to view the auxiliary view. Notice how the Paper Space icon tilts perpendicular to the edge of the auxiliary view. Pick a location for the auxiliary view and establish a viewport. The result is illustrated in Figure 22–14.

Command: **SOLVIEW**
Enter an option [Ucs/Ortho/Auxiliary/Section]: **A** *(For Auxiliary)*
Specify first point of inclined plane: **End**
of *(Pick the endpoint at "A" in Figure 22–13)*
Specify second point of inclined plane: **End**
of *(Pick the endpoint at "B" in Figure 22–13)*
Specify side to view from: *(Pick a point inside of the viewport at "C" in Figure 22–13)*
Specify view center: *(Pick a point to locate the view–see Figure 22–14)*
Specify view center <specify viewport>: *(Press ENTER to place the view)*
Specify first corner of viewport: *(Pick a point at "D" in Figure 22–14)*
Specify opposite corner of viewport: *(Pick a point at "E" in Figure 22–14)*
Enter view name: **AUXILIARY**
UCSVIEW = 1 UCS will be saved with view
Enter an option [Ucs/Ortho/Auxiliary/Section]: *(Press ENTER to exit this command)*

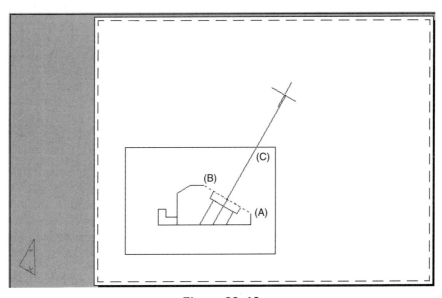

Figure 22–13

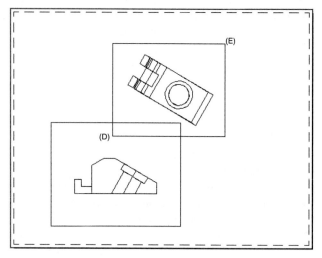

Figure 22–14

Run the SOLDRAW command and turn off the VPORTS layer. The finished result is illustrated in Figure 22–15. Hidden lines display only because this linetype was previously loaded.

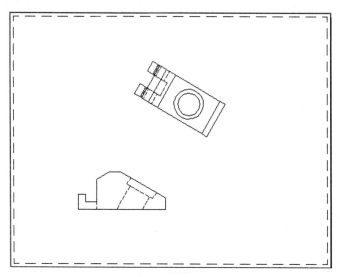

Figure 22–15

CREATING A SECTION VIEW

The SOLVIEW and SOLDRAW commands can also be used to create a full section view of an object. This process will automatically create section lines and place them on layer (*-HAT) for you.

 Try It! - Open the drawing file 22_Section. From the model in Figure 22–16, create a Top view based on the current User Coordinate System as in Figure 22–17.

 Command: **SOLVIEW**
Regenerating layout.
Enter an option [Ucs/Ortho/Auxiliary/Section]: **U** *(For Ucs)*
Enter an option [Named/World/?/Current] <Current>: *(Press* ENTER*)*
Enter view scale <1.0000>: *(Press* ENTER*)*
Specify view center: *(Pick a point to locate the view—see Figure 22–17)*
Specify view center <specify viewport>: *(Press* ENTER *to place the view)*
Specify first corner of viewport: *(Pick a point at "A" in Figure 22–17)*
Specify opposite corner of viewport: *(Pick a point at "B" in Figure 22–17)*
Enter view name: **TOP**
UCSVIEW = 1 UCS will be saved with view
Enter an option [Ucs/Ortho/Auxiliary/Section]: *(Press* ENTER *to exit this command)*

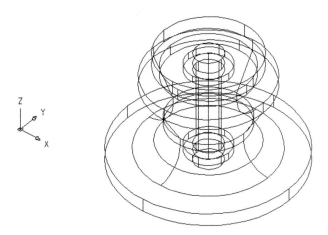

Figure 22–16

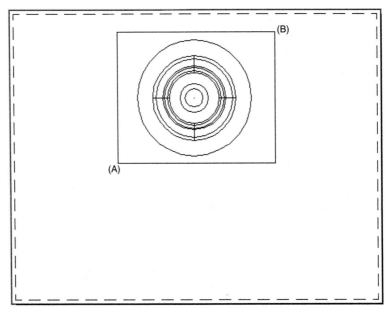

Figure 22–17

Begin the process of creating the section. You must first establish the cutting plane line in the Top view as in Figure 22–18. After the cutting plane line is drawn, you select the side from which to view the section. You then locate the section view. This is similar to the process of placing an auxiliary view.

 Command: **SOLVIEW**
Enter an option [Ucs/Ortho/Auxiliary/Section]: **S** *(For Section)*
Specify first point of cutting plane: **Qua**
of *(Pick a point at "A" in Figure 22–18)*
Specify second point of cutting plane: *(Turn Ortho on, pick a point at "B" in Figure 22–18)*
Specify side to view from: *(Pick a point inside of the viewport at "C" in Figure 22–18)*
Enter view scale <1.0000>: *(Press* ENTER*)*
Specify view center: *(Pick a point below the top view to locate the view–see Figure 22–19)*
Specify view center <specify viewport>: *(Press* ENTER *to place the view)*
Specify first corner of viewport: *(Pick a point at "D" in Figure 22–19)*
Specify opposite corner of viewport: *(Pick a point at "E" in Figure 22–19)*
Enter view name: **FRONT_SECTION**
UCSVIEW = 1 UCS will be saved with view
Enter an option [Ucs/Ortho/Auxiliary/Section]: *(Press* ENTER *to continue)*

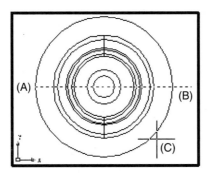

Figure 22–18

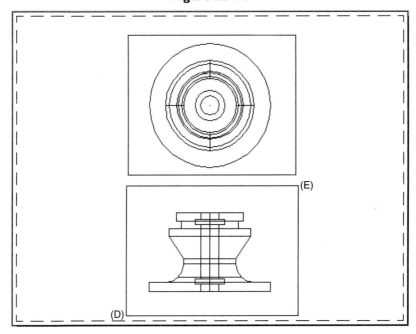

Figure 22–19

Running the SOLDRAW command on the viewports results in the image in Figure 22–20. You can also activate the viewport displaying the section view and use the HATCHEDIT command to edit the hatch pattern. In Figure 22–21, the hatch pattern scale was increased to a value of 2.00 and the viewports turned off.

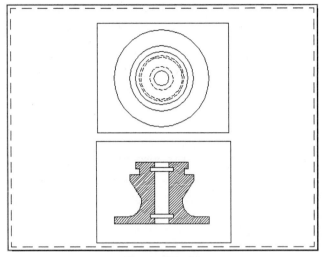

Figure 22–20

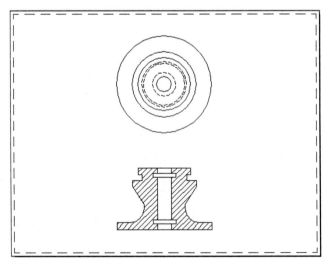

Figure 22–21

CREATING AN ISOMETRIC VIEW

Once orthographic, section, and auxiliary views are projected, you also have an opportunity to project an isometric view of the 3D model. This type of projection is accomplished using the UCS option of the SOLVIEW command and relies entirely on the viewpoint and User Coordinate System setting for your model.

Try It! - Open the drawing file 22_Iso. This 3D model should appear similar to Figure 22–22. To prepare this image to be projected as an isometric view, first define a new User Coordinate System based on the current view. See the prompt sequence below to accomplish this task. Your image and UCS icon should appear similar to Figure 22–23.

 Command: **UCS**
Current ucs name: *WORLD*
Enter an option [New/Move/orthoGraphic/Prev/Restore/Save/Del/Apply/?/World]
<World>: **N** *(For New)*
Specify origin of new UCS or [ZAxis/3point/OBject/Face/View/X/Y/Z] <0,0,0>: **V** *(For View)*

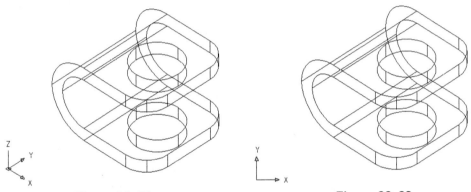

Figure 22–22   **Figure 22–23**

Next, run the SOLVIEW command based on the current UCS. Locate the view and construct a viewport around the isometric as shown in the sample layout in Figure 22–24. Since dimensions are placed in the orthographic view drawings and not on an isometric, you can tighten up on the size of the viewport.

 Command: **SOLVIEW**
Regenerating layout.
Enter an option [Ucs/Ortho/Auxiliary/Section]: **U** *(For Ucs)*
Enter an option [Named/World/?/Current] <Current>: *(Press ENTER)*
Enter view scale <1.0000>: *(Press ENTER)*
Specify view center: *(Pick a point to locate the view—see Figure 22–24)*
Specify view center <specify viewport>: *(Press ENTER to place the view)*
Specify first corner of viewport: *(Pick a point at "A" in Figure 22–24)*
Specify opposite corner of viewport: *(Pick a point at "B" in Figure 22–24)*
Enter view name: **ISO**
UCSVIEW = 1 UCS will be saved with view
Enter an option [Ucs/Ortho/Auxiliary/Section]: *(Press ENTER to exit this command)*

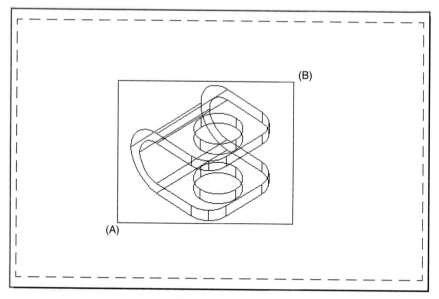

Figure 22–24

Running the SOLDRAW command on the isometric results in visible lines as well as hidden lines being displayed in Figure 22–25. Generally, hidden lines are not displayed in an isometric view. The layer called ISO-HID, that was created by SOLVIEW, contains the hidden lines for the isometric drawing. Use the Layer Properties Manager dialog box to turn off this layer. This results in the image shown in Figure 22–26.

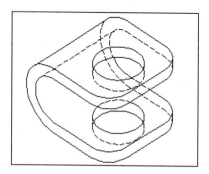

Figure 22–25

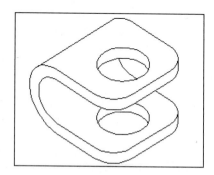

Figure 22–26

TUTORIAL EXERCISE: 22_SOLID DIMENSION.DWG

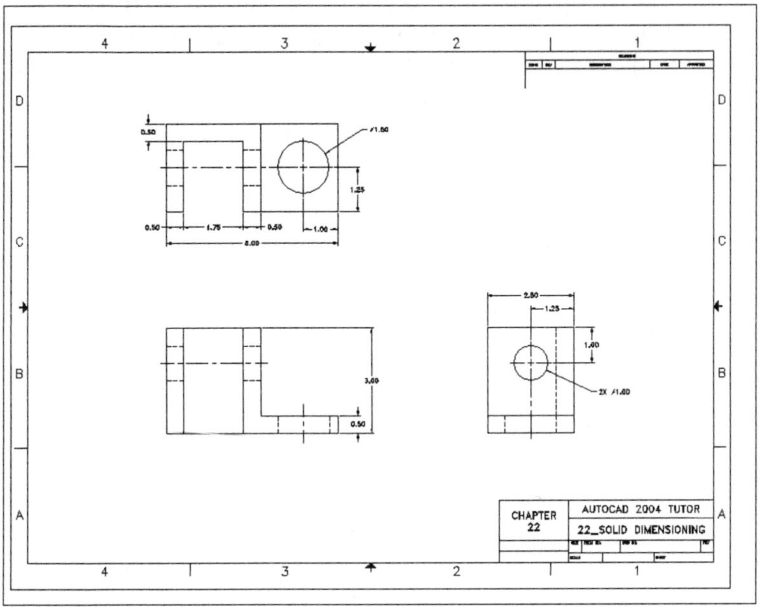

Figure 22–27

Purpose

This tutorial exercise is designed to add dimensions to a solid model that has had its views extracted using the SOLVIEW and SOLDRAW commands. See Figure 22–27.

System Settings

Drawing and dimension settings have already been changed for this drawing.

Layers

Layers have already been created for this tutorial exercise.

Suggested Commands

To start this tutorial, activate the front viewport and make the Front-DIM layer current.

Then add dimensions to the front view. As these dimensions are placed in the front view, the dimensions do not appear in the other views. This is because the SOLVIEW command automatically creates layers and freezes layers in viewports that do not display the dimensions. Next, activate the top viewport, make the Top-DIM layer current, and add dimensions to the top view. Finally, activate the right side viewport, make the right side-DIM layer current, and add the remaining dimensions to the right side views.

Whenever possible, substitute the appropriate command alias in place of the full AutoCAD command in each tutorial step. For example, use "CP" for the COPY command, "L" for the LINE command, and so on. The complete listing of all command aliases is located in chapter 1, table 1–2.

STEP 1

Begin this tutorial by opening up the drawing *22_Solid Dimension.* Your display should appear similar to Figure 22–28. Viewports have already been created and locked in this drawing. This means that as you activate the viewport and accidentally zoom, the image inside of the viewport will not zoom; rather the entire drawing will be affected by the zoom operation. Centerline layers have also been created and correspond to the three viewports. Centerlines have already been placed in their respective views.

Be sure Osnap is turned on with Endpoint being the active mode.

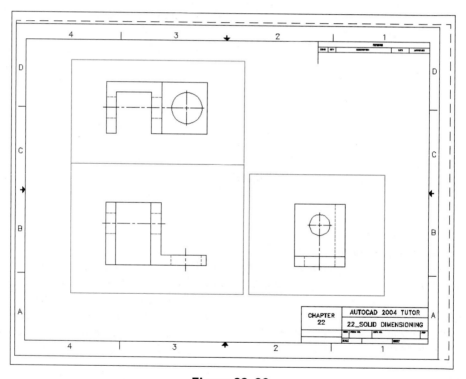

Figure 22–28

STEP 2

Activate the viewport that contains the Front View by double-clicking inside of it. You know you have accomplished this if the floating model space icon appears. Then make the Front-DIM layer current as shown in Figure 22–29.

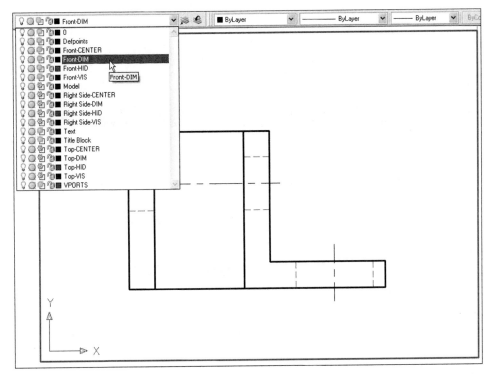

Figure 22–29

STEP 3

Most of the layers illustrated in Figure 22–29 were created by the SOLVIEW command. Notice how they correspond to a particular viewport. For example, study table 22–1 below regarding the layer names dealing with the front viewport:

Table 22–1

Layer Name	Purpose
Front-CENTER	Center lines
Front-DIM	Dimension lines
Front-HID	Hidden lines
Front-VIS	Visual (Object) lines

Notice the Layer Front-DIM in Figure 22–30. Because the viewport holding the front view is current, this layer is thawed, meaning the dimensions placed in this viewport will be visible. Notice that the other dimension layers (Right Side-DIM and Top-DIM) are frozen. The SOLVIEW command automatically set up the dimension layers to be visible in the current viewport and frozen in the other viewports.

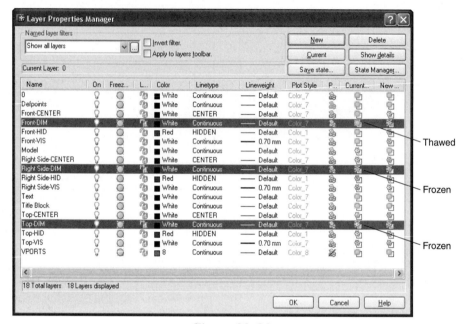

Figure 22–30

STEP 4

Add dimensions to the front view using Figure 22–31 and the following prompt sequences as guidelines.

⊞ Command: **DLI** *(For DIMLINEAR)*

Specify first extension line origin or
   <select object>: *(Pick the Endpoint at "A")*

Specify second extension line origin: *(Pick the Endpoint at "B")*

Specify dimension line location or
[Mtext/Text/Angle/Horizontal/Vertical/
   Rotated]: *(Locate the dimension in Figure 22–31)*

Dimension text = 0.50

⊞ Command: **DBA** *(For DIMBASELINE)*

Specify a second extension line origin
   or [Undo/Select] <Select>: *(Pick the Endpoint at "C")*

Dimension text = 3.00

Specify a second extension line origin or
   [Undo/Select] <Select>: *(Press ENTER)*

Select base dimension: *(Press ENTER)*

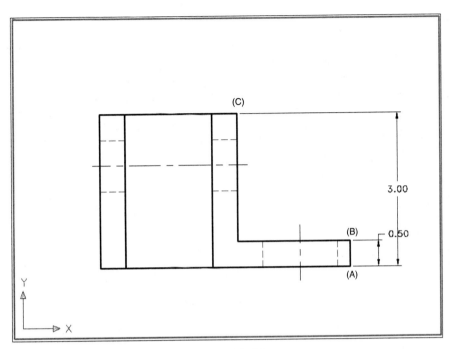

Figure 22–31

Using grips, stretch the text of the 0.50 dimension so that it is located in the middle of the extension lines as shown in Figure 22–32. This completes the dimensioning of the front view.

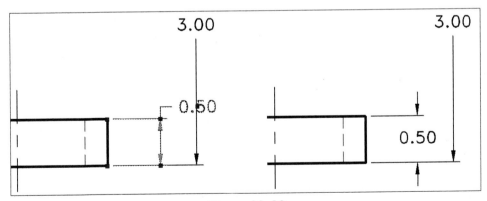

Figure 22–32

STEP 5

Activate the viewport that contains the Top View by double-clicking inside of it. Then make the Top-DIM layer current. This layer is designed to show dimensions visible in the Top viewport and make dimensions in the Front and Right Side viewports frozen (or invisible). See Figure 22–33.

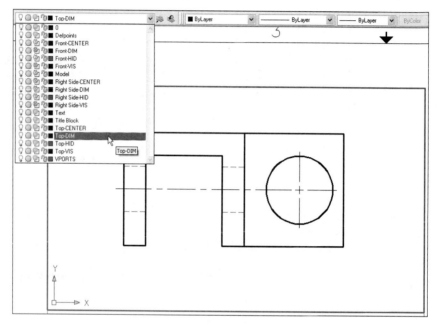

Figure 22–33

STEP 6

Add dimensions to the top view using Figure 22–34 and the following prompt sequences as guidelines.

 Tip: Because the viewports are locked, use the zoom and pan operations freely while dimensioning in floating model space.

Command: **DLI** (*For DIMLINEAR*)
Specify first extension line origin or
  <select object>: (*Pick the endpoint at "A"*)
Specify second extension line origin: (*Pick the endpoint at "B"*)

Specify dimension line location or
[Mtext/Text/Angle/Horizontal/Vertical/
  Rotated]: (*Locate the dimension in Figure 22–34*)
Dimension text = 1.75

⊞ Command: **DLI** (*For DIMLINEAR*)

Specify first extension line origin or
   <select object>: *(Pick the endpoint at "B")*
Specify second extension line origin: *(Pick
   the endpoint at "C")*
Specify dimension line location or
[Mtext/Text/Angle/Horizontal/Vertical/
   Rotated]: *(Locate the dimension in Figure
   22–34 – try picking the endpoint of the
   dimension line for the 1.75 dimension to
   align it exactly)*
Dimension text = 0.50

⊞ Command: **DLI** (*For DIMLINEAR*)

Specify first extension line origin or
   <select object>: *(Pick the endpoint at "A")*
Specify second extension line origin: *(Pick
   the endpoint at "D")*
Specify dimension line location or
[Mtext/Text/Angle/Horizontal/Vertical/
   Rotated]: *(Locate the dimension in Figure
   22–34)*
Dimension text = 0.50

⊞ Command: **DLI** (*For DIMLINEAR*)

Specify first extension line origin or
   <select object>: *(Pick the endpoint at "E")*
Specify second extension line origin: *(Pick
   the endpoint at "F")*
Specify dimension line location or
[Mtext/Text/Angle/Horizontal/Vertical/
   Rotated]: *(Locate the dimension in Figure
   22–34)*
Dimension text = 0.50

Using grips, stretch the text of the 0.50
dimension and locate it in the middle of
the extension lines in Figure 22–35.

⊞ Command: **DLI** (*For DIMLINEAR*)

Specify first extension line origin or
   <select object>: *(Pick the endpoint at "G")*
Specify second extension line origin: *(Pick
   the endpoint at "H")*
Specify dimension line location or
[Mtext/Text/Angle/Horizontal/Vertical/
   Rotated]: *(Locate the dimension in Figure
   22–34)*
Dimension text = 1.00

⊞ Command: **DLI** (*For DIMLINEAR*)

Specify first extension line origin or
   <select object>: *(Pick the endpoint at "H")*
Specify second extension line origin: *(Pick
   the endpoint at "J")*
Specify dimension line location or
[Mtext/Text/Angle/Horizontal/Vertical/
   Rotated]: *(Locate the dimension in Figure
   22–34)*
Dimension text = 1.25

⊞ Command: **DLI** (*For DIMLINEAR*)

Specify first extension line origin or
   <select object>: *(Pick the endpoint at "D")*
Specify second extension line origin: *(Pick
   the endpoint at "H")*
Specify dimension line location or
[Mtext/Text/Angle/Horizontal/Vertical/
   Rotated]: *(Locate the dimension in Figure
   22–34)*
Dimension text = 5.00

⊘ Command: **DDI** (*For DIMDIAMETER*)

Select arc or circle: *(Pick the edge of the
   circle)*
Dimension text = 1.50
Specify dimension line location or [Mtext/
   Text/Angle]: *(Locate the dimension in
   Figure 22–34)*

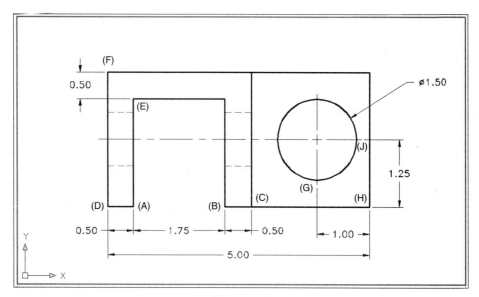

Figure 22–34

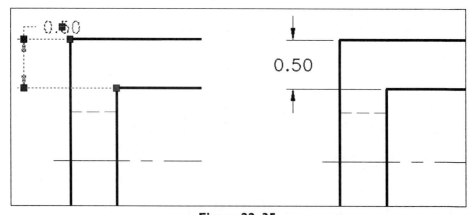

Figure 22–35

STEP 7

Activate the viewport that contains the Right Side View by clicking inside of the viewport. Then make the Right Side-DIM layer current. This layer is designed to make dimensions visible in the Right Side viewport and make dimensions in the Front and Top viewports frozen (or invisible). See Figure 22–36.

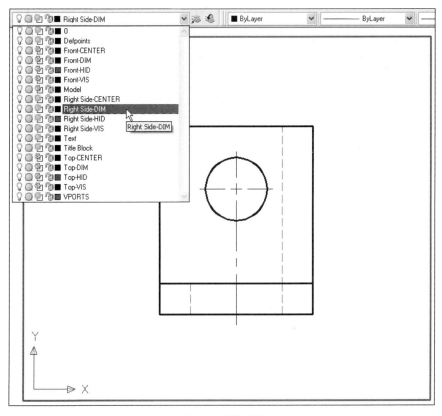

Figure 22–36

STEP 8

Add dimensions to the right side view using Figure 22–37 and the following prompt sequences as guides.

⊟ Command: **DLI** *(For DIMLINEAR)*
Specify first extension line origin or
 <select object>: *(Pick the endpoint at "A")*
Specify second extension line origin: *(Pick the endpoint at "B")*
Specify dimension line location or
[Mtext/Text/Angle/Horizontal/Vertical/
 Rotated]: *(Locate the dimension in Figure 22–37)*
Dimension text = 1.25

⊟ Command: **DLI** *(For DIMLINEAR)*
Specify first extension line origin or
 <select object>: *(Pick the endpoint at "B")*
Specify second extension line origin: *(Pick the endpoint at "C")*
Specify dimension line location or
[Mtext/Text/Angle/Horizontal/Vertical/
 Rotated]: *(Locate the dimension in Figure 22–37)*
Dimension text = 1.00

Command: **DLI** (For DIMLINEAR)

Specify first extension line origin or <select object>: (Pick the endpoint at "B")

Specify second extension line origin: (Pick the endpoint at "D")

Specify dimension line location or [Mtext/Text/Angle/Horizontal/Vertical/Rotated]: (Locate the dimension in Figure 22–37)

Dimension text = 2.50

 Command: **DDI** (For DIMDIAMETER)

Select arc or circle: (Pick the edge of the circle)

Dimension text = 1.00

Specify dimension line location or [Mtext/Text/Angle]: (Locate the dimension in Figure 22–37)

 Note: The "2X" will be added to the dimension text in the next step

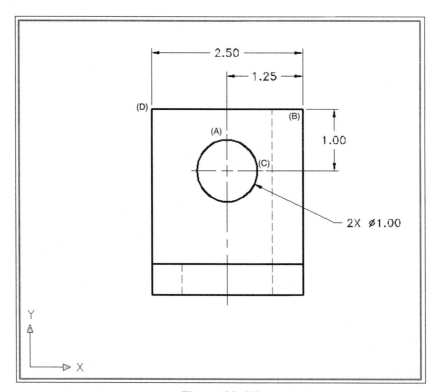

Figure 22–37

STEP 9

Edit the diameter dimension text to reflect two holes. Use the DDEDIT command to accomplish this. This will activate the Multiline Text Editor box illustrated in Figure 22–38.

Command: **ED** (For DDEDIT)

Select an annotation object or [Undo]:
(Select the 1.00 diameter text. The Multiline Text Editor box displays in Figure 22–38)

When the Text Formatting dialog box appears, type "2X" from the keyboard and click the OK button.

Select an annotation object or [Undo]:
(Press ENTER to exit the command and return to the drawing)

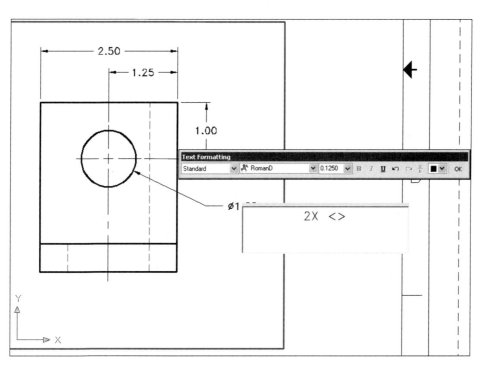

Figure 22–38

STEP 10

The completed dimensioned solid model
is illustrated in Figure 22–39.

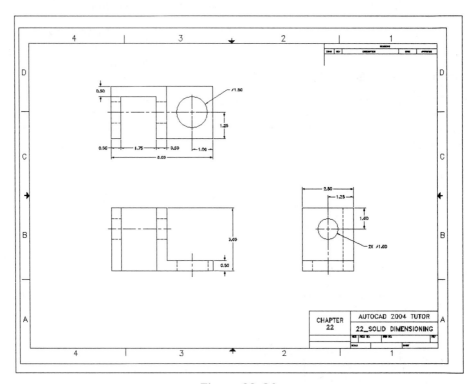

Figure 22–39

 Open the Exercise Manual PDF file for Chapter 22 on the accompanying CD for more tutorials and exercises.

 If you have the accompanying Exercise Manual, refer to Chapter 22 for more tutorials and exercises.

CHAPTER 23

Producing Photorealistic Renderings

This chapter introduces you to renderings in AutoCAD and how to produce realistic renderings of 3D models. The heart of the rendering operation is the Render dialog box. Most of its options will be discussed in detail. Adding lights will be the next topic of discussion. This important feature is where you produce unlimited special effects depending on the location and intensity of the lights. The Point, Distant, and Spotlights features will all be discussed. Once you have enhanced your drawings with lights, you can group these lights with existing named views in what are called Scenes. You will be able to attach materials to your models to provide additional realism. Materials can be selected from a library available in AutoCAD, or you can even make your own materials.

AN INTRODUCTION TO RENDERINGS

Engineering and architectural drawings are able to pack a vast amount of information into a 2D outline drawing supplemented with dimensions, some symbols, and a few terse notes. However, training, experience, and sometimes imagination are required to interpret them, and many people would rather see a realistic picture of the object. Actually, realistic pictures of a 3D model are more than just a visual aid for the untrained. They can help everyone visualize and appreciate a design, and can sometimes even reveal design flaws and errors.

Shaded, realistic pictures of 3D models are called renderings. Until recently, they were made with colored pencils and pens or with paint brushes and airbrushes. Now they are often made with computers, and AutoCAD comes with a rendering program that is automatically installed by the typical AutoCAD installation and is ready for your use. Figure 23–1 shows, for comparison, the solid model of a bracket in its wireframe form, as it looks when the HIDE command has been invoked, and when it is rendered.

This chapter is designed to give you an overview of the Rendering topic, which will allow you to create pleasing, photorealistic renderings of your 3D models.

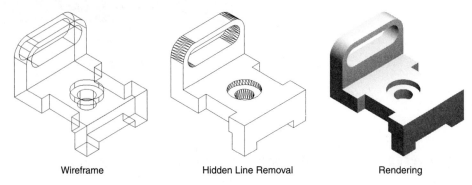

| Wireframe | Hidden Line Removal | Rendering |

Figure 23–1

Illustrated in Figure 23–2 is the Render toolbar. All rendering commands can be accessed from this toolbar.

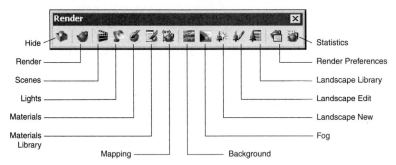

Figure 23–2

Rendering commands can also be selected from the View pull-down menu illustrated in Figure 23–3.

Figure 23–3

The commands covered in this unit are:

RENDER—Displays the Render dialog box and performs the render.

RPREF— Displays the Rendering Preferences dialog box, which allows you to preset the options for a rendering.

LIGHT—Displays the Light dialog box designed for placing lights in your drawing.

SCENE—Displays the Scene dialog box designed for collecting lights and views and placing them in named scenes.

RMAT —Displays the Materials dialog box designed for attaching materials to your models.

MATLIB— Displays the Materials Library dialog box designed for loading new material definitions into your drawing.

BACKGROUND—Displays the Background dialog box designed for placing a color or image in the background of your rendering for extra special effects.

FOG—Displays the Fog dialog box designed to create the effect of fog.

SAVEIMG—Designed to save a rendering to a file.

THE RENDER COMMAND

The RENDER command is a more flexible and advanced version of the SHADE command (see chapter 20). The RENDER command creates much more realistic images than the SHADE command does. For example, RENDER smoothes out all of the facets you see on a 3D sphere, creating the image of a round ball.

As with the SHADE command, you can create a rendering by simply typing the RENDER command. Or you can set many options to create a complex rendering: you can set up many lights, apply materials to objects, place a background image, and save the rendering in one of several common file formats.

However, RENDER requires that your graphics card be capable of displaying at least 256 colors. For the most realistic renderings, your computer system should be set up with a graphics card that displays 16.7 million colors (also known as "24 bits" or "true color"). To display that many colors, your graphics card needs 64MB of RAM, a feature that is becoming common and inexpensive by today's standards.

Before continuing with this unit, you should use the Windows Control Panel to check that the Display is configured to True Color in order to achieve the most realistic renderings possible.

AN OVERVIEW OF RENDERING

Begin your study of rendering on the 3D model of a cylinder head. Open the drawing 23_Valve Head.Dwg and activate the Render dialog box shown in Figure 23–4 by clicking on the Render button located in the Render toolbar. If this is the first time that the render command has been used, AutoCAD will take a few seconds to load this application.

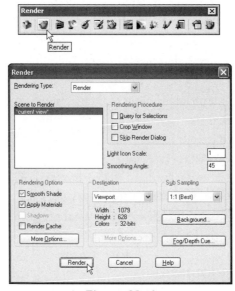

Figure 23–4

With the Render dialog box displayed, ignore all options for now and click on the OK button. Within a few seconds (depending on the speed of your computer) you should see a rendered image of 23_Valve Head, as shown in Figure 23–5.

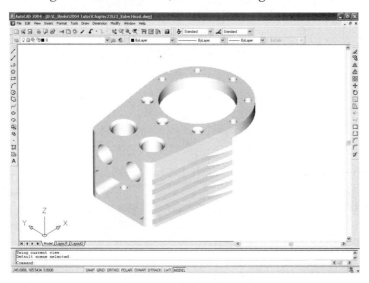

Figure 23–5

RENDERING PREFERENCES

For most renderings, you probably won't need to know anything more about the render command. For total control, though, it is useful to know all about the options of the Render dialog box. To manipulate these values, click the Rendering Preferences button in the Render toolbar to activate the Rendering Preferences dialog box displayed in Figure 23–6 The RPREF command (Rendering Preferences) allows you to preset options for the Render dialog box .

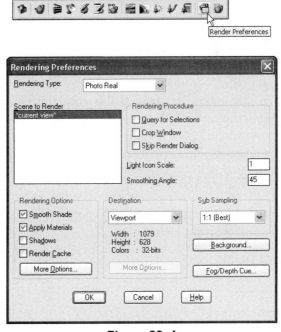

Figure 23–6

RENDERING TYPE

Three options are available in this drop-down box, namely Render, Photo Real, and Photo Raytrace. Each option is progressively more realistic in the results it displays, but takes longer to render.

SCENE TO RENDER

With your first rendering, there is only one scene defined: the current view. Later, we will use the scene command to create named scenes whose names appear in this part of the Render dialog box.

RENDERING OPTIONS

Rendering options are the core of the render command. For the most part, you will leave the options as they are set by default. If you want to change any of them, here is a summary of the effect they will have on the final rendering:

Smooth Shade—When this is turned on, AutoCAD creates a smooth transition through faceted areas of the 3D model. This makes the model look more realistic; the cutoff angle for smoothing facets is specified by the Smooth Angle option (default = 45 degrees). You would only turn off the Smooth Shading option for faster renderings.

Apply materials—When this is turned on, AutoCAD renders the objects using the materials defined by the RMAT (Materials) command. This helps the object look more realistic. You would turn off this option for faster renderings.

Shadows—This option is only available with the Photo Real and Photo Raytrace rendering types. When this is turned on, shadows are generated by direct, spot, and point lights. Turning on this option slows down the rendering process but makes the image more realistic.

Render Cache—When this is turned on, this option speeds up the renderings after the first one, by only rendering those parts that have changed from the previous rendering. Turning on this option makes the first rendering slower but speeds up subsequent ones.

RENDERING PROCEDURE

AutoCAD's Render module gives you three choices when performing each rendering.

Query for Selections—When this is turned on, AutoCAD prompts you to select specific objects for rendering. This is useful for creating a faster rendering, since not every object in the drawing is rendered.

Crop Window—When this is turned on, AutoCAD prompts you to window a portion of the drawing to render. Once again, this speeds up rendering by rendering a smaller portion of the overall drawing.

Skip Render dialog—Use this option when you have all rendering parameters already set up and you don't want to see the Render dialog box every time you enter the RENDER command. You will have to use the RPREF command to deselect this option if you want the Render dialog box to be available again.

LIGHT ICON SCALE

The lights that you place in a rendering scene are actually AutoCAD blocks. This option lets you specify a different size from the default scale factor of 1.0 units.

DESTINATION

AutoCAD gives you three destinations to which you can have the rendered image sent: these are the Current Viewport; an Independent Rendering Window; and a File on Disk. When you select File, there are additional options you will need to decide on, such as the file format, resolution, interlacing, and so on.

SUB SAMPLING

This option determines the number of pixels that are rendered. The default 1:1 renders all pixels in the scene. The setting of 8:1 renders one pixel out of eight, which results in a very chunky looking rendering (see Figure 23–7). When placing lights in the drawing and adding materials to the model, use higher sub sampling values, such as 4:1, to preview the results and speed up the rendering process. When you are satisfied with your light and material placements, switch to a sub sampling of 1:1. The rendering will take longer to process but will provide the best results.

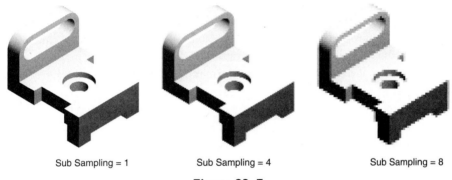

Sub Sampling = 1   Sub Sampling = 4   Sub Sampling = 8

Figure 23–7

BACKGROUND

Use this button to display the Background dialog box, which lets you select the kind of background you want to appear behind the rendered image. For example, the background could be a raster image of a grassy field, clouds, or cityscape. You will learn more about using backgrounds later in this chapter.

FOG/DEPTH CUE

Clicking on this button displays the Fog/Depth dialog box, which allows you to blur or fog the foreground or background in your rendering. This feature will be discussed in greater detail later in this chapter.

MORE OPTIONS

Clicking on this button displays the Render Options dialog box. When the rendering type is set to AutoCAD Render, the dialog box illustrated in Figure 23–8 is displayed. There are two Render Quality options you can choose from:

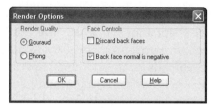

Figure 23–8

Render Quality—Gouraud and Phong are the names of the algorithms used by AutoCAD Render to create its images. Each is named after the computer scientist who came up with the rendering algorithms. Phong is more realistic when your model has lights assigned, but it takes longer to render than Gouraud does.

Face Controls—AutoCAD uses the concept of *face normals* to determine which parts of the model are the front and which are the back. A *face* is the 3D polygon that makes up the surface of the model; a *normal* is a vector that points out at right angles from the face. AutoCAD determines the direction of the normal by applying the right-hand rule to the order in which the face vertices were drawn. A *positive* normal points toward you; a *negative* normal points away from you. To save time in rendering, negative face normals are ignored, since they won't be seen in the rendering. You would turn on Discard Face Normals for faster rendering; turn off the option when mistakes appear in the rendering.

PHOTO REAL RENDER OPTIONS

When the Rendering Type is Photo Real, the following dialog box is displayed (see Figure 23–9). This dialog box gives you four rendering options you can adjust:

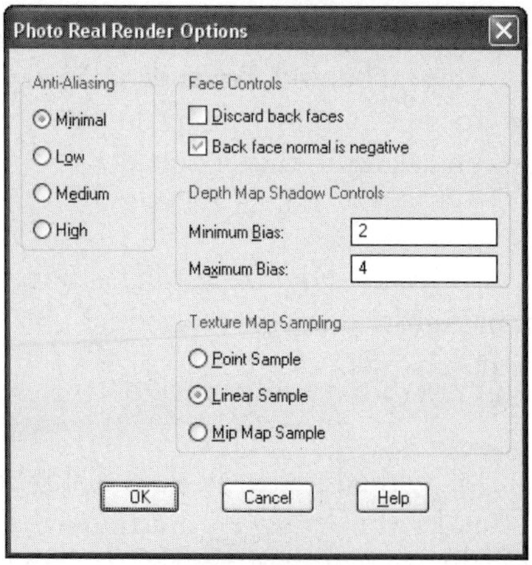

Figure 23–9

Anti-Aliasing—This is a software technique for reducing the stairstep-line edges of a rendered image to make the edge look smoother. The stair-step effect or "jaggies" are due to the monitor's limited resolution of about 72 dpi (dots per inch). That is a much coarser number than a typical 600 dpi laser printer. Thus, the Render software adds subtle pixels of gray and other colors to make the edges look smoother. Technically, AutoCAD uses the following styles of anti-aliasing:

> Minimal—Analytical horizontal anti-aliasing
>
> Low—Use 4 shading samples per pixel
>
> Medium—Use 9 shading samples per pixel
>
> High—Use 16 shading samples per pixel

The drawback to anti-aliasing is that the rendering process takes longer and the edges can look fuzzy. The higher the anti-aliasing setting, the smoother the look, but the longer it takes for the rendering to process.

Face Controls—The same issues apply for Photo Real as for the AutoCAD Rendering options.

Depth Map Shadow Controls—This feature helps prevent erroneous shadows, such as a shadow that casts its own shadow or a shadow that doesn't connect with its object. Autodesk recommends that the Minimum Bias should range between 2 and 20, while the Maximum Bias should range between 4 and 10 greater that the minimum. (Note that additional shadow controls are available when defining lights.)

Texture Map Sampling—A texture map is usually a raster image that is applied to a 3D object to make it look more realistic. For example, you could apply a picture of bricks onto a fireplace or a picture of a square of carpet onto the floor. This controls how a texture map is handled when it is applied to an object smaller than the texture map.

> Point—The nearest pixel in the bitmap.
>
> Linear—Averages 4 pixel neighbors.
>
> MIP Map—Pyramidal average of a square sample.

RAYTRACE RENDER OPTIONS

When the Rendering Type is Raytrace, the following dialog box is displayed (see Figure 23–10). This dialog box gives you two more options to play with:

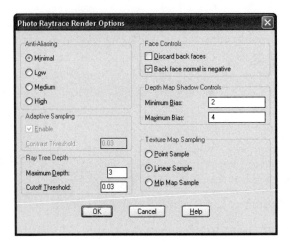

Figure 23–10

Adaptive Sampling—This option is only available when you turn on Low, Medium, or High anti-aliasing. Adaptive sampling allows for a faster anti-aliased rendering by ignoring some pixels. The Enable checkbox toggles this option. The Contrast Threshold adjusts the sensitivity between 0.0 and 1.0. For values closer to 0.0, AutoCAD takes more samples; for values closer to 1.0, AutoCAD takes fewer samples for faster rendering speed but possibly lower quality rendering.

Ray Tree Depth—In ray tracing, AutoCAD follows each beam of light as it reflects (bounces off opaque objects) and refracts (transmits through transparent objects) among the objects in the scene. This control allows you to speed up rendering by limiting the amount of reflecting and refracting that takes place. The Maximum Depth is the largest number of "tree branches" AutoCAD keeps track of as the light bounces and transmits through the scene; the range is 0 to 10. The Cutoff Threshold determines the percentage that a Raytrace must affect a pixel before ray tracing stops; this range is 0.0 to 1.0.

Try It! – This exercise consists of an overview of rendering. You will be attaching a material to one of the 3D elements and using the three different render types to observe the results. (Materials and the Materials Library will be covered in greater detail later in this chapter.) Open the drawing file 23_Render Types. You will notice a container and two glasses resting on top of a flat platform. Click on the Materials Library button in the Render toolbar shown in Figure 23–11, which will activate the Materials Library dialog box shown in Figure 23–12.

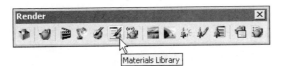

Figure 23–11

With the Materials Library dialog box active, browse through the various material types along the right side of the dialog box and pick CHECKER TEXTURE. Then click the Import button to load this material from the library into the 23_Render Types drawing (see Figure 23–12).

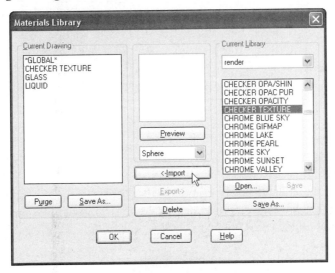

Figure 23–12

Next, click on Materials in the Render toolbar illustrated in Figure 23–13. This will launch the Materials dialog box shown in Figure 23–14.

Figure 23–13

With the Materials dialog box active, click on the material CHECKER TEXTURE; then click the Attach button as shown in Figure 23–14. When you return to the drawing editor, pick the edge of the rectangular base. This will apply the CHECKER TEXTURE to this object.

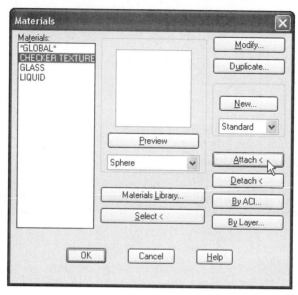

Figure 23–14

Now experiment with the various rendering types and observe the results. The first render type to sample is Render. Picking this type and clicking on the Render button at the bottom of the Render dialog box produces the results illustrated in Figure 23–15. The Render type is the lowest form of rendering. As a few highlights are picked up (liquid in the container and glasses), the base does not display the CHECKER TEXTURE.

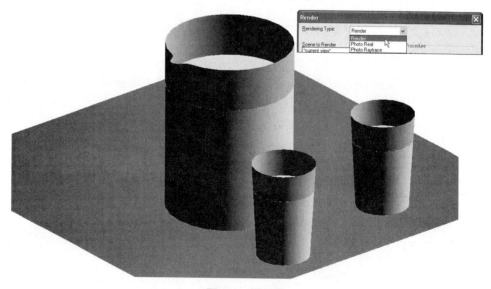

Figure 23–15

Return back to the Render dialog box. This time, pick the Photo Real rendering type. Then pick the Render button at the bottom of the Render dialog box. This will produce the results illustrated in Figure 23–16. Notice that the CHECKER TEXTURE is visible along with transparent glasses and shadows cast.

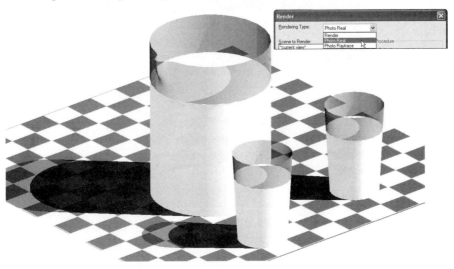

Figure 23–16

The highest form of rendering is the Photo Raytrace type. In addition to shadow casting, notice the reflection of the CHECKER TEXTURE in the glasses and container as shown in Figure 23–17.

Figure 23–17

Return back to the Render dialog box and pick the Background button (located in the lower-right corner of the dialog box), as shown in Figure 23–18. You will apply a background image to the rendering.

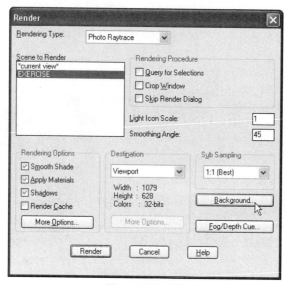

Figure 23–18

Clicking the Background button displays the Background dialog box illustrated in Figure 23–19. Of the four background types (Solid, Gradient, Image, and Merge), pick the Image type. Then pick the Find File button located in the lower-left corner of the dialog box.

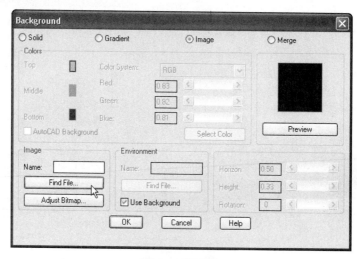

Figure 23–19

A number of textures are available in the Background Image dialog box. Click the file "Cloud.tga" as shown in Figure 23–20. Click the Open button. This will return you to the Render dialog box. Check to see that Photo Raytrace is selected and click the Render button located at the bottom of the dialog box.

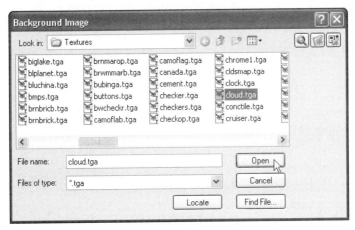

Figure 23–20

The result of applying a background is illustrated in Figure 23–21. Now, even the cloud and sky background appears to be reflected in the glasses and container, along with the checker texture pattern.

Figure 23–21

CREATING LIGHTS FOR RENDERING PURPOSES

Whereas the SHADE command is limited to a single light source, the LIGHT command has four different kinds of light sources. These are point, spot, distant, and ambient lights. Except for ambient light, you can place as many of each of these lights in your drawing as you want. All lights can emit any color at any level of brightness. The names of the lights have special meanings:

> **Point Light**—This type emits light in all directions with varying intensity. The best example of a point light is the light bulb typically located in the ceiling in a room.

> **Spotlight**—This type emits light beams in the shape of a cone. The best example of a spotlight is a high-intensity desk lamp or a vehicle-mounted spotlight. When you place spotlights in a drawing, you specify the *hotspot* of the light (where the light is brightest) and the *falloff* (where the light diminishes in intensity).

> **Distant Light**—This type emits light of parallel beams and constant intensity. The best examples of distant lights are the sun and moon. Typically, you want to place a single distant light to simulate the sun. To simulate a setting sun, you would change the color of the light to orange-red.

> **Ambient Light**—This type is an omnipresent light source that ensures that the object in the scene has illumination. There is a single ambient light source in every rendering. You would turn off the ambient light to simulate a nighttime scene.

 ## PLACING LIGHTS IN A DRAWING

The LIGHT command allows you to place and adjust settings for lights in the drawing. To place lights in a drawing, select Lights from the View pull-down menu or the Render toolbar in Figure 23–22. This will display the Lights dialog box illustrated in Figure 23–23. Initially no lights are defined other than a single default light source located at your eye.

Before placing the light, you need to decide on the type of light you wish to use: Point Light, Distant Light, or Spotlight. For now, select Distant Light from the list box, because a sun-like light is the easiest to work with (Point Lights and Spotlights are more challenging to place.)

With Distant Light selected, click on the New button to give the light a name and to specify its direction and other parameters. The dialog box shown in Figure 23–24 for each of the three light types is roughly similar, depending on their characteristics.

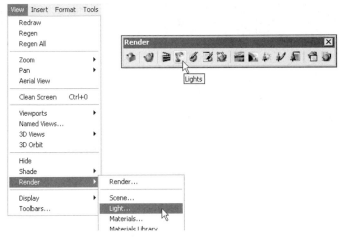

Figure 23–22

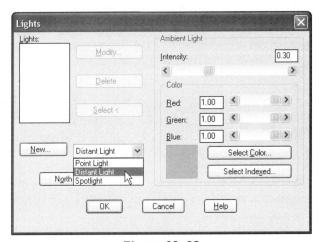

Figure 23–23

The following are common to all three light types:

Name—Give the light a convenient name, such as "SUN".

Intensity—The brighter the light, the higher the intensity value. An intensity of zero turns off the light.

Color—You can select any color for each light. Clicking on Select Custom Color displays the Windows Color dialog box. Select from the ACI (the AutoCAD Color Index), the easiest option to use. Select the color yellow for the color of the light and click the OK button.

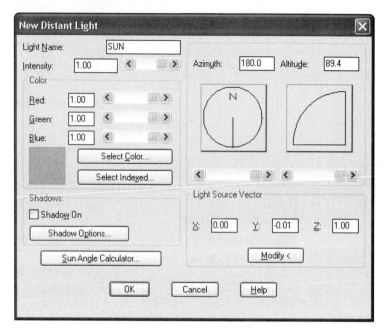

Figure 23–24

The New Distant Light dialog box in Figure 23–24 also lets you define the sun's azimuth (how far around in the day) and altitude (height in the sky). You do not have to select a position for the light, since the light's intensity doesn't change with distance. It doesn't matter how far the sun is from the object.

The Shadows area lets you specify the type of shadow to use. Volumes are faster with sharp borders; Map is slower but has soft borders (see Figure 23–25).

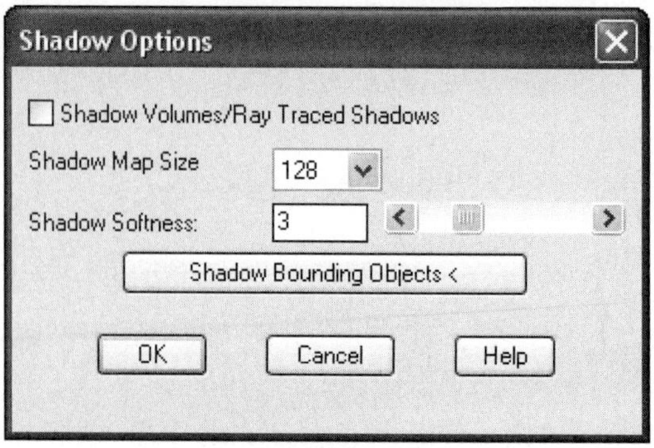

Figure 23–25

Clicking on the Sun Angle Calculator... button activates the Sun Angle Calculator dialog box shown in Figure 23–26, which allows you to select the sun's position by date, time, and geographic location.

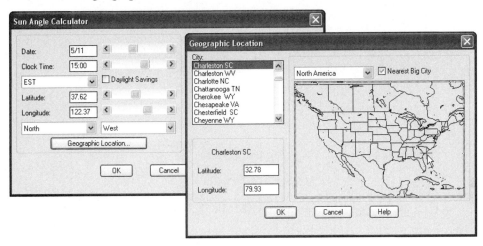

Figure 23–26

Exiting all light-related dialog boxes places an image of the distant light in the drawing.

When placing lights in the drawing, AutoCAD places a block on a special layer named "ASHADE". Except for the ambient light type, each type of light has a unique block shape, as shown in Figure 23–27.

Figure 23–27

Try It! – Open the file 23_Light Exercise.dwg. This drawing file, shown in Figure 23–28, contains the solid model of a machine component that we will use for a rendering. For general illumination of the model, you will use a single distant light that is aimed perpendicular to the model's center line and is pointed down at a fairly steep angle so that the shaded side is on the bottom.

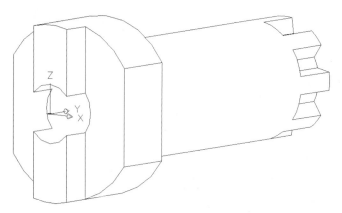

Figure 23–28

To install this light, initiate the LIGHT command. Make sure that ambient light intensity is set to its default value of 0.30, and all of its color values are 1.00 (so that its color is white). Then select Distant Light from the light type pull-down list, and click the New button as shown in Figure 23–29.

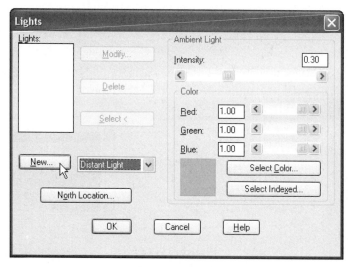

Figure 23–29

In the New Distant Light dialog box, assign any name you want to this light. Make sure that the light's intensity is 1.00, all of the three color values are 1.00, and the Shadow On checkbox is turned off (there is no surface under the model, so shadows will have little effect). Also, set the distant light's azimuth to 90°, and set its altitude to 60°. Your display should appear similar to Figure 23–30.

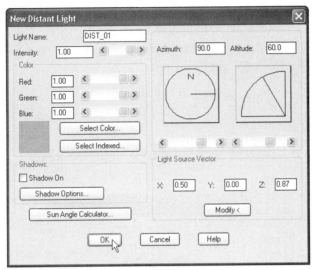

Figure 23–30

Click the OK button to return to the main Light dialog box, and click its OK button to complete the LIGHT command. Then start the RENDER command. In the Render dialog box, select Photo Real as the rendering type, then click the Render button. The distant light did a good job with the cylindrical surfaces, as shown in Figure 23–31, but the front of the model is almost completely in the shade. We will add a spotlight to illuminate it.

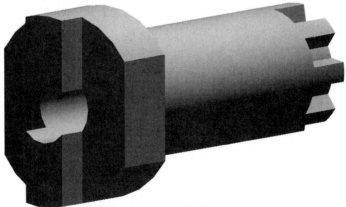

Figure 23–31

Invoke the LIGHT command again. Select Spotlight as the light type and click the New button. In the New Spotlight dialog box illustrated in Figure 23–32, assign a name of your choice to the light. Make certain that the light color is white and that the Shadow On checkbox is cleared.

To make the three separate but parallel surfaces stand out from each other we will use inverse square attenuation to make the more distant surfaces darker than closer surfaces. We also want the light to be relatively dim and to have wide cone angles. Therefore, set the following light parameters:

Attenuation:	Inverse square
Intensity:	3.0
Hotspot:	160.00
Falloff:	160.00

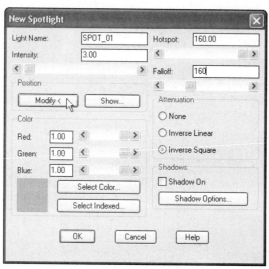

Figure 23–32

Click the Modify Position button. AutoCAD will temporarily dismiss the dialog box and issue command line prompts for the light's target and location points. Respond as follows:

Enter light target <current>: **0,0,0**
Enter light location <current>: **1,-2,0**

Click OK to exit the New Spotlight dialog box, and then click OK to end the LIGHT command.

Invoke the RENDER command again and click the Render button to render the current view. Your rendering should look similar to the one shown in Figure 23–33. Although the results are not dramatic, you can now see the detail of the front surface.

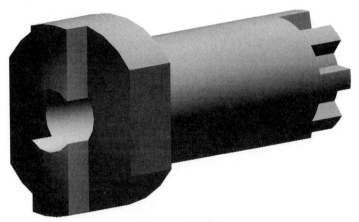

Figure 23–33

MORE WORK WITH LIGHTS

This next topic demonstrates how to make modifications on lights already placed in a drawing through the Properties Palette.

 Try It! – Open the file 23_Piston Render. This drawing consists of an exploded model of a piston assembly. Notice in Figure 23–34 the existence of three point lights labeled LT1, LT2, and LT3.

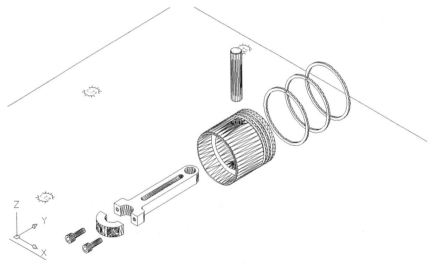

Figure 23–34

When placing lights, you have a better idea as to their location and orientation to the model if you switch the model to plan view. This will align your view of the model perpendicular to the XY plane of the User Coordinate System, as shown in Figure 23–35.

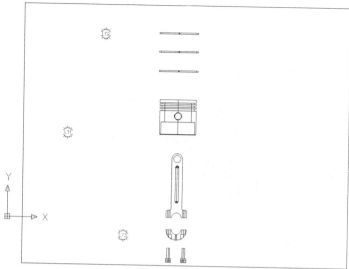

Figure 23–35

With the lights placed, activate the Render dialog box and perform a rendering operation on this assembly using the Photo Realistic rendering type. The results are illustrated in Figure 23–36. Whenever placing lights in plan view, while it is easy to orientate them to the model in the XY plane, all lights have a Z-axis of 0. In the figure, only portions of the assembly are illuminated while the base has areas of hot spots due to the lights being placed directly on top of the surface.

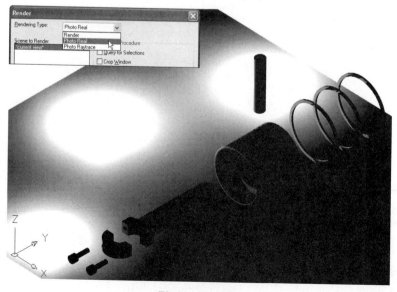

Figure 23–36

To give elevation to each light, select the light and activate the Properties Palette. In Figure 23–37, LT1 is selected. When the Properties Palette displays, change the Z-coordinate entry from 0 to 8 units. This will elevate LT1 8 units up from the top of the base surface.

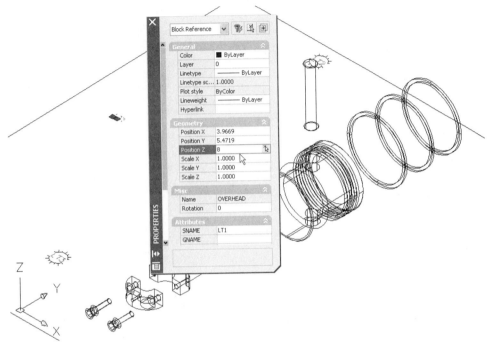

Figure 23–37

Continue using the Properties Palette to change the elevation of LT2 and LT3 to 6 units up from the base surface. This operation is illustrated in Figure 23–38 on LT3.

 Note: Whenever using the Properties Palette on an inserted light, the light icon will change; however, the text identifying the light number will remain in its original location. This will not affect the final outcome of the rendering.

With all three lights elevated to different Z-coordinate values, activate the Render dialog box and perform a rendering operation based on the Photo Realistic type. The results are shown in Figure 23–39. A more realistic image of the assembly is shown now that the lights have been moved from right on top of the surface to hovering above the assembly. The casting of shadows adds a dramatic affect.

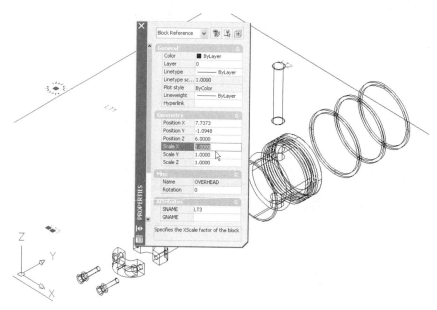

Figure 23–38

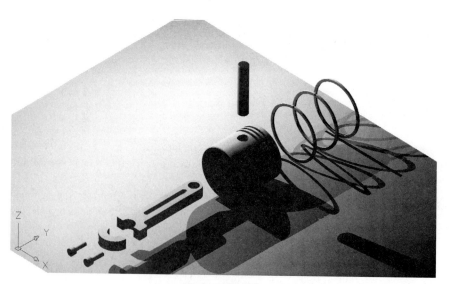

Figure 23–39

COLLECTING LIGHTS INTO SCENES—THE SCENE COMMAND

The SCENE command does just two things: First, it lets you decide which lights should be used in a rendering, and second, it lets you specify the name of the view for that rendering. The lights and the view are collected into a named scene. Let's see how the SCENE command works:

 Try It! - Open the drawing file 23_Piston Scenes. The same piston assembly used in the previous exercise displays. Select Scene from the Render area of the View pull-down menu or the Render toolbar, as shown in Figure 23–40. This will activate the Scenes dialog box shown in Figure 23–41.

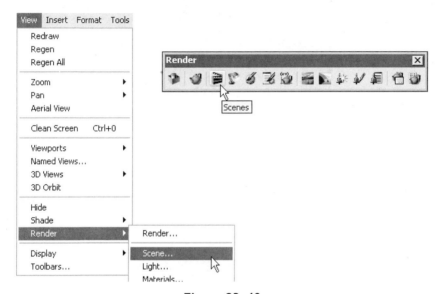

Figure 23–40

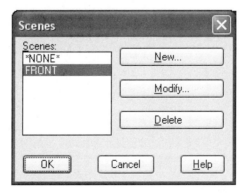

Figure 23–41

An existing scene called FRONT already exists. Clicking on the New button displays the New Scene dialog box illustrated in Figure 23–42. Type a name for your new scene, such as "TOP". Along the left side of the dialog box, select the Top view. Along the right side, select "*ALL*" and click the OK button to dismiss this dialog box. Then click the OK button in the main Scenes dialog box to return to your drawing.

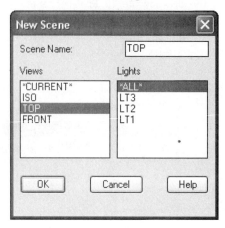

Figure 23–42

Activate the Render dialog box. This time, the RENDER command's dialog box displays the name of the scene you just defined: TOP, as shown in Figure 23–43. Select the TOP scene, then click on the Render button.

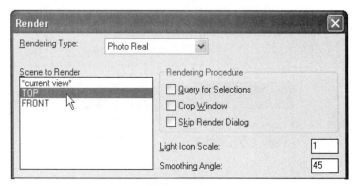

Figure 23–43

The results are displayed in Figure 23–44. Even though you were looking at an isometric view of the assembly, the rendering display switches to the view created in the scene. Regenerating your drawing screen will return you back to the isometric view of the assembly.

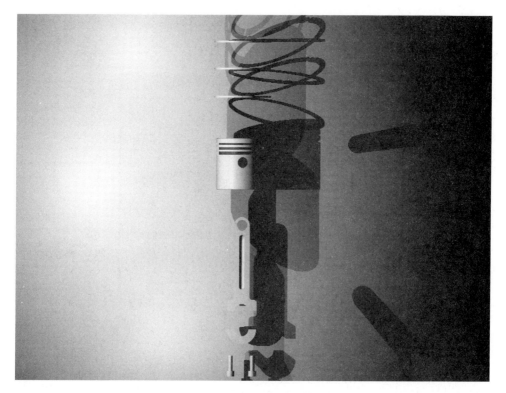

Figure 23–44

📝 LOADING MATERIALS WITH THE MATLIB COMMAND

Before a material can be used, it must first be loaded. This process is similar to loading and using linetypes in layers. AutoCAD comes supplied with a library of 145 predefined material definitions stored in a file called Render.MLI. Let's use the MATLIB command to load a material definition.

Clicking on the Materials Library Button in the Render toolbar will launch the Materials Library dialog box illustrated in Figure 23–45. You can browse through all the sample materials listed along the right side of the dialog box. Clicking on a material and then clicking the Preview button will give you an idea of what the material will look like. By default, the Preview shape is a sphere. You also have the option of previewing the material using a cube.

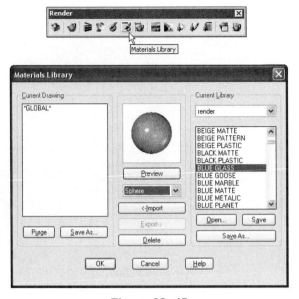

Figure 23–45

 APPLYING MATERIALS

Once a material or group of materials have been loaded into a 3D model drawing, the next step is to attach these materials to drawing shapes and components. This is accomplished through the RMAT (Materials) command.

Try It! – Open the drawing file 23_Connecting Rod. From the View menu, select Materials... from the Render area or pick the Materials button found in the Render toolbar (illustrated in Figure 23–46). This displays the Material dialog box shown in Figure 23–47, which presently lists no materials.

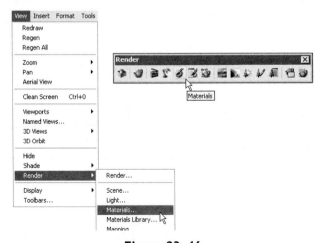

Figure 23–46

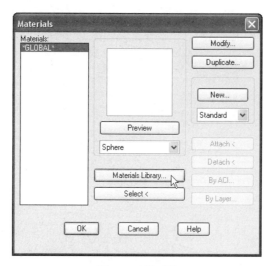

Figure 23–47

As mentioned earlier, before you can attach a material, you have to load its definition into the Materials dialog box. Clicking on the Materials Library button in the Rendering toolbar displays the Materials Library dialog box in Figure 23–48. (You could also have gone directly to this dialog box with the MATLIB command.)

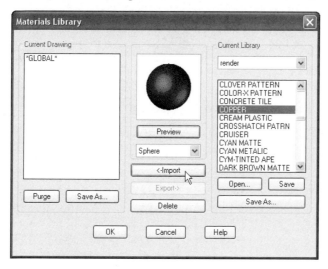

Figure 23–48

Find and click on the material called COPPER, then click on the Import button to add this to the materials list. For a quick preview of how a material looks, click on the Preview button. AutoCAD quickly renders a sphere using the selected material (see Figure 23–48). (Note: This does not work when more than one material type is selected.) Click on the OK button to return to the Materials Library dialog box.

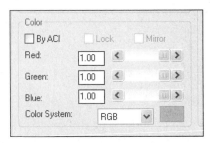

Figure 23–49

Copper is listed in the Materials list box. To attach the COPPER material to the 3D model, click on the Attach button as shown in Figure 23–49. AutoCAD clears the dialog box and returns you to the drawing screen where you are prompted to select objects (see Figure 23–50):

Select objects to attach "COPPER" to: *(Pick the edge of the Connecting Rod – see Figure 23-50)*
I found
Select objects: *(Press ENTER)*
Updating drawing . . . done

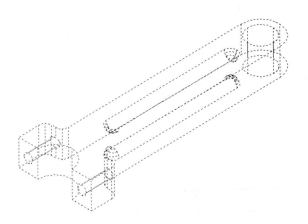

Figure 23–50

Note: In addition to attaching the material to individual objects, you can attach a material definition to all objects of a specific color by clicking on the By ACI button, or to all objects on a layer by clicking on the By Layer button.

The Materials dialog box returns. Click the OK button, and activate the Render dialog box to render the 3D model and check that the Apply Materials option is turned on. The newly rendered model looks browner than before due to the definition of the copper material, as shown in Figure 23–51.

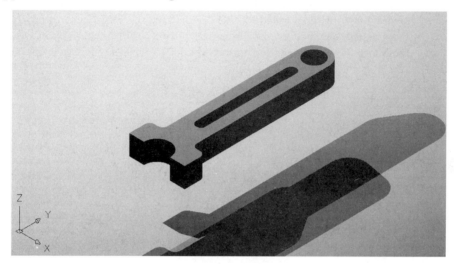

Figure 23–51

Experiment with applying other material definitions to see the difference they make in color and shininess. Note that the GLASS material definition illustrated in Figure 23–52 makes the object transparent and could take a longer amount of time to render.

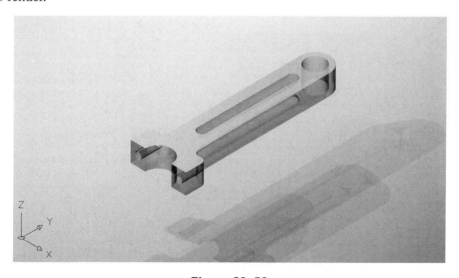

Figure 23–52

CREATING YOUR OWN MATERIALS

One way to render detailed reflections is to use ray tracing, which gives the most realistic reflections.

 Try It! – Open the file 23_Piston Mirror, as shown in Figure 23–53. You will create a new material called MIRROR and attach this material to the piston in the assembly file. Because of the nature of this custom material, you will use the Photo Raytrace rendering option in order to achieve reflective properties.

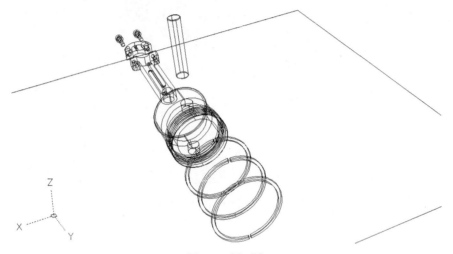

Figure 23–53

Activate the Materials dialog box and make sure that Standard is the material type displayed below the New button. Click the New button, as illustrated in Figure 23–54.

Figure 23–54

In the New Standard Material dialog box, enter MIRROR as the Material Name at the top of the dialog box illustrated in Figure 23–55.

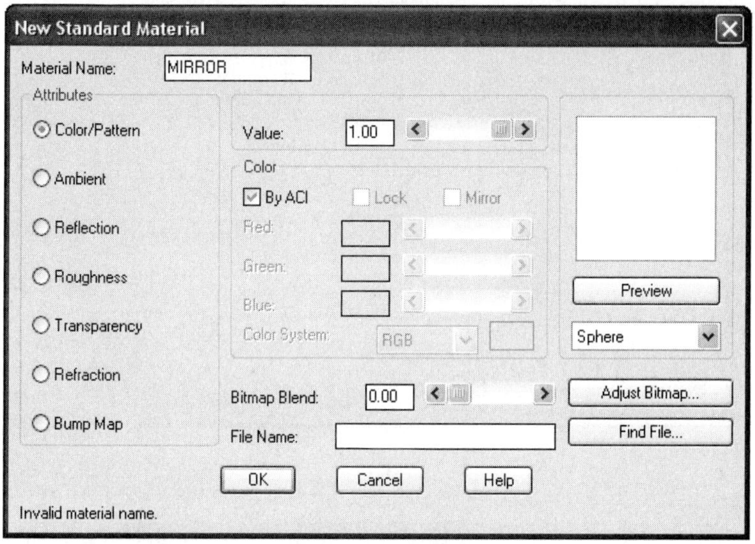

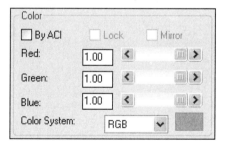

Figure 23–55

Begin making the next series of changes in this dialog box in order to create the Mirror material. With the Color/Pattern rendering attribute selected, click on the By ACI toggle to turn it off, as shown in Figure 23–56. Then make the color white by sliding the Red, Green, and Blue components all to a value of 1.00 (see Figure 23–56).

Figure 23–56

Select the Reflection rendering attribute, and then click the Mirror checkbox to turn it on, as shown in Figure 23–57.

Select the Roughness rendering attribute, and then use the Value slider to set it to 0.00 as shown in Figure 23–58. When you have finished making all of these changes, click the OK button.

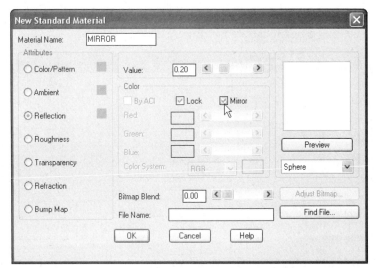

Figure 23–57

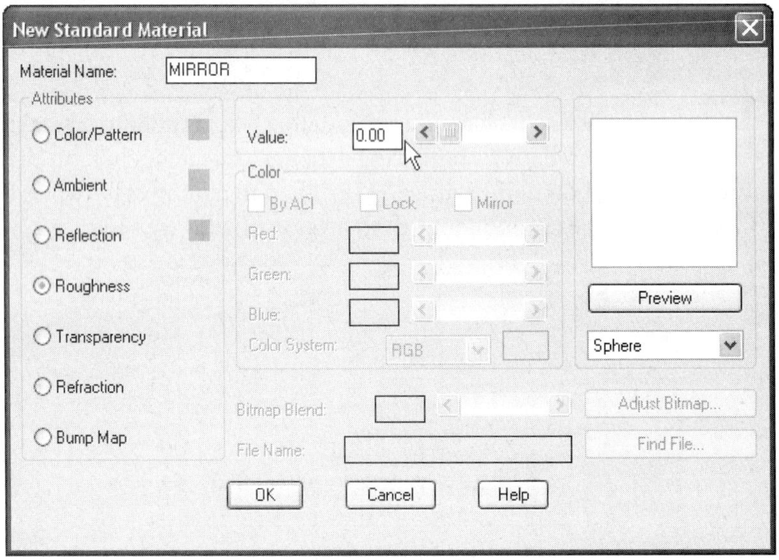

Figure 23–58

You will return back to the Materials dialog box. In the Materials dialog box, notice the appearance of the MIRROR material type. Click the Attach button, as shown in Figure 23–59.

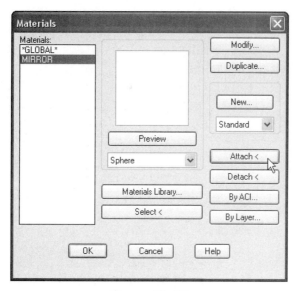

Figure 23–59

On the AutoCAD graphics screen, pick the piston in Figure 23–60 then press ENTER. Click OK to close the Materials dialog box.

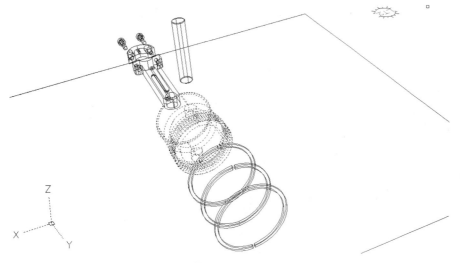

Figure 23–60

Activate the Render dialog box and be sure that the Photo Raytrace rendering type is selected, as shown in Figure 23–61. Then click the Render button to start the rendering operation.

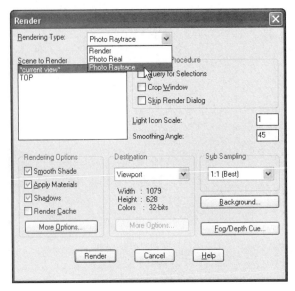

Figure 23–61

The results are illustrated in Figure 23–62. Notice that the piston rings reflect in the top portion of the piston. Notice also how the shadows of other parts reflect in the piston.

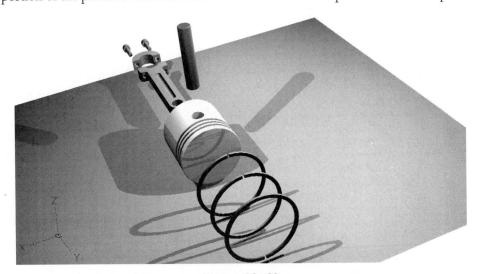

Figure 23–62

 APPLYING A BACKGROUND WITH THE BACKGROUND COMMAND

Now let's place an image behind an object with the BACKGROUND command. This is useful for enhancing the rendering. For example, if you have designed a 3D house in

AutoCAD, you could place a landscape image behind the rendering. To place a background image, the image must be in raster format. AutoCAD comes supplied with many sample background images. To load a background image, select Background… from the Render area of the View pull-down menu or click the Background button found in the Render toolbar, both shown in Figure 23–63.

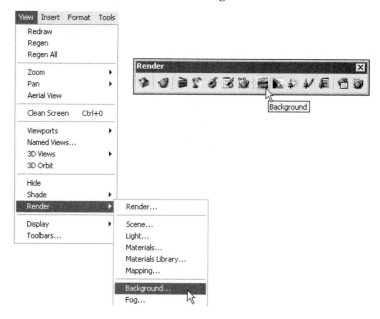

Figure 23–63

AutoCAD displays the Background dialog box illustrated in Figure 23–64.

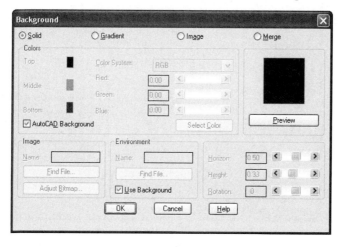

Figure 23–64

With the Background dialog box displayed, notice along the top of the dialog box how AutoCAD gives you a choice of four different kinds of backgrounds. They are:

Solid—A solid background means that AutoCAD replaces the default white (or black) background of the drawing screen with another color. You choose the color from the Colors section of the dialog box.

Gradient—A gradient means the color changes from one end of the screen to the other, such as from red at the bottom to light blue at the top (to simulate a sunset). AutoCAD gives you three ways to control a linear gradient—look carefully: they are tucked into the lower-right corner. Horizon specifies where the lower color ends; a value closer to 0 moves the lower color lower down. When Height is set to 0, you get a two-color gradient; any other value gives you a three-color gradient. Rotation rotates the gradient.

Image—You select a raster image for the background. The image can be in BMP (Bitmap), GIF, TGA (Targa), TIF (Tagged Image File Format), JPG (JPEG File Interchange Format), or PCX (PC Paintbrush) formats. As mentioned earlier, AutoCAD supplies 142 TGA files, some of which are suitable as background images. To adjust the positioning of the image, click the Adjust Bitmap button. AutoCAD displays the Adjust Background Bitmap Placement dialog box, shown in Figure 23–65.

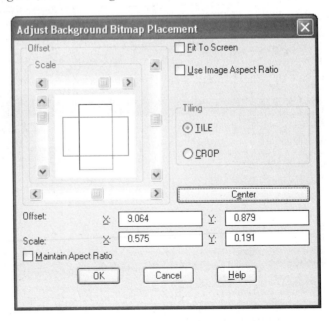

Figure 23–65

The Scale section shows the relative placement of the background image in relation to the viewport. The red rectangle represents the extents of the objects that will be rendered; the magenta rectangle represents the background image (turn off the "Fit To Screen" feature).

Several check boxes and radio buttons let you automatically position the image, such as stretching it to fit the entire viewport, retaining the image's aspect ratio (for no distortion), and whether to tile or repeat the pattern.

Merge—The current image is used as the background.

 Try It! – Open the drawing file 23_Piston Background. After rendering your display should appear similar to Figure 23–66.

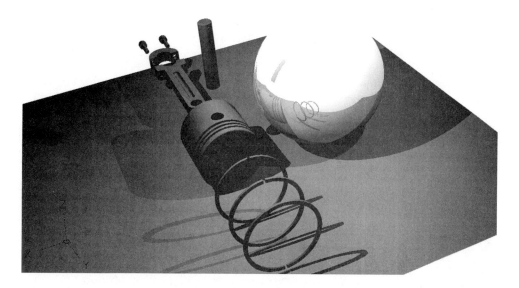

Figure 23–66

Enter the BACKGROUND command. Picking the Image radio button activates the Image Name edit box located in the lower-left corner of the Background dialog box, shown in Figure 23–67. When you select the Find File button, a number of sample raster files in TGA (Targa) format appear. Clicking the Preview button found along the right side of the Background dialog box will allow you to see the currently-selected color, gradient, or raster image.

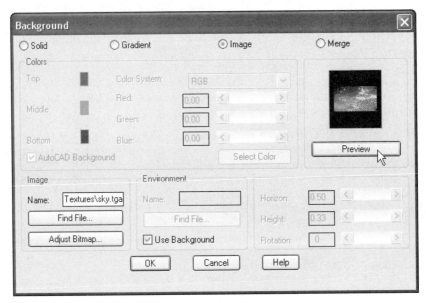

Figure 23–67

After selecting the sky.tga file, render the 3D Model through the Render dialog box. Auto-CAD renders the object on top of the background image, as shown in Figure 23–68.

Figure 23–68

FOG AND DISTANCE EFFECTS—THE FOG COMMAND

AutoCAD's Fog effect simulates fog by applying an increasing amount of white with distance. The further away the object, the more dense the level of white.

The color need not be white. The subtle use of black, for example, can enhance the illusion of depth, since objects further away tend to be darker in color. A limited application of yellow could create the illusion of a glowing lamp. You access the fog effect from the Render dialog box's Fog/Depth Cue button, or directly with the FOG command, which is selected from the Render area of the View pull-down menu or from the Render toolbar, as shown in Figure 23–69.

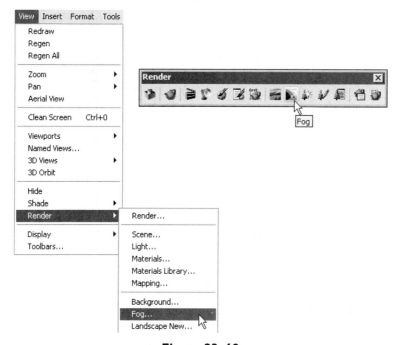

Figure 23–69

The FOG command displays the Fog/Depth Cue dialog box, shown in Figure 23–70.

The following components of the Fog/Depth Cue dialog box will now be explained in detail:

> **Enable Fog**—This toggle lets you turn the fog effect on and off without affecting any other parameters in this dialog box.

> **Fog Background**—This toggle determines whether the fog affects the background. For example, if the background color in your rendering is normally white but you choose black for the fog color, then the background becomes black.

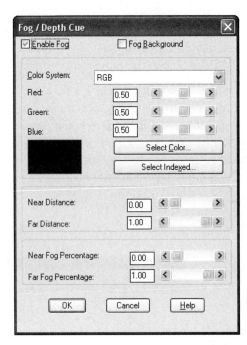

Figure 23–70

Color—Select a color for the fog. White is the default. Move all sliders to 0.0 for black, or choose a color from the Windows Color and AutoCAD Color dialog boxes.

Near Distance—This slider lets you position where the fog begins. The range is from 0.0 (the default) to 1.0. This slider can be tricky to understand, since it represents a relative distance from the camera to the back clipping plane. You may have to adjust the fog's Near Distance a number of times—each time you use the RENDER command—until the effect looks correct.

Far Distance—This slider lets you position where the fog ends. Like the Near Distance slider, Far Distance (default = 1.0) represents the percentage of distance from the camera to the back clipping plane.

Near Fog and Far Fog Percentages—These two sliders determine the percentage of fog effect at the near and far distances. Normally, these are 0.0 and 1.0, respectively. You should increase the value of Near Fog for a stronger effect; you should reduce the value of Far Fog for a more subtle fog effect.

 Try It! - Open the file 23_House Plan Fog. Rendering this image displays the results shown in Figure 23–71.

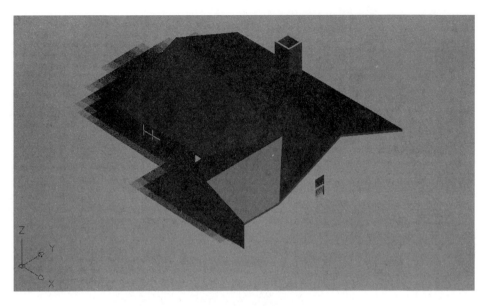

Figure 23–71

Activate the Fog/Depth Cue dialog box shown in Figure 23–72. Be sure that Enable Fog is checked. Also, near the bottom of the dialog box, set the slider bar adjacent to Near Fog Percentage: to a value of 50.

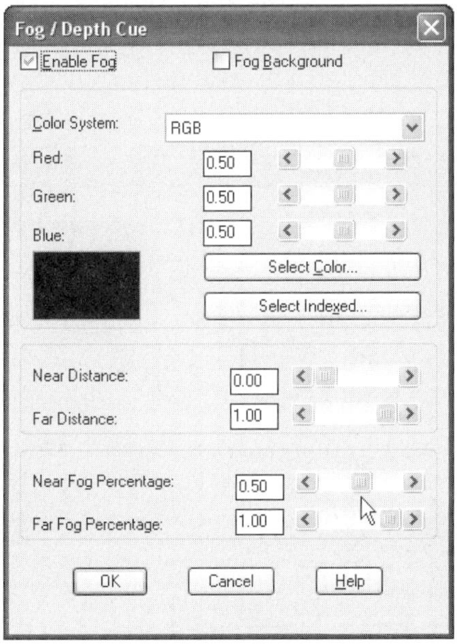

Figure 23–72

The results are displayed in Figure 23–73.

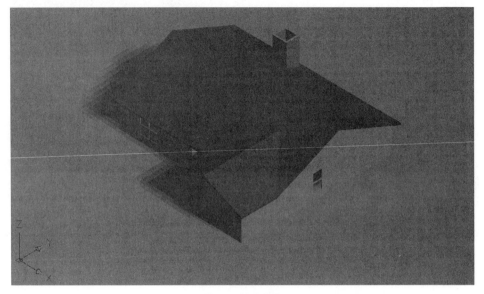

Figure 23–73

SAVING RENDERINGS TO A FILE—THE SAVEIMG COMMAND

There are a number of ways to save the results of a rendering. One method occurs when the Render dialog box appears. Select File from the Destination area and then click on the More Options button. This will display the File Output Configuration dialog box illustrated in Figure 23–74.

Figure 23–74

A number of file formats and resolutions are available to you, as shown in Figure 23–75.

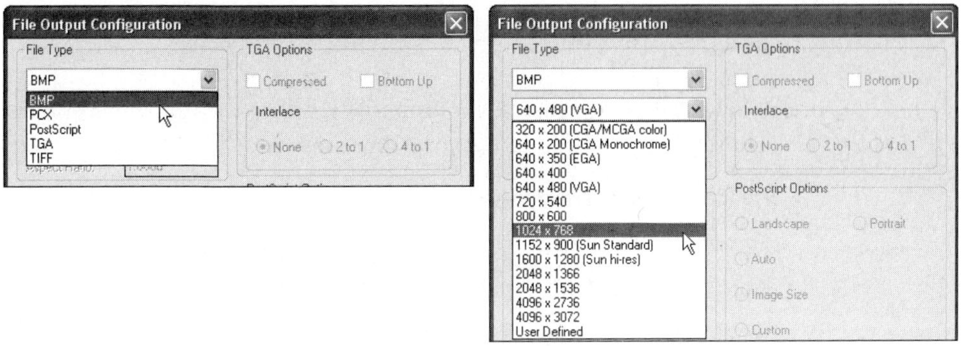

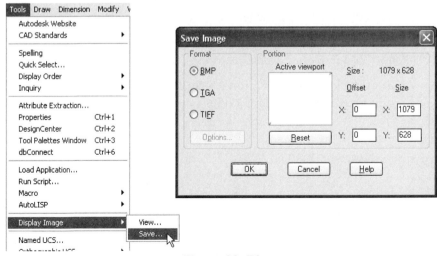

Figure 23–75

Here you can choose to save the rendering in one of five file formats and adjust a variety of parameters, such as the number of colors and compression schemes. There is one drawback to using the File option of the RENDER command, however; you do not see the rendered image before it is saved.

Use the SAVEIMG command to save the rendering to a file. This command is used after the RENDER command has created the rendering in a viewport. Select this command by clicking on Save... in the Display Image area of the Tools pull-down menu. As you can see from the illustration of the Save Image dialog box in Figure 23–76, you have only three choices of output file format to save in, compared with the File option of the RENDER command—namely BMP, TGA, and TIFF. Clicking on the Options button of this dialog box lets you select the type of image compression.

Figure 23–76

RENDERING TIPS

The time it takes for AutoCAD to create a rendering depends on many factors. Use the following hints as a means of speeding up the rendering process.

Options selected for the RENDER **command**—Setting the following options results in a rendering that is at least two times faster:

Rendering type: **Render**

Smooth shading: **Off**

Apply materials: **Off**

Rendering quality: **Gouraud**

Discard back faces: **On**

Destination: **Small viewport or 320 x200 file**

Lights: **None**

Render cache: **On**

Shadows: **Off**

Sub sampling: **4:1**

Enable Fog: **Off**

Background: **Solid**

Speed and memory of the computer—The faster the computer's CPU and the larger the computer's RAM, the faster the rendering.

Complexity of the model—A complex model with many parts takes longer to render than a simple model with only a few parts.

Amount of the model being rendered—Rendering all of the model takes longer than rendering a small portion.

Size of the viewport being rendered—Rendering to a very small viewport greatly decreases the rendering time.

TUTORIAL EXERCISE: SUNLIGHT STUDY

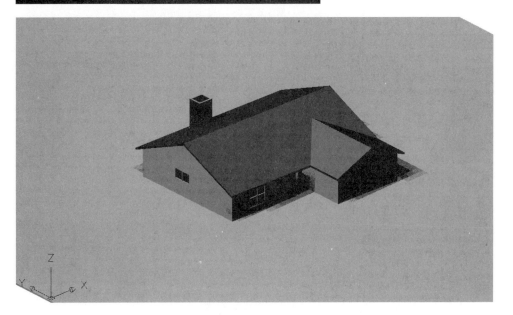

Figure 23–77

Purpose

This tutorial is designed to simulate sunlight on a specific day and time, and to observe the shadows that are cast by the house shown in Figure 23–77.

System Settings

Since this drawing is provided on the CD-ROM, all system settings have already been made.

Layers

The creation of layers is not necessary.

Suggested Commands

Begin this tutorial by opening up the drawing 23_House Plan Rendering. You will be performing a study based on the current location of the house and the position of the sun on a certain data, time, and geographic location. Shadow casting will be utilized to create a more realistic study.

Whenever possible, substitute the appropriate command alias in place of the full AutoCAD command in each tutorial step. For example, use "CP" for the COPY command, "L" for the LINE command, and so on. The complete listing of all command aliases is located in chapter 1, table 1–2.

STEP 1

Open the file 23_House Plan Rendering drawing illustrated in Figure 23–78.

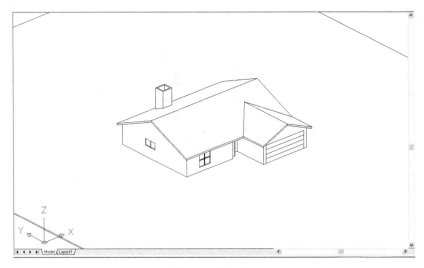

Figure 23–78

STEP 2

Activate the Lights dialog box in Figure 23–79. Select the Distant Light option from the drop-down list next to the New button, then click the New button.

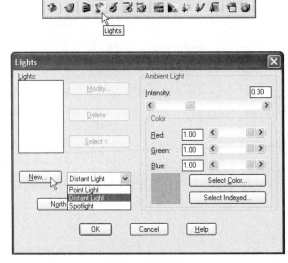

Figure 23–79

STEP 3

When the New Distant Light dialog box appears, enter SUN in the Name field. While in this dialog box, click in the checkbox to turn Shadow mode On.

Your dialog box should appear similar to Figure 23–80. Then click on the Sun Angle Calculator.

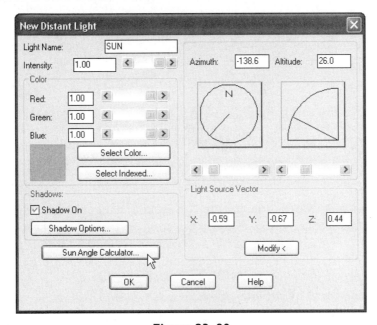

Figure 23–80

STEP 4

When the Sun Angle Calculator dialog box appears (as shown in Figure 23–81), change the following items:

> **Date:** 5/16 (the Sun Angle Calculator shows the date as month/day)
>
> **Time:** 10:00 (in the Sun Angle Calculator, you specify time with a 24-hour clock; for example, 15:00 is 3P.M.)

In the popup list below the Clock Time, make sure that EST—for Eastern Standard Time—is selected.

Your display should appear similar to Figure 23–81. When you are finished entering information into this dialog box, click the Geographic Location button.

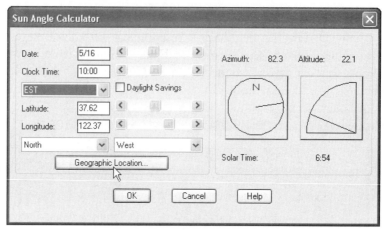

Figure 23–81

STEP 5

When the map of North America displays in the Geographic Location dialog box, check to see that "Charleston SC" appears above the Latitude and Longitude controls, as shown in Figure 23–82. You could also experiment by clicking on the map and changing the location. Other maps from throughout the world are also available. When you are satisfied with the location, click the OK button to leave the Geographic Location dialog box.

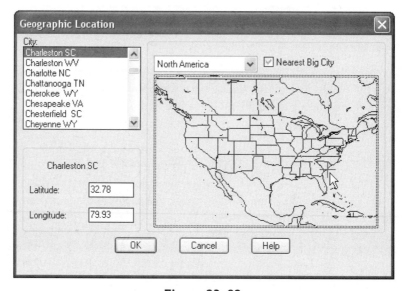

Figure 23–82

STEP 6

Continue clicking the OK button to accept the settings in the New Distant Light dialog box, and click the OK button one last time to leave the Lights dialog box and return back to the display of the house. Activate the Render dialog box.

Be sure that Photo Real is selected as the Rendering Type, as shown in Figure 23–83. Also be sure that the Shadows toggle is turned On. Finally, click the Render button.

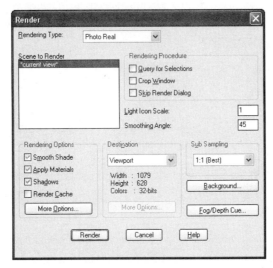

Figure 23–83

Your display should appear similar to Figure 23–84.

Figure 23–84

Note: You can improve the rendering speed by turning off the Smooth Shading and Apply materials toggles.

Experiment with other locations and times. Figure 23–85 illustrates the Geographic Location dialog box and the other continents available when clicking the down arrow.

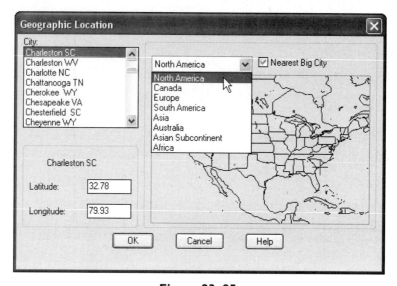

Figure 23–85

 Open the Exercise Manual PDF file for Chapter 23 on the accompanying CD for more tutorials and exercises.

 If you have the accompanying Exercise Manual, refer to Chapter 23 for more tutorials and exercises.

Note: Entries in SMALL CAPS indicate keyed-in commands; page numbers in **bold** indicate material in

LICENSE AGREEMENT FOR AUTODESK PRESS
A Thomson Learning Company

Educational Software/Data

You the customer, and Autodesk Press incur certain benefits, rights, and obligations to each other when you open this package and use the software/data it contains. BE SURE YOU READ THE LICENSE AGREEMENT CAREFULLY, SINCE BY USING THE SOFTWARE/DATA YOU INDICATE YOU HAVE READ, UNDERSTOOD, AND ACCEPTED THE TERMS OF THIS AGREEMENT.

Your rights:

1. You enjoy a non-exclusive license to use the enclosed software/data on a single microcomputer that is not part of a network or multi-machine system in consideration for payment of the required license fee, (which may be included in the purchase price of an accompanying print component), or receipt of this software/data, and your acceptance of the terms and conditions of this agreement.

2. You own the media on which the software/data is recorded, but you acknowledge that you do not own the software/data recorded on them. You also acknowledge that the software/data is furnished "as is," and contains copyrighted and/or proprietary and confidential information of Autodesk Press or its licensors.

3. If you do not accept the terms of this license agreement you may return the media within 30 days. However, you may not use the software during this period.

There are limitations on your rights:

1. You may not copy or print the software/data for any reason whatsoever, except to install it on a hard drive on a single microcomputer and to make one archival copy, unless copying or printing is expressly permitted in writing or statements recorded on the diskette(s).

2. You may not revise, translate, convert, disassemble or otherwise reverse engineer the software/data except that you may add to or rearrange any data recorded on the media as part of the normal use of the software/data.

3. You may not sell, license, lease, rent, loan, or otherwise distribute or network the software/data except that you may give the software/data to a student or and instructor for use at school or, temporarily at home.

Should you fail to abide by the Copyright Law of the United States as it applies to this software/data your license to use it will become invalid. You agree to erase or otherwise destroy the software/data immediately after receiving note of Autodesk Press' termination of this agreement for violation of its provisions.

Autodesk Press gives you a LIMITED WARRANTY covering the enclosed software/data. The LIMITED WARRANTY can be found in this product and/or the instructor's manual that accompanies it.

This license is the entire agreement between you and Autodesk Press interpreted and enforced under New York law.

Limited Warranty

Autodesk Press warrants to the original licensee/ purchaser of this copy of microcomputer software/ data and the media on which it is recorded that the media will be free from defects in material and workmanship for ninety (90) days from the date of original purchase. All implied warranties are limited in duration to this ninety (90) day period. THEREAFTER, ANY IMPLIED WARRANTIES, INCLUDING IMPLIED WARRANTIES OF MERCHANTABILITY AND FITNESS FOR A PARTICULAR PURPOSE ARE EXCLUDED. THIS WARRANTY IS IN LIEU OF ALL OTHER WARRANTIES, WHETHER ORAL OR WRITTEN, EXPRESSED OR IMPLIED.

If you believe the media is defective, please return it during the ninety day period to the address shown below. A defective diskette will be replaced without charge provided that it has not been subjected to misuse or damage.

This warranty does not extend to the software or information recorded on the media. The software and information are provided "AS IS." Any statements made about the utility of the software or information are not to be considered as express or implied warranties. Delmar will not be liable for incidental or consequential damages of any kind incurred by you, the consumer, or any other user.

Some states do not allow the exclusion or limitation of incidental or consequential damages, or limitations on the duration of implied warranties, so the above limitation or exclusion may not apply to you. This warranty gives you specific legal rights, and you may also have other rights which vary from state to state. Address all correspondence to:

AutodeskPressExecutive Woods5 Maxwell DriveClifton Park, NY 12065Albany, NY 12212-5015